D0022955

The Hunters or the Hunted?

R. A. Dart, pioneer in the interpretation of australopithecine-bearing bone accumulations, photographed at Makapansgat.

The Hunters or the Hunted?

An Introduction to African Cave Taphonomy

C. K. Brain

The University of Chicago Press
Chicago and London

C. K. BRAIN is director of the Transvaal Museum, Pretoria.

The University of Chicago Press, Chicago 60637
The University of Chicago Press, Ltd., London

Printed in the United States of America
85 84 83 82 81   5 4 3 2 1

Library of Congress Cataloging in Publication Data

Brain, Charles Kimberlin.
The hunters or the hunted?

Bibliography: p.
Includes index.
1. Australopithecines. 2. Animal remains (Archaeology)—Africa, Southern. 3. Caves, Africa, Southern. I. Title.
GN772.22.S6B7   573   79-28104
ISBN 0-226-07089-1

Behold, I will cause breath to enter into you, and ye shall live: and I will lay sinews upon you, and will bring up flesh upon you, and cover you with skin, and put breath into you, and ye shall live.

Ezek. 37:5–6

R. F. Ewer

With this biblical quotation, Griff Ewer prefaced one of her papers on saber-toothed tigers (1954*a*); it summed up the aim of much of her paleontological work.

More than twenty years ago Griff Ewer first introduced me to the pleasures of a career in which the study of animal behavior mingles freely with that of paleontology. Her influence has molded the form of this book, and it is to her memory that the work is gratefully dedicated.

The University of Chicago Press
gratefully acknowledges a subvention
from the National Science Foundation
in partial support of the costs
of production of this volume.

Contents

Acknowledgments

Paleontological work is usually time-consuming, and this project has been no exception. In my particular circumstances I have been able to complete it only through prolonged misuse of time intended for recreation; the person who has borne the brunt of this misuse is my wife Laura, who has supported me beyond measure with her particular blend of loyalty and laughter. Our four children, Rosemary, Virginia, Timothy, and Conrad, have all helped me in many ways, from sorting innumerable bones to surveying the caves they came from. In addition, Virginia has drawn many of the diagrams and charts that illustrate the text.

The investigation described in this book has been, for me, an adventure of the mind, prompted by the imaginative concepts of Professor R. A. Dart. On this adventure my almost daily companion has been Dr. Elisabeth Vrba, who has freely given me the benefits of her lucid mind and unbounded enthusiasm. She had also added a good deal to the significance of this book through the results of her own research.

At the Transvaal Museum I would not have been able to remain scientifically active without the help of Mrs. M. C. Erasmus, who has willingly shouldered much of what I would normally have been expected to do.

Other museum colleagues have also aided me in many ways, particularly Mrs. Elizabeth Voigt, who helped with the analysis of bone accumulations during the early stages of the project and is now continuing her studies of faunal remains from archaeological sites. Dr. Alan Kemp has aided me with research in the Kruger National Park, and the companionship of Mr. O. P. M. Prozesky was appreciated during fieldwork in South-West Africa. Ms. Imogen Chesselet has prepared many of the drawings in this book; Mrs. Ronel Goode has tracked many obscure library references for me, and Mrs. Elsa Kirsten typed the manuscript with patience and precision.

For seven years my field team at Swartkrans has been supervised by Mr. George Moenda, and the services of Mr. Absalom Lebelo and Mr. Jack Sepeng have not gone unnoticed.

The Swartkrans site was acquired by the University of the Witwatersrand in 1968 and, since then, the Board of Control of the Bernard Price Institute for Palaeontological Research has generously allowed me to continue my investigations there. I am much indebted to Professor S. H. Haughton and Professor S. P. Jackson for their interest in this work.

In the Anatomy Department of the University of the Witwatersrand Professor P. V. Tobias remains an esteemed colleague and warm friend, and he and Mr. Alun Hughes have been my companions during many days of fruitful discussions at the caves. On some such occasions we have been joined by Dr. James Kitching, Mr. Brian and Dr. Judy Maguire, and Dr. Tim Partridge. I have benefited greatly from the experience and kindness of these people.

It would not have been possible to initiate the research described in this book without the personal interest of Mrs. Lita Osmundsen and the generosity of the Wenner-Gren Foundation. The symposium this foundation sponsored in Austria during 1976 helped to crystallize the new science of taphonomy and to chart its future course. My appreciation is due to fellow taphonomists Dr. Andrew Hill, Dr. Kay Behrensmeyer, and Dr. Alan Walker for their active guidance in this venture.

On geological aspects of this project I have had the benefit of Professor Karl Butzer's wide experience and critical appraisal. My research has been the better for it. I have also appreciated the wise counsel of Professor F. Clark Howell and the fruitful cooperation of Professor Richard Klein. At the South African Museum, Dr. Brett Hendey has generously helped me in a variety of ways.

It is a pleasure to acknowledge the help and hospitality of Mr. Attila Port and his wife Karen during productive periods of fieldwork in South-West Africa. I am also grateful to Mr. C. K. Cooke, who was my companion and guide during many happy days spent in Rhodesian caves. Col. J. Scott has generously allowed me access to his Uitkomst nature reserve where so many of my observations have been made.

Many thanks are due to the following friends and colleagues without whose help the work described in this book would not have been completed: the late Professor W. W. Bishop, Professor C. S. Churcher, Dr. R. J. Clarke, Mr. C. G. Coetzee, Professor H. B. S. Cooke, Dr. D. H. S. Davis, Professor H. J. Deacon, Mrs. J. Deacon, Dr. N. J. Dippenaar, Mr. W. du Plessis, Professor L. Freedman, Dr. C. E. Gow, Professor R. F. Holloway, Professor G. Ll. Isaac, Professor T. Jenkins, Dr. M. D. Leakey, Mr. R. E. F. Leakey, Professor A. E. Mann, Professor R. J. Mason, Professor H. McHenry, Mr. M. G. L. Mills, Dr. U. de V. Pienaar, Dr. I. L. Rautenbach, Professor J. T. Robinson, Dr. B. H. Sandelowsky, Dr. M. K. Seely, Miss V. Scott, Professor J. D. Skinner, Mr. F.

van den Broek, Dr. W. E. Wendt, Professor M. H. Wolpoff, and Professor A. Zihlmann.

Finally my thanks must go to the Board of Trustees of the Transvaal Museum, under chairmanship of Professor F. C. Eloff, the University Research Division of the Council for Scientific and Industrial Research, headed by Mr. W. J. Weideman, and the National Monuments Council for their sympathetic support of my research.

Drawings initialed I. M. C. were done by Imogen Chesselet. All photographs are by the author except those otherwise acknowledged in the legends.

Part 1
A Guide to the Interpretation of Bone Accumulations in African Caves

1 Introduction

This is a detective story, but a rather odd one. The clues are bones, and the aim of the investigation is to establish causes of death, but the evidence is ancient and no witnesses survive to relate their experiences. More normal detective stories often show some confident and efficient inspector systematically unraveling the evidence, with the professional expertise of Scotland Yard at his elbow; in this case the investigator is a zoologist who found himself in an uncharted field where guidelines were few and ill defined. What was needed was some kind of paleodetective's handbook—a guide to the interpretation of bony clues found in African caves. Part 1 of this book forms the rudiments of such a guide, and in Part 2 evidence from the caves of Sterkfontein, Swartkrans, and Kromdraai is presented and interpreted in terms of the guide's criteria.

The title of this book, *The Hunters or the Hunted?*, has been used before, in the same context. After attending the Third Pan-African Congress on Prehistory at Livingstone in July 1955, S. L. Washburn visited the Wankie Game Reserve to study baboons. Here it was not only baboons that claimed his attention; he also made observations on bones left by a variety of carnivores, his interest having been sparked by an extremely provocative paper delivered at the Livingstone congress. It was called "The Makapansgat Australopithecine Osteodontokeratic Culture," and in it R. A. Dart (1957*a*) drew some remarkable conclusions about early hominid behavior. From an analysis of more than 7,000 bones from Makapansgat, Dart concluded theat the collection represented food remains of *Australopithecus*, who had apparently been a highly effective hunter, capable of killing the largest and most dangerous animals of the times. The unusually high proportion of cranial remains among the fossils was taken to indicate that australopithecines had been headhunters, sometimes practicing their art on their own kind.

Looking at the remains of kills in the Wankie Game Reserve, Washburn pondered Dart's far-reaching claims. He observed that many of the kills retained their skulls long after other parts had disappeared; could it be that hyenas had taken such residual remains to the caves? That the australopithecines had in fact been the hunted, rather than the hunters? Such a suggestion had already been made by K. P. Oakley (1954*a,b*) of the British Museum after his study tour of fossil sites in southern Africa. From the point of view of understanding early man, the question was significant, as Washburn explained in his paper "Australopithecines: The Hunters or the Hunted?" (1957, p. 612):

> The taste for meat is one of the main characteristics distinguishing man from the apes, and this habit changes the whole way of life. Hunting involves cooperation within the group, division of labor, sharing food by adult males, wider interests, a great expansion of territory, and the use of tools. It is therefore important to date the beginning of hunting in order to interpret the origin of human behaviour. Did man take to the grasslands because he was a hunter, or did he become carnivorous long after leaving the forests? The answers to these questions may lie in the earliest australopithecine deposits, those at Makapan and Sterkfontein.

Washburn was not the only delegate at the Livingstone congress to be stirred by Dart's challenging claims. In subsequent years I followed the development of his thesis concerning "the predatory transition from ape to man" with great interest and in 1965 returned from Rhodesia to the Transvaal Museum specifically to pursue the topic further. My intention was to analyze bone accumulations from the other australopithecine sites of Sterkfontein, Swartkrans, and Kromdraai and to see if they could shed more light on man's predatory beginnings. It soon became apparent that interpreting these collections would be possible only after a good deal of background research on bone-accumulating agencies in African caves. Some of this work had now been done and is reported on in Part 1. Results of the bone accumulation analyses themselves, with an attempted interpretation, are presented in part 2. That the product of my ten years' research on this topic appears under the same title as Washburn's original paper is deliberate. Our motives for writing were the same and I gratefully acknowledge the impetus his thoughts have given to my work.

Over forty years, many paleontologists have described fossils from the cave breccias of Sterkfontein, Swartkrans, and Kromdraai, and the fossil animals are now reasonably well known. In the past, however, it has not been possible to view the collection from each site as a complete assemblage, to estimate minimum numbers of all the animals whose remains are involved, or to draw overall interpretive conclusions. This is partly because the bones typically occur in solid rock that has the consistency of concrete. They must be partially or wholly freed from this enclosing matrix before they can be identified as to skeletal part or animal taxon. This is an extraordinarily tedious business, so that the analysis that can now be presented, involving 19,487 bones from a minimum of 1,331 animals is the result of an enormous amount of painstaking work by my paleontological colleagues, my helpers, and myself.

Dart's Predatory Hypothesis

> The fossil animals slain by the man-apes at Makapansgat were so big that in 1925 I was misled into believing that only human beings of advanced intelligence could have been responsible for such manlike hunting work as the bones revealed. . . . These Makapansgat protomen, like Nimrod long after them, were mighty hunters.
>
> They were also callous and brutal. The most shocking specimen was the fractured lower jaw of a 12-year-old son of a manlike ape. The lad had been killed by a violent blow delivered with calculated accuracy on the point of the chin, either by a smashing fist or a club. The bludgeon blow was so vicious that it had shattered the jaw on both sides of the face and knocked out all the front teeth. That dramatic specimen impelled me in 1948 and the seven years following to study further their murderous and cannibalistic way of life. [Dart 1956*b*, pp. 325–26]

In a remarkable series of thirty-nine papers published between 1949 and 1965, Dart developed his hypothesis of predatory, cannibalistic australopithecines, who practiced an osteodontokeratic (bone, tooth, horn) culture. Looking back over his research in 1962, Dart (1962*e*) concluded that the evolution of his concept had passed through seven distinct stages.

The first stage involved the realization that many of the primate skulls from the australopithecine-bearing caves showed damage that, in Dart's estimation, could have resulted only from purposeful predatory behavior on the part of the early hominids. The sample consisted of 58 baboon skulls, endocranial casts, or other specimens—21 from Taung, 22 from Sterkfontein, and 15 from Makapansgat—many of which showed depressed fractures of their cranial vaults. In addition to the baboons, 6 australopithecine specimens were selected as showing evidence of interpersonal violence. Dart went so far as to conclude that 64% of the skulls had received blows delivered directly from the front and 17% from the left side; only 5% appear to have received blows delivered from the right side. The australopithecines were, it seems, mainly right-handed. But with what weapon were these lethal blows delivered? In Dart's opinion it was a bludgeon consisting of the shaft and distal end of an antelope humerus, whose double-ridged extremity caused characteristic depressions in the skulls of the prey. Recognition of these bludgeons represented the second stage in the evolution of Dart's hypothesis, and the concepts were presented in two papers. "The Predatory Implemental Technique of *Australopithecus*" (1949*a*), and "The Bone-Bludgeon Hunting Technique of *Australopithecus*" (1949*b*).

Stage three of the process consisted of "running the hyena myth to earth." Ever since Dean Buckland (1822) attributed the very large bone accumulation in the Kirkdale cave, Yorkshire, to the activities of spotted hyenas, it had been usual to regard hyenas as important collectors of bones in caves. Such a concept has been vigorously challenged by Dart in his paper "The Myth of the Bone-Accumulating Hyena" (1956*a*), his monograph on the osteodontokeratic culture (1957*b*), and elsewhere. Some views on the current status of hyenas as bone accumulators are given in chapter 4.

Stage four in the evolution of Dart's hypothesis involved the recognition that "the transition from apehood to manhood" had been conditioned by hunting behavior. In his paper "The Predatory Transition from Ape to Man" (1953*a*, p. 209), Dart wrote: "On this thesis man's predecessors differed from living apes in being confirmed killers: carnivorous creatures, that seized living quarries by violence, battered them to death, tore apart their broken bodies, dismembered them limb from limb, slaking their ravenous thirst with the hot blood of victims and greedily devouring livid writhing flesh."

The fifth stage in the progress of Dart's appraisal of *Australopithecus* came with the statistical analysis of the osteodontokeratic fragments from Makapansgat (1957*a,b*). Here some startling facts were brought to light through the analysis of 7,159 fossils extracted from the gray breccia: 91.7% of the specimens were of bovid origin, while 4.0% were from non-bovid ungulates and 4.3% were from other animals. Fossils in the last category were almost exclusively cranial, leading Dart to suggest that the australopithecines were "headhunters." That verte-

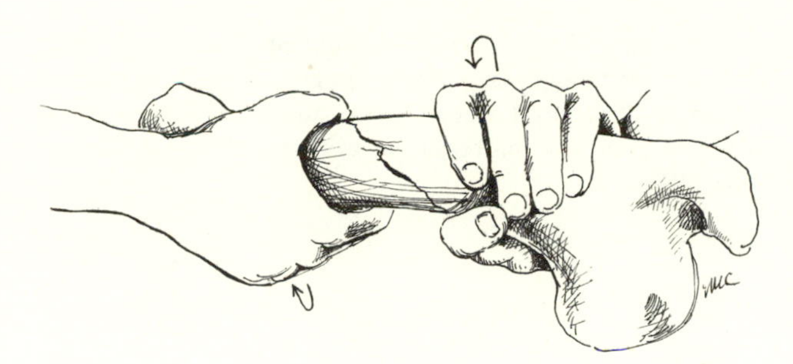

Seven basic techniques thought by Dart (1964*a*) to have been practiced by the Makapansgat australopithecines.

1. The "crack and twist" technique, in which a long bone was given a blow on the shaft and the two ends were then twisted apart. This resulted in a spiral fracture of the shaft, providing a perforating tool or blade as well as solid ends that could be used as pounders (1957*a*,*b*, 1959*g*, 1960*c*,*e*, 1961*b*, 1962*a*, 1964*b*,*c*).

2. Splitting of bones longitudinally by forcing one bone into the cavity of another. One example was a gazelle horn that had been thrust into the cavity of a spirally broken femur of a larger antelope (1957*c*). Other examples have also been described (1965*b*) and compared with an Iron Age skin-preparing tool from the western Transvaal (Mason 1964), which consisted of a proximal radius of a large bovid, surrounding a distal tibia shaft into which had been forced a rib fragment as the cutting blade.

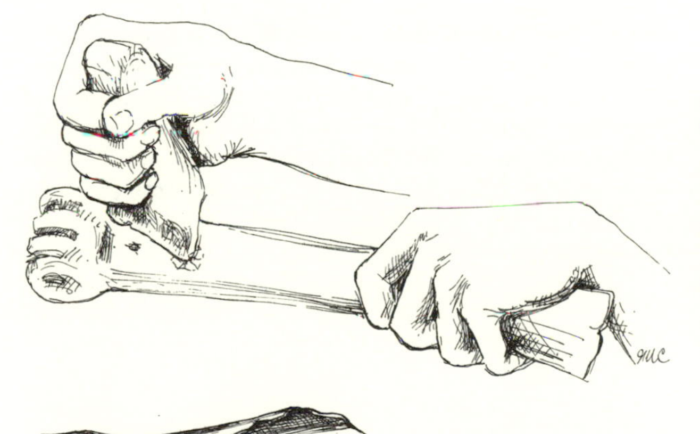

3. Localized or precision battering to produce scoops or spatulate tools. The bone most frequently used was a bovid metapodial, which was subjected to repetitive battering on one surface with a pointed object (1957*a*,*b*, 1959*e*,*f*, 1960*a*, 1962*a*,*f*, 1964*c*). Experiments were described in which as many as 140 blows were required (1959*e*,*f*, 1960*c*) to produce a metapodial "scoop," similar to "apple-corers" that were still in use during the past century (Campbell 1959) and are known as paleolithic fossils from Egypt (Arkell 1957).

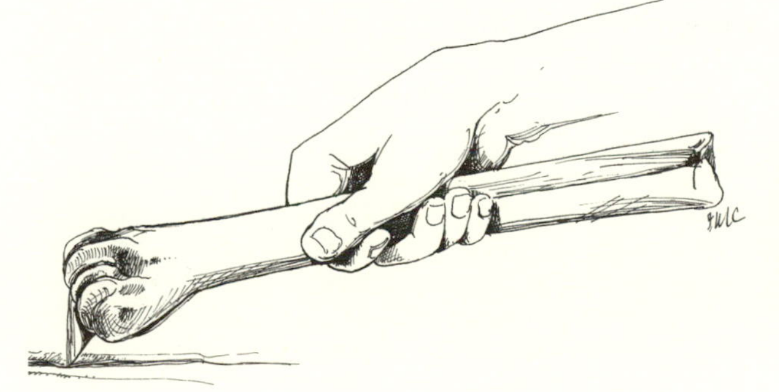

4. Making of composite tools, in which the cleft between the two articular processes at the distal end of a bovid cannon bone was used as a slot into which a tooth, bone, or stone fragment could be wedged, serving as a replaceable cutting blade (1957*b*, 1959*f*, 1960*a*,*c*, 1962*a*, 1964*c*).

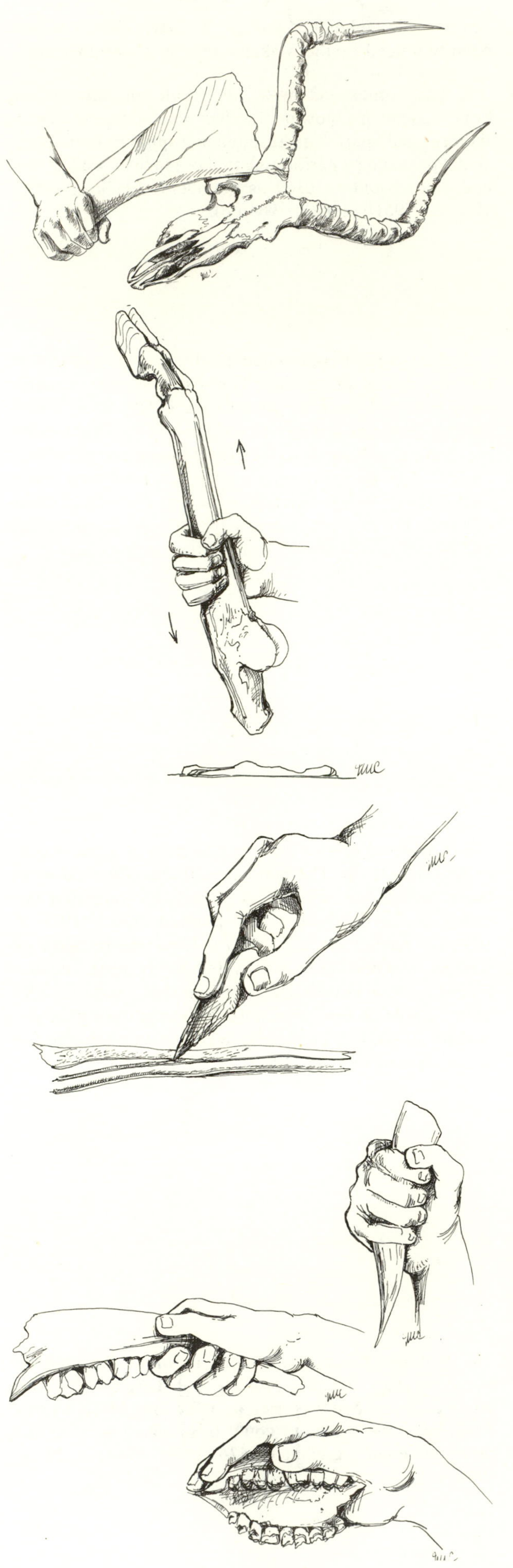

5. Splitting of antelope skulls transversely into anterior and posterior halves by striking them with a blade such as a scapula. The skull pieces were thought to have served as bowls, mortars, platters, and saucers (1957*a*, *b*, 1961*b*, 1962*f*, 1964*c*).

6. Detaching of the tibia at the talotibial joint by a blow that severed the malleolus and gave access to the distal tibia, talus, and calcaneus—bones thought to have been of cultural significance to the australopithecines (1957*a*, *b*, 1961*b*, 1962*d*).

7. Splitting of tendons and cutting of skins to form thongs. The tools thought to have been used for this purpose were specially prepared bone flakes and sharp ends of distal tibial pieces. Smooth indentations on the ends of long-bone shafts and flakes were thought to have resulted from the preparation of thongs of varying widths. Transverse grooves have also been observed on antelope limb bones, resulting, according to Dart, from abrasion by leather thongs (1957*b*, 1962*e*, 1964*b*).

In addition to these specific techniques, Dart visualized the australopithecines as using many other parts of skeletons as tools or weapons—mandibular toothrows as saws, maxillae as scrapers, horn-cores as points and daggers, and so on. In fact, a use could be visualized for each of the many thousands of bones preserved in the gray breccia at Makapansgat.

brae were almost entirely absent suggested that the hominids had been "professional decapitators." The bulk of the fossils were found to have come from 293 individual antelopes, the remains of which showed remarkable skeletal disproportion. Vertebrae again were scarce, but most of those present came from the upper neck region, suggesting that the antelope heads had been severed from the bodies high up. Dart interpreted the almost total absence of tail vertebrae to mean that the tails had been systematically employed as signals or flags and whips outside the cave." Interesting disproportions were also present among limb bones; while 336 humeri were found, for instance, only 56 femurs were present. Within limb bones themselves, disproportions between proximal and distal ends were marked: 336 distal ends of humeri were found associated with only 33 proximal ends. Dart interpreted these disproportions as meaning that the australopithecines purposely selected certain parts of the skeleton for use as tools. These were brought back to the cave while others were left at the sites of the kills. Some supplementary information about these disproportions is given in chapter 2 of this work.

Stage six of Dart's appraisal involved the "direct and detailed comparison between indisputably human and australopithecine bone splitting, and through eliminating the porcupine gnawing confusion." The most detailed comparison was between the Makapansgat bone assemblage and that from Kalkbank, an open site 64 km northwest of Pietersburg in the central Transvaal (Mason 1958; Dart and Kitching 1958). An excavation in this old pan produced 3,619 animal bones associated with 50 stone artifacts and 38 waste fragments, dated at about 15,000 years B.P. The bones, 25% of which had been gnawed by porcupines, were interpreted as human food remains and were compared directly with specimens from Makapansgat.

An assessment of the role of porcupines as bone collectors in caves was made in Dart's osteodontokeratic monograph (1957*b*) and in his paper "Bone Tools and Porcupine Gnawing" (1958*b*). Further observations on this topic are presented in chapter 5.

The seventh phase in Dart's interpretive process was reached in 1962 and consisted of "finding out more about what the man-apes did with the bones." In his review of the Makapansgat investigations from 1925–1963, Dart (1964*a*) listed seven basic techniques he visualized the Makapansgat australopithecines as having practiced. (See pp. 5–6.)

Analysis and Interpretation of Bone Assemblages

Dart's study of the Makapansgat fossils was a pioneering project in that it represented the first analysis and interpretation of a bone assemblage from an African cave. Before this, a great deal of paleontological work had been done on fossils from caves, but this did not involve complete bone assemblages; it was concerned with evaluating specific fossils isolated from such assemblages.

The analysis and interpretation of complete bone assemblages or representative samples of them has come to be known as taphonomy. The term was coined by Efremov (1940) in a paper entitled "Taphonomy: A New Branch of Palaeontology" and means literally the "laws of burial." It concerns itself with what happens to animal remains between death and fossilization. The aims of taphonomic work are often very similar to those of paleoecological reconstruction—making use of the fossil assemblage to draw conclusions about the living community that gave rise to the remains and about the interaction of that community with the environment of the time. Unfortunately, the fossilized remains, or death assemblages, are very different from the living communities that produced them. Between death and fossilization a variety of biases operate, and the evaluation of these biases preoccupies a good deal of taphonomic research (Hill and Walker 1972; Hill 1975).

The aims of taphonomists are varied. Fossil assemblages may be used to reconstruct the faunal composition of the original community; the nature of the environment in which the community lived; the process of community succession; or the determination of relative ages of different communities. All these are valid objectives, but they do not concern us particularly here. My taphonomic aims in this study are quite specific: *to analyze the fossil assemblages from the caves of Sterkfontein, Swartkrans, and Kromdraai in order to decide how these bones may have found their way into the caves and to draw conclusions about the behavior of the hominids and other animals that interacted with them.*

The Caves and Their Fillings

The three well-known cave sites of Sterkfontein, Swartkrans, and Kromdraai occur within 3 km of one another in the Bloubank River Valley, 9 km north-northwest of Krugersdorp in the Transvaal. The country rock is Precambrian dolomitic limestone of the Transvaal supergroup, dipping north at approximately 30°. Although the form of the caves and their fillings may be complex, as described in subsequent chapters, the sites have presumably passed through the same generalized stages of formation, as is shown by the idealized vertical sections in figure 1 (Brain 1975*a*). In stage 1 the cavern has formed by solution of the dolomite in the phreatic zone beneath the water table; its contours may well have been determined by planes of weakness in the country rock.

By stage 2 the water level has dropped through valley incision at some point in the general area, and the cavern is now air filled. It may be ventilated by a distant and indirect connection with the surface, and secondary cave travertines such as stalactites and stalagmites have started to form within it. Stage 3 shows aven formation in the dolomite overlying the cavern. The avens represent joints, or other planes of weakness in the rock, enlarged by rainwater passing down through them. The passage of this water is accelerated by the presence of the cavern, since the water can drip freely from the roof of the cavern as rapidly as it percolates down through the joint.

By stage 4, one of the avens has broken through to the surface, providing the first direct link between the cavern and the outside. A talus cone begins to form beneath the aven, containing bones of animals living on the surface. This cone may be calcified by lime-bearing solutions dripping from the roof, in which case the resulting deposit is known as a cave breccia.

Stage 5 shows the cavern almost filled with breccia and the roof further dissected by aven formation. In stage 6, surface erosion has removed much of the roof, exposing the bone-bearing breccia on the surface. The sites of Sterkfontein, Makapansgat, Swartkrans, and Kromdraai are currently at this stage, though in some cases the cave histories have been complicated by floor subsidences or secondary decalcification of the breccia masses.

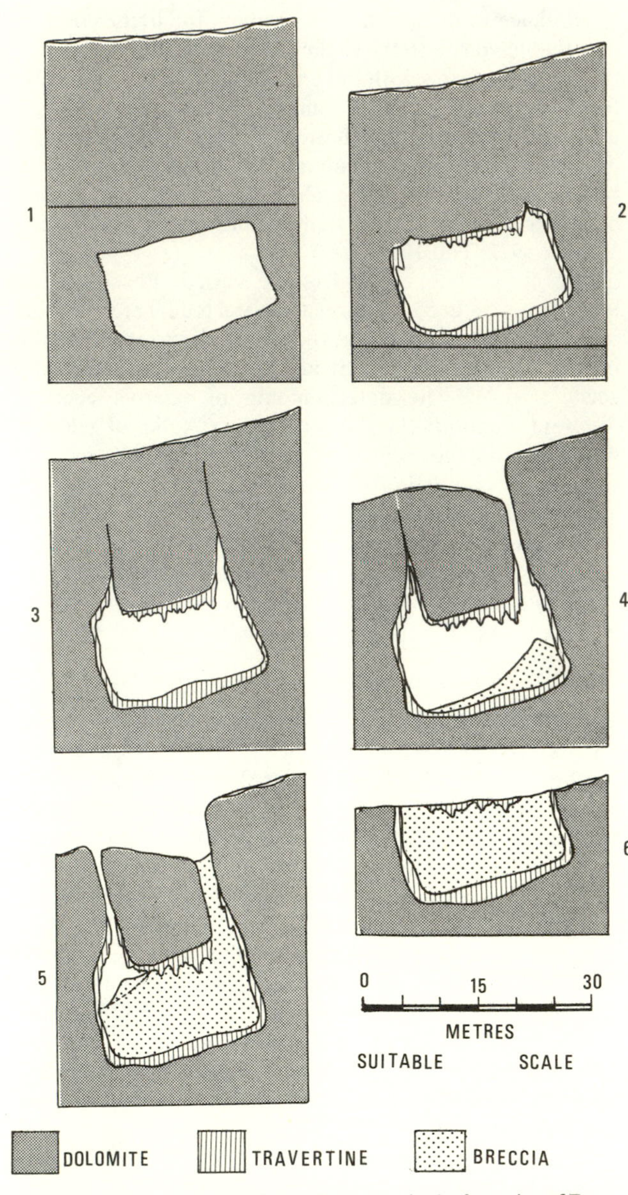

Fig. 1. Vertical sections showing typical stages in the formation of Transvaal dolomitic caves and their fossiliferous deposits. Details are given in the text. From Brain 1975*a*.

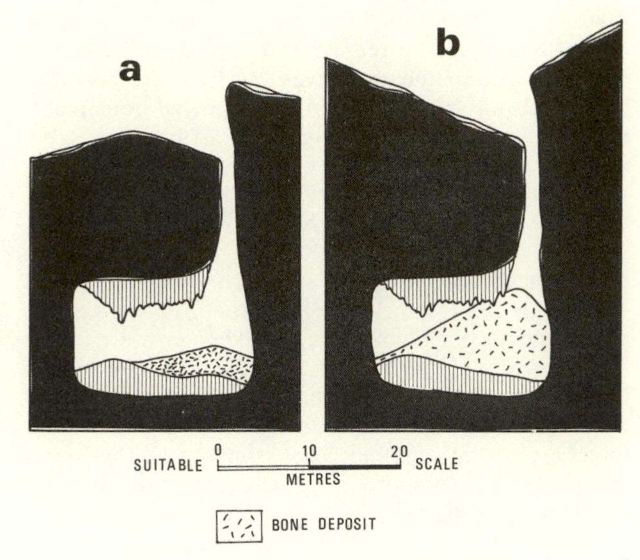

Fig. 2. Vertical sections through a hypothetical cave where the rate of influx of bones is equal in each case. In *(a)* the inflow of sediment is minimal, resulting in a dense bone concentration; in *(b)* the bones are much diluted with sedimentary matrix. From Brain 1975*a*.

The concentration of bones in the cave breccias varies greatly and is regulated by the rate of bone inflow relative to that of the enclosing matrix. Figure 2 shows two cave situations where the number of bones passing down the two avens per unit of time is the same. In case (*a*) the rate of sediment accumulation is low, as a result of the small catchment area round the cave entrance, so that the resultant deposit shows a high concentration of bone. By contrast, the sediment accumulation in case (*b*) was comparatively rapid, "diluting" the bone breccia and resulting in a low concentration of the bones themselves. With the exception of the gray breccia at Makapansgat Limeworks, where the bones show remarkable concentration, the other cave breccias referred to in this study show varying degrees of sediment dilution.

The Bone Assemblages

As will be outlined in subsequent chapters, the bone accumulations from the Sterkfontein valley caves fall naturally into two groups: the microvertebrate component, consisting of bones of small animals such as rodents, insectivores, birds, and reptiles that were almost certainly derived from the regurgitations of owls roosting in the cave (chap. 6), and the macrovertebrate component made up from the bones of larger animals whose remains found their way into the caves in a variety of ways (chaps. 3, 4, 5).

The main concern of this book is with the macrovertebrates, although samples of the microvertebrate fraction have been prepared and analyzed wherever possible.

Stratigraphic work at the sites has resulted in subdivision of the deposits into discrete members or other site units. The following is a summary of how the macrovertebrate bone assemblages used in this study are related to such subdivisions:

Sterkfontein	Member 4	(ST 4)	1,895	specimens
Sterkfontein	Member 5	(ST 5)	1,202	specimens
Sterkfontein	Member 6	(ST 6)	454	specimens
Swartkrans	Member 1	(SK 1)	2,372	specimens
Swartkrans	Member 2	(SK 2)	5,895	specimens
Swartkrans	Channel fill	(SK C)	837	specimens
Kromdraai A, faunal site		(KA)	1,847	specimens
Kromdraai B, australopithecine site		(KB)	4,985	specimens
		Total	19,487	specimens

Sterkfontein

As will be described in chapter 9, this cave deposit has been divided into six members. Although fossils are known to occur in the lower levels, collections from only the upper three members were available for this analysis.

All the specimens from Member 4 came from lime-mining and paleontological excavations in the Type Site area that occurred between 1936 and 1948. The recent work of P. V. Tobias and A. R. Hughes at Sterkfontein has produced a wealth of other fossils from this and other breccia members, and their evaluation of this material is anticipated with much interest. Microvertebrate remains in Member 4 breccia are not abundant, and acetic acid preparation of the other fossils failed to provide me with an adequate sample for analysis.

Member 5 was explored paleontologically after stone artifacts were found within it during 1956. Two seasons of excavation (1957–58) by J. T. Robinson provided the bone assemblage analyzed here. The member contains a remarkably rich microvertebrate concentration, and Robinson sampled this extensively in the course of his excavation. Subsequently I dissolved a quantity of these blocks in acetic acid and analyzed their bone content. This sample contained remains from a minimum of 644 individual animals, details of which are given in chapter 9.

Member 6 was formerly known as Robinson's "upper breccia." It is an insignificant unit appreciably more recent than the others. Fossil bones are present, but microvertebrate remains are rare.

Swartkrans

The original fieldwork conducted by the late Robert Broom and J. T. Robinson between 1948 and 1952 led to the recovery of 3,600 fossils, including a very large and significant hominid collection. My subsequent work at the site between 1965 and 1975 added 5,504 macrovertebrate fossils to the total. Formal separation of the Outer Cave filling into Members 1 and 2 has occurred fairly recently (Butzer 1976; Brain 1976*a*), and the fossil sample has since been divided between these members on the characteristics of the enclosing matrix, as is described in chapter 10.

In addition to the two breccia members, channel fills are present in the Outer Cave, varying a great deal in age and degree of calcification. The bone sample from the Swartkrans channel fill was excavated in 1973 and is appreciably younger than the main mass of Member 2 breccia.

Microvertebrate bones are very abundant in the Member 1 sediment, and two samples were prepared and analyzed in the course of this study. A minimum of 527 animals was found to have contributed to the two samples.

Since I completed the manuscript of this book, new excavations at Swartkrans have uncovered a hitherto unrecorded earlier component of Member 1. This extensive deposit is older than that previously designated Member 1 and is now being investigated.

Kromdraai

Australopithecine remains from Kromdraai are known only from site B, but it has been customary in the past to associate the fauna from Kromdraai A (the "faunal site") with the hominids from Kromdraai B (the "ape-man site"), although this association, in strict temporal terms, is now known to be invalid. For historical reasons, therefore, Kromdraai A is included in this study.

The macrovertebrate fossil assemblage from site A came from Broom's excavation of 1947; the microvertebrate sample, containing bones from a minimum of 273 individual animals, was obtained by dissolving blocks of breccia, recently collected at the site, in acetic acid.

The type specimen of *Paranthropus robustus* came from Kromdraai B in June 1938; it was followed in 1941 by a small number of other remains, but the bulk of the sample analyzed here was derived from an excavation I undertook there between March 1955 and May 1956. Most of the bones came from the decalcified fringe of the breccia mass along the north wall; they included an extremely large microvertebrate collection that is being studied by D. H. S. Davis and T. N. Pocock.

Some Methods and Procedures

My objectives in this study of bone accumulations were to establish what animals had contributed to the assemblage and by what skeletal parts these individuals were represented. The first step in such an analysis of a fossil assemblage involves removing all bone pieces whose species can be identified with certainty. These form the basis of the species list and generally consist of cranial pieces or other skeletal parts with diagnostic characteristics. For this stage of the work, a complete and well-organized osteological reference collection is indispensable, for it is on this, as well as on the competence of the investigator, that the reliability of the investigations will depend. If specialists on particular groups of animals are available, problematic specimens will naturally be referred to them.

After removal of specifically identifiable specimens, a second sorting is aimed at removing bone pieces referable to broader taxonomic categories such as "suid," "carnivore," or "bovid." Special mention should perhaps be made of the bovid or antelope groupings, since southern African bone accumulations from Quaternary sites are often dominated by antelope remains. Where these are fragmentary, or where many species are involved, it is generally not possible to do more than group the antelopes from which they came in size classes. For this purpose, four antelope size classes have been proposed (Brain 1974*a*), based on the liveweights of the animals whose remains are present in the bone accumulations.

Thirty-four extant species of antelope are currently recognized in the southern African region, and in table 1 these are arranged in order of increasing weight from dikdik to eland. The list has been divided into four arbitrary, partly overlapping categories as follows:

Antelope class I, 0–23 kg liveweight, with the upper limit represented by a large female common duiker.

Antelope class II, 23–84 kg: upper limit, large male blesbok.

Antelope class III, 84–296 kg: upper limit, large wildebeest or roan antelope.

Antelope class IV, more than 296 kg: very large animals such as eland or buffalo.

Various investigators have used variations of this scheme of bovid size classes to suit their particular needs. In his study of bones from southern Cape caves, Richard Klein (1976*a*) has employed a fifth size class to accommodate the extinct giant buffalo, *Pelorovis antiquus*, an animal appreciably larger than the living Cape buffalo or eland. In a recent consideration of southern African Bovidae relative to habitat and carnivore predation patterns, Elisabeth Vrba (1976*b*) has used four weight classes separated at liveweights of 27, 125, and 343 kg. These figures have cube roots of 3, 5, and 7, which aids graphic representation.

On completion of these sorting procedures, the sample will have been reduced to a residue of fragments that cannot be placed with confidence in any taxonomic category. One often finds that a large part of this residue consists of pieces from the shafts of long bones, particularly of bovid origin. In primitive human food remains, the long bones will generally have been smashed to extract marrow, resulting in characteristic bone fragments. These are referred to as *bone flakes* if they conform to the following requirements: that they come from the shafts of long bones such as the femur, radius, or metapodial; that they lack complete articular ends; and that they do not preserve more than half the circumference of the long-bone shaft. In cases where more than half the circumference of the shaft has been preserved, the specimens are called *shaft pieces*. Fragments showing recognizable articular ends are classified according to anatomical parts and would therefore not form part of the residue.

After removal of the bone flakes and shaft pieces, the remainder of the residue is listed as consisting of *miscellaneous fragments*.

In any interpretation of a bone assemblage, it is important to know how many individual animals of various kinds have contributed to the sample. The estimation of individual animal numbers on the basis of a bone assemblage has been attempted in a variety of ways by different workers (Clason 1972). The usual method is to estimate the minimum number of individuals from a count of the skeletal element occurring most frequently. If, for instance, the assemblage includes 50 left distal humeri of a particular species, 45 right ones, and all other elements in lesser numbers, then the minimum number of individuals for that species is 50. As Chaplin (1971) emphasized, skeletal elements should be matched against all others in the sample by age, sex, and size in an attempt to decide which bones could belong to one individual. Reservations about the reliability of the "minimum number of individuals" estimate have been expressed by Perkins (1973), who suggested instead that the "relative frequency" of each species should be determined. I have little doubt, however, that for the purposes of this study the minimum number method is the more appropriate.

In the course of the sorting procedures outlined here, the bone pieces should be examined for special features.

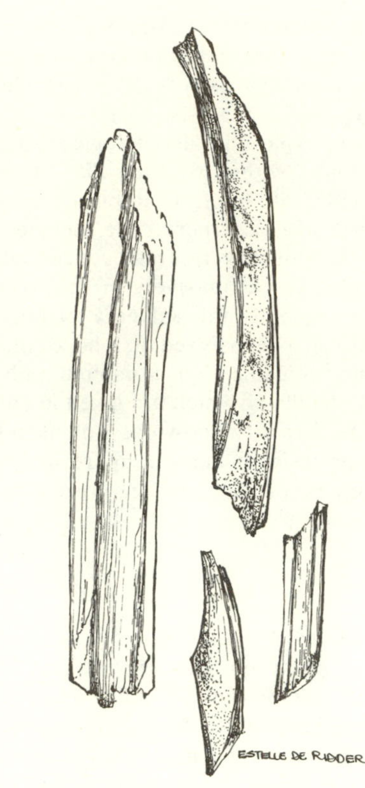

Bone flakes

These include evidence of use as tools, surface abraision, cut-marks, carnivore damage, porcupine gnawing, and pathology. The incidence of such features in an assemblage will greatly influence the result of any interpretation that is attempted.

Finally, it is remarkable how much information can be obtained from the study of bone accumulations—often from those parts of the assemblage that, in the past, have been ignored or discarded by paleontologists. It is desirable that *all* bone fragments from an excavation be retained, as the seemingly uninteresting fragments often provide clues vital to the interpretation.

2 Parts of the Skeleton: Survival and Disappearance

Occasionally, to the delight of paleontologists, the entire skeleton of a long-dead animal is preserved, with every bone present and in place. Such an event occurs only in special circumstances: the body of the animal must have come to rest in a place where it could lie undisturbed for a great many years; it must have been rapidly covered with a suitable enclosing matrix, and it must have been preserved and strengthened by percolating solutions of appropriate chemical composition. More usually, the animal's body is subjected to the destructive influences that characterize any natural environment. The skeleton becomes disarticulated—broken down into its individual parts, each of which has then to contend with the attention of carnivores and with the forces of decay and disintegration.

A typical mammalian skeleton consists of parts that differ greatly in size, shape, and ability to withstand destructive treatment. Compare, for instance, the astragalus and the scapula of an antelope shown in figure 3. One is a compact, subspherical object, enormously resistant to damage, while the other is essentially a thin, bladelike structure, fragile and delicate, apart from its head, which is somewhat more robust.

Such differences may not interest all paleontologists—some specialists are concerned with individual fossils only—but when a study is made of a fossil assemblage and an explanation is sought for how the bones found their way into a cave, then the presence or absence of certain parts of the skeletons can provide vital interpretative clues.

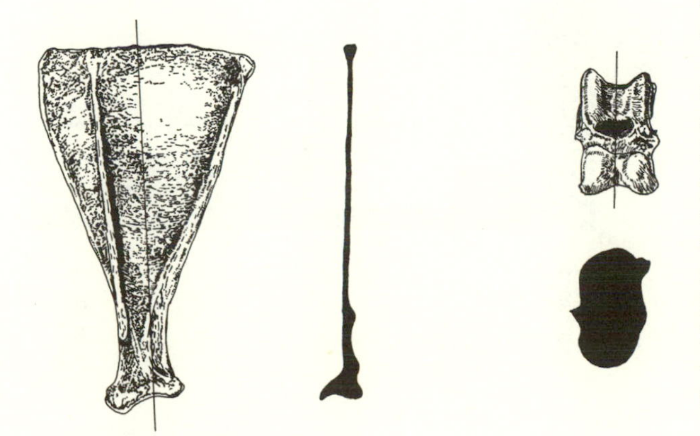

Fig. 3. Variety of form in mammalian skeletal parts: a fragile bovid scapula and a robust astragalus.

Very early in this study it became apparent to me that the consistent absence of certain skeletal parts from the Sterkfontein valley fossil assemblages could well be related to their original delicacy and inability to survive destructive influences. Direct observation suggested that some skeletal parts were more robust than others, but an experimental situation was clearly required to assess such robusticity objectively. At that time, in 1965, an interesting account, "Skeletal Durability and Preservation," had just been published by Chave (1964). It described the results of laboratory experiments aimed at testing the resistance of various skeletal materials to destructive forces, but it unfortunately restricted itself to invertebrate animals. The calcareous skeletons of various marine invertebrates were tumbled in a porcelain barrel containing water and either chert pebbles or sand. From time to time the fossils were examined and weighed. It was concluded that the major factors controlling the resistance of shells to physical destruction were the microarchitecture of the shells themselves and the disposition of organic matrix among the crystals of carbonate. Dense, fine-grained skeletons like those of *Nerita, Spisula,* and *Mytilus* proved most resistant to destruction—surely a result to be expected.

During 1965 I was planning a similar series of experiments on bones from mammalian skeletons, and part of the project involved observations of the effects of weathering on bones in various environments. To this end I visited the Namib Desert Research Station, 96 km inland from Walvis Bay, in South-West Africa, to set up a bone-weathering experiment in that extremely arid environment. After ten years, the experiment has yielded some interesting results (see fig. 120 in chap. 5).

While in this part of the Namib, I happened to visit several Hottentot villages and was struck by the abundance of goat bone fragments that lay about among the huts. Making a small collection of these, I laid them out at the research station and sorted them into skeletal parts as an exercise in osteology. It was immediately obvious that certain parts were well represented, while others were rare or absent. Distal humeri, for instance, were common, but, search as I might, I could not discover a single proximal humerus. The explanation was not difficult to find—the sample represented the resistant residue of goat skeletons able to survive the treatment they had received.

But what was this treatment? Inquiries and observations during the following week showed that goats were virtually the only source of meat for the Hottentots; when a goat was slaughtered, its skeleton was treated in a traditional manner and those parts the Hottentots considered inedible were tossed to the dogs. When the dogs in their turn had finished, a residue of parts unchewable by Hottentot or dog was left to bleach on the desert surface. Here recovery was easy, since the ground was devoid of vegetation.

For someone in search of an experimental situation where the comparative robusticity of mammalian skeletal parts could be tested, the activities of the Hottentots and their dogs proved a godsend. Only goat bones were involved, and these were being subjected to treatment that could still be observed. Here was, in fact, an experimental situation that could hardly have been bettered—short, perhaps, of substituting australopithecines for the Hottentots.

After the initial reconnaissance in 1965 (Brain 1967*b*, 1969*b*, 1976*b*), I returned in March 1966 to finish collecting all available bones and to investigate the circumstances in greater detail. On this occasion I was accompanied by Trefor Jenkins of the South African Institute for Medical Research, who undertook a thorough genetic study of the Hottentot population (Jenkins and Brain 1967).

The Kuiseb River Environment

The Kuiseb is one of several rivers that arise in the central plateau region of South-West Africa and make their way across the Namib plain to the Atlantic Ocean. On the eastern fringe of the Namib plain the Kuiseb passes through a spectacular canyon before continuing almost 160 km to its mouth in the vicinity of Walvis Bay. The lower part of the Kuiseb course runs through highly arid countryside where no more than a few centimeters of rain can be expected in a year. In the south is the Namib sand sea, with some of the highest dunes in the world; to the north are gravel plains with occasional outcrops of the Precambrian country rock. Except for a few water holes, the Kuiseb river course is normally dry, although subsurface flow sustains a dense growth of trees, some of them extremely large. Once or twice a year, rains on the escarpment bring the Kuiseb down in flood, scouring the riverbed of the dune sand that has encroached upon it from the south.

The Hottentots and Their Villages

At the time of our 1966 study, the total population of the lower Kuiseb valley was 133 persons, living in eight separate villages, all on the north bank of the Kuiseb, as shown in figure 4. Eighteen villages are known to have existed in recent years, but ten of these are now deserted, the people having moved to other centers in South-West Africa. Table 2 lists the numbers of people, dogs, and goats living at each of the eight occupied villages. A total of 61 males and 72 females were counted, together with 40 dogs and 1,754 goats. The histogram in figure 5 shows the age- and sex-distribution of all human inhabitants of the villages when our study was made. Very few individuals in the 20–40-year age group were found because in their late teens many leave home to seek employment in Walvis Bay and elsewhere. Some girls marry while they are away, but many become pregnant, and in this event the baby is brought home to grow up in the grandparents' care while the mother returns to her work.

A village is composed essentially of a family group, and figure 6 shows a typical village genealogy, representing the inhabitants of Soutrivier. In the past marriage partners have been found in neighboring villages, but as migration to Walvis Bay has increased the selection of partners farther afield has been possible.

Spacing of the villages along the riverbank is determined by the number of goats kept at each, because grazing is restricted to the riverbed and the extent of a

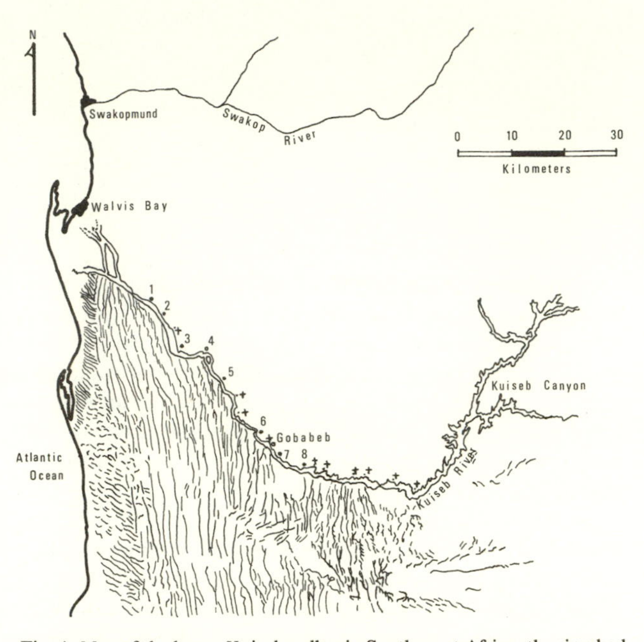

Fig. 4. Map of the lower Kuiseb valley in South-west Africa; the riverbed forms the division between sand dunes to the south and bare gravel plains to the north. Positions of Hottentot villages are shown scattered along the north bank; occupied villages are numbered, deserted village sites are marked with crosses. 1, Rooibank; 2, Ururas; 3, Itusib; 4, Swartbank; 5, Klipneus; 6, Soutrivier; 7, Natab; 8, Ossewater.

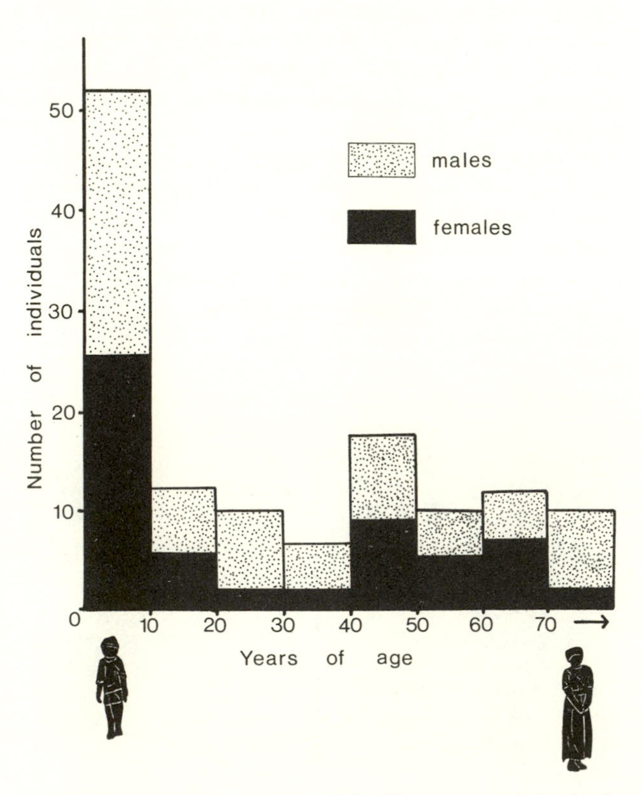

Fig. 5. The age and sex structure of the Hottentot population in the Kuiseb River villages.

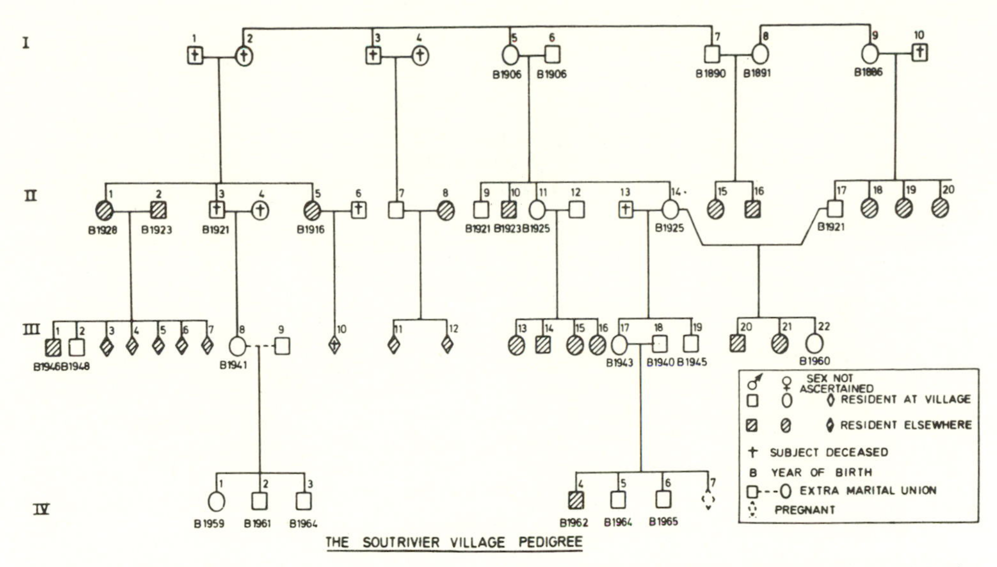

Fig. 6. Genealogy of the people of the Soutrivier village, lower Kuiseb valley (after Jenkins and Brain 1967).

village's pasturage is measured linearly along the Kuiseb bed. The main stockowner at the Ossewater village explained how, as the size of his original goat herd at Soutrivier increased, he was obliged to move his home to a new village to avoid disputes over pasturage.

The villages themselves are typically situated on the banks overlooking the riverbed, exposed to the glaring sun. The cooler, shady riverbed is avoided, for fear of the occasional floods that sweep down the valley. The layout of a typical village, in this case Soutrivier, is shown in figure 7. Three wells, each about 3 m deep, have been dug close to the riverbank; the upper part of each is lined with tree trunks, and the water is raised in a bucket suspended from the end of a long counterbalanced pole running over a fulcrum. Living quarters consist of beehive-shaped huts covered with long strips of bark from *Acacia albida* trees, which grow in the riverbed (fig. 8). Occasionally the individual huts are connected to form clusters with interleading doors.

Goats and lambs are kept in separate enclosures made of dry tree trunks and branches embedded upright in the sand. A similar palisade had been constructed around a small tobacco garden on the riverbank and also protected a few maize plants—the only attempt at agriculture found in any of the villages when the study was made.

The goats were found to subsist very largely on the fallen pods of *Acacia albida* trees (fig. 9), and herding is simple because the only water usually available to the animals is at the wells that have been dug at the villages. The goats therefore return to the village for water and, after their evening drink, are herded into kraals for the night.

Most of the people in the villages regard themselves as Topnaar Hottentots, and a smaller number consider themselves Bergdamara. One claimed to be a Herero, and two men were of mixed origin—one had had a Hottentot mother and an English father, the other a Bergdamara mother and a German father. The Hottentots were of slight build and had light brown-yellow skin, in contrast to the Bergdamara, who were more robust, darker, and more negroid.

Sir Francis Galton, the famous English traveler and anthropologist, traveled a short distance up the Kuiseb River at the start of his journey from Walvis Bay to Damaraland in 1850. He stayed for a while with missionaries at Schleppmansdorf, now called Rooibank (Galton 1889), then moved up the Swakop River to Barmen, where he described the physical characteristics of the Hottentots he saw there. In a letter to his cousin, Charles Darwin, he wrote on 23 February 1851:

> I have just left the land of the Hottentots, I am sure that you will be curious to learn whether the Hottentot ladies are really endowed with that shape which European milliners so vainly attempt to imitate. They *are* so, it is a fact, Darwin. I have seen figures that would drive the females of our native land desperate—figures that could afford to scoff at Crinoline, nay more, as a scientific man and as a lover of the beautiful I have dexteriously even without the knowledge of the parties concerned resorted to actual measurement. Had I been a proficient in the language I should have advanced, and bowed and smiled like Goldney, I should have explained the dress of the ladies of our country, I should have said that the earth was ransacked for iron to afford steel springs, that the seas were fished with consummate daring to obtain whalebone, that far distant lands were over-run to possess ourselves of caotchouc—that these 3 products were ingeniously wrought by competing artists, to the

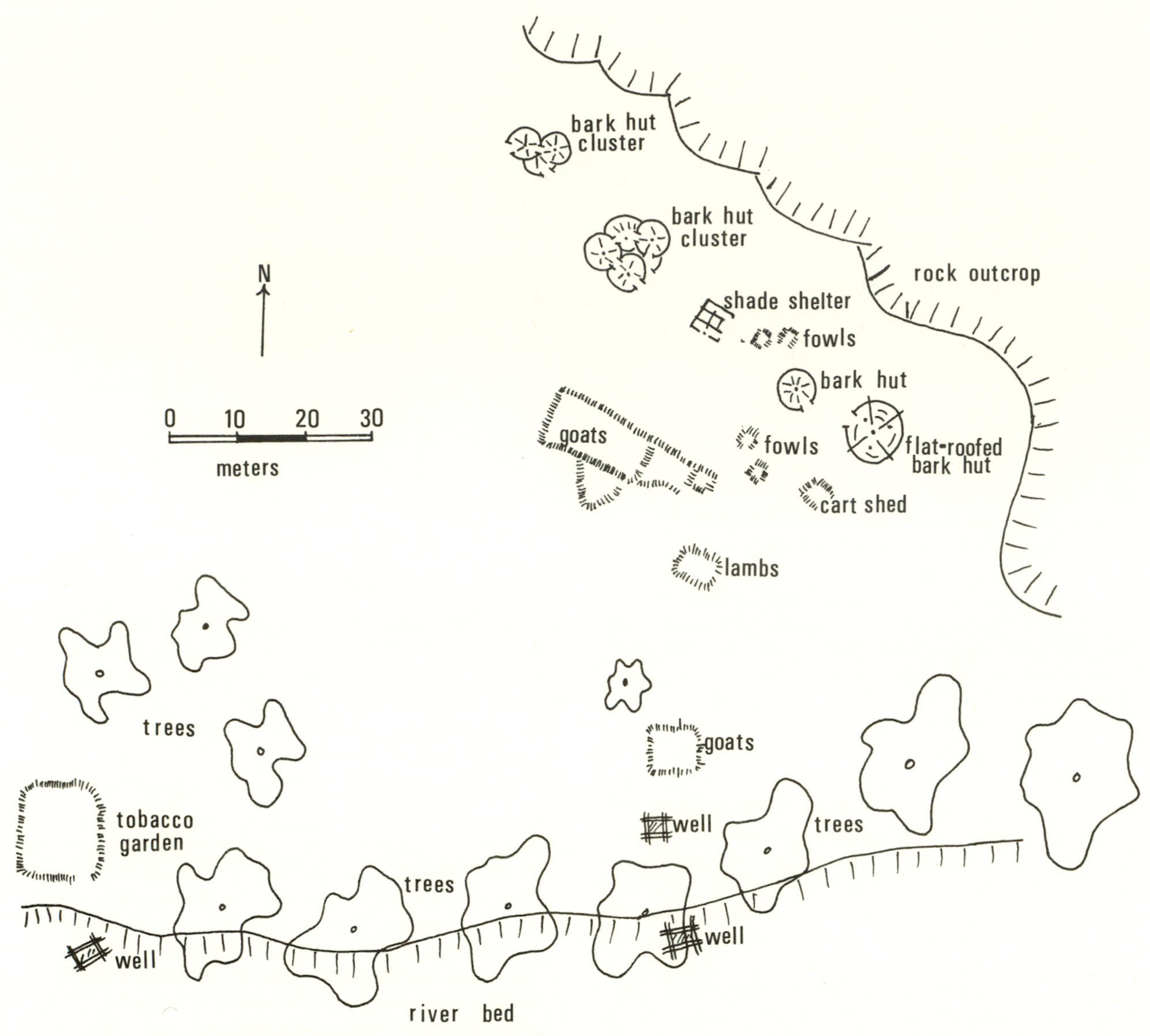

Fig. 7. Plan of the Soutrivier village as it was in 1967.

Fig. 8. Part of the Natab village showing characteristic beehive huts and the Kuiseb River bed in the background.

utmost perfection, that their handiwork was displayed in every street corner and advertised in every periodical but that on the other hand, that great as is European skill, yet it was nothing before the handiwork of a bounteous nature. Here I should have blushed, bowed and smiled again, handed the tape and requested them to make themselves the necessary measurement as I stood by and registered the inches or rather yards. This, however, I could not do—there were none but Missionaries near to interpret for me, they would never have entered into my feelings and therefore to them I did not apply—but I sat at a distance with my sextant, and as the ladies returned themselves about, as women always do, to be admired, I surveyed them in every way and subsequently measured the distance of the spot where they stood—worked out and tabulated the results at my leisure. [Galton, in Pearson 1914]

A short while before Galton's visit, the Hottentots in the Walvis Bay area had been described by the early traveler J. E. Alexander (1838) as the tallest and best-built of the Namaquas; but when Mrs. Hoernlé studied them in 1923 she described them as "the most miserable remnant of all," ascribing their decline to the fact that they had sold their stocks to sailors in Walvis Bay, in return for beads, brandy, and "other worthless articles" (Hoernlé 1923).

Processes Involved in Building up the Bone Accumulation

Fairly detailed information on Hottentot butchering techniques and eating habits became available from direct observations and from questioning of local people. An apparently typical goat-processing procedure will be described, based on observations made in the Soutrivier village during March 1966. The goat, a young male estimated to be one year old, in which the second molars were about to erupt, was led to a particular tree where slaughtering is normally carried out (fig. 10). Several Hottentots held the goat down on its left side while another cut its throat with a pocketknife. The blood was caught in an enamel basin and fed to two waiting dogs, who lapped it avidly. Once dead, the goat was suspended by its hind feet from an overhanging branch and the skin was removed complete, being split along the midventral line, along the insides of the limbs, and around the neck just behind the horns. It was salted and pegged out in the shade. The abdominal cavity was opened next, and the viscera were removed; the stomach was slit open, its contents emptied out, and its lining washed. This, together with the liver and kidneys, was said to be a delicacy. The intestine, once the contents had been squeezed out, was kept for making sausage. Other abdominal organs were fed to the dogs.

The front legs were then removed complete with the scapulae; the hind limbs were taken off with the innominate bones attached, by cutting through both the pubic symphysis and the sacroiliac joints. The feet were severed from the legs at the metapodial/phalangeal joints and were taken by some children, who cooked them themselves over a fire.

The ribs on one side of the carcass were separated at their vertebral articulations, and finally the head was removed, a knife being used to sever the axis from the third cervical vertebra. The atlas and axis vertebrae remained attached to the occiput.

All meat is normally cooked before it is eaten, either by boiling in large metal pots or by direct roasting over the fire. The head was dealt with as follows: the horns were broken off at their bases by sharp blows from an ax and were discarded. The dogs chewed the horn-core bases before rejecting them. The complete head was then boiled for several hours in a pot standing over the fire. All edible meat was picked from it and eaten, after which the brain case was smashed in the occipital region with a hammerstone for removal of the brain. The skull and mandibles were then passed to the dogs.

As eating progressed, all marrow-containing bones were broken. They were held on a rock anvil and hammered with another stone. Neither the anvil nor the hammerstone is an artifact in the usual sense of the word—they are simply suitable pieces of rock that happen to be at hand (fig. 11). The Hottentots seemed to habitually eat while squatting on the ground, and their utensils were pocketknives, rock anvils, and hammerstones. Their feeding behavior seemed to be a mixture of longstanding tradition and European influence.

Once discarded by the Hottentots, the goat bones were gnawed sporadically for many days by the dogs, all of which were jackallike in size (see fig. 10). Jackals themselves were extremely rare in the vicinity of the villages at the time of the study, and spotted hyenas had not been seen in the area for some years. It therefore seems likely that these scavengers do not enter the picture.

Pied crows are fairly common along the Kuiseb River and, when they can, will carry off scraps of meat, sometimes with bones adhering to them. On one occasion in 1966 a crow was seen flying from the Soutrivier village with most of a goat's tail in its bill.

When lying fully exposed on the gravel surface, bone fragments become bleached and degreased within about three months. Exposure to the sun weathers the bone surface, and a soft, chalky superficial layer develops. Gnawing of the bones by gerbils of the genus *Desmodillus,* whose burrows are often concentrated around old goat kraals, is not uncommon.

While collecting bone fragments from the vicinity of the Hottentot villages, I was surprised to find many pieces that appeared to be bone tools. They tapered to points (fig. 12) and showed wear and polish that had surely resulted from human use. In reply to my queries, the Hottentots denied that they made use of bone tools at all, and I had to find a different explanation for the remarkably suggestive appearance of these "pseudotools." Further observations showed that the worn and polished bones were specially abundant in areas regularly used by men and animals, such as around the Ossewater water hole, where 460 goats converge daily to drink (fig. 13), in the immediate vicinity of goat kraals, and along paths used by the Hottentots and their goats in the riverbed. In protected areas among rocks, for instance, the bones would develop their characteristic chalky surfaces but would lack signs of wear and polish. The mechanism of pseudotool production was therefore clearly related to the disturbance of the sand in which the bones lay by the feet of animals and men (Brain 1967*c*). The process may therefore be summarized as follows: bones come to rest on the sand, and their surfaces weather to a chalky consistency. Regular disturbance of the sand by the feet of animals abrades the chalky surface as it forms, leading to bones that are both worn and polished. If the whole piece of bone is lying in the disturbed sand zone, it is likely to

Fig. 9. Goats eating the fallen seeds of *Acacia albida* in the Kuiseb River bed.

Fig. 10. Slaughtering a goat in the Soutrivier village. A characteristic Hottentot dog is shown.

Fig. 11. A Hottentot in the Soutrivier village breaking a goat limb bone with a quartz hammerstone. The bones are broken to extract the marrow before being tossed to the dogs. From Brain 1976*b*. Courtesy C. K. Brain: copyright © 1976 by W. A. Benjamin, Inc.

acquire wear and polish on all surfaces (fig. 12*a*), but if some part of it is buried deeper this will remain protected, and only a part of its surface will be converted into a pseudotool. Selective abrasion of this kind has been observed on a number of metapodial and other limb-bone pieces that had been buried with their long axes vertical, or at least inclined. This meant that parts of such bones were buried too deep to be affected by superficial sand movements, so that wear and polish occurred on one end only (fig. 12*b*). Pseudotool production is not restricted to arid environments like that of the Kuiseb River, and the mechanism should be borne in mind when any interpretation of a bone assemblage is undertaken. Some of the worn bones showed numerous shallow striations, usually at right angles to their long axes. I suspect that these were made either by the hooves of goats or by chewing by goats. It may seem unlikely that noncarnivorous ungulates should regularly chew bones, but they certainly do, as Sutcliffe (1973*b*, 1977) has documented in detail (see also chap. 7).

It seemed advisable to be able to separate the damage done to goat bones by the Hottentots themselves from that caused by their dogs. Consequently I bought a goat from one of the inhabitants of the Soutrivier village and then gave it back to the people of the community. Over two days they consumed, in their traditional manner, all that was edible of the goat and returned the bones to me without allowing their dogs access to them. The goat was a subadult male, and the following is a summary of damage to its skeleton as a result of Hottentot feeding:

Skull: The 18 cm horns were broken off at their bases to allow cooking of the head; the occiput was smashed to allow removal of the brain; snout and palate were broken off as a unit; the mandible, in two halves, was undamaged.

Vertebrae: The head was removed by chopping through the axis; the atlas and part of the axis remained attached to the occiput. Very little damage was done to the other cervicals. Thoracic vertebrae suffered fairly extensive damage to their dorsal spines and transverse processes. Lumbars showed slight damage to their transverse processes; the sacrum was undamaged. Of the caudals, only the first survived, the rest having been chewed and eaten.

Ribs: Slight damage was suffered by the distal ends only.

Scapulae: Undamaged.

Pelvis: This had been chopped through the pubic symphysis and across the acetabulums. There was no other damage.

Humeri: Both shafts were broken transversely for the extraction of marrow. One proximal end was completely chewed away, the other was complete; both distal ends were undamaged.

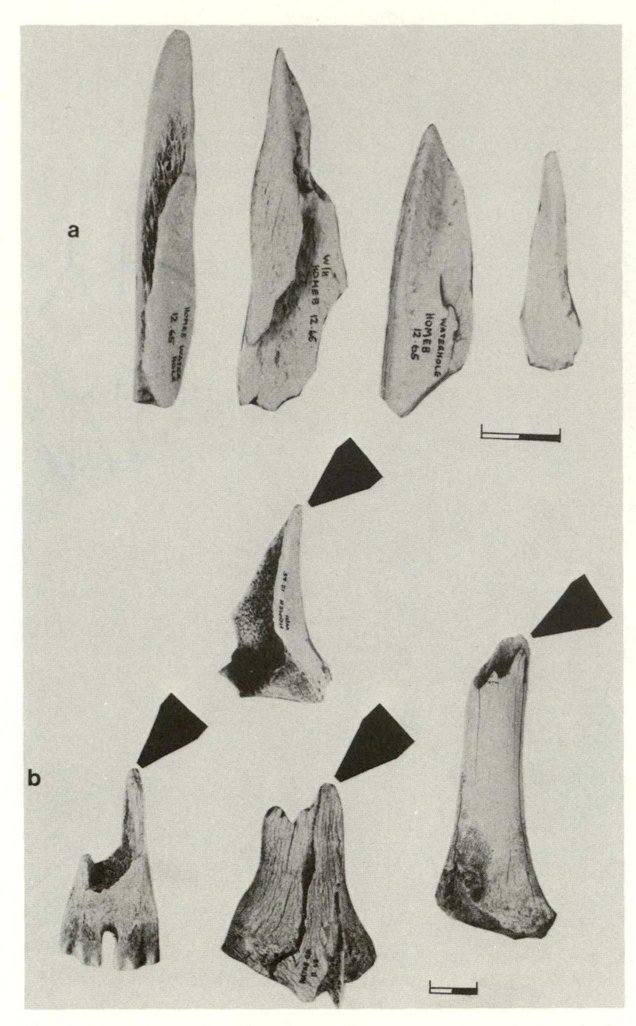

Fig. 12. "Pseudotools" collected around the Ossewater water hole in the Kuiseb River bed: *(a)* specimens showing overall wear and polish; *(b)* bones with localized wear at points indicated by the arrows.

Radii and ulnae: Both were severely shattered by stone impact.

Femurs: Heads and trochanters had been removed and proximal shaft ends chewed; both shafts were broken through the middle; both distal epiphyses were removed and the distal shaft ends were chewed.

Tibiae: Both shafts were broken through the middle, and there was some damage to each end.

Metapodials: All four proximal ends were complete, but all the distal epiphyses had been removed and the shaft ends severely chewed back.

Carpal and tarsal bones: Undamaged.

Phalanges: Undamaged.

Apart from the results of stone impact, it was surprising to find that the Hottentots were capable of inflicting considerable damage on bones with their teeth. Fifteen tail vertebrae were chewed and swallowed, and limb bones such as femurs and metapodials suffered severely at their ends. It is doubtful that the condition of Hottentot teeth is as good as that of hunter-gatherer peoples. The staple Hottentot diet, apart from occasional meat, is mealie-meal porridge, which very likely results in accelerated dental decay. It is to be expected that Stone Age people would have done even greater damage to bones with their teeth than do Kuiseb River Hottentots (this topic is mentioned again in chap. 7).

The Composition of the Bone Accumulation

The collection made in the Hottentot villages during 1965 and 1966 consisted of 2,373 goat bone pieces; the composition in terms of body parts is shown in table 3. The minimum number of individual goats that contributed to the sample, estimated on horns, is 190. Since the bone accumulation was originally described, however, I have found that this figure is deceptively high. The reason is as follows: in the extreme aridity of the Namib environment, horn is almost indestructible and lasts for many years after the last trace of bone has disappeared. Part of the original sample came from two deserted village sites that had not been occupied for more than ten years. These yielded horns to the almost complete exclusion of other skeletal parts. The average annual rainfall in the Kuiseb study area is less than 2.5 cm per year; but in more normal environments, with more than 25 cm of rain annually, horn disappears rapidly, exposing the core, which is composed of easily destructible spongy bone. It is now apparent that if data from the Kuiseb sample are compared with those for other bone accumulations, the incidence of horns will appear deceptively high. In this discussion, therefore, horns will be omitted.

After horns, the most numerous single skeletal parts are mandibles. I found that the 188 fragments could be divided into 53 left and 64 right half-mandibles. This indicates that a minimum of 64 individual goats contributed to the sample. Initially, I used aging criteria quoted by Cornwall (1956) for sheep, but J. P. White kindly pointed out (pers. comm.) that these figures were almost certainly low. Revised age estimates, based on Silver's (1969) figures for "rough goats" have therefore been made; they allow three months for first and second molars to come into wear after eruption. The aging criteria are therefore as follows, even though these estimates may be on the low side according to recent information on known-age goats given by Noddle (1974):

First molar unerupted: under 6 months
First molar in use, second unerupted: 9–12 months
Second molar in use, third unerupted: 15–30 months
Third molar in use: more than 30 months

On this basis, indicated ages for the left and right half-mandibles are given in table 4. The figures suggest that there was one goat in the sample under 6 months of age, 23 between 9 and 12 months, 7 between 15 and 30 months, and 35 more than 30 months of age. The goats had therefore been slaughtered largely when either just under a year of age or when fully mature. The Hottentots confirmed that this was their usual practice, the yearlings usually being the surplus males.

The combined feeding action of Hottentots and dogs has resulted in the disappearance of some parts of the skeletons and the survival of others. It has also resulted in some very characteristic damage to certain parts. Such damage may be summarized as follows:

Skull: The braincase has been broken open by stone impact to allow removal of the brain. In most

Fig. 13. The Ossewater water hole in the Kuiseb River bed, where "pseudotools" have been collected. Here bones have been worn and polished by the movement of sand around them. The movement has been caused by the feet of 460 goats that converge on the water hole daily.

cases the occiput or floor of the skull has been broken out, producing a receptaclelike fragment (fig. 14). In most cases the palates have been detached from the braincases complete, and damage to mandibles is usually confined to their lower margins, angles, and ascending rami.

Vertebrae: These show damage particularly on their spines and processes.

Ribs: These have generally been chewed at both ends.

Scapulae: Extensive damage is usually present on the blades.

Pelvises: These have characteristically been gnawed down to little more than their acetabular portions.

Damage to limb bones is best reflected by the presence or absence in the sample of their ends, to be discussed shortly. Shafts have typically been broken through by hammerstone impact, and spiral fractures are common (fig. 14). Carpal, tarsal, and phalangeal bones, when they occur, are generally undamaged.

Figure 15 diagrams the way different ends of the long bones are numerically represented in the sample. In the humerus, for instance, 82 distal ends were found, but not a single proximal end has survived. Again, the proximal end of the fused radius/ulna is represented by 62 pieces, compared with 19 distal ends. In the femur, the proximal end occurs twice as commonly as the distal end, but the position is reversed for the tibia, where the distal end is about six times as common as the proximal end. With both metacarpal and metatarsal, proximal ends are more commonly found than distal ends.

Survival and Disappearance of Skeletal Parts

The survival of parts of these goat skeletons is clearly based on their durability. Certain elements disappear when subjected to the chewing of Hottentots and their dogs, others do not. The percentage survival of different parts is therefore a measure of their resistance to this kind of destruction.

Working on a minimum number of 64 individual goats,

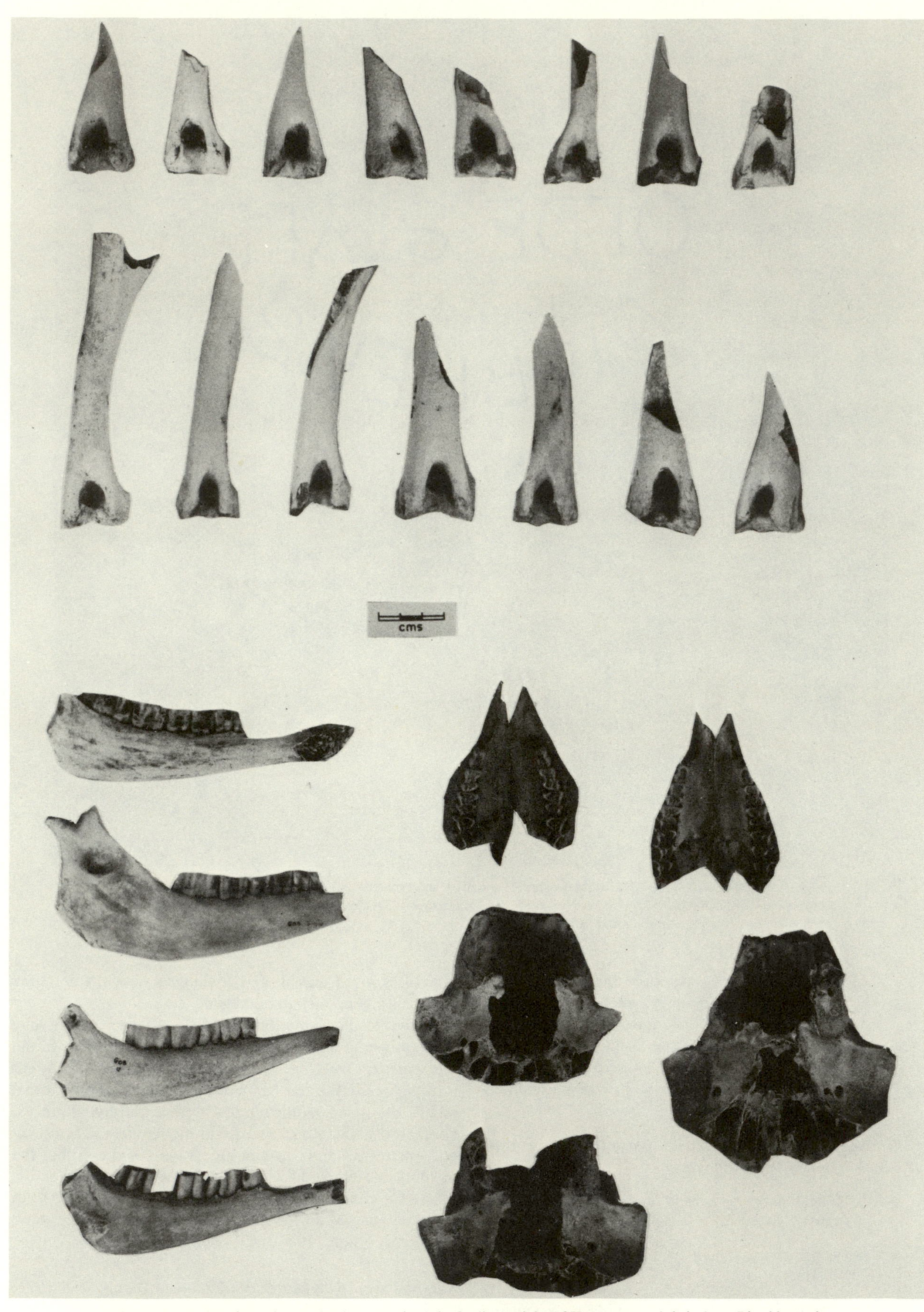

Fig. 14. Examples of goat bones that have survived the feeding activity of Hottentots and their dogs. Distal humeri are shown above, cranial pieces below.

it is possible to calculate the original number of each skeletal part that must have existed, and from this one may estimate the percentage survival of the part in the sample. In the case of ribs for, instance, 26 of which are found in a single goat skeleton, the original number contributed by 64 goats must have been 1,664. Only 170 have been found, indicating a 10.2% survival. Table 5 shows different parts of the goat skeleton arranged in descending order of survival, and these results are plotted in figure 18*a*. The parts most resistant to destruction are mandibles and distal ends of humeri, and these are the most numerous. Proximal ends of humeri and caudal vertebrae have proved so vulnerable as to have disappeared completely.

The Predictable Pattern of Survival in Limb Bones

It is clear that the parts of the goat skeletons that survive best are the unchewable ones. Nevertheless, for limb bones, percentage survival can be related in quantitative terms to particular qualities. In the course of this study it occurred to me that, in a sample derived essentially from immature animals, the survival of limb-bone ends could be related to the time at which the epiphysis of that bone fused to the shaft. Consider the humerus for instance, in which survival of the proximal end is nil, but that of the distal end amounts to 64%. The proximal epiphysis is likely to fuse to the shaft at about 36 months, whereas the distal epiphysis is fully fused by 12 months. An unfused epiphysis is linked to its shaft by a cartilaginous interface that is easily broken, making the shaft vulnerable to damage. This means that, when a year-old goat is eaten, the distal end of humerus will be fully ossified and unchewable, while and proximal end remains cartilaginous. Epiphyseal fusion times are known for domestic animals, but at the time of my earlier writing about the Kuiseb River goats (Brain 1967*b*, 1969*b*) data on goats were not available, and so figures for sheep were used. In the meantime, a study on ages of epiphyseal closure in feral and domestic goats has been carried out by Noddle (1974), whose figures are used here and are listed in table 6.

In addition to fusion times, structural considerations are very important. The proximal end of the humerus is wide, thin-walled, and filled with spongy bone; the distal end is comparatively narrow and compact. Such qualities may be quantitatively expressed in terms of specific gravity of each end of the bone. The experimental procedure was as follows: the shaft of a dry, defatted humerus was cut through at right angles to its axis, midway along the length of the bone. Each end was weighed individually, and the cut ends of the hollow shaft were filled with Plasticine. Any other openings were similarly filled. The volume of each end was then measured by submersion in water, and specific gravities were calculated. I found that the proximal end of a goat humerus had a specific gravity of approximately 0.6, and that of the distal end was about 1.0. There is a clear and direct relationship between the specific gravity of the end of a long bone and its percentage survival.

Table 7 gives figures for percentage survival, specific gravity, and fusion time for each end of the goat limb-bones listed. These figures are plotted in figure 16. Percentage survival of a part is related *directly* to the specific gravity of that part, but *inversely* to the fusion time expressed in months.

The conclusion to be drawn is simply that survival is not haphazard but is related to the inherent qualities of the parts.

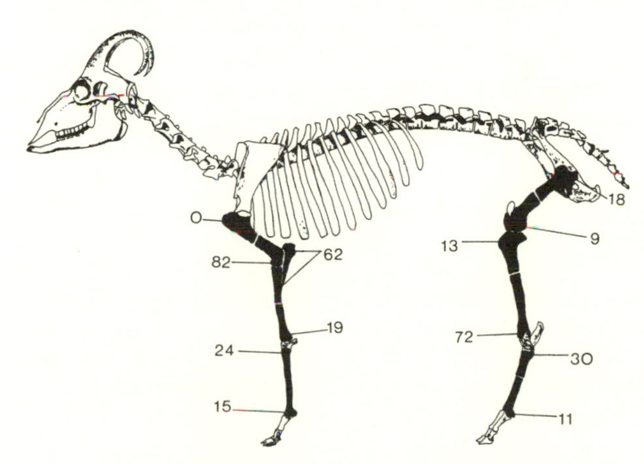

Fig. 15. Diagram of a goat skeleton showing numbers of each end of the principal limb bones found in the Kuiseb River sample.

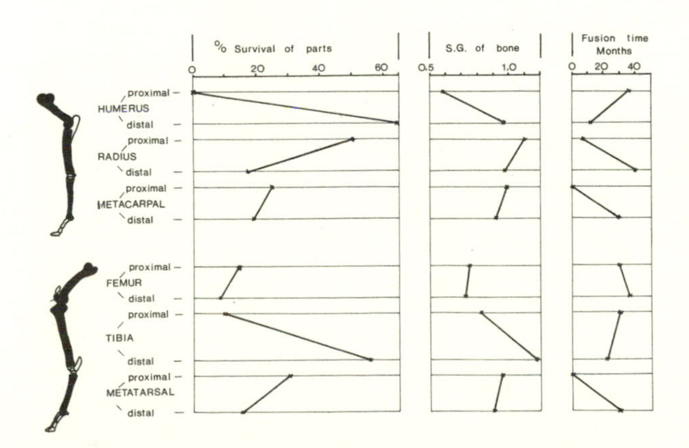

Fig. 16. The percentage survival of proximal and distal ends of goat limb bones correlated with the specific gravities of the bone pieces and times of fusion of the epiphyses to the shafts.

Survival of Parts in the Makapansgat Bone Sample

Dart's (1957*a*) analysis was undertaken on remains from 293 antelopes. His estimation of minimum numbers of individual animals of different sizes was as follows: large antelopes, based on 74 radius fragments, 39 individuals; medium antelopes, based on 238 humeral fragments, 126 individuals; small antelopes, based on 191 mandible pieces, 100 individuals; and very small antelopes, based on 53 mandible fragments, 28 individuals. Using the total number of 293 individuals, it has been possible to calculate the percentage survival of different parts of the skeleton as was done for the Kuiseb River goat bones. Skeletal parts, lised in descending order of survival, are given in table 8 and plotted graphically in figure 17.

The Makapansgat/Goat Comparison

If we compare the plots of the Kuiseb bones (fig. 18*a*) with those of the Makapansgat fossils (fig. 17), we see that the form of the two histograms is similar. In both, parts with the highest percentage survival are mandibles, followed by distal humeri. At the lower end of the survival curve in both collections are such parts as thoracic and caudal vertebrae. But, in spite of a broad similarity of the two histograms, the detailed order of survival of parts differs. For direct comparison, the percentage survival figures for the Makapansgat sample are replotted in figure 18*b* so that they follow the order laid down by the goat bones. Although the two histograms are not identical, we can see that the trends in survival order are broadly similar.

When comparing these results one should bear in mind that the Makapansgat sample is made up of bones from animals ranging in size from eland to steenbok. They have almost certainly been subjected to a variety of destructive treatment, including feeding action of primary predators and scavengers. By contrast, the goat-bone sample is made up of bones from one species of small bovid, subject only to feeding by men and domestic dogs. In view of this, the overall similarity in composition of the bone collections is remarkable. It reflects the predictable pattern of survival that manifests itself when whole bovid skeletons are subjected to destructive treatment.

The differential survival of anatomical parts in archaeological contexts has been further explored in an important contribution by Binford and Bertram (1977), whose aim was to obtain information on assemblages of

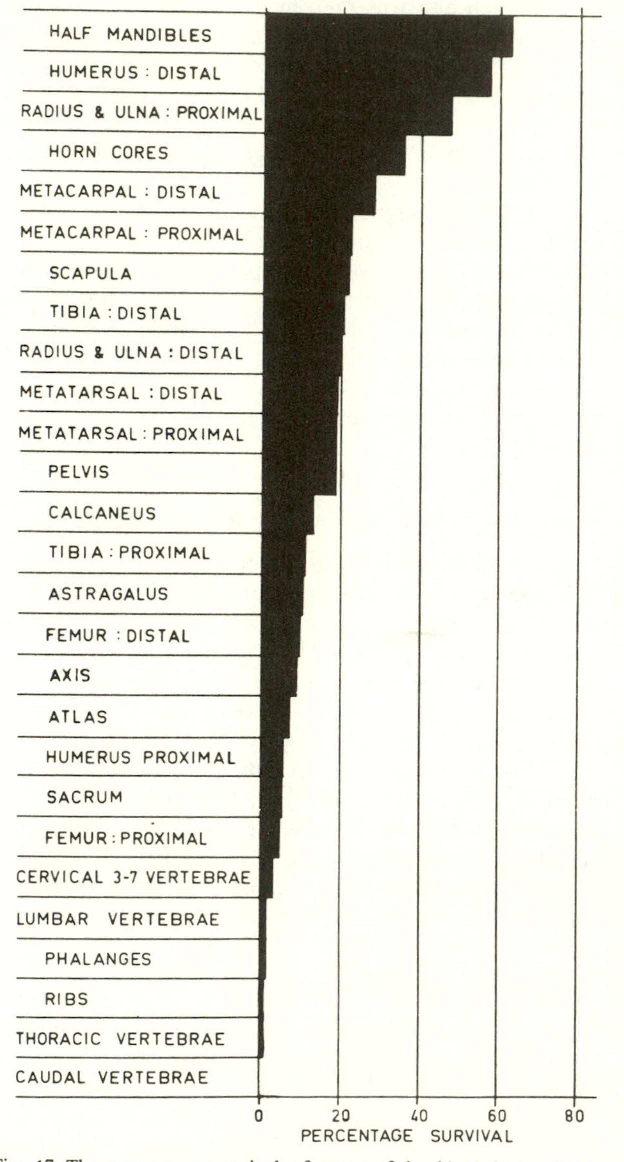

Fig. 17. The percentage survival of parts of bovid skeletons from Makapansgat. The sample consisted of bones from a minimum of 293 individuals. From Brain 1976*b* (data reworked from Dart 1957*b*). Courtesy C. K. Brain: copyright © 1976 by W. A. Benjamin, Inc.

bones left after a known attritional agent had acted upon them. They also aimed to measure characteristics of bones relevant to their survival. Significant observations were made on the food remains of Nunamiut Eskimos and their dogs as well as on those of Navajo Indians in Arizona. Binford and Bertram were able to confirm that bone density was a vital factor in survival, but that bone densification was not an allometric growth process and that bones changed their density at different rates with age. Applying their results to the Makapansgat assemblage, they wrote (1977, p. 148): "We conclude that, based on the survivorship of anatomical parts at Makapansgat, there is absolutely no basis for the assumption that the hominids present played a behavioural role in the accumulation of the deposit."

R. R. Inskeep (pers. comm.) has drawn my attention to an interesting, and much earlier, parallel to the Kuiseb River goat-bone study. In his book *Early Man in Europe,* Rau (1876, p. 111) described the contents of Stone Age refuse dumps, or kitchen middens, in Denmark, pointing out that the bones were the discarded food remains of the people and their domestic dogs. To verify this assumption, Rau described how Professor Steenstrup locked up some dogs, restricting them to a diet of bones, and thereby "ascertained that all the bones rejected by the dogs were the same that are present in the kitchen-middens, while the bones or portions of bones devoured by them are correspondingly missing there."

These observations were also reported on by Lubbock (1865). Quite clearly, the characteristics of bones described here as influencing the survival or disappearance of such parts are not the only significant ones. The strength of attachment of one skeletal element to another will also affect the potential survival of each part. The strength and durability of interpart attachments will be indicated by the natural sequence of disarticulation of a skeleton after the animal's death. Detailed information on such disarticulation sequences is now available from observations made by Andrew Hill (1975) on skeletons, particularly those of topi (*Damaliscus korrigum*), in northern Kenya. Hill has described twenty-one stages in the disarticulation process, starting with the first, where the forelimb, including the scapula, separates from the rib cage. Second to detach are the caudal vertebrae, and thereafter all the other parts of the skeleton separate in an order determined by the nature of their attachments. Last to separate are the cervical vertebrae, whose survival doubtless is enhanced by the durability of their ligamentous connections.

It is not unusual for the separate specimens in a bone accumulation to have been transported by water before their deposition, and a factor of importance here is the ease with which different skeletal parts can be carried in an aqueous current. A remarkable fossil bone concentration of Pliocene age has been studied in the Verdigre quarry of northeastern Nebraska by Voorhies (1969). In an attempt to understand the reasons for the presence and absence of different skeletal parts there, Voorhies undertook stream-table experiments that suggested that current sorting was probably responsible for the scarcity of elements such as ribs, vertebrae, sacra, and phalanges compared with rami, metapodials, and tibiae.

Important taphonomic studies have recently been made at the East African hominid localities by Kay Behrensmeyer (1975*a,b*). She has pointed out that bones are

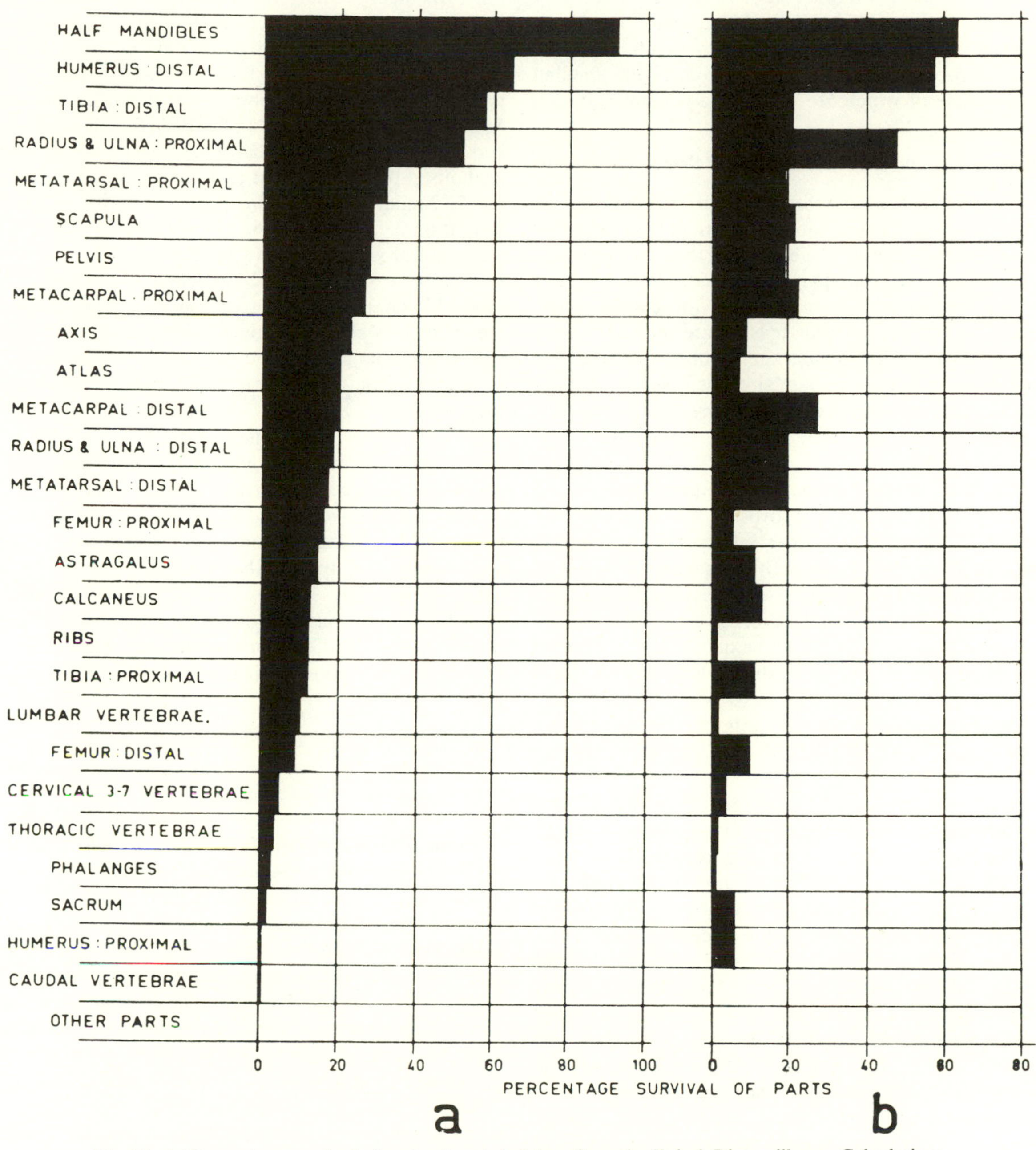

Fig. 18. *(a)* Percentage survival of parts of goat skeletons from the Kuiseb River villages. Calculations are based on a minimum of 64 individuals. *(b)* Percentage survival of parts of bovid skeletons from Makapansgat, arranged in the same order as for *(a)*. From Brain 1976*b*. Courtesy C. K. Brain: copyright © 1976 by W. A. Benjamin, Inc.

not necessarily buried where an animal died, and that they are even less likely to have been buried where the animal lived. In fact, it is important to be able to distinguish between *autochthonous* and *allochthonous* assemblages: the former consist of bones that have not been moved far from the *general* environment where the animals died, whereas the latter have been significantly transported. In the course of water transport, certain parts of hominid skeletons are transported more readily and farther than others, as is indicated by flume experiments (Boaz and Behrensmeyer 1976). Complete craniums proved to be the fastest-moving elements, while cranial fragments and isolated teeth fell into the lag group, to become buried in aggrading parts of a fluvial system. Behrensmeyer (1975*b*) suggested that such taphonomic sorting of skeletal parts could have accounted for the large concentration of hominid teeth recorded from the Shungura Formation of the Omo basin in Ethiopia by Howell and Coppens (1974). Selective transport of different stone artifact types in water is also an important factor, as experimental work by Glynn Isaac (1967*b*) has shown.

When assemblages of hominid skeletal remains from Olduvai beds I–IV are compared with those from the Shungura and Koobi Fora formations (Behrensmeyer 1975*b*), it is found that the composition of the collections relate to the degree of taphonomic sorting to which each has been subjected. The Omo fossils appear to have suffered the most from transport, sorting, and reworking, as is to be expected in a large, active river situation. Olduvai fossils have been affected the least, having been buried in lake-margin and deltaic environments, while the fossils from the Koobi Fora Formation indicate effects intermediate between these two extremes.

There are at least two other ways in which skeletal-part disproportions are often introduced into bone assemblages. These are human butchery practices, to be discussed in chapter 3, and selective carnivore feeding patterns, some of which are mentioned in chapter 4.

Comparative Vulnerability of Bovid and Primate Skeletons

A striking feature of the Sterkfontein valley fossil assemblages is that antelope are represented by a far wider range and abundance of skeletal parts than are primates. In fact, postcranial remains of hominids, baboons, and monkeys are rare in comparison with cranial ones, whereas bovid postcranial fossils are comparatively common. Dart (1957*a*) made a similar observation for the Makapansgat assemblage, and, since he assumed that the bones were hominid-collected, he likewise inferred that the australopithecines had been "professional decapitators" and headhunters.

I will now discuss some observations relevant to bovid/primate skeletal disproportions, made incidentally when I had intended to gather information on another theme. Early in 1966 I decided to examine the kind of damage done to prey skeletons by cheetahs, in the hope that this might throw some light on the way saber-toothed cats treated bones. The postcanine dentition of a cheetah is specialized for slicing meat and includes a very small bone-crushing component. In this respect cheetahs and sabre-toothed cats are similar, although the former are smaller than most of the known southern African machairodonts.

During February 1966 the Natal Parks Board was acclimatizing six adult cheetahs in a large enclosure in the Umfolosi game reserve, before their release there. With the kind cooperation of the Natal Parks Board staff, particularly R. C. Bigalke and John Clarke, I was able to make detailed observations of the cheetahs' feeding behavior and food remains. It was soon apparent that the cheetahs did very little damage to the skeletons of the antelopes provided as food. The remains of an adult female bushbuck, *Tragelaphus scriptus,* are shown in figure 19*a*. As can be seen, the whole skeleton is there, even though six adult cheetahs had been feeding on it for most of one day and part of the succeeding night. Damage was restricted to the distal ends of the ribs, the blades of the scapulae, and the vertebral processes.

Fieldwork in the Kruger National Park during 1967–68 allowed further observations on cheetah food remains in natural circumstances, although these were continually hampered by interference from jackals and hyenas. In fact, it was extremely difficult to find remains that had not been modified by scavengers. I was fortunate to have the help of A. C. Kemp, who was working in the field close to Satara at that time, studying hornbills. He took me to the few cheetah kills we were able to locate. Remains of an adult male impala that had been killed and eaten by several cheetahs in the bed of the Timbavati River during September 1968, which we studied before scavengers arrived, are shown in figure 19*b*. As with the Umfolozi bushbuck, the skeleton was virtually undamaged.

The problem of interference by scavengers in the cheetah field observations suggested to me that controlled feeding experiments would, after all, provide the most satisfactory results, and the opportunity to undertake these presented itself at Valencia Ranch, in South-West Africa, through the kindness and cooperation of Attila F. Port, the ranch owner. The Valencia Ranch environment is described in some detail in chapter 4, since studies were also made there of leopard food remains. As is shown in figure 20, two enclosures had been constructed close to the Valencia Ranch farmhouse. The smaller enclosure, 55 m by 28 m, contained a six-year-old female cheetah that had been in captivity for four years; the larger, 55 m by 37 m, housed three cheetahs: two adult males caught on 20 and 23 January 1968 and a female, somewhat smaller and caught on 25 January 1968. These three had formed part of a natural group and had been trapped at a particular tree close to the Hakos River leopard breeding lair described in chapter 4. This tree had an inclined trunk, and the cheetahs visited it regularly, climbing up the trunk and defecating from an elevated position. I am not sure if this is characteristic cheetah behavior, but the regular visits the cheetahs made to the tree certainly aided their capture. In fact the two remaining members of the group, both adult males, were caught on 13 and 20 May 1968 and added to the three already in the enclosure.

The cheetahs were fed largely on karakul sheep, and between March and September 1968 many observations were made, confirming that cheetah damage to adult sheep skeletons was minimal. On 22 March, detailed observations were made as the three cheetahs in the large enclosure fed on the complete carcass of a freshly shot springbok, the following being an extract from my field notes at the time:

> *22 March 1968:* an adult male springbok weighing 103 lb was shot on the farm Friedental and placed in the three-cheetah enclosure at 2 P.M.—a sunny and rather hot day. It was immediately dragged into the shade by the two males while the female threatened us. All three cheetahs started to feed on the ventral surface of the body (fig. 21*a*): one worked from there onto the inner parts of the hind legs, one chewed away the sternal rib attachments, and the third entered the thorax. The intestines were partly eaten and partly discarded a few meters to one side. The two male cheetahs were obviously dominant over the female, who was threatened several times. One male left after 1 hr of feeding but returned 15 min. later. All activity had ceased by 5 P.M.
>
> The remains were left in the enclosure until Sunday, 24 March, when they were removed and photographed (fig. 21*b*). A good deal of the skin was still intact, but this was cut away before the photo was taken. Damage was restricted to the ends of the ribs and the vertebral processes: the right scapula and humerus had been damaged when the springbok was shot.

The pattern of damage to this skeleton was typical of what I had observed previously, but two days earlier I had obtained some very different results in a feeding experiment. Although we had spent a considerable time searching for a suitable antelope to feed to the cheetahs, on that occasion we found none and instead shot a male baboon who had been taunting us from the top of a nearby cliff. The baboon was offered to the three cheetahs, with quite unexpected results:

> *20 March 1968:* The body of an adult male baboon weighing 29.5 kg was placed in the enclosure at 9:05 A.M. It was immediately taken by the two male cheetahs and carried by its arms to the shade of a tree. All three cheetahs started to feed on the ventral sur-

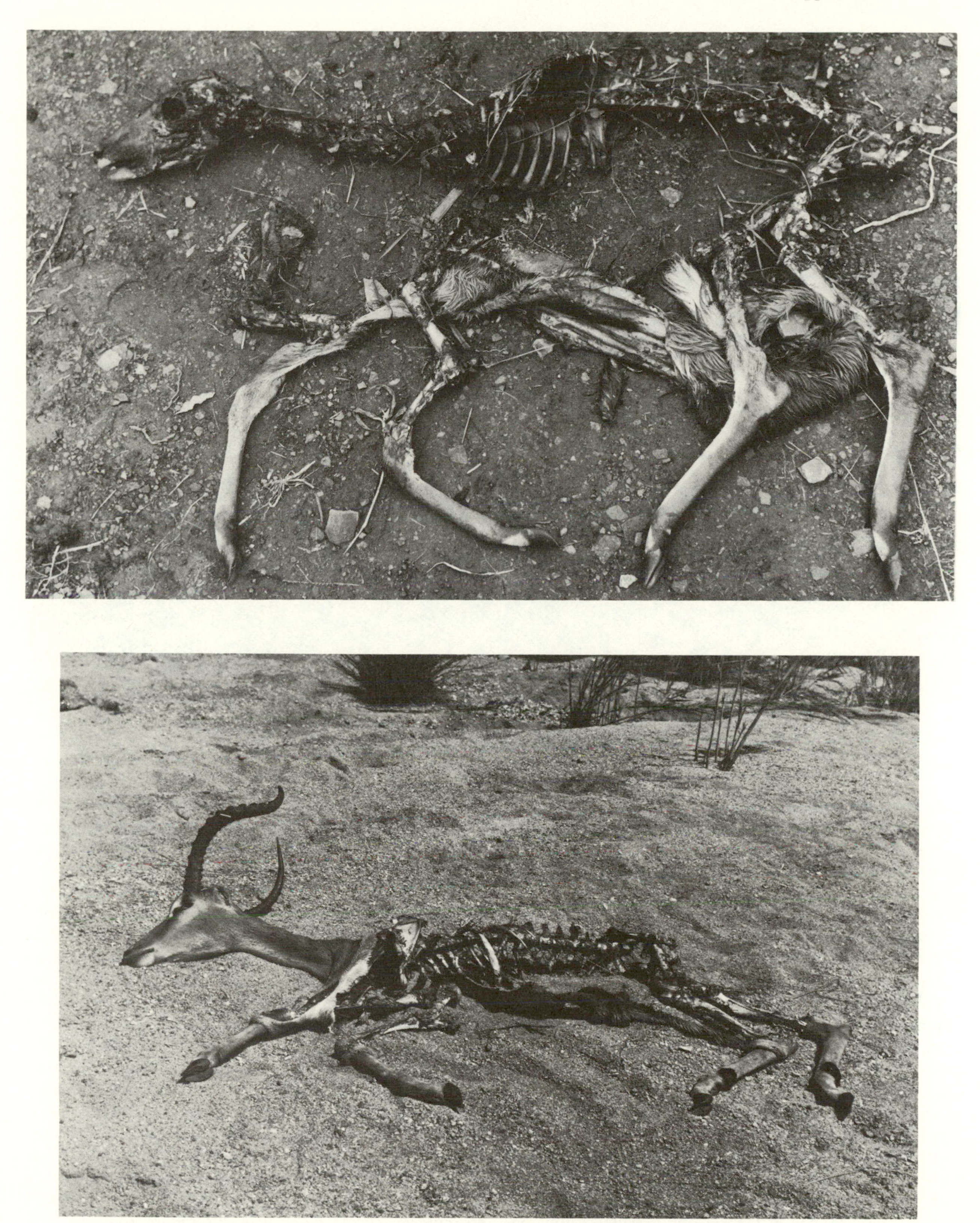

Fig. 19. *(a)* Remains of an adult female bushbuck in the Umfolosi game reserve, Zululand, after feeding by six cheetahs (from Brain 1978). *(b)* Remains of an adult male impala left by cheetahs in the Timbavati River bed, Kruger National Park.

Fig. 20. The Valencia Ranch environment in South-West Africa where the cheetah-feeding experiments were undertaken. The cheetahs were confined in the enclosures visible in the middle foreground.

face of the abdomen; the viscera were removed and part of the intenstine eaten. The rib cage was quickly chewed away and the vertebral column simply crunched up and swallowed—quite unlike the antelope situation. As the vertebral column was destroyed, the pelvis and both hind limbs were removed by one cheetah and carried a short distance away. The sacrum was eaten so that the femurs, still articulated into the innominates, were separated. One cheetah left the baboon after 1 hr, 10 min, the others remained 15 min longer, then left, but all three returned intermittently throughout the day.

The remains were removed and photographed the next day (fig. 22*a*); the entire vertebral column, from atlas to first caudal, had disappeared, as had most of the ribs. The innominate bones showed damage round the edges, and both knee joints had been disarticulated and chewed.

The disappearance of the vertebral column in this baboon carcass came as a complete surprise and suggested that a primate backbone was less resistant to carnivore chewing than its bovid counterpart. To test this suspicion, an adult sheep of almost precisely the same liveweight as the baboon was fed to the cheetahs when they showed equivalent signs of hunger. The result is shown in fig. 22*b;* although the innominates had been separated from the sacrum, the vertebral column was intact except for the tail, which had been consumed.

During 1968, many controlled feeding tests were done with the cheetahs on baboons and bovid prey. A characteristic feeding sequence is shown in figure 23*a–c*, in which the prey was a subadult male baboon whose vertebral column was also entirely eaten except for the tail.

I do not intend to suggest that cheetahs would ever have contributed food remains to bone accumulations in African caves—they are simply amenable carnivores on which experimental work can be done. Under controlled conditions they are able to do far more damage to the skeleton of a baboon than to that of a bovid of equivalent liveweight. Most striking is the fact that a primate's vertebral column is far less resistant to carnivore damage than is an antelope's. The reason is clearly the comparative structure and robusticity of the two backbones—individual bovid vertebrae are comparatively resistant structures with more robust spines and processes. This is of the greatest significance when bone accumulations composed of primate and bovid remains are interpreted. *Primate vertebral columns will suffer more from whatever carnivore action they are subjected to than will bovid ones.* In the same way, primate hands and feet are far more palatable to a carnivore than are bovid equivalents.

In view of these observations, it is in no way surprising that antelopes should be represented in a fossil assemblage by more abundant and more complete skeletal parts than are primates of similar size. The key lies in different degrees of resistance to equivalent carnivore action.

Fig. 21 *(a)* Three cheetahs feeding on a springbok in controlled circumstances. *(b)* Remains laid out after the cheetahs had left them. Damage is restricted to the ribs and vertebral processes. From Brain 1978.

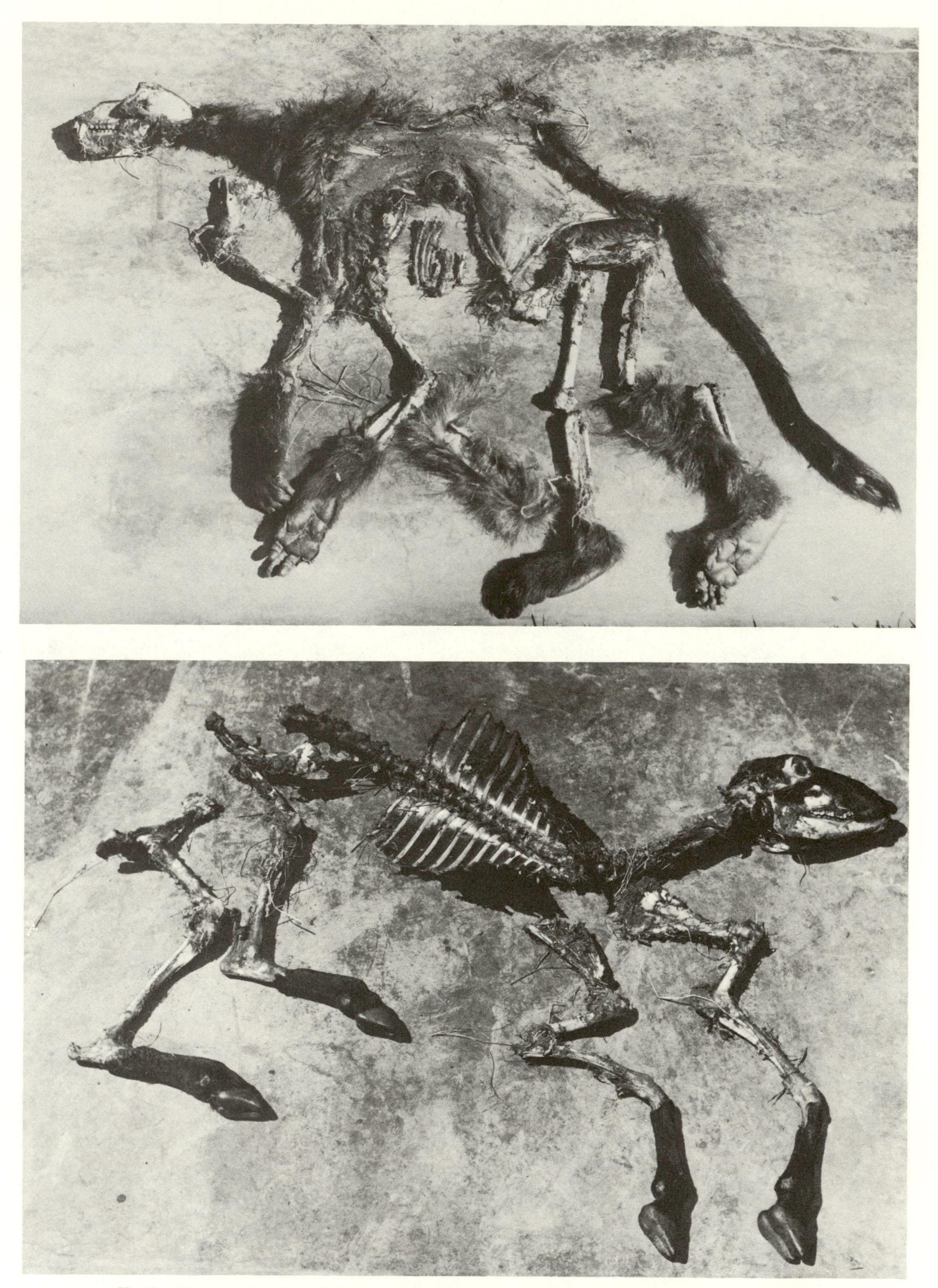

Fig. 22. *(a)* Remains of an adult male baboon after feeding by three cheetahs in controlled circumstances. *(b)* Remains of a karakul sheep of the same liveweight as the baboon after feeding by the three cheetahs. The pattern typical of cheetah damage to bovid skeletons is apparent. From Brain 1978.

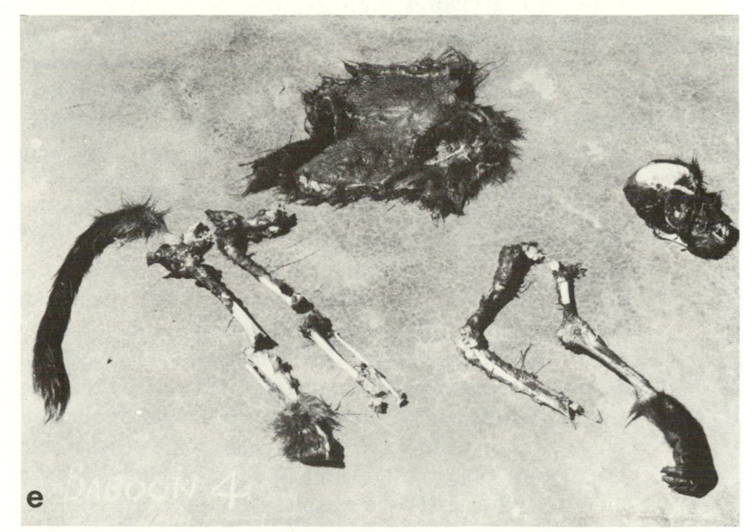

Fig. 23. Three cheetahs feeding on a subadult male baboon in controlled circumstances. *(a)* The cheetahs approach. *(b)* Two of them investigate the freshly killed baboon. *(c)* They carry the baboon off between them. *(d)* Feeding takes place in the shade. *(e)* The baboon remains, laid out after all feeding had ceased; almost the whole vertebral column has disappeared.

3 Food Remains of Primitive People in Southern African Caves

> To date, the hunting way of life has been the most successful and persistent adaptation man has ever achieved. Nor does this evaluation exclude the present precarious existence under the threat of nuclear annihilation and the population explosion. It is still an open question whether man will be able to survive the exceedingly complex and unstable ecological conditions he has created for himself. If he fails in this task, interplanatary archaeologists of the future will classify our planet as one in which a very long and stable period of small-scale hunting and gathering was followed by an apparently instantaneous efflorescence of technology and society leading rapidly to extinction. "Stratigraphically," the origin of agriculture and thermonuclear destruction will appear as essentially simultaneous. [Lee and Devore 1968, p. 3]

Hominids are changeable creatures and, in comparison with most carnivores, are very tiresome. A leopard of two million years ago not only looked like one of today, but probably behaved in the same way. Given equivalent prey, it would presumably have left the same food remains then as now. Not so the hominids; *Homo habilis* was a very different animal from *Homo sapiens,* not only in the flexibility of its behavior, but also in the traces of its former presence. From the point of view of this investigation, the characteristics of early hominid food remains are more relevant than those of their late descendants, but in paleontology one has to be content with what one can get. Food remains resulting from the early phases of human existence are understandably less common than those from later phases, and so it is that the traces of Middle and Later Stone Age people are the better documented. In an attempt to gain some insights into the nature of primitive human food remains, I have analyzed extensive bone samples from four caves or rock shelters that, on archaeological grounds, appear to have been intensively occupied by Stone Age peoples. The sites have a wide geographic scatter in southern Africa, one being in Zimbabwe, one in the Transvaal, one in the Cape, and one in South-West Africa. Some details of these sites and bone samples will now be considered in an attempt to discern consistent patterns between them. They will then be examined in the context of results from other sites, obtained by other investigators.

Pomongwe Cave

The cave is in the Matopo Hills south of Bulawayo, an area of rugged granite hills that will be discussed further in connection with leopards and black eagles (chap. 4) and Cape eagle owls (chap. 6); the cave has formed near the base of a very large granite dome, whose converging sides form a natural tree-filled amphitheater at the cave's mouth (fig. 24). The name Pomongwe appears to be of Karanga origin, perhaps derived from the word *mamongwe,* the term for a small wild melon of whose shape the granite dome is reminiscent (C.K. Cooke 1963).

The cave, which has resulted from negative spheroidal weathering of the granite, is dome-shaped and dry—a place exceptionally suitable for human habitation and one that has been used for many thousands of years. In plan (fig. 25*a*) it is roughly triangular with a gently sloping floor (fig. 25*b*) descending into a delightful tree-filled glade. Two springs arise on the granite hill and form pools close to the cave's entrance almost throughout the year. With the abundance of wildlife in the area, I can think of few more desirable places to live.

Three separate excavations were conducted by C. K. Cooke and his helpers, the first commencing in June 1960 (C. K. Cooke 1963). Bedrock was reached at a depth of 13 ft, 6 in, or 4.12 m, and the profile was found to be made up to twenty-seven definable layers, the upper ones rich in ash of various colors, while the lower part of the section consisted of granitic sand with a certain amount of bone breccia. A striking feature of the profile was a layer rich in granite spalls, typically occurring 1 m below the surface. The spalls had become detached and had fallen from the roof and walls of the cave, perhaps in response to some changing conditions of temperature or humidity; according to currently available dates, this happened between 16,000 and 20,000 years ago.

On cultural grounds the profile was originally divided into ten stratigraphic units (C. K. Cooke 1963). Reading from the bottom up, these were Proto-Stillbay, layers 27–21; Stillbay, 20–13; Scraper 1, or Magosian, 12–10; Scraper 2, 9–7; and Wilton, 6–surface. The cultural phases have undergone many changes in nomenclature (e.g., C. K. Cooke 1966, 1968, 1969), which I am not competent to assess, and the subdivisions should probably now read: Bembesi, layer 27; Charama, 26–22; Bambata, 21–13B; Tshangula, 13A–10; Pomongwe, 7; and Matopo, 6–1. Figure 26 shows subdivisions of the profile into levels and cultural units, linked to radiocarbon dates and to the occurrence of the 17,756 bone fragments and 39,032 stone artifacts recovered from the excavation.

In recent years, radiometric dating has suggested that many of the cultural phases in the southern African Pleistocene are far more ancient than had been previously assumed. In his important paper on the ecology of early man in southern Africa, Klein (1977) expressed the view that the Middle Pleistocene ranged between one million and 130,000 years ago and that its characteristic stone culture included hand axes, cleavers, and associated flake tools, which are referred to the Acheulean Industrial Complex (see also Deacon 1975). This was followed by the Middle Stone Age industries that lack hand axes and cleavers but are characterized by a variety of flake tools, especially scrapers, points, and denticulates. The time span is 130,000 to 40,000 years. Post-Middle Stone Age industries were practiced between 40,000 years ago and the present, although they are best known from the past 20,000 years, as there was often a nonoccupational hiatus in southern African caves between 40,000 and 20,000 years B.P. This was probably brought about by population movements in response to low temperatures and environmental changes during the middle and later parts of the last glacial.

In the framework just mentioned, levels 1–8 at Pomongwe would fall within the Post-Middle or Later Stone Age, while levels 7–10 should be regarded as Middle Stone Age. However, I have reservations about the reliability of the radiocarbon dates for the Pomongwe samples and, for this discussion on food remains, I group levels 1–4 as Later Stone Age, 5–6 as Middle Stone Age, and 7–10 as Pre-Middle Stone Age.

The excavated area of the cave floor is not sufficient to allow a full reconstruction of how the space was used by the Stone Age people. Cooke concluded that most of the artifacts were made near the cave entrance and also in certain areas close to the walls where groups of rocks

Fig. 24. The large granite dome of Pomongwe hill in the Matopos of Zimbabwe. The cave is at the base of the dome behind the screen of trees.

appear to have served as convenient seats. The main fire-making areas were in the central area of the cave floor.

The most striking feature of the bone sample from Pomongwe was the extreme fragmentation of the individual pieces, there being virtually no complete bones (fig. 27). Of the 17,756 pieces, 9,549 were found to consist of long-bone flakes, as defined in chapter 1, and there were 981 fragments too incomplete to be placed skeletally or taxonomically. As listed in table 9, 5,364 fragments could be identified, with varying degrees of confidence, as to skeletal part; there were 1, 357 pieces of tortoise carapace and plastron; 366 ostrich eggshell pieces, 136 land-snail fragments, and 3 mussel shells.

The fragmentation of the bones at Pomongwe can be attributed largely to two factors: purposeful breakage by Stone Age people to extract marrow, and trampling by people walking around in the cave. I am inclined to think that the influence of trampling would be less for the Pomongwe deposit than for others, such as Wilton Shelter, since the soft Pomongwe ash layers would presumably have cushioned and protected the fragments. The overall size distribution of bone fragments in an accumulation will indicate fragmentation but will of course be influenced strongly by the initial sizes of the complete bones and by the sizes of the animals that contributed them. More meaningful as an index of fragmentation are the sizes of bone flakes derived from the limb bones of the prey, particularly antelope. Body size of the original animals will of course also constitute a factor here, but to a lesser extent than with the whole bone sample.

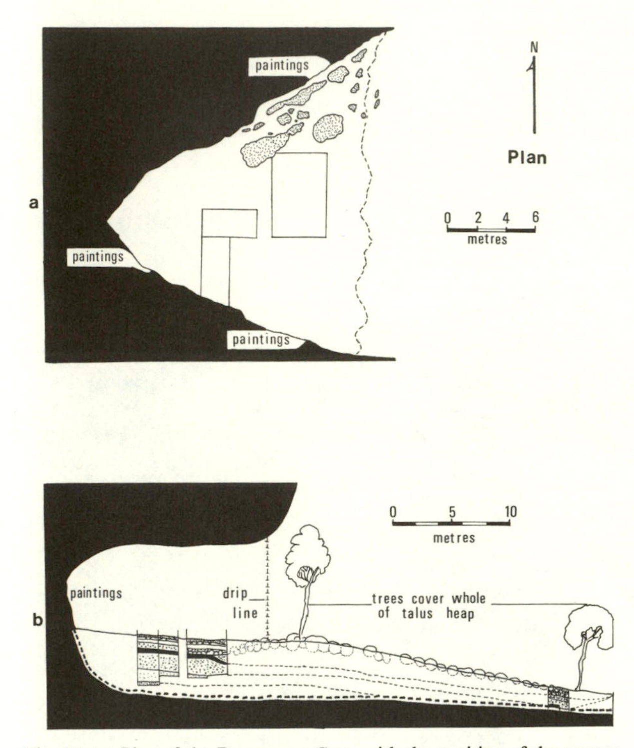

Fig. 25. *(a)* Plan of the Pomongwe Cave with the position of the excavations shown. *(b)* Section through the cave and talus heap. Redrawn after C. K. Cooke (1963).

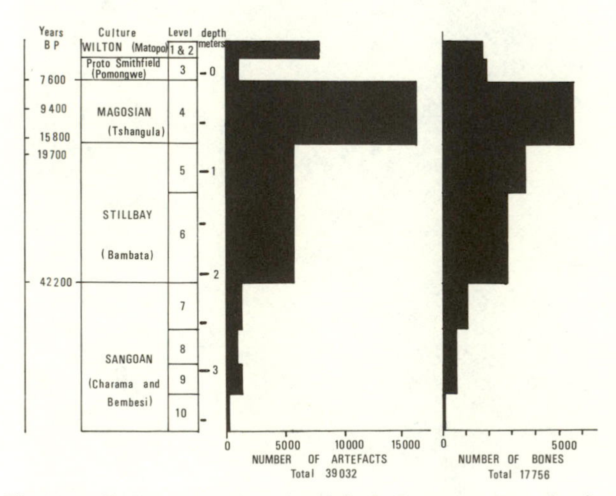

Fig. 26. Profile through the floor deposit in the Pomongwe Cave showing the relationship of artifact numbers per level to those of bones.

As mentioned earlier, of the 17,756 bone pieces in the entire sample, 9,549, or 53.8%, consisted of bone flakes. Each of these was individually measured, and the results are given in table 10. As figure 28 shows, the length distributions of the bone flakes from the various Pomongwe layers are remarkably constant, though the samples were broken by different people whose combined lives spanned an immense period of time. It seems clear that the efficient use of prey skeletons as food by Stone Age people was not random and haphazard. Long bones for instance, were systematically broken with hammerstones so that no marrow was wasted. This procedure at Pomongwe has resulted in an abundance of bone flakes between 1 in and 2 in (2.5–5.2 cm) long. As will be discussed in chapter 7, the production of these bone flakes causes extremely few marks on the bone pieces that indubitably indicate human action—the same is true of bone flakes that I have made experimentally with a stone hammer and anvil. However, their association with very large numbers of stone artifacts and their burial in ash from artificial hearths suggests that these are indeed food remains of people.

Although many of the bone fragments from Pomongwe were charred to some extent, the actual frequency of charring was not determined. To do this, each piece would have had to be broken so that a fresh surface could be examined. This was not thought feasible or justifiable.

On the basis of all bone fragments that could be placed taxonomically, minimum numbers of individual animals that must have contributed to the sample were estimated for each level. The list for the combined Pre-MSA levels is given in table 11, for the MSA levels, in table 12, and for the Later Stone Age horizons, in table 13. Obviously the various animals on these lists, depicted in figure 29, varied considerably in the quantity of edible meat they would have provided. An attempt has therefore been made to estimate the percentage contribution each would have made if the sample were a true reflection of the overall meat diet of the hunters. Quite clearly, such an assumption is dangerously untrue, owing to particular butchery practices, portability of kills, and similar factors that will be discussed shortly. But, accepting the existence of these biases, we can gain some rough indication of what the people ate *inside* the cave. Mean liveweights for each of the animal taxa listed are given in the tables, and the assumed contribution to the diet is then calculated, working on the assumption that 70% of each animal's weight would have been edible. In his paper "A Method of Calculating the Dietary Percentage of Various Food Animals Utilized by Aboriginal Peoples." White (1953*a*) concluded that 50% by weight of animals such as bison, elk, deer, and antelope would have been edible, as opposed to 70% for others like pigs, raccoons, and badgers. In view of the intense fragmentation to which the Pomongwe people subjected their prey skeletons, I have taken 70% edibility for all animals on the lists.

Table 14 gives figures for the percentage contributions to the diet of the people at Pomongwe made by the various animal groups, and this information is presented graphically in figure 30. Most of the meat eaten came from medium-sized antelopes (class III) and nonbovid mammals, particularly dassies, zebras, and warthogs. Although many tortoises were collected, their contribution to the overall diet was very small.

In an analysis of this sort, information on the body parts by which each kind of animal is represented would be useful in reconstructing something of the behavior of the hunters. Unfortunately, the extreme fragmentation of the remains makes this impracticable at Pomongwe for all the larger animals. With dassies, however, the situation is more hopeful. Two species, *Procavia capensis* and *Heterohyrax brucei*, occur very commonly among the granite boulders in the Matopos, and both were obtained in large numbers by the Stone Age people, either by hunting or by snaring. The two forms may be separated easily by cranial remains, but this is not true for post-cranial bones. For the present discussion the remains are considered together.

The Pomongwe sample was found to contain 1,192 pieces of dassie skeleton, as listed in table 15. On the basis of the most commonly occurring part, the distal humerus, at least 96 individual animals are represented in the sample, but the minimum number demanded by parts separated into discrete levels is 114 animals. It appears that the dassie bones were not broken up with stone hammers as were those of larger prey, but that they were simply chewed by the people, presumably after the whole carcass was roasted over a fire. The chewable parts were

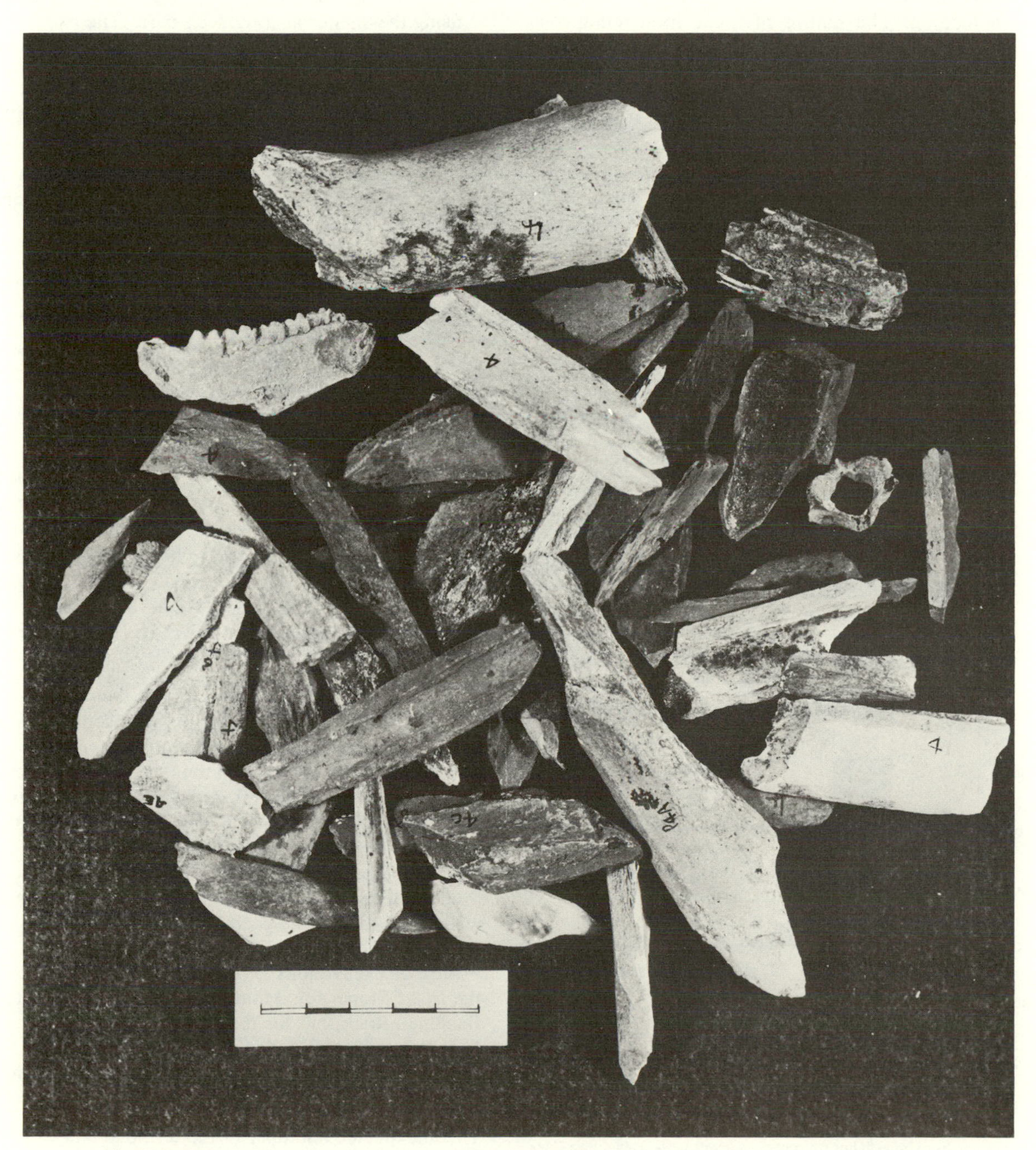

Fig. 27. Bone fragments representing Stone Age human food remains from Pomongwe Cave. The extreme fragmentation appears typical of bones broken up with hammerstones for the extraction of marrow.

chewed and swallowed, and the unchewable pieces were discarded into the ash, where many became charred. The Pomongwe sample therefore appears to consist of those parts of the dassie skeletons that the people rejected as inedible. The nature of these parts is further discussed in chapter 7, but it is clear from the table that those parts most commonly rejected were maxillae, mandibles, and distal humeri. These show consistent patterns of damage (fig. 31). The cranium typically has been broken away at the back for the removal of the brain, and this damage could have been done either with a stone tool or with the teeth. The angles of the mandibles and ascending rami have frequently been damaged during removal of the tongue, and the proximal end of the humerus has simply been chewed away. The distal humerus, perhaps often articulated with the almost equally resistant proximal radius and ulna, was then discarded. In addition to these commonly recurring parts, a scatter of other skeletal fragments, as indicated in table 15, is likely to result from human feeding. The nature of these food remains is therefore very different from those that typically result when leopards and other large felids feed on dassies, as is discussed in chapter 4. Such consistent differences in food remains are valuable indicators as to collectors of food remains in caves.

The Pomongwe bones suggest that, throughout the period of cave occupation, the people were successfully hunting large and small game and were also collecting many slow-moving animals such as tortoises. Some of the animals, like the occasional leopards represented, were perhaps hunted for their skins rather than their meat,

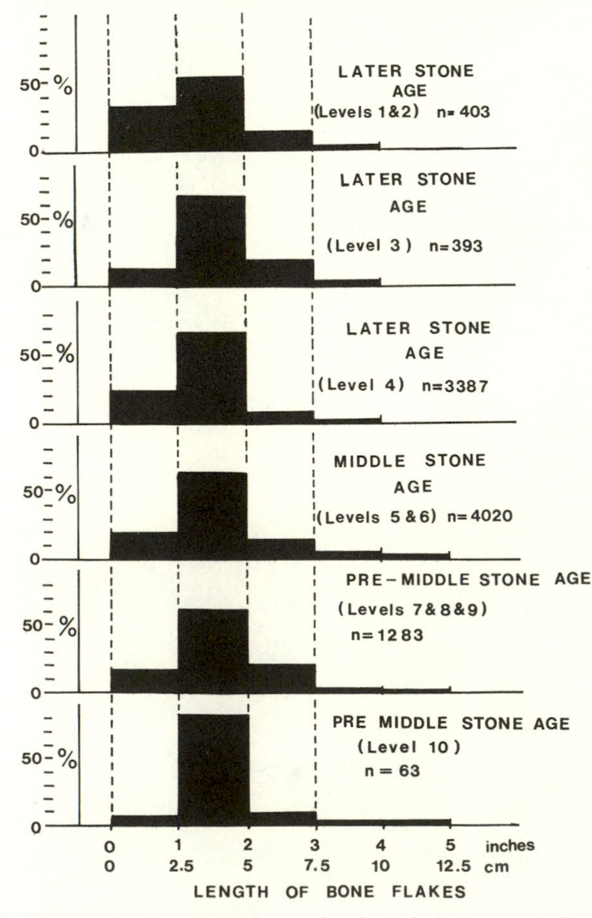

Fig. 28. Lengths of bone flakes from various levels in the Pomongwe Cave deposit. As shown in the histograms, the majority of flakes from all levels are between 2.5 and 5 cm in length.

although what is eaten or rejected appears to depend on the traditions of a particular people. At the same time, we must remember that the staple diet of African hunter-gatherers has probably always been vegetable—a matter to be discussed shortly—and that meat has represented a much sought-after, but often unessential, bonus.

The Bushman Rock Shelter

The site is in a south-facing dolomite ridge (fig. 32) close to the Echo Caves, 30°38′E, 24°35′S, in the Ohrigstad District of the eastern Transvaal. A gentle slope leads up to the shelter from the valley floor, as shown in the section (fig. 33), and the shelter itself is roughly triangular in plan (fig. 33), about 52 m wide, 23 m deep, and 14 m high. A few paintings in poor condition still decorate its walls.

The first systematic excavation of the floor deposit was undertaken in 1965 by A. W. Louw after the discovery of Middle Stone Age artifacts in pits dug by the farmer who was using the shelter as a tobacco barn. The excavation covered an area of 15 ft. by 5 ft. (4.6 m by 1.5 m) and descended to a depth of 8 ft., or 2.4 m. On the basis of lithology, Louw (1969) divided the profile into forty-three layers, 1–2 resulting from Bantu occupation, 3–26 from Later Stone Age, and 27–43 from Middle Stone Age habitation. Charcoal was found to be abundant in the sediment, and seven radiocarbon dates were obtained (Vogel 1969). These read:

Layer 9: 9,510 ± 55 B.P.
12: 9,940 ± 80 B.P.
21: 12,090 ± 95 B.P.
27: 12,160 ± 95 B.P.
28: 12,510 ± 105 B.P.
38: 51,000 B.C.
41: 45,500 B.C.

The samples from the two lowest layers contained so little radioactive carbon that only a minimum age could be given, but the measurements indicated that MSA cultures existed before 51,000 B.C.—an unexpectedly early date at that time. According to Mason's (1969) interpretation of the evidence, the cave was unoccupied for a considerable period at the end of the Middle Stone Age before Later Stone Age peoples again made use of it. In this regard he wrote (1969, p. 57).

> Layers 27 and 28 are apparently at the contact between Later and Middle Stone Age occupations of the cave. Layer 27 is the bottommost layer containing L.S.A. artefacts, but also contains M.S.A. artefacts. Charred wood GrN 4815 from layer 27 dated 10210 B.C. like the wood GrN 4816 from layer 28 dated 10560 B.C. may come from Later Stone Age occupants of Bushman Rock who walked into the surface of Middle Stone Age gravel layer 28 and disturbed it, adding the wood GrN 4816 to the gravel and also churning up the M.S.A. tools found in layer 27.
>
> Further excavations should try to find evidence on the Later Stone Age–Middle Stone Age contact at Bushman Rock, but for the moment it seems that M.S.A. occupation ended in layer 28 at about 51000–45500 BC and L.S.A. occupation began in layer 27 at c. 10560 B.C.

The unexpected antiquity of the terminal phases of the Middle Stone Age, indicated for the first time by the Bushman Rock evidence, has since been confirmed by

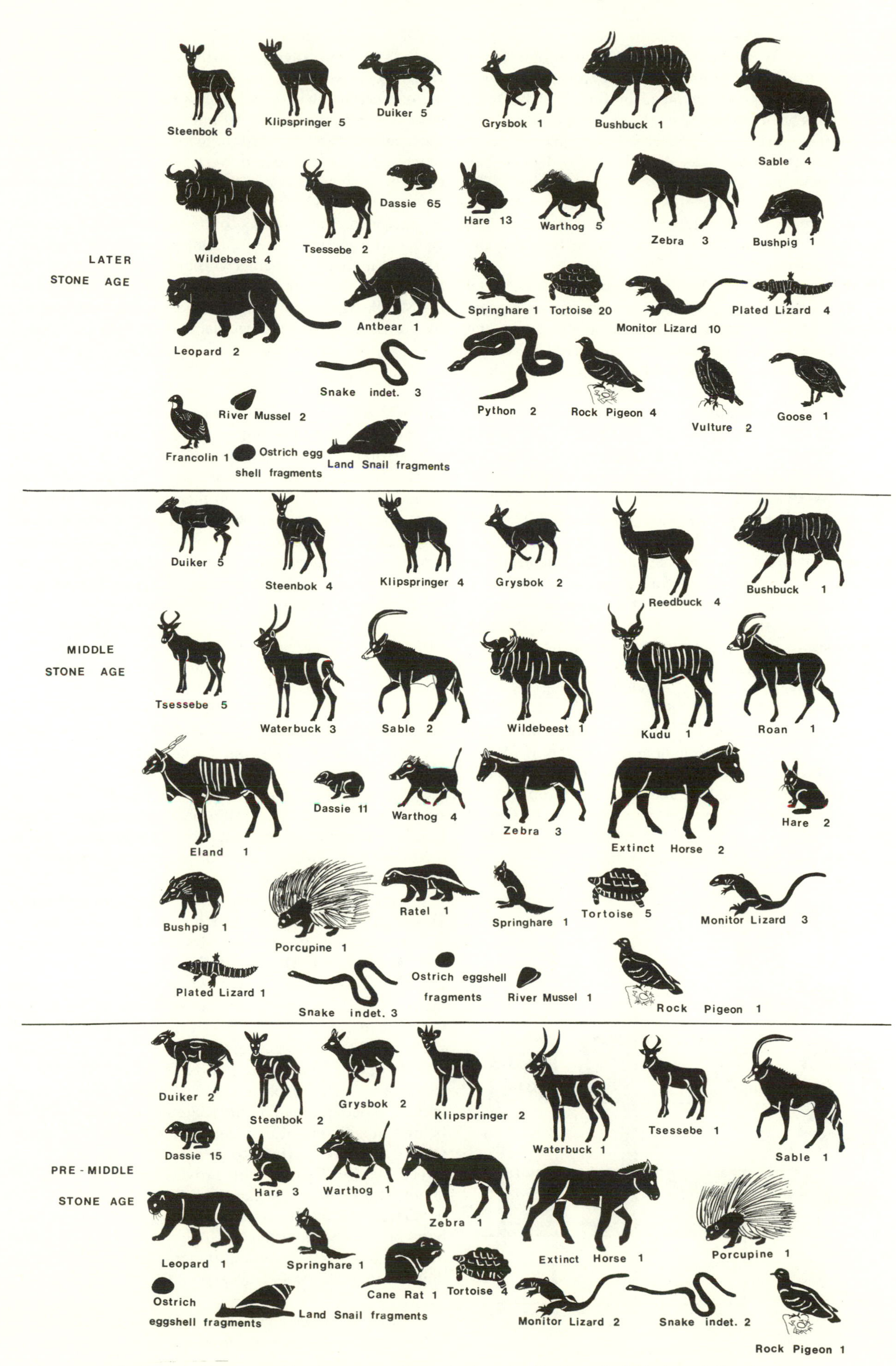

Fig. 29. Animals represented by the Pomongwe human food remains. Numbers of individuals are indicated.

many dated sequences elsewhere in southern Africa. This matter has already been touched upon in the discussion of Pomongwe Cave. After the preliminary work by Louw, a new excavation was commenced in July 1967 by J. F. Eloff and has been continued every year since then. After two seasons' work J. F. Eloff (1969) was able to report that his new excavation was already slightly deeper than that of Louw, but that bedrock was not in sight. The Middle Stone Age material appeared to represent a southeastern expression of the Pietersburg Industrial Complex, with close resemblances to artifacts from Bed 4 at the Cave of Hearths.

Faunal remains from the new excavations have been analyzed by Mrs. I. Plug, who reports (pers. comm.) that the hiatus between the end of the MSA occupation and the beginning of the LSA occupation, observed in Louw's section, is not as marked as had been expected. Publication of the results of Plug's study, and of Eloff's excavation, is awaited with interest.

A sample of 4,819 bone pieces came from Louw's initial excavation, and this was kindly made available to me for study by Louw and Mason. Each piece of bone was individually marked as to its level before sorting proceeded.

It was found that, of the 4,819 fragments, 1,775 consisted of recognizable skeletal parts, albeit broken, and there were also 2,723 bone flakes and 341 indeterminate fragments. As at Pomongwe, fragmentation of the bones was extreme, there being very few unbroken skeletal elements except those of the smallest animals. As with the Pomongwe sample, an attempt was made to assess fragmentation by measuring the lengths of each of the 2,723 bone flakes. Results by individual level are given in table 16, and these are combined into MSA, LSA, and Bantu horizons in table 17, with the figures presented graphically in figure 34. Fragmentation in the MSA and LSA horizons is very similar, with the majority of flake lengths falling between 1 in and 3 in. In the uppermost Bantu layer fragmentation is more extreme, but this could have resulted from a great deal of modern activity when the shelter was in use as a tobacco barn.

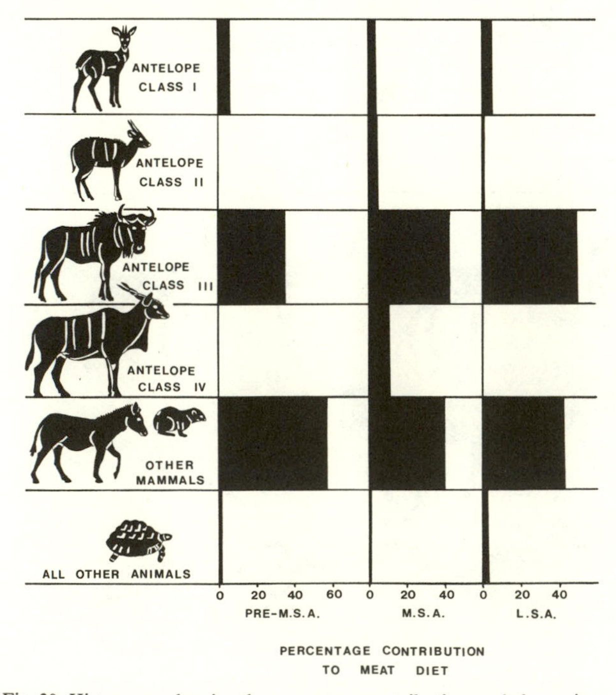

Fig. 30. Histograms showing the percentage contribution made by various animal groups to the diet of Stone Age people in Pomongwe Cave.

Table 18 lists the animal taxa identified from the bone, tooth, and shell fragments, with an indication of occurrence per level. The minimum numbers of individual animals occurring in the Middle and Later Stone Age horizons are given in tables 19 and 20 respectively, while these animals are depicted in figure 35. In contrast to the Pomongwe Cave situation, very few dassies were being eaten at Bushman Rock, although tortoises were gathered in fair numbers. As indicated in tables 19 and 20 and figure 36, the main contributions to the meat diet of the people came from medium-sized antelopes (class III) and non-bovid mammals, particularly zebras and warthogs. Remains of smaller antelopes are surprisingly absent from the Middle Stone Age sample, although some were hunted by the Later Stone Age people.

The Wilton Rock Shelter

The site is known as the Wilton Large Rock Shelter to distinguish it from the Wilton Cave, a deeper shelter occurring on the same farm, Wilton, 33°19′S, 26°5′E, in the Alicedale area of the eastern Cape. The two sites are significant in that it was from them that the Wilton Culture, one of the expressions of the Later Stone Age, was originally described by John Hewitt (1921), at that time director of the Albany Museum in Grahamstown. His description followed the discovery of pygmy crescents in the large rock shelter by C. Windsor Wilmot, postmaster of Qumbu.

During 1921 Hewitt, assisted by the Reverend Mr. Stapleton, the Reverend Mr. Kilroe, and W. W. Wilmot, owner of the farm Wilton, spent five days on an excavation in the rock shelter. They found that the deposit consisted largely of ash, devoid of stratification, and reached a depth of 4 ft (1.2 m). Four burials were uncovered, in each case the skeleton having been covered by flat stones, painted on their undersides with red ocher. The burials proved to be of flat-faced "shortheaded bushmen," rather different in appearance from those found in the nearby "Wilton Cave," which were of prognathous people and were thought to be more recent.

In view of the significance of the remains recovered by Hewitt, a subsequent excavation at the large rock shelter was undertaken by Dr. and Mrs. H. J. Deacon during 1966 and 1967 (J. Deacon 1972). The positions of their excavations relative to the plan of the shelter are shown in figure 37, and the appearance of the shelter at the time of my visit in April 1969 is shown in figure 38. The rock shelter has formed in a steep-sided valley composed of Witteberg quartzite cut by a tributary of the New Year's River. The valley itself is heavily wooded, but areas of grassland occur above it, and there are areas of bushveld below. Inhabitants of the shelter therefore had access to several habitats in which to gather food and hunt.

The Deacons' excavation provided a lithic sample of about 34,000 pieces, which included 1,353 tools, a few potsherds and fragments of ostrich eggshell, and 43,629 bone pieces and invertebrate fragments. The profile was divided into four layers, numbered from the surface downward as follows (J. Deacon 1972, p. 13):

> Layer 1: Layer 1 varied in thickness from 25 mm in the south to 130 mm in the west of the grid. Leaves, twigs and other organic debris accumulated largely from natural sources were variably preserved. The only plant-food remains were some corm scales of *Watsonia*

sp. One or possibly two oval areas more twiggy in composition than the leaf fall were noted as possibly representing decayed bedding heaps, but there is no other evidence to support this suggestion.

Layer 2: Below a transitional interface (2A), Layer 2 is developed as a white compact horizon (2B). This white ash covers the whole of the excavated area and varies in thickness from 150 mm to 300 mm, becoming thicker towards the dripline. Although the deposit has the superficial appearance of ash, grading analysis shows it to be a medium-grade sand, much the same as Layer 3.

Layer 3: Layer 3 is a somewhat oxidised, red-brown deposit 50 cm to 70 cm thick. Oxidisation has not been uniform and towards the back wall the colour grades to grey-black. Subdivision of Layer 3 was partly on the basis of thin, lenticular, interdigitating hearths and partly on arbitrary spits. None of the subdivisions is more than 8 cm thick. Layer 3D marks a general but not a defined change between the deposit above (3C–3A) and below (3E–3I), and includes three major white ashy lenses (Hearths 1, 2, and 3) which interleave, making vertical separation difficult. For purposes of analysis, the lithic material from the hearths was grouped together with the material from the rest of the layer, but the faunal samples from the hearths were kept separate.

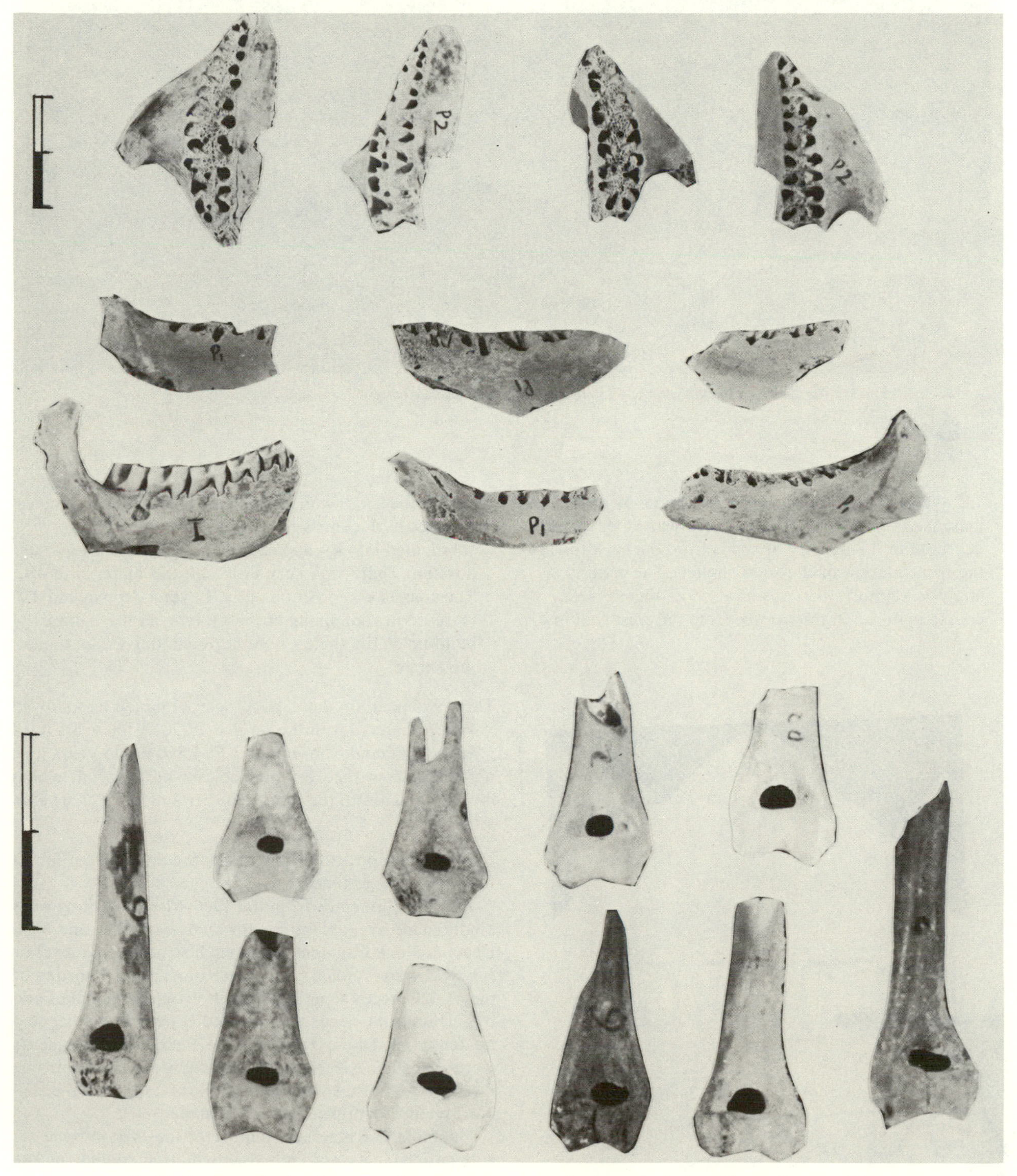

Fig. 31. Dassie remains from Pomongwe Cave, apparently representing human food refuse. The maxillae, mandibles, and distal humeri appear to represent unchewable remnants of the dassie skeletons.

Fig. 32. The Bushman Rock Shelter in a dolomite ridge near Ohrigstad, eastern Transvaal.

Layer 4: The top of Layer 4 is marked by an abrupt increase in roof rock spalls. There was an observed decrease in the density of spalls from the back towards the front margin of the rock shelter. The grading analysis supports this observation as there is less coarse material in the sample from the west wall in 4A than in the south wall (ie the back of the shelter). Layer 4 varies in thickness from 35 cm to 50 cm and is subdivided into 4A above and 4B below a dark marker horizon. There was very little cultural material in 4B. The coarser composition of the Layer 4 deposit and the increase in roof spalls probably reflects the natural build-up on the shelter floor during initial occupation of the shelter.

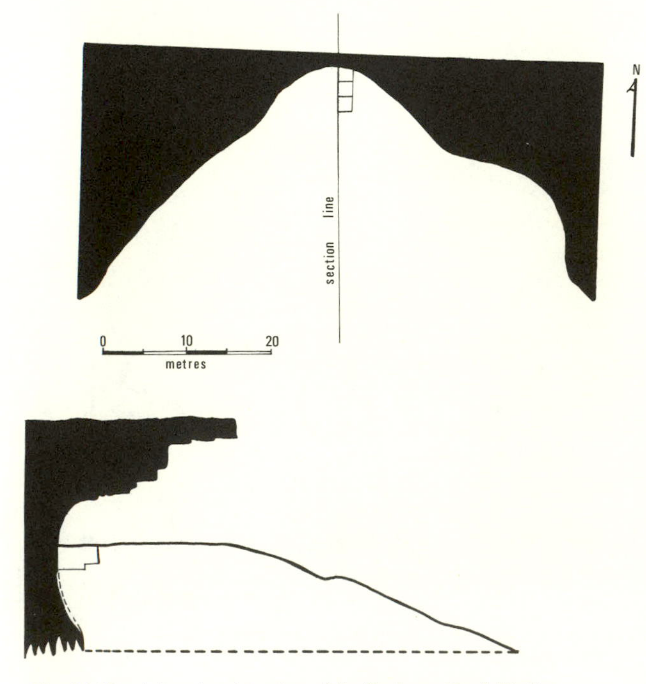

Fig. 33. Plan *(above)* and section of the Bushman Rock Shelter.

Three radiocarbon dates have been obtained as follows:

Wood charcoal from Layer 2B: 2,270 ± 100 years B.P.

Wood charcoal from Layer 3 F: 4,860 ± 115 years B.P.

Sample from the vertebral column of a burial, which probably relates to the base of Layer 3: 8,260 ± 720 years B.P.

The whole deposit appears to have accumulated during Later Stone Age times.

The faunal remains from the 1966–67 excavations were kindly made available to me for analysis by Dr. and Mrs. Deacon. Bone fragments from each stratigraphic level or sublevel were studied as discrete units. Initial sorting of the 43,629 pieces separated the unrecognizable flakes and fragments from pieces that showed at least some diagnostic features. The unrecognizable flakes and fragments were placed in size groups and counted. Recognizable pieces were sorted into skeletal parts and, where possible, specific identifications were made.

The most characteristic feature of the Wilton bone accumulation is its extreme fragmentation, which makes identificaton of skeletal parts and animal taxa difficult. Of the 43,629 pieces, 33,891, or 77.6%, consisted of short bone flakes or unrecognizable fragments less than 1 in (2.5

cm) long. The length distributions of these, by stratigraphic unit and by layer, are given in tables 22 and 23. In view of the very short lengths of many of the bone flakes, it was not practicable to separate these pieces from the miscellaneous unrecognizable fragments, as has been done in analyses of the other three site assemblages. The fragmentation at Wilton is more extreme than at the other sites and may reflect more intensive trampling of the cave floor, superimposed upon the usual breakage for the extraction of marrow. The degree of fragmentation is remarkably consistent in each of the four layers, as is shown in figure 39.

I found that, 6,702 pieces in the collection were complete enough to allow identification of the skeletal part involved. After sorting into parts, these were allocated to the following groups, according to the kind of animal from which they came: bovid, mammalian other than bovid, bird, reptile, fish, and invertebrate remains. Results of this analysis are given in table 24, but it must be stressed that very few of the pieces listed consisted of complete bones; rather, most were fragments with enough diagnostic features to allow recognition.

It will be seen that 1,454 pieces came from bovid antelopes, and, as would be expected, the most resistant parts of the skeletons were best represented. Tooth fragments were numerous, and pieces of the lower legs such as carpals, tarsals, metapodials, and phalanges were well represented. The reason such parts survived is doubtless

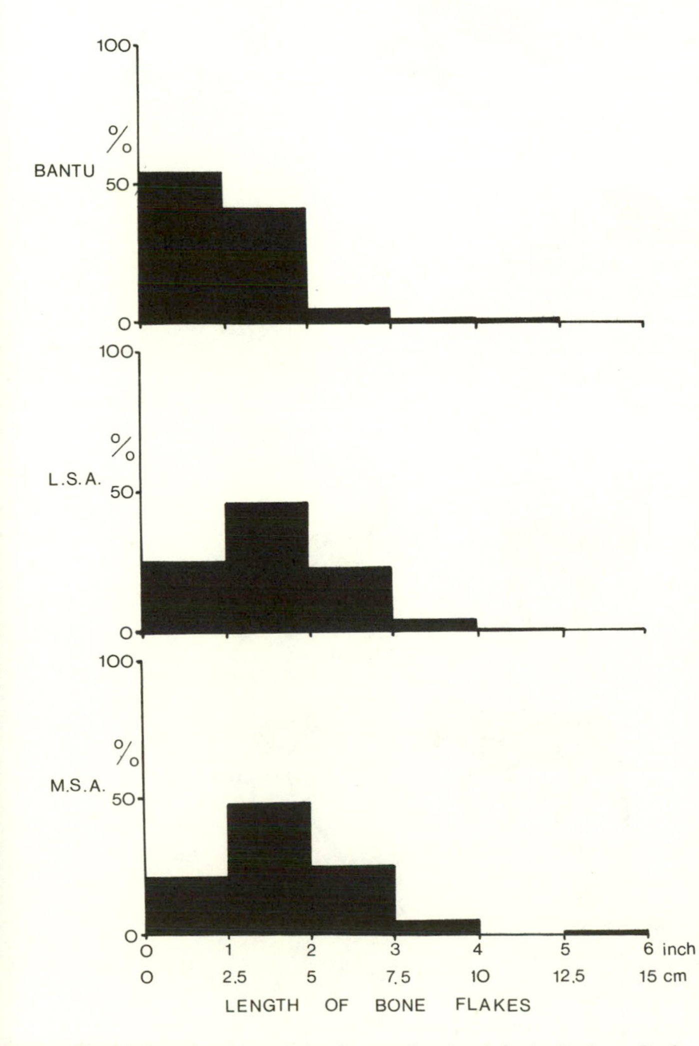

Fig. 34. Lengths of bone flakes from various levels in the Bushman Rock Shelter deposit.

that they contain little marrow; nor is much meat to be had from bovid feet.

Nonbovid mammalian remains were not numerous, except for microfaunal fragments such as bones and teeth of small rodents and shrews. These have almost certainly come from owl pellets, and evidence of barn owls in two levels suggests that the shelter was sometimes used as a roosting place, as one would expect.

Among the reptile remains, tortoise fragments are abundant in all layers. The tortoises were probably collected and eaten by the Wilton people and appear to have been a favored food throughout the occupation period. The heads and necks of the tortoises have largely disappeared, suggesting that these parts may have been consistently chewed and swallowed. Two forms, *Chersine angulata* and *Geochelone pardalis,* appear to have been involved.

Snakes are well represented by isolated vertebrae in many layers. Some of these may have come from owl pellets, but they are usually abundant in relation to the rodent remains, suggesting that snakes may have featured regularly in the diet of the Later Stone Age people or that they were used for other purposes, as discussed below.

There is strong evidence that a freshwater stream existed close to the shelter throughout the Wilton occupation period. Widespread crab and mollusk fragments occur, as well as fish vertebrae at two levels.

Wherever possible, I have tried to identify the species of animals involved and to estimate the minimum number of individuals that must have contributed to the bone accumulation; results are given in table 25 and presented pictorially in figure 40. Table 26 provides an inventory of skeletal parts upon which identifications were based.

Blue duiker has been positively identified in five layers on the basis of isolated teeth, horn-cores, and a metatarsal fragment. The species occurs in the area today and has clearly done so throughout the Wilton time span. There is an indication of gray duiker in the uppermost layer only, the diagnosis being based on a single lower molar.

Gray rhebuck has been identified in two layers on the basis of a few teeth only. Although these conform well to the teeth of *Pelea* in the comparative collection, a firm identification requires more complete material. The same is true of the bushbuck, identified from Layer 3B on the basis of an isolated tooth. Large bovids were poorly represented at the site. A domestic cow is tentatively identified from the uppermost layer on the basis of a single terminal phalanx, and a large alcelaphine, almost certainly a wildebeest, is present in Layer 4.

Remains of common dassies were found in nine layers and came from a minimum of 12 individuals. These are represented by the same skeletal parts as were the dassies in the Pomongwe Cave deposit. Almost as numerous are remains of scrub hares, likewise represented by a scatter of resistant skeletal parts. Three small carnivore individuals are represented by isolated teeth only, one of which was positively identified as coming from a striped skunk. These animals were perhaps not hunted for food, though their skins may well have been used. Baboons have been identified in two layers on the basis of worn incisors. It is not known whether these primates were eaten. The same can be said of the single puff adder, identified from Layer 3D, on the basis of a maxillary bone complete with fang. The possibility exists that the Wilton people used puff adder venom as an arrow poison; during historic times Bushmen in the eastern Cape are reported to have used this venom on their arrows, perhaps mixed with plant poison such as *Euphorbia* juice (Shaw, Woolley, and Rae 1963). Part of a marine limpet (*Patella* sp.) was found in Layer 3A and is certainly indicative of human transport. The shell may have had some cultural significance.

As with remains from Pomongwe Cave and Bushman Rock Shelter, I have attempted to estimate the percentage contributions to the diet of the people made by each animal taxon. Results are given in table 27, and the figures for broader animal groupings are listed in table 28 and depicted graphically in figure 41. It is immediately obvious that, in terms of meat yield, class II antelope have been the most important resource, followed by larger and smaller bovids, other mammals, and finally reptiles, birds, fish, and invertebrates. Although crabs were eaten in large numbers, they constituted a delicacy rather than a staple food.

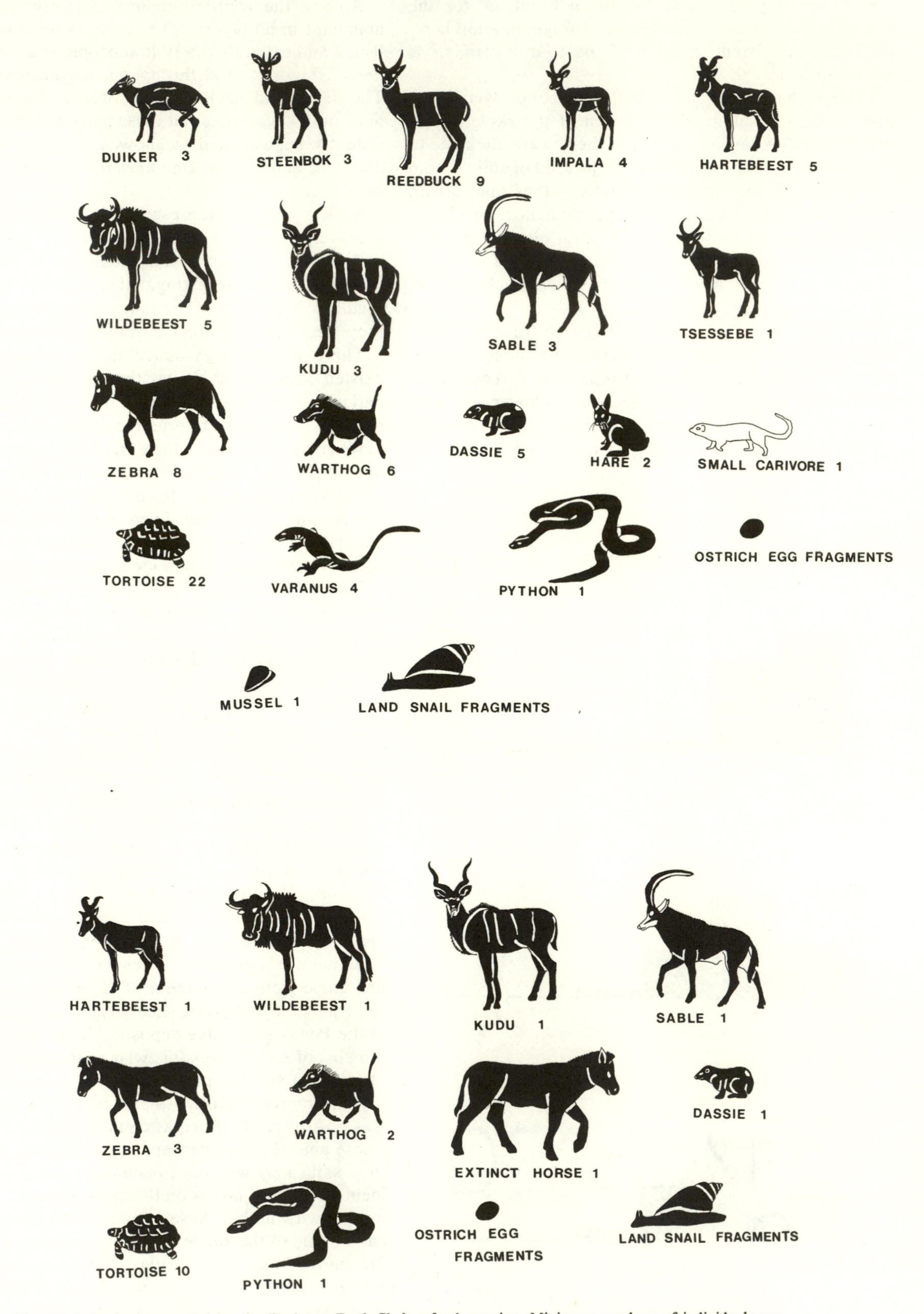

Fig. 35. Animals represented by the Bushman Rock Shelter food remains. Minimum numbers of individuals are indicated.

Fackelträger Shelter

The site is on the farm Omandumba West 137, in the Omaruru District of South-West Africa, 21°34′S, 15°32′E. The country rock in this northwestern part of the Erongo Mountains is Precambrian granite that forms prominent bare domes and piles of boulders (fig. 42) in an arid environment of sparse thorn scrub.

As is shown in the plan and section (fig. 43), the shelter consists of an extremely large granite boulder resting on five smaller ones. The undersurface of the roof block is decorated with many paintings, and the shelter was one of several selected by W. E. Wendt for excavation as part of his research program, conducted between 1968 and 1971, on archaeological deposits associated with rock art (Wendt 1972). The bone sample from this excavation has very kindly been made available to me for study by Dr. Wendt.

The excavated area extended over 22 m^2, as indicated on the plan, and reached a maximum depth of 165 cm. At the base of the excavation, in some of the innermost squares, a layer of coarse granite grit was found, containing Middle Stone Age artifacts but no trace of bone. Above this was a deposit of brownish gray, gritty sediment without clear stratification except for some ash lenses and pockets of charcoal. This upper layer, which can be regarded as Later Stone Age, yielded 26,707 stone artifacts as well as 82 grinding stones. The artifacts included 663 formal tools, a number of bone tools, including points, awls, a scoop, and various disks, pendants, and tubular bone beads. In addition, there were several pendants of mica schist, some ostrich-eggshell beads and pendants, a complete ostrich eggshell worked into a container, a fragment of engraved ostrich eggshell, and a possible rattle made from a cocoon.

Two carbon dates are available from charcoal from the upper part of the deposit: from 28–36 cm, 2,980 ± 120 years B.P., and, from 9–11 cm, 2,770 ± 120 years B.P. Since all the bones are derived from the upper part of the deposit, the assemblage was certainly built up within the last few thousand years.

The bone sample was found to consist of 3,098 fragments, among which were 1,588 bone flakes, whose lengths are listed in table 29 and whose distribution is shown graphically in figure 44. It is apparent that the great bulk of the flakes have lengths less than 2 in (5 cm). The fragmentation is likely to be due partly to deliberate breakage for extraction of marrow and partly to the trampling of people's feet on the rather hard and gritty granitic sand surface. Fragmentation of the other skeletal parts is also extreme, so much so that many of the identifications are necessarily tentative. A list of animal taxa and the parts by which they have been identified is given in table 30, and the animals involved are shown pictorially in figure 45.

When we consider the numbers of individual animals represented in the bone accumulation, we find that the sample is dominated by small antelopes, hyraxes, hares, tortoises, monitor lizards, and birds. The inhabitants of the shelter clearly hunted small and medium antelopes as well as dassies and hares. Many tortoises were gathered, and monitor lizards were favored food, as were rock pigeons and larger birds such as francolins and geese.

Tooth fragments of a single black rhino were found to occur in square B5 at a depth of 15 cm. These remains highlight an effect inherent in any attempt to calculate the contributions made by individual animals in the diet of primitive people. If the meat yield of the rhino is included in the calculations, its contribution to the diet of the people, as indicated by the available bone sample, will outweigh that of the numerous smaller animals that presumably constituted a more regular and predictable component of the meat supply. Apart from the tooth fragments, no other parts of the rhino's skeleton were found among the excavated remains, and it is highly likely that the rhino was largely eaten where it lay rather than being transported to the shelter. The dietary contributions of the various animals and animal groups, expressed as percentages of diet indicated by the remains, with and without the single rhino included, are listed in tables 31 and 32 and shown graphically in figure 46. When the rhino is included, the contribution of its group of "other mammals" outweighs that of the various antelopes, whereas if the rhino is excluded from the calculation, most of the meat is found to have come from steenbok, springbok, and gemsbok.

Regrettably, little information can be derived from the Fackelträger sample about butchery practices or about which parts of prey skeletons were brought back to the cave. One gains the impression, however, that parts of the entire skeletons of steenbok and springbok are present among the fragmented remains, including lower leg elements. It is likely that these smaller bucks were carried back whole to the shelter. Remains of large antelopes are too sparse to be informative.

A minimum of 18 dassies, one of which was a juvenile, were represented by 68 bone pieces. The skeletal parts involved here are very similar to those found at Pomongwe, there being 9 maxillary pieces, 19 mandibular pieces, 16 distal humeri, and a scatter of other parts, consistent with the pattern of unchewable remains built up from the Pomongwe sample.

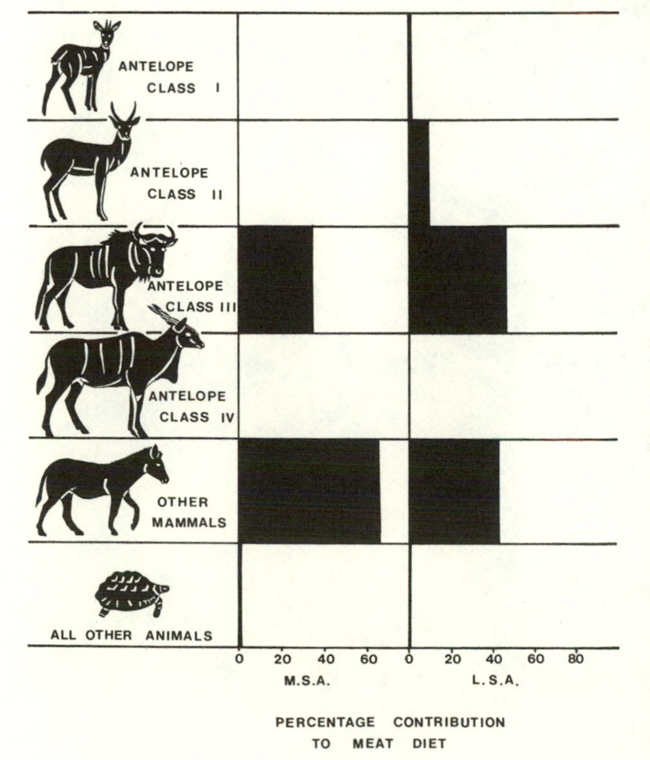

Fig. 36. Histograms showing the percentage contribution made by various animal groups to the diet of Stone Age people in the Bushman Rock Shelter.

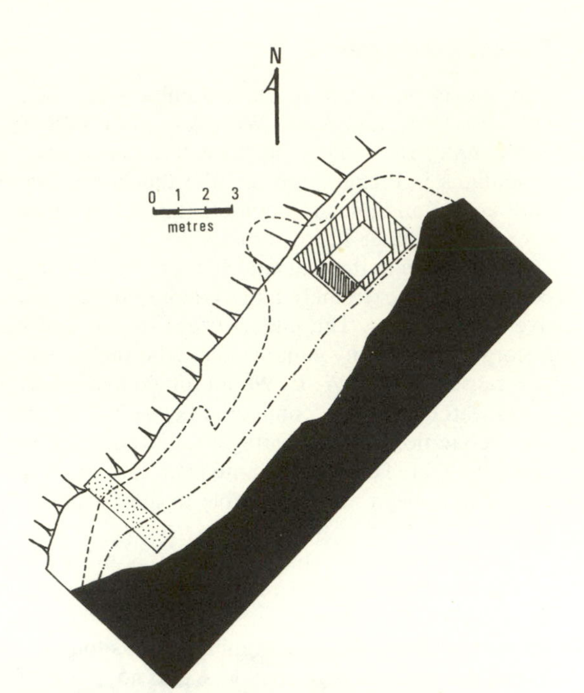

Fig. 37. Plan of the Wilton Rock Shelter showing the position of the excavations and dripline *(outer dotted line)*.

Fig. 38. A view along the length of the Wilton Rock Shelter showing the overhanging cliff of Witteberg quartzite.

Very little in the way of recognizable remains has survived to attest the presence of at least 10 hares and the same number of monitor lizards. Apart from teeth and jaws, the skeletons of these animals appear to have been highly vulnerable to human feeding action. As at the other sites, tortoises are represented by abundant fragments of carapace and plastron and a few limb bones.

Discussion

It would be very naive to imagine that the results of the four analyses presented here could do more than provide a glimpse of the remains of the animals eaten at those specific sites. Collections from other areas will give a very different picture and will serve to emphasize that *Homo* is a remarkably adaptable animal, making use of whatever food resources happen to be available. Fortunately, the results of some excellent studies on food remains from other caves in southern Africa have been published within the last ten years, and these place my results in perspective and provide detailed insights I was unable to obtain. A severe drawback of the collections I studied was that the bone pieces had been very heavily fragmented, and as a result I could deduce little about the butchery practices of the Stone Age people or about the parts of animals brought back to the caves. Fragmentation seems to be an attribute of bones from Stone Age cave or shelter sites that have been intensively occupied. Fortunately, remains from the Klasies River Cave I, analyzed by Richard Klein (1975*b*, 1976*a*), proved more complete than usual and have provided some extremely valuable data.

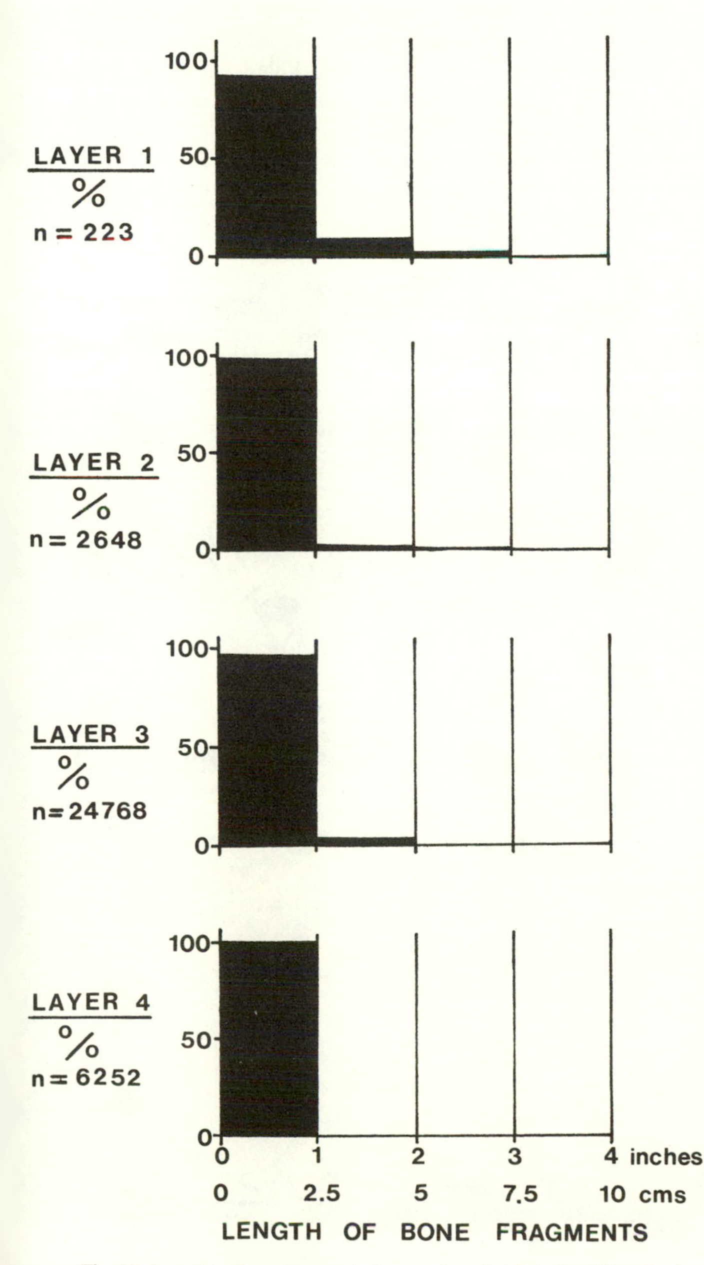

Fig. 39. Lengths of bone fragments from various levels in the Wilton Rock Shelter deposit.

Information on Skeletal Disproportions from Southern Cape Caves

Seven caves and shelters have been excavated in the Klasies River Mouth complex, on the southern Cape coast about 130 km west of Port Elizabeth. Excavations were organized by R. Singer and undertaken by J. J. Wymer between December 1966 and July 1968. The faunal remains have been studied in detail by Klein, who was able to discuss three aspects of the body-part data the bone samples provided (Klein 1976*a*, pp. 85–97).

Effects of Differences in Bone Density and Epiphyseal Fusion Times

Although the bovid remains from Klasies River Mouth came from animals differing widely in size and in age at death, Klein found that the conclusions reached on the basis of Kuiseb River goats, described in chaper 2, concerning the relationship between bone density and survival held good for those bones as well. He wrote: "As in the case of the Kuiseb goats, those ends of the bovid long bones at Klasies which are denser and fuse earlier tend to be more common than their less dense opposite ends that fuse later" (p. 86).

Effects of Species Size

The Klasies bovids were placed in five size groups, varying from "small" to "very large," and it was apparent that the pattern of body-part representation differed according to the size class being considered. To clarify this situation, Klein drew up a table in which the minimum numbers of animals in the "small bovid" category demanded by each skeletal part were arranged in descending order of abundance. Thus, at the top of the column 54 small bovids were demanded by scapulae, followed by 46 on the basis of mandibles, right down to nil on the basis of carpals and tarsals. Following the body-part order determined for the small bovids, the same was done for animals in the four other size classes, and histograms were plotted for each. The form of these histograms was found to vary a great deal; for instance, it was only in the smaller animals that scapulae were the most abundant parts—in the larger ones mandibles took over the position at the top. Various statistical tests that Klein applied confirmed that the pattern of body-part representation in large and very large bovids was very similar, but that it was strikingly different from that in the small and medium bovids. In general, as the size of the bovid decreased, the ratio of cranial to postcranial parts increased, whereas the ratio of limb bones to foot bones decreased. What do these differences mean? Klein suggested that they are related to the portability of the prey—to what Perkins and

Daly (1968) have called the "schlepp effect." In this connection, Klein wrote (1976*a*, pp. 87–88):

> Basically they postulated that hunters were likely to bring home smaller animals intact, but they would probably bring back only selected parts of larger animals. This is because larger animals would be butchered at the place of the kill and the less useful parts would be left there. In documenting the operation of the "schlepp effect" at the early Holocene ("Neolithic") hunters' site of Suberde in Turkey, Perkins and Daly showed specifically that larger bovids tended to be represented disproportionally well by their foot-bones versus leg-bones, just as at Klasies. They postulated that the Suberde people discarded many larger bovid limb bones at the kill sites, but brought back the feet either as handles in the skins (used as carrying containers for the meat?) or because the feet were particularly valued, perhaps as sources of sinews for sewing.

It seems that heads of larger bovids were selectively brought back and that mandibles have survived better than maxillae as a result of their structural strength.

Effects of Site Type: "Occupation Sites" versus "Kill-Butchery Sites"

Klein has pointed out that, if the "schlepp effect" had been in operation, it is reasonable to expect that the skeletal parts of antelope prey found at "base camps" or "occupation sites" like Klasies would differ from those encountered at "kill or butchery sites." He has compared the body-part information from Klasies with that from an open-air kill and butchery site he investigated at Duinefontein 2, near Melkbosstrand in the southwestern Cape (Klein 1976*b*). Those parts, such as vertebrae, that were underrepresented at the Klasies "base camp" were abundant at the Duinefontein 2 "butchery site." Klein has pointed out that a rather similar effect may be observed for bison remains at American Indian village and hunting sites. He compared body-part information from four village sites described by T. E. White (1953*b*, 1954*a*, *b*) with that for the Casper kill site investigated by Frison (1974) and found that, although many differences existed, it was reasonable to assume that the "schlepp effect" had been responsible for the low axial/appendicular ratio in the Klasies remains.

Observations on contemporary hunter-gatherers such as those of Yellen (1977) are very valuable in showing what kinds of activities led to the fossil traces as we know them. Working with the !Kung Bushmen in northwestern Botswana, Yellen was able to observe butchering of various large antelopes that had been killed with poisoned arrows. He described twenty-eight typical steps in the process and found that the cannon bones were characteristically cooked and eaten at the site of the kill, as were some of the ribs. Such parts would then be missing from "base camp" food debris.

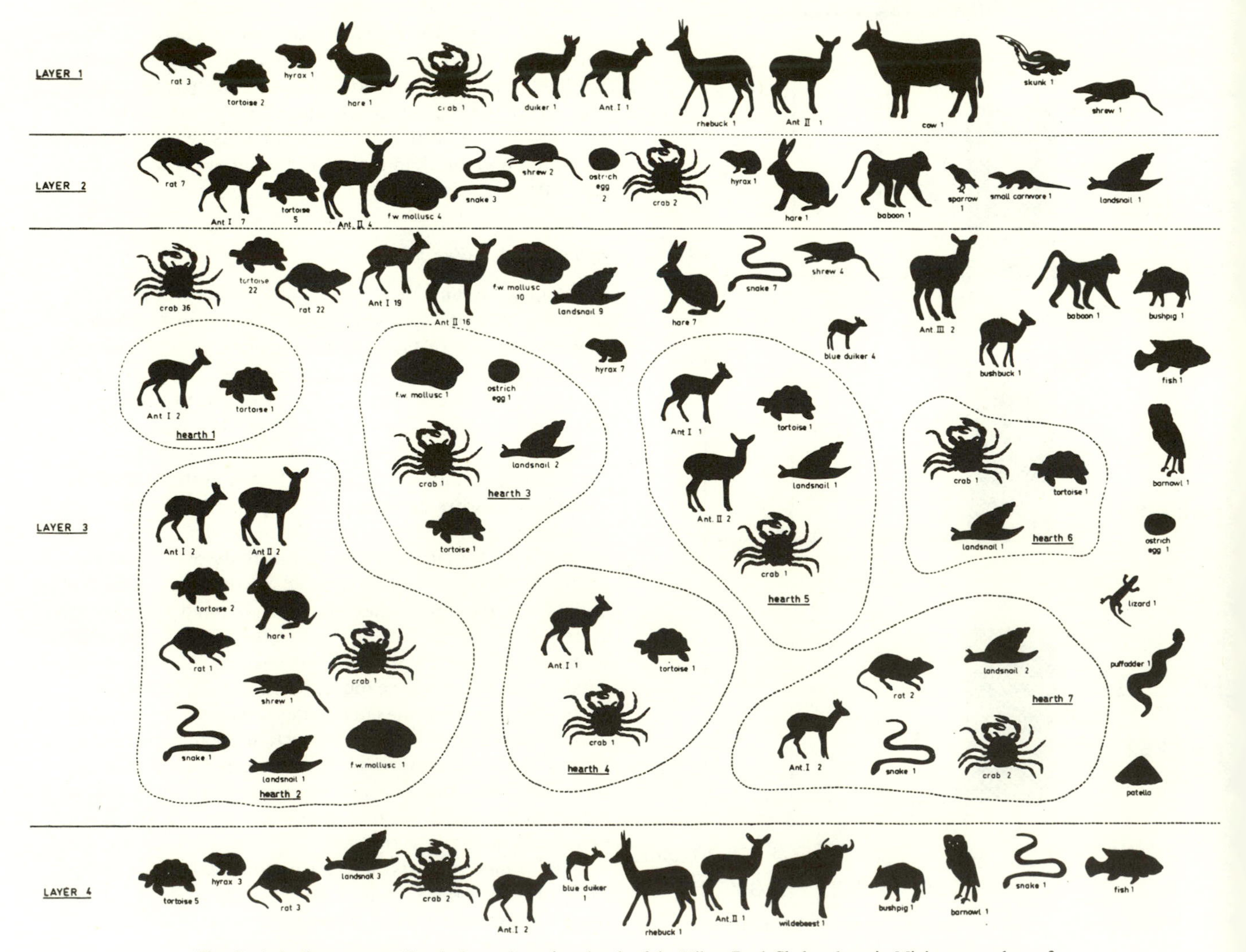

Fig. 40. Animals represented by the bones in various levels of the Wilton Rock Shelter deposit. Minimum numbers of individuals are indicated.

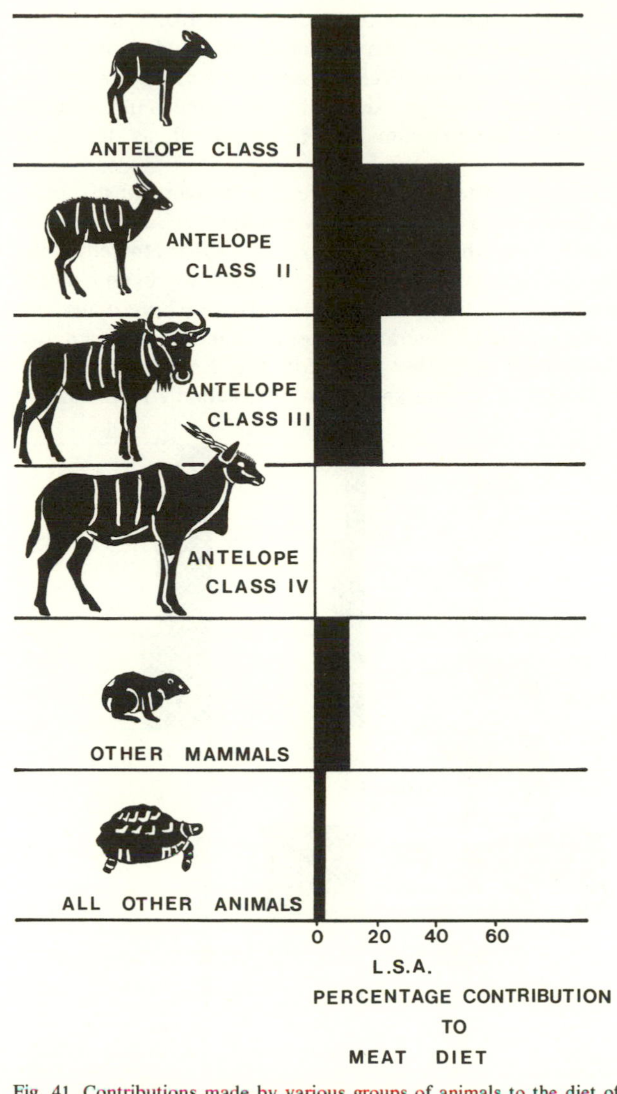

Fig. 41. Contributions made by various groups of animals to the diet of Stone Age inhabitants of the Wilton Rock Shelter.

Fig. 42. A view of the Fackelträger Shelter under a large granite boulder in the Erongo Mountains. Photo by W. E. Wendt, from Wendt 1972.

A Classification of Bone-Bearing Archaeological Sites

In his paper on the diet of early man, Isaac (1971) proposed a classification of bone-bearing archaeological sites that is particularly relevant to this discussion. It is based on the relative abundance of stone artifacts and bone refuse as indicated in figure 47, and is properly applicable only to undisturbed sites where preservation has been good. The diagonal defined by increasing artifact and bone densities also represents increasing intensity of occupation.

Transitory camps, characterized by a low density of stone artifacts *and* of bone refuse, where the traces have resulted from brief occupation—perhaps overnight—of a moving human band. Examples suggested by Isaac are Koobi Fora site FxJj1 (Isaac, Leakey, and Behrensmeyer 1971), parts of Olorgesailie (e.g., Isaac 1968), and Peninj (Isaac 1967*a*).

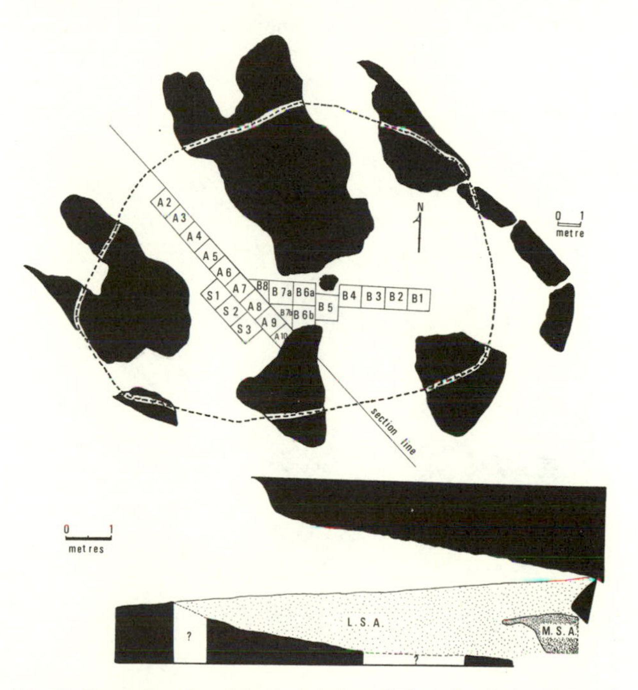

Fig. 43. Plan and section of the Fackelträger Shelter. Granite boulders and bedrock are shown in black.

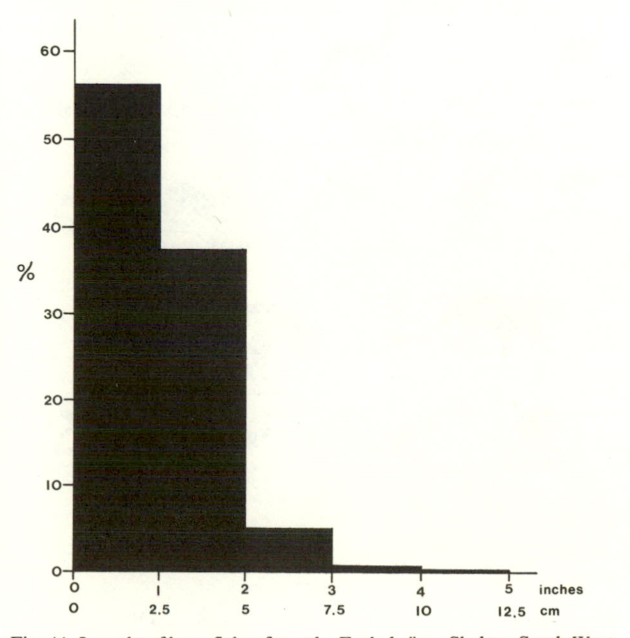

Fig. 44. Lengths of bone flakes from the Fackelträger Shelter, South-West Africa.

Kill or butchery sites, characterized by a high density of bone but small numbers of artifacts, these sites may center on the remains of a single animal or contain remains of many that were attracted to some resource, such as a water hole, and were killed there.

A well-documented instance of a single elephant skeleton, surrounded by Acheulean tools that had clearly been made on the spot for butchery, has been described from Mwanganda's Villge in Malawi (Clark and Haynes 1970). A rather similar situation, but later in time and involving two elephants, has been reported from the zoo grounds in Windhoek, South-West Africa (Sydow 1961, 1963; MacCalman 1967). Other elephant butchery places have been recorded from Olduvai Gorge in the top of Bed 1, level 6, of site FLK N1, where the skeleton of an *Elephas recki* was associated with 123 artifacts. Nearby, in the base of Bed II at FLK, the skeleton of a *Deinotherium* was preserved with 39 artifacts (M. D. Leakey 1971). Farther afield, some fine examples of elephant butchery places have been described from the middle Acheulean sites of Torralba del Moral and Ambrona in Spain (Howell 1966; Freeman and Butzer 1966), where, in alluvial beds from the Rio Ambrona, remains of elephants and other large mammals were preserved, as they were left, by the Stone Age hunters.

Butchery sites of hippos are also known, for instance, at Koobi Fora FxJj3 and Olorgesailie (see above) and at Isimila in southern Tanzania (Howell 1961; Howell, Cole, and Kleindienst 1962). Mention has already been made of the Duinefontein 2 site near Cape Town where many animals, including several buffalo, appear to have been systematically butchered by Middle Stone Age peoples.

Finally, it is remarkable how much detailed information

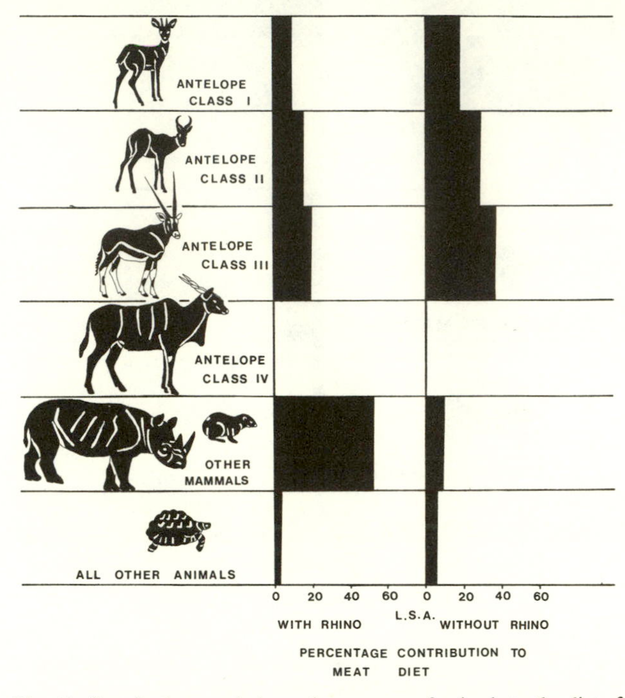

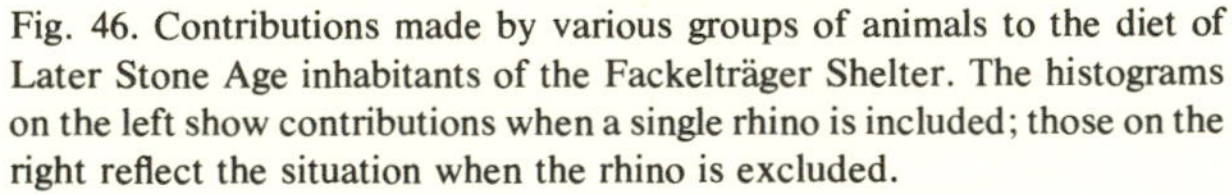

Fig. 46. Contributions made by various groups of animals to the diet of Later Stone Age inhabitants of the Fackelträger Shelter. The histograms on the left show contributions when a single rhino is included; those on the right reflect the situation when the rhino is excluded.

STEENBOK 9
cf. STEENBOK 6
SPRINGBOK 1
cf. SPRINGBOK 8
cf. GEMSBOK 2
DASSIE 18
HARE 10
ROCK RAT 6
ELEPHANT SHREW 3
MONGOOSE 2
LARGE CARNIVORE 1
RHINO 1
TORTOISE 17
VARANUS 10
LIZARD 2
SNAKE 2
cf. ROCK PIGEON 8
INDET. BIRDS 7

Fig. 45. Animals represented by the bones from the Fackelträger Shelter. Minimum numbers of individuals are indicated.

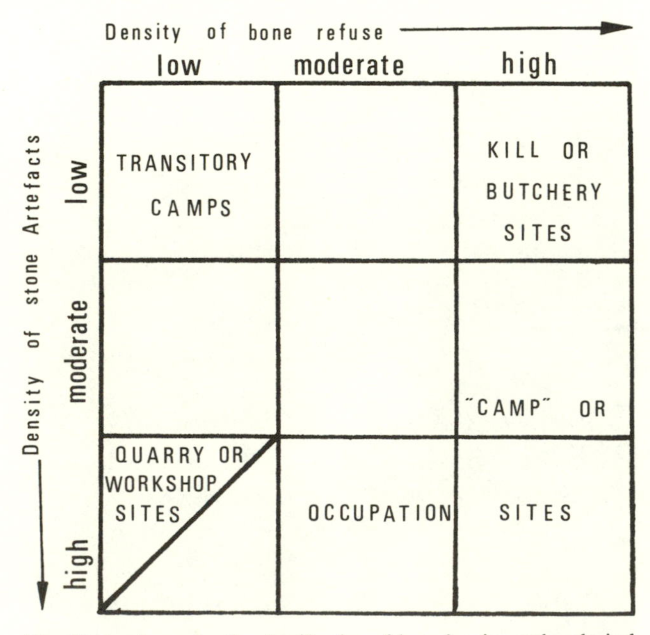

Fig. 47. A scheme for the classification of bone-bearing archaeological sites, proposed by Isaac (1971).

about the behavior of Stone Age hunters may be derived from meticulous excavation of butchery sites, as has been pointed out by Desmond Clark (1972). For instance, work at the Olsen-Chubbuck site in Colorado has allowed the reconstruction of a bison hunt that took place about 8,500 years ago (Wheat 1967). It seems that about two hundred bison were stampeded and driven into a valley, where their remains have been found. Investigation of these remains has made it possible to reconstruct, with reasonable certainty, the month in which the hunt took place, the wind direction on that particular day, and the direction of the hunters' drive, along with details of the butchery process and of which cuts were eaten on the spot.

Quarry or workshop sites, characterized by an abundance of stone artifacts but a low density of bone refuse. Isaac mentions the "chert workshop site" from lower Bed II at Olduvai as a possible Lower Pleistocene example (Stiles, Hay, and O'Neil 1974), as well as several more recent ones, including parts of the Cave of Hearths accumulation at Makapansgat (Mason 1962*a*).

Camp or occupation sites, typified by high densities of stone artifacts *and* bone. The cave sites discussed earlier all qualify for this definition, as do open-air localities like the "Zinj" floor at Olduvai FLK 1, the DK floor, some levels at FLK N1 and the DE/89 horizon B at Olorgesailie, where the remains of at least 40 adult and 13 juvenile baboons, of the large extinct species *Theropithecus oswaldi,* together with 15 kg of bone pieces, were found among very numerous Acheulean artifacts (Isaac 1968, 1969).

As mentioned earlier, evolutionary changes in the *Homo* lineage during the course of the Pleistocene led to marked behavioral changes that must inevitably have been reflected in the traces left by the people. Some information on progressive changes in the nature of such traces from Bed 1 at Olduvai through upper Bed II has been collected by M. D. Leakey (1971) and further discussed by Isaac (1976). It is especially significant that, with the passage of time, a rise in the ratio of artifacts to bone refuse is discernible, while the fragmentation the bones have suffered has likewise increased. Larger animals are slightly better represented in the later levels, and the early preponderance of bovids is replaced by a wider prey spectrum, including equids and hippos. In upper Bed II, at Localities SHK and BK, the earliest evidence of herds of antelopes having been successfully hunted is found. *Antidorcas recki* remains are found at the former and *Pelorovis* at the latter.

Any interpretation of food remains associated with the southern African australopithecines should take into account the fact that these early hominids probably did not break up bones as intensively as did subsequent members of the genus *Homo*.

The Gathered Component in the Diet of Stone Age Peoples

The people we have been discussing were, presumably, *hunter-gatherers,* who derived part of their sustenance from *gathered* foods. Some of the bone refuse at the sites I have considered, such as remains of tortoises, came from animals that were gathered rather than hunted, but the main gathered component in the diet surely came from plant foods, except at certain coastal sites where gathered shellfish were extremely important. In fact, in his survey of the diet of living hunter-gatherers, Richard Lee wrote (1968, pp. 42–43):

> Although hunting is rarely the primary source of food, it does make a remarkably stable contribution to the diet. Fishing appears to be dispensable in the tropics, and a number of northern peoples manage to do without gathered foods, but with a single exception, *all* societies in all latitudes derive at least 20 per cent of their diet from the hunting of mammals. Latitude appears to make little difference in the amount of hunting that people do. Except for the highest latitudes, where hunting contributes over half of the diet in many cases, hunted foods almost everywhere else constitute 20 to 45 per cent of the diet. In fact, the mean, the median, and the mode for hunting all converge on a figure of 35 per cent for hunter-gatherers at all latitudes. This percentage of meat corresponds closely to the 37 per cent noted in the diet of the !Kung Bushmen of the Dobe area. It is evident that the !Kung, far from being an aberrant case, are entirely typical of the hunters in general in the amount of meat they consume.

From his study of the !Kung Bushmen in northwestern Botswana, Lee concluded that these people "eat as much vegetable food as they need, and as much meat as they can." Gathered plant food constituted the stable dietary base, and, in comparison with hunting, gathering proved more productive. In the Dobe area, Lee found that one man-hour of hunting typically produced about 100 edible calories, while gathering brought in 240. In fact, hunting proved to be a *high-risk, low-return* subsistence activity, while gathering represented a *low-risk, high-return* one. This may well have been true for African Stone Age hunter-gatherers as well.

In the Dobe area, Lee observed that the Bushmen made use of a wide range of plant foods, but that mongongo nuts from the tree *Richinodendron rautanenii* were the most important dietary item, the average daily per capita consumption being 300 nuts, which yielded about 1,260 calories and 56 g of protein. Such a helping of nuts, weighing about 213 g, contained the caloric equivalent of 1.1 kg of cooked rice or the protein equivalent of 397 g of lean beef. Although the habitat in northwestern Botswana was harsh and arid, Lee established that a Bushman's life

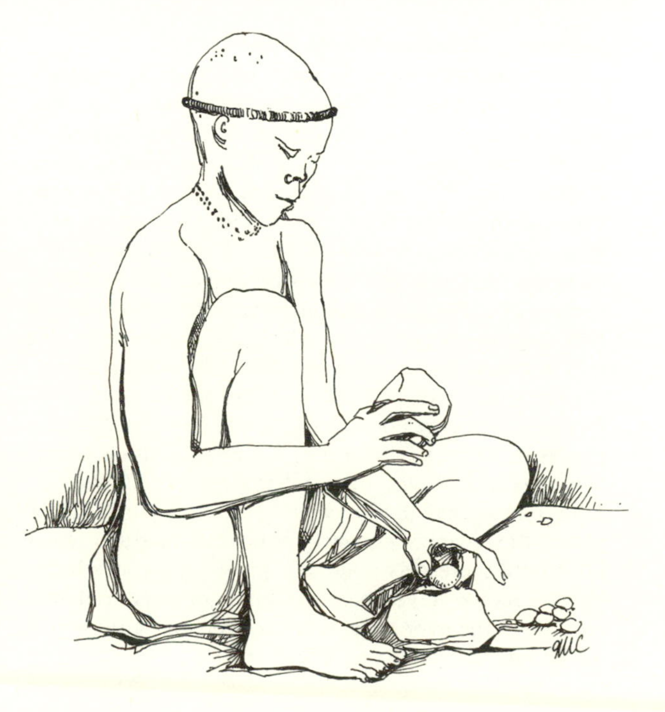

was not the precarious and arduous struggle for existence that most people thought was the unavoidable lot of hunter-gatherers. The Bushmen of Dobe devoted only twelve to ninteen hours a week to food-getting; the rest of the time was spent in leisure and relaxation. Despite this, they were well-nourished, consuming an average of 2,140 calories and 93 g of protein per day—quantities greater than those known to be required by people of small stature participating in vigorous activity.

The mongongo nut tree does not occur everywhere, and in other areas hunter-gatherers are known to depend on a variety of different food plants. After his study of the G/wi Bushmen in central Botswana, Silberbauer (1972) listed thirty-five species of plants regularly used as food sources. These included four kinds of melon, seven other fruit species, two types of seeds, eight types of leaves, and fourteen root or storage-organ forms. Of these, twelve species represented food sources of major importance in the economy of the people. A rather smaller number of plant foods is regularly used by the Hadza hunter-gatherers of the Lake Eyasi area of Tanzania (Woodburn 1968*a*). The bulk of their vegetable food comes from only ten species of plants the edible part being the root of four species, the berry of five, and the fruit and seeds of the rest.

Over most of southern Africa, hunter-gatherers have disappeared from the scene, and their activities can no longer be observed. But when the first Europeans started to explore the interior of the subcontinent, they frequently encountered bands of people who practiced neither agriculture nor animal husbandry. In the western Cape such people were known as the Sonqua or San, and a particular study of them has been made by J. E. Parkington. In addition to his archaeological studies, Parkington (Parkington and Poggenpoel 1971) has investigated the accounts of early travelers who described contacts with hunter-gatherer bands. One source of such information is Simon van de Stel's journal of his expedition to Namaqualand, 1685–86, translated and edited by Waterhouse (1932). The following extracts from this journal indicate the kind of information available from such historical sources (Waterhouse 1932, pp. 117–18, 128):

> September. Tuesday the 4th, 1685.
>
> They said they were Sonquas or Obiquas, and they had come here to look for an eland which they had shot with poisoned arrows the day before. They carried arrows, bow, and assegai. They have no cattle, living on honey and the wild animals they shoot. Their skin is very rough and scurvy owing to the frequent hunger they endure and the lack of fat with which to smear themselves. The honourable Commander made them a present of a sheep and although they are not men of any breeding they were nevertheless polite enough to give him in return three tiger-cat skins. They immediately cut the sheep's throat, flayed it and then cut off both forelegs, allowing nothing to go waste except four glands in the legs, which they cut out and threw away. When we asked why they did so, they could give no other reason than that they never ate such things. They placed the meat under ashes for half an hour and feasted on it until there was none left, gnawing the legs like beasts. As we stayed here the whole day and took our altitude in the evening and discovered our latitude to be 32°58′ and our longitude 38°57′.
>
> October, Tuesday the 2nd, 1685.
>
> They are lean and slight of frame, due to the great hunger and hardship which they endure. They eat nothing but the bulbs of flowers which they [we?] call *uyentjes,* tortoises and a certain large kind of caterpillar, together with locusts which are found here in abundance.

The *uyentjes* referred to here appear to have been an important item in the diet of hunter-gatherers in the Cape. Parkington (in Parkington and Poggenpoel 1971) has suggested that the chief task of the women in Cape hunter-gatherer bands was collecting roots, corms, bulbs, and tubers with a digging stick and kaross. Early Euro-

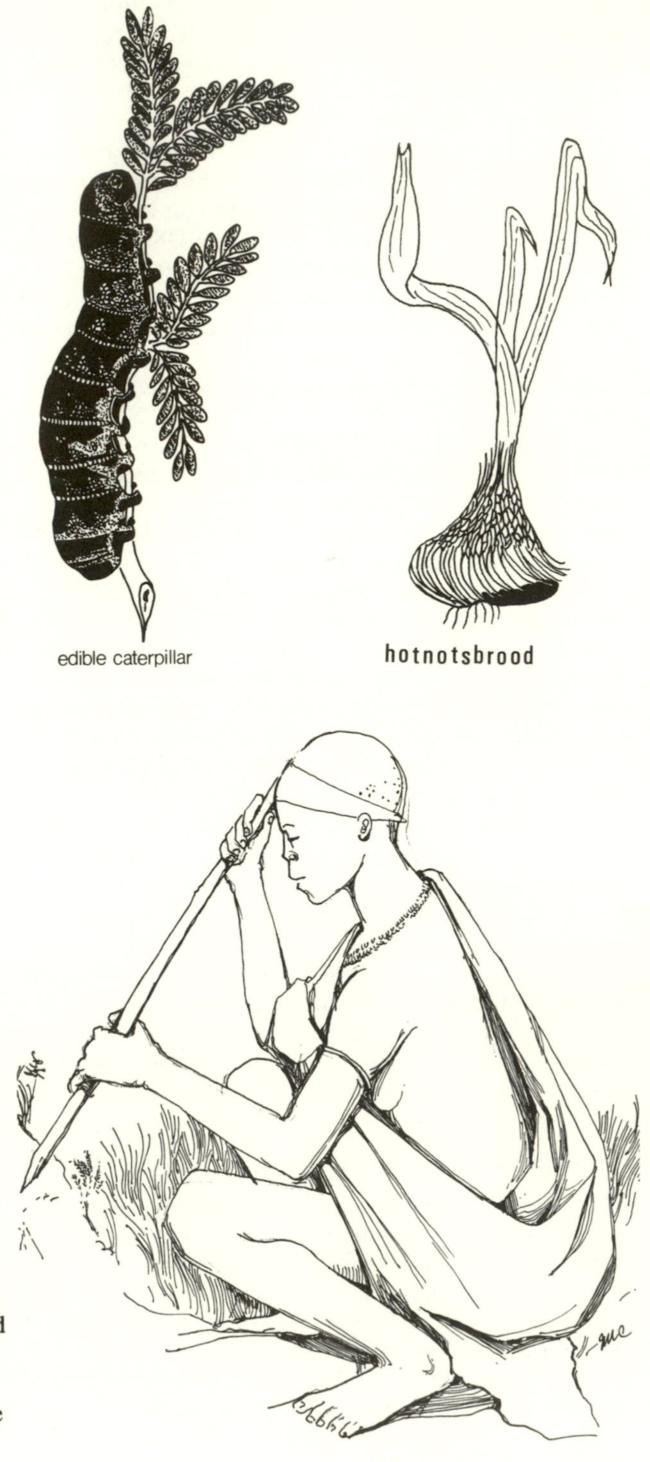

pean settlers used the name *uintjiestok,* "onion stick," for the digging tool and *uintjiesak,* "onion sack," for the kaross. The records suggest that the corms were ground and roasted before being eaten, the common name for several species of *Watsonia* being *Hotnotsbrood* or Hottentot's bread.

The importance of the corms of *Watsonia* and similar plants in the Stone Age human diet has been confirmed by plant remains preserved in various southern Cape caves. The first such evidence came to light at the Melkhoutboom Cave, Alexandria District, in the course of an excavation by Hewitt (1931) during May 1930. The site was reexcavated in 1967 by Deacon; one of his aims was to recover further plant remains for analysis (H. J. Deacon 1969, 1970). Hewitt's original list of plant species was extended, and the main food plants were shown to be *Hypoxis* sp., *Watsonia* sp., *Freesia* sp., *Moraea* sp., *Bulbine alooides, Oxalis* sp., *Cyperus usitatus, Schotia afra,* and *Harpephyllum caffrum.* By the time H. J. Deacon (1972) wrote his review of the Postpleistocene in South Africa, information on food plants was available from five more rock shelters in the eastern and southern Cape. Farther afield, sites such as Border Cave (Beaumont 1973), Kruger Cave (Mason, Friede, and Pienaar, 1974), and Mirabib Shelter (Sandelowsky 1974) have also been yielding valuable insights into the plant foods of their Stone Age inhabitants.

In addition to plants, Stone Age inhabitants of the southern African coast had been using the resources of the sea for a very long time. In fact the Middle Stone Age layers in the Klasies River Mouth Cave, ranging in time from about 125,000 to 55,000 years B.P., contain clear evidence for the earliest systematic exploitation of marine resources known anywhere in the world. These layers are packed with mollusk shells and also contain bones of seals and penguins that the people had discarded. Klein (1977) has pointed out that at more recent coastal sites such as Nelson Bay (Klein 1972*a,b*) and Elands Bay (Parkington 1976), where the levels with marine food remains are all younger than 12,000 years, remains of seals and penguins, comparable to those from Klasies, are accompanied by numerous bones of fish and flying birds, which are rare at Klasies. It seems that fishing and the hunting of airborne birds were beyond the technological capabilities of the Middle Stone Age people, but that Later Stone Age people had acquired these skills.

The evidence of shell middens around the southern African coast, and of shell-packed deposits in coastal caves, testifies to the importance of gathered shellfish in the Stone Age diet. The first analysis of the faunal content of a southern African shell midden was carried out at the Bonteberg Shelter on the western side of the Cape Peninsula (Maggs and Speed 1967), where excavations conducted between 1962 and 1964 revealed a midden deposit 104 cm deep consisting of two layers dated at 2,050±95 and 4,505±100 years B.P. (Grindley, Speed, and Maggs. 1970). In the report on the Bonteberg material, Elizabeth Speed (now Voigt) laid down some guidelines for the analysis of molluscan shells in middens and subsequently (Voigt 1975) further defined criteria for the estimation of minimum numbers of individual mollusks in an assemblage. Using such methods, it was possible to show that in the shell midden at Bonteberg Shelter the lower level contained 55 percent limpets (*Patella* sp.) as well as large numbers of welks (*Burnupena* sp.), while in the upper level limpets decreased and whelks and winkles (*Oxystele* sp.) increased.

The analysis of mollusks from the Klasies River Mouth Caves (Voigt 1973), showed equally clear variations in dietary preference over an enormous span of time. The representation of limpets in the shellfish diet of the people varied from 71% to 17% at different phases of the accumulation, and winkles also fluctuated in popularity. By her layer-by-layer analysis, Voigt was also able to

cm

Painted burial stone

Robberg Cave

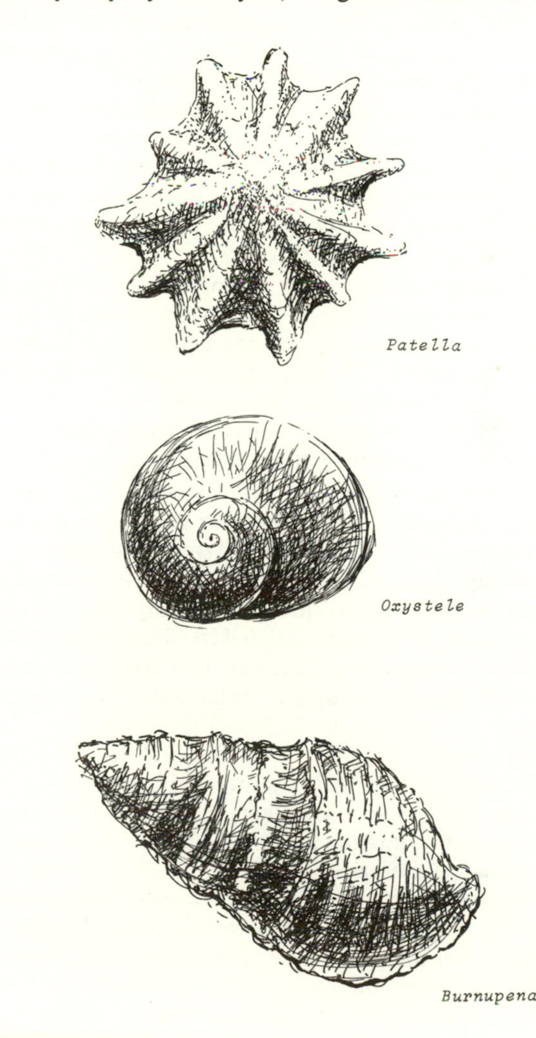

show which parts of the intertidal zone were being exploited at each cultural phase.

The extent and thickness of some of the southern African coastal shell middens is remarkable, and it would be interesting to know how many people were involved in their accumulation. Most of the authentic "strandloper" bands have disappeared, but coastal tribesmen of the Transkei still exploit shellfish in a traditional manner (Bigalke 1973), and so direct observations can be made on the rate of midden accumulation. For instance, Voigt (1975) has shown that a shell midden covering an area of 24 m² was accumulated by four persons over a period of 40 years.

This discussion has made it clear that the *gathered* component has always been extremely important in the diet of Stone Age peoples. In a recent review entitled "Gathering and the Hominid Adaption," Adrienne Zihlman and Nancy Tanner (n.d.) have gone so far as to suggest that gathering rather than hunting exerted the greater influence on the development of typically human behavior and thus on the course of human evolution.

Some Evidence for Seasonal Occupation of Southern African Caves

It is clearly important for hunter-gatherer peoples to be knowledgeable about the availability of natural foods throughout the year, and it is by moving about on a regular seasonal round that the bands will make the fullest use of their resources. Hilary Deacon (1969) has recorded an oral tradition concerning the movements of the last Bushman band in the Long Kloof of the southeastern Cape. This group, of about sixteen hunter-gatherers, was said to have occupied a series of three caves—one on the coast for about two months in summer, another in Baviaanskloof during the winter, and a third in a tributary of the Kouga River for the rest of the year. In this way they presumably made use of both marine and terrestrial resources to their best advantage.

Evidence, some of it ingenious and interesting, is steadily accumulating on the seasons during which Cape caves were occupied, and the evidence seems to be at variance with the spoken tradition about the Long Kloof Bushmen. For instance, plant foods preserved in the inland Melkhoutboom Cave deposit and excavated by Hilary Deacon (1969) include inflorescences of *Themeda triandra* and *Helipterum milleflorum,* which could have been gathered only in spring or early summer, and indications provided by shells in Wilton middens at the Nelson Bay Cave indicate that the mollusks were collected only in the colder months, so that this particular coastal cave must have been occupied in winter. The conclusion is based on the ratios of oxygen-16 to oxygen-18 in the last growth increments of the shells concerned (Shackleton 1973).

Mention has already been made of the importance of the corms of *Watsonia* and other Iridaceae in the diet of Stone Age people in the Cape. Parkington (1976) has provided histograms for flowering times of Iridaceae in the southwestern Cape that show that August, September, and October are the months when most flowers are present. It is known that the corms of plants like *Watsonia* attain their maximum size after flowering in about October; thereafter the leaves wither and the plant becomes much more difficult to find. This strongly suggests that a cave such as De Hangen in the Clanwilliam area (Parkington and Poggenpoel 1971), where abundant remains of Iridaceae corms are preserved, was occupied in early summer. The De Hangen deposit also contains a good deal of plant material that the prehistoric inhabitants brought in for bedding. This was made up of four main components, *Helichrysum* sp., *Ehrharta* sp., *Pennesetum* sp. and a sedge, either *Cyperus* sp., or *Mariscus* sp. All these plant remains had florescences, and all are known to flower between October and December.

Faunal remains from De Hangen confirm the summer occupation of the cave. The bones include pieces from at least 313 tortoises that had presumably been gathered in summer. During the winter, the tortoises in this part of the country tend to hibernate. Parkington also recovered remains of at least 64 dassies, many of which were immature, but which could be aged on the basis of tooth eruption, since it is known that the young are born there between September and November. This study showed that all the dassies at De Hangen could have been killed between November and February, but that very few could have died in the winter.

An equally detailed study has been carried out by John Parkington (1976) at sites such as Elands Bay on the adjacent west coast. Here the evaluation of seal bones discarded by the cave's inhabitants appears to have great potential for establishing the season of the year when the hunting took place. At Elands Bay Cave the sizes of seal mandibles, compared with those of known-age animals, strongly suggest occupation between July and October.

Inhabitants of these west coast caves and shelters ate very large numbers of marine mollusks, which are frequently toxic in summer months owing to contamination by "red tide" dinoflagellates (Grindley and Nel 1970). The people presumably avoided being poisoned by eating the shellfish only during the winter.

The accumulating evidence certainly seems to suggest that Stone Age hunter-gatherers living within reach of the southern African coast made seasonal rounds whereby summer months were spent inland and the winters were spent on the seashore.

In the interior, Stone Age bands presumably moved seasonally in response to migrations of the game herds. A model for Later Stone Age seasonal movements between the Drakensberg escarpment and the Natal coastal areas has been suggested by Carter (1970), who pointed out that during spring and early summer the Highland Sourveld provides attractive grazing. In late summer the grazing value declines, and in winter these highland areas experience very low temperatures. Game herds used to respond to such seasonal changes by migrating up and down the escarpment edge. They were presumably followed by the hunter-gatherer bands who were responsible for the spectacular Drakensberg rock paintings. On this topic Carter wrote (1970, p. 57):

> Acceptance of the model would have important repercussions on the interpretation of the many rock-paintings that occur in the Drakensberg. The virtual restriction of painting to the summer months could be interpreted as evidence of periodic, possibly annual, ceremonies in which painting was an integral and important part. Such an interpretation would be in accord with findings in which the ritual aspects of the paintings are emphasised. It is tempting to suggest that the summer occupation of the Drakensberg was a time of plenty for the hunter-gatherers of the Late Stone Age. With

adequate supplies of animal and vegetable food it would be economically possible for band size to be larger and ceremonial gatherings to take place. During the warm, dry, winter months in the Thorn Veld, band size would of necessity be reduced, camp sites would have been in the open and ceremonial gatherings less likely to occur.

Observations on living hunter-gatherers (e.g., Silberbauer 1972; Lee 1972; Woodburn 1968*b*) emphasize the loose social structure of bands and their constantly changing size. In view of this there could well have been times when large numbers of people came together for ceremonial purposes. In his paper on group size during the Later Stone Age, Maggs (1971) has drawn evidence from various sources, including rock paintings, and has concluded that bands of eight to twelve members or of twenty to twenty-five members were most common, but that much larger aggregations were by no means unusual. Figures for human group sizes in Drakensberg rock paintings have also been given by Pager (1971); these varied from two to thirty-two.

The evidence discussed here comes very largely from the closing phases of the Stone Age, but it is likely that seasonal movements characterized the way of life of hunter-gatherer bands throughout the Pleistocene. An attempt has recently been made to document seasonal occupation of Olduvai Gorge living sites through comparisons of food remains left on these floors with those of contemporary Bushmen bands. In this attempt, Speth and Davis (1975) listed animals eaten by !Kung and G/wi Bushmen in Botswana and correlated these with the seasons when they were hunted. They worked out percentage representations of three groups of animals—Bovidae (the various antelopes), Carnivora (largely jackals and foxes), and Chelonia (tortoises and terrapins) and compared these with percentages derived from Mary Leakey's data on bone food refuse from various floors and levels in the Olduvai sequence (M. D. Leakey 1971). A complication in the comparison is that the percentages of Bushman prey were based on numbers of individual animals, whereas those from Olduvai were calculated on numbers of bones identified in each taxon. Serious discrepancies could result, particularly for tortoises and terrapins, since when their carapaces are broken up very numerous fragments result. Nevertheless, Speth and Davis did find an interesting correspondence between the proportions of the three animal taxa in the Bushman prey and the Olduvai remains. They were able to establish that Bushmen ate the tortoises and terrapins almost exclusively during the summer rainy period, since they were hibernating in the dry winter months. Assuming that climatic seasonality was comparable between the Olduvai and Botswana habitats, Speth and Davis concluded that all but four of the twenty-two Olduvai collections had resulted from dry-season occupation. The four levels that suggested wet-season occupancy were DK (all levels), FLK NN level 3, FLK NN level 1, and the MNK skull site. Three of these are among the earliest occupations in the gorge and may have coincided with a period of slightly reduced aridity.

If there is validity in these interesting deductions, it appears that Olduvai Gorge, or rather the lake that occupied the position of the present gorge, was one of the stopping places on an early hominid seasonal round that took in the adjacent plains and mountains.

It seems likely that the early hominids whose remains are preserved in the Sterkfontein valley caves would also have made regular seasonal movements to exploit the resources of the highveld grassland and adjacent bushveld areas. The Sterkfontein valley is close to the northern edge of the highveld plateau, and within 20 miles to the north, other habitats are available that would have been more productive in winter. The open highveld was almost certainly a rich and desirable habitat in spring and summer but may well have been largely avoided in winter. Hunter-gatherers, unlike their sedentary successors, were locked into an ecological framework that they could influence but little. Like other animals, they would have tried to make optimal use of a patchy environment by exploiting each part of it during its most productive season (MacArthur and Pianka 1966).

Some Consistent Features in Primitive Human Food Remains

In the following brief discussion I will restrict my comments to the animal component of Stone Age human diet, represented by skeletal remains.

The Nature of the Prey

Study of food remains from habitation sites in southern Africa has made it abundantly clear that the people were opportunists, feeding on whatever animal protein source was available to them, although from time to time and from place to place their diet may have been restricted by customs and taboos. For instance, in his study of the Hadza hunter-gatherers, Woodburn (1968*a*) found that these people rejected civets, monitor lizards, snakes, and terrapins, although they would eat lions, leopards, servals, wild cats, hyenas, jackals, and vultures. Reptiles such as monitor lizards, snakes, and terrapins are certainly eaten by contemporary Bushmen, and they also featured regularly in the diet of Stone Age people, as did tortoises. In fact, I would go so far as to say that abundant fragmented remains of tortoises in a bone accumulation strongly suggest human involvement.

Evidence from the southern Cape caves suggests that regular and effective fishing was beyond the capabilities of Middle Stone Age (or earlier) peoples, although the technique was mastered during the Later Stone Age. The same appears to be true for the hunting of flying birds, although there are some bird remains in the bone accumulations from almost all the sites studied.

Invertebrates such as land snails, freshwater mollusks, and crabs appear to have been eaten whenever they were available, and large-scale exploitation of marine mollusks goes back well into the Middle Stone Age, if not further.

Comprehensive data are available on the animals that contributed to bone accumulations in thirteen southern African cave sites known to have been occupied by Stone Age peoples. The relevant collections come from Pomongwe, Bushman Rock, Fackelträger, and Wilton, which I have discussed in this chapter, from Nelson Bay, Die Kelders 1, Klasies River Mouth caves, and Redcliff, described by Klein in various publications, from Scott's Cave, described by Klein and Scott (1974), from Andrieskraal 1 (Hendey and Singer 1965), from De Hangen (Parkington and Poggenpoel 1971), and from Elands Bay (Parkington 1976). Information from other important sites such as Beyeneskrans, Boomplaas, and Border Cave will become available shortly.

In table 33 I have listed the contributions made to each collection by individual animals of different mammalian orders and have calculated percentage representations of each order. The ranges and mean percentage representations per order are given in table 34 and plotted graphically in figure 48.

Primates Other Than Homo. Human remains, occasionally derived from burials in the caves, have been excluded. Sparse remains of baboons were found in eight of the thirteen collections studied, suggesting that these animals were occasionally hunted. Elsewhere in Africa, baboons may have been hunted on a larger scale from time to time. This is suggested by evidence at locality DE/89 of Olorgesailie, where remains of at least 50 adult and 13 juvenile *Theropithecus oswaldi* baboons were preserved along with Acheulean tools. Glynn Isaac (1968) suggests that these baboons may have been hunted, in the same way that the contemporary Hadza people occasionally hunt them, by surrounding a sleeping troop at night, dislodging them with arrows, and clubbing them to death.

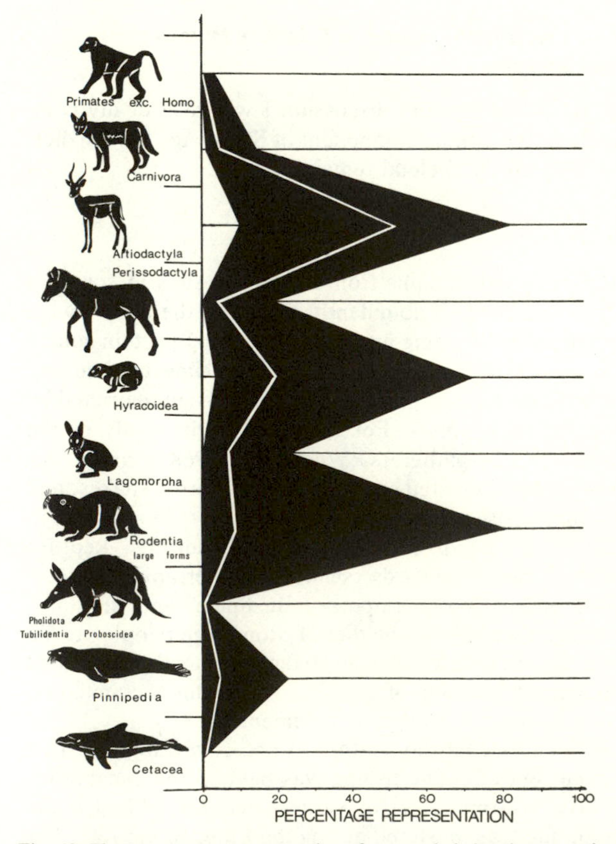

Fig. 48. The percentage representation of mammals belonging to various orders in the remains from thirteen Stone Age cave-habitation sites in southern Africa. The range for each order is indicated in black, and the mean value is shown as a white line.

Carnivora. Small quantities of carnivore remains are a consistent feature of the bones from all thirteen sites; they come typically from fairly small animals, although leopards are represented at some sites. It is difficult to be sure whether the carnivores had invariably been eaten by the people, or whether they had been hunted for their skins. In two collections, those from Scott's Cave and De Hangen, several mongooses and genets appear to have died naturally at the sites; their remains were therefore excluded from the calculations.

Artiodactyla. These even-toed ungulates, the bucks, pigs, giraffes, and hippos, were unquestionably the most important meat source for the Stone Age people at most of the thirteen sites. As was shown earlier in this chapter, where meat yields were calculated for prey animals from Pomongwe, Bushman Rock, Wilton, and Fackelträger, medium-sized antelope contributed most of the protein to Middle and Later Stone Age diets. The species hunted vary with habitat and age.

Perissodactyla. Remains of zebras and rhinos are far less prominent in the food refuse than are those of antelopes. When large animals such as rhinos or the extinct Cape horse, *Equus capensis,* were eaten, it is probable that the hunter-gatherer band moved to where the carcass lay, rather than transporting much of the meat back to the cave. Zebra remains were particularly abundant in the Redcliff deposit, perhaps owing to local abundance of equids there during the Rhodesian Stone Age.

Hyracoidea. As was discussed earlier in this chapter, dassies were much favored Stone Age prey and were hunted or snared in large numbers. At a site like De Hangen they were brought back to the shelter more frequently than were other mammals, although they were far outnumbered by tortoises.

Lagomorpha. Remains of hares have been found at all the sites under review. These animals apparently made a small but consistent contribution to the Stone Age meat diet.

Large Forms of Rodentia. Occasional remains of porcupines, springhares, and cane rats are found in the food refuse at the sites, but at a cave like Die Kelders 1, on the southern Cape coast, remains of incredible numbers of dune mole rats, *Bathyergus suillus,* have been found. Klein (1975*b*) showed that their remains characteristically lacked foot bones, suggesting that the pelts had been used and that the feet were left attached to the skins.

If it were not for the large numbers of mole rats represented at Die Kelders and Klasies River Mouth, large rodents would have featured insignificantly in the prey spectrum.

Pholidota, Tubulidentata, and Proboscidea. Pangolins and ant bears were apparently eaten on rare occasions.

Some meager traces of elephants are present, but these large animals were probably eaten at the kill sites.

Pinnipedia. The evidence of the coastal sites indicates that seals were regularly exploited. At Elands Bay, and possibly elsewhere, seal-hunting appears to have been a regular seasonal activity.

Cetacea. Occasional remains of dolphins and whales are found in the coastal caves. They presumably came from stranded animals that were used by the people.

On account of the apparently extremely catholic tastes in meat of Stone Age people in southern Africa, I am doubtful if human food remains could be recognized solely on the basis of the prey species involved. Richard Klein (1975*a*, 1977) has suggested that hominid food refuse might be separated from carnivore refuse by the proportional representation of *carnivore remains themselves* within them. The assumption is that carnivores typically feed on other carnivores more frequently than do hominids. Klein has proposed (1975*a*, p. 284) that the *carnivore-ungulate* ratio may be used as indicator of the bone-accumulating agency. Higher figures would suggest carnivore involvement, lower figures, hominid activity. I am inclined to think that the term "ungulate" needs to be somewhat restricted to be really useful in this context. Its conventional usage embraces five orders of mammals—Artiodactyla, Perissodactyla, Hyracoidea, Proboscidea, and Sirenia (Wender 1948). In my opinion it would be expedient to exclude the Hyracoidea, since the presence of dassies at a southern African site is usually determined by the presence of a rocky habitat nearby. Inclusion or exclusion of elephants and sea cows would be immaterial to the present discussion. I suggest that, in this context, "ungulate" might be taken to mean the artiodactyl plus perissodactyl component. On this basis I have calculated carnivore-ungulate percentages for the thirteen sites under discussion:

Bushman Rock, 1:59	1.7
Wilton, 2:72	2.8
Pomongwe, 4:95	4.2
Redcliff, 35:643	5.4
Klasies (excluding Cave 1), 13:223	5.8
Scott's Cave, 3:42	7.1
Klasies Cave 1, 38:424	9.0
Die Kelders 1, 28:254	11.0
Fackelträger, 3:27	11.1
Nelson Bay, 38:347	11.1
De Hangen, 2:15	13.3
Elands Bay, 18:120	15.0
Andrieskraal 1, 10:31	32.2

The mean of these thirteen percentages is 10.0. The ratio for Andrieskraal 1 is 32.2—a figure appreciably higher than those for the other sites. It is conceivable that from time to time this site was a carnivore lair in addition to being a human occupation site.

The Nature of the Damage

It will have become abundantly clear, from the discussions in this chapter, that a striking feature of food remains from human occupation sites is the damage the bones have suffered, both during marrow extraction and from trampling by people's feet. Extracting marrow from antelope limb bones has typically involved smashing the bone shafts with stone choppers, which produces very numerous bone flakes. The ratio of bone flakes to other pieces in the Middle and Later Stone Age collections studied is surprisingly constant. At Pomongwe, Bushman Rock, and Fackelträger, the percentages were 53.6, 56.9, and 51.2 respectively. At Wilton, fragmentation from trampling makes it difficult to distinguish bone flakes from other fragments; here 77.6% of the sample falls into this nondescript category. As discussed earlier, lengths of bone flakes in the collections tend to be remarkably consistent, probably because efficient extraction of marrow typically results in bone flakes with a mean length of about 5 cm. Lengths of flakes from the four sites where measurements have been made are shown in figure 49.

As will be described in chapter 4, hyenas produce bone flakes when breaking up long-bone shafts with their premolars. I am not confident of my ability to distinguish these hyena-made bone flakes from flakes produced with stone choppers. This matter will be touched upon again in chapter 7 (see also figures 146 and 147).

Stone Age people were undoubtedly capable of chewing and swallowing a good deal of bone, and the pattern of unchewable skeletal parts from animals such as dassies is highly diagnostic, as mentioned earlier and in chapter 7.

Chopping and cutting with sharp stone tools produce occasional marks on bones that could not have been caused by the teeth of carnivores or by other agencies. When these are found, human involvement is confirmed. A study I carried out on bone food refuse from the Iron Age site of Zimbabwe (Brain, 1974*c*) suggested that cut marks and chop marks are far more common on bones when iron, rather than stone, tools have been employed. Once primitive people acquired containers in which to cook meat, the pattern of carcass-preparation and of bone damage appears to have changed in various ways. For instance, horns were often chopped from antelope calvariae so the head could be boiled in a pot. Long bones were also smashed into very small fragments for boiling to prepare "bone grease"—a practice described among

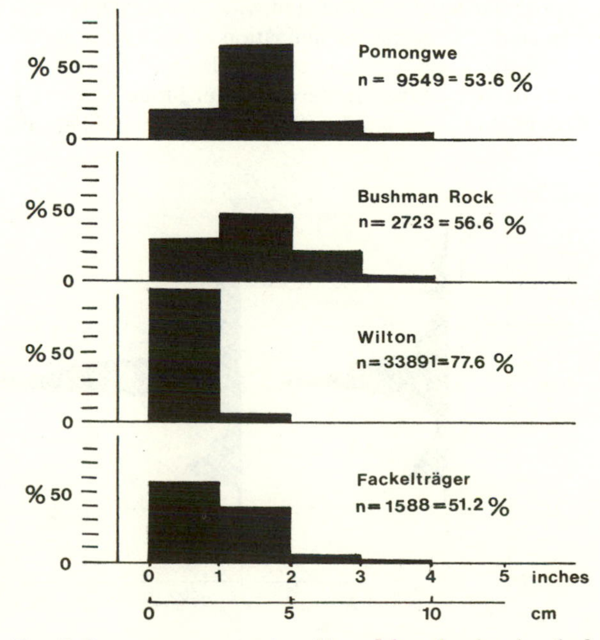

Fig. 49. Percentage representation of bone flakes of various lengths from the four Stone Age cave sites discussed in the text.

North American Indians (Leechman 1951) but probably employed by African Iron Age peoples as well.

A good deal has been written about the human "crack and twist" technique for breaking long bones. The shaft frequently breaks in a spiral manner, and sharp-pointed bone pieces result. It is true that no animal other than man is capable of grasping a long bone in its two hands and twisting the shaft in opposing directions. If spiral fractures resulted *only* from this process then the presence of spirally fractured bones in an assemblage would be excellent evidence for human involvement. Regrettably, this is not true (see fig. 144). Such bones are found among food remains of hyenas and leopards, and the fact that shafts have broken in a spiral manner is more the result of inherent properties of the bones than anything else (fig. 145). This is further discussed in chapter 7.

The Presence of Biases in the Bone Sample

It is obvious that the feeding action of people, like that of other carnivores, will result in the consistent disappearance of certain skeletal elements and the persistence of others. The relevant factors here are the dental performance of the eaters and the robusticity of the bones being chewed. Other more typically human biases likely to be introduced into collections of food remains result from culturally determined butchery practices and traditions. The disappearance of foot bones from mole-rat skeletons at Die Kelders suggests, for instance, that the people used the pelts of these animals and typically left the foot bones attached to the skins. Likewise, the paucity of large antelope vertebrae at human occupation sites is probably because the axial skeletons of such prey were characteristically left at the kill sites.

Association with Artifacts

Perhaps the surest and most direct evidence of human involvement in a bone accumulation is the association of artifacts with the remains. Studies at caves like Pomongwe indicate that, at least for Middle and Later Stone Age assemblages, an abundance of bone pieces in any particular level is associated with an equivalent richness in artifacts. A further indication of such an association is given in figure 50, where weights and numbers of artifacts in each of the Middle and Later Stone Age levels at Redcliff are correlated with weights and numbers of bone pieces found in association (Brain and Cooke 1967; Brain 1969*a*). Such associations were probably characteristic of human occupational debris throughout the Pleistocene, although evidence from Olduvai, discussed earlier, suggests that the very early occupation sites were typified by fewer artifacts per unit of bone than the later ones.

Association with Evidence of Fire

Bone fragments from the Middle and Later Stone Age occupation sites I have discussed were frequently preserved in layers rich in ash, and charring of the pieces themselves was not uncommon. Some simple experiments with bones in the ashes of campfires has demonstrated to me that there are two distinct stages in the charring of bone. The bone starts white then, as the collagen within it is carbonized, turns black. Finally, with continued heating, the black carbon is oxidized and the bone reverts to a white color and a chalky consistency. It is quite possible for a single piece of bone to show all three stages, as does the one shown in figure 51.

Before the introduction of cooking pots, meat presumably was cooked directly over the fire or beneath the coals. This could have charred exposed bones, but I imagine charring occurred more frequently when inedible bone pieces were discarded into the hot ash or when a new fire was made over the refuse from a previous meal.

In open sites, bone could be charred by natural veld fires; for instance, in the Pliocene deposit at Langebaanweg, tortoise remains frequently show the effects of fires that may not have been deliberately started (Hendey 1974*a*). However, consistent evidence of fire in layer after layer of a cave site is highly suggestive of purposeful hominid activity.

There is apparently no acceptable evidence of fire in southern African caves predating the late Acheulean, where undoubted traces occur in the Cave of Hearths at Makapansgat and at Montagu Cave in the Cape (Oakley 1956). At one time it seemed likely that the blackened bones in the gray breccia at Makapansgat Limeworks showed evidence of having been burned, and more than fifty years ago Professor Dart wrote (1925*b*, p. 454):

> As the deposit seemed to be of the cave variety and some of the bone fragments had a blackened, charred appearance, the agency of man in its formation was

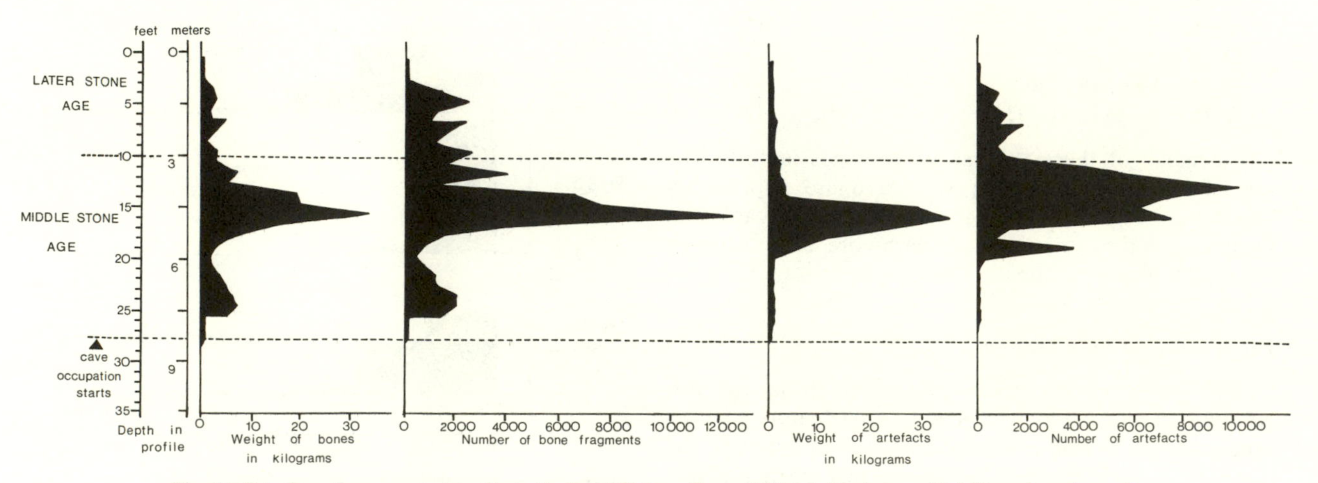

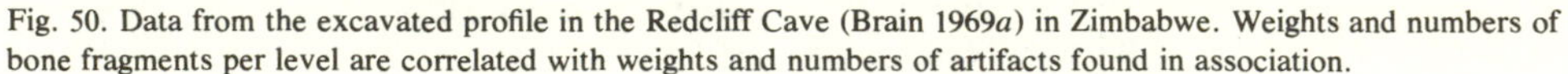
Fig. 50. Data from the excavated profile in the Redcliff Cave (Brain 1969*a*) in Zimbabwe. Weights and numbers of bone fragments per level are correlated with weights and numbers of artifacts found in association.

highly probable. For this reason some promising bone fragments were taken for chemical examination to determine if any free carbon were present in the bone. This examination was brought to a successful conclusion through the courtesy of Dr Moir, of the Government Chemical Laboratories, and Dr Fox, of the South African Institute for Medical Research. After the bone had been treated with acid and the soluble material had been washed away, a residue of yellowish dirt containing numerous black particles was revealed. These particles had an appearance similar to that of carbon particles under the microscope, and by means of transformation into carbon dioxide a considerable percentage of the element was demonstrated chemically.

This evidence certainly seemed reasonable enough, and it prompted Dart (1948*a,b*) to name the first australopithecine remains from Makapansgat *Australopithecus prometheus*. Regrettably, the first indications no longer seem as unambiguous as they did, and in 1956 Kenneth Oakley wrote (1956, p. 103):

> First one naturally looks for confirmation that there really is free carbon in the bed containing the remains of *Australopithecus*. Unfortunately it is not forthcoming. A number of new samples have been tested and no free carbon can be found in any of them. The blackness of the bone fragments in all the samples tested recently proves to be due to manganese dioxide. Yet the fact remains that competent chemists determined considerable quantities of carbon in the pieces collected at the same site in 1925. One cannot avoid entertaining a doubt whether the carbon in the original samples was indigenous. There is always the possibility to be considered that material from recent fires had, by some strange chance, infiltrated, or been introduced through blasting. So long as no carbon can be found in samples of the Australopithecine bed collected under test conditions, one feels bound to say that there is no valid evidence that *Australopithecus* was a fire-user.

Food Remains of Precultural Hominids

The recognition of hominid food remains in the absence of both artifacts and traces of fire would be extremely difficult, and I would certainly not be competent to make such a diagnosis. It presumably would depend on the damage done to bones by hominid teeth—a subject requiring a good deal of further study.

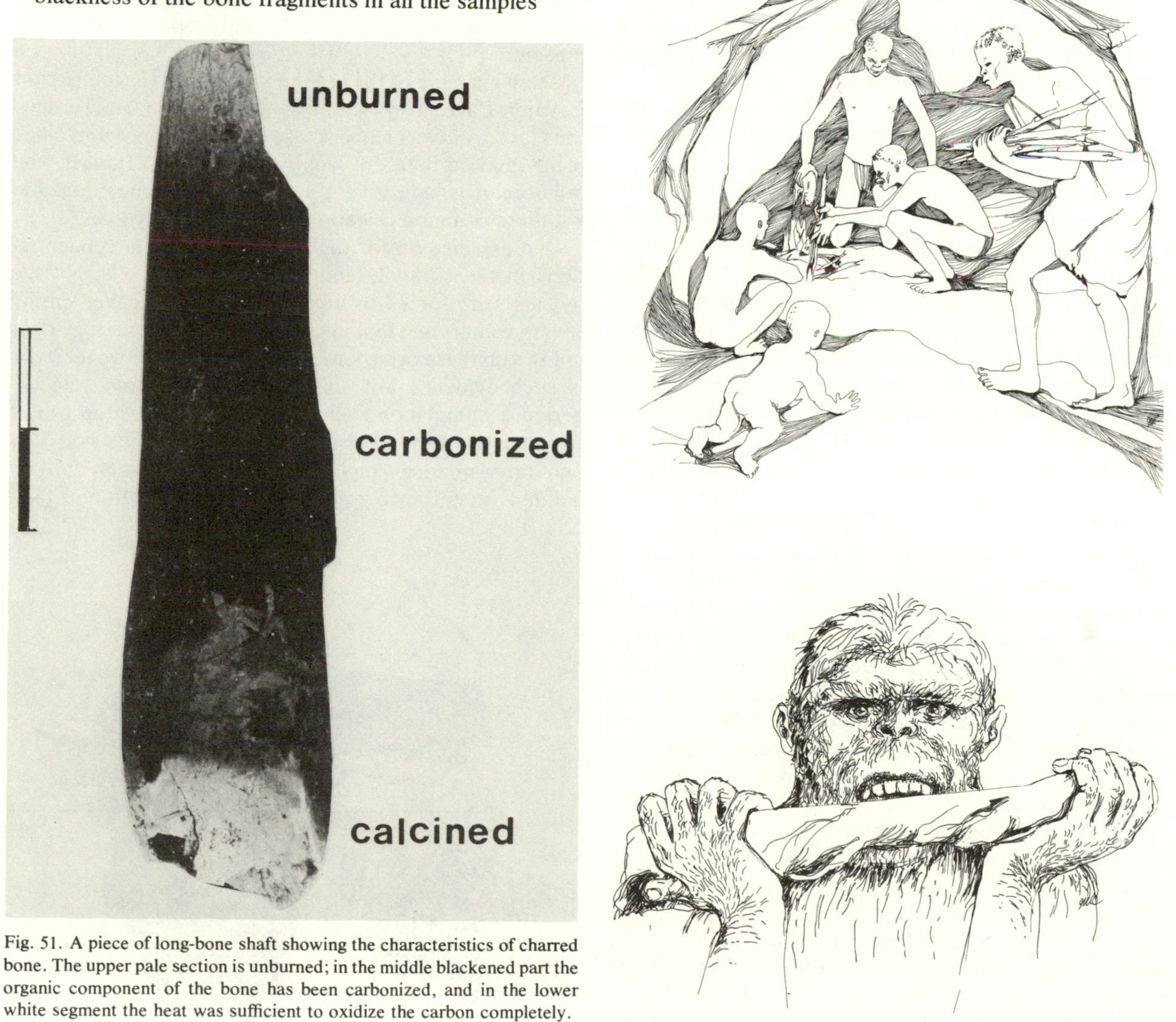

Fig. 51. A piece of long-bone shaft showing the characteristics of charred bone. The upper pale section is unburned; in the middle blackened part the organic component of the bone has been carbonized, and in the lower white segment the heat was sufficient to oxidize the carbon completely.

4 Food Remains of Carnivores in African Caves

> If one had to give a general answer to the question, "What do the Carnivora eat?" it would be a very simple one—"What they can get." [Ewer 1973]

A variety of carnivorous animals use caves as retreats, feeding places, and breeding sites. In such places bony food remains may accumulate, and in certain circumstances these may become fossilized. From the taphonomic point of view it would be useful if bone accumulations built up by different carnivores could be distinguished. This chapter will consider mainly living carnivores other than hominids that may take bones to African caves; the role of certain extinct carnivores that perhaps were important in this respect will be touched upon briefly.

The living carnivores that appear to be of special significance are the three species of hyena and the leopard. Several other species may contribute occasional bones to an accumulation in a cave; these include the civet, ratel, wildcat, wild dog, serval, and caracal, as well as certain species of mongoose, all of which rear their young in holes or rock crevices that may fall within the catchment area of a cave entrance. Their possible contributions should be noted but will not be considered here. Mention will also be made of the black eagle as an accumulator of hyrax bones.

The Spotted Hyena, *Crocuta crocuta* (Erxleben)

As will be mentioned in chapter 8, there is some argument about the geographic origin of the spotted hyena, which today has a wide African distribution (fig. 52). During the Pleistocene the species was also abundant in Europe and Asia. According to Kurtén (1968), *C. crocuta* first appeared in Europe in the I–Günz II at Süssenborn and Gombasek and then steadily increased in body size to become the true cave hyena *C. crocuta spelaea* Goldfuss. At the end of the last glaciation it became extinct both in Europe and in Asia.

The first suggestion that spotted hyenas may have been responsible for collecting bones in caves was made well over a century ago. An excavation conducted in 1821 by the Reverend (later Dean) William Buckland at Kirkdale Cave, Yorkshire, revealed vast numbers of hyena remains, together with teeth of hippopotamus and other animals. He concluded that the cave had served as a hyena lair during antediluvian times and that the layer of mud that covered the remains had been laid down in the cave by the waters of the biblical Deluge (Buckland 1822, 1823).

Sutcliffe (1969) has suggested that there may have been two kinds of hyena lairs in Britain. The first type is exemplified by Kirkdale Cave and by Tornewton Cave in Devon, in which the bone contents consisted mainly of splintered and gnawed remains of the spotted hyenas themselves, especially teeth and foot bones. The hyenas had been of all ages, including many juveniles. They were associated with remains of other animals, including hippopotamus, which indicated an interglacial age for the deposit, and very numerous hyena droppings were also present.

As an example of the second kind of lair, Sutcliffe cites Levaton Cave, Torbryan, less than a mile from Tornewton Cave. A low chamber close to the entrance contained many gnawed bones of woolly rhinoceros, reindeer, and other animals indicative of last glaciation times, together with less common remains of the spotted hyenas.

One explanation for the remarkable hyena concentrations in the Kirkdale and Tornewton caves is that the hyenas were forced to take refuge in such caves during severe winters and that many died and were eaten by their fellows on these occasions. If this were so, however, why is such evidence not also forthcoming from a glacial-period accumulation such as that in the Levaton Cave? As an alternative explanation, Sutcliffe suggested that two different zones might have existed within an individual den: an entrance zone with its concentration of food

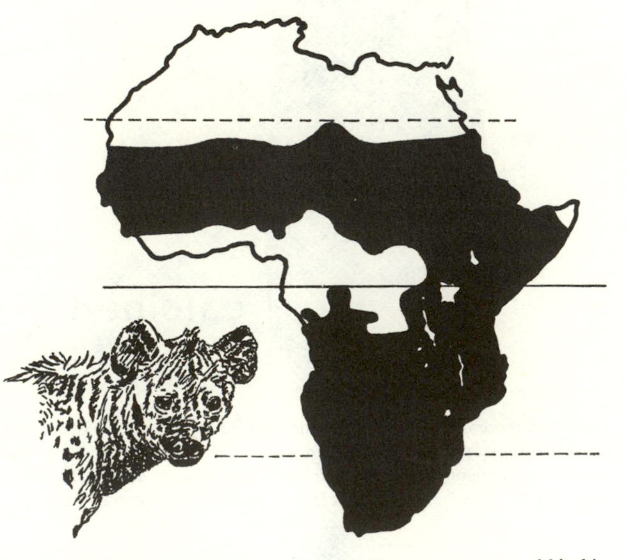

Fig. 52. Distribution of the spotted hyena, *Crocuta crocuta*, within historic times.

debris, and an inner residential zone with abundant hyena remains. The former zone could have been represented by Levaton Cave, the latter by Kirkdale and Tornewton.

Very shortly after Buckland's assertion that spotted hyenas were responsible for bone accumulations in caves and that they ate one another in their lairs, the concept of the bone-collecting, cannibalistic hyena was challenged by Robert Knox, who traveled extensively in South Africa between 1812 and 1817. He published his remarks under the title "Notice Relative to the Habits of the Hyaenas in Southern Africa" (Knox 1822), and his African research has been further remarked upon by Kirby (1940). Concerning hyena behavior he wrote (Knox 1822, p. 383): "yet I do not wish it to be understood that I deny that hyaenas ever drag their prey, including the bones, to the caverns, or wild mountain tracts, they inhabit. Many instances occurred, however which indicate that they do not . . . the young of these animals follow them early into the field, so that I much question if they ever carry a portion of their prey, on any occasion whatever, to their dens."

Despite doubts expressed by Knox, the concept of the bone-accumulating, cannibalistic hyena persisted. Dawkins (1874) interpreted the very extensive bone accumulation in the Wookey Hole Cave near Wells in Somerset as representing a spotted hyena den, and hyenas were probably generally regarded as the most important bone collectors in caves until this concept was challenged by R. A. Dart after the discovery and description of the first australopithecine remains from Makapansgat Limeworks (Dart 1948*a–c*). Dart was attempting to explain the origin of the extensive bone accumulation in the Makapansgat gray breccia, which included the australopithecine remains, and he reacted strongly both to Buckland's early hyena hypothesis and to more recent writers, such as Broom (in Broom and Schepers 1946), Von Koenigswald (1953), and Oakley (1954*a,b*), who expressed the opinion that the bones, including the australopithecine ones, in the well-known Transvaal caves were probably taken there by carnivores.

In searching for a solution to the Makapansgat bone accumulation problem, Dart enlisted the aid of A. R. Hughes, who undertook a study of modern spotted hyena lairs. He visited the farm Malamala, 26 km north of Skukuza, close to the Kruger National Park, where he thoroughly investigated two lairs (Hughes 1954*a,b*). One consisted of a series of nine ant-bear holes, the vicinity of which had been virtually cleared of vegetation by the trampling of the hyenas. Hughes excavated one of the tunnel systems and found that it covered a surface area of 13 m by 5 m, descending to a depth of almost 2 m. Although the tunnels had clearly been used as a hyena breeding lair, they were empty except for a single tortoise carapace. Outside the entrance to the lair were four chewed bones and one set of hyena droppings.

The second lair at Malamala consisted of a low shelter under an outcrop of exfoliating granite. It too was empty, although a few broken bones, a tortoise shell, and two fecal evacuations were found outside it.

In his search for additional evidence, Hughes (1958) also investigated four spotted hyena lairs in the Kalahari National Park. These consisted of two hyena resting places beneath calcrete outcrops along the east bank of the Auob River, both of which were devoid of bones or droppings, and two breeding lairs. The latter took the form of low recesses beneath calcrete outcrops, one in the east bank of the Nossob River at Saint John's Dam, the other in the bank of the Auob River, 10 km upstream from Twee Rivieren. No bones were found inside either lair, although two chewed springbok horns were found outside the first and eighteen bones and horns outside the second.

On the basis of this evidence, Dart (1956*a*, 1957*b*) vehemently rejected earlier claims that hyenas could be responsible for extensive bone accumulations in caves. He proposed instead that such accumulations were the work of hominids—either australopithecines or early men—and in his paper "The Myth of the Bone-Accumulating Hyena" Dart (1956*a*) concluded that the abundant spotted hyena remains in caves such as Kirkdale had been taken there by primitive men; that the whitened droppings that others had interpreted as hyena feces could have been produced by bone-crunching hominids or, if they were in fact hyena feces, could have been collected by people for therapeutic purposes. Dart pointed out that in more recent times such whitened droppings have been widely used medicinally, being termed *album graecum*. One prescription is as follows:

> Take Album Graecum, 1 ounce; Pulp of the Conserve of Roses, 2 ounces; Syrup of white Poppies, as much as is sufficient. This is to be spread pretty thick, and apply'd to the Throat, from Ear to Ear; and renewed every 6 or 7 hours, or oftener, if dry. A Poultess against Quinseys. [*The Pharmacopoeia Officinalis et Extemporanea; or, A Complete English Dispensatory* (1736), p. 681]

Despite Dart's forceful and eloquent rejection of hyenas as bone-collecting agents, evidence has been steadily building that spotted hyenas do take bones to breeding lairs and that these may accumulate either inside the lair or within the catchment area of its mouth. It has now been clearly established that breeding lairs of two kinds may be used: either burrows in the earth or shelters—recesses or caves among rock outcrops. Such lairs will now be considered in greater detail.

Crocuta Lairs in Burrows

I have already mentioned the breeding den Hughes examined at Malamala; it took the form of tunnels originally dug into friable granite subsoil by ant bears *(Orycteropus afer)*, which had then been used by warthogs *(Phacochoerus aethiopicus)* before the spotted hyenas moved in (Hughes 1954*a*).

A rather similar system of burrows was studied in 1935 by Harrison Matthews (1939) on the Balbal Plains in Tanzania. An area of about 30 m by 15 m was honeycombed with burrows and was bounded on the west side by a donga about 6 m deep, with steep earth banks. The first burrows had evidently been dug as adits into the side of donga, but as they approached the surface the roofs collapsed so that the breeding dens opened into steep-sided craters. Matthews concluded that the warrens were the products of some years' work by the hyenas.

After observations on a variety of *Crocuta* breeding dens in Ngorongoro and on the Serengeti plains, Kruuk was able to give this general description (1972, p. 242):

> Hyena dens usually have a number of entrances, sometimes a dozen or more. They are, with few exceptions, on flat ground; only rarely do hyenas use holes of an underground-river system or tunnels dug out in soft tuffs in a vertical surface. The tunnels are

mostly oval in section, more wide than high, and they narrow down from an entrance width of ½ to 1 m to sometimes less than 25 cm; they are 1 ½ to 3 m long. Sometimes large tunnels extend considerably farther; they may go on for several meters, usually horizontally, about ½ to 1 m under the surface. Most of them join underground.

One of the Ngorongoro hyena breeding dens was studied by Sutcliffe (1973*a*) during 1967 and 1970 when he succeeded in photographing a juvenile *Crocuta,* about three months old, inside one of the tunnels. Two similar burrows were also excavated in alluvial sediments in the Queen Elizabeth Park, Uganda (Sutcliffe 1970). The first was simple in form, with a single terminal chamber about a meter below ground level, opening to the surface by a trifurcating tunnel. The second was more complicated, with a central chamber and a series of tunnels leading to at least ten openings situated around it.

Disused ant-bear burrows that have been taken over by spotted hyenas as breeding dens have also been described from Zululand by Deane (1962). On an exposed and windswept ridge called "Nkwankwa" in the Hluhluwe Game Reserve, two colonies are situated about 180 m apart, each consisting of about six burrows. Around the warrens the grass had been trampled out by the hyenas and the warthogs that shared the breeding dens (see below).

Use of ant-bear burrows by spotted hyenas has been recorded in the Kalahari National Park by Mills and Mills (1977) and in the Kruger National Park by Stevenson-Hamilton (1934).

It seems that, in areas where ant-bear burrows are available, these are greatly favored by spotted hyenas as breeding lairs. They are frequently enlarged and modified by the cubs themselves.

Crocuta Lairs in Caves and Recesses

Mention has already been made of spotted hyena lairs in British caves such as Kirkdale, Tornewton, and Levaton, and similar sites are known from continental Europe. In Africa it has become clear that in habitats where deep soil exists, spotted hyenas favor ant-bear burrows for their lairs; in rocky or mountainous habitats, however, they make use of recesses or caves both for resting places, retreats, and breeding lairs.

In the course of his study on the bionomics of the spotted hyena, Harrison Matthews described lairs of both kinds. Concerning the cave lairs he wrote (1939, p. 49): "Near the Serronea river the country is dotted with small granite boulders. Amongst the rocks the hyaenas make their dens, which are natural cavities and are not enlarged or dug out by the animals."

The use of caves as lairs in rocky country of East Africa has been recorded by Percival (1924). This reference is quoted by Kruuk (1972), though I have not read the original account myself.

An interest in hyena lairs in Britain led Sutcliffe (1969, 1970, 1973*a*) to investigate caves in East Africa and to record the bones they contained. Although *Crocuta* food remains and droppings were found in caves on both Mount Elgon and Mount Suswa, Sutcliffe found no evidence that these caves had been used as lairs. He wrote (1973*a*, p. 46): "The nearest approach to a hyaena lair cave which we found was on the shore of Lake Langano, Ethiopia, where an impressive cave opening with scattered bones and droppings on its floor closed down to two low tunnels, apparently occupied by hyaenas at its innermost extremity. The openings of these tunnels were nevertheless still within sight of daylight."

In addition to this cave, Sutcliffe recorded two lairs near Kajiado Hospital in Kenya. They were in natural horizontal cracks in lava that apparently extended some distance into the rock but were too low to allow investigation beyond about 4.6 m. Bones were visible inside these lairs for a distance of about 7.6 m.

The Kalahari National Park in the northern Cape is an area of duneland traversed by two riverbeds, those of the Auob River and Nossob River, which are normally waterless. These are bounded by low calcrete scarps in which many solution cavities occur; one is described in chapter 5 as an example of a porcupine lair. Many of the solution cavities, which are often low but extensive recesses, are used by spotted hyenas. Hughes (1958) recorded two resting lairs and two breeding lairs in such places, and Mills and Mills (1977) described bones collected from a breeding lair at Urikaruus on the Auob River. This lair, which has also been used by porcupines (see below), has three entrances, one of which is shown in figure 53.

In the arid regions of the northern Cape and South-West Africa, I have investigated five reputed spotted hyena lairs; but in each case I found that porcupines had also been tenants of the sites. Bones present in the lairs could have been collected by either agent. In fact, the use of hyena lairs by more than one species must be taken into serious consideration when any interpretation of bones in lairs is made.

A den that is likely to provide valuable evidence on spotted hyena bone-collecting has recently been described by Andrew Hill (1978). It is on the dried-up flats of Lake Amboseli in Kenya and consists of tunnels beneath a calcrete crust, opening to the surface through a trench in the calcrete. According to Hill, porcupines have not been involved in collecting the numerous bones that litter the entrance trench. Hill's full description of the bone assemblage, and of the damage suffered by individual pieces, is awaited with interest.

Use of Lairs by More Than One Species

The interpretation of bone accumulations in hyena lairs can be very seriously complicated by collecting by porcupines. It appears that the kind of sites spotted hyenas prefer as breeding lairs coincide closely with those that porcupines select as diurnal retreats and, presumably, breeding sites. Both animals select burrows, caves, or recesses that are low-roofed, with lighting subdued or virtually absent.

As is discussed in chapter 5, porcupines are extremely active collectors of bones, and this must be borne in mind whenever a bone accumulation from a hyena lair is studied unless porcupine involvement can be definitely excluded. It may seem strange that spotted hyenas will tolerate porcupines in a lair they are currently using, but this does seem to happen. At the Urikaruus den in the Auob River, referred to above, Mills and Mills (1977) saw two porcupines enter the lair while the hyenas were there, but the species ignored each other. Another lair, known as Wright's den, on the Nossob River was found to be used successively by spotted hyenas, brown hyenas, and porcupines. On the basis of other observations in the Kalahari National Park, F. C. Eloff (1975) reports peaceful coexistence of spotted hyenas and porcupines in the hyenas' breeding lair.

Sutcliffe (1973*a*) has likewise stressed that some of the spotted hyena lairs he studied in East Africa showed evidence of multiple use. He pointed out that the hyenas may take over burrows from warthogs that may in turn have taken them from the ant bears that originally excavated them. He found that a lair in the Ngorongoro crater that had been used by spotted hyenas in 1967 was used by bat-eared foxes in 1970.

From the taphonomic point of view, the multiple use of lairs is of the greatest significance. The lack of aggression by spotted hyenas toward other tenants of the same lairs is surprising but well documented. A detailed description of interactions between spotted hyenas and warthogs at a series of communal breeding burrows in the Hluhluwe Game Reserve has been given by Deane (1962, pp. 34–35):

> The warthogs arrive anything up to two hours before sunset, slowly feeding their way towards the warrens and showing no signs of fear of the hyaenas and, in fact, will graze within a few feet of them. On the other hand the hyaenas, particularly the younger animals, often display a degree of uncertainty at the arrival of the warthogs. The very young whelps watch them intently, often bolting into one of the warrens if they approach too close. The warthogs arrive singly or in family parties and as many as seven have been observed arriving together and entering one of the warrens.
>
> On one occasion four mature hyaenas were standing close together when a large warthog female appeared and approached to within a distance of twelve feet from them. They stood watching the warthog intently and were obviously uneasy about her attitude. After they had eyed each other for a few seconds, the warthog suddenly trotted straight at them with the result that the hyaenas scattered. The warthog then stopped where the four hyaenas had been standing and after having satisfied herself that they had run off to a respectable distance, she approached one of the warrens and entered without further ado. As soon as she had entered the warren, the four hyaenas returned and lay down in the sun.
>
> On another occasion a noise was inadvertently made and two hyaena whelps, who were playing outside the warrens, took fright and immediately dashed down the nearest warren, which two warthogs had entered only a few minutes previously. After about fifteen seconds they appeared at the entrance of the hole and after peering around cautiously they came out again, apparently having met no hostility from the warthog below.

Fig. 53. A spotted hyena taking the skull of a gemsbok into its den beneath a calcrete bank in the Auob River bed, Kalahari National Park. Photo by M. G. L. Mills, from Mills and Mills 1977.

Transport of Bones by Spotted Hyenas to Secluded Feeding Places and Lairs

The complication introduced because more than one species of animal may use a spotted hyena lair means we cannot assume that all bones found in such a lair were brought there by the hyenas. Thus any direct observations of hyenas actually carrying bones to their feeding places and lairs are especially significant. Such observations have been made by a number of zoologists interested in hyena behavior.

As Kruuk (1972) has pointed out, the spotted hyena, with its strong, well-developed neck and forequarters and relatively weak hindquarters, is well adapted to carrying heavy objects in its jaws. He has observed a *Crocuta* running at considerable speed while carrying a 15 kg wildebeest head well clear of the ground. Far heavier objects may be alternately carried and dragged for appreciable distances. The early hunter F. C. Selous (1908) was of the opinion that a spotted hyena may carry a heavy object by throwing it over its shoulders while galloping off. He relates that when he was camping one night in western Matabeleland near the Gwaai River, he tethered a fat goat that weighed 23 kg (50 lb) to one of the wheels of his wagon. In the night, before Selous was asleep, a spotted hyena approached, seized the goat by the back of the neck, broke the tethering thong, and ran off, apparently with the goat thrown over its shoulders.

Stevenson-Hamilton (1934) also relates that two spotted hyenas in the Transvaal lowveld killed more than a hundred goats over a three-year period. It was found that some were being carried 10 km to a den consisting of several old ant-bear holes at the side of a dry torrent bed. Stevenson-Hamilton wrote that, judging from the bones and other indications, the den must have been used by successive generations of hyenas.

The ability of hyenas to carry off large parts of a carcass became very clear to me during 1967 and 1968 when I made observations on spotted hyenas in the Kruger National Park. At that time, A. C. Kemp of the Transvaal Museum was undertaking ornithological fieldwork in the vicinity of Satara and kindly located forty-one lion kills that I was then able to keep under observation for varying periods up to fifteen months. The objective was to study the scattering of bones and the survival of skeletal parts. Two of the kills were of particular relevance here because they were made very close to the Satara rest camp in situations where the hyenas did not feel at ease. Both skeletons were broken up and carried away with remarkable rapidity, as the following descriptions indicate.

Kill H28. A subadult zebra *(Equus burchelli),* of unknown sex and weighing an estimated 180 kg, was killed by a single male lion in the early hours of 10 August 1968 within sight of the east fence of the Satara rest camp. Soon after dawn the lion was seen feeding on the rib cage of the zebra, but apart from damage to the ribs the skeleton was virtually intact. One spotted hyena and three black-backed jackals were waiting about 30 m away. With the daylight, the lion, hyena, and jackals retreated, and the carcass remained untouched until the evening. The lion did not return, but five spotted hyenas did; they were clearly nervous at being so close to the rest camp, and they made a good deal of noise. The following morning, 11 August, the vicinity of the kill site was searched with great care, but only one bone could be found—the left patella shown in figure 54. The rest of the skeleton had been dismembered and carried away, and, although we searched game paths for several kilometers around from the site, no parts of the zebra carcass could be found.

Kill H40. A medium-sized adult giraffe of unknown sex was killed by several lions in the early hours of 13 July 1968, on the bank of a small tributary of the Nwanedzi River, 0.8 km east of Satara rest camp. Soon after daylight, A. C. Kemp saw three lions feeding on the carcass. At 6:45 the next morning Kemp again visited the kill site but found only the thorax (rib cage with articulated thoracic vertebrae) and right front leg at the kill site. Vultures were feeding, a black-backed jackal was nearby, and tracks of several hyenas were discernible. In the course of the day the left hind leg was found by game guards west of the Satara rest camp, having been dragged by hyenas for 1.4 km.

The following day, 15 July 1968, Kemp and I again visited the site. The thorax, seen the previous day, had completely disappeared except for bone fragments, and it was clear that feeding had taken place at two spots, on the north and south sides of the streambed (fig. 55). The right front leg had been moved about 46 m to the northwest and was intact and articulated except for the proximal end of the humerus, which had been chewed away, and the olecranon process of the ulna, which had been removed (fig. 56*b*). One hooded vulture was feeding on the leg when we arrived.

Besides the leg, the remains consisted of the following:

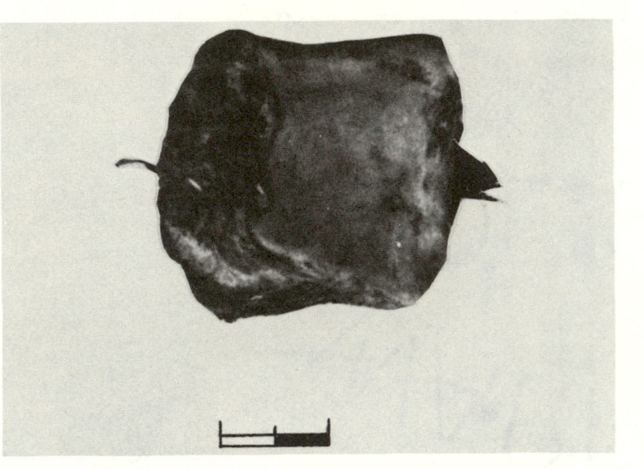

Fig. 54. The left patella of a subadult zebra—all that could be found at the Satara kill site after one night of spotted hyena activity.

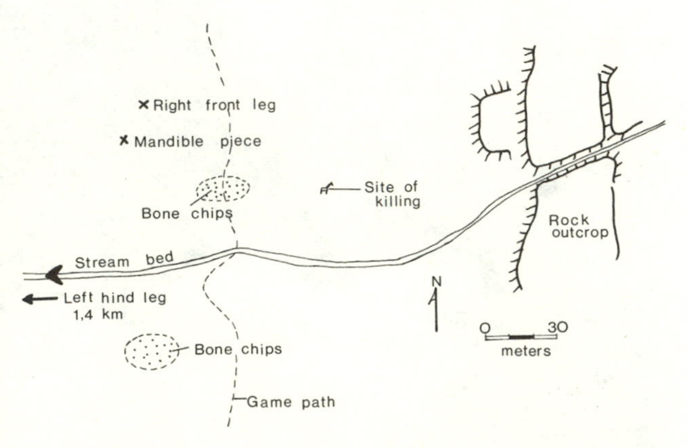

Fig. 55. Sketch map showing the distribution of remains two days after a giraffe was killed by lions close to the Satara rest camp, Kruger National Park. Spotted hyenas were responsible for disarticulating and scattering the skeletal parts.

north bank area, 128 chewed fragments of rib, 1 piece of mandible with a full set of incisors; south bank area, 148 chewed rib fragments, 1 neural spine from a thoracic vertebra, and 4 sesamoid bones (fig. 56*a*). Apart from these fragments, the giraffe skeleton had been disarticulated and scattered by spotted hyenas in two nights. A search of the surrounding countryside failed to reveal where the hyenas had taken the parts, including such weighty objects as the head, neck, right hind limb, and left forelimb.

There seems little doubt that the rapidity with which the hyenas dismembered this giraffe carcass was influenced by the proximity of the kill site to the Satara rest camp. A second giraffe had been killed four months earlier farther away from the rest camp, and its skeletal elements were not scattered to the same degree. Details follow.

Kill H39. An adult giraffe of unknown sex was killed by lions close to the Nwanedzi road about 4 km southeast of Satara rest camp on 11 March 1968. The kill was visited on the first day by Kemp, who also observed feeding by lions, spotted hyenas, and vultures over the next three days. The feeding process was not disturbed by the proximity of humans and was far more leisurely than with the two kills close to the rest-camp fence. There was very little activity of any kind after the first seven days, and I finally made a sketch plan of the distribution of the remains on 18 July 1968 (fig. 57). Certain skeletal elements had disappeared, but the greater part of the skeleton was present, although scattered as much as 23 m in various directions from the original kill site. The distribution was rechecked on 1 September 1968 and found to be essentially the same. The pelvis was still attached to the five lumbar vertebrae and last four thoracic vertebrae. Damage was restricted to the edges of the ilia, the right ischium, and the transverse processes of the vertebrae. The skull was almost complete, except for damage to the nasals and left ossicone.

Hyena activity at this kill was probably more typical than at the first two, in that most of the feeding took place close to the kill site. However the first two kills are of interest because they indicate that, when disturbed and uneasy, spotted hyenas are capable of dismembering a carcass with surprising rapidity and carrying off the parts to secluded feeding places.

An instance of transport of bones to a lair was recorded by Sutcliffe (1970) at Kajiado, Kenya, where, among other bones, three human craniums were found in a low

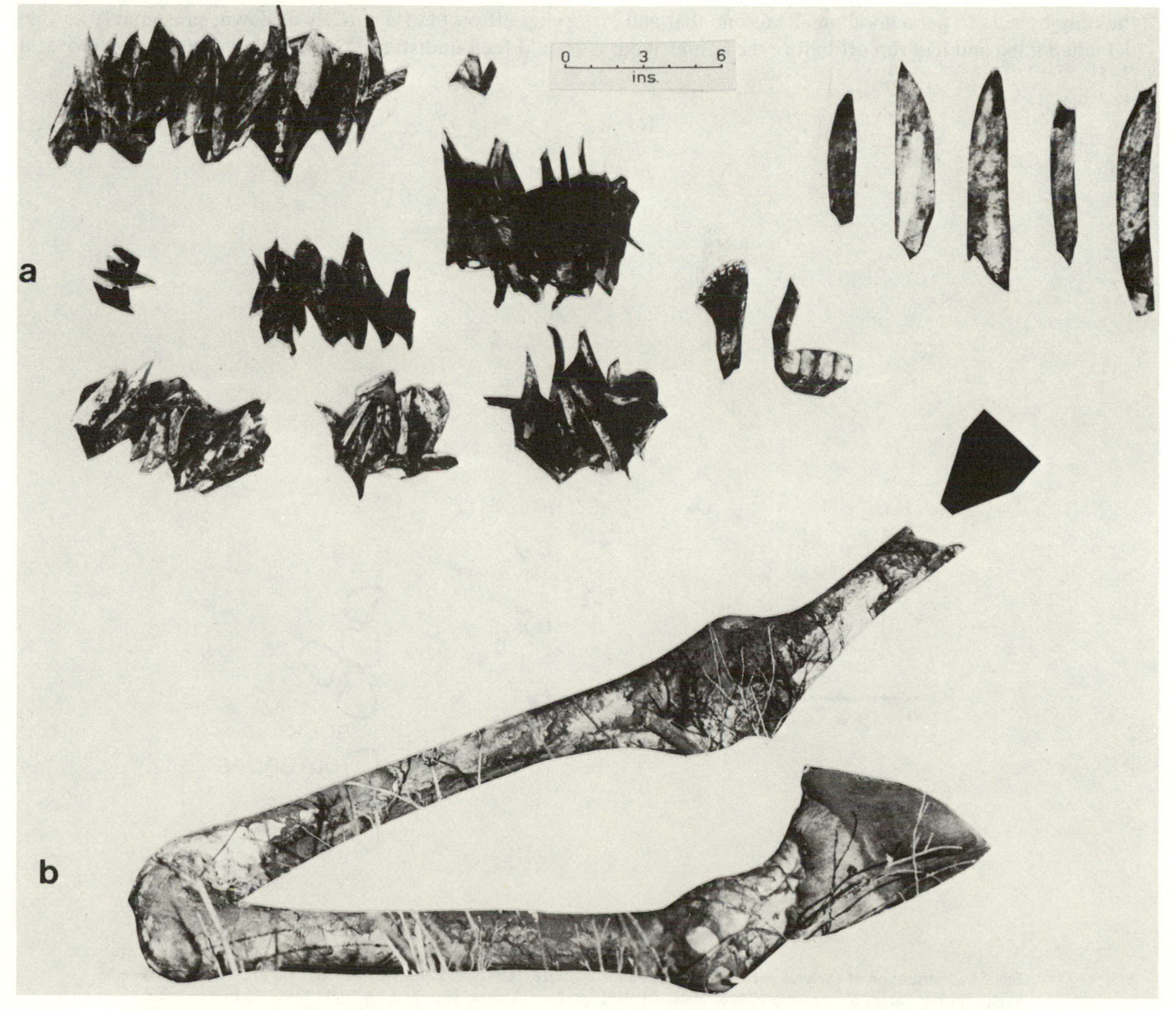

Fig. 56. *(a)* Remains of a giraffe skeleton, H40, collected at the kill site (see fig. 55) after two nights of spotted hyena action. The hyenas were uneasy at the close proximity of the rest camp and scattered the remains more rapidly than usual. *(b)* The right front leg of the giraffe showing damage caused by hyenas chewing the proximal end of the humerus.

recess used by spotted hyenas. The skulls were found to have been dug up by the hyenas at Kajiado hospital cemetery 4 km away and carried to the lair.

On the basis of his detailed study of spotted hyenas, Kruuk was able to remark as follows (1972, p. 244):

> Hyenas also do not carry any substantial amount of food to their cubs; they may occasionally take a bone or a head along, but this seems comparable to taking it to their lying-up place, where they can chew in peace. Cubs may chew these bits, but only after the original owner has finished with them. Older cubs are more likely to carry bones to their den than adults; they sometimes take them down their holes, but more usually these are left in front of the den, or in the entrance. Some dens accumulate quite a collection of bones around the entrance, contrary to the belief expressed, for example, by Hughes, but this does not often happen; it may be more common among brown hyenas.

Working in collaboration with Kruuk, Sutcliffe (1973*a*) made further observations on bone-carrying. He concluded that, in the Ngorongoro crater, where spotted hyenas often kill their own prey only to lose the carcass to scavenging lions,

> the only hyaena to get a good meal was one that had detached a leg and had run off before the lion arrived. There is also competition between hyaenas, so that in a very short time parts of carcasses, by being passed from one hyaena to another, may be carried a distance of several miles. We tried placing a head of a wildebeest which we had taken from a hyaena, outside one of the Ngorongoro lairs. In a short time a half grown hyaena came out and, walking backwards, pulled the head into the entrance of its lair where it jammed by the horns. The hyaena repeated this when we carried the head away again. Its motive for doing this was apparently to take the head somewhere where it could eat quietly, without competition from other hyaenas or lions or jackals or vultures. [Sutcliffe (1973*a*), p. 49]

In another study of spotted hyenas in the Ngorongoro crater, van Lawick-Goodall and van Lawick-Goodall (1970) described how a female carried a large piece of antelope backbone back to the den where she had left her twins. Finding that the cubs were no longer there, she moved from den to den with the bone, calling at each, until she relocated her young. She then put the bone down next to them and lay down to rest.

In the course of his study of spotted hyenas in the Transvaal lowveld, Bearder (1977) observed hyenas carrying off parts of a carcass at dawn, presumably so they could feed undisturbed in a secluded place. Likewise, in

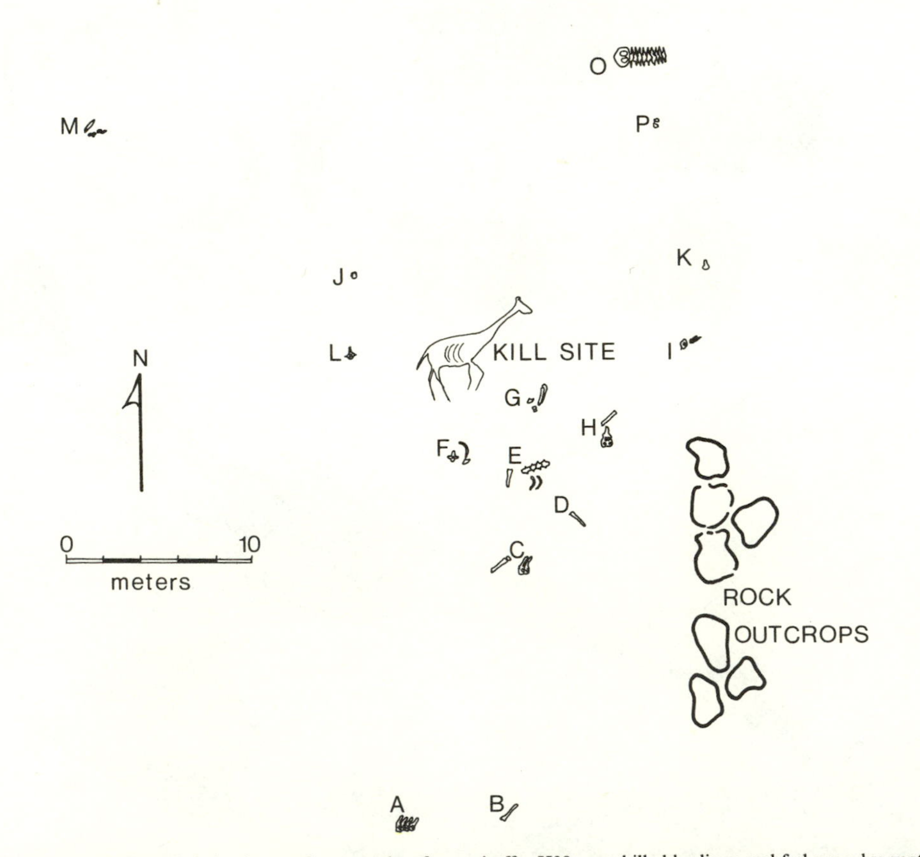

Fig. 57. Distribution of skeletal parts four months after a giraffe, H39, was killed by lions and fed upon by spotted hyenas in the Satara area of the Kruger National Park: *A*, four thoracic vertebrae; *B*, right tibia; *C*, two thoracic vertebrae, left metatarsal and tarsal; *D*, right metatarsal; *E*, articulated cervical vertebrae, two ribs, and left tibia; *F*, thoracic vertebrae, rib, and hoof; *G*, hoof, right astragalus, and right radius/ulna; *H*, skull and right metacarpal; *I*, atlas vertebra and hyena droppings; *J*, right tarsal; *K*, proximal humerus fragment; *L*, thoracic vertebra; *M*, bone flake and hyena droppings; *N*, bone flake; *O*, pelvis with nine articulated vertebrae; *P*, two carpal bones.

the Kalahari National Park F. C. Eloff (1975) has observed a female spotted hyena dragging parts of a springbok carcass to a den where there were two sets of cubs of about three and eight months old, and Mills has photographed a spotted hyena taking a gemsbok skull into the Urikaruus den on the Auob River (fig. 53).

Bones and Other Animal Remains Recorded at Spotted Hyena Lairs

Although many observations have been made of food transport by spotted hyenas, particularly to lairs, relatively few analyses have been undertaken of bone collections in and around such lairs. I will now discuss the few that are currently available.

During his early observations in the Kalahari National Park, Hughes (1958) listed bones from two spotted hyena breeding dens. From the Saint John's lair on the Nossob River he recovered two pieces of springbok horn, and from Site 2, another breeding den in the bank of the Auob River, he retrieved 18 pieces as follows: 4 springbok horns, 7 antelope bone fragments, 4 medium-sized vertebrae, 1 ulna, 1 innominate, and 1 sacrum from a small antelope.

More recently, Mills and Mills (1977) have studied three more dens in the Kalahari National Park: Kaspersdraai, used only by spotted hyenas; Urikaruus, used by spotted hyenas and porcupines, and Wright's den, used by both spotted and brown hyenas as well as porcupines. Wright's den yielded only 9 bone pieces that will not be considered further.

Mills recovered 16 bone pieces from the Kaspersdraai den, as follows:

Antidorcas marsupialis, springbok: 3 adults, based on 3 right mandible pieces, 1 left mandible piece, 1 right and 1 left horn sheath.

Alcelaphus buselaphus, hartebeest: 1 old adult, based on 1 mandible piece.

Antelope class III: 1 right tibia, 1 calcaneus piece, 1 right metatarsal, and 1 right metacarpal; 1 indeterminate rib piece and 4 bone flakes.

The 71 bone pieces collected at the Urikaruus den were as follows:

Oryx gazella, gemsbok: 1 adult, based on 1 cranial piece and 1 horn sheath; 2 juveniles, based on 2 cranial pieces; 1 young juvenile, based on 1 mandible piece.

Antidorcas marsupialis, springbok: 1 adult, based on 1 articulated forefoot.

Antelope class II: 2 scapula pieces.

Antelope class III: 1 metatarsal piece

3 indeterminate pieces and 59 bone flakes. There is no doubt that many of the bone flakes in the Urikaruus sample were derived from regurgitations—this will be discussed further.

From the Queen Elizabeth Park in Uganda, Sutcliffe (1970) has described the bone content of two spotted hyena burrows, though unfortunately he has not provided detailed analyses. One burrow had a simple form with a single terminal chamber about 1 m below ground. The access tunnel was found to contain a few fragments of antelope bone and some juvenile hippo bones; the terminal chamber yielded the complete skeleton of a juvenile *Crocuta* and a single dropping. The second burrow consisted of a central chamber, opening to the surface by at least ten tunnels that contained numerous bones including skulls of a juvenile hippo, warthog, and kob, bones of a buffalo, a juvenile elephant humerus, and skull fragments of a juvenile hyena. Most of these remains had been damaged by hyenas.

In the Ngorongoro crater, a spotted hyena burrow investigated by Sutcliffe (1970) contained many bones, mainly of zebra and wildebeest, together with the skull of an adult hyena; and two natural horizontal cracks in lava, used by spotted hyenas at Kajiado, Kenya, contained substantial quantities of bones. These included remains of domestic donkey, cow, and dog, bones of an ostrich and parts of an ostrich eggshell, and also human remains such as 3 calvariae and a mandible.

In the course of his study of *Crocuta* in the eastern Transvaal, Bearder (1977) collected 409 bone pieces from six breeding dens in the Timbavati reserve. His analyses of these pieces is presented in table 35. Of the 409 pieces, 123 could be assigned to species, as shown in the last column of table 36. The animals involved were zebras, giraffes, impalas, wildebeests, and kudus.

Regurgitations as a Source of Bones at Lairs

It has been established, it seems, that spotted hyenas do not regurgitate food for their cubs (Kruuk, 1972; Bearder, 1977), but it has been equally firmly established that they regularly regurgitate indigestible residues of recent meals. As Kruuk has pointed out (1972, p. 244):

> Adults do regurgitate near the den, but what comes out is almost invariably a large slimy ball of hair and bone slivers with nothing edible about it. The reaction of any hyaena (also a cub) to the sight and sound of another one regurgitating is to run up, sniff the results of this action, and then immediately roll in it, shoulder first, then with the whole back, legs bent and head lifted. Often the regurgitating animal joins in.

Regurgitation and rolling were likewise observed among Ngorongoro spotted hyenas by van Lawick-Goodall, who also recorded that the hyenas will pick out pieces of undigested bone from the fresh regurgitations and eat them again. Despite this habit, accumulations of regurgitations may build up near spotted hyena dens; in the Timbavati reserve, Bearder (n.d.) found that up to 35 regurgitations accumulated at well-used den sites in the course of a month, and 200 complete regurgitations were collected in the vicinity of four such dens during his study period. Bearder was able to identify the animals that had contributed hair and other undigested residues to the regurgitations; his list is given in the first column of table 36. He showed that, although bone pieces, chips, and slivers occurred in both regurgitations and scats, they tended to be larger in the former than in the latter.

Bone pieces that have undergone corrosion in spotted hyena stomachs before regurgitation show recognizable characteristics. Corrosion of bones, almost certainly caused by spotted hyenas' gastric juices, has been observed on very numerous specimens from the Pin Hole Cave in the Creswell Crags on the border between Derbyshire and Nottinghamshire in England. It was here that Magins Mello and Boyd Dawkins worked from 1875 onward and demonstrated for the first time that man had been a contemporary of the mammoth. In his account of the mammalian faunal remains from the cave, Dawkins (1877) concluded that the numerous corroded bones had suffered through the action of carbonic acid in the cave soil. This explanation seemed unlikely to James Kitching

(1963), who had samples of the cave earth analyzed and found that they were highly alkaline. Kitching concluded that the corroded bones were from spotted hyena regurgitations or from the stomachs of hyenas dismembered in the cave. His illustrations show that corroded bones from the Pin Hole Cave compare very closely with bone fragments from the stomachs of spotted hyenas shot in the Kalahari National Park. Kitching concluded that 2,946 bone flakes showed signs of partial digestion, the same being true for the following specimens: 32 eroded juvenile incisors of *Equus*, 101 deciduous upper and 68 lower molars of *Rangifer*, and 45 deciduous molar plates of *Elephas*, as well as 15 astragali, 29 phalanges, and 142 mastoid, carpal, and tarsal bones of reindeer.

One of the more striking of the corroded bones from Pin Hole Cave was part of a bison's rib with numerous holes in it. This specimen had been regarded by A. L. Armstrong (1936), excavator of the cave, as a human artifact; he termed it a bull-roarer. Kitching, however, has argued that it is simply a rib fragment that had suffered corrosion in a hyena stomach.

The similarity of hyena-corroded bone fragments to human artifacts is sometimes striking. This should be borne in mind when interpreting bone accumulations to which hyenas may have contributed.

Droppings of Spotted Hyenas

The abundant presence of hyena droppings in certain British caves has already been mentioned; such droppings may also be found in African caves, such as at Redcliff in Rhodesia, where their abundance in an excavated profile has been plotted (Brain, 1969*a*). It was found that concentrations of hyena coprolites coincided with layers in which quantities of human cultural material were low, the inference being that hyenas used the cave at times when the human inhabitants were elsewhere.

Although spotted hyenas may defecate inside their lairs or in the immediate vicinity, they also tend to have specific "latrine areas" where they indulge in social defecation. The existence of such areas has been known for a long time—under the heading "scatology," Harrison Matthews wrote as follows (1939, p. 47):

> The faecal masses of the Spotted Hyaena resemble those of a large dog in shape and size, and consist almost entirely of mineral matter derived from bones, with sometimes a few hairs or feathers. The lower part of the intestine in all specimens exmined contained a green paste, with some lumps, which turned white on drying in the air. When freshly deposited the faeces are green in colour, but as soon as they are thoroughly dry they become quite white, and are then the classical "album graecum." The animals' habit is to defaecate at regular latrine areas, spaces of limited extent covered with a plentiful deposit of album graecum. These places are conspicuous at a distance when the sun is shining on the chalky white faeces. Some of the latrine areas are very large: one seen near the Serronea river must have covered quite a quarter of an acre. A resident of Arusha was heard to describe such a deposit as "like a fall of snow"; nor is the simile greatly exaggerated. It is uncertain whether the animals consciously repair to these areas to defaecate, but it seems more probable that the sight of the deposit, when casually encountered, provides the necessary stimulus to the individual.

Hyena droppings (fig. 58) can provide valuable information about the diet of the animals that produced them. In the course of his East African *Crocuta* study, Kruuk (1972) collected 810 fecal samples, ground them up, and extracted hair for identification of the prey eaten. His analysis provided information about hyena diet during the wet and dry seasons as well as in various habitats. In an attempt to assess the reliability of fecal analysis as an indicator of actual diet, Kruuk plotted the percentages of total occurrence of wildebeest, zebra, and Thompson's gazelle in feces from different ranges in Ngorongoro and from Serengeti areas, at different times of the year, against percentages of actual observations of hyenas feeding in those areas at the same time. A close correlation was found between the two sets of observations, confirming the usefulness of fecal analysis as an indicator of diet.

Conclusions on spotted hyena diet in the eastern Transvaal have been drawn up by Bearder (1977), largely through analysis of 527 droppings and 200 regurgitations. Results are given in table 36. Of particular interest is the difference in representation of large and small prey species in the scats and regurgitations. The percentage occurrence of giraffe hair in the scats was 38.3 as opposed to 16.3 in the regurgitations. Impala hair, on the other hand, was present in 36.8% of the scats and 60.2% of the regurgitations. Bearder's explanation for this and similar discrepancies is that a hyena feeding on a small animal such as an impala is likely to consume more hair relative to a given weight of meat than when eating from a large animal such as a giraffe.

In this connection Bearder observed that, in addition to the usual whitened droppings, brown, organic-rich ones could sometimes be found. These were apparently produced by hyenas that had fed on meat to the exclusion of bone, which they periodically do when a large carcass is available. Although Kruuk (1972) estimated that the average food intake for Ngorongoro spotted hyenas in the dry season was about 2 kg per day, he also pointed out that a single hyena will readily eat a third of its body weight at one time. This has been confirmed by Bearder (n.d.), who reported that a three-year-old *Crocuta* ate 18 kg of elephant meat in one night and then returned for more the following evening. On another occasion, two adults ate most of a 50 kg impala in 1 hr, 20 min.

In conclusion, the presence of characteristic coprolites in a cave deposit provides valuable evidence that hyenas had made use of the site. The chances are good that they contributed to any bone accumulation that may be present there.

The Diet of Spotted Hyenas

Detailed information about what spotted hyenas eat is relevant to this study in that it indicates the range of bony objects the hyenas may take back to a cave lair. It has been amply demonstrated, in many parts of the African continent, that spotted hyenas are active and effective predators as well as being scavengers.

Kruuk (1972) has collected a wealth of information on *Crocuta* diet from direct observations in the Serengeti and Ngorongoro areas. Results based on 513 carcasses in the former area and 297 in the latter are given in table 37. Wildebeest forms the most important dietary item in both areas, followed by Thomson's gazelle in the Serengeti and by zebra in Ngorongoro. Surprising as it may seem, the hyenas were found to successfully hunt animals as large as adult waterbuck, eland, and buffalo.

Information about spotted hyena kills in the Kruger National Park over a thirty-year period has been compiled by Pienaar (1969). As is shown in table 38, impala forms 58.8% of the kills, followed by waterbuck, wildebeest, and kudu.

In the nearby Timbavati reserve, Bearder (1977) showed that impala was also the most frequent prey item in regurgitations, although giraffe hair appeared with greater regularity in scats. These two species were followed in percentage representation by wildebeest and zebra. No actual kills were observed during Bearder's study period.

In other areas, spotted hyenas clearly feed on whatever they can get. In the Kalahari National Park, F. C. Eloff (1964) recorded that three adult gemsbok and two juveniles were found to have been killed by spotted hyenas between 1958 and 1961, and remains collected from the Kalahari Park dens, referred to earlier, came from gemsbok, springbok, and hartebeest.

In Zululand, Deane (1962) has recorded *Crocuta* predation on wildebeest and warthog, as well as on an old and sick black rhino. Other odd dietary items are fish (Stevenson-Hamilton 1947), tortoises and lions (Pienaar 1969), hippo calves (Cullen 1969), and young elephants (Bere 1966).

Many accounts have been published of spotted hyenas attacking sleeping people and attempting to drag them off. In the Mlanje District of Malawi, several attacks have been described by Balestra (1962); during September 1955 an adult African man, who happened to be the village idiot, was killed and eaten on a path between two villages. It is estimated that at least four spotted hyenas participated, leaving nothing but some patches of blood and shreds of clothing. Seven days later an old woman was dragged from her hut by a *Crocuta* and lost an arm before help could reach her. She died the next day. The same year a six-year-old child was dragged from a veranda where she was sleeping and eaten; only her head was left. This too would probably have been eaten had the hyenas not been driven off by village dogs.

The habit, which persisted till within historical times among various African tribes, of putting aged, feeble, or sick individuals outside villages at night so that they could be disposed of by scavengers must have meant that spotted hyenas regularly fed on human flesh. The early traveler and hunter F. C. Selous related an incident in which the body of a man was eaten by spotted hyenas. During 1872 Selous was camped near the Jomani River in eastern Matabeleland; at that time one of the residents of the area was shot by a Hottentot, who was then clubbed to death by the friends of the murdered man. The Hottentot's body was left lying in the bush about 275 m from Selous's wagon. The first night several hyenas created a good deal of noise around the corpse from dark to daylight; on the second night they once again left it alone, but on the third night they devoured it (Selous 1908).

Fig. 58. Characteristic whitened droppings of a spotted hyena: Satara area, Kruger National Park.

I am not aware of any accounts providing detailed information about bony remains resulting from hyena action on human bodies, nor is such information easy to obtain experimentally. The closest I have come to it was a feeding experiment I undertook involving a baboon. In October 1967 a large male baboon weighing 27 kg was shot at Satara in the Kruger National Park while making a nuisance of itself in the rest camp. The body of the baboon was securely tied down to the ground 50 m from the rest-camp fence, and several hyenas were heard to approach soon after darkness fell. The following morning the area was carefully searched for remains, but, apart from a patch of blood where the body had been eaten, some tufts of hair, and a piece of intestine 20 cm long, nothing was left.

The early observations of Buckland (1822) on the bone accumulation in Kirkdale Cave, Yorkshire, led to the conclusion that the spotted hyenas that lived there had been cannibalistic, feeding upon one another inside the cavern. The concept of cannibalistic hyenas was vehemently rejected by Dart (1956*a*, 1957*a*) and Hughes (1954*a*,*b*, 1958) in their consideration of the possible role of hyenas in accumulating bones at Makapansgat. Despite this, however, evidence has been coming to light that clearly shows that spotted hyenas *do* eat one another upon occasion, and that *Crocuta* remains in a cave could well be food remains of their conspecifics.

When Hughes (1958) wrote "Some Ancient and Recent Observations on Hyaenas," he quoted M. Cowie, warden of the Royal National Park in Kenya, as follows: "With spotted hyaenas I have seen many willing to eat their own kind, but only after a dead one has become almost putrefied. In other words although I have seen spotted hyaenas killing each other, either in fighting for a lady or over a kill, I have never seen them eat each other until the dead one has been lying for something like four days, and has become what to us would be a most offensive object."

The tendency of spotted hyenas to wait until the body of one of their fellows has partially decomposed before they are willing to eat it is borne out by two observations I made in the Satara area of the Kruger National Park. Details of the incidents are as follows:

Kill H44. An adult *Crocuta* of undetermined sex was found about 90 m from the Satara rest-camp gate on the bank of the Shitsakana stream on 28 May 1968. It had apparently been killed by lions five days before and, although decomposing, was complete. Two days later, that is, seven days after death, it was eaten by other spotted hyenas, the only remains being the left and right scapulae, shown in figure 59*a*. These remains were left until 27 June 1968 when, there being no change, they were collected.

Kill H45. An adult *Crocuta* of undetermined sex was found to have been killed by lions 0.8 km west of Satara on 20 March 1968. Four days later it was beginning to rot but was still complete. Within the next two days it was consumed by one or more spotted hyenas, except for a number of ribs and chips of bone. These were left where they lay till 30 April 1968 when, there being no further activity, they were collected. As is shown in figure 59*b*, the remains consisted of 13 complete ribs, 6 sternal bones, 2 sternal cartilages, 2 proximal tibiae, 1 patella and 29 bone flakes.

Another incident of cannibalism in the Kruger National Park was recorded by Pienaar (1969). On a night in August 1966, two spotted hyenas were seen feeding on the hindquarters of a younger one along the Nahpe Road. Whether this carcass was fresh or putrid was not recorded.

Spotted hyenas do not appear to have a strong inhibition against feeding on young of their own species. F. C. Eloff (1975) reports a case in the Kalahari National Park when an adult was seen feeding on the remains of a very young cub, "which was probably its own," and Kruuk (1968) records an incident in which his tame *Crocuta* cub, Solomon, was carried off one night by a wild adult. The cub was outside Kruuk's house at Seronera one moonless night when, at 2 A.M., it was grabbed by a wild adult. The cub's cries caused Kruuk to give chase and recover the cub, which had suffered a gashed throat, punctured windpipe, and broken jaw. With penicillin treatment the cub recovered.

Two other incidents of spotted hyena cannibalism have been reported from the Ngorongoro crater in Tanzania. The first, reported by Kruuk (1972), occurred in September 1967 when a hyena from the Mungi clan killed a wildebeest within the territory of the adjacent Scratching Rocks clan. A number of Mungi clan hyenas started to feed on the carcass, but they were challenged by the Scratching Rocks residents, and one was severely injured in a fight. The following day it was partly consumed by several hyenas that, at the time of Kruuk's observation, had already eaten about a third of the body.

The second incident, recorded by van Lawick-Goodall and van Lawick-Goodall (1970), took place in the same area, on top of the Scratching Rocks hill. Here one of the hyenas was killed by a lion; the body was left untouched throughout the following day but was then consumed by members of the clan to which it belonged, a number of cubs participating in the feeding.

The conclusion to be drawn from all these observations is that cannibalism is frequent among spotted hyenas; both adults and juveniles may be involved, and it may take place within a single clan. It is highly probable that hyena bones would be added to a bone accumulation in any long-term *Crocuta* lair through cannibalism.

Food Storage in Water: A Possible Source of Bones in Caves

In dolomitic countryside it is not unusual for a pool of water to exist either inside the entrance to a cave or within its catchment area. This is particularly true of situations where streams either disappear underground through a cave entrance or emerge again from a cavern. Spotted hyenas have a habit of storing surplus food in water for future consumption, and this can result in an accumulation of bones at the bottom of a water hole. If the water hole happens to be close to a cave entrance, this strange hyena habit could contribute to a bone accumulation inside the cave.

In connection with the food-storage habit of *Crocuta*, Kruuk has written (1972, p. 119):

> Hyenas leaving a kill often carry a large chunk of meat or bone away with them, which is usually eaten quietly some distance off. Occasionally, however, they cache it by dropping it into 30–50 cm of standing water and collect it later. In Ngorongoro, I have seen them do this three times, always in the central lake,

and in the Serengeti both Schaller and I once saw a hyena leave part of a carcass in a small waterhole. On several occasions I found fresh pieces of carcass partly or wholly submerged in Serengeti waterholes, without hyenas nearby; but from the way the food was partly eaten, it was all but certain that the hyenas had put them there. In the crater lake, hyenas had to wade up to 10 m from the shore to reach the proper depth. They usually returned to a cache within a day, but the many bones that were exposed when part of the Ngorongoro lake dried up probably meant that several caches had been left unused. A cache can most likely be recovered only by visual relocation, and this is not always effective enough. I twice saw a hyena wade into the lake and repeatedly plunge its head deep under water in a small area, obviously

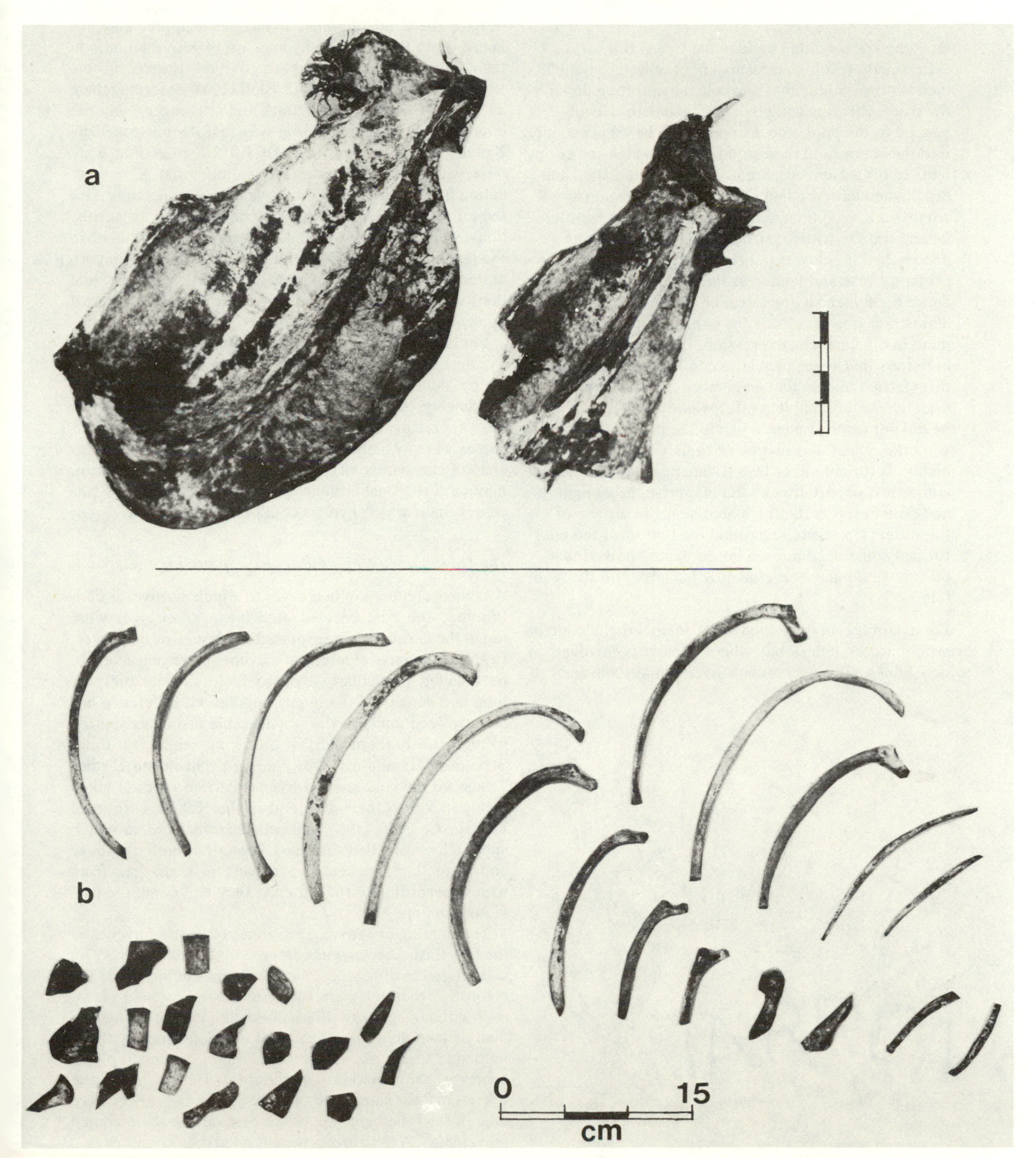

Fig. 59. *(a and b)* Skeletal remains of two spotted hyenas collected after their decomposing bodies had been eaten by other spotted hyenas in the Satara area of the Kruger National Park.

> searching but eventually leaving without finding anything. But I also twice saw a hyena successfully recover a cache in this way, once the head of a wildebeest, and the second time the remains of a dead lion; neither of these was in any way visible from above the surface of the extremely muddy water.

An earlier but rather similar observation was made by a former warden in the Kruger National Park and published by Wolhuter in his *Memories of a Game Ranger* (1948, pp. 178–79):

> One day I was resting under a big tree at the 'Mbeamede spruit near a small pool of water about 18 inches deep. Suddenly I saw a hyaena trotting down the bank, carrying something in his mouth. He advanced to the pool, and dropped what he was carrying into the water, and then stood back to watch the effects of his action. Apparently he was astonished and dissatisfied with the fact that the piece of meat (or whatever it was) floated on the surface for presently he entered the water, seized the meat in his mouth, and pushed it below the surface once more. It promptly rose and floated again, and this seemed to cause the hyaena a great deal of concern. I could almost imagine that I saw the expression of astonishment in his face. His expression, in fact, was so ludicrous that I was unable to control a chuckle, and this startled him so that he glanced nervously in my direction before ambling off. I waited a while, but as he did not return I went down to the pool to find out what the object was that he obviously valued so highly. It turned out to be a fragment of lung, of either wildebeest or waterbuck, and of course, being light and bouyant, it naturally floated near the surface of the water. On scratching round the bottom of the pool I found some decomposed bones, which proved that this unusual place was evidently his larder for the lean times.

Food storage of this kind may seem insignificant in transient water holes, but where a cave is involved in which bones may accumulate over some thousands of years, the water-storage procedure could be of great importance.

The Importance of Scavenging as Opposed to Hunting

Spotted hyenas have obvious scavenging adaptations in the form of hammerlike premolars mounted in robust jaws that are powered by very large muscles. They are able to crack remarkably large bones, thus making a meal of remains left by much larger predators such as lions. But, despite these adaptations, it has been appreciated for many years that spotted hyenas are effective hunters in their own right. Referring to spotted hyenas in the Kalahari National Park, F. C. Eloff (1964) described them as "a curious mixture of hunter and scavenger," but the effectiveness of their hunting was first demonstrated by Kruuk's study in Tanzania. Of 1,052 hyenas that were observed feeding, Kruuk (1966) found that 82% were eating from animals killed by hyenas, whereas only 11% fed on animals killed by other predators, such as jackals, lions, wild dogs, leopards, and cheetahs; cause of death in the remaining 7% was doubtful. In fact, in the Ngorongoro crater the traditional relationship between hyenas and lions was reversed: the lions obtained most of their meat by scavenging from hyena kills.

For interpreting a bone accumulation in a cave to which hyenas may have contributed, it would be useful to know which bones resulted from hunting and which from scavenging. Active hunting by spotted hyenas is likely to have been more frequent in environments where few carcasses were available for scavenging. The abundant presence of carnivores larger than hyenas, fossil or modern, may be a reasonable indicator of the availability of carcasses upon which hyenas could have scavenged.

The Hunting/Scavenging Range of Spotted Hyenas

A bone accumulation in a cave, to which carnivores contributed, can best be evaluated if we know over what range the carnivores transported the bones or carcasses. In spotted hyenas, hunting or scavenging range is likely to be affected particularly by two factors: availability of food and density of the hyena population. Reference has already been made to the considerable distances spotted hyenas in southern Africa travel in search of food: Stevenson-Hamilton (1934) recorded that in the eastern Transvaal lowveld spotted hyenas carried parts of goats about 10 km to their den, and in the Kalahari National Park F. C. Eloff (1964) described an incident in which spotted hyenas killed some goats near the warden's house and were then tracked 40 km back to their den, from which, according to their tracks, they had come the previous evening.

The hunting or scavenging ranges of spotted hyenas in the two East African areas studied by Kruuk (1966, 1972) were found to differ a good deal. Hyenas living inside the Ngorongoro crater were found to belong to eight clans, each containing 10 to 100 individuals; each clan had its own defined area, in which most of the feeding by those individuals took place. On the adjoining Serengeti plains however, the situation was different. Here some of the hyenas had the same kind of clan system, but others were migratory, following the wildebeest during their annual movements. A third group were termed the "commuters"; they retained the same dens on the plains throughout the year, from which they made long excursions, perhaps lasting several days, to where the wildebeest happened to be. Such excursions might involve a distance of 80 km each way.

It is difficult to judge what system would have been followed by spotted hyenas living in the Sterkfontein valley during lower Pleistocene times. The habitat is known to have been fairly open, and a Serengeti type of arrangement could well have operated.

Some Notes on the Mechanics of Crocuta Chewing Action and Resulting Damage to Bones

Spotted hyenas are remarkable among carnivores for their exceptional ability to crack bones with their teeth—an ability that allows them to benefit from remains other carnivores have discarded. In this bone-crushing adaptation the premolars, especially P^3 and P_3, together with the jaw muscles that power them, are of particular significance. The anterior and posterior cusps of these premolars have been reduced and the central cusps enlarged and widened, so that the teeth have been converted from bladelike structures into heavy conical hammers (Ewer 1973).

Powerful closure of the jaws is made possible by three sets of muscles; the action of two of the sets is shown in figure 60. The temporalis muscle arises from the lateral surface of the braincase and from the ligament behind the eye, whence its fibers run down to insert on the upper part of the coronoid process. Its anterior fibers (T1) pull directly upward, while the more posterior ones (T2 and T3) pull upward and backward. The fact that the coronoid process is situated lateral to the braincase means that the temporalis fibers also tend to pull the jaw mediad.

Fibers of the masseter muscle originate along the lower border of the zygoma, running down to insert on the angle of the mandible and in the masseteric fossa. Those fibers arising from the anterior part of the zygoma (M1) are the most superficial ones and wrap around the angular process of the mandible. They tend to pull the jaw forward and upward. Fibers of the intermediate masseter layer arise from the middle of the zygomatic arch (M2) and run straight down to insert on the outer surface of the angular process; their action is a simple closing one. The deepest masseter fibers arise from the posterior part of the zygoma (M3) and run forward to insert in the masseteric fossa. They pull the mandible backward and upward. In opposition to the temporalis muscle, the masseter exerts a lateral pull on the mandible, since the zygoma is laterally placed. The zygomaticomandibularis muscle is particularly effective in this respect but is not easily separated from the deep layers of the masseter (Ewer 1973).

A third set of muscles involved in jaw closure, but not shown in figure 60, is the pterygoideus, which originate on the side of the skull below the orbit and run back to the inner surfaces of the mandible. The pull is anteriad and mediad.

Observations I have been able to make on spotted hyenas in the Kruger National Park, and on remains they left there, have made it clear to me that the chewing action on bones falls into two distinct categories: cracking bones between the premolars, and gnawing uncrackable pieces with the incisors and canines.

Bone Cracking with Premolars. As is shown in figure 61, the main teeth involved are P^3 and P_{3-4}, the former acting as a hammer, the latter as a cradle or anvil. The most effective bone-cracking appears to occur when the jaws are about half open, that is, when there is a gap of about 4 cm between the crowns of P^3 and P_3, as shown in figure 60*b*. With the aid of a model based on the skull of an adult *Crocuta*, I have been able to measure the percentage change in length of each of the muscle components when the jaws are half open (fig. 60*b*) and fully open (fig. 60*c*).

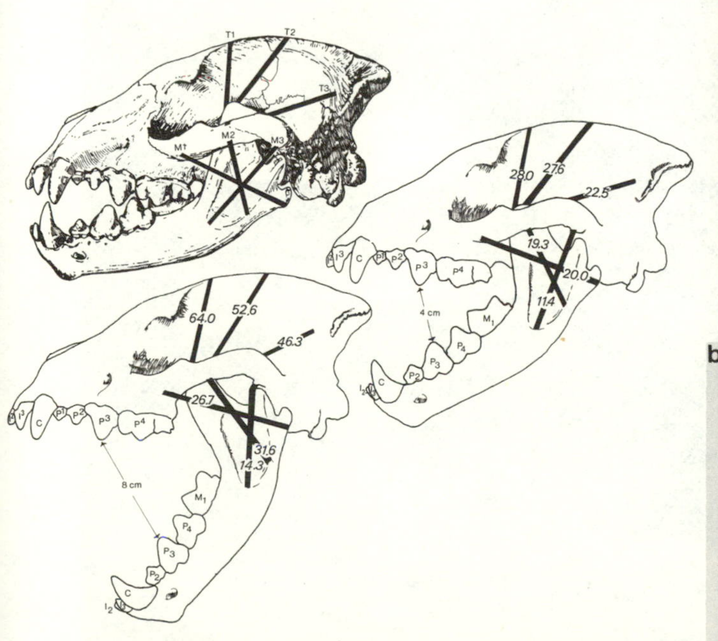

Fig. 60. Diagrams of a spotted hyena skull showing some of the muscles involved in closing the jaw. T1–3 are the anterior, middle, and posterior fibers of the temporalis muscle, and M1–3 are the superficial, intermediate, and deep fibers of the masseter. The lower jaw is shown at rest *(a)*, half open *(b)*, and fully open *(c)*; the percentage changes in length of each muscle component are indicated for the different jaw openings.

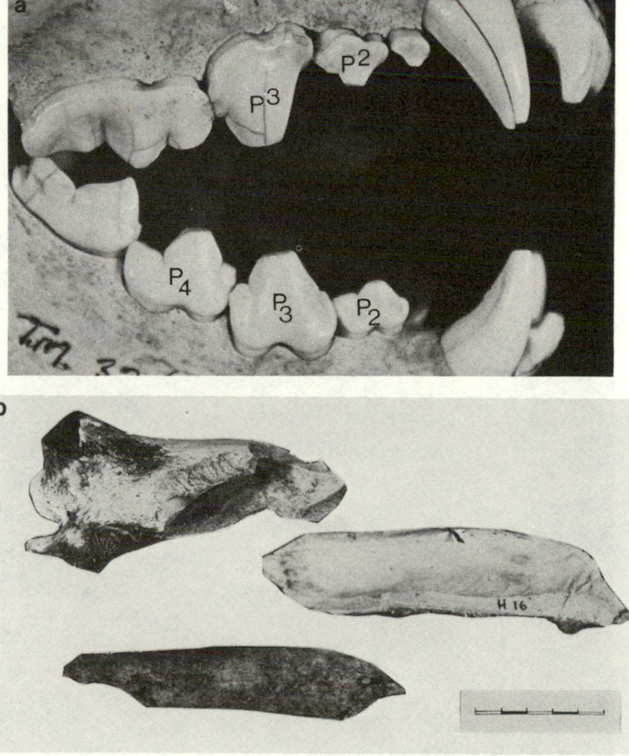

Fig. 61. *(a)* Premolars of a spotted hyena, which are of special significance as bone-crackers. *(b)* Bone flakes produced from bovid limb bones by spotted hyenas with the help of their premolars.

The lengths of the relevant muscle fibers and the percentage change in length per component are given in table 39.

I do not know the relationship between the percentage change in muscle fiber-length and the effectiveness of that particular muscle component. But it is clear that all muscle components, with the exception of M3, show appreciable changes in length when the jaw is moved from the closed to the half-open position.

With the jaws fully open, as shown in figure 60*c*, length changes on all components of the temporalis muscle are considerable; they are appreciably greater than masseter length changes. It seems that the combination of muscles *Crocuta* has allows powerful jaw closure, to a marked degree, over a wide range of jaw openings. A good deal of force clearly is exerted on the premolars when cracking limb bone shafts, which may be up to 6 cm in diameter. The typical result of such shaft-cracking is the production of bone flakes, as shown in figure 61*b*. These are very similar in form to those produced by hominid-wielded hammerstones. A comparison of hyena- and human-produced bone flakes is given in chapter 7, together with a discussion of the form of the fractures that bone shafts are likely to suffer.

The form of other premolar-shattered bone pieces will depend on the anatomy of the parts concerned. For instance, the blades of scapulae are likely to be severely shattered, as are the ascending rami of mandibles. Damage to horizontal rami is consistently characteristic, as is shown in figure 62. The lower margins of the mandibles are typically broken away by a series of premolar crunches, exposing the tooth roots and the marrow around them. Very similar damage has been described for hyena-chewed bones from Ethiopia by Shipman and Phillips (1976).

Gnawing of Bones with the Incisors and Canines. Many of the bones in which spotted hyenas interest themselves are too large or cumbersome to be cracked between the premolars. They are therefore worked upon by the incisors and canines (fig. 63*a*), often at wide jaw openings.

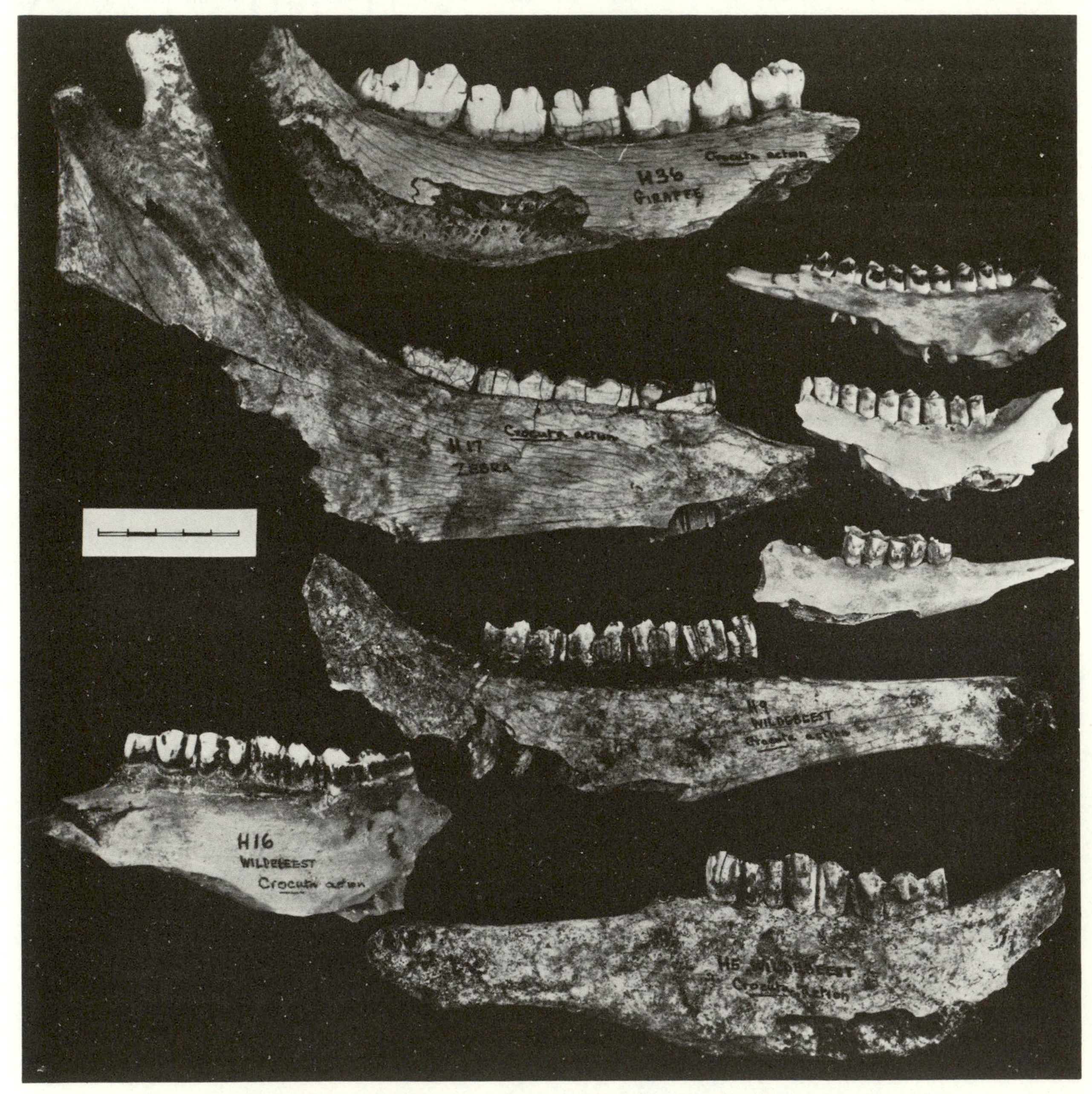

Fig. 62. A series of artiodactyl mandibles from the Satara area of the Kruger National Park showing characteristic damage caused by the premolars of spotted hyenas.

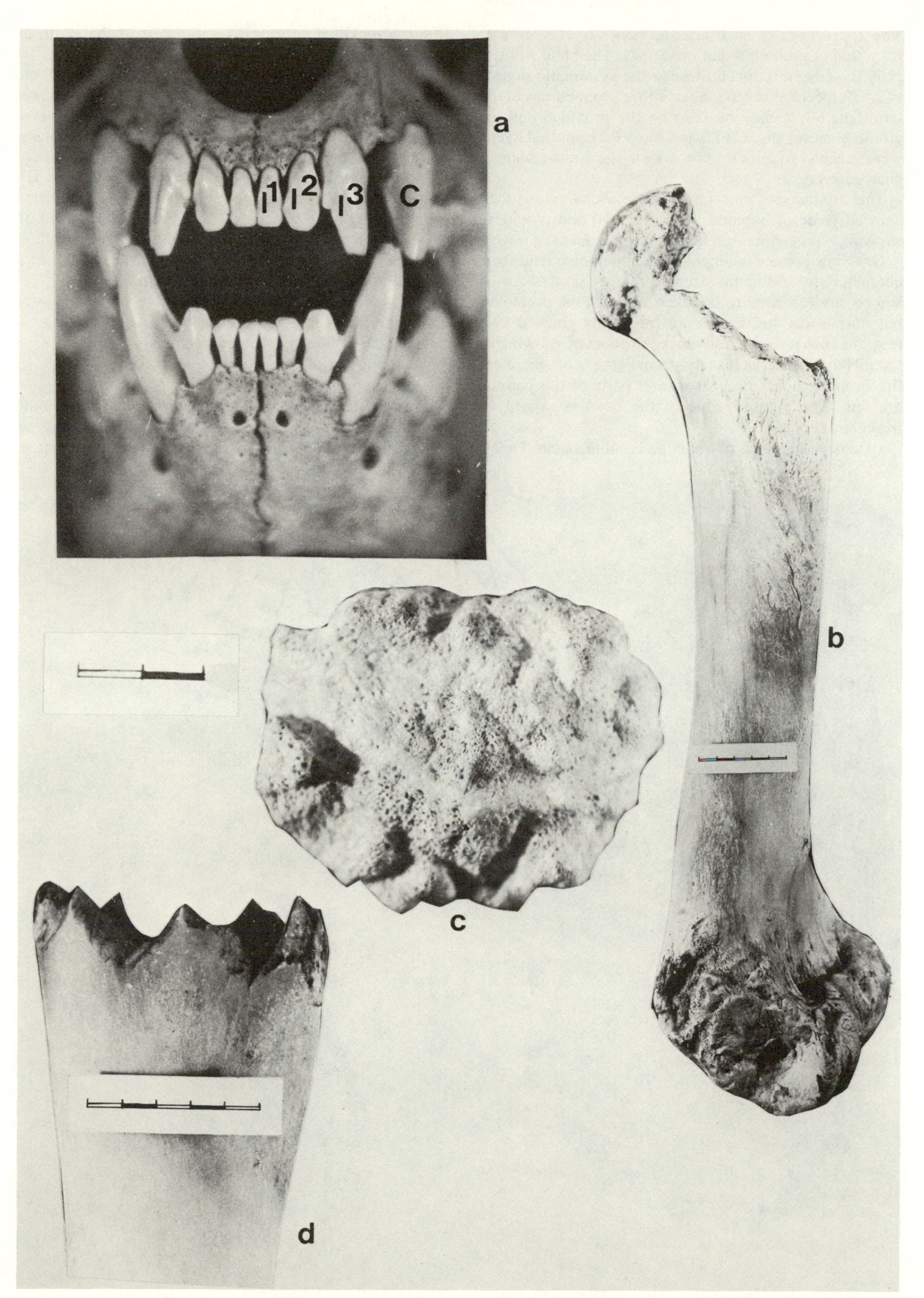

Fig. 63. *(a)* The incisor-canine battery of a spotted hyena, used in gnawing uncrackable bones. *(b)* A giraffe humerus showing hyena gnawing around the proximal epiphysis. *(c)* Part of a giraffe's humeral head, characteristically gnawed by hyenas. *(d)* A giraffe's humerus, the proximal end of which has been gnawed off by spotted hyenas, leaving a typically ragged end to the shaft.

The appearance of such gnawed bones is very different from that of premolar-shattered ones. The bone will typically have been reduced through the systematic removal of small pieces that have been either scooped or chipped out. This effect may be seen on the proximal end of a giraffe humerus (fig. 63*b*) that Kruger Park spotted hyenas were unable to crack, but which they systematically gnawed away.

The products of this process are a characteristically gnawed piece of humeral head (fig. 63*c*) and a humerus showing a ragged margin at its proximal end (fig. 63*d*).

Gnawing damage on ungulate craniums also tends to be characteristic. Medium-sized antelope skulls, for instance, are too large to be taken between the premolars, but horn-cores and nasals are frequently gnawed away (e.g., fig. 64*a*). A more advanced product of gnawing is a "skull bowl" such as that shown in figure 64*b*, where only the frontals, horn-core bases, and parts of the parietals are left. The gnawed edge of the bowl is ragged and irregular.

Extensive gnawing of antelope craniums also tends to free the maxillae entirely from the rest of the skull, as shown in fig. 64*c*.

Various skeletal parts, such as vertebrae of large mammals, may show damage caused both by the premolars and by the incisors. Neural spines and transverse processes are typically cracked off with the premolars, and the vertebral bodies are extensively gnawed.

In his paper "Spotted Hyaena: Crusher, Gnawer, Digester and Collector of Bones," Sutcliffe (1970) described four types of bone damage he observed to be caused by *Crocuta* in East Africa and compared them with the damage reported by Zapfe (1939) when bones were fed to a spotted hyena at the Schönbrunn Zoo in Vienna.

The types were:

1. Splintering of bones by adult animals.
2. Gnawing of bones by juvenile animals.
3. Scooping out of cancellous bone.
4. Damage by partial digestion.

Concerning type 2, Sutcliffe concluded that young hyenas with milk teeth were unable to splinter bones as adults do but simply gnawed them, leaving striations at

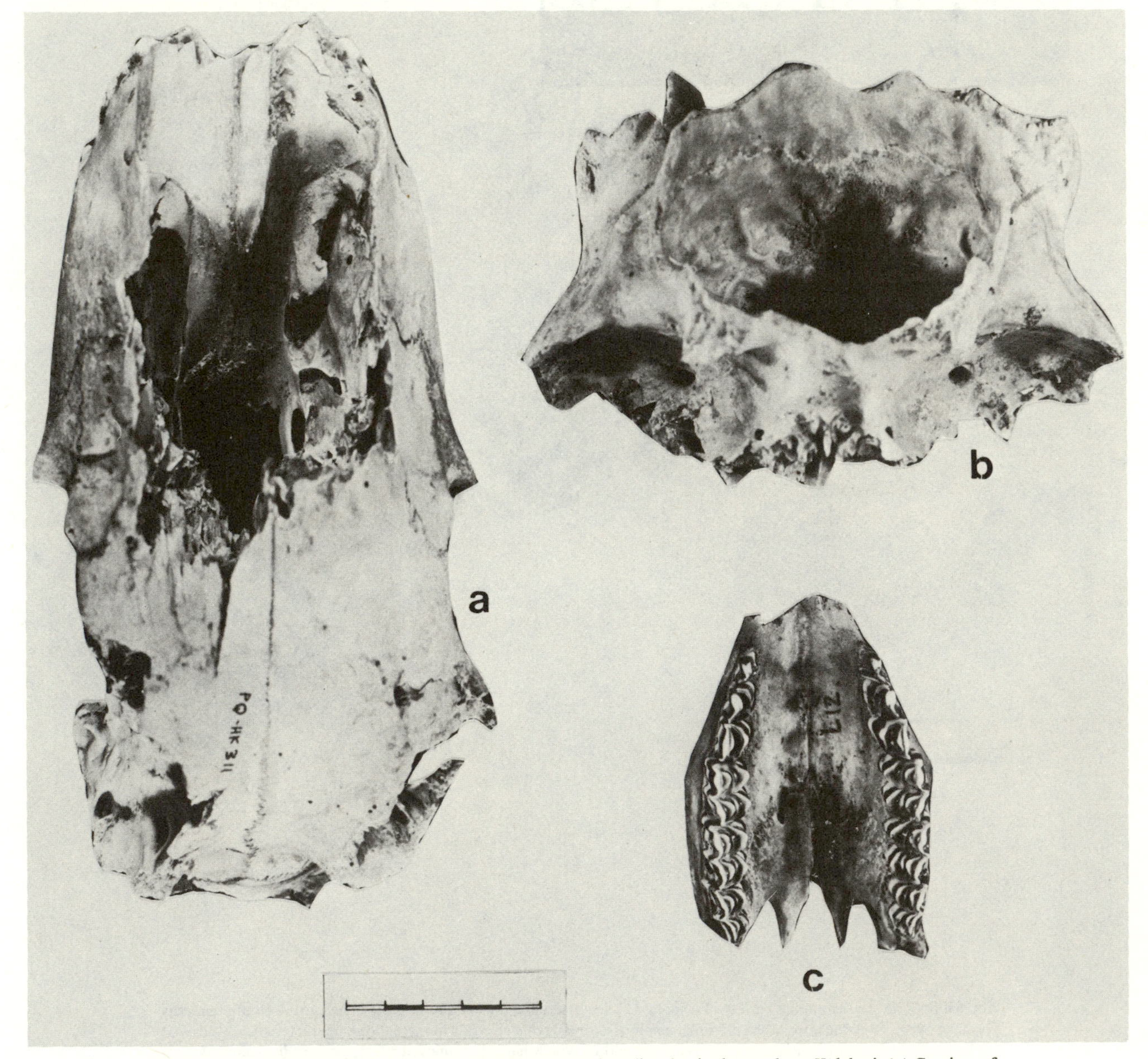

Fig. 64. Examples of skulls gnawed by spotted hyenas at a breeding den in the southern Kalahari: *(a)* Cranium of a juvenile gemsbok showing characteristic damage to the nasals and horncores. *(b)* A gemsbok "skull bowl." *(c)* A maxilla isolated by gnawing.

right angles to the long axes of the bones. Such marks would be indistinguishable from those caused by other small carnivores but are very different from the effects, classified under types 1, 3, and 4, characteristic of adult spotted hyenas.

Sutcliffe's first category would certainly be equivalent to my "premolar-cracked" class, and his third could be equated with incisor-canine gnawing. Damage to regurgitated bone pieces by partial digestion has already been touched upon.

The Brown Hyena, *Hyaena brunnea* Thunberg

As will be discussed in chapter 8, the earliest recorded representative of the *Hyaena brunnea* stock appears to have been the mid-Pliocene form, *H. pyrenaica,* from Europe, which entered Africa from the north and, in the southern part of the continent, replaced the striped hyena, *H. hyaena.* The latter species has not been recorded in southern Africa since the accumulation time of Makapansgat Limeworks. The distribution of brown hyenas as represented in the last decade is shown in figure 65, although within historic times brown hyenas still ranged throughout the Cape, to the shores of Table Bay (Skinner 1976).

The brown hyena is typically rather smaller than *Crocuta,* though its jaws may be as powerful. It is characterized by a sloping back, shaggy hair, and characteristic banding on the front legs (fig. 66).

Brown Hyena Lairs in Burrows

In the Kalahari National Park and adjacent areas, Mills and Mills (1977) have shown that brown hyenas regularly use ant-bear holes as breeding dens. These may be enlarged and modified through digging by the hyenas themselves. At the "Botswana den," three series of disused ant-bear holes within an area of 1 km^2 were modified and used; at Kaspersdraai, seven ant-bear burrows in an area of 2 km^2 were used successively by the same litter of cubs; the Kannaguass den consisted of a single burrow; at Rooikop, five ant-bear burrows were used within an area of 1 km^2 and at Kwang two ant-bear burrows 200 m apart were used.

On the banks of the Sabi River, near Skukuza in the Kruger National Park, Stevenson-Hamilton (1947) found a brown hyena den consisting of a series of ant-bear holes with about six entrances, "covering some thirty feet of open space." The site was screened by thick bush and reeds and was surrounded by an appreciable bone accumulation that will be discussed shortly.

During his study of brown hyenas in the Transvaal, Skinner (1976) reported on two dens in modified ant-bear holes; these were on the farms Leeufontein and Tweeputkoppies. He provided a plan and sections of the Leeufontein den, which had three entrances; the internal dimensions of the tunnels varied in height from 25 to 35 cm and in width from 35 to 53 cm. Brown hyenas were seen entering the tunnels, and Skinner suggests that the rather small dimensions helped keep larger predators away from the cubs.

Brown Hyena Lairs in Caves and Recesses

During 1955 a cave was pointed out to me by the late J. Templeton on the farm Uitkomst, north of the Sterkfontein valley; it was used as a breeding lair by brown hyenas and opened into the side of a steep dolomitic valley (fig. 67*a*). Inside, in almost total darkness, were numerous bones among which were scattered 395 whitened hyena droppings. Some of the bones and droppings are shown in figure 67*b*. According to Templeton, the cave had been used as a brown hyena breeding site for a number of successive years before 1955.

Another low cave used as a breeding lair occurs on the farm Boekenhoutkloof, near Cullinan, northeast of Pretoria. The site and resident hyenas have been described by Skinner (1976), and I first visited the place on 6 September 1975 in Skinner's company. I subsequently returned and surveyed the site, drawing up the plan shown in figure 68.

As is shown in figure 69, the site is close to the top of a wooded ridge of Waterberg sandstone, facing east. It consists of a low cliff, about 5 m high, forming a corner on the south side; several large sandstone blocks are scattered in front, as is shown in the plan. The lair consists of a low recess, generally about 0.5 m high, beneath the sandstone cliff, descending at an angle, with a substrate of fine, loose dust. At the time of the first visit, spoor of adult and juvenile brown hyenas was visible both outside and inside

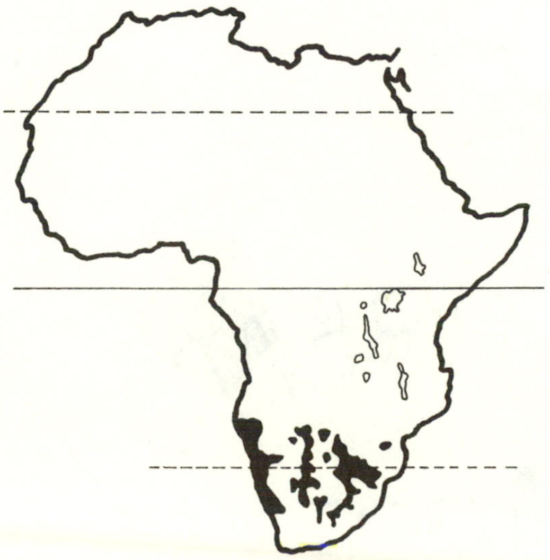

Fig. 65. Current distribution of the brown hyena, *Hyaena brunnea.*

Fig. 66. A mounted specimen of a brown hyena from the Transvaal. The sloping back, shaggy coat, and banding on the forelimbs are clearly shown.

the lair. On either side of the entrance to the lair were concentrations of hyena droppings; 316 were collected.

It is clear that brown hyenas, like spotted hyenas, use breeding lairs of two types: either modified ant-bear burrows or natural caves and recesses.

Use of Brown Hyena Lairs by Other Species

It appears that, like *Crocuta,* brown hyenas share their dens with other animals, which they appear to tolerate. The den figured and excavated by Skinner (see above) was found to house a porcupine as well as brown hyenas, and "Wright's den" described by Mills and Mills (1977) from the Kalahari National Park was used, perhaps in rotation, by spotted and brown hyenas, as well as by porcupines.

The Composition of Bone Accumulations at Brown Hyena Dens

At the time of my first visit to the Boekenhoutkloof lair, a partly eaten lynx, *Felis caracal,* was lying about 12 m from the cave entrance, the position being indicated in figure 68. The lynx is shown in figure 70: its forequarters were intact, but the abdomen, including the vertebral column between the thoracic region and the sacrum, was missing. The pelvis showed carnivore damage, and the hind legs, except for the head of the right femur, were absent.

Fig. 67. *(a)* Entrance to a dolomitic cave used by brown hyenas as a breeding lair on the farm Uitkomst. *(b)* Bones and hyena droppings on the floor of the inner chamber in total darkness.

A short while before this observation was made, the local farmer found the hindquarters of a large domestic dog slightly closer to the entrance of the lair. He returned a few days later to find that these remains had been further eaten, so that only fragments of the hind limbs remained. The only other bones found subsequently at the lair were the head, shoulders, and thorax of a domestic calf, although a number of other species were identified by means of hair and feathers (Skinner 1976).

In the course of his brown hyena study in the northern Transvaal, Skinner collected 39 bone pieces from the vicinity of five dens on the farm Tweeputkoppies. These were found to come from 15 individual mammals belonging to seven species as listed in table 40A. From the breeding lair on the farm Leeufontein 6 bone pieces were collected; they came from 5 individuals of four species as listed in table 40B.

The bone accumulation from the brown hyena den described by Stevenson-Hamilton from the banks of the Sabi River in the eastern Transvaal is of great interest. Concerning it, Stevenson-Hamilton wrote as follows (1947, p. 210):

> The vicinity was littered with bones, and it was quite evident that most if not all the animals had been seized alive and killed by these hyaenas. It was to me rather a remarkable discovery, since hyaenas, as a tribe, have never been regarded as primarily hunters. In the present instance there was no doubt about the matter. The heads of fourteen full-grown impala rams, all quite recently killed, the skulls of several baboons, and of two chitas (one of them a full-grown animal), remains of guinea fowls, and a large tree snake ("boomslang") partly chewed, were among the exhibits. The carcasses of the animals had seemingly been dragged down the holes, including any hornless females, for only heads of fully grown horned males were found outside; the prevailing odour was fairly good proof of what had happened. These hyaenas must have developed a sound hunting technique to be able to catch and kill

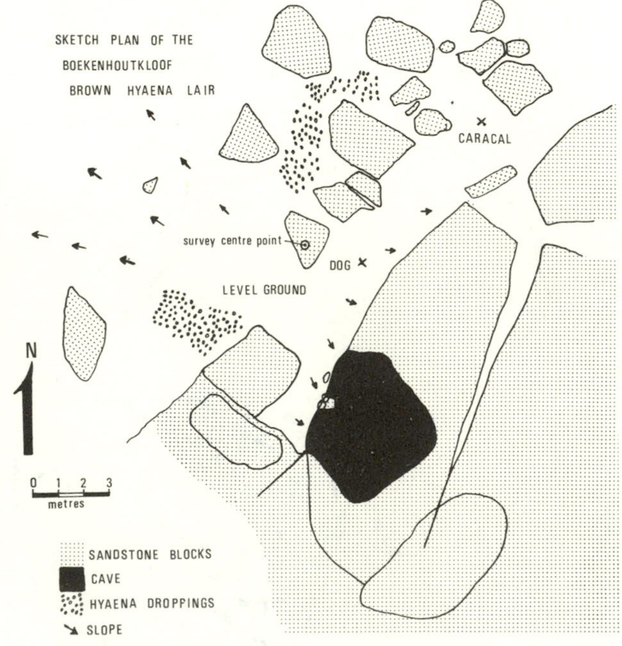

Fig. 68. Sketch plan of a brown hyena breeding lair on the farm Boekenhoutkloof in the Cullinan district of the Transvaal.

Fig. 69. The Boekenhout brown hyena lair; the entrance is indicated by the arrow at the base of the sandstone cliff.

Fig. 70. The partly consumed body of a caracal close to the entrance of the Boekenhoutkloof brown hyena lair.

so wary and quick an animal as an impala, and one so relatively formidable as a chita, or a baboon.

These observations were made in 1941, but Stevenson-Hamilton gave further details of the site in his annual reports to the National Parks Board for 1942 and 1943. These have been made available by Pienaar (1969, p. 139):

> 1942—The family of brown hyaenas again bred in the north bank of the Sabi a few miles east of Skukuza, but changed their breeding holes for others nearer the Sand River junction, their old home having been to a great extent destroyed by some lions which had dug into it extensively to reach the meat which they could smell inside. Established at their new site, these hyaenas continued to kill considerable numbers of animals, and skulls of impala, bushbuck and baboons were found scattered outside, evidently hunted and killed by the occupants of the den.
>
> 1943—The family of brown hyaenas on the north bank of the Sabi continued in their new abode, to take toll of other animals. In addition to the remains of full-grown impala, several baboons and one more cheetah skull was found.

Unfortunately, Stevenson-Hamilton did not provide a detailed list of the bones found at the Sabi River dens, but the following is a summary of animals recorded: 1941, 14 adult impala rams, several baboons, 2 cheetahs (1 adult, 1 subadult), several guinea fowls, water tortoises, and a tree snake; 1942, several impalas, baboons, and a bushbuck; 1943, impalas, several baboons, and 1 cheetah.

The Uitkomst cave, used as a brown hyena breeding lair (fig. 67*a,b*), which I investigated in 1955, was found to contain, besides 395 hyena droppings, 90 identifiable bones, of which 60 belonged to hares and hyraxes, as well as 128 bone flakes. I recently reexamined this collection and found that, though many of the bones showed clear carnivore damage, others had been gnawed by porcupines or charred in fires. The cave has obviously been used by porcupines and Iron Age people in addition to brown hyenas, and all three agents will have contributed to the bone accumulation. For our present purposes, therefore, the composition of this bone accumulation is of limited relevance.

Also on the farm Uitkomst, the brown hyenas made use of a spot between some large chert boulders in the dry bed of the "python valley" as a resting and feeding place. When I made a routine visit in this place on 6 June 1969, I found that the hyenas had brought back the right hind leg of a donkey, still articulated to the innominate bone. All meat had been eaten from the leg, and damage had been done to the pelvic bone, the greater trochanter, and the calcaneus. In addition, there was the complete palate of a calf and the right mandibular ramus of a blesbok, showing damage that appears very characteristic of hyena action (see fig. 76). These were the only bones found at the spot, although hair and droppings have been regularly seen there. The hind leg of the donkey had been brought by the hyenas from beyond the boundaries of the Uitkomst reserve, probably over a distance of 5 to 7 km.

The most detailed and significant information on bones around brown hyena dens comes from the Kalahari National Park, where M. G. L. Mills has been conducting his study of *H. brunnea* (Mills 1973, 1974, 1976; Mills and Mills 1977). Mills has kindly made his collection of bones found around the Kalahari brown hyena dens (fig. 71)

Fig. 71. Bones accumulated by brown hyenas around the entrance to their breeding lair, Botswana den, Kalahari National Park. Photo by M. G. L. Mills, from Mills and Mills 1977.

available to me. The sample consists of 235 pieces from the following sites: Botswana dens (53 pieces), Kaspersdraai dens (14 pieces), Kannaguass den (66 pieces), Rooikop dens (22 pieces), and Kwang dens (80 pieces). Particulars of the dens have already been discussed and are given by Mills and Mills (1977).

The 235 bones are found to come from a minimum of 75 individual animals, belonging to sixteen taxa as depicted in figure 72. The bones by which each taxon is represented are listed in table 41.

The most remarkable feature of this bone accumulation is that carnivores are represented in exceptional abundance. Of the 75 individual animals whose remains are found, 35, or 46.7% of the total, are other carnivores, black-backed jackals and bat-eared foxes being the most numerous, followed by lynx, ratel, and aardwolf. As will be discussed shortly, these animals are frequently represented by surprisingly complete skulls. It is now well established that brown hyenas, at least when rearing cubs, deliberately hunt other carnivores (up to the size of cheetahs, on the Kruger Park evidence) and take them to the breeding dens as food for the cubs. Although most of the skeletons of these prey items may be eaten, skulls are typically left. This carnivore-hunting habit of brown hyenas has, I suspect, been of the greatest importance in the accumulation of bones in southern African caves.

Of the 3 animals represented by bony remains at the Boekenhoutkloof lair, discussed above, 2 were carnivores; the proportion is appreciably lower in the 45 bones collected by Skinner at the Tweeputkoppies and Leeufontein lairs in the northern Transvaal, but this may reflect the fact that small carnivores are locally scarce there, or that brown hyena cubs were not actually being reared in the dens at the time the collections were made.

An upper Pleistocene bone accumulation containing abundant carnivore remains has been described by Klein (1975*a*) from the southwestern Cape. It ocurs in an irregular fissure hollowed out in calcareous aeolianites forming a steep cliff flanking False Bay at Swartklip, 30 km east-southeast of Cape Town. The site shows no evidence of porcupine or human involvement, and Klein has suggested that the bones were collected by carnivores, probably hyenas. Since the collection includes remains of 2 adult and 2 juvenile *H. brunnea,* it seems conceivable that Swartklip was a brown hyena breeding lair. Bones from a total of 32 species have been recorded at the site; of these, 13, or 40.6%, are of other carnivores.

Other bone-filled burrows of upper Pleistocene age have been described from Duinefontein, 50 km north of Cape Town (Klein 1976*b*). These are still being investigated but probably result from hyena action.

Diet of H. brunnea as a Guide to the Potential Bone Content of Lairs

In addition to the species represented by the bones already listed from the various lairs, further information is available from analysis of hair in scats and regurgitations (Skinner 1976) and from direct observations in the Kruger National Park. Identified from the hair analysis were the following species besides those already listed: *Proteles cristatus,* aardwolf; *Suricata suricatta,* suricate; *Herpestes sanguineus,* slender mongoose; *Lepus saxatilis,* scrub hare; and *Pronolagus crassicaudatus,* red rock hare.

For the Kruger National Park, Pienaar (1969) gives the following list of 239 animals found killed by brown hyenas during the periods 1936–46 and 1954–66: kudu 78, impala 49, waterbuck 44, zebra 8, baboon 7, guinea fowl 7, water tortoise 6, wildebeest 4, sable 3, tsessebe 3, buffalo 2, reedbuck 2, warthog 1, roan antelope 1, eland 1, duiker 1, bushbuck 1, ant bear 1, and tree snake 1. In addition, Pienaar states that brown hyenas prey on lion cubs and ostriches.

From these observations it is clear that brown hyenas are able to hunt a wide variety of animals, including the largest antelopes. When rearing cubs they also hunt the full range of carnivores that they are able to overpower, up to the size of cheetahs.

Cannibalism appears to occur among brown hyenas, as it does in *Crocuta*. In his list of food items eaten by brown hyena cubs in the Kalahari National Park, Mills (1974) notes cannibalism on two occasions, without specifying whether the animals eaten were other cubs or adults.

I have made one personal observation on brown hyenas eating a dead *Crocuta* close to the Nossob rest camp in the Kalahari Park. On 16 August 1975, in the company of M. G. L. Mills, I examined the remains of an adult spotted hyena (fig. 73) that had died as a result of porcupine quill injuries and had been partly consumed by brown hyenas. On the other hand, Mills (1974) has recorded two incidents in which spotted hyenas ate brown hyenas. In the first, a 14-month-old brown hyena was killed near a water hole; in the second, an adult that was recovering from the effects of an immobilizing drug was killed by six spotted hyenas.

Brown Hyena Predation on Primates

I am not aware of any authenticated reports of brown hyena attacks on people, although Stevenson-Hamilton (1947) quoted Africans in Gazaland as saying that brown hyenas sometimes took small children from their huts at night or attacked sleeping adults.

Predation on baboons appears to be fairly common, there being evidence of it both in the Kruger National Park (7 baboon skulls were among the remains at the Sabi River breeding den) and at the northern Transvaal dens investigated by Skinner. Baboons are probably one of the characteristic prey species brought back to lairs when cubs are being raised.

I was able to arrange one feeding experiment involving a baboon on the farm Uitkomst. An adult male baboon weighing 25 kg was shot and securely tied to a tree, close to the brown hyena feeding place in Python valley referred to earlier. It was eaten two days later on 30 May 1969 by two brown hyenas, and the remains were then collected. They consisted of 2 pieces of mandible and 3 fragments of the cranial vault (fig. 74). The maxillary toothrows were swallowed by one of the hyenas.

Relative Importance of Hunting and Scavenging

It is well established that brown hyenas both hunt and scavenge, but little has been published thus far on the relative importance of each food-getting means. The research currently being undertaken by M. G. L. Mills will provide valuable information of this kind.

Where the brown hyena's range extends to the coast, the animal has become known as the "strandwolf," from its habit of scavenging from dead marine animals washed up on the beach. It is worth remembering that bones of marine vertebrates in coastal caves could very well have been brought there by brown hyenas.

The Hunting or Scavenging Range of Brown Hyenas

With the aid of radio telemetry, Mills (e.g., 1974) is establishing the movement patterns of *H. brunnea* in the Kalahari National Park. He reports that a foraging distance of 50 km per night is not exceptional for a brown hyena. Such trips are generally undertaken singly, and when the cubs first start to forage they do so on their own.

Distances traveled in the course of a night will certainly be influenced by the availability of food and will therefore vary greatly from region to region. The home range of 340 to 544 km² estimated by Mills (1976) for the Kalahari Park brown hyenas is a good deal larger than the range Skinner (1976) estimated for six adult and two juvenile brown hyenas living within an area of 20 km² at Boekenhoutkloof in the Transvaal.

It would probably be reasonable to suggest that brown hyenas might carry bones back to a lair over a distance of at least 10 km.

Food Storage by Brown Hyenas

The storage of surplus food in water practiced by spotted hyenas has not been observed in *H. brunnea*. Instead, brown hyenas simply hide food in long grass or take it

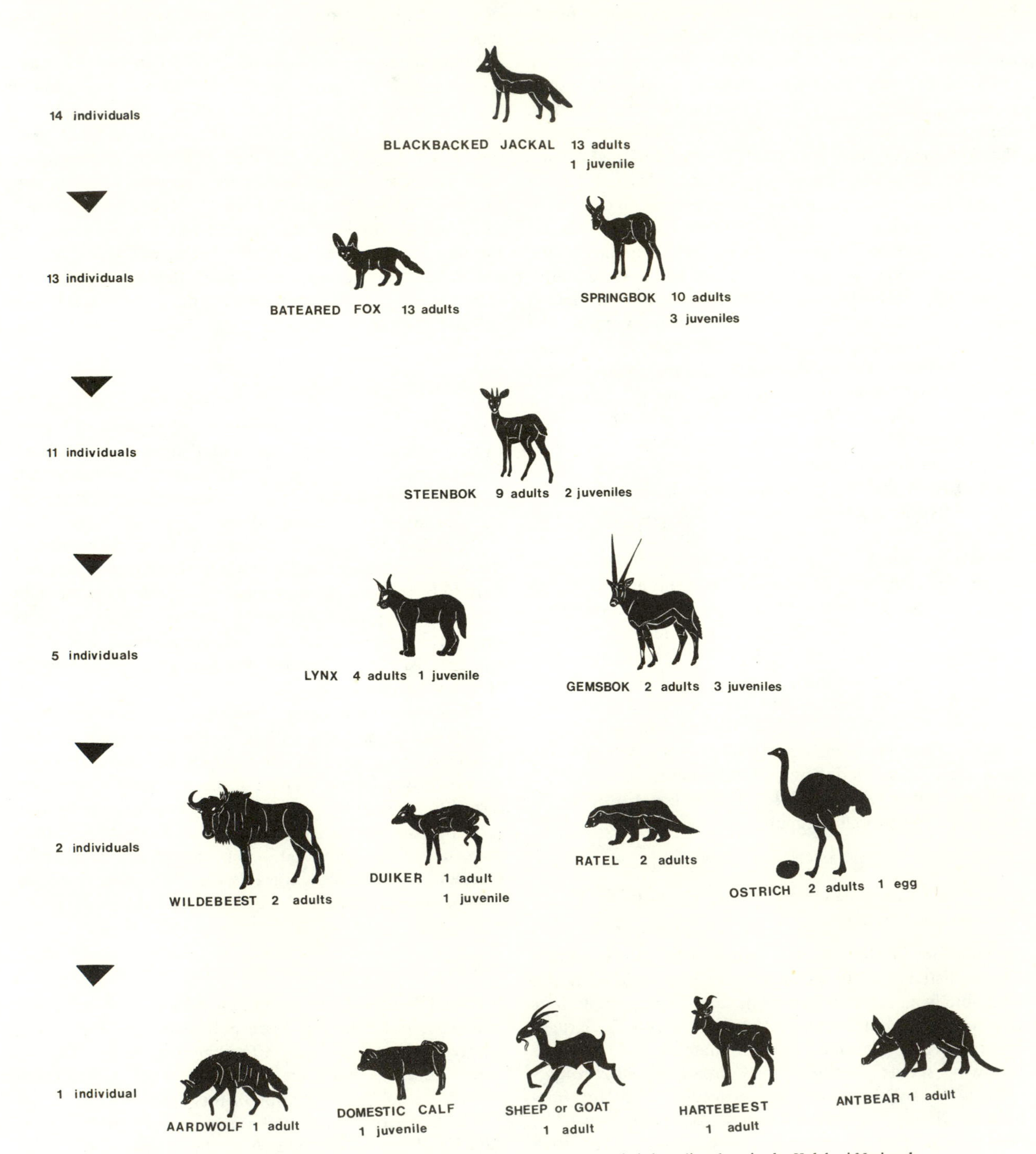

Fig. 72. Animals represented by remains collected by brown hyenas at their breeding dens in the Kalahari National Park. Minimum numbers of individuals are indicated.

Fig. 73. Remains of a spotted hyena partly consumed by brown hyenas near Nossob Camp, Kalahari National Park.

down a hole (Mills 1973, 1974). One attempt by a brown hyena to bury surplus food at Boekenhoutkloof has been reported by Skinner (1976).

Some Characteristics of Brown Hyena Food Remains

The skull structure and dentition of *H. brunnea* are basically similar to those of *Crocuta*, although there are adaptational differences. According to Ewer (1973), *Hyaena* dentitions are characterized by less efficient carnassial shear, so that P_4 and the anterior part of P^4 are also modified for bone-crushing. In fact, third molars of *Hyaena* are less dominant than they are in *Crocuta*.

Bony food remains of the two species of hyena appear remarkably similar, except for those of soft-bodied prey animals, which brown hyenas take to their dens when rearing cubs. These are highly diagnostic. At brown hyena feeding places, chewed bones indicate that *H. brunnea* follows the *Crocuta* pattern of *cracking* manageable bones with its premolars and *gnawing* larger ones with its incisors and canines. *Hyaena*-cracked limb bones and mandible pieces are shown in figure 75. The damage is indistinguishable from that inflicted by spotted hyenas, as portrayed in figures 61 and 62. The characteristics of *gnawed* bones are probably also very comparable. Figure 76*a* shows a calf palate, representing the gnawed remnant of a complete cranium, and figure 76*b* shows a "skull bowl" that resulted from brown hyena gnawing of a springbok skull.

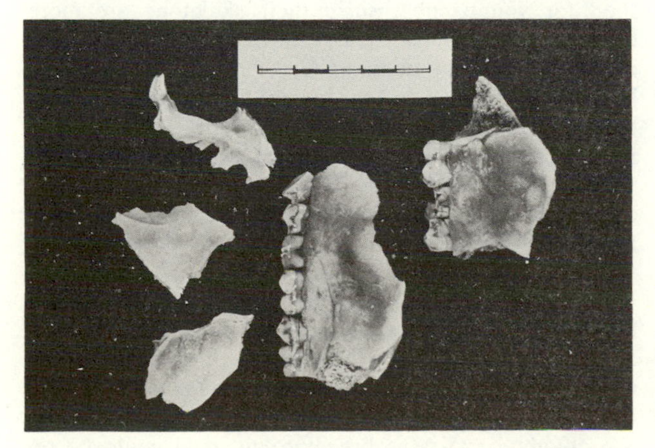

Fig. 74. Remains of an adult male baboon consumed by brown hyenas on the farm Uitkomst, Transvaal.

An important difference between *H. brunnea* and *Crocuta*, from the point of view of this discussion, is that the former actively kills small animals, particularly other carnivores, and brings their bodies to the cubs. The most striking aspect of the carnivore remains found around the Kalahari brown hyena breeding lairs is the undamaged state of many of the craniums. On the basis of skulls, the following numbers of individual carnivores are represented by the collection: black-backed jackal, 13 adults, 1 juvenile; bat-eared fox, 13 adults; caracal, 4 adults, 1 juvenile; ratel, 2 adults; aardwolf, 1 adult. Yet, despite the abundant craniums of these animals, postcranial bones are almost unrepresented. For instance, the black-backed jackals are represented by a single proximal humeral piece; bat-eared foxes by a single articulated forefoot and most of a vertebral column; and the only other carnivore remains are 5 isolated vertebrae, a piece of innominate, and a tibia. It is obvious that the hyenas have consistently eaten the bodies of the carnivores, leaving only the heads, or parts of them. This is surprising, since a single crunch of an adult brown hyena's

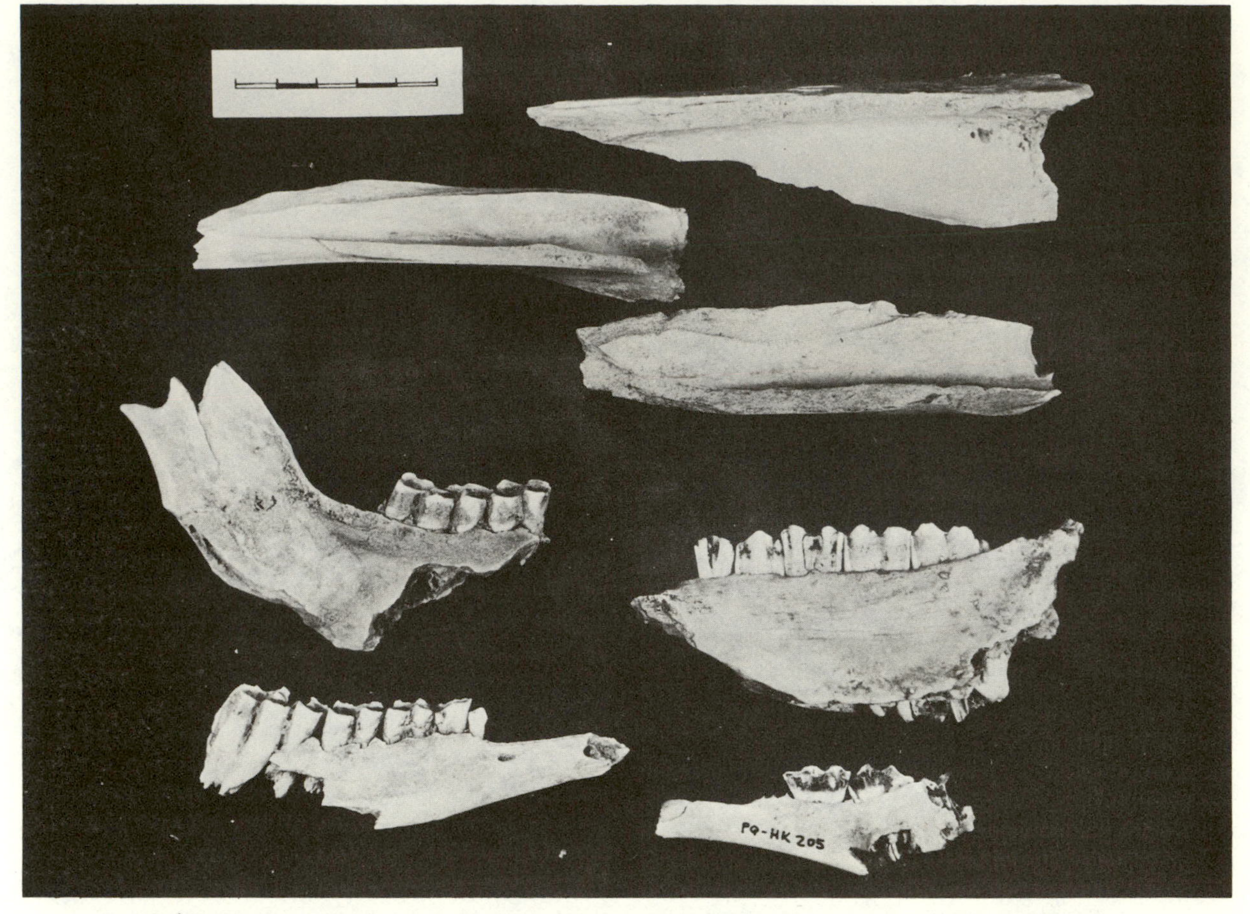

Fig. 75. Examples of bones cracked by the premolars of brown hyenas. The specimens are from the Kalahari National Park and from Uitkomst, Transvaal.

jaws would shatter the braincase of a jackal, fox, or caracal, making the brain immediately accessible. Yet the adult hyenas have clearly not done this. I can only suggest that cub-rearing brown hyenas have an inhibition against feeding on the animal they bring as food for the cubs and that the cubs are unable, or disinclined, to break up the skulls.

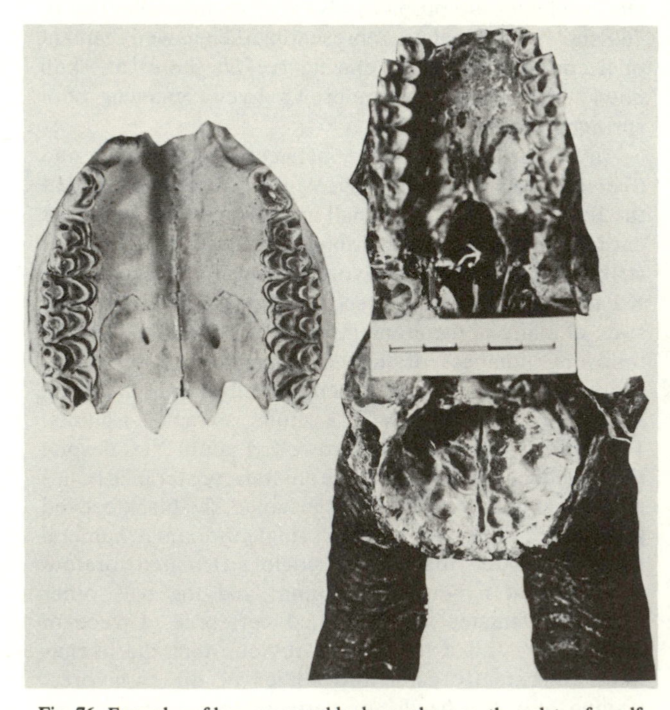

Fig. 76. Examples of bones gnawed by brown hyenas: the palate of a calf from Uitkomst, Transvaal, and a springbok "skull bowl" from the Kalahari National Park.

The parts by which each species of carnivore is represented are listed in table 41, but it can also be said that, of the black-backed jackal skulls, 6 had undamaged braincases (see fig. 77); 11 of the bat-eared fox calvariae were complete, and all the braincases of the caracal and ratel skulls were undamaged. Mandibles were typically present and reasonably undamaged, except for the bat-eared foxes, where all but one of the mandibles had disappeared. It appears that the brown hyena cubs were able to chew up and swallow the rather weak fox mandibles but typically rejected those of the other carnivores.

I suspect that cub-rearing brown hyenas purposely select small carnivores and, in some areas, primates as food for young cubs, since their skeletons are more crunchable than those of bovids of the same size (see chap. 2). When the cubs are larger they seem to be fed largely on antelopes that the hyenas hunt or scavenge. By this time the cubs are probably able to cope with the more robust bovid bones. The easily chewed carnivore and primate bones would provide the cubs with a readily accessible source of calcium, but why brown hyenas should feed their young in this way while spotted hyenas do not, I am unable to explain.

The brown hyenas' habit of bringing the bodies of carnivores (and baboons, on Kruger Park evidence) to their lairs and of often leaving the heads of these prey animals undamaged is of great significance to our understanding of bone accumulations in caves. Brown hyenas undoubtedly use dolomitic caves as breeding lairs and could be responsible for extensive bone accumulations made over many years.

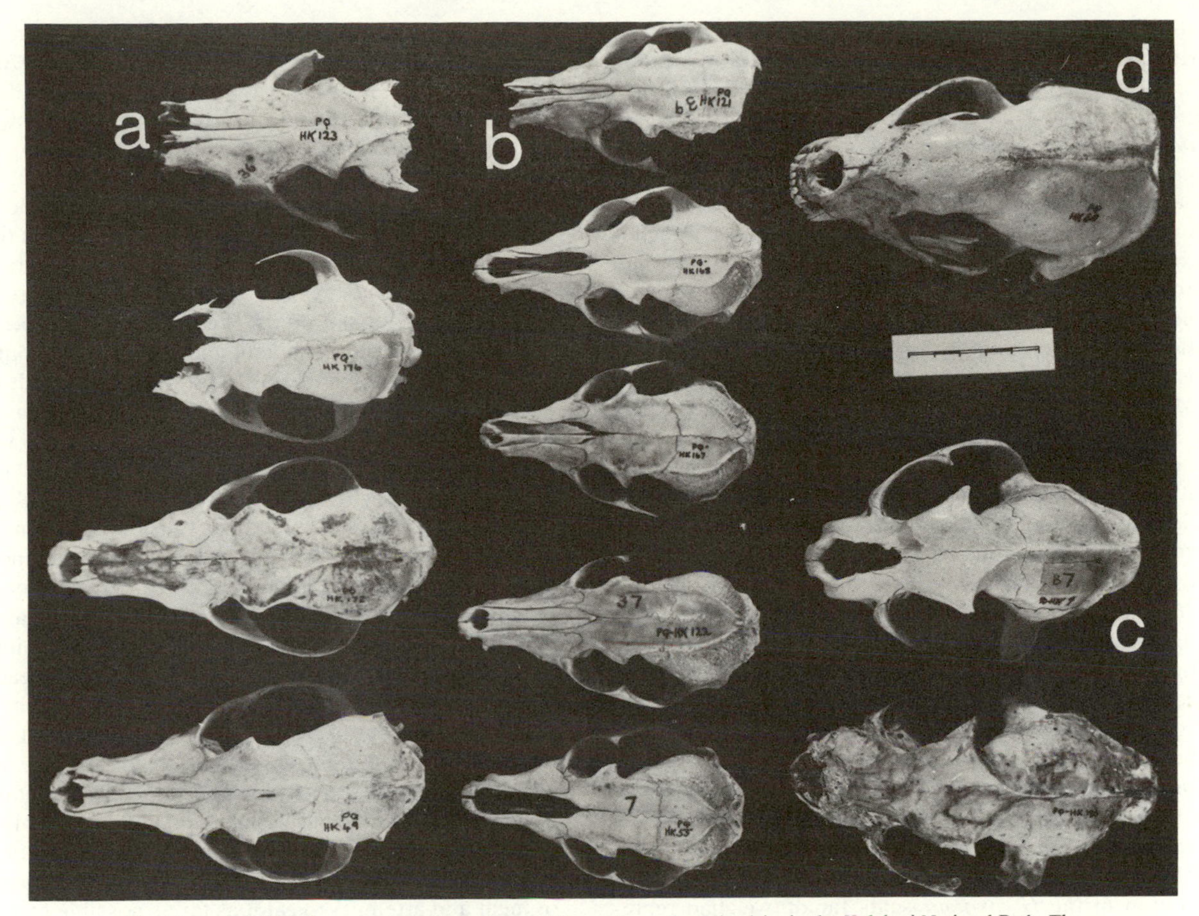

Fig. 77. Skulls of small carnivores collected around brown hyena breeding lairs in the Kalahari National Park. The carnivore bodies served as food for the hyena cubs: *(a)* black-backed jackal; *(b)* bat-eared fox; *(c)* caracal; *(d)* ratel.

The Striped Hyena, *Hyaena hyaena* Linnaeus

As will be mentioned in chapter 8, it appears that *Hyaena hyaena* was derived from the late Pliocene form *H. abronia,* known from Langebaanweg in the southwestern Cape. The lineage was a rather conservative one and underwent little change in the course of the Pleistocene, the fossil form *H. h. makapanensis* from Makapansgat Limeworks, more than three million years old, being only subspecifically different from the living striped hyena.

As is shown in figure 78, the striped hyena is currently distributed in north and northeast Africa, overlapping in range extensively with the spotted hyena, but nowhere with the brown hyena. Outside Africa, striped hyenas are found from the Gulf of Bengal westward throughout southern Asia and the Middle East and then as far north as southern Siberia and the Caucasus (Kruuk 1976).

The striped hyena is appreciably smaller than *Crocuta* but has a large head and long, pointed ears; it has a prominent erectile mane of long hair, and its general color varies from pale brown to white, with sharply defined stripes on the body and transverse barring on the legs. It is generally solitary and nocturnal.

I have made no personal observations on striped hyenas, but the animals are of significance here because they are known to carry bones to cave lairs. I will therefore present a brief review of some relevant information from the literature.

Striped Hyena Lairs and Breeding Sites

Referring to the striped hyena in India, Pocock (1941) remarked: "By day it lies up in enlarged porcupine-burrows, caves, or in crevices under boulders." In East Africa, Kruuk (1976) has photographed a den under overhanging boulders on a rocky hill about 10 km northeast of Seronera in Tanzania. It therefore appears that *H. hyaena,* like *H. brunnea* and *Crocuta,* makes use of either burrows or caves and recesses as lairs and breeding sites.

There is unfortunately not much information available about the composition of bone accumulations at striped hyena lairs. The late L. S. B. Leakey once told me that he had seen striped hyena cave lairs in East Africa well stocked with bones; but, to my knowledge, further information was not published. In search of such infor-

Fig. 78. Distribution of the striped hyena, *Hyaena hyaena,* within historic times.

mation, Hughes (1961) paid a visit to East Africa; unfortunately, in Uganda he was not successful in locating striped hyenas or their lairs, but at Olorgesailie in Kenya he saw a number of holes and overhangs used by *H. hyaena*. At the time of his visit, two sites in the area were occupied by porcupines, but Hughes decided to excavate a hole from which striped hyenas were seen emerging. From the front of this hole, and in the entrance, Hughes recovered 61 bones that he described as "old, dry and porcupine-gnawed"; he concluded that these had been collected by porcupines, and thus his East African observations served to confirm his conviction that hyenas do not transport bones to their lairs.

More recently, Kruuk (1976) has listed 9 items near a *H. hyaena* den in the Serengeti; these came from wildebeest, kongoni, impala, Thomson's gazelle, ostrich, vulture, tortoise, and dung beetle. From regurgitated hair balls near the den he was able to add the following food items: zebra, lion, spotted hyena, hare, hedgehog, fruit, and domestic rubbish.

An interesting comparison between the diets of striped and spotted hyenas has been provided by Kruuk (1976), who analyzed the contents of 50 feces of the former species and 42 of the latter from Ngare Nanyuki, in the Serengeti. Remains of large mammals such as zebra, wildebeest, kongoni, and topi featured in only 26% of the *H. hyaena* scats. But in 74% of the *Crocuta* scats, among medium-sized mammals such as gazelle and impala, the percentages were equivalent (68 and 71 respectively), and in the small and very small mammal categories, representation in the *H. hyaena* scats far outweighed representation in the *Crocuta* scats (28% : 2%). Likewise, reptiles, birds, insects, and vegetable foods were eaten in large quantities by striped hyenas but were largely ignored by spotted hyenas. Moreover, Kruuk found that 20 droppings of striped hyenas had a mean calcium content of about 12% as opposed to 25% for spotted hyena scats from the same area. These figures indicate that *Crocuta* was eating more bone, relative to other foods, than was *H. hyaena*.

To summarize Kruuk's results, it appears that the diet of striped hyenas in the Serengeti consists mainly of smaller vertebrates, particularly mammals, either hunted or scavenged, in addition to substantial quantities of fruit and insects.

Striped hyenas are still found in Israel, where various aspects of their natural history have been described by Ilani (1975). Of particular relevance here are Ilani's comments on the use of caves by *H. hyaena* (1975, pp. 12–13):

> At times of inactivity and when the cubs are growing up, hyenas stay in caves. These usually have narrow entrances and widen out considerably inside. Hyenas are good at digging and if the interior of the cave does not suit their requirements they do their best to improve and enlarge it. In regions where there are no natural caves, hyenas make their homes in the lairs of porcupines or badgers. It has even been claimed that hyenas occasionally share their lairs with porcupines, but this may stem from a misunderstanding of the living habits of the hyenas. My observations tend to indicate that hyenas make use of several caves at one time and do not remain permanently in any one lair. Nevertheless it is worthwhile to examine this theory more closely. Hyenas do indeed share their lairs with bats and invertebrates common to dark caves and their skins are consequently infested with ticks.
>
> Let me describe a typical lair located some 3 km. south of Arad that I investigated in November, 1970. The entrance to the cave was on a cliff about 4 meters above a dry wadi bed and was accessible by a narrow rock ledge. The diameter of the opening was about 40 cm. and we had to crawl on our stomachs for four meters before the tunnel widened out and we found ourselves in a roomy chamber 1.40 m. high. From the cave walls five openings led to smaller cave-chambers. In the center of the big room there was a pile of broken bones and animal skulls. We checked some of this material by electric torch-light and found that among others there were hundreds of dog skulls, hundreds of skulls and other skeletal fragments of domestic animals such as donkeys, camels, goats and sheep as well as two human skulls that were apparently dug out of a Bedouin cemetery in the vicinity. There were also several gazelle skulls, about ten skulls of ibex, bones of foxes and badgers, hedgehog skins and porcupine quills. The floor of the cave was of dry, powdery clay in which were innumerable hyena spoor. We managed with great difficulty to crawl into some of the adjoining smaller caves and, to our surprise, discovered in each one of them additional but smaller accumulations of bones. We wondered at what seemed to be a habit of dragging food to remote corners of the lair and it was only two years later when I watched the eating habits of older hyena cubs that I could advance a plausible explanation for this. Food brought by the adults is not eaten together but each cub apparently seizes a piece of meat and attempts to eat it as far as possible from his siblings leaving separate piles of bones in various places. Other caves that I examined were on the whole similar in size, layout and contents.
>
> In front of the cave openings, one usually finds hyena droppings. These are easy to identify because of their white color that is due to the calcium salts that remain from the bones eaten by the hyena. Less common, are wads of undigested hair that are exgurgitated outside the lair opening. These wads usually indicate that the cave has not only been used as a shelter but that cubs were raised in it.

More recently, J. D. Skinner undertook a study of striped hyenas in Israel from December 1977 to January 1978. He inspected five dens and referred to one in particular—a maternity den near Arad in northern Negev, in the same area as the one described by Ilani. Skinner wrote (n.d., p. 16):

> At one particular maternity cave we crawled through a passage some ten metres long to enter a large cavern where hyaenas had been feeding over a very long period in a special area. The den was occupied and the fresh remains of a goat *Capra hircus* were in evidence on top of a layer of bones covering the floor of the cave for 40 m². Hundreds of skulls and bone fragments covered the floor and two square metres were sampled ten metres apart.

The sample of bones was analyzed by S. Davis at the University of Jerusalem and was found to consist of 267 pieces from a minimum of 57 individual animals. These included 14 camels, 14 sheep or goats, 12 dogs, 11 donkeys, and single individuals of fox, cow, porcupine, gazelle, pig, and man. If one assumes that the layer of bones was uniformly dense over the 40 m² of floor area, one may estimate that the cave contained 5,340 pieces from a minimum of 1,140 individuals.

Although information is not yet available on the skeletal parts present in the sample, Skinner (n.d.) discounts the activities of porcupines in the collecting process.

H. hyaena is well represented among fossils from the gray breccia at Makapansgat Limeworks, and the observations made in Israel could be, to my mind, of the greatest significance in the interpretation of the Makapansgat fossil accumulation. This matter will be discussed again in chapter 13.

The Possible Role of Extinct Hyenas as Collectors of Bones in Southern African Caves

Extinct hyenas, whose remains have been found in the Sterkfontein valley caves, belong to four genera: *Crocuta, Hyaena, Hyaenictis,* and *Euryboas*. Although several species of *Crocuta* have been described on the basis of Transvaal fossils (see chap. 8), these are currently regarded as subspecies of *Crocuta crocuta,* the well-known spotted hyena. Morphologically, African crocutas of two or three million years ago were very similar to those of today, and there is no reason to believe that their behavior, with particular relevance to bone-collecting in caves, was any different.

Remains of *Hyaena brunnea* virtually identical to those of the contemporary brown hyena are known from Swartkrans Members 1 and 2, and a less firm identification has been made from Kromdraai B. Remains of a very similar hyena, *H. bellax,* have been described from Kromdraai A. Again, I am inclined to assume that these animals behaved much like their modern brown hyena counterparts.

The situation with the other two genera, *Hyaenictis* and *Euryboas,* is very different. Together they constitute the "hunting hyenas" to be described in chapter 8. They were predacious, fast-running animals, and representatives of both genera were perhaps derived from a Pliocene *Hyaenictis* stock.

It is not easy to reconstruct the behavior of mammals whose living representatives no longer exist, but a few speculations may be ventured. Since *Hyaenictis* and *Euryboas* were true hyaenids, they very probably reared their cubs in burrows or caves as the three contemporary forms of hyena do; their dentitions and limb bones indicate that they killed their own prey, almost certainly running it down in a cheetahlike fashion, although, like other hyenas, they probably scavenged as well; they were very probably social, hunting in packs like spotted hyenas or wild dogs, and, finally, they very probably brought back parts of their kills as food for their cubs.

That remains of these hunting hyenas are found in abundance in the Sterkfontein valley cave breccias strongly suggests that *Hyaenictis* and *Euryboas* frequented caves, at least when they were rearing cubs. If this was so, their contribution to the bone accumulations at such sites could have been considerable. An important difference between food remains of spotted and brown hyenas at breeding sites is that the latter include many skulls of small carnivores, from prey that the brown hyena parents fed to their cubs. It is not known whether hunting hyenas fed their cubs other carnivores or not, but the fact that hunting hyenas appear to have been active predators suggests that their food remains would consist more predominantly of *hunted* than of *scavenged* components. Furthermore, those *hunted* components are likely to have included bones from medium or large bovids and equids that were pursued and dismembered by hunting hyena packs.

The Leopard, *Panthera pardus* Linnaeus

Leopards have a remarkably wide distribution both in Africa and in Asia, and, although they have been exterminated from many areas, they are usually the last of the large predators to disappear under pressure from human encroachment. They are clearly very successful animals, able to adapt to a variety of circumstances and environments. Furthermore, they have been present in the Sterkfontein valley since Swartkrans Member 1 times—a probable span of 1.5 million years, showing negligible change in form over the period.

Within the living species, however, there is a good deal of variability in size, body conformation, and color pattern. In his taxonomic review of the leopards of Africa, R. I. Pocock (1932) regarded seventeen subspecies as valid and considered that these could be placed in five groups linked to the environment in which they occur: the "savanna or veldt type," the "desert type," the "mountain type," the "forest type," and various "dwarf forms," perhaps adversely affected by their harsh surroundings. Only thirteen of the former subspecies were admitted by Smithers (1971*b*) in the most recent taxonomic evaluation; he remarked that there seems to be a tendency for the unusual specimens to find their way into the hands of taxonomists rather than representative samples of the populations concerned.

Leopards are particularly relevant to this study in that they are secretive predators, making use of caves as retreats, feeding places, and breeding lairs. In addition, they have the habit of storing food in trees, which, in the case of the Sterkfontein valley caves, may have had special local significance from the taphonomic point of view. Although frequent passing reference is made in the literature to the use of caves by leopards, there are very few specific descriptions of caves that have served as habitual lairs or feeding places. In the course of his study of caves in the East African Rift Valley, Sutcliffe (1973*a*) found

Fig. 79. Distribution of the leopard, *Panthera pardus,* within historic times.

evidence of leopard occupation in two caves on Mount Elgon. The Kitum Cave contained a scatter of broken bones that appeared to be food remains of a resident leopard, and, a kilometer away, the Makingnen Cave was of special interest for this. The cave opens at the head of a small valley, behind a crescent-shaped cliff fall passable only by way of an elephant path between the rockfall and one of the cave walls. The path is extremely steep, but droppings along it and within the cave show that elephants are able to negotiate it. It may seem strange that elephants should wish to enter a cave, but they do, in fact, regularly visit some of the Mount Elgon caverns in search of salt, mainly mirabilite, which crystalizes on the cave walls. Various other animals are similarly attracted, and it is on these that the leopards appear to prey.

In the case of the Makingnen Cave, Sutcliffe located some habitual leopard feeding places on the right-hand side of the cave entrance, while scattered behind the entrance rockfall in the north chamber were numerous broken bones of antelopes, monkeys, and forest hogs—apparently food remains of the leopard. Unfortunately, Sutcliffe did not fully complete his investigation: "The last intended stage of the investigation of this cave was to have been a photographic session of the north chamber where the leopard eating place was situated. On this occasion Una and I were alone at the cave mouth, the light was beginning to fail (it gets very dark very quickly in East Africa) and an animal of unknown identity made a noise at us from the back of the cave. We confess to taking fright, abandoning the photographic session and concluding the investigation prematurely!" (Sutcliffe 1973*a*, p. 56).

Dr. and Mrs. Sutcliffe have my sympathy. I have several times experienced the apprehension of entering a cave knowing a leopard was inside; perhaps ape men felt the same apprehension on nights long ago when leopards prowled around their sleeping sites.

Use of the Mount Suswa Lava Caves by Leopards

Some of the best evidence for the use of caves by leopards in East Africa has come from Mount Suswa or Oldoinyo Nyokie, "the Red Mountain," a dormant volcano on the floor of the Rift Valley about 48 km northwest of Nairobi. The central part of the mountain consists of a wide, shallow caldera, about 11 km across, the floor of which lies at an altitude of 1,830 m. On the northeast slope of the mountain is a series of about forty-five holes, roughly circular in outline and varying from 1.5 m to 60 m in diameter. These consist of collapses into underlying lava tunnels or tubes that were formed when the phonolitic or trachytic lava of which the mountain is composed was still fluid (Glover et al. 1964).

In the course of her initial study of bone remains from these lava tunnels, Coryndon (1964) concluded that a number of collecting agents had been at work (see also Isaac 1967*b*), including leopards, hyenas, owls, and men. In addition, some of the surface holes concealed by vegetation had acted as natural traps, evidenced by the skeletons of 3 rhinos, a giraffe, and various antelopes. Concerning leopard involvement, Coryndon wrote: "Predatory animals prefer certain definite conditions in which to take their food—thus the leopard seems to prefer a very dark recess where no light can penetrate. One such cave had a floor strewn with the remains of many baboons and a few small antelopes, presumably brought in by leopards (Coryndon 1964, p. 61).

A more specific study of leopard involvement in the bone accumulations of the Mount Suswa caves has been made by J. W. Simons (1966), who studied both the skeletal remains of the leopards themselves and those of their prey. Remains of 8 or 9 leopards were found in various parts of the lava cave system; 3 leopard skeletons lay on a talus cone beneath the vertical-sided, 21 m-deep, "caldera blowhole," together with the dehydrated remains of 3 adult and 1 juvenile female baboons. Simons concluded that there was little doubt that the leopards fell to their death while trying to catch baboons, which regularly make use of the blowhole as a sleeping place. The baboons may also have fallen while being chased by leopards.

Another leopard skeleton, associated with that of a baboon, was found beneath collapse hole 6, which has a drop of about 12 m. In the leopard skeleton, the shafts of the right humerus and left tibia were broken, damage that Simons suggests might have resulted from a fall while attempting to catch the baboon. Scattered remains of 3 or 4 other leopards were found in other parts of the lava cave system.

Concerning the prey of the leopards, Simons estimated that there were parts of at least 37 baboons in eleven different areas of the cave system. Most of the remains were concentrated in the tunnels linking surface holes designated 6, 36, 37 and holes 18a,b,c,d. Certain very characteristic damage appeared to have been done to the baboon remains; such damage to the skulls may be summarized as follows: lateral pterygoid plates on the ventral surface chewed; tooth holes in the borders of the orbits; zygomatic arches sometimes chewed away; tooth holes in the sides of the muzzle, and brow ridges frequently chewed. Vertebrae and ribs were found to be underrepresented, and a good deal of damage had been done to the other skeletal parts.

The Suswa lava cave situation is fairly clear as regards the origin of its bone accumulation: baboons regularly use the holes in the lava slope as sleeping sites, and they are preyed on there by leopards that then feed on them in the subterranean tunnels. Occasionally the leopards and their intended victims fall to their deaths below the cliff faces. No actual leopard hunts have been observed at Suswa, but Simons has seen a baboon troop coming to sleep on ledges around collapse hole 18. Mounds of baboon droppings below the collapse holes indicate that this has been a long-standing baboon habit.

During June 1970 I was able to examine all the bones collected up to that time from the Suswa caves and housed in the Centre for Prehistory in Nairobi. It was soon apparent that, as Coryndon (1964) and Isaac (1967*b*) had observed, the bones owed their presence in the caves to a variety of agents. Some were charred and showed damage clearly attributable to humans; others appeared to have been cracked by hyenas; and still others had been gnawed by porcupines. Only one area in the cave system seemed to represent a reasonably uncontaminated leopard feeding site. It is the tunnel running from collapse hole 36, where live baboons are frequently observed, toward hole 37. Here, in almost total darkness, a steep passage leads upward for about 6 m from the main tunnel to a low cavern in which baboon remains were scattered in great profusion. Apart from leopard-inflicted damage, the bones showed a small amount of porcupine gnawing, and it is not impossible that porcupines added some bones to the leopard food remains.

The bone collection from this particular spot, designated "leopard lair 36E," consists of 176 pieces that were collected in March 1964. Particulars of the individual bones and the damage they have suffered are given in table 42. The remains come from a minimum of 10 anubis baboons as follows: 1 juvenile male, 1 subadult male, 2 adult males, 5 adult females, and 1 juvenile of undetermined sex. There are also remains of 1 adult leopard, 1 adult male klipspringer, and 2 unidentified birds.

Practically all the baboon remains show extensive carnivore damage, including numerous punctate marks presumably caused by the leopard canines. Skulls show the least damage of all the skeletal elements, and in no case was a braincase broken in such a way that the leopards could have gained access to the brain, although the damage suggests that they fed on the jaw muscles, tongues, and eyes.

The scarcity of vertebrae, and the almost total absence of ribs, shows that the baboon body from neck to tail was typically eaten. This is true of the hands and feet as well, and the limbs were clearly disarticulated, each segment being chewed from both ends. It is of interest that many of the long bones show transverse breaks in their shafts similar to the "spiral fractures" described by Dart (1957*a*) as resulting from hominid activity. In the case of these Suswa bones there does not seem to be any reasonable doubt that the breaks were caused by leopards. Some examples of characteristic damage are shown in figure 80.

A Study of Leopard Lairs in South-West Africa

Between 1968 and 1970 I was able to investigate a number of leopard lairs in South-West Africa through the kind cooperation of Attila F. Port, a well-known game conservationist and farmer in the area. At that time Mr. Port was owner of Valencia Ranch, and the adjoining farm known as Portsmut, in the Hakos Mountains, about 160 km southwest of Windhoek, an area where leopards were formerly extremely abundant and where they still occur in reduced numbers. It was in this vicinity that the observations described here were made (see also chap. 2).

The country rock in the Valencia Ranch area is Precambrian mica schist, which gives rise to rugged terrain (fig. 81) traversed by watercourses leading westward toward the Namib plain. Average annual rainfall is about 175 mm, falling in summer, and the vegetation is an arid thorn savanna. In the absence of a well-developed soil cover, the mica schist outcrops extensively, giving rise to numerous overhangs and shelters but seldom to caves of any size. Mr. Port was able to show me eight overhangs or holes, as well as one cave, where leopards had been seen; four of these could be regarded as habitual leopard lairs, used either as breeding or as feeding sites. One was situated on Valencia Ranch, two on Portsmut, and one on Verloren, as indicated on the map (fig. 82). I will give details of two breeding and two feeding lairs and then discuss some of their consistent features.

The Portsmut Breeding Lair. The site is in the wall of a steep-sided tributary of the Hakos River about 1 km northeast of the Portsmut homestead and consists of a tunnel less than 1 m in diameter (fig. 83) extending 8 m into the mica schist wall. As is shown in the plan (fig. 84), the tunnel opens into a dark chamber about 3 m wide. Both tunnel and chamber contained a shallow floor deposit of loose micaceous dust.

The recognition of this site as a leopard breeding lair occurred during 1927 in rather dramatic circumstances. Attila Port's father, first owner of the farm Portsmut, was a military man who believed his son should grow up tough and fearless. Suspecting that a leopard was rearing cubs in the hole, Mr. Port provided his five-year-old son with a thorny branch for protection and induced him to crawl into the tunnel. Equipped with a torch and a .450-caliber revolver, the father followed. As was to be expected, the leopard charged the intruders, and Mr. Port shot it over his son's head. After they dragged the dead leopard from the lair, the boy crawled back in and brought out the two young cubs.

The site has apparently been used as a breeding lair on several occasions since 1927, and in this interval 8 leopards have been trapped at the water hole a few hundred meters downstream.

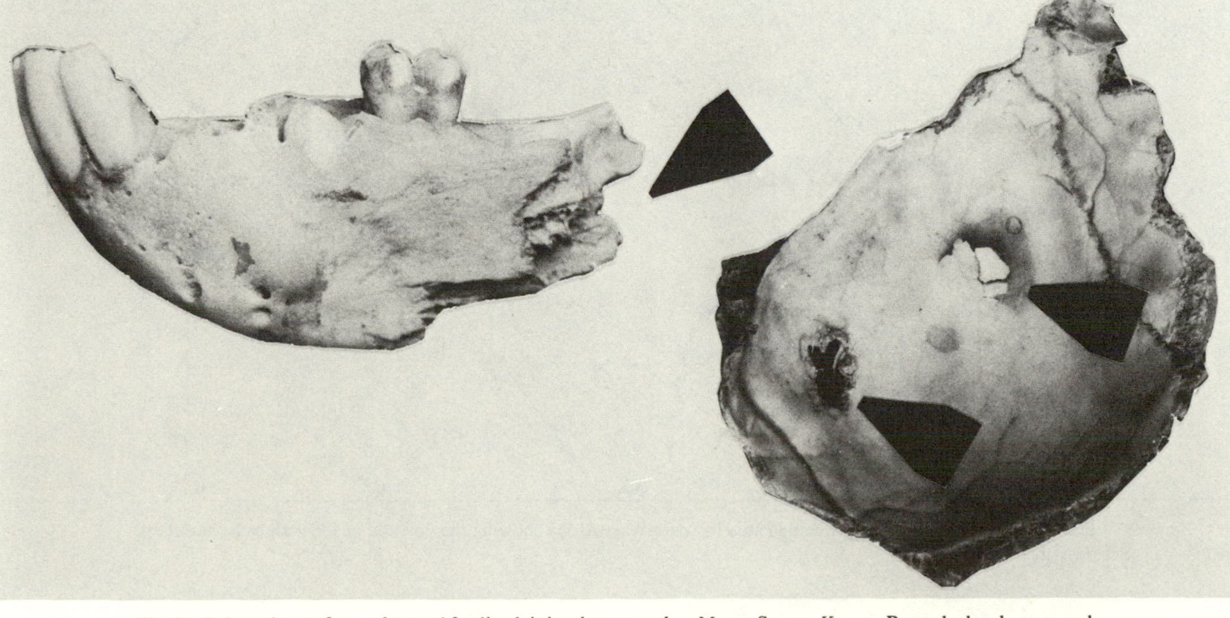

Fig. 80. Baboon bones from a leopard feeding lair in a lava tunnel on Mount Suswa, Kenya. Ragged edge damage and punctate marks are shown.

Fig. 81. An aerial view of the Hakos Mountains in South-West Africa, where studies of leopard lairs were undertaken. The arrow indicates the Quartzberg lair in an outcrop of white, crystalline quartz.

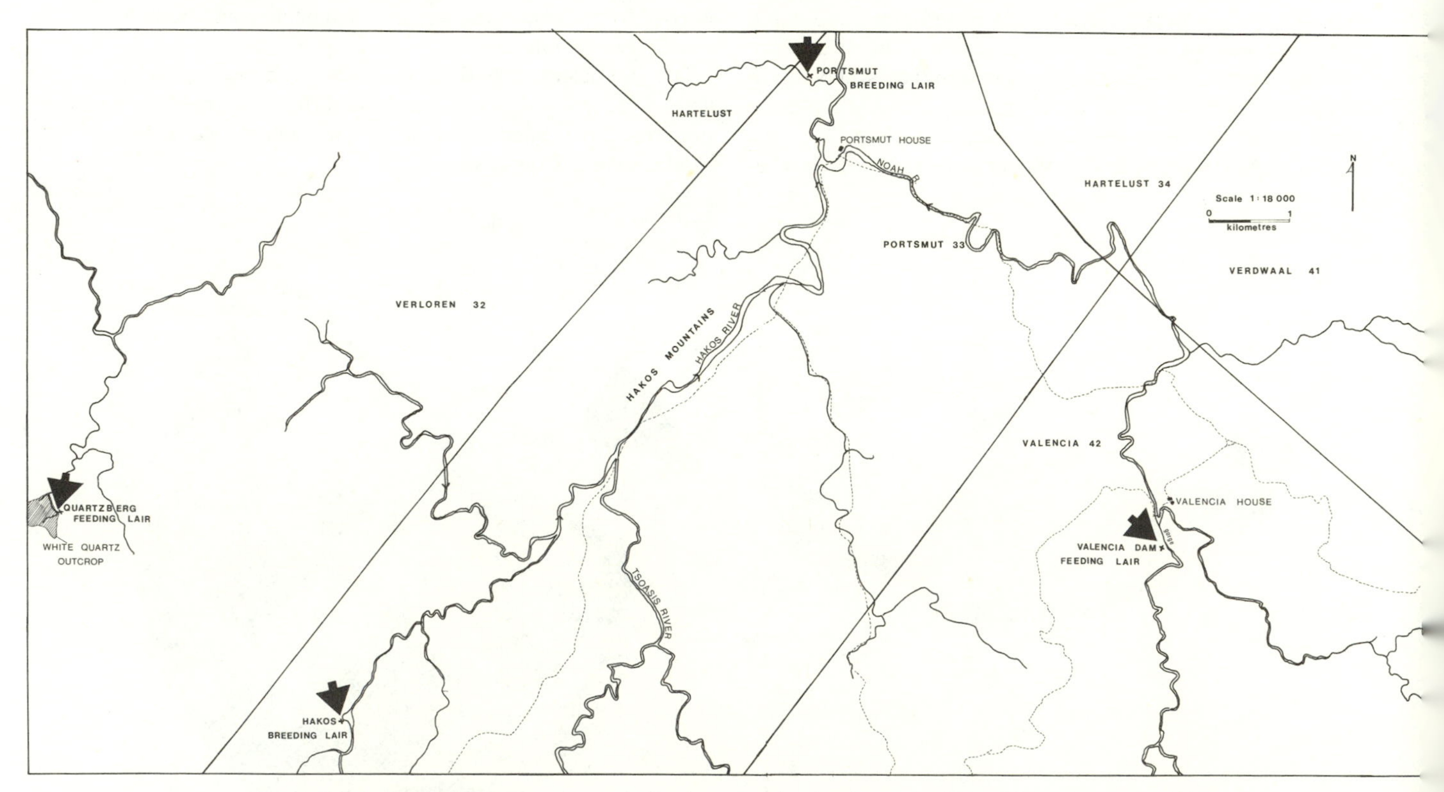

Fig. 82. A map of the Hakos Mountain area in South-West Africa showing the position of the various leopard lairs described in the text.

Fig. 83. The entrance to the Portsmut leopard breeding lair. Mr. Attila Port and helpers are shown sieving the micaceous dust removed from the floor of the lair.

My first visit to the lair was on 20 March 1968, when leopard tracks were visible in the dust at the entrance and the stomach of a freshly killed dassie was found there. I was anxious to investigate the interior of the lair but apprehensive about my reception by its possible occupant. Nor was Mr. Port's attitude—doubtless hardened by his own experiences at the place—particularly reassuring. In response to my query about what to do if the leopard were to attack me, he handed me the tibia of a zebra that happened to be lying there and suggested I thrust it into the creature's mouth! After crawling some distance into the tunnel, I came upon the intestines of a dassie and heard ominous noises from the dark chamber, suggesting it would be inadvisable to proceed farther. I therefore retreated and returned a few days later with more suitable equipment. On the second visit the lair was unoccupied; I undertook a survey and removed and sieved the floor deposit.

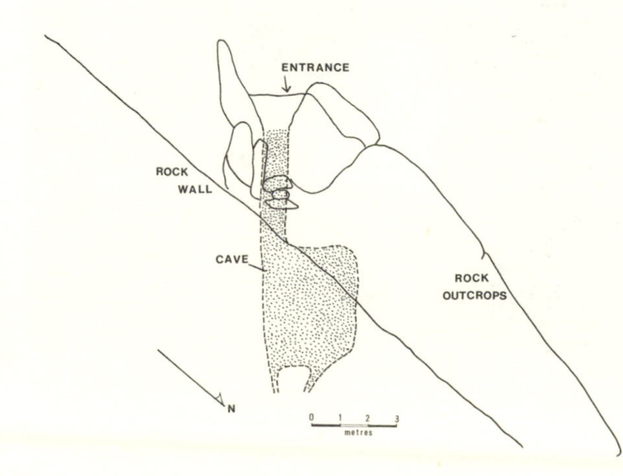

Fig. 84. Sketch plan of the Portsmut leopard breeding lair.

The resulting collection consisted of 192 bone pieces and 17 cultural objects apparently left in the cave by a hunter-gatherer. The presence of these latter objects raises the possibility that the cave was used for shelter by primitive people and that bones found there might represent human food remains. I have doubts about this, however, in view of the low roof of the access tunnel and the darkness of the inner chamber. The grooved stones, grooved pottery piece, pestles, and hematite and quartz pieces were found together in the innermost corner of the cave, suggesting that a hunter had stored them there but had failed to come back for them.

Porcupine gnawing was discernible on 19 of the bones, in particular on the weathered cranium and mandible of a baboon and the equally weathered zebra limb bones. These and some of the other bones were very probably brought to the cave by porcupines.

As is detailed in table 43, the remains were found to have come from an adult male baboon, an adult leopard, 2 juvenile klipspringers, 2 class II bovids—probably sheep or goats—1 class III antelope, 2 adult mountain zebras, 4 dassies, and a tortoise. At least some of these animals are likely to have been leopard prey, though the remains of the leopard itself probably came from an animal that died there.

There is no way of telling from these remains whether the prey was brought to feed cubs in the lair or whether the site was used as a feeding retreat between litters.

The Hakos River Breeding Lair. The site consists of three intersecting tunnels in the north bank of the Hakos River (figs. 85 and 86) about 9 km west of the Portsmut homestead. The tunnels open just above the level of the riverbed, which is normally dry, and have formed in a soft band of inclined mica schist; they contained a deposit of micaceous sand varying in depth from 15 cm to 40 cm, which was carefully removed and sieved to recover all bones.

According to Attila Port's records, the site was last used as a breeding lair by leopards in October 1967, when the female was not disturbed. Before this, 104 leopards were trapped in the riverbed outside the lair between 1910 and 1950.

Details of the 339 bones recovered from the lair are given in table 44; 65 of the pieces showed evidence of porcupine gnawing, and so it is to be expected that some of the collection was made by porcupines. Five of the bones had been partially burned—these may have been picked up by the porcupines at a human fire place. As at the Portsmut lair, bones of leopards were encountered in the assemblage. These came from a young cub and an adult. It therefore appears not unusual for remains of the leopards themselves to be found in habitual lairs.

It was somewhat unexpected to find remains of an adult wild dog among the bones. Whether this animal had been killed by a leopard or whether it had used the lair itself I am unable to say. Its skeletal parts showed no carnivore damage.

Remains of the various bovids and mountain zebras, many of them from young animals, could well represent leopard prey. As in the Portsmut lair, however, it is not possible to decide whether the prey was taken to the lair at times of cub-rearing.

The indications from these two small caves, which have certainly been used as leopard breeding places, is that such places may be expected to contain remains not only of leopards themselves but of their prey.

The Quartzberg Feeding Lair. Although this cave is on the farm Verloren, which adjoins Portsmut to the north, the countryside in this vicinity is extremely broken, and no tracks or roads link the two farms. Access to the Quartzberg lair is therefore gained by the Walvis Bay Road at the foot of the Gamsberg pass. The surroundings of the cave are easily recognized, since they are pure white quartz, the only outcrop of its kind in a vast area of drab micaceous schist (fig. 81). Here a small stream, fed by a perennial spring, makes its way through a spectacular gorge with glistening walls of white crystalline quartz. As one of the few water sources in these barren mountains, the spring attracts a variety of animals, some of which fall prey to the resident leopards.

In the side of the gorge, just below the spring, is a cave of considerable size, its rounded entrance partially obscured by a fig tree and by large fallen blocks of rock (fig. 87). The cave has resulted from the dissolution of a calcite seam in the quartz and has a complex form, as shown in the sketch plan. Passing between the fallen entrance blocks, one enters a dimly lit chamber that becomes progressively darker toward the back (fig. 88), where a low passage leads into a second cavern hung with stalactites on the right-hand side. Beyond these a tunnel continues, decreasing in height and width.

At the time the plan of the cave was drawn, on 25 March 1968, a leopard sunning itself on the rocks at the cave entrance was disturbed by our approach and retreated into the darkness. We followed its tracks across the floor of the outer cave, through the low passage into the dark inner cavern, and thence into the tunnel on the right-hand side. Here we left it, deciding that peaceful coexistence was preferable to confrontation. My companions guarded the entrance to the tunnel with their guns while I surveyed the cave and collected the bones from the inner chamber.

Fig. 85. Entrance to the Hakos River leopard breeding lair.

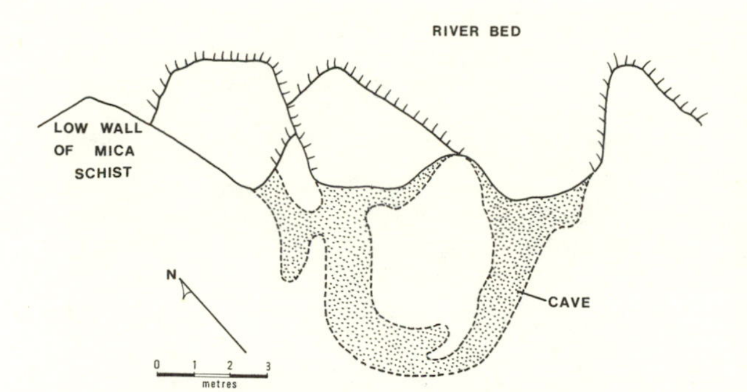

Fig. 86. Sketch plan of the Hakos leopard breeding lair.

The Quartzberg cave is of interest in this investigation because it has served both as a habitual leopard feeding place and as a retreat when resident leopards are disturbed. It is likely that it has also been used as a breeding lair, but its inaccessability means few people have visited it, and cub-rearing has not been observed there.

That porcupines also make use of the Quartzberg cave complicates any interpretation of the bones found there. Fortunately, porcupines prefer dry and naturally defatted bones (see chap. 5 for discussion on this topic) that have been lying in the open for some time, and it is these that they carry into their lairs. They would probably have ignored any fresh leopard food remains until these were dry and defatted. All the bones collected in the Quartzberg cave seemed to fall into two groups: unweathered remains with tissue adhering, derived from animals that were killed or died in the cave, and second, clean, defatted bones clearly derived from outside the lair, showing abundant signs of gnawing and almost certainly carried in by porcupines.

The cave was divided into three zones, the light, the twilight, and the dark, as shown on the plan; bones collected from each zone were separately marked. Table 45 lists the bones found in the Quartzberg cave. Of the 211 pieces, 147 are regarded as probably porcupine-collected and 64 as probably leopard-collected. As is shown in table 46, it is estimated that the porcupine component came from a minimum of 6 individuals, while the minimum number of individuals in the leopard component is 15; 80.2% of the bones in the presumed porcupine component were gnawed by porcupines, and 23.4% in the presumed leopard component showed such marks, suggesting that the porcupines may have picked up some of the dry remains from leopard kills at the cave entrance and gnawed them in their part of the lair.

Table 47 gives some figures on the distribution of bones in the Quartzberg cave relative to the three lighting zones. Only 3% of the bones in the presumed porcupine component were found in the light zone, whereas 36% of the leopard component came from this zone. These data suggest that, in caves, porcupines prefer the twilight and dark zones, avoiding the well-lit area, whereas leopards appear equally at home in light, twilight, and dark zones. They probably do most of their undisturbed feeding at the entrance to the cave, but they do not hesitate to retreat into the darkest interior when necessary.

Leopard damage to bones at the Quartzberg lair showed some interesting features. Lower limb bones of klipspringers were characteristically left undamaged and articulated (fig. 90), while the remains of 6 individual dassies were almost exclusively cranial. The skulls had been damaged in a consistently typical fashion that will be discussed further.

The Valencia Feeding Lair. This small cave is of interest because it was a feeding site of the leopard upon which feeding experiments—shortly to be described—were undertaken. After its escape from a cage at the Valencia homestead in October 1968, this large male leopard made its headquarters for approximately two weeks at the Valencia dam, on a tributary of the Noah River, 0.7 km west of the house. Here it killed an adult karakul sheep and a subadult goat, dragging both of them into a small cave in the mica schist a short distance above the dam wall (fig. 91). After eating parts of these, as well as most of a subadult dassie, the leopard moved away and was never seen again.

The cave consists of a triangular recess, 15 m wide, in the steep rocky hillside. As shown in the plans and sections (fig. 92), the outer overhang leads by way of a low divide into an inner chamber with a maximum height of 1.5 m. It was on the floor of this inner chamber, littered with dassie droppings, that the remains of the leopard's prey were found. All the bones were removed for cleaning and study; the skeleton of the karakul sheep was virtually complete, suggesting that the leopard may have been disturbed and left the area before finishing its meal. The damage to the goat skeleton was more extensive; the axis vertebra had been chewed through and all vertebrae between this and the sacrum consumed. The pelvis showed tooth marks; both scapulae were extensively damaged, one with a characteristic punctate mark; the proximal ends of both humeri had been chewed off, and both knee joints were chewed through, resulting in the loss of the distal femurs and proximal tibiae. The skeleton was otherwise complete.

The dassie remains consisted of a characteristically damaged skull with no postcranial bones. The nature of leopard damage to dassie prey will be considered further.

Leopard Damage to Dassie Prey Skeletons

Remains of 6 dassies were found in the Quartzberg leopard lair and of 1 at the Valencia feeding site. At Quartzberg, only 3 postcranial bones were found associated with the 6 craniums; at Valencia there were none at all. Furthermore, it appeared in each case that the braincase and posterior part of the mandible had been sheared off, leaving ragged, tooth-marked edges. Examples from these two feeding sites are shown in figure 93, together with a similarly treated skull from the Suswa lava caves in Kenya, a leopard feeding situation discussed earlier. It seemed highly likely that when leopards fed on dassies they typically consumed the whole postcranial skeleton and that, working up toward the head from the back, they sheared off the braincase and ascending mandibular rami to remove the brain and tongue. The remaining unchewable maxillary and mandible parts were rejected.

The only way to confirm this would be by direct observation of a feeding free-ranging leopard, or by controlled feeding experiments on a captive one. The possibility of observing feeding behavior of a wild leopard in the Valencia area seemed remote owing to their extremely secretive behavior there, but Attila Port kindly offered to catch one for experimental purposes. The trap was consequently set on 20 March 1968 in one of the tributaries of the Hakos River near the Portsmut homestead. It consisted of a substantial rectangular cage with falling doors at each end activated by a foot treadle. The cage was positioned on the floor of the steep-sided valley, and a

Fig. 87. Entrance to the Quartzberg leopard lair. The large blocks of rock at the entrance are used by the resident leopards as feeding places.

Fig. 88. Inside the Quartzberg leopard lair cave: the dimly lit chamber leading from the entrance.

barricade of thorn branches was built on each side of it so that, we hoped, any leopard making its way along the valley would pass through it. During the first week the cage was sprung, and we approached it with great anticipation but found, to our surprise and disappointment, that we had caught nothing more spectacular than a large mountain tortoise. But finally, on 26 April, a leopard entered the trap and captured itself (fig. 94). This proved to be the largest male I had seen, weighing 96 kg (211 lb) and

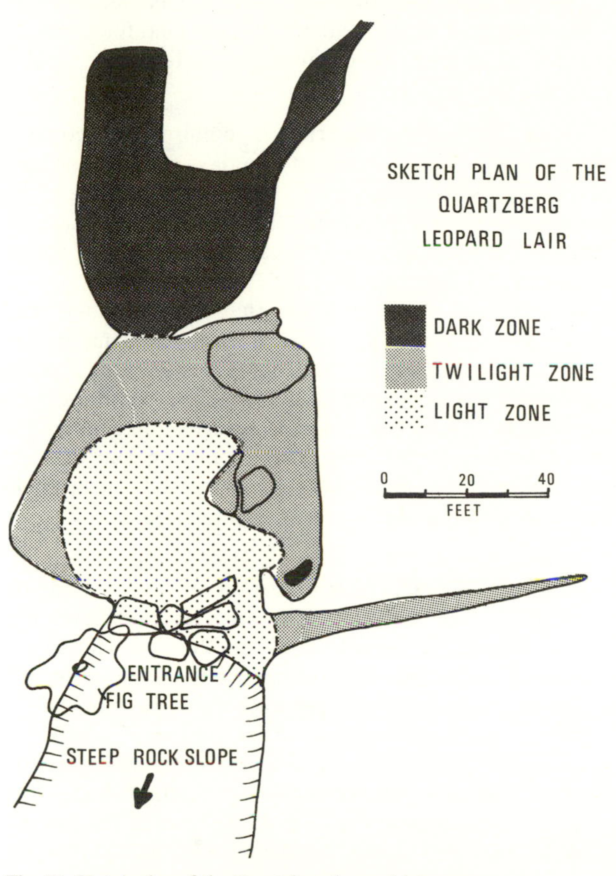

Fig. 89. Sketch plan of the Quartzberg leopard lair cave.

was so ill tempered that it could be handled only with the greatest difficulty. Most authorities quote the maximum weight for adult leopards at about 82 kg (180 lb), so this was an unusual specimen; *Roland Ward's Records of Big Game* states that leopards weighing more than 200 lb have occasionally been recorded (Best 1973), and Turnbull-Kemp (1967) in his book *The Leopard* admits seven such records.

The captive leopard was taken to the Valencia homestead and transferred to a specially constructed cage. Handling was difficult, since any entry to the cage precipitated an uncontrollable attack; to overcome this problem, the cage was divided into two components with a lockable door between them. Food could be placed in one compartment while the leopard was in the other and this door then opened.

Records kept by Attila Port over many years showed that food preferences of wild leopards in the Valencia area were, first, domestic calf; second, kudu calf; third, young mountain zebra; and fourth, gemsbok calf. Sheep were taken occasionally, but goats rarely. Typically no more than 10 kg of meat would be eaten from the prey, and the leopard usually did not return to the carcass, but would kill again within the next few days. In the absence of hyenas in the Valencia area, the leopards never used trees to store their food, although they would often drag prey to a habitual feeding site where disturbance was not likely. The relatively easy life of Valencia leopards may have permitted the exceptional size of some individuals.

In captivity, the male leopard ate even less from the carcasses provided for it than had been observed in the wild state. Weights eaten from three kudu calves and three baboon carcasses were as follows:

Kudu calf 1: 3 lb 14 oz or 1.8 kg eaten
Kudu calf 2: 4 lb 2 oz or 1.8 kg eaten
Kudu calf 3: 2 lb 3 oz or 1.0 kg eaten
Baboon adult female: 4 lb 7 oz or 2.0 kg eaten
Baboon adult female: 4 lb 1 oz or 1.8 kg eaten
Baboon adult male: 2 lb 8 oz or 1.1 kg eaten

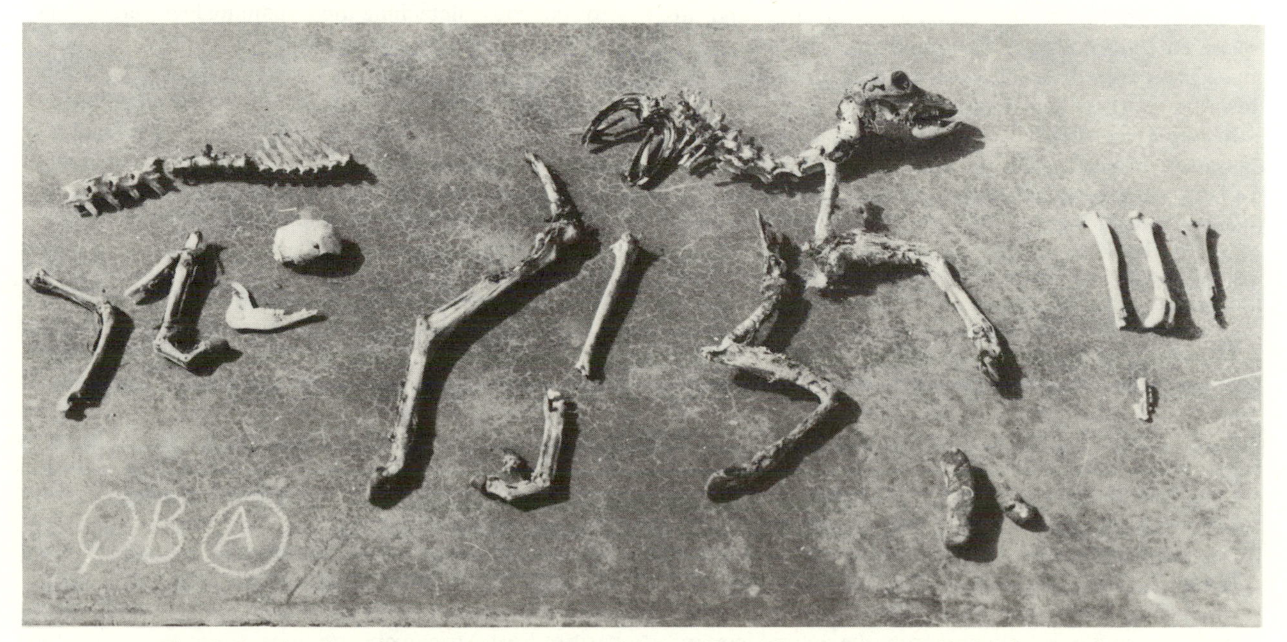

Fig. 90. Remains of two klipspringers and a baboon, left by a leopard that used the large blocks of rock at the entrance to the Quartzberg cave as a feeding place.

Fig. 91. The Valencia leopard lair, a small cave in the mica schist hillside, used as a feeding retreat.

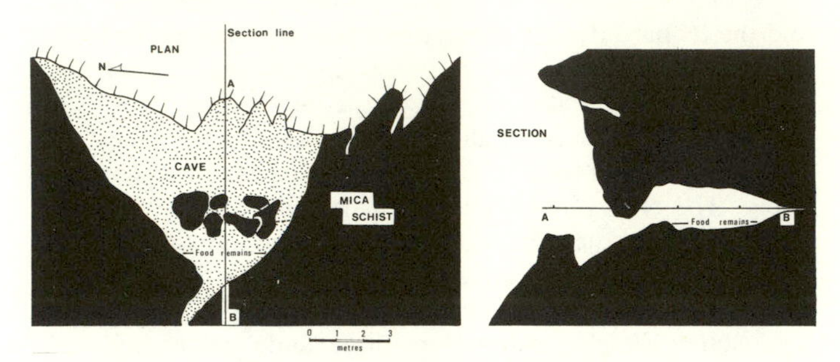

Fig. 92. Plan and section of the Valencia leopard lair.

The mean weight of meat eaten from these six carcasses was 1.6 kg, and on no occasion did the leopard return to the remains to eat a second time. The purpose of catching the leopard had been to allow it to eat dassies in controlled circumstances so that the remains could be compared with those found in the natural leopard lairs. Unfortunately, only two dassies had been fed to the leopard before it escaped one night, killed a young mountain zebra in an adjoining enclosure, and disappeared. Subsequently we found that the leopard had made use of the Valencia dam cave as a feeding lair, as described earlier, before moving away altogether.

In September 1968, when the feeding experiments on dassies were performed, the captive leopard was still extremely aggressive and secretive about its feeding. It refused to eat during the day or when an observer was in sight. What observations were possible were made at night from a concealed position. Some particulars of the two feeding experiments are as follows:

Dassie 1:
Adult female, shot 21 September 1968, weight 6 lb, 4 oz, or 2.8 kg; eaten during the night, a total weight of 5 lb, 0 oz, or 2.3 kg, being eaten. A good deal of fur was first plucked from the ventral surface of the dassie, the body cavity was opened, and the stomach and intestines were removed and placed to one side. The entire body and skeleton was then eaten, the only part remaining being the snout. As is shown in figure 95, the damage to the skull is remarkably similar to that observed in dassie craniums from the natural leopard feeding lairs. The braincase has been chewed away, and a punctate mark is present below the right orbit. All of the right and most of the left ascending mandibular rami have been chewed away, leaving typical ragged edges. No postcranial bones remained.

Dassie 2:
Subadult male shot 23 September 1968, weight 5 lb, 8 oz, or 2.5 kg, eaten during the night, a total of 4 lb, 0 oz, or 1.8 kg, being eaten. As before, the fur was plucked out, the alimentary canal was removed and rejected, and the entire body, with the exception of the front part of the head and the left hind foot, was eaten. Damage to the cranium was virtually identical to that of dassie 1.

These two feeding experiments confirm that leopards tend to leave the anterior part of a dassie skull, damaged in a characteristic manner, but that little or nothing of the postcranial skeleton remains. Thus, in a fossil context, leopard food remains might reasonably consist of dassie craniums showing "sheared-off" characteristics.

It is perhaps worth recording that when cheetahs feed on dassies they employ almost exactly the same procedure, except that on no occasion have I seen them pluck fur from the body before starting to feed. Working with the captive cheetahs at Valencia described in chapter 2, I was able to make detailed observations on four dassies, each eaten by a single cheetah. Extracts from my field notes concerning one of the experiments read:

19 March 1968, 2 P.M. Shot an adult dassie on the cliff below the Valencia farmhouse and offered the body to an adult female cheetah, who took it immediately. She first chewed off the right front leg, then opened the chest from the side and below, eating the lungs, liver, and heart. She then worked backwards past the diaphragm and removed most of the alimentary canal complete. This was dumped under a tree, and the rest of the body was moved to another patch of shade.

The cheetah next chewed off the head of the dassie at the neck and worked on the back of the head for 10 min (fig. 96), eating the brain but leaving the snout with the fur on it. The dentition of the dassie appeared undamaged. She then ate away all the flesh and bones of the body, detaching the remaining legs one by one, chewing and swallowing each.

After 1 ½ hr a patch of back skin about 8 in by 6 in was left, together with the snout and the alimentary canal. The cheetah moved away, and so I entered the enclosure to collect the remains. As I was leaving, the cheetah attacked me and put on an impressive intimidation display. She slowly withdrew when I stood still.

Records of the other three dassie feeding experiments were very similar, and in each case the snout and alimentary canal remained. It is fairly clear why the dassie stomach is rejected by both leopards and cheetahs. In most specimens I have examined, the stomach, a bean-shaped organ up to 15 cm long, was packed to capacity with green leaves and other vegetable matter. This was so tightly compacted that it is surprising digestion could proceed.

The reason for these apparently overfilled dassie stomachs has to do with the animals' feeding behavior, which has been described by Sale (1965, 1966). On the basis of East African observations, the average daily food intake of dassies is not excessive relative to their body

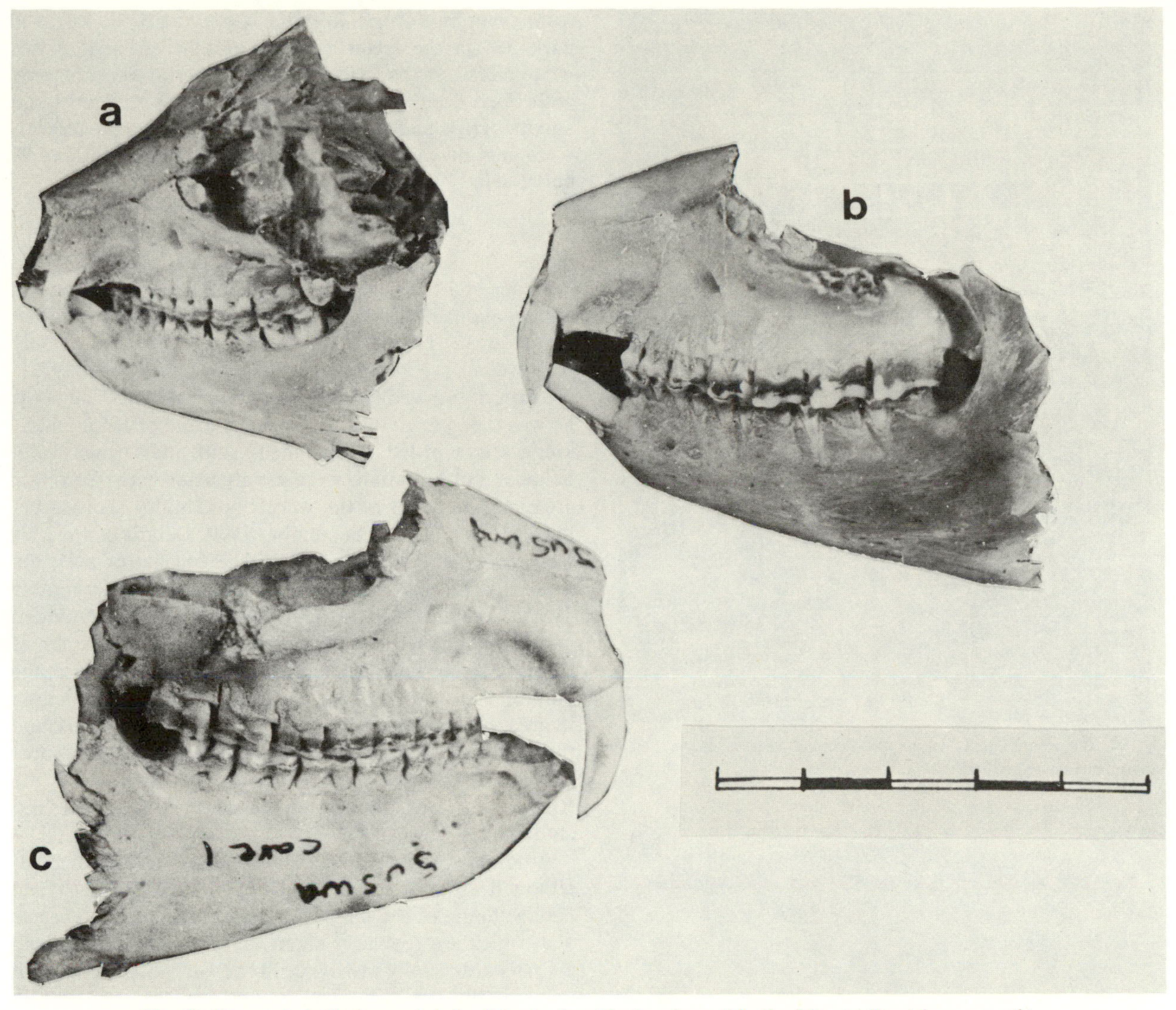

Fig. 93. Characteristically damaged skulls of dassies found in three leopard feeding lairs: *(a)* Quartzberg cave; *(b)* Valencia lair; *(c)* Suswa leopard lair, Kenya.

weight, but the mode of ingestion is unusual. The incisors are not used and appear unsuitable for biting off shoots and leaves. Instead, the head is turned sideways, at right angles to the body, so that the entire premolar-molar toothrow may be used in shearing the vegetation, allowing remarkably rapid ingestion. In fact feeding, which is usually performed in groups, usually occupies less than an hour a day, split into morning and afternoon sessions. During these sessions the dassies' stomachs are rapidly packed with vegetation. The briefness of the feeding sessions appears to be an antipredator adaptation, for it is when foraging away from the protection of their rocky retreats that dassies are most vulnerable to predation.

The Diet of Leopards as Reflected in Observed Kills

There are two obvious ways of establishing what leopards eat: direct observations of their kills and analysis of their droppings. The two methods are likely to produce strikingly different results, as will be discussed shortly. Information on leopard diet is relevant to the present study of bone assemblages in caves because it will indicate the range of animals that could be consumed in feeding lairs. Five detailed studies of leopard kills are currently available, one from the Transvaal, one from Rhodesia, one from Zambia, and two from Tanzania. Some particulars of the areas in question are now provided.

Kruger National Park, Transvaal. Situated along the eastern boundary of the Transvaal, this park had an area of 19,084 km² and includes a variety of veld types, the main ones being lowveld, sour bushveld, arid lowveld, mopani veld, and mixed bush veld (Edwards 1974). Details of kills have been recorded by park staff over many years, and Pienaar (1969) collated them for the periods 1933–46 and 1954–66. The recorded kills totaled 1,940 for the first period and 5,525 for the second, a remarkable total of 7,465.

Rhodes Matopos National Park, Rhodesia. This national park covers an area of about 43,200 ha situated 48 km southwest of Bulawayo and consists of broken granite hills and gorges surrounded by woodland containing patches of grassland and savanna woodland. The area was mentioned in chapter 3 when I discussed the Pomongwe Cave. A study of leopards in the Matopos has been made over a number of years (Wilson 1969; Wilson and Grobler 1972; Smith 1977), and information is available both from records of kills and from analyses of scats. In the most recent paper, Smith provides data on 38 kills observed between 1960 and 1974.

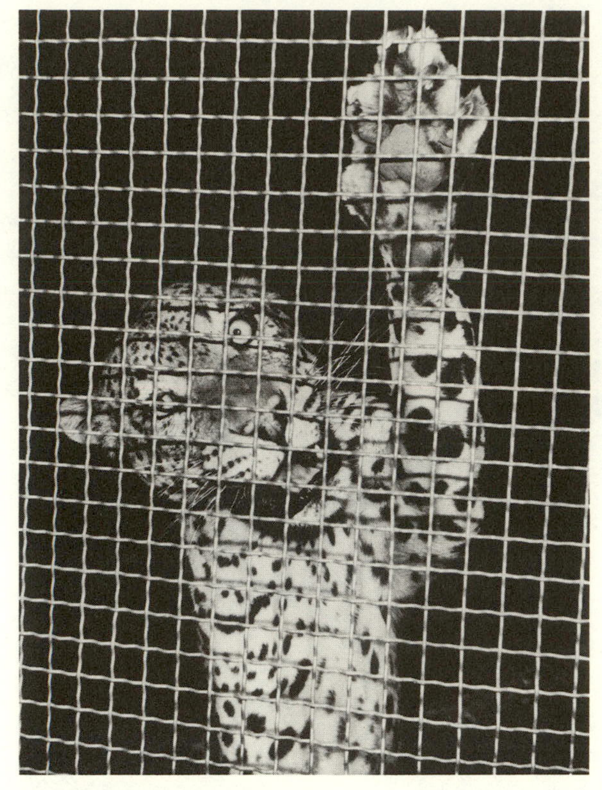

Fig. 94. The exceptionally large male leopard used in dassie feeding experiments described in the text.

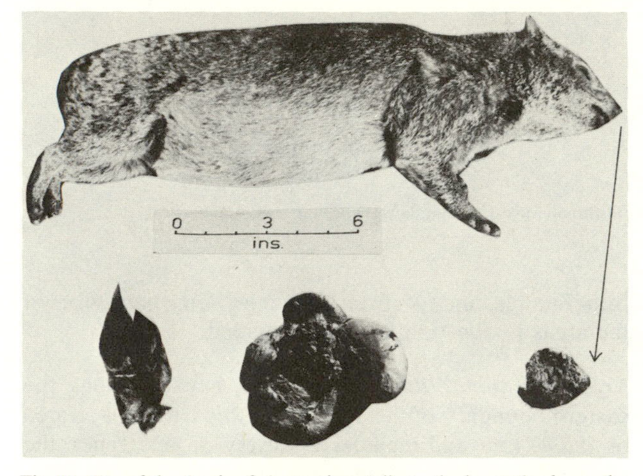

Fig. 95. One of the dassies fed experimentally to the leopard: *above*, the entire animal; *below*, discarded remains—the snout, tufts of fur, and alimentary canal.

Fig. 96. A captive cheetah on Valencia Ranch eating a dassie. The whole body, with the exception of the stomach and snout, is crunched up by the cheetah's carnassials. Damage to the skull is similar to that observed in leopard feeding.

Kafue National Park, Zambia. Covering 8,600 mi², the park lies in the basin of the Kafue River and carries woodlands of the *Brachystegia-Julbernardia* complex with open drainage lines supporting grass savannas and aquatic grasslands in the wetter patches. Kill records were kept between June 1960 and May 1963, a total of 96 being listed by Mitchell, Shenton, and Uys (1965).

Serengeti National Park and Surrounding Areas, Tanzania. A vast area of plains and hills lying between latitudes of 1°S and 3°30′S, the national park itself has an area of about 13,250 km², supporting two main vegetation types: grassy plains and wooded grasslands with widely scattered trees, creating a parklike aspect (Schaller 1972).

Turner observed leopard predation from 1957 till 1964; he was then joined by Kruuk until the end of 1965, and 55 kills were recorded in this time (Kruuk and Turner 1967). Schaller's observations were mostly made in the Seronera area and the edges of the woodlands, in dry seasons between June 1966 and September 1969. Details of 164 kills were recorded (Schaller 1972). The full list of prey observed in the course of these five studies is given in table 48, and a summary of the main prey categories is provided in table 49 and depicted graphically in figure 97. Each of the five studies reveals that bovids in size class II form the bulk of the observed kills, the species varying from area to area, depending on the habitat. In the Kruger Park and the Matopos, for instance, impala kills are the most numerous, whereas reedbuck and puku dominate in the Kafue area and Thomson's gazelles in the Serengeti studies.

Antelopes in other size classes are less prevalent, although duikers are favored prey in the Matopos and Kafue. Some of the larger antelopes such as wildebeests and sables are regularly taken, but in such cases young animals are usually involved. In the very large species, it is invariably calves that are killed.

Among nonbovid ungulates, zebras and warthogs feature fairly prominently; among primates, kills of baboons and vervet monkeys have been recorded; and a wide range of carnivores is known to have been killed by leopards. The literature contains many references to the strong taste these cats have for domestic dogs and to their disregard for danger in obtaining them. I will cite two instances. The first, related by Major C. Graham, occurred in the Mongalla Province of the Sudan. It concerned the late C. H. Stigand, who was in civil charge of the region (Graham 1953, pp. 244–45):

> one night he and his wife were woken up by a commotion in the room next to their bedroom. They thought their dog was attacking a small duiker kid they had inside the house, so they jumped up and rushed to the rescue. You can imagine the shock they got when a leopard dashed past them out of the room and into their bedroom! He had come right into their solidly built stone house, forcing his way through a rough but stout wooden door, and then, passing through a small entrance hall, had gone for the dog in the room opposite. The dog was only badly mauled and they shot the leopard through a window of their bathroom.

The second incident took place in Angola while the Benguella railway line was under construction. One of the locomotive drivers named Jones had a fox terrier dog which stayed with him in the gangers' cottage. One evening the driver had been sitting at a small table in his lantern lit room, eating his supper. His dog had been lying on the floor at his side with its back to the door. Suddenly a huge leopard sprang into the room, but owing to the smooth cement floor, it slid across the room knocking Jones and his table over. Jones, the table, the dog and the furniture, skidded in a circular movement around the room. Unfortunately the leopard regained its balance first, and fleeing towards the door, caught the dog in its jaws and disappeared from sight. [Ryan 1961, p. 68]

Certain leopards also have a preference for jackals—one was observed by Estes (1967) in the Ngorongoro crater to kill eleven in three weeks. Some of the larger carnivores are also preyed upon: Pienaar (1969) provides a photograph of an adult male cheetah weighing 45 kg that was strangled by a leopard and hoisted 3.6 m into the fork of a marula tree in the Kruger National Park.

Birds seen to have been killed by leopards include a number of European storks. Schaller (1972) suggests that these may be rather naive about large feline predators that are absent from their nesting areas in the northern hemisphere.

The Overall Diet of Leopards

It is well known that leopards frequently feed on small prey that would seldom feature in any lists of observed kills. For this reason, information derived from such kill observations is likely to be strongly biased in favor of large prey items. The analysis of leopard feces can, however, provide very valuable information about their total diet. For some years leopard droppings have been regularly collected in the Matopos National Park, and an analysis of 358 samples as given by Smith (1977) is listed in table 50. It is not usually possible to determine from fecal analyses how many of each kind of animal have been eaten, but the percentage occurrence of remains from a particular prey item in the leopard scats must surely bear a close relationship to the frequency with which these items were eaten.

In the Matopos study area it is possible to compare results of kill observations directly with those of fecal analysis. On the basis of observed kills, the sequence of importance of prey items, as listed in table 48, was, first, impala (32%); second, reedbuck (13%); third, duiker and sable, both with 11%; and, fourth, wildebeest (8%). Data from fecal analysis provides a very different picture; first is dassie, with 46% occurrence; second, klipspringer (10%); third, hare (8%); fourth, duiker (7%); then rats and mice (5%) (fig. 98). It is likely that when many of the smaller prey animals are eaten by a leopard in a feeding lair, there would be no bony food remains.

Leopard Predation on Primates

From the point of view of the interpretation of the Sterkfontein valley bone assemblages, leopard predation on primates, particularly baboons and hominids, could have been very significant. Neither in the lists of observed kills nor in the results of fecal analysis do primates feature as important leopard food sources. Yet in certain circumstances, such as on Suswa mountain, where baboons coming to their sleeping sites are preyed upon, predation on primates can assume important proportions.

Baboons as Leopard Prey

Although a single baboon cannot contend with a hungry leopard, a baboon troop in an aggressive mood is a different proposition. Concerning leopard/baboon interactions

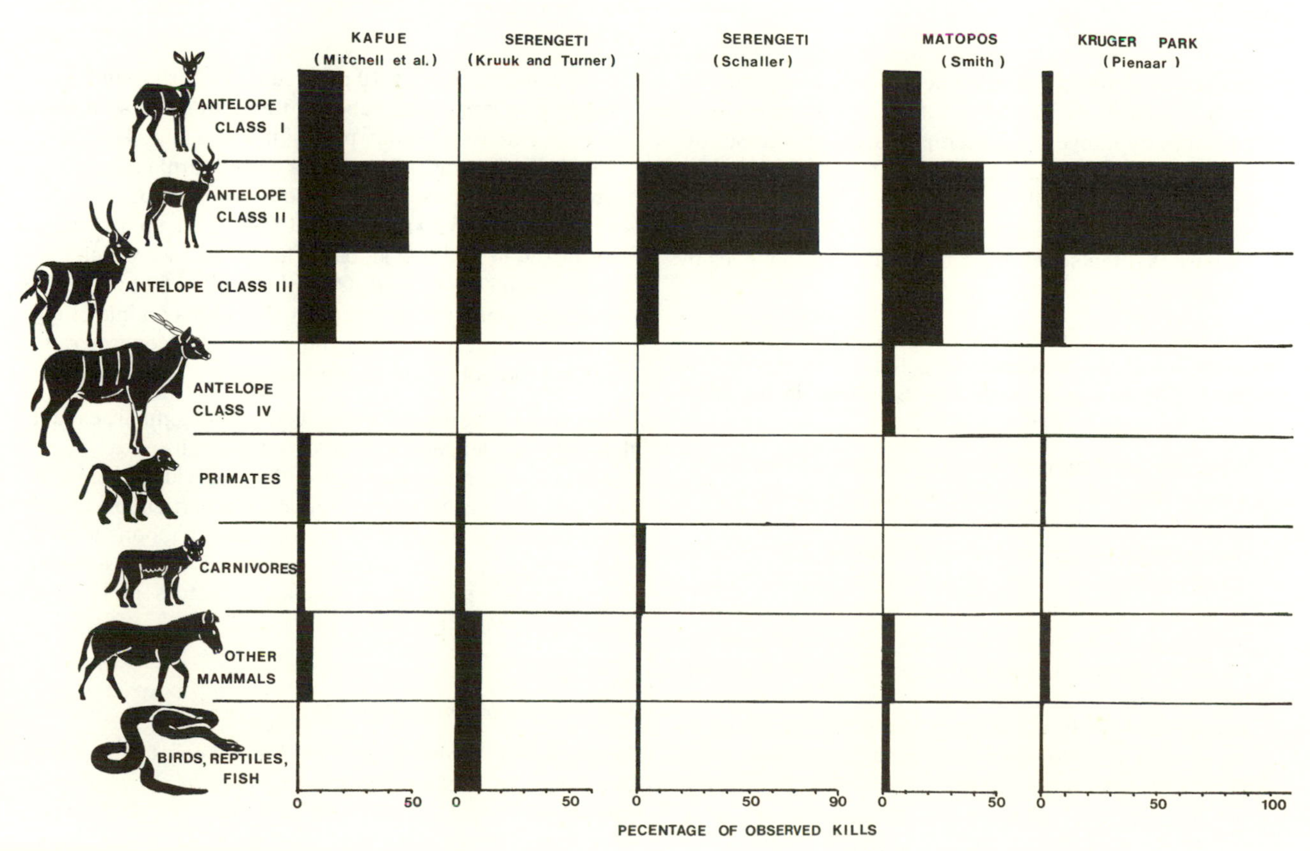

Fig. 97. Broad categories of leopard prey as determined by five field studies in various parts of Africa. See tables 48 and 49.

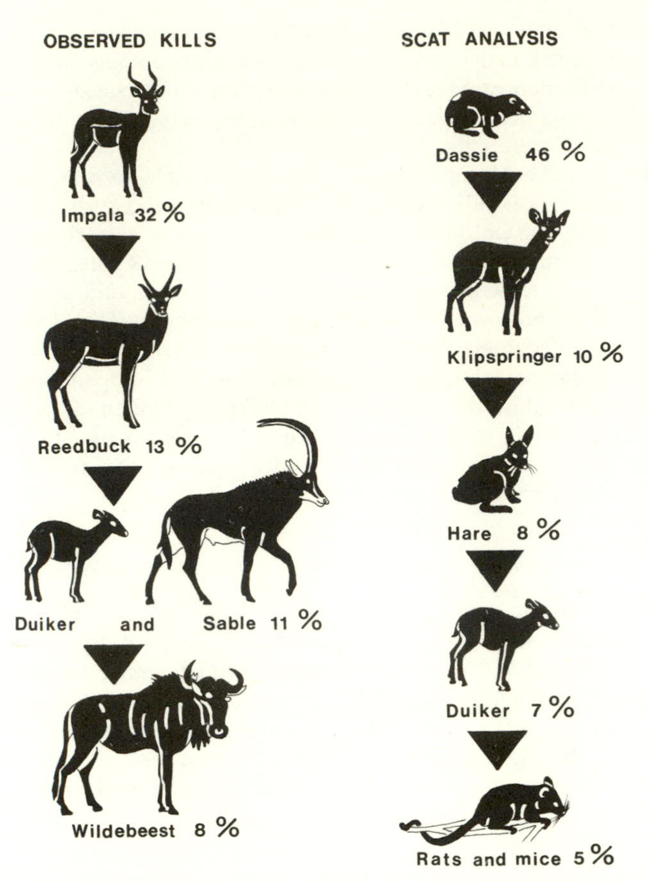

Fig. 98. On the basis of fieldwork done in the Matopos of Zimbabwe (Smith 1977), it is possible to compare results of leopard kill observations with those of fecal analysis. The sequence of importance of prey items differs considerably in the two methods, as is shown here.

in the Kruger National Park, Pienaar has written (1969, p. 123):

> Leopards kill many baboons when they are stealthily stalked or pounced upon at their roosts during the night. Occasionally however, a leopard is driven to rashness by hunger and attempts to snatch one out of a troop in broad daylight. In such cases a number of big male baboons will usually come to the aid of the shrieking victim and in the ensuing free-for-all the leopard is often severely wounded or torn to pieces by the enraged primates.

Leopards seem fully aware of the danger posed by a baboon troop, as evidenced by the following two incidents. The first was quoted by Smithers, in his *Mammals of Botswana* (1971*a*, p. 116):

> While the predation of leopards on baboons has been widely stressed in literature, an adult leopard is no match for the co-operative efforts of a troop of baboons during daylight hours. At Wankie, Rhodesia, an incident of this type was observed, the leopard lying up near a waterhole and, being spotted by a large baboon, took to its heels. When the baboon charged it, the troop including many juveniles then followed until the leopard took to a high tree where it was harried by members of the troop, the remainder sitting around on the ground at the base of the tree until sundown, when they straggled off leaving the leopard still marooned in the high branches. Baboons do, however, figure among the prey species but are probably taken from the fringes of the troop resting after dark or, if during daylight hours, solitary males or stragglers away from the troop.

The second incident, which took place in the Kruger National Park, was reported among readers' letters to the journal *African Wildlife* (Bates 1971, p. 154):

> We witnessed a rather exciting episode during July 1966 at Manurge Kop. It was about 5 pm and we were taking a slow drive home before returning to Pretoriuskop rest camp when from the rocks which make up Manurge Kop there arose a great clamour.
>
> Out of the bush burst a leopard followed by five large male baboons. The leopard leapt into the tree at the roadside followed by the baboons, who were barking and making a great deal of noise. The leopard proceeded upward and outward onto the thinnest branch—whilst his tummy worked furiously—the result falling onto the roadside in front of us!
>
> The baboons sat still for a few minutes and then with a great bark pounced at the leopard, who with great dexterity managed to avoid them and from at least 20 ft. high leapt out of the tree to disappear into the grassland alongside the road.
>
> The baboons, with great chattering, returned to the safety of Manurge Kop.

The records of these incidents serve to emphasize the effectiveness of coordinated action within the baboon group. Successful hunting of baboons will therefore depend on stealth and surprise attack, most of which is probably carried out at night.

It appears that leopards, when hungry, will eat most or all of a baboon body, leaving the skull comparatively intact. V. J. Wilson, former curator of the National Museum at Bulawayo, has kindly sent me extracts from his field notes concerning three baboon kills he found near Chipangali in eastern Tanzania. The description of one is as follows:

> 17 January, 1960, 5.10 pm. Adult leopard found feeding on large male baboon. Leopard spent 45 minutes at carcass and only part of it was eaten. The carcass was then carried up into a large Marula tree growing at the base of a granite rocky outcrop.
>
> 18 January, 6.20 am. When I visited the spot I was surprised to find two leopards at the carcass which was now on the ground. Only the skull in fact remained together with pieces of skin and a couple of leg bones. As a result of disturbing them the leopards ran off and did not return to the carcass.

Events at the other two kills were rather similar, except that only one leopard was involved in each.

Several observations suggest that individual leopards may develop a taste or preference for a particular kind of prey. In certain situations they may hunt baboons almost exclusively, and in various areas of the Transvaal leopards are regarded as regulating factors in baboon population numbers (Stoltz 1977).

Leopard Predation on Monkeys

Leopards are known to prey on both species of monkey that occur in southern Africa—*Cercopithecus aethiops*, the vervet, and *C. mitis*, the samango, but few direct observations have been published on these interactions. Wilson (pers. comm.) has sent me notes on one observation made at Kalichoo, eastern Zambia, when on 6 October 1962 he encountered a leopard feeding on a large vervet monkey.

> It spent quite some time licking the hair off the animal and then ate the skin and complete carcass, leaving only the skull together with the lower jaw. In fact the entire head was chewed off and left intact together with the eyes, skin, etc.

That leopards tend to leave the heads of their monkey prey intact was confirmed by observations on the feeding behavior of two young captive leopards, caught and reared in eastern Zambia (Wilson, pers. comm.; Wilson and Child 1966). Wilson (pers. comm.) has expressed the opinion that skulls of monkeys and baboons frequently found among rock outcrops in rugged country of eastern Zambia are food remains of leopards.

Leopard Predation on Homo

Evidence of leopard predation on people is relevant to this study for the light it might throw on the possibility that leopards may have preyed on early hominids. The ancient and modern situations are, however, very different. Through their technology, men have become highly dangerous adversaries, and leopards learn early either to avoid contact with people or to treat them with great circumspection.

Cases of man-eating by leopards are not uncommon and show at least one aspect that is important to us here. When a leopard takes to eating people, it generally continues to do so, perhaps to the virtual exclusion of other prey sources. It is frequently suggested in the literature that man-eaters, once they acquire the taste for human flesh, become exclusive predators of people. The evidence confirms that this is at least partly true, but such behavior may not be as unusual as it sounds. Leopards appear to develop strong food preferences, which vary from place to place and perhaps from time to time in an individual leopard's life. The leopards I was able to study in the Satara area of the Kruger National Park seemed to have a virtually exclusive preference for impala, whereas in other parts of the park certain individuals showed a strong preference for reedbuck or waterbuck. Turnbull-Kemp (1967) described feeding preferences among leopards marooned on islands that formed in the Kariba dam during its filling process. One leopard would go to great lengths to catch agile duikers on an island, when catching baboons, enfeebled by malnutrition, would have required hardly any effort. That particular leopard's habitual diet was duiker, and substitute foods were not favored as long as a single duiker was available. Likewise, the bone accumulation in the Suswa feeding lair suggests that the leopards involved there had developed a preference for baboon meat that was almost exclusive of other food sources.

So it is, it seems, with man-eaters. Human flesh may first have been sampled as carrion or through a chance encounter. The well-known man-eating leopard of Rudraprayag in India started preying on humans in 1918 after an influenza epidemic claimed many lives in that area. The Hindu custom is to cremate the dead, but because the epidemic was killing so many people, full cremation was impossible. Hindu tradition thus was satisfied by placing a live coal into the mouth of the deceased person and tipping the body down one of the steep valleys that abound there. Leopards are known to be attracted to carrion, and this may have started the man-eating habit of this particular Indian leopard (Corbett 1954). Despite concerted efforts by the authorities in the Garhwal region of India, as well as by various sportsmen, the leopard killed at least 125 people between 9 June 1918 and 14 April 1926, when it was finally shot by Col. Jim Corbett. During this period it obviously ate other prey, but a number of goats used as bait were killed but not eaten, the leopard trying instead to kill members of the party hunting it. As an indication of the strong preference the Rudraprayag leopard showed for human flesh, the following quotation from Corbett's book *The Man-Eating Leopard of Rudraprayag* is interesting and relevant (Corbett 1954, pp. 11–12):

> A boy, an orphan aged fourteen, was employed to look after a flock of forty goats. He was of the depressed—untouchable—class, and each evening when he returned with his charges he was given his food and then shut up in a small room with the goats. The room was on the ground floor of a long row of double-storied buildings and was immediately below the room occupied by the boy's master, the owner of the goats. To prevent the goats crowding on him as he slept, the boy had fenced off the far left-hand corner of the room.
>
> This room had no windows and only the one door, and when the boy and the goats were safely inside, the boy's master pulled the door to, and fastened it by passing the hasp, which was attached to a short length of chain to the door, over the staple fixed in the lintel. A piece of wood was then inserted in the staple to keep the hasp in place, and on his side of the door the boy, for his better safety, rolled a stone against it.

> On the night the orphan was gathered to his fathers his master asserts the door was fastened as usual, and I have no reason to question the truth of his assertion. In support of it, the door showed many deep claw marks and it is possible that in his attempts to claw open the door the leopard displaced the piece of wood that was keeping the hasp in place, after which it would have been easy for him to push the stone aside and enter the room.
>
> Forty goats packed into a small room, one corner of which was fenced off, could not have left the intruder much space to manoeuvre in, and it is left to conjecture whether the leopard covered the distance from the door to the boy's corner of the room over the backs of the goats or under their bellies, for at this stage of the proceedings all the goats must have been on their feet.
>
> It is best to assume that the boy slept through all the noise the leopard must have made when trying to force open the door, and that the goats must have made when the leopard had entered the room, and that he did not cry for help to deaf ears, only screened from him and the danger that menaced him by a thin plank.
>
> After killing the boy in the fenced-off corner, the leopard carried him across the empty room—the goats had escaped into the night—down a steep hillside, and then over some terraced fields to a deep boulder-strewn ravine. It was here, after the sun had been up a few hours, that the master found all that the leopard had left of his servant. Incredible as it may seem, not one of the forty goats had received so much as a scratch.

This particular leopard was one of many that, in recent times, had become man-eaters in India. Perhaps the most notorious was the man-eater of Panar, which was reputed to have killed four hundred people. Incidents continue to be reported. G. B. Schaller worked in the Kanha National Park of central India between 1963 and 1965 and reported (1967) that three people were killed there by a leopard in 1961. Then, between 2 August 1964 and 19 March 1965 when the leopard was shot, there were six attacks on people, four of which were fatal. The leopard proved to be a male weighing 97 lb (44 kg) with its lower right canines missing and showing evidence of having received a gunshot wound in its face some time previously.

In Africa, incidents of man-eating have been reported from many places. Swayne (1899) related that for several years before 1899 a leopard had haunted the Mirso ledge of the Golis Range, inland from Berbera in Somalia, and was reputed to have killed more than a hundred people. It had the habit of lying in wait at a corner of a dark, rough path overhung by large rocks. The Somalis pointed out a boulder, a meter from the path, from the top of which the leopard was said to spring on unsuspecting travelers.

Between 1936 and 1937 a leopard was reported to have killed sixty-seven people on the Chambezi River in eastern Zambia and to have met its end when it jumped on a man who happened to be carrying a large fishing spear (Brelsford 1950). The same author reported that eight people were killed by a leopard in the Isoka area of the Luangwa valley during 1938, and an African woman's baby was snatched from her back by a leopard at Luwingu in 1943.

Finally, the activities of a man-eater at Kanganga station during the construction of the Benguella railway line have been described by Ryan (1961). This particular leopard appears to have started preying on humans when it came across a two-year-old African child lying on a skin rug where its mother had placed it while she was working in a field. The leopard grabbed the child and disappeared into the undergrowth. Thereafter it attacked five other people, three of them fatally, before it was shot in a pit trap.

It would be interesting to know whether man-eating leopards had any particular characteristics in common. For instance, the Rudraprayag animal proved to be a male, somewhat past its prime, with one canine tooth broken. Turnbull-Kemp (1967) states that, of 152 recorded man-eating leopards, only 9, or 6%, were females. He provides further information about 78 man-eaters, showing that, of these, 80% were mature individuals without injuries; 12% were aged; and 4% were immature leopards. It seems safe to assume that the man-eating habit is acquired typically by mature male leopards, unaffected by injuries.

Leopard Hunting Ranges

It would be useful, when interpreting a bone accumulation in a leopard feeding lair, to know over what geographic range the prey had been hunted. Fortunately, some information is available on home ranges and hunting areas of leopards. In the Seronera area of the Serengeti, Schaller (1972) found that seven leopards occupied an area of 200 km^2 and that the minimum ranges of two females there were 40 and 60 km^2. He wrote (1972, pp. 284–85):

> The ranges of resident leopards overlapped considerably, although each animal tended to focus its activity in an area little used by others at the time. Two adults were occasionally within .5 km of each other, and during one period three females and a subadult male hunted along the same 5 km-long stretch of river, yet I only once saw two adults together when they were not courting. This indicates a strong mutual avoidance probably based both on direct visual contact and on such indirect methods as marking with scent. There was no evidence that these females actively defended a territory, but the fact that only one adult male used the area suggests that he possibly did so.

Further information has been forthcoming from a remarkably detailed study carried out in the Matopo Hills of Rhodesia (Smith 1977): 730 sets of leopard tracks were followed until they were lost, and when plotted they showed definite clumping. Judging from differences in spoor size, it was concluded that each of the areas showing spoor clumping (fig. 99) represented the home range of an adult male and one or two adult females, there being seven home ranges involved with an average area of 18 km^2 and a variation of from 10 to 19 km^2. These ranges were occupied by about twenty resident leopards as well as by a number of transitory individuals that wandered over much wider areas.

As the figure shows, there is a certain amount of overlap between the home ranges, and Smith estimated that the overall leopard density in that part of the Matopo Hills was one animal per 4.5 to 5 km^2. Although this sounds high, it was found in the Wilpattu National Park of Ceylon that leopard home ranges did not exceed 10 km^2 and that those of the males were virtually exclusive but overlapped ranges of females (Eisenberg 1970; Eisenberg and Lockhart 1972).

There are some indications that, where the home range of a leopard is fairly extensive, a regular routine is discernible in its movements. Astley Maberley (1953) reported that, on his farm Narina in the Letaba District of the northeastern Transvaal, a leopard made predictable, near monthly, visits to the valley where the farmhouse was situated. The following dates indicate the regularity of the process:

Arrival	*Departure*
28 June	4 July
12 August	20 August
5 September	12 September

Evidence collected by Corbett (1954) suggests that the man-eating leopard of Rudraprayag operated over an exceptionally large area. If all the human kills are included, an area of 500 mi² (1,296 km²) is involved. However, as is shown in figure 100, almost all the kills occurred in a smaller area, namely 150 mi² (388 km²).

In its search for human victims, this man-eater clearly trespassed on other males' home ranges, as evidenced by a vicious fight that took place on 14 April 1926 between it and a territorial male leopard. The man-eater suffered minor injuries and was forced to retreat, covering 29 km in the course of that night.

Transport of Prey by Leopards

When interpreting bones in a leopard feeding lair, it would also be useful to know over what distances they are likely to have been transported. In completely undisturbed circumstances a leopard may feed where it kills, but this seems rare, and the prey is usually carried or dragged some distance to a feeding site. My own observations on this are scanty. In the Satara area of the Kruger National Park, where I examined twelve impala kills, all but one were placed in trees, probably very close to where the actual killings took place. The abundance of suitable trees in this area meant that it was probably never necessary to drag the carcass more than 100 m.

I have been fortunate in having access to a motion picture sequence of a leopard kill taken at a dam in the central area of the Kruger National Park during the winter of 1966. The incident was observed by B. Stapelberg of White River, who filmed it in 8 mm color film, from which I have made black and white prints and selected single frames. As is shown in the prints (fig. 101), a leopard approached three juvenile impalas drinking at the edge of the dam and managed to overpower one of them while the others escaped. After killing the impala, it dragged it for about 100 m to a rocky kopje, where it fed on it. The latter part of the film is indistinct but shows that the leopard licked the hair from the ventral body surface before starting to feed on the abdomen. During the dragging, the impala was gripped by the neck and straddled by the front legs of the leopard.

Evidence from the Valencia feeding lair, discussed earlier, indicated that a mature karakul sheep weighing 30 kg and an immature goat weighing rather less were killed when drinking at the dam and were dragged about 300 m up a steep and rocky slope to the cave where they were eaten.

Valuable information on transport distances for leopards in the Matopo Hills has been provided by Smith (1977); data for 28 kills is given in table 51. It is of particular interest that the prey was moved a good deal farther in the wet season than in the dry season. Fifteen

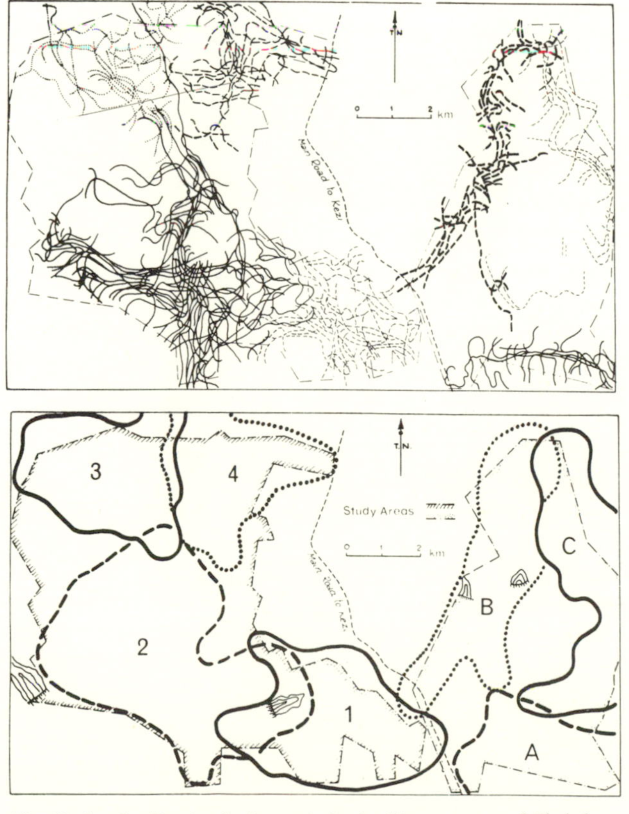

Fig. 99. Land utilization by leopards in the Matopos area of Zimbabwe (from Smith 1977). *Above*, the clumped distribution of leopard tracks, giving an indication of the home range pattern of individual animals shown below. From Smith 1977.

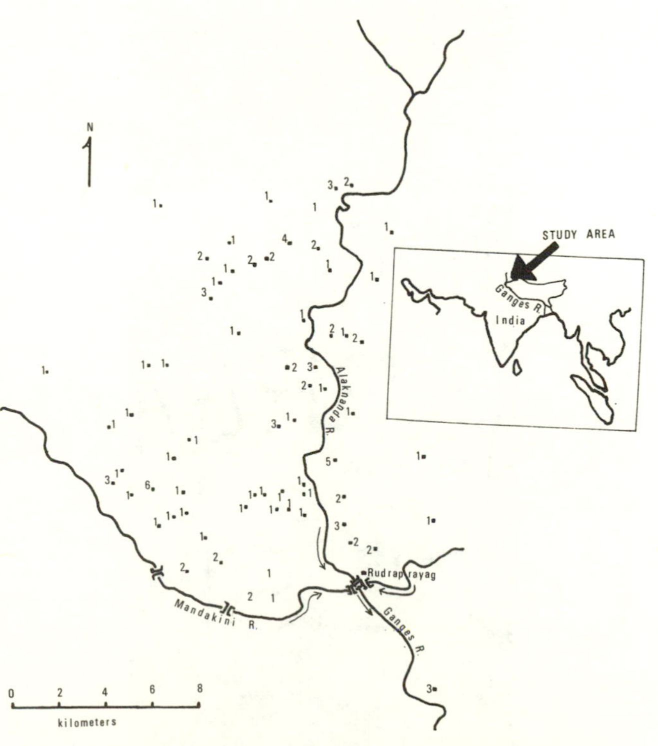

Fig. 100. The area in northern India where the "man-eating leopard of Rudraprayag" killed its victims between 1918 and 1926. Figures on the map indicate the number of people killed at each locality. Data from J. Corbett (1954).

Fig. 101. Single frames from an 8 mm motion-picture record of a leopard kill at a dam in the Kruger National Park.

1. A leopard *(arrow)* emerges from behind some large rocks and approaches the dam; it is partly concealed by a loose boulder.

2. Three juvenile male impalas at the water's edge take fright and start to turn. The stalk and final rush of the leopard occupied 74 frames, or 4.1 sec.

3. The leopard springs onto the back of the nearest impala, an event covered by 2 frames, or 0.1 sec.

4. The leopard puts its head around the right side of the impala's neck and clamps its teeth into the throat.

5. Leopard and prey fall to the ground while the last impala makes its escape.

6. The leopard lies on its left side, gripping the impala's throat; the impala is on its back with its feet in the air.

7. The leopard maintains its throat grip for about 5 min, suffocating the impala. Two minor changes in the leopard's position occur during this time.

8. The dying impala rolls over onto its left side.

9. Maintaining the same grip, the leopard lifts the impala's body by the throat.

10. Straddling the impala's body, the leopard starts to drag it away.

11. The leopard drags the impala about 100 m to a rocky kopje.

12. The leopard starts to feed on the impala among the rocks.

kills in the dry season were moved an average of 120 m, the range being 0–400 m, whereas during the wet season eleven kills were moved an average of 260 m, the range being 50–1,000 m. It is clear that leopards prefer to feed in dry circumstances, so that during the rains they dragged their prey up into the granite kopjes where they found shelter and better drainage. The longest drag distance was for a sable calf that was transported 1,000 m. According to Smith, large prey was dragged straddled between the leopard's front legs, and it was often eviscerated along the way, the viscera being covered by brush or plant litter.

Smith has recorded a single observation of a dassie being killed by a leopard; it was carried 1,600 m before being eaten.

Corbett (1954) has mentioned the distances over which six of the Rudraprayag man-eater's victims were transported. All but one were moved no more than a few hundred meters, the exception being an adult man who was dragged 2 mi (3.2 km) up the slope of a well-wooded hill and then the same distance down the other side through dense scrub jungle, a total of about 6.4 km. It is of interest that on two occasions the leopard took its human prey down 3.6 m high walls, suggesting that in the past leopards would not have had difficulty in taking their kills down into somewhat inaccessible caves.

In conclusion, what evidence we have suggests that prey would normally not be transported more than a few hundred meters to a feeding lair, although distances of up to 6 km are not impossible.

Food Storage by Leopards

In many parts of their geographic range, leopards store surplus food. The habit seems best developed in areas where hyenas and other scavengers are troublesome, and my own observations, as well as those of others (e.g., Pienaar 1969), suggest that leopards are frequently subordinate to spotted hyenas in competition for a carcass. This being so, unless a leopard can promptly take its prey into a tree or other inaccessible spot, it risks losing it to hyenas. In the Satara area of the Kruger National Park, all but one of the twelve impala kills I studied were stored in trees, while one or more spotted hyenas waited hopefully below (fig. 102). In the one exception, the impala was dragged into a thicket but was stolen by hyenas before the leopard could eat it.

A variation in this pattern is found in the Valencia Ranch area, where hyenas are absent and leopard food is abundant. Here, at the time of my observations, leopards never used trees for food storage, although they did use secluded feeding lairs. Frequently they failed to return to prey carcasses, simply killing again when the were hungry.

The habit of storing food in protected places and returning to it on several occasions is clearly economical. The weight of leopard prey frequently exceeds what can be consumed at a single sitting. Turnbull-Kemp (1967) quotes fifteen cases in which the meat consumed by a single leopard in a 12-hr period has been measured. Meat weights eaten were found to range from 18 to 39 lb (8.2–17.7 kg), with a mean of 28 lb (12.7 kg). Depending on the size of the prey, therefore, feeding duration will vary. In the Matopos, Smith (1977) found that an adult leopard would consume a full-grown impala in four days, and a yearling wildebeest lasted six days.

A well-documented case of a leopard returning to its kill on successive nights was provided by Schaller (1967) from observations in the Kanha National Park of India. The prey was a 72 lb (32.7 kg) goat, killed on the night of 9 July 1964 and eaten during four consecutive nights. On the first night the leopard fed for 60 min and ate 17.0 lb (7.7 kg); on the second night 9.5 lb (4.3 kg) was eaten in 40 min; on the third night, 10.5 lb (4.7 kg) in 65 min; and on the fourth night 2.0 lb (0.9 kg) in about 20 min. This means that a total of 39.0 lb (17.7 kg) was eaten in about 185 min by a leopard estimated to weigh 100 lb (45 kg).

Food storage in trees above a cave, or within the cave itself, may have been of special significance in the case of Swartkrans, to be discussed in chapter 10.

Leopard Damage to Impala Prey Skeletons

Observations on eleven impala carcasses, stored and fed on by leopards in the Satara area of the Kruger National Park, suggested a pattern of skeletal damage that may characterize antelope prey of this size subjected to repeated leopard feeding sessions. The following extract from my field notes describes the feeding process at one of the kills:

> *16 July 1968.* A freshly killed female impala was seen in a tree on the east bank of the Timbavati River, 14 km upstream from the Timbavati rest spot. A large adult leopard had just started to feed when discovered during the late afternoon.
>
> *17 July, 8:50* A.M.: The kill has been moved to another tree, 30 m south of the first one, and is being eaten by a second leopard, considerably smaller than the first. The leopard is eating from the back of the head. No vertebral column or body remains, only a large area of skin linking the head with one front leg (the other has fallen) and both hind legs, hanging down on strips of skin. In the front legs the shafts of the radius/ulna have been chewed through, as have both tibia shafts in the hind limbs.
>
> *9:30* A.M.: The leopard has just eaten the ears and is busy on the throat, but the skull and mandible are apparently undamaged.
>
> *10:30* A.M.: The head has just fallen and been taken by a waiting hyena. The leopard has climbed down the tree and departed, leaving only three legs of the impala hanging from strips of skin over a branch about 10 m up. These were collected and photographed.

These remains are shown in figure 103. The feeding process described here is atypical in that two leopards, instead of the usual single individual, fed on the carcass, thereby reducing the typical consumption time considerably.

In undisturbed circumstances the leopard's feeding in all the observed cases was neat and orderly, resulting in the "eaten-out" appearance of the carcass that various authors (e.g., Turnbull-Kemp 1967) have remarked upon. The body, neck, and upper limbs are eaten, so to speak, out of the skin, which is left surprisingly intact, linking the rejected skeletal parts—the head and lower leg segments. Ideally, it should be possible to collect these remains as a unit at the end of the third or fourth day, when the leopard finally leaves the carcass. In practice, the head and some of the limbs typically fall, partly as a result of putrefaction

Fig. 102. A leopard stores its prey in a tree. When this photograph was taken by A. C. Kemp, in the Satara area of the Kruger National Park, the leopard was asleep on a branch below its partly consumed impala prey. A spotted hyena was hopefully waiting for pieces to fall to the ground.

of the linking skin, and are snatched up by waiting hyenas. In one case we decided to remove and examine the carcass before the head fell and was lost to hyenas. The kill was a juvenile impala that, in September 1968, had been stored 8 m high in the fork of a marula tree near Satara. At the end of the second day we climbed the tree, put the leopard to flight, and recovered the remains shown in figure 104. The linking skin was artificially removed before the photograph was taken to show that the vertebral column had disappeared between the axis and the lumbar vertebrae, while both humeri had been chewed through just above their distal extremities. Had we left the carcass for another night, the rest of the vertebral column and upper limb bones would almost certainly have been eaten by the leopard, but the head would very probably have fallen to a waiting hyena.

The Possible Role of Extinct Cats as Collectors of Bones in Southern African Caves

Cats that may be significant from this point of view, and whose remains are preserved in the Sterkfontein valley cave breccias, are these: the leopard; a related extinct feline, *Panthera crassidens;* and the false and true saber-toothed cats.

Remains of leopards from Swartkrans have been described as a separate subspecies, *Panthera pardus incurva* (see chap. 9), slightly smaller than typical living leopards. However, there is a good deal of size variation in *P. pardus* over its wide geographic range. There is no reason to suspect that the leopards that lived in the Sterkfontein valley two million years ago behaved differently from their contemporary descendants.

Remains of *P. crassidens,* a cat that seemed to have had characteristics of both a leopard and cheetah (see chap. 9) are known from Kromdraai A and also from various localities in East Africa. What little information there is suggests that we may be dealing with a leopardlike feline in which cursorial adaptations were emphasized. Whether the "crassidens cat" was a regular frequenter of caves is not known. If it was, then it was presumably far less common than leopards were. Food remains of the two cats would probably have been similar, although the crassidens cat may have actively pursued its prey rather than taking it by stealth as is usual with leopards.

False saber-toothed cats of the genus *Dinofelis* were probably of singular importance as accumulators of bones in the Sterkfontein valley caves. As is described in Chapter 9, remains from three species are known from southern Africa: *D. diastemata* from Langebaanweg, *D. barlowi* from Sterkfontein, Makapansgat, and Bolt's Farm, and *D. piveteaui* from Kromdraai A. The three appear to have formed an evolutionary sequence, with *D. diastemata* showing the most primitive dental characters and *D. piveteaui* the most advanced. The indications are that *Dinofelis* was a heavy-bodied feline with a short tail; as is shown in figure 105, the craniums of both *D. barlowi* and *D. piveteaui* are appreciably larger than those of large male leopards.

The frequency with which *Dinofelis* remains are found in the Transvaal cave breccias suggests that these were reasonably common, cave-frequenting cats. Their bodily conformation suggests that they would have relied on stealth, rather than sustained speed, to capture their prey; their behavior was probably similar to that of leopards. However, their appreciably larger size means they would have been able to overpower larger prey—class III antelopes would have been more readily accessible to them than they are to leopards.

For what purpose did *Dinofelis* use its enlarged canines? They were clearly highly effective killing weapons and, in combination with the powerful forelimbs, would have enabled *Dinofelis* to hunt not only *large,* but also dangerous prey. I am tempted to suggest that *Dinofelis* would have been a very efficient hunter of primates such as baboons and early hominids. Individually, a baboon or australopithecine may not have been a particularly dangerous adversary, but an attack on an individual would probably have precipitated retaliation by the entire group. In such circumstances, silent, efficient killing would have been a great advantage to the predator. One may visualize *Dinofelis* biding its time in a concealed position while a troop of baboons or hominids passed; selecting a straggler, the cat would then overpower its prey, holding it down with its powerful forelimbs while killing it silently and quickly by a throat bite.

The suggestion that *Dinofelis* frequented caves gains some support from fossils discovered in Pit 23 of Bolt's Workings (see chap. 9). Here remains of three *Dinofelis* individuals were preserved alongside the remains of several baboons. It seems that the cave represented a natural trap into which the baboons had fallen. The cats followed and probably fed upon the baboons, only to discover that they could not escape either.

Remains of true saber-toothed cats fall naturally into two groups—those with crenulate sabers and highly specialized carnassials *(Machairodus* and *Homotherium)* and those with smooth-edged canines and less specialized cheek teeth *(Megantereon).* These animals are further discussed in chapter 9. From the regularity with which their remains are found in Transvaal cave deposits, saber-toothed cats appear to have been both fairly common and prone to frequent caves. Their size and formidable canines suggest that they probably preyed on large animals, though the structure of their cheek teeth makes it clear that they would have done minimal damage to bones. It is quite conceivable that Transvaal saber-toothed cats used caves as feeding and breeding lairs and that their food remains should have accumulated in such places. Such remains would be characterized, I suspect, by the large size of the animals involved and by the lack of damage to the bones themselves.

The Black Eagle, *Aquila verreauxi* Lesson

Black eagles probably never enter caves, but they do consume very large numbers of dassies, whose remains accumulate below their nests and habitual perches. If either the nest or the perch is within the catchment area of a cave mouth, black-eagle food remains will form part of the bone accumulation in the cave.

The black eagle is a most spectacular bird, coal black with a white back that forms a white V when the wings are closed. According to Leslie Brown (1970), its grace in flight is surpassed only by that of the Lammergeier. In many ways it is very like the golden eagle, *Aquila chrysaetos,* which it replaces in Africa south of the Sahara. Its usual habitat includes rocky gorges and kopjes where dassies abound, and in Kenya and Ethiopia it breeds at altitudes of up to 13,500 ft (4,117 m).

The biology of black eagles has been studied in great detail in various parts of the birds' geographic range.

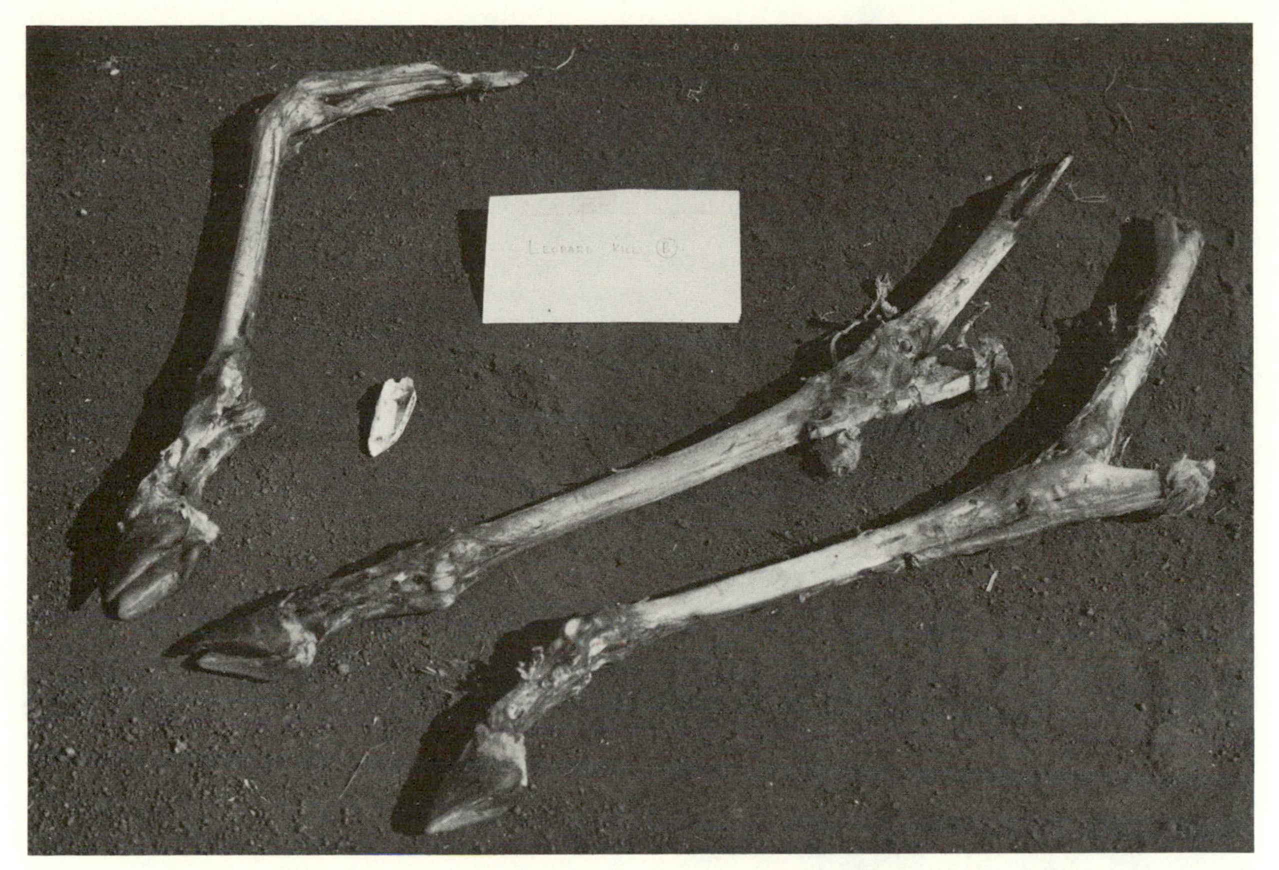

Fig. 103. Legs of an impala after two leopards had fed on the carcass in the Kruger National Park. The entire body was eaten, but the fourth leg and the head fell to waiting hyenas.

Fig. 104. Remains of a juvenile impala eaten by a leopard in the Kruger National Park. The leopard was disturbed before it completed its feeding on the hindquarters.

After some early breeding observations in South-West Africa by Hoesch (1936), Rowe (1947) located a conveniently placed nest on a rocky hill near Mbulu in north-central Tanzania. A crevice on the nesting cliff allowed him to observe at a range of 21 m, and he watched the eagles for a total of 1,013 hr. The first egg, laid on 13 June 1944, hatched on 27 July; the second, laid on 16 June, hatched on 29–30 July, and the surviving eaglet flew on 1 November, which meant a laying-fledging period of 140 days. Rowe described the process whereby the second chick to hatch was pecked and harried by the first until it disappeared from the nest on 6 August; he ascribed its death to maltreatment, starvation, and exposure. Rowe speculated about the biological significance of this "Cain and Abel" process and suggested that, since the first chick is stimulated by the presence of the second to take more food than it needs, a more robust eaglet results.

During Rowe's observation of the nest, 41 carcasses were brought—24 mammals and 17 birds. Among the mammals were 22 dassies, 1 hare, and 1 mongoose, and the birds were francolins, guinea fowl, and poultry. About half the mammals had been decapitated and eviscerated before being brought to the nest, a process that may be related to the adults degree of hunger. Rowe observed that the eagles ate most of the bones of the prey, so that only skulls and larger skeletal elements accumulated below the nest.

Further observations on black eagle nests were made in the Embu District on the eastern slopes of Mount Kenya and elsewhere in East Africa (Brown 1952, 1955, 1966), at Jansenville in the Orange Free State (Visser 1963), in the Matopo Hills of Rhodesia (to be discussed shortly), and elsewhere, allowing Siegfried (1968) to collate 99 breeding records for southern Africa. These showed that, over a considerable geographic range, most eggs were laid in May and June, the usual clutch being two, though clutches of one and three were occasionally encountered.

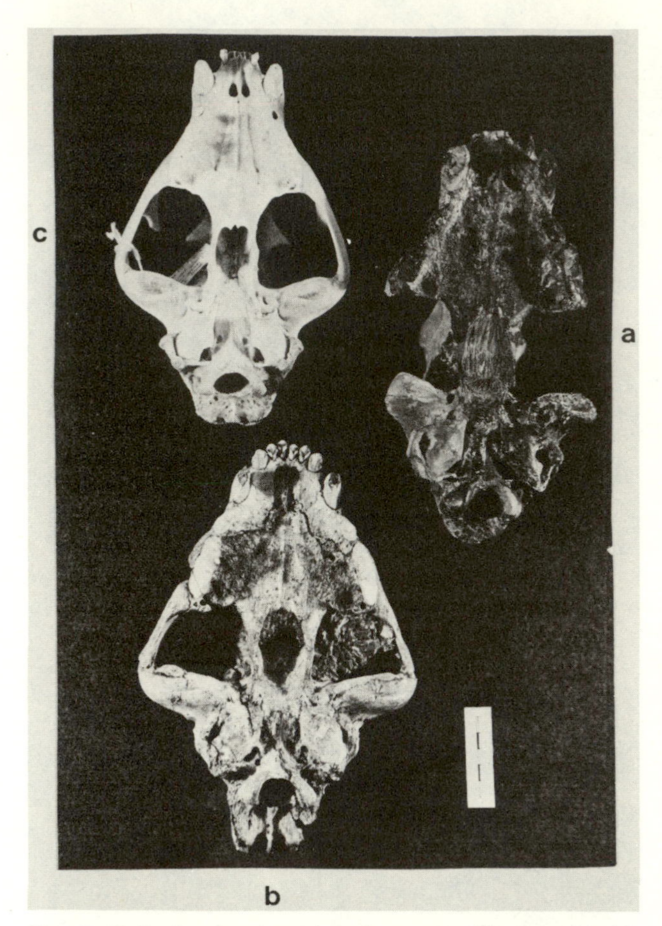

Fig. 105. Skulls of two species of the extinct cat *Dinofelis* compared with the skull of a large male leopard. *(a) D. barlowi* from Bolt's Farm; *(b) D. piveteaui* from Kromdraai A; *(c)* a modern *Panthera pardus*.

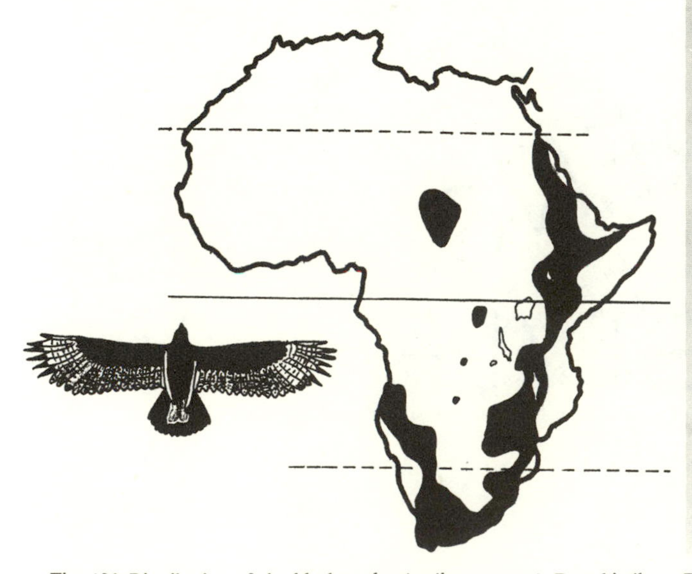

Fig. 106. Distribution of the black eagle, *Aquila verreauxi*. Data kindly provided by D. Snow, British Museum of Natural History.

Fig. 107. A black eagle, photographed in the Hakos Mountains of South-West Africa by A. C. Kemp.

Almost invariably the first-hatched chick killed the second or subsequent ones, and about this Siegfried wrote (1968, p. 144): "the apparent biological waste involved in the two-egg clutch, followed by the killing of the one chick by its sibling partner, is not understood."

Some experiments designed to throw light on the "Cain and Abel" struggle were carried out at black eagle nests in the Matopos by V. Gargett (1967, 1970*a*). She concluded that it was *not* shortage of food that led to the survival of one rather than two chicks; it appeared that the proximity of a second chick elicited an aggressive response as marked at six, seven, and eight weeks as it was in the first few days after hatching. The experiments demonstrated that there was no reason the second chick should not grow to maturity if shielded from the aggression of the first.

The Matopo Hills of Rhodesia, where the experiments were carried out, are remarkable for black eagles, probably supporting an eagle population more dense than any other in the world. For many years it has been known that black eagles are particularly abundant in the Matopos (e.g., Mouritz 1915), a spectacular area of granite hills south of Bulawayo (see also chap. 3). By 1960 the positions of 35 black-eagle aeries had been determined in the Matopos National Park, and this information was the basis for a survey of black eagles in the park (Vernon 1965). From this survey, conducted in 1964, Vernon was able to conclude that there were then 41 nest sites in 160 mi^2 of the Matopos, that one breeding pair occurred in every 5.3 mi^2, although the actual population density was even greater, and that this density was limited by the need for each pair of eagles to have a territory large enough to display in. The suitability of nest sites was determined more by security considerations than by aspect, although it was found that adjacent nests that overlooked one another had to be more than a mile apart or one pair would inhibit the breeding of the other.

This 1964 survey has been continued by an ornithological team under the leadership of V. Gargett and has resulted in an impressive body of information on the biology and population dynamics of the eagles (Gargett 1965, 1967, 1970*a,b,* 1971, 1972*a,b,* 1975, 1977). Within the 620 km^2 study area, and over the thirteen-year study period, 442 breeding attempts were recorded and 339 young eagles seen to be fledged, a reproductive success of 0.52 young per pair per year. Territory size varied from 585.4 ha to 1,437.6 ha, depending on the nature of the terrain and the availability of food, and territory shape in the airspace appeared to be an inverted truncated cone, which resulted in airspace overlapping above ground-level boundaries.

It is of particular relevance to the present study of bone accumulations that about nine months of a black eagle's year would normally be taken up with some part of the breeding cycle, centered on the nest. Apart from this, habitual perches are used, whose positions, relative to the nests in eighteen adjacent territories, are shown in figure 108. Food remains could be expected to accumulate mainly below nests, but also beneath these perches. Prey was almost exclusively dassies, many individuals having been decapitated by the eagles at a perch before their bodies were taken to the nests. Gargett (1971) quoted several observations of actual kills. It appears that the eagles often fly low around the rocky outcrops, exploiting surprise, and knock the prey off the rocks with their outstretched talons. They then circle and pick up the bodies.

Through the cooperation of J. H. Grobler of the Department of National Parks and Wildlife Management in Zimbabwe, I have been able to study a sample of black eagle food remains collected below nests and perches in the Matopos. Details of the 111 bones are given in table 52; they came from a minimum of 8 adult common dassies, *Procavia capensis,* 40 yellow-spotted dassies, *Heterohyrax brucei* (39 adult and subadult, 1 juvenile), and 1 hinged tortoise, *Kinyxis belliana*. The three plastron pieces of this tortoise were in a regurgitation, surrounded by dassie fur, and it is remarkable that the largest of these pieces was 7.5 by 8.5 cm—larger, I would have thought, than an eagle could swallow and regurgitate. Yet one certainly did so.

I have been able to examine two other collections made below black eagle perches. The first was from the Magaliesberg at Wonderboom, within the municipal limits of Pretoria. The collection was made in 1969 by O. P. M. Prozesky and W. Spofford; it consisted of 56 pieces, as listed in table 53. Common dassies, *Procavia capensis,* dominated the prey and included 8 adults, 5 subadults, and 4 juveniles. Parts of 2 hares, a mountain tortoise, and 2 francolin-sized birds were also present.

The second collection, consisting of 131 bone pieces, listed in table 54, was made by A. C. Kemp beneath a black eagle perch close to its nest on the Portsmut Farm in South-West Africa—a locality already described in connection with leopard lairs—in June 1969. Represented by these remains were 9 adult, 1 subadult, and 2 juvenile *Procavia capensis,* as well as 1 hare.

The three collections of food remains clearly confirm earlier observations that the main prey of black eagles is dassies, either *Procavia* or *Heterohyrax*. Other mammals,

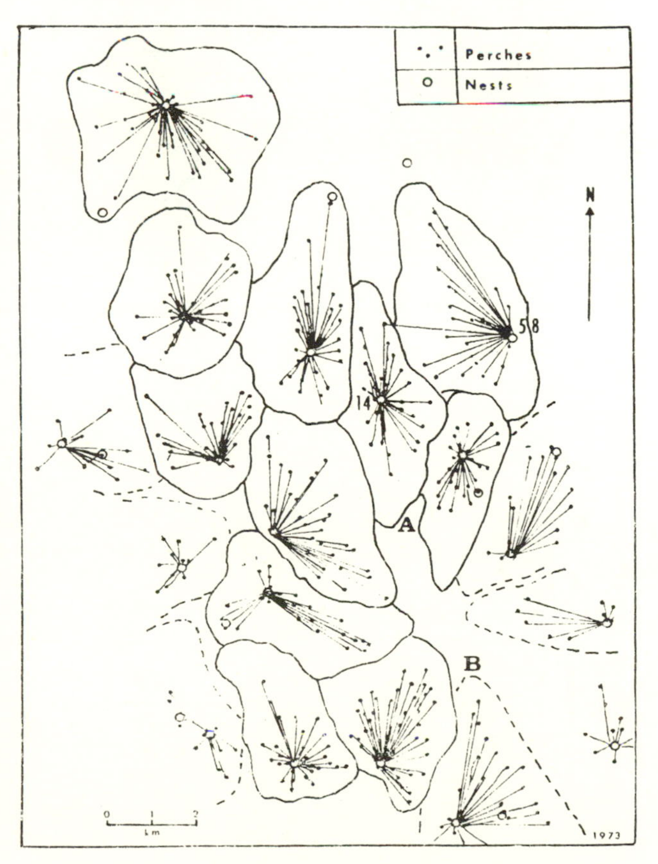

Fig. 108. Territories of eighteen pairs of black eagles in the Matapos area of Zimbabwe. In each case, habitual perches are linked to nest sites. From Gargett (1977).

birds, and tortoises are of limited importance. The food remains of dassies are mainly cranial parts, together with pelvises and a few of the larger limb bones. The skulls show very characteristic damage: in most instances the braincases have been opened from the back or the side to remove the brain. In some cases the entire calvaria has been eaten away. Damage is also common on the angles of the mandibles, probably inflicted when the tongue was removed from behind or below.

The damage to skulls, as shown in figure 109, has been inflicted by the points of the black eagles' sharp recurved bills. Frequently a round opening is made in the braincase—an interesting difference from the "sheared-off" damage to dassie skulls caused by leopard or cheetah feeding.

If dassie skulls that had been discarded by black eagles were found in a fossil bone accumulation, it should be possible to separate them, by the nature of the damage they have suffered, from remains left both by felid predators and by human hunters.

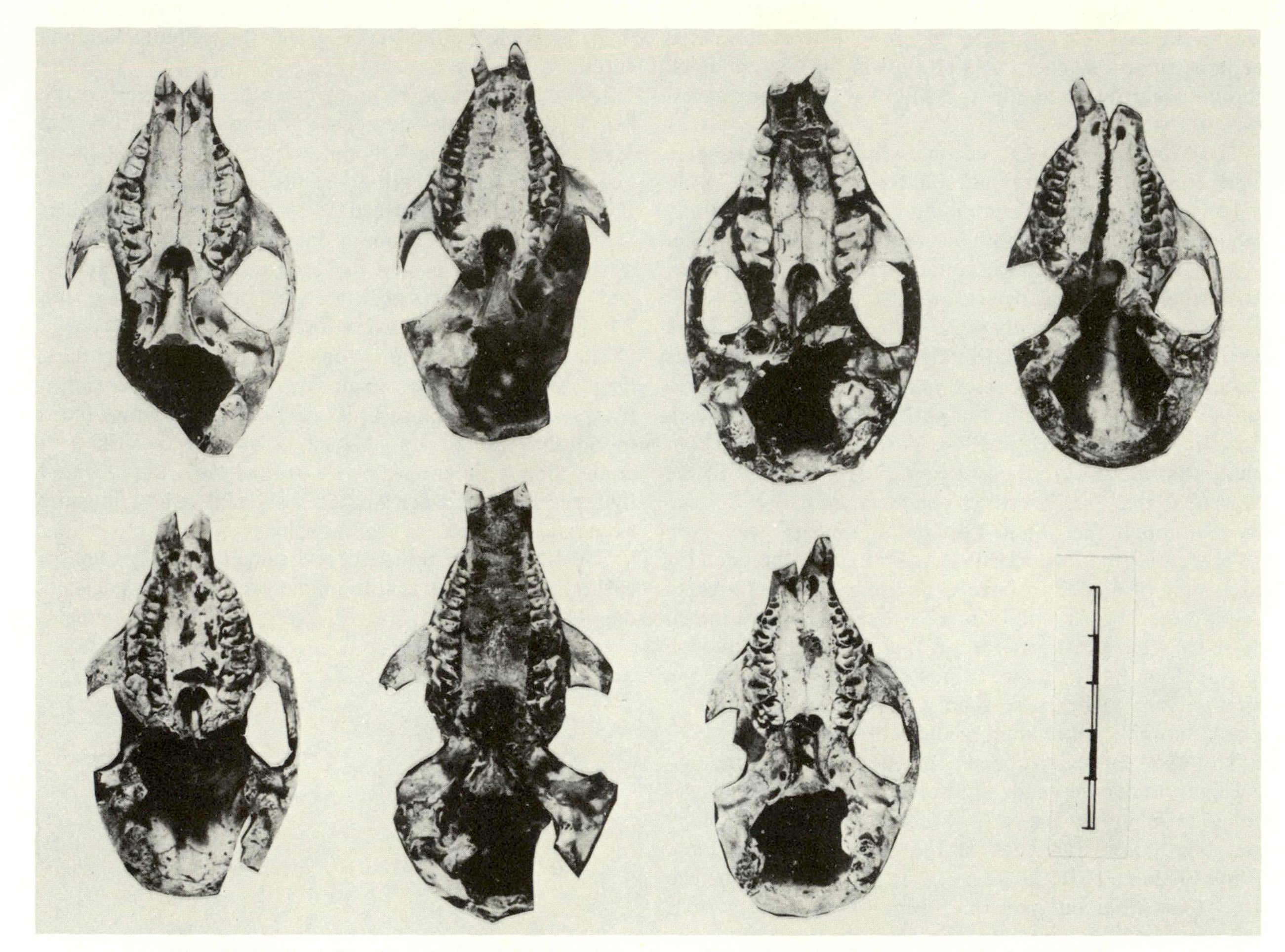

Fig. 109. Dassie skulls collected beneath black eagle perches and nest sites, showing characteristic damage to the braincases.

5 Porcupines as Bone Collectors in African Caves

It is rather surprising that a vegetarian rodent like *Hystrix africaeaustralis* should prove to be an important collector of bones; yet I suspect porcupines carry more bones to African caves than does any other species. As long ago as 1934, Shortridge wrote in his account of the mammals of South-West Africa: "The porcupine has a well-known habit of gnawing bones, accumulations of which often betray the whereabouts of a warren." He went on to quote earlier accounts by Lydekker and Pitman about the gnawing abilities and collecting habits of African porcupines.

Attention was specifically focused on this strange behavior of porcupines during the 1950s when R. A. Dart and A. R. Hughes were trying to make sense of the remarkable bone accumulation in the gray breccia at Makapansgat Limeworks (Hughes 1954*a;* Dart 1957*a*). About that time Hughes started his observations of contemporary animal lairs, including those of porcupines. In September 1954 he visited the Kalahari Gemsbok National Park in the northern Cape and examined several lairs in the banks of the dry Auob and Nossob rivers there. Two of these were clearly inhabited by porcupines, and from them Hughes (1958) took 147 bones and other objects that the porcupines had apparently hoarded. Both sites were in the east bank of the Auob River, 17 and 26 km upstream from the main camp at Twee Rivieren.

Some direct information on porcupine hoarding behavior came to light about the same time. In January 1956 a young male porcupine was caught near Grahamstown and taken to the Zoology department of Rhodes University, where it was observed over a considerable period (A. J. Alexander 1956). The porcupine, named Aristotle (fig. 111), made its lair in the photographic darkroom and wandered out freely at night. Before long it started to stock its lair with objects collected during its nocturnal wanderings—bones, a tortoise carapace, "a wicker basket, an enamel dish, a duster, a piece of flex, a drain pipe, a buck horn and a piece of softboard." Manageable objects it simply carried in its mouth, and larger ones it dragged, walking backward all the way to its lair.

Of particular interest was Anne Alexander's observation that the porcupine showed no interest in fresh bones with meat on them, though dry bones were immediately taken to the lair and gnawed there. The purpose of the gnawing was soon apparent: it had to do with wearing down the incisors rather than with nutrition. Like other rodents, the African porcupine has open-rooted incisors that grow throughout its life (fig. 112); they require regular attrition to keep them at a usable length. So it seems that porcupines have developed a behavior pattern that requires them to collect dry bones and other hard objects and hoard them in their lairs. When resting during the day, the porcupines select some of their favored objects and gnaw them. The collecting behavior appears to have become a compulsion—they will bring back far more objects than they can possibly use and do not get around to gnawing anywhere near all the treasures they collect.

Marks on bones caused by porcupine incisors are highly characteristic (fig. 113), as are the end products of prolonged gnawing on various parts of bovid skeletons. In her dissertation on various aspects of fossil and living porcupines, Judy Maguire (1976) has considered the nature and distribution of gnawing marks on different skeletal elements. Her conclusions may be summarized thus:

Craniums:
Gnawing tends to be random and widespread.
Mandibles:
The symphyseal region and ascending ramus tend to be gnawed first; thereafter the horizontal ramus is gnawed, which may cause the teeth to fall out. The angle of the mandible tends to remain till last.
Vertebrae:
Neural spines and transverse processes are gnawed

Fig. 110. Distribution of the African porcupine, *Hystrix africaeaustralis.*

Fig. 111. A tame African porcupine, Aristotle, photographed in the Zoology Department of Rhodes University during 1955. This animal provided the first clear evidence of bone-hoarding and bone-gnawing.

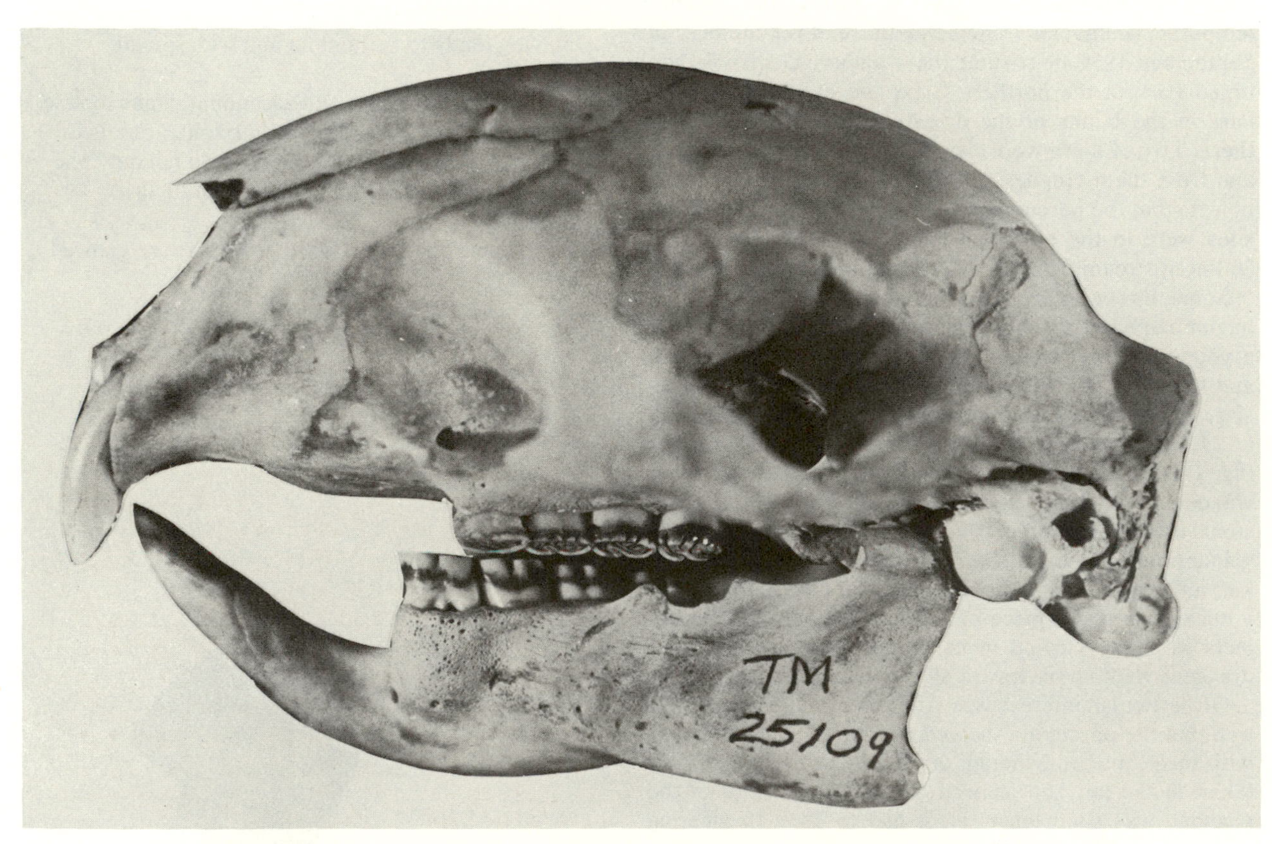

Fig. 112. The skull of an African porcupine showing its prominent incisors, which grow throughout life and require regular attrition.

first, thereafter the zygapophyses and centra. An irregular and unrecognizable fragment may result.
Ribs:
The ends and body are prone to gnawing. Damage along the anterior and posterior edges causes ribs to split longitudinally, producing thin slivers of bone in which the porcupines show little further interest.
Scapulae:
The compact head tends to disappear first, and thin pieces of blade, of little interest to the porcupines, are left.
Pelvises:
Gnawing proceeds from all sides, usually leaving the acetabular region till last.
Long bones:
Proximal and distal ends of the various limb-bone elements tend to be destroyed first, and the gnawing then proceeds down the shafts from each end, to result in characteristic pieces shaped like napkin rings, with gnawing scars around each circumference.
Astragali, calcanei, and phalanges:
Gnawing tends to occur on all surfaces of these bones.

An interesting comparison between a porcupine collection and a bone assemblage from a human occupation site has been made by Hendey and Singer (1965). Construction of a dam and canals on the Kougha River in the Gamtoos valley near Port Elizabeth necessitated the de-

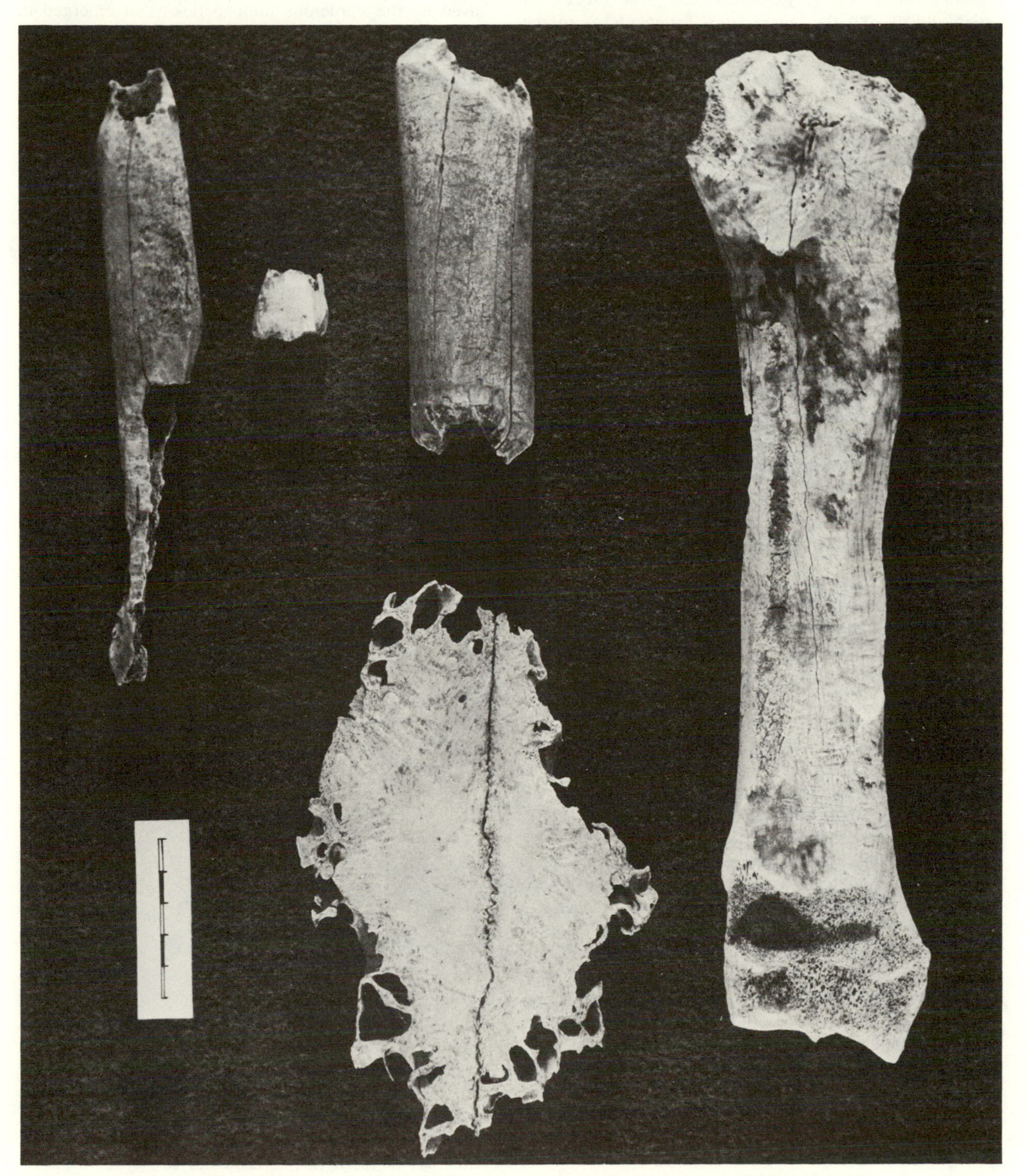

Fig. 113. Examples of porcupine-gnawed bones. The marks caused by the chisellike incisors are highly characteristic.

molition of three rock shelters in 1961 and 1962. Two of these, relevant here, were on the west bank of the Gamtoos valley at Andrieskraal. The first, designated Andrieskraal 1, consisted of a small rock shelter containing a shallow deposit rich in stone artifacts and bone fragments (J. Deacon 1965). The second, Andrieskraal 2, a few hundred meters downstream but at the same height above the river, consisted of a crevice in the Enon conglomerate, too low to have been occupied by man but containing a remarkable collection of gnawed bones. The significance of these two bone accumulations from adjacent sites is clear: a direct comparison can be made between a bone assemblage resulting from Later Stone Age human feeding and one collected by porcupines. The "human site," AK 1, yielded 11,056 bone pieces, of which 8,887 (80.4%) were too fragmentary to be identified; the "porcupine lair," AK 2, produced 1,105 pieces, of which only 465 (42.1%) were unidentifiable. There was therefore a striking difference in fragmentation of the bones—this is also reflected in the weight of each sample. The 11,056 bone pieces from AK 1 weighed 20.9 kg, giving an average weight per piece of 1.9 g; the 1,105 pieces from AK 2 weighed 85.3 kg, an average weight per piece of 77.2 g. Likewise, only 0.3% of the identified bones from AK 1 showed signs of porcupine gnawing, whereas 60.0% of those from AK 2 did so. The striking differences between these two bone accumulations are particularly significant because the range of animal species represented in the two collections is very similar, as is reflected in table 55.

A porcupine lair I have been able to study in some detail is in the east bank of the Nossob River, 12.9 km upstream from Twee Rivieren in the Kalahari National Park. Analysis of its bone accumulation has provided insights into a number of aspects of collections made by porcupines.

The Nossob Porcupine Lair and Its Evidence

The site was first visited by A. R. Hughes in 1956; on 21 May of that year he removed from it 1,420 bones and other objects that appeared to have been collected by porcupines. At the time, he made the following notes (Hughes, pers. comm.):

> The lair was situated in one of a number of intercommunicating solution cavities in the calcrete at the top of the east bank of the Nossob R. 8 miles from the base camp at Twee Rivieren in the south of the Kalahari Gemsbok National Park. In the front of the entrance was a 72 sq. ft. area of loose ground 6 ins.–1 ft. in depth which contained 411 porcupine quills and many very small hairlike quills, gnawed bones, skulls and horns of large and small animals, 9 tins, pieces of iron, bottles, 105 pieces of wood up to 3 ft. in length and other articles collected by porcupines. The inside of the lair was about 425 sq. ft. in extent and tapered away from the entrance for 36 ft. to a small opening in the calcrete surface superadjacent to the river bank. On the inside of the lair, especially the sides near the entrance, were scattered the bones, skulls and horns of animals which had died during the last 25 years as was indicated by the presence of the horns of domestic rams and cows; domestic cattle had not lived in the neighbourhood since 1932 when the park was proclaimed. Inside the lair was a circular raised piece of clean ground well consolidated by the porcupines that had lain there sleeping.

I have been able to visit the site twice—in March 1968, when all bones and other objects were again removed from the lair, and in June 1969, when the sketch plan shown in figure 114 was drawn. The site itself and the abundance of bones inside its entrance are shown in figures 115 and 116. Besides the main entrance at A on the plan, the lair has three openings to the surface. The roof is low, varying from 45 cm to 60 cm, a situation that makes retrieval of bones difficult. Two porcupines were visible in the farther recesses of the lair at the time of my visits—they shook their quills in a threatening manner but were not otherwise troublesome. More discomfort was caused by the voracious tampan ticks that emerged in large numbers from the dusty substrate of the lair while we were creeping around inside it.

The 1956 collection, made by Alun Hughes and kindly put at my disposal, consisted of 1,328 pieces as listed in table 56. By 1968 380 more objects were available for collection, making a total of 1,708. These were all bone, horn, or tortoise shell, except 71 pieces of wood and 17 metal objects (iron bars, rusty cans, enamel mugs, and the like). As is reflected in table 57, the remains in the whole collection came from a minimum of 106 individuals; bovids such as springbok, gemsbok, hartebeest, and wildebeest were particularly well represented on the basis of their horns.

The study of the bone assemblage from the Nossob lair has provided information on a number of questions:

What Is the Rate of Accumulation of Objects in a Porcupine Lair? The Nossob lair was cleared by Hughes in

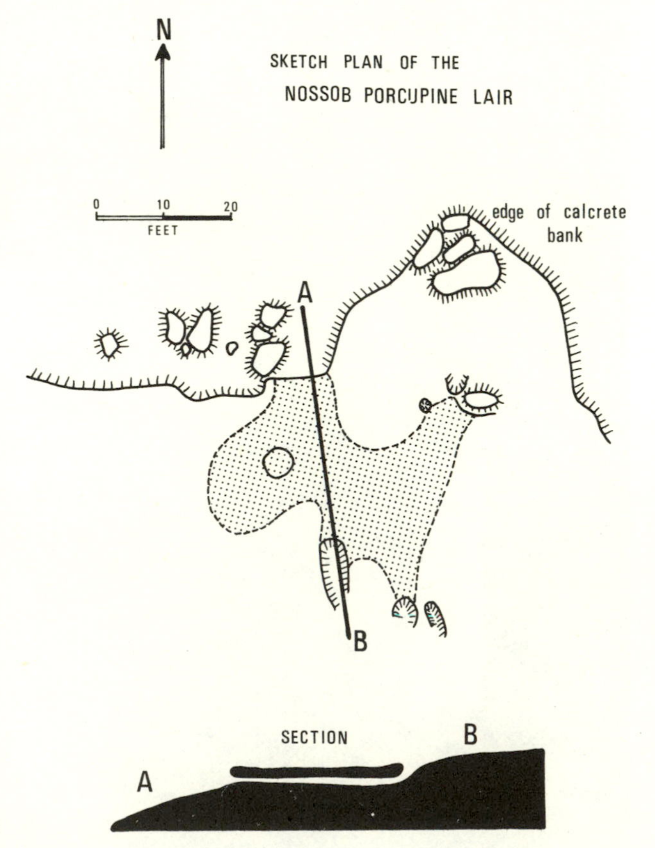

Fig. 114. Plan of, and section through, the Nossob porcupine lair. This extensive but low-roofed cave has formed in the calcrete bank of the Nossob River, Kalahari National Park.

Fig. 115. The main entrance to the Nossob porcupine lair.

Fig. 116. The interior of the Nossob lair showing abundant bones that had been collected by the resident porcupines.

May 1956 and again by me in March 1968, giving a time lapse of almost twelve years. During this period 380 objects were taken to the lair, a rate of about 32 per year. I was able to observe two porcupines in the lair on two occasions, but there may well have been others in the deeper recesses that were out of sight. The rate of collection of bones would depend on their availability; I suspect that the Nossob locality would generally provide an abundance of bones for porcupines to collect.

The only other estimate of accumulation rate of which I am aware was made by Maguire (1976) at the Hartebeesthoek porcupine lair west of Pretoria. The lair, carefully cleared and reexamined after precisely one year, was found to contain 44 gnawed bones, 5 gnawed branches, 6 gnawed mealie cobs, and 2 gnawed stones—a total of 57 objects, which is slightly higher than the figure for the Nossob lair.

Do the Bones in the Porcupine Collection Reflect the Natural Abundance of Animals from Which They Come? It is fortunate that the Nossob lair is situated in the Kalahari National Park—an area for which census data on the contemporary bovid populations are available. These unpublished data have been kindly provided by G. De Graaff (De Graaff, Bothma, and Moolman, unpublished) of the National Parks Board. The census figures for springbok, gemsbok, hartebeest, wildebeest, eland, steenbok, and duiker are given in table 58 and may be compared with the minimum numbers of individuals reflected in the porcupine collection. The relevant numbers for these species are springbok, 40; gemsbok, 15; hartebeest, 9; wildebeest, 5; steenbok, 5; duiker, 2. Eland has not been identified from cranial remains, but at least 3 individuals of an eland-sized bovid are represented in the postcranial parts.

Percentage abundance of the various species is plotted in figure 117, from which it is evident that the minimum numbers of individual animals represented by the porcupine-collected remains do indeed mirror the actual abundance of the antelope species.

An interesting point here is that, although more springbok are represented by their horns than are larger antelopes, the postcranial remains of the larger bovids are more numerous than are those of their smaller counterparts. For convenience, southern African bovids have been divided into four size classes (table 1) according to weight: class I, up to about 23 kg (50 lb); II, 23–90 kg (50–200 lb); III, 90–295 kg (200–650 lb); and IV, above 295 kg (650 lb). Springbok fall into class II; gemsbok, hartebeest, and wildebeest, into class III.

Table 59 lists the bovid skeletal parts in the 1968 collection from the Nossob lair. On the basis of horns, the 1968 collection contains parts of 15 class II antelopes (springbok) and 9 from class III (gemsbok, hartebeest, and wildebeest); nevertheless, class II antelopes are represented by 132 pieces and class III by 183. The reason for this discrepancy is probably that the horns of almost every antelope killed survive and may be collected by porcupines, but that much less is left of the skeleton of a smaller antelope after predator and scavengers are finished.

What Parts of Bovid Skeletons Are Represented, and Are These the Most Resistant Elements in the Skeleton? Table 60 lists the parts of bovid skeletons represented in the two Nossob collections. There is wide representation of almost all skeletal parts. On the basis of horns, at least 81 individual antelopes appear to have contributed to the sample. Working with this total, it is possible to calculate how many individual parts of the skeletons there should have been if none was lost and all were collected. These figures are given in the third column of table 61. Surprisingly enough, following pelvis pieces, atlas and axis vertebrae have the highest survival figures, whereas caudal vertebrae have the lowest. Following through the vertebral column, the sequence of percentage survival from highest to lowest is as follows: atlas (30.9%); axis (28.4%); other cervical (18.8%); lumbar (15.6%); sacral (8.6%); thoracic (5.0%); caudal (0.1%).

These survival figures very probably reflect the robusticity of the different vertebral types but may also have been influenced by the porcupines' selective preference. Perhaps porcupines simply prefer a substantial-looking, chunky cervical vertebra to a scrawny-looking thoracic one with its long neural spine.

The individual parts of a bovid skeleton vary considerably in strength and in resistance to destructive treatment. Under a given destructive regime, the individual parts will survive in proportion to their robusticity. Study of the survival of goat bones subjected to Hottentot and dog feeding (chap. 2) has shown which parts of the skeleton are best able to withstand destructive treatment. Table 5 lists the various parts of the goat skeletons and provides figures on their percentage survival. Horns had the highest survival, followed by mandibles, maxillae, and distal humeri. Proximal humeri and caudal vertebrae were found to have a nil survival value. Survival may be correlated with the compactness of the bone, expressed as specific gravity, and with fusion times of the epiphyses of the long bones. In fact, the survival of parts will follow an entirely predictable pattern if the destructive influences are known.

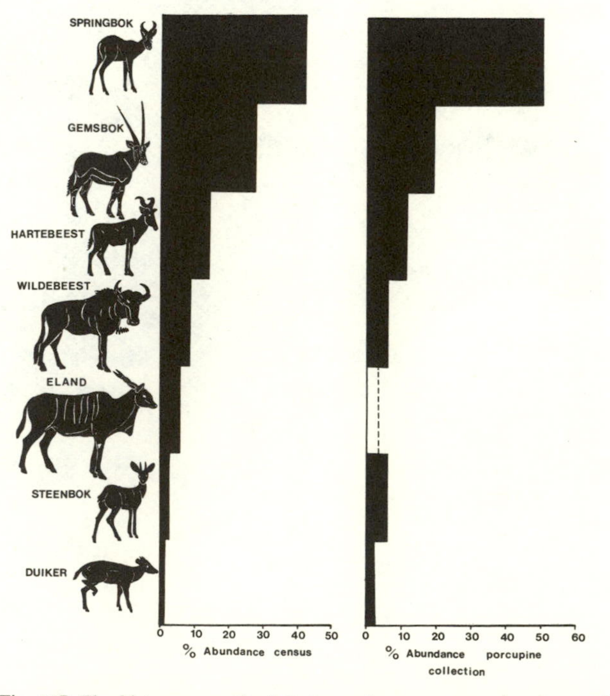

Fig. 117. The histogram on the left shows the percentage abundance of various mammals in the Kalahari National Park, according to National Parks Board census data; that on the right reflects the percentage abundance of the same species as represented by bones collected by the porcupines.

In figure 118 the percentage survival of bovid skeletal parts in the Nossob sample is plotted and contrasted with that for the same parts in the Hottentot goat-bone sample. The Nossob survival sequence does not follow the Hottentot pattern; to my mind it does not follow the pattern that could be predicted for skeletons subjected to a good deal of destructive treatment. For instance, the ratio of proximal to distal humeri in the Hottentot sample is 0 : 82; in the Nossob sample it is 9 : 20. It is perfectly clear that the bovid skeletons in this part of the Kalahari Park have not been subjected to extreme scavenging pressure; complete bones frequently survive, as do delicate elements. The residue left for porcupines to collect therefore does not necessarily consist of only the most resistant elements.

Do Porcupines Preferentially Collect Objects of a Particular Size and Weight? To answer this question, each individual object in the Nossob collection was weighed and measured—the greatest length per piece being recorded in inches. The results given in tables 62 and 63 are plotted in histogram form in figure 119. As far as weight is concerned, the great bulk of the sample falls in the 0–50 g category, though some objects up to 750 g are included in the sample. In length, the bulk of the objects have maximum dimensions of between 1 and 6 in, though some up to 36 in (mostly gemsbok horns and pieces of wood) do occur.

Do Porcupines Preferentialy Gnaw Objects of a Particular Size? Tables 64 and 65 give the percentages of gnawed bones in each weight and length class. It is immediately apparent (fig. 119) that the percentages of gnawed bones in the smallest weight and length classes are *lower* than those in the larger categories. The conclusion, therefore, is that, although porcupines collect large quantities of small bone pieces, they prefer to gnaw the larger ones. It is probably more difficult for a porcupine to hold a small object between its forepaws while gnawing than to hold a larger one.

Are the Collected Bones Generally Fresh or Weathered? Any curator of an osteological collection will have found that, unless bones are defatted before being stored, they will continue to exude grease indefinitely. An interesting point about the Nossob porcupine collection is that only a very small number of bones (no more than 15 out of a total of 1,620) showed appreciable traces of fattiness. I have little doubt that if the bones had been brought to the lair in a fatty condition, this fattiness would have remained indefinitely in the shaded conditions of the lair. I formed this conviction as a result of a bone-weathering experiment I set up more than ten years ago at the Namib Desert Research Station in South-West Africa. Two series of fresh bones were placed in a wooden frame; one series was fully exposed, protected only by wire netting. The other series was shielded from direct sun by surrounding screens of ⅛-in pegboard with holes 1 in apart. The frame was set up in a fully exposed situation on 1 December 1965; effects were observed at yearly intervals, and some specimens were removed on 21 March 1973 for photography, after a time lapse of seven years. Two halves of a pig mandible showed striking differences (fig. 120). The

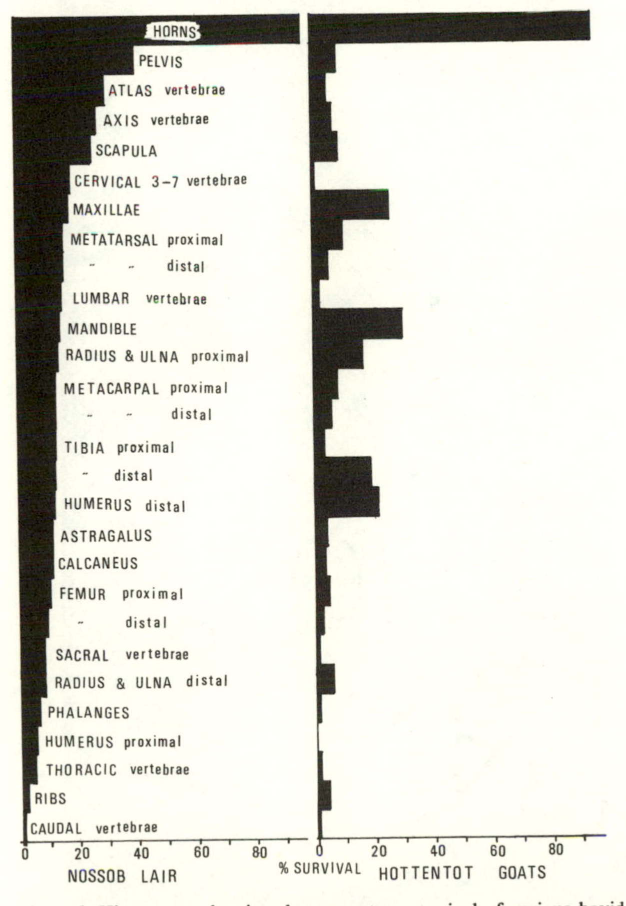

Fig. 118. Histograms showing the percentage survival of various bovid skeletal parts in the porcupine-collected sample from the Nossob lair. The figures are contrasted with those for the Hottentot goat bones described in chapter 2.

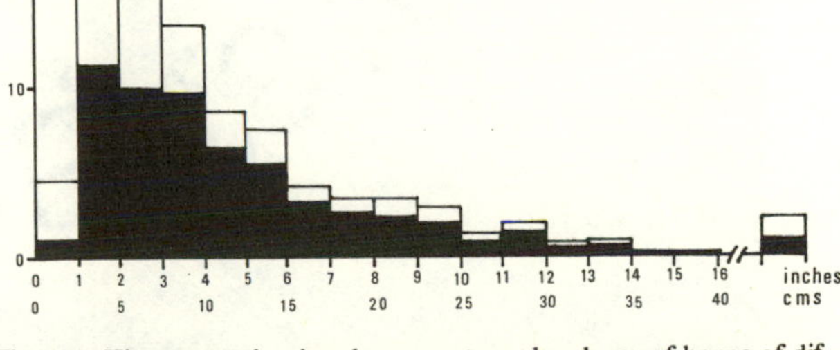

Fig. 119. Histograms showing the percentage abundance of bones of different weight and length classes in the porcupine-collected sample from the Nossob lair. The proportions of gnawed and ungnawed bones in each class are also shown.

half that had been exposed to full sun was bleached and cracked, while the other, which had been shaded from the sun in a well-ventilated situation, retained enough fat after seven years' storage for dust to be adhering to it. The defatting process of the fully exposed bone was complete within one year, and I would suspect within three months. In a shaded situation, defatting may take decades.

On the basis of these observations, I have no doubt that the bones generally brought to the Nossob lair by porcupines are ones that have been naturally defatted through surface exposure. In fact, the porcupines show a marked preference for bleached, defatted bones.

Maguire (1976) reached a similar conclusion after an experiment in which 13 ungnawed bones, varying in freshness, were placed 5 m from the entrance of a porcupine-occupied ant-bear hole near Klaserie in the eastern Transvaal. Gradually the porcupines removed all these bones, but they left the undefatted, greasy ones till last. After three months these were found to have been naturally degreased, and they were then attractive to the porcupines.

Conclusion

The most reliable indication that a bone accumulation in a cave has been built up by porcupines is the presence of typical gnawing marks on defatted and frequently weathered bones falling within the size and weight range

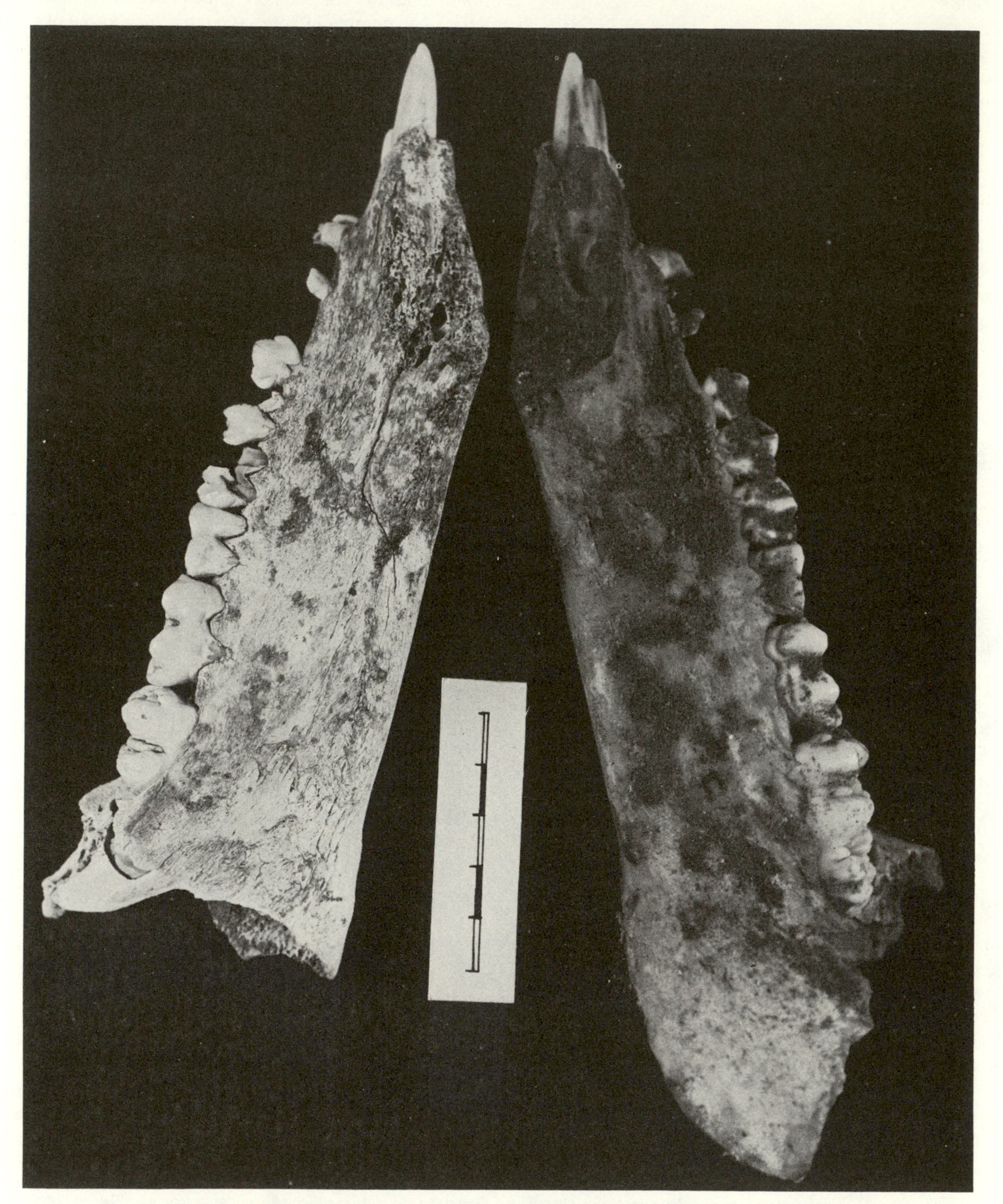

Fig. 120. The long-term persistence of fat in bones is demonstrated by these two halves of a pig mandible, both of which were removed from a bone-weathering experiment after seven years. The half on the left was subjected to full sun and was fully defatted within one year; that on the right was stored in a shaded, well-ventilated situation and retained its fat after the full seven years.

suggested in this chapter. The incidence of gnawed bones in the whole collection from any site will depend on the abundance of bones available to the porcupines at the time. Where bones were readily available collections will show less gnawing than where bones were scarce. Several bone accumulations attributable to porcupines have now been studied. Percentage abundances of gnawed bones in these collections varied as follows:

- 22.0% gnawed. Winkelhoek lair, northern Natal, 463 objects (Maguire 1976).
- 60.0% gnawed. Andrieskraal 2 lair, eastern Cape, 640 identified bones (Hendey and Singer 1965).
- 61.1% gnawed. Nossob lair, 1,708 objects (Brain, this study).
- 69.0% gnawed. Auob lair, Kalahari National Park, 287 objects (Brain, unpublished).
- 76.0% gnawed. Rand van Tweespruit lair, western Transvaal, 106 objects (Maguire, 1976).
- 80.0% gnawed. Kimberley lair, northern Cape, 22 objects (Maguire, 1976).
- 96.0% gnawed. Hartebeesthoek lair, central Transvaal, 479 objects (Maguire, 1976).
- 100.0% gnawed. Hanglip lair, northern Transvaal, 51 objects (Maguire, 1976).

It is perhaps of interest that a bone accumulation remarked upon many years ago by Peringuey (1911) in his "The Stone Ages of South Africa as Represented in the Collection of the South African Museum," from a low cave at Hawston on the Cape coast, shows clear signs of being a porcupine collection. Of the 115 bones I have been able to examine from the site, 44 showed the imprint of porcupine incisors.

6 The Contribution of Owls

> Though I have devoted comparatively little attention to systematic zoology, I have from time to time made pretty extensive collections of the bones of small mammals, and the best hunting grounds I have invariably found to be the haunts of owls. In the disgorged pellets often found in great abundance in rock clefts the small mammal skulls are usually preserved uninjured, and the owls frequently obtain specimens which the collector of skins will not readily come across. [Broom 1907, p. 262]

The habit of certain southern African owls of roosting and nesting close to the entrances of caves is of considerable taphonomic significance. For it is during daytime rest periods that the owls may regurgitate pellets containing the indigestible remains of their prey. These pellets accumulate on the cave floor in considerable numbers and, in certain circumstances, become fossilized as a breccia rich in microvertebrate remains (figs. 121 and 122).

In southern Africa, the owl species most important from this point of view is undoubtedly the barn owl, *Tyto alba* (Scopoli), although certain species of eagle owl may also play a role in the accumulation of microvertebrate remains within caves.

The Mechanism of Pellet Production

As early as 1927 an American study was made of pellet formation in the great horned owl, *Bubo virginianus*, by Reed and Reed (1928). Seven nestlings were fluoroscoped after ingesting barium paste or food mixed with paste. It was found that pellets were ejected from 12 to 20 hr after feeding and that the formation of the pellets was made possible by structural peculiarities of the stomach. The pyloric opening proved to be only about 1 mm in diameter and, in the normal position, was found to lie on a level with the opening of the esophagus so that practically the whole stomach, even when greatly distended, lay below the level of the pylorus. The small size of the pyloric opening provided a mechanical bar to anything but finely divided or liquefied food.

Very little free acid was detectable in the gastric contents at any stage of digestion, or in extracts from the pellets. Total acidity of a sample taken 5 hr after eating was 0.16%; 2.5 hr after a meal it was 0.44%; and at 4 hr it was 0.28%. Despite the lack of acidity, peptic activity was found to be generally about three times as potent as that of dog gastric juice at comparable stages of digestion.

Summing up the situation, Reed and Reed concluded that pellet formation involved three factors: first the high placement and small size of the pyloric opening; second, feeble gastric motility that would preclude stirring up or freeing of hair or feathers enmeshed in the whole mass; and, third, potent peptic activity that would digest flesh free from hair, feathers, and bones and liquefy it so that pyloric passage would be possible. The actual ejection of a pellet was similar to vomiting—the owls showed evidence of nausea for 15 to 20 min before ejection.

It seems likely that other species of owl may show variations in the action of their gastric juices. A pellet of the wood owl, *Ciccaba woodfordi*, from Zimbabwe contained fur and feathers but no bones at all (Brooke 1967), suggesting that digestion of bones may occur in this species, as it apparently does in birds such as the black-headed heron, *Ardea melanocephala*, which characteristically ejects bone-free pellets (Crass 1944). A digestive oddity in the barn owl, *Tyto alba*, recorded in California is that these birds appear unable to digest eggshell or yolk (Banks 1965).

A number of studies (e.g., Clark 1972) have suggested that hawks digest bone more extensively than do owls. This has, in fact, been demonstrated in a series of experiments on gastric digestion in raptors (Duke et al. 1975), in which pH and proteolytic activity of gastric juices in several hawks and eagles (falconiformes) was compared with that in owls (strigiformes). The basal, or preprandial, pH in falconiform gastric juice was found to range from 1.3 to 1.8, with a mean of 1.6, while that of the strigiform equivalent ranged from 2.2 to 2.5 with a mean of 2.35. The greater acidity of hawk gastric juice could be expected to damage bones more extensively than happens in owls, and this has now been experimentally demonstrated (Cummings, Duke, and Jegers 1976). Mouse bones were incubated in solutions simulating gastric juices with pH levels of 1.6 and 2.35, and it was shown that the former corroded bones significantly more, at all pepsin concentrations, than the latter. Proteolytic activity proved slightly greater in solutions with a lower pH.

In an important paper on the role of avian predators as accumulators of fossil mammal material, Mayhew (1977) emphasized that mammal remains in pellets of diurnal predators showed consistent differences from those in pellets of nocturnal predators. He concluded that diurnal birds of prey were able to digest bones more completely than were nocturnal ones, but that the former sometimes

Fig. 121. Modern owl pellets on the cave floor beneath the Bolt's Farm owl roost.

Fig. 122. A block of Member 1 breccia from Swartkrans, partly etched in acetic acid. The abundant microvertebrate remains came, almost certainly, from owl pellets.

built up assemblages of mammalian teeth, as in the Cromerian Upper Freshwater Bed at West Runton in England.

It is also to be expected that some parts of prey skeletons should suffer more from digestion than others, and this has been shown to be true for artificially fed prey of three species of European owls (Raczynski and Ruprecht 1974). Determinations were made, by analyzing pellets, of elements missing from skeletons of birds and mammals fed to the owls. It was found that the pelvic girdle was most often missing (up to 80%), followed by craniums (up to 35%) and mandibles (25%). The largest number of missing skeletal elements occurred in the tawny owl, *Strix aluco* (51%), with fewer in the long-eared owl, *Asio otus* (46%), and barn owl, *Tyto alba* (34%). Furthermore, the quantity of missing skeletal elements showed an inverse relationship to the age of the owls and of their prey.

Further information on pellet formation has resulted from a laboratory study on the short-eared owl, *Asio flammeus,* a diurnal species common in Scotland and Wales (Chitty 1938). An experimental owl was housed in a specially constructed cage that allowed a record to be kept of the time of feeding and ejection of pellets. It was found that the interval between eating and pellet production varied from 1.5 to 13 hr, depending on the weight of the meal. A food item weighing 4.8 g led to ejection of a pellet after 1.5 hr, whereas one weighing 37.8 g took 13 hr. Likewise, the size of a pellet depended on the nature of the meal; pellet lengths varied from 10 to 65 mm, the largest recorded pellet resulting from digestion of a 54.2 g vole. This 65 mm-long pellet had a volume of 13.5 cc and a wet weight of 10.9 g. Typically, the weight of a dry pellet is one-half to one-third that of the fresh pellet.

The timing of pellet production is clearly important from the point of view of microfaunal accumulations in caves. Only pellets ejected during rest periods in the cave roost will contribute to the deposit. It seems likely that pellet-ejection times will vary a great deal in natural circumstances. A study of twenty pairs of tawny owls, *Strix aluco,* over a period of eight years in Britain (Southern 1954) revealed that this species typically casts all its pellets before going to roost.

After a study of barn owls in France, Guérin (1928) concluded that there are characteristically two pellet ejections in each 24-hr period. The first results from prey caught early in the evening and takes place away from the roost at the "station nocturne." Thereafter, the owl tends to feed again before retiring to its roost, or "station diurne." Here, in the course of the day, the second pellet is produced. If this procedure is typical of barn owls in general—and it is confirmed by Wallace's study (1948) in Michigan—it is clear that only part of the diet will be reflected in the pellets that accumulate under a day roost. In addition, the consumption may be unexpectedly high; Guérin showed that a wild barn owl captured a total of 1,197 animals in a 90-day observation period; of these, it ate 837.

Various attempts have been made to assess the food requirements of owls by collecting and analyzing their pellets. Montgomery (1899) studied the short-eared owl, *Asio flammeus,* from this point of view but concluded that many pellets would have been disgorged at casual feeding perches that could not be located. The same is thought to be true of the barn owl in the Netherlands (Honer 1963), which was found to hunt over fixed routes, using high points as lookouts and pellet-casting posts.

The rate of pellet-casting by barn owls at a diurnal roost in the Transvaal has been established by Dean (1973*a*). He found that the rate varied from 0.6 to 1.5 pellets per day, with a mean of 1.10 per owl per day. This involved a mean daily consumption of vertebrate biomass ranging from 41.7 to 81.9 g per day, with a mean of 55.3 g per day per owl. Consumption of food was highest in late autumn and winter, the months when young are reared. The figures given by Dean were based only on pellets cast at the roost. They therefore represent only part of the owl's total diet.

Southern African Owls Responsible for Microvertebrate Collections

As I mentioned earlier, the species of importance here are the eagle owls and barn owl. The former are perhaps of lesser significance than the latter, but they certainly warrant mention. Three species of eagle owl occur in southern Africa: the spotted eagle owl, *Bubo africanus;* the Cape eagle owl, *B. capensis,* and the giant eagle owl, *B. lacteus.*

Spotted Eagle Owls, *Bubo africanus* (Temminck)

The spotted eagle owl is the commonest large owl in southern Africa and has a wide distribution farther to the north. It tends to rest during the day among rocks or in large trees, emerging at dusk, when it may be seen regularly on vantage points in the hunting area. It is frequently killed by cars on country roads at night.

Though less important than barn owls as accumulators of bones in caves, spotted eagle owls cannot be ignored, since they certainly roost and nest in caves from time to time. While prospecting for cave deposits in South-West Africa, I came across the nest of a spotted eagle owl inside a dolomitic cave on the farm Chicago, in the Otavi District, on 23 May 1954. The cave consisted of a vertical shaft about 9 m deep, overhung by a large fig tree whose roots descended the shaft and aided entry. The nest was on the bare floor of the cave, in the twilight zone, and contained three eggs and two downy owlets. The vicinity of the nest was littered with pellets, which I unfortunately failed to collect, but a freshly killed rodent close to the nest was identified as an African bush rat, *Aethomys chrysophilus.*

Fig. 123. Distribution of the spotted eagle owl, *Bubo africanus.*

Spotted eagle owls, like many other predators, are opportunists, taking whatever prey is to be had at the time, whether rodents or invertebrates. The stomachs of 7 adult owls killed on roads between 1956 and 1961, examined by Benson (1962), contained nothing but insects and one snail; similarly, an owl killed near Trelawney in Zimbabwe (Carnegie 1961) had eaten 11 beetles and a locust.

On the other hand, analyses of pellet accumulations below spotted eagle owl roosts have revealed a variety of vertebrate prey species. Nel (1969) identified a total of 565 individual animals from a collection of 252 pellets that had accumulated on the floor of an old building at Sossus Vlei in the Namib of South-West Africa. The prey sample was dominated by desert gerbils, *Gerbillurus vallinus* (62%), followed by golden moles, *Eremitalpa granti,* various rodents (10%), and geckos and birds (7%). A rather similar prey composition was found in pellets collected below a cliff roost at Homeb, on the Kuiseb River in the Namib Desert Park (Brain 1974*b*). Other rodents feature prominently in the prey of spotted eagle owls from other southern African localities. In the sandveld area of the southwestern Cape, the gerbil, *Tatera afra,* was found to be the commonest prey (Siegfried 1965), whereas in a Johannesburg sample the vlei rat, *Otomys irroratus,* dominated (Welbourne 1973).

A few years ago I had an opportunity to compare the faunal content of spotted eagle owl pellets with those of barn owls hunting simultaneously over the same area. During September 1972 a pair of eagle owls roosted in some large *Celtis* trees against a dolomitic cliff 300 m southwest of Swartkrans, while a barn owl was using the Inner Cave at Swartkrans as its day roost. In the course of the month I collected pellets beneath both roosts; thereafter the eagle owls left, and the barn owl continued to make intermittent use of the Swartkrans cave roost until July 1975. Prey animals identified from the two pellet collections are listed in table 66, and the major prey groupings are depicted in figure 125. The contributions made by the various prey animals to the diets of the two species of owl were surprisingly similar.

Fig. 124. A spotted eagle owl, *Bubo africanus,* photographed by A. C. Kemp.

The tentative conclusion is that microvertebrate bone accumulations built up in any particular cave by spotted eagle and by barn owls would probably not differ much in species composition. In fact, in these terms it would not be possible to separate one set of food remains from the other.

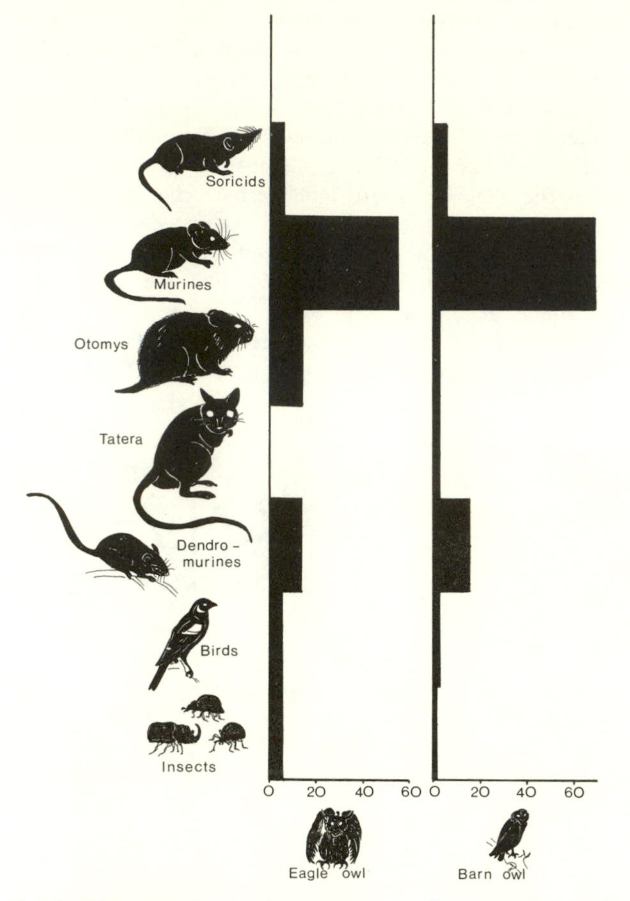

Fig. 125. Histograms showing the percentage contributions made by various animals to the diet of spotted eagle owls and barn owls, hunted simultaneously over the same range at Swartkrans. Prey selection by the two owl species proves to be remarkably similar.

Cape Eagle Owls, *Bubo capensis* Smith

In South Africa this is a rare species, not easy to distinguish from the spotted eagle owl, but with very much larger feet that allow it to overpower larger prey. Three subspecies are currently recognized (Benson and Irwin 1967):

B. c. capensis, confined to the Republic of South Africa, in the mountainous area from Cape Town eastward to Natal and Zululand.

B. c. makinderi, found at scattered localities from Zimbabwe to the highlands of Kenya and west to Mount Elgon. Larger than *capensis,* with less distinct barring on the lower abdomen and thighs.

B. c. dillonii, an owl intermediate in size between the last two, from the highlands of Ethiopia and Eritrea.

As figure 126 shows, the present distribution of the owl is highly discontinuous, but the currently isolated relict populations were probably linked during cold periods in the Pleistocene.

Very little is known about the South African subspecies, although Clancey (1964) states that these birds can overpower "quite large mammals such as the springhare, *Pedetes caffer,* the fawns of antelopes, as well as quite large birds." The northernmost subspecies, *B. c. dillonii,* has also received scant attention from ornithologists, although Brown (1970) has observed it in the Bale Mountains of Ethiopia. He states that in the cedar forests there it nests at the bases of low cliffs, and the numerous pellets he collected contained the remains of fruit bats and mole rats in about equal proportions.

Most of the information available about *Bubo capensis* is based on the subspecies *mackinderi,* named after Sir Halford Mackinder, the first man to climb Mount Kenya where, incidentally, the owl occurs up to an altitude of 13,500 ft (4,117 m). The first detailed study of *B. c. mackinderi* was made by Sessions (1966, 1972) at Lengetia Farm on the Mau Narok plateau close to the west wall of the Central Rift Valley, 56 km west of Gilgil in Kenya. Here ten nest sites were found, each breeding pair having a territory of about 2 km^2. The nests were on bare ground, beneath suitable cover, and invariably close to water. Concerning feeding behavior, Sessions wrote (1972, p. 4):

> Typical of many eagle owls, Mackinder's Owl is not entirely nocturnal and is often seen hunting before sunset and again in the early morning; sometimes one may see it sunning itself by day on an exposed rock or branch.
>
> The usual routine is, at sundown, the owl gives a preparatory hoot from its daytime roost, sometimes giving a few answering calls to its mate. It then flies up the valley in short stages, pausing *en route* on suitable posts, until it finally reaches its hunting ground on the ridge top. Here it will utilise a favourite perch such as a tall tree or telephone pole, or even a chimney top, from which to spot its prey, which may be eaten on the ground or on the perch. After a couple of hours' hunting, the birds return to the roost site in the valley, where they appear to spend a considerable part of the night settled on the ground near the water. The purpose of this habit is not clear, but it may be in order to catch frogs and crabs. Before dawn they are back hunting again on top of the ridge, returning to roost as the sun comes up.

From an analysis of the contents of pellets, Sessions concluded that the main prey of the owls was mole rats, *Tachyoryctes splendens,* which weigh up to 250 g, and a single owl may eat three each night. Second in importance

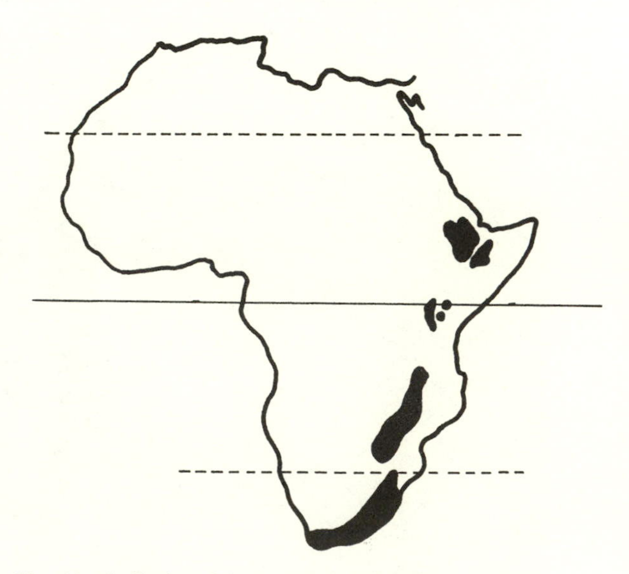

Fig. 126. Distribution of the Cape eagle owl, *Bubo capensis.*

are freshwater crabs; it is not unusual to find a whole carapace up to 25 mm across in a pellet, or a claw 39 mm long. Crabs are also brought to the nestlings, and on one occasion Sessions found a pile of their carapaces on a feeding block about 5 m from the nest. The crabs had not been eaten whole as they would have been by an adult owl; the meat had been picked out and the shell discarded.

At the top of the Teleki valley on Mount Kenya, Sessions also found numerous roosts of Mackinder's owl and, from an examination of the food remains beneath them, concluded that dassies had been the main prey. The heads of the dassies had not been swallowed, and the skulls were simply discarded alongside the pellets.

Bubo capensis was not known from Zimbabwe until specimens unexpectedly came to light from Inyanga cave and Bambata cave in the Matopos (Benson and Irwin 1967). The first information about the owls' diet there was published by Brooke (1973), who listed the contents of a collection of pellets from a small cave on the Shotshe Kopje in the Matopos. Prey included three hares, confirming a chance observation by Jackson (1973*a,b*), who found that a female owl collected in Mozambique near the Zimbabwean border had the leg of a red rock hare in its esophagus. A number of breeding and roosting sites of *B. c. mackinderi* have now been located in the Matopo Hills (Gargett and Grobler 1976; Gargett 1976; Gargett, n.d.). These show that the owls are sparsely distributed in comparison with, say, black eagles. Seven breeding sites have been located, indicating that the owls occur at a density of one pair to 88.6 km^2. Since 1972, regular visits have been made to the nest and roost sites in the Matopos, and all pellets and other food remains have been collected and analyzed. A remarkable list of 925 prey items has now been compiled (Gargett and Grobler 1976), a simplified version of which is given in table 67 and figure 128. Working with the liveweights of the various prey species, it has become clear that hares, principally the red rock hare, make up 70% of the owls' diet. Second in importance are the two forms of dassie occurring in the Matopos, followed by the two species of cane rat and a variety of mammalian prey including springhare, hedgehog, squirrel, mongoose, genet, and juvenile civet. Birds, lizards, and invertebrates also contributed, albeit insignificantly, to the diet.

Concerning treatment of the prey skeletons, Gargett and Grobler reported that the hare craniums showed characteristic damage to the calvariae and nasals. Skulls of prey of this size had not been swallowed but occurred side by side with the smaller skeletal elements regurgitated in the pellets.

I have been able to observe a pair of *B. capensis* roosting in a small cave 15 m up on a cliff face adjacent to the Skeerpoort River on the farm Uitkomst, about 30 km northeast of Swartkrans. Remains of the owls' prey in the cave, collected on 29 September 1972, came from 2 adult and 2 juvenile hares, probably *Pronolagus randensis* and from 1 hedgehog, *Erinaceus frontalis*. The hares had been dismembered and eaten piecemeal, but the remains of the hedgehog were contained in a single pellet shown in figure 129*a*. The hedgehog had obviously been swallowed whole, despite its many spines, and at the time this struck me as very odd. Subsequently I found that the eating of hedgehogs by eagle owls is not as uncommon or as unlikely as it seemed. Hedgehogs were recorded among the food remains of *B. capensis* from the Matopos (Gargett

Fig. 127. Young of the Cape eagle owl, photographed in the Matopo Hills of Zimbabwe by Mrs. V. Gargett. The comparatively large feet of the species are well shown.

and Grobler 1976), and giant eagle owls also prey upon them, as discussed below.

A second visit was made to the Uitkomst cave roost on 18 April 1978; the owls were not there but had left a large number of bony food remains on the cave floor. These were found to come from 18 hares, of which at least 10 were immature, from 4 birds, 1 crab, 1 dassie, 1 elephant shrew, and 26 rodents, of which 17 were vlei rats of the genus *Otomys*. A striking feature of the hare skulls was the characteristic damage they had suffered to both nasals and calvariae, as shown in figure 129*b*. This was the same as that described and figured from a Cape eagle owl roost in the Matopos by Gargett and Grobler (1976).

In conclusion, it appears that Cape eagle owls are likely to contribute significantly to bone accumulations in caves. They have the habit of roosting and nesting in suitable caves, and they leave uneaten food remains and regurgitated pellets there. In contrast to the barn and spotted eagle owls, their mammalian prey is often comparatively large, and their diet also includes freshwater crabs.

Giant Eagle Owls, *Bubo lacteus* (Temminck)

By far the largest of the southern African owls, with a wingspan of up to 1.6 m, this species has a very wide distribution from the Cape to Ethiopia and Senegal, wherever its preferred habitat of riverine forest, broad-leaved woodlands, or forests interspersed with grassy glades occurs. It is the only eagle owl in Africa in which the eyes are dark brown rather than brilliant yellow or orange as in the other species. It's eyelids, however, are bright orange or pink and are doubtless used in social signaling. According to McLachlan and Liversidge (1970) and Brown (1970), giant eagle owls tend to breed in disused nests of eagles and vultures, incomplete nests of hamerkops, or occasionally in hollow trees. They probably make little regular contribution to bone accumulations in caves, but could well do so if their roost or nest was situated within the catchment area of a cave entrance. Like the Cape eagle owl, this species is known to feed on hedgehogs. Benson (1962) recorded an instance of one of these owls swallowing a complete adult hedgehog at Nyamandhlovu near Bulawayo, while Vernon (1971) found hedgehog remains below a *B. lacteus* roost on Tsumis Estate in South-West Africa. Other prey items recorded here were ground squirrel, *Xerus inauris,* suricate, *Suricata suricatta,* yellow mongoose, *Cynictis penicillata,* a gerbil, *Tatera* sp., a raptorial bird, and a korhaan.

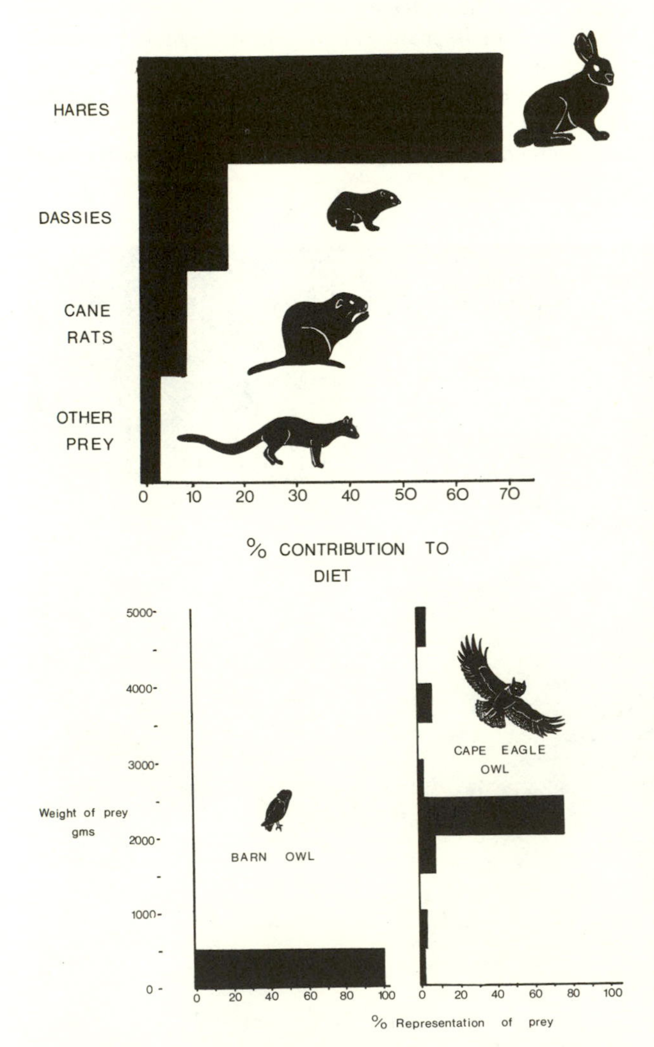

Fig. 128. *(a)* Percentage contributions to the diet of Cape eagle owls in the Matopo Hills of Zimbabwe by animals of various kinds (data from Gargett and Grobler 1976). *(b)* The weights of prey items in Cape eagle owl diet compared with those in the diet of barn owls.

Hedgehog prey of *B. lacteus* has also been recorded by Brown (1965) at a nest 21 km from Nairobi. Here hedgehogs proved to be the most common prey species; 4 skulls and at least 11 skins were collected below the nest, and many other skins were not picked up. The adult owls had skinned the hedgehogs before offering them to the young. Other prey remains came from rodents, fruit bats, a bushbaby, various birds including a barn owl, a house snake, and numerous unidentified frogs.

Of possible interest here is Broom's (1937*b*) record of a fossil hedgehog, *Atelerix major,* from Bolt's Farm. The specimen came from the same small cave deposit which yielded abundant remains of a game bird, possibly a francolin, as well as bones of a bird of prey. It is conceivable that the fossils represented the food remains of a large owl such as an ancestral *B. lacteus*.

Barn Owls, *Tyto alba* Scopoli

These familiar birds are of great significance as accumulators of microvertebrate bones in southern African caves, a fact emphasized by Davis (1959) in his paper "The Barn-Owl's Contribution to Ecology and Palaeoecology."

As figure 132 shows, the barn owl has a remarkably wide distribution throughout Africa, Europe, America, the Far East, and Australia, feeding on whatever prey happens to be available in the particular habitat. Even within the southern African region, there is a good deal of variation in principal prey species, as has become apparent from no fewer than eighteen published analyses of barn owl pellets from seventy-seven separate localities in southern Africa. References to these studies are given in table 68 along with particulars of the species occurring most abundantly in each pellet collection. Insectivores dominated in three collections, rodents in sixty-eight, birds in five, and reptiles in one. Among the rodents, the multimammate mouse, *Praomys natalensis,* proved to be the dominant species in by far the largest number of collections: Smithers (1971*a*) gave the mean weight of 14 adult male *P. natalensis* from Botswana as 64.5 g, and that of 16 adult females as 56.8 g. These figures suggest that barn owls, over much of their southern African range, typically feed on prey weighing approximately 60 g per item.

Some interesting information is available on chick-rearing by barn owls in circumstances where food is superabundant. The observations were made by Wilson (1970) at the eland research station (21°28′S, 29°42′E) in Zimbabwe, between January and December 1967, at the time of a *Praomys natalensis* population explosion that, in that area, peaked during July and August and then gradually subsided. Barn owls were nesting in a hollow baobab tree and, during the eleven months of observations, four clutches varying from seven to nine eggs were laid and the young raised. Young owls were found to fly for the first time at about forty-five days of age and, since the eggs were laid and hatched in sequence, young were still in the nest when the eggs of the next clutch were laid. In fact, the young birds probably served to incubate the new batch of eggs. The rapidity of the breeding sequence observed here was doubtless a response to the abnormally abundant food supply, but it does show the reproductive potential of barn owls in favorable circumstances.

Dean (1973*a*) made a study of pellets collected at monthly intervals throughout 1971 at a roost near Warmbaths, Transvaal. The prey was also strongly dominated by *Praomys natalensis:* there were at least 431 *Praomys* individuals out of a total of 647 animals. Each *Praomys* specimen was placed in one of six age classes, depending on the degree of wear on the first upper molars (Dean 1973*b*), and it was established that a substantial part of the sample consisted of subadult individuals. Dean estimated that the mean liveweight for the *Praomys* individuals in the pellet sample would have been 38.8 g. Particulars of the species present in the Warmbaths sample, together with mean liveweight estimates, are given in

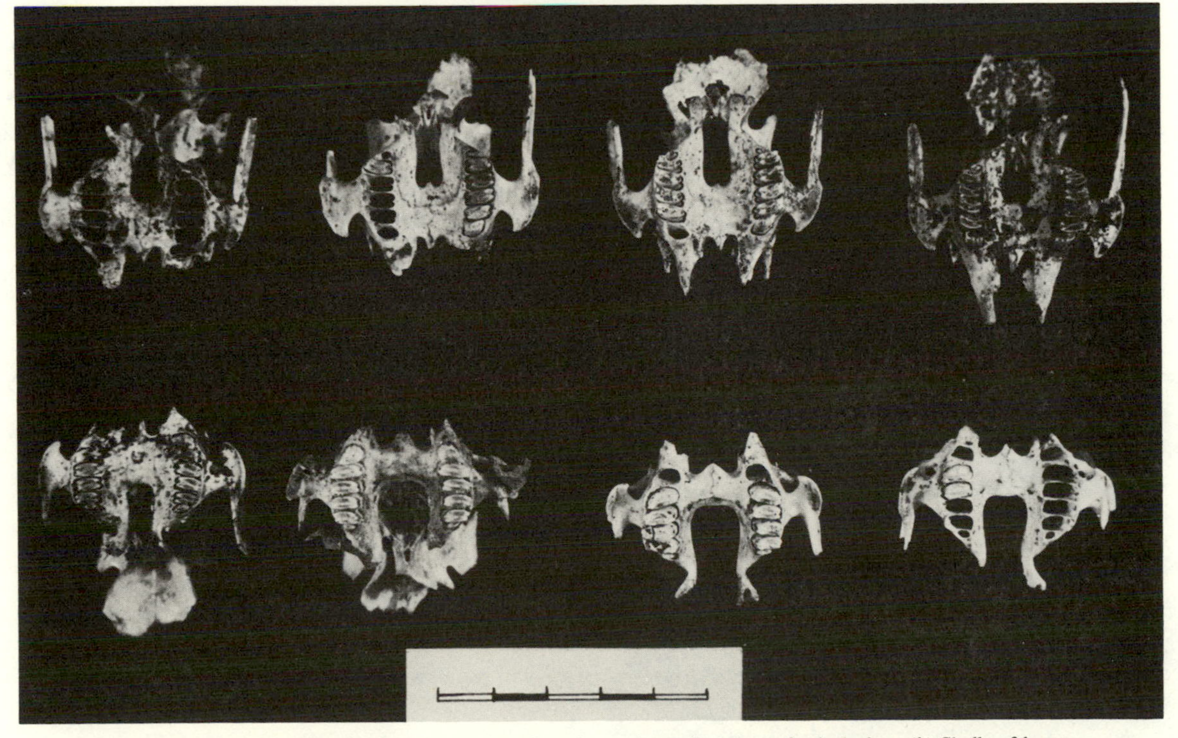

Fig. 129. *(a)* A regurgitated pellet of a Cape eagle owl containing the spines of a hedgehog. *(b)* Skulls of hares discarded by Cape eagle owls in a cave roost, Uitkomst, Transvaal. Damage to the nasals and calvariae is characteristic.

table 69. On the basis of these data, it has been possible to count the number of individual animals in each weight class; figures are given in table 70. As is apparent in figure 134, the great majority of prey items in this sample fall in the 20–40 g weight range.

It is abundantly clear that, although barn owls may have some dietary preferences, they will feed on the most readily available food source as long as it is palatable and of manageable size. This means that, over the owls' wide geographic range, dominant prey will vary from one area to another depending on the species composition of the potential prey fauna. Apart from this, fluctuations from one dominant prey taxon to another may occur in a particular area with the changing seasons. A monthly analysis of barn owl diet based on owl pellets collected in Bryanston, Johannesburg, during 1955 and 1956, enabled de Graaff (1960*b*) to conclude that the owls concentrated mainly on birds in summer and rodents in winter. A similar situation had previously been observed in the Transvaal by Kolbe (1946) and Davis (1959). Generally, however, in favorable habitats a barn owl's diet is very diverse, so that clear-cut alternations between prey taxa may be difficult to isolate.

Fig. 130. Distribution of the giant eagle owl, *Bubo lacteus*.

Fig. 131. A giant eagle owl, *Bubo lacteus*, photographed at Lake McIlwaine in Zimbabwe.

A particularly uncomplicated example of prey alternation came to light recently in the Namib of South-West Africa. The Mirabib Rock Shelter in the Namib Park contains ledges along its back wall that are currently used as barn owl roosts (fig. 135) and apparently have been for at least six thousand years (Brain 1974*b*). Bones from the resulting pellets have been incorporated in the stratified floor deposit, which also contains vegetable matter and human occupation debris. The floor deposit has been extensively excavated by Sandelowsky (1974).

Recently an analysis of microvertebrate remains in the stratified floor deposit has been undertaken (Brain and Brain, 1977), providing a record of owl diet over a six-thousand-year period. It was found in all layers that a very large proportion of this diet was made up of two prey categories: desert gerbils *(Gerbillurus vallinus* and *G. paeba*) and geckos (probably largely *Pachydactylus bibroni*). Together, individuals of these two categories often made up 90% of the total prey items taken. Inevitably, therefore, when the percentage abundance of one category's individuals, for example, gerbils, increased, that of the other, the geckos, declined. This reciprocal relationship is shown graphically in figure 136.

The key to this striking relationship lies in the fact that the gerbil population on the plains surrounding the Mirabib Hill is prone to remarkable fluctuations. After rain, which in this part of the Namib is sporadic and sparse, annual grasses appear in great abundance. In response to the grass cover, the gerbil population builds up within a few months to a relatively high density. If no rain falls in the following year, the population again declines to a residual level.

When gerbils are abundant, the barn owls prey on them to the virtual exclusion of all other prey sources. When they become rare, however, as in poor rain years, the barn owls fall back on the standing gecko resource, mainly on the rocky hillsides, which is little utilized when the gerbils are readily available. It seems reasonable to

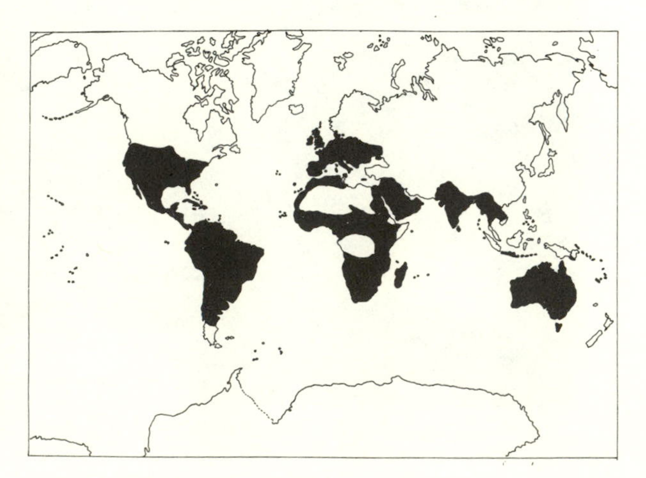

Fig. 132. Distribution of the barn owl, *Tyto alba*.

correlate high gerbil abundance in the owl's diet with summers in which adequate rain fell; gecko abundance, on the other hand, tends to suggest times of drought.

This Namib example provides an indication of how barn owl diet may reflect environmental changes in the hunting area.

Species Composition of Barn Owl Prey as a Reflection of the Habitat

It would be very valuable from the paleoecological point of view if microvertebrate bone accumulations derived from owl pellets could be used as indicators of the habitat over which the owls hunted. For any such attempt it is clearly necessary to know what the hunting range of an owl from its roost might be. The range is likely to be greatly affected by the availability of food; when prey is abundant, the owl's range will be smaller than in times of food scarcity.

Reliable information on hunting ranges will not be forthcoming until radio telemetric tracking of owls has been carried out. Meanwhile we have little more than tentative indications. As a result of a study of barn owls in Israel, Bodenheimer (1949) concluded that "it can be assumed that their hunting grounds do not exceed an area of 25 sq. klm. and are probably much more restricted." Another indication has come from a study of barn owls on the west coast of North America. Owls roosting in a small cave on the northernmost of of the Islas Los Coronados, Baja California, included in their pellets prey that had been obtained on the mainland. Forays to the mainland for food would have involved flights of 10 mi, or 16 km, each way (Banks 1965).

Evidence from the Mirabib owl roost in the Namib of South-West Africa (Brain 1974*b*) clearly indicates that barn owls in those circumstances do not fly the 25 km to the nearest dune fields where they may obtain dune-adapted prey, in the form of Namib golden moles. The meager evidence currently available suggests that barn owls may radiate as much as 16 km from their roost on nightly forays, but not as much as 25 km. Thus, from the point of view of environmental reconstructions, owl prey should be regarded as coming from within a few kilometers of the roost. When prey is normally abundant, some recent evidence suggests that owl hunting in the Transvaal may be extremely localized. I have been able to

Fig. 133. Two young barn owls, from Brits in the Transvaal, at different stages of development. Barn owl eggs are typically laid, and hatch, in sequence, so that young of different sizes are found in one nest.

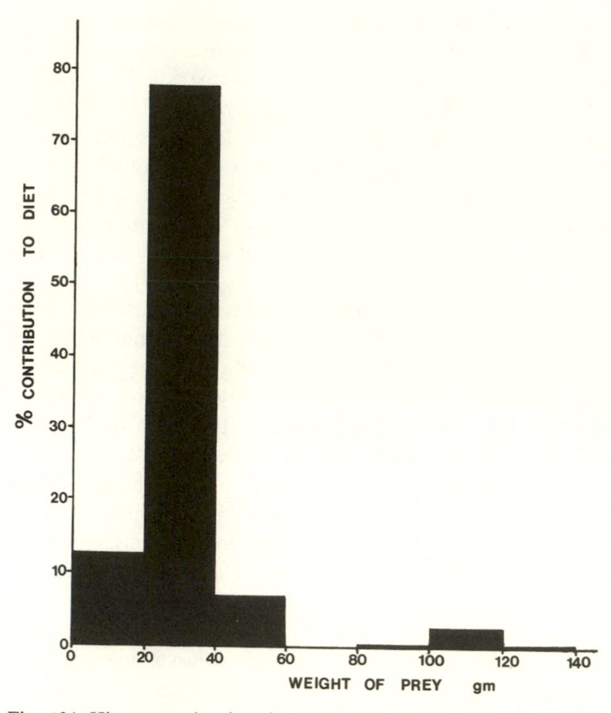

Fig. 134. Histogram showing the percentage representation of various weight groups in the prey of barn owls roosting at Warmbaths in the Transvaal. Data reworked from Dean (1973*a*).

make pellet collections at two pairs of adjacent barn owl roosts, covering the same time periods, and the results show interesting local variation.

The first pair of roosts were the Swartkrans Cave itself and the Bolt's Farm roost, 1,470 m to the southwest. Both roosts are on the north side of the Bloubank River valley, Swartkrans being 250 m from the streambed and Bolt's 350 m away (see fig. 137). The habitat looks very similar. Analyses of pellet collections from each site made in 1975 are given in tables 66 and 71. Although the same prey species occur in both samples, owls operating from Swartkrans were exploiting a local concentration of pygmy mice, *Mus minutoides,* which constituted 27.2% of the Swartkrans prey animals but only 1.8% of those at Bolt's Farm.

The second pair of roosts were in two quarries close to the village of Irene, south of Pretoria. The grassland habitat on dolomite country rock is very similar to that in the Sterkfontein valley. The new quarry is 1,500 m east-southeast of the old one; both are in similar surroundings, except that the former is 1,300 m from the Hennops River and the latter is only 120 m away. The Hennops River is a perennial stream with fringing trees and a fairly extensive alluvial floodplain. Pellet collections were made at each site in July 1975, and results are given in table 71. As is to be expected, the same species occur in both collections, although there are interesting differences. The New Quarry collection, as a result of its proximity to the alluvial plain, contains larger numbers of vlei rats, *Otomys irroratus/angoniensis:* 50.6% as opposed to 32.0% at the Old Quarry. On the other hand, the owls based at the Old Quarry were exploiting a *Tatera* colony nearby, so that these gerbils constituted 26.0% of the prey there in contrast to 5.1% at the other roost.

Taking it into account that extremely localized variations in prey composition are to be expected at any particular roost, it would still be useful if the composition of prey could be used as an indicator of habitat type. On a gross basis across southern Africa this is certainly possible. Davis (1962) has divided southern Africa into four biotic zones, the main features of each being these:

The southwestern Cape. A distinct area climatically and biotically. The southwestern portion has a Mediterranean climate and winter rainfall, but along the coast toward the east the rainfall becomes more regularly distributed throughout the year. The zone corresponds to the Cape Macchia vegetation type.

The southwest Arid. The area consists of true desert, the Namib, and semidesert, the Kalahari and Karroo, all of it receiving, on average, less than 50 cm (20 in) of rain a year. It is divided by the Orange River and contains the western and southwestern portions of the great escarpment. It extends across the Kunene River to about latitude 12°S.

Southern savanna. This zone occupies the eastern part of southern Africa. It includes montane and highveld grasslands, bush and lowland woodlands, and the tropical Mozambique plain.

Forest. Isolated patches of montane and subtropical evergreen forest are distributed in the savanna zone and southwestern Cape, mainly below the escarpment. Endemic mammals are extremely rare, and no rodents of the family Muridae can be regarded as true forest forms.

Davis (1962) has listed murid rodent species that are endemic to each of the biotic zones. The two zones of particular relevance to us here are the southern savanna and southwest arid; species occurring in these zones that Davis regards as endemic or nearly endemic are as follows:

Southwest arid zone: endemic: *Zelotomys woosnami, Aethomys granti, Petromyscus collinus, Parotomys littledalei, Gerbillurus vallinus;* near endemic: *Malacothrix typica, Otomys unisulcatus, Parotomys brantsi, Desmodillus auricularis,* and *Gerbillurus paeba.*

Southern savanna zone: endemic: *Thamnomys dolichurus, Pelomys fallax, Acomys* spp., *Cricetomys gambianus, Dendromys mystacalis, D. nyikae, Otomys angoniensis, Tatera inclusa;* near endemic: *Aethomys chrysophilus, Thallomys paedulcus, Praomys natalensis, Aethomys namaquensis, Rhabdomys pumilio, Mus minutoides, Dasymys incomtus, Lemniscomys griselda, Saccostomus campestris, Steatomys* spp., *Dendromys mesomelas, D. melanotis, Otomys laminatus, O. irroratus, O. saundersiae, O. sloggetti, Mystromys albicaudatus, Tatera brantsi, and T. leucogaster.*

Abundant remains of any of these species in microvertebrate accumulations would indicate a habitat typical of one or other of the two biotic zones. Owl pellet collections from the Kalahari National Park (e.g., Davis 1958; Nel and Nolte 1965) or Namib (e.g., Brain 1974*b*) that are dominated by *Gerbillurus vallinus/paeba* remains are, in fact, indicative of true arid zone habitats, whereas those from the very numerous Transvaal localities, often dominated by *Praomys natalensis* bones, clearly indicate savanna zone habitats.

The differences between arid and savanna zone habitats are often extreme, whereas paleoecological reconstructions generally aim at more subtle habitat characterization. Within the southern savanna zone are habitats as diverse as open grassland and woodland; it would be interesting to know if the species composition of microvertebrate accumulations could be used as indicators of such vegetation types. To test such a possibility, comparable collections of barn owl pellets have been analyzed from an open grassland and a savanna woodland habitat.

Fig. 135. Mirabib Rock Shelter in the Namib desert, with the owl roost indicated by an arrow. The floor deposit contains abundant remains derived from owl pellets.

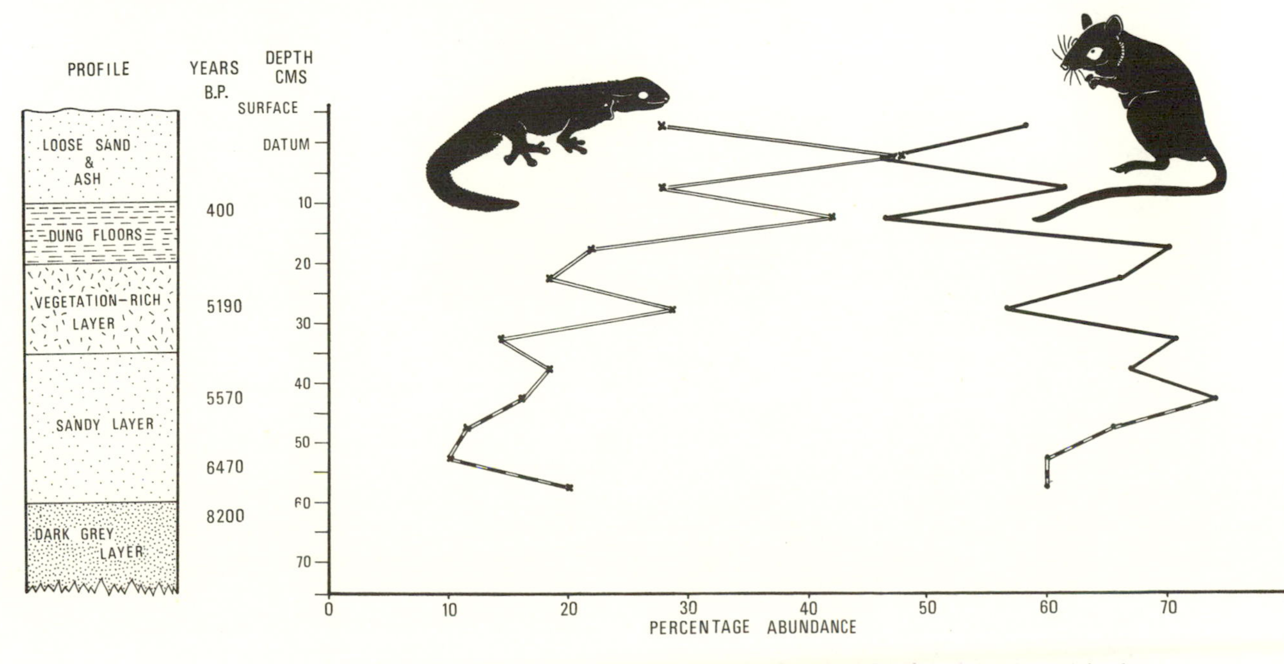

Fig. 136. Profile of the Mirabib floor deposit with percentage abundance of gerbils and geckos plotted for each level.

The roost in the open grassland habitat is one on Bolt's Farm referred to previously. It is 1.5 km southwest of Swartkrans, as shown in figure 137. The whole of the Sterkfontein valley is within the grassveld type designated as "Bankenveld" by Acocks (1953) (fig. 139). He states that this rather open habitat results from severely frosty winters, combined with regular burning. Although the main veld type is described as a "particularly sour, wiry grassveld, virtually ungrazable in winter," the rocky hills carry woody vegetation dominated by *Protea caffra, Acacia caffra,* and *Celtis kraussiana*. Sheltered valleys or sinkholes retain remnants of temperate or transitional forest.

The selected savanna zone owl roost was in the collapsed cone area of the Makapansgat Limeworks, near Potgietersrust. The cave site is in the veld type designated "sourish mixed bushveld" by Acocks. This is described as a rather open savanna with *Acacia caffra* the dominant tree, in a fairly tall and dense grassveld dominated by *Cymbopogon plurinodis, Themeda triandra, Elyonurus argenteus,* and *Hyparrhenia* spp.

Circles with scale equivalents of the 2 km radius were drawn on enlarged aerial photographs of the countryside surrounding the Bolt's Farm/Swartkrans and Makapansgat localities. In each case the center of the circle was taken as the position of the owl roost; each

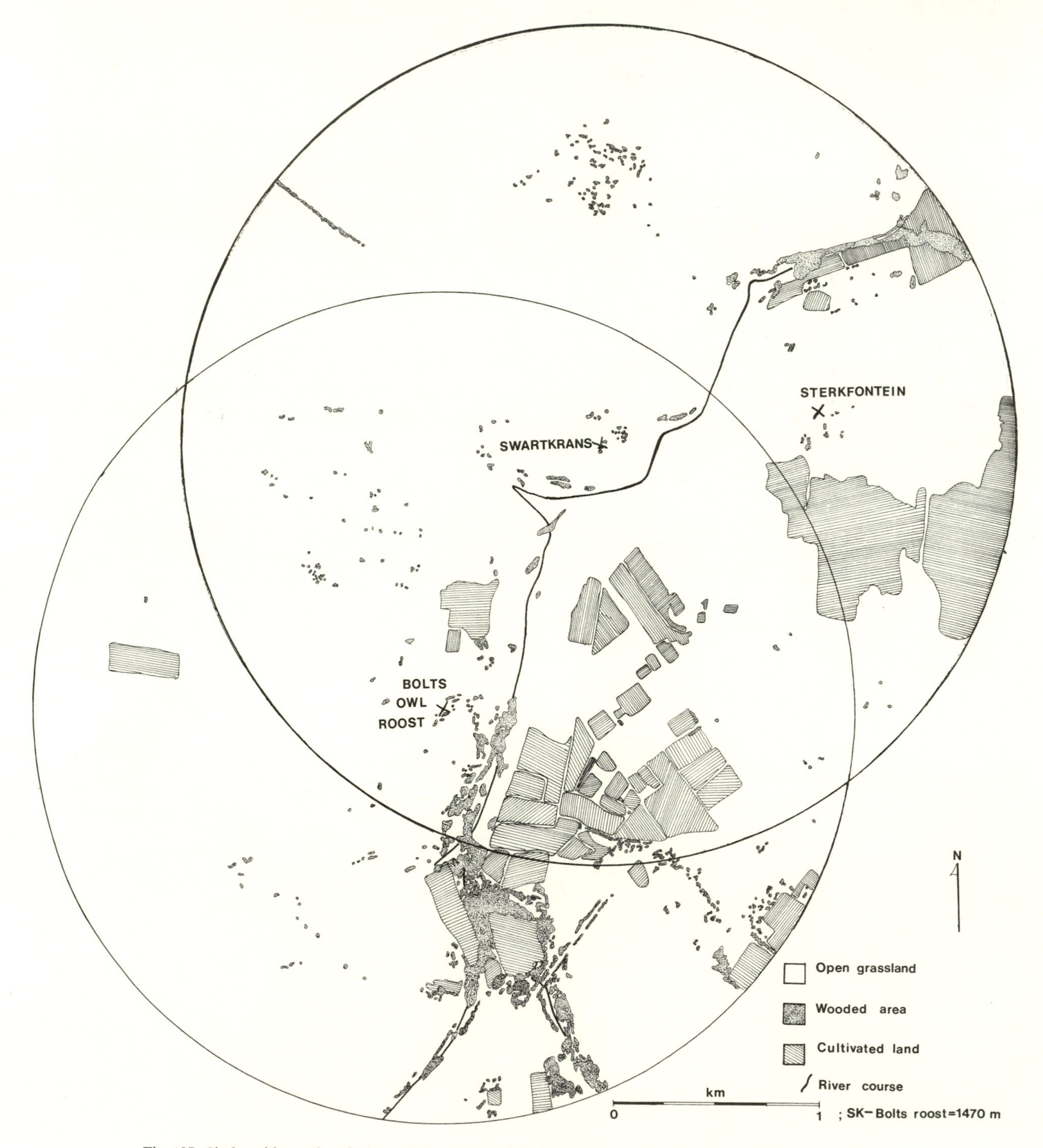

Fig. 137. Circles with a radius of 2 km each drawn around the Swartkrans and Bolt's Farm owl roosts. As is indicated in table 72, 88% of the area of the circles is occupied by open grassland.

circle enclosed an area of 12.6 km^2, which would certainly have covered the core area of the barn owls' hunting territory, if not its entire extent.

Tracings from the two aerial photographs depicting cultivated fields, wooded areas (including patches of exotic vegetation, but excluding the mixed bushveld, of Makapansgat), open grassland, mixed bushveld, and evergreen forest were prepared (figs. 137 and 138). From these the areas occupied by each vegetation type were calculated. The figures are given in table 72.

The Bolt's Farm/Swartkrans habitat is dominated by *open grassland,* occupying 88.3% of the total area. By contrast, 90.5% of the Makapansgat area is taken up with *sourish mixed bushveld*—a rather open savanna. Cultivated fields, largely planted in mealies, are present to about the same extent in both circles. About 1% to 2% of the Bolt's Farm/Swartkrans habitat is wooded, some of this covered by exotic trees, while 1% of the Makapansgat area is taken up by evergreen forest in the upper reaches of the valley.

As the two photographs show (figs. 139 and 140), the general appearance of the two areas is very different. The Sterkfontein valley countryside is essentially open and treeless, while that of the Makapansgat valley is comparatively well wooded. Despite this obvious difference, the species composition of barn owl prey from the two areas (table 71) is remarkably similar, and it would not be possible, by species composition, to tell which sample came from which habitat. Both samples (see fig. 141) are dominated by multimammate mice, *Praomys natalensis,* with lesser proportions of vlei rats *(Otomys),* gerbils *(Tatera),* and the other taxa shown in the figure. The remarkable similarity of the two samples could mean that barn owls may select similar "microhabitats" in which to hunt, despite the differences in "grosshabitat" appearances. It is extremely likely that there *are* some microvertebrate species that occur in the Makapansgat habitat but not at Bolt's Farm, but the owls (in these limited samples at least) have failed to catch them. For instance, tree rats of the genus *Thallomys* are not uncommon in the *Acacia* trees of the Makapansgat valley—in fact, they would be good indicators of thorn trees—yet they escape predation by barn owls. Likewise, the white-tailed rat, *Mystromys albicaudatus,* may be regarded as a good indicator of the

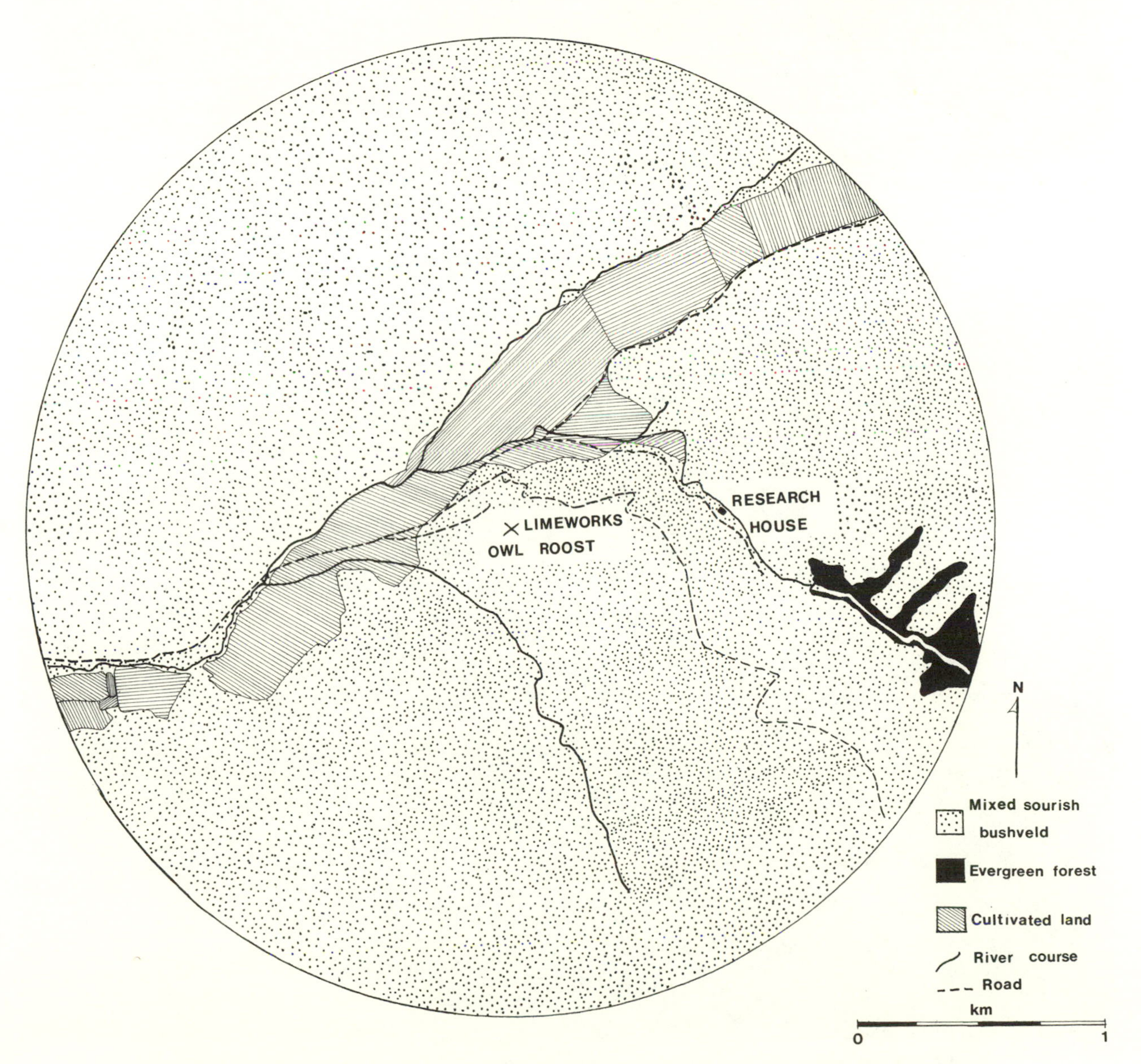

Fig. 138. A circle with a radius of 2 km drawn around the Makapansgat Limeworks owl roost. As is detailed in table 72, 90% of the area is occupied by "sourish mixed bushveld."

Fig. 139. A view of the Sterkfontein valley habitat, looking east from Swartkrans. The habitat is predominantly an open grassland.

Fig. 140. The Makapansgat valley immediately above the Limeworks. The dolomite hillsides are covered with mixed sourish bushveld.

open grasslands in the Sterkfontein valley habitat, but it is currently so rare there that the barn owls often fail to catch it.

There is clearly a need, and a good deal of scope, for further study on the habitat preferences of small southern African mammals and other animals commonly found in microvertebrate accumulations. Several small mammals have clear requirements—knowledge of such requirements could be very useful in reconstructing past habitats. An example is the shrew genus *Myosorex*, in which both extant southern African species, *M. varius* and *M. caffer*, require moist and densely vegetated microhabitats (Brain and Meester 1964).

The Presence of Barn Owl Roosts as Indicators of Past Cave Form

A large collection of owl-pellet-derived microfauna in a dolomite cave clearly implies that a suitable roost site once existed within the catchment area of the cave entrance. If any consistent features of roost sites could be found, these would be of value in reconstructing cave form where the relevant sturctures of ancient caves have been destroyed by erosion.

Roost sites selected by barn owls are extremely variable. They may be church spires, lofts, disused nests of the hamerkop *(Scopus umbretta)*, caves, or even gutter pipes. And yet, despite this diversity of chosen sites, the favored locations of roosts in natural dolomite caves are often rather similar. The roost is very often on a ledge beneath an overhang in the twilight zone of the cave. Subdued light is definitely favored, provided the owls have reasonably direct access to the cave entrance.

In the course of this study, pellet accumulations beneath three roosts in natural dolomite caves have been investigated. These were at Swartkrans, Bolt's Farm, and Makapansgat Limeworks. Sectional profiles have been drawn through each cave at the position of the relevant roosts; these are shown in figure 142. Despite differences in the size of the caves, selected roost positions have a basic similarity.

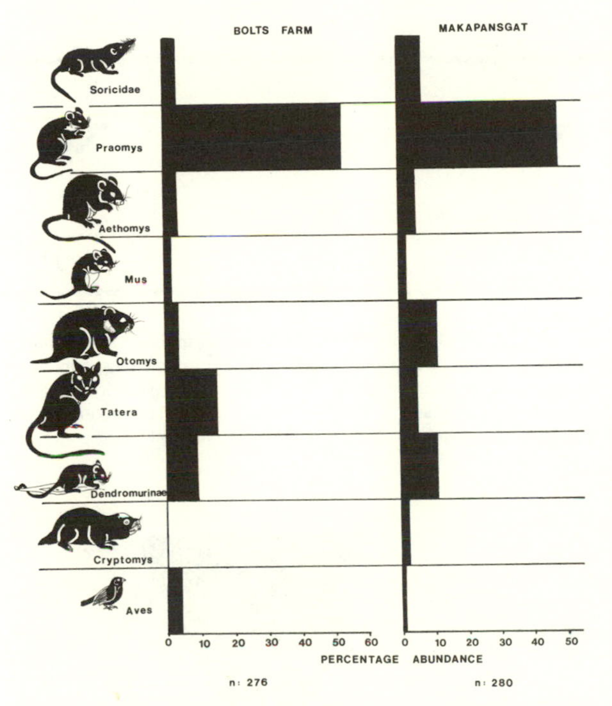

Fig. 141. Percentage representation of various animals in the prey of barn owls hunting from the Bolt's Farm and Makapansgat Limeworks owl roosts. Despite the difference in habitat of the two areas, prey-selection is remarkably similar.

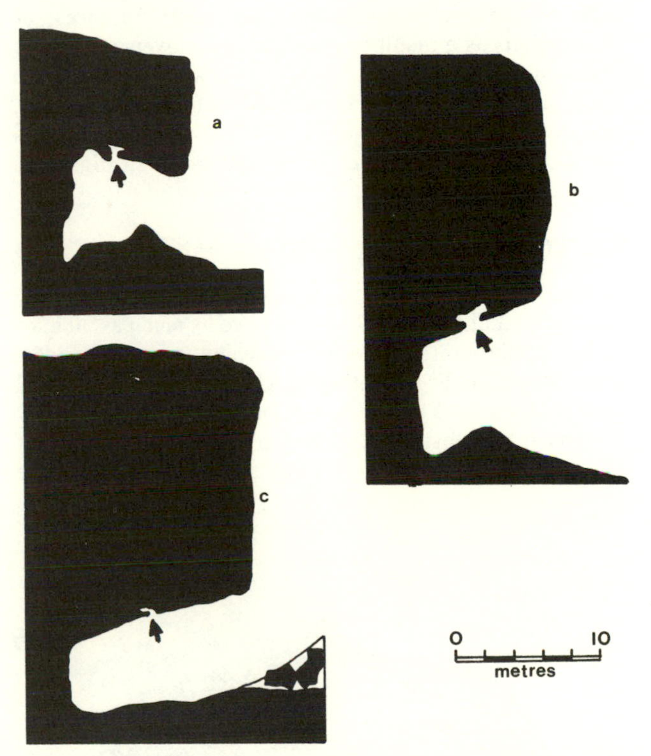

Fig. 142. Sectional profiles through three caves where barn owl roosts have been studied. Arrows indicate the roosts: *(a)* Swartkrans Inner Cave; *(b)* Bolt's Farm; and *(c)* Makapansgat Limeworks.

7 Some Compressional Effects on Bones Preserved in Cave Breccia

Although the preservation of bones in cave breccias may be extremely good, fossils are often found that have suffered the most remarkable distortions, fractures, or dislocations as a result of pressure and movements in the sediment. It may not be easy to separate such effects from injuries the bone sustained before fossilization, particularly when the fossil has been freed from its matrix and can no longer be placed in its stratigraphic context.

The Effects of Pressure

A fossil buried close to the bottom of a deep cave filling such as that at Sterkfontein may have as much as 30 m of sediment above it, the weight of which is very considerable. But, in addition to this weight, the overburden load on a fossil buried within a dolomitic cave may be vastly increased by subsidence of the cave roof. Denudation of the hillside above a cavern results in the breakup of the

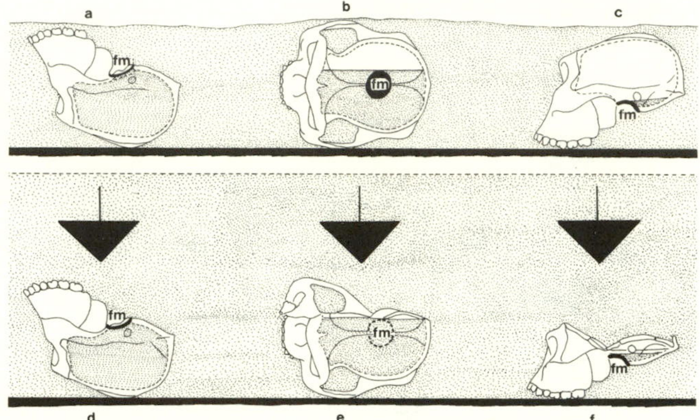

Fig. 143. Compressional effects on skulls buried in different positions in cave earth: *(a)* foramen magnum uppermost; *(b)* foramen magnum to one side; *(c)* foramen magnum below. When overburden pressure is applied, the filled skull *(d)* retains its shape; the partly filled cranium *(e)* collapses partially; and the empty one *(f)* is completely flattened.

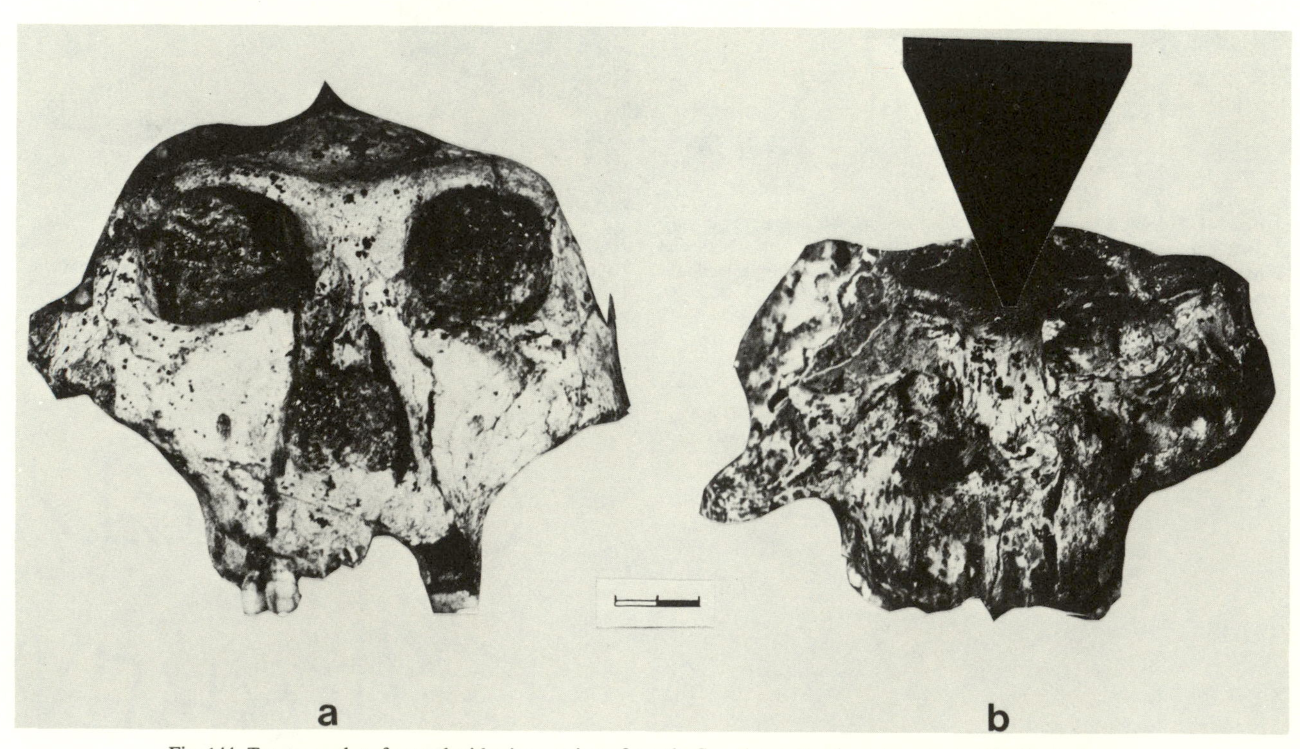

Fig. 144. Two examples of australopithecine craniums from the Swartkrans Member 1 breccia. *(a)* The filled calvaria of SK 48 has suffered very little distortion. *(b)* The almost empty cranial vault of SK 79 has collapsed completely under the weight of the overburden.

dolomite spanning such a cavern. Roof blocks collapse or settle onto the fossiliferous sediment, introducing compressional stresses of enormous magnitude.

The ability of buried bones to withstand such pressures will depend on the structural characteristics of the bones and the degree to which their cavities are infiltrated by sediment. I will restrict my comments here to craniums, although effects on other skeletal parts may readily be visualized.

From the taphonomic point of view, a skull consists essentially of a hollow, subspherical capsule whose walls will collapse inward if sufficient pressure is applied to them. The capsule, or calvaria, is breached by a number of orifices, of which the foramen magnum is the largest and, for us, most significant.

Skulls embedded in cave breccia have generally been covered by sediment brought into the cavern through the combined action of stormwater and gravity. Whether the hollow capsule of the calvaria will fill with sediment depends on the position in which the skull comes to rest on the cave floor. If it is upside down (fig. 143*a*), the foramen magnum is uppermost and total filling of the endocranial space is likely; if it is on its side, the space will fill to the level of the foramen magnum's upper edge (fig. 143*b*), and if it lies right way up, with the foramen magnum beneath, the endocranial space typically remains empty (fig. 143*c*), except for a probable frosting of calcite crystals. When overburden pressure is applied to these skulls, the degree of compressional collapse will be related to the amount of sediment filling. The totally filled calvaria will distort but little under pressure (fig. 143*d*); the half-filled skull may suffer collapse of the unfilled portion of its vault (fig. 143*e*), and the empty calvaria will probably collapse completely.

Figure 144 shows two examples of australopithecine skulls that demonstrate the extremes of the process. The endocranial space of the well-known cranium SK 48 from Swartkrans (fig. 144*a*) had been almost completely filled

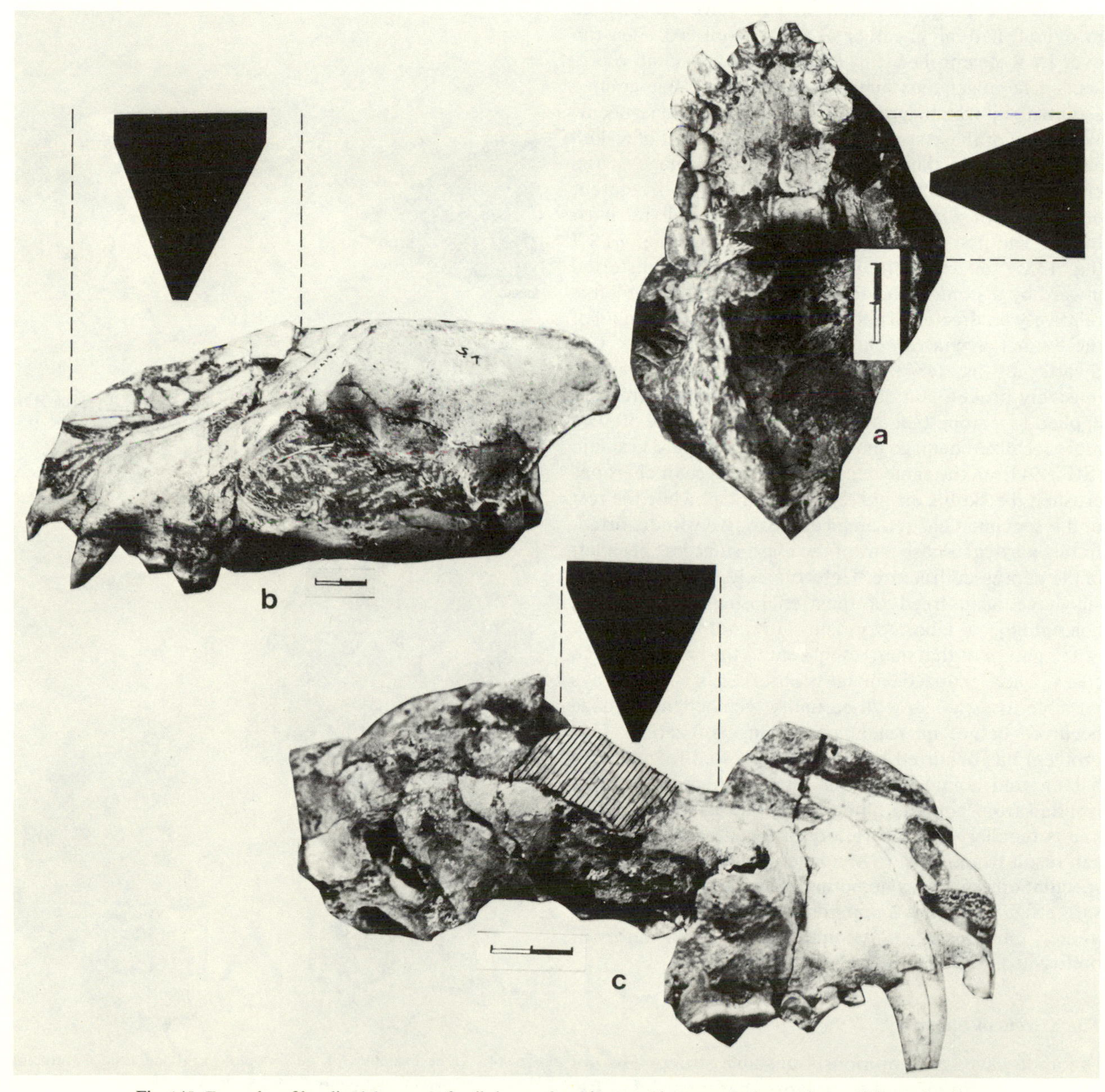

Fig. 145. Examples of localized damage to fossils in cave breccia caused by stones in the matrix to which overburden pressure has been applied. *(a) Australopithecus* mandible (STS 7) showing localized damage to right corpus only. *(b) Hyaenictis* skull (SK 314) with damage restricted to its snout. *(c) Panthera* skull (SK 349) with part of a stone *(stripes)* still embedded in its crushed orbital region.

with sediment and, although the specimen was clearly subjected to pressure, retained its shape except for some collapse in the occipital area. By contrast, the endocranial space of SK 79 (fig. 144*b*) remained empty and collapsed completely under pressure from above.

It occasionally happens that a cranium survives intact and undistorted throughout the burial and fossilization process. Such an example is the well-known skull of "Mrs. Ples" (STS 5) from Sterkfontein Member 4, whose endocranial cavity remained empty except for a calcitic encrustation. It happened that this skull had originally come to rest in a part of the Sterkfontein cave close to the back wall, where the sediment surface was almost in contact with the underside of the roof (see fig. 157). The result was that the specimen was not subjected to overburden pressure or the effects of roof collapse. Fossils preserved in the same matrix but a few meters away from the back wall show very marked effects of pressure, probably induced by subsidence of roof blocks.

The effects I have mentioned are to be expected in a reasonably homogeneous matrix. They have typically led to overall flattening, rather as if the skull had been run over by a steamroller. But a cave breccia seldom represents a homogeneous matrix—rather, it is a fine-grained sediment in which large objects such as stones occur. When one of these stones is in contact with part of a skull and pressure is then applied, localized damage is the characteristic result. Part of the skull may be fractured, depressed, or distorted. Three examples will be illustrated. The first is an *Australopithecus* mandible, STS 7 (fig. 145*a*), in which the right ramus has been distorted inward by a stone in the adjacent matrix, while the left is relatively unaffected. The second example is a cranium of the extinct hyena *Hyaenictis forfex* from Member 1 at Swartkrans (fig. 145*b*), in which the muzzle has been extensively broken and depressed by overburden pressure applied to a stone that lay against it. An example of even more localized damage may be seen on a leopard cranium (SK 349) from the same deposit. Here a piece of chert has crushed the skull's orbital region (fig. 145*c*) while the rest of the specimen has remained comparatively undistorted. In this particular case part of the chert stone has been left in the depressed fracture. Unfortunately, many other fossils have been freed of their enclosing matrix in the paleontological laboratory, and no record has been kept of the nature of that matrix adjacent to the bones. In such cases, when localized damage is observed, it is not always possible to establish with certainty whether the damage occurred before interment or during fossilization. This problem has occurred relative to depressed fractures on baboon and hominid craniums that Dart speculated had resulted from hominid bludgeoning (see chap. 10). Until one is familiar with the variety of localized damage that can result from pressure in a heterogeneous matrix, such speculations are very tempting. Having observed pressure effects on many hundreds of fossils, I am now extremely cautious when attempting to isolate instances of prefossilization bone injury.

The Effects of Shearing

Dolomite caves are notoriously unstable structures—not only do their roofs tend to collapse, but their floors frequently drop out as other cavities develop below them. When the latter happens with a filled cave, the filling itself becomes unstable and changes its shape to accommodate the new situation. A common result of such shape adjustments is shearing deformation of fossils that happen to be enclosed in the matrix. Figure 146 shows an example of a baboon skull, *Dinopithecus ingens,* SK 603, from Swartkrans Member 1, that has been sheared by forces acting in the direction of the arrows.

Some Effects on Rotted Bone

Fossil bones are found in the Sterkfontein valley caves in various states of weathering, indicating that they lay around for various lengths of time before they were interred and calcified. But, in addition to such weathering effects, it is not unusual to find fossils in which the structure of the bone has broken up into discrete spicules that have lost their original orientation (fig. 147). For some time I was at a loss to explain the circumstances in which such drastic effects might occur until, by chance, I happened to encounter a modern baboon skull lying in a perennial pool deep in the subterranean portion of Peppercorn's Cave at

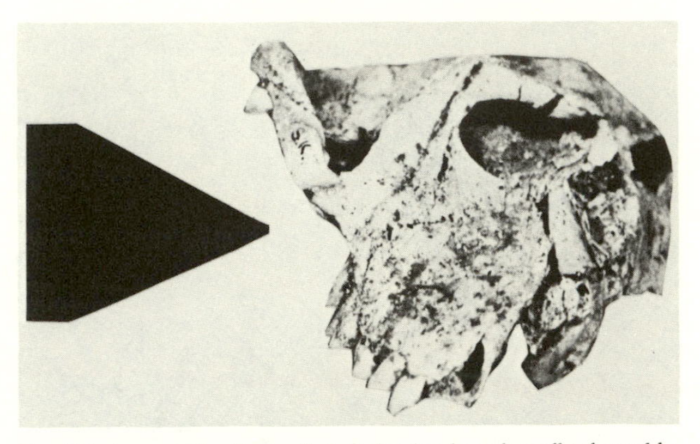

Fig. 146. An example of a fossil skull that has been laterally sheared by movements in the enclosing matrix: a baboon, *Dinopithecus* (SK 603) from Swartkrans Member 1.

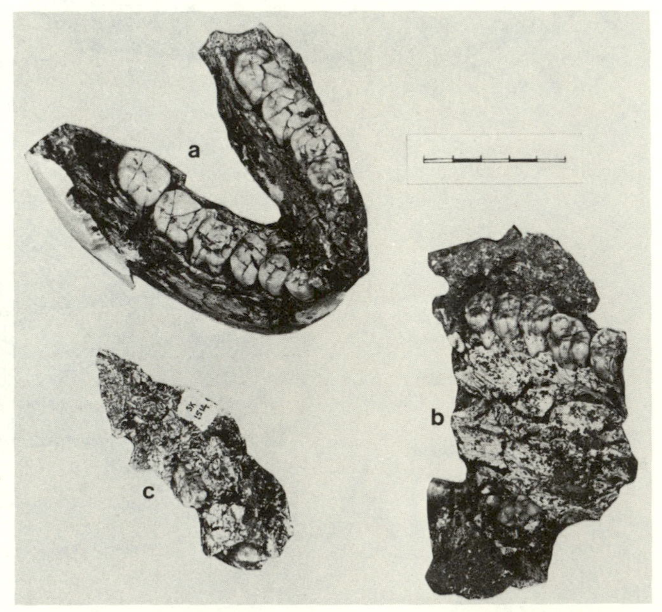

Fig. 147. Three examples of *Australopithecus* bone from Swartkrans Member 1 that had apparently been extensively rotted in water before fossilization. *(a)* Mandible (SK 876) showing cracking of the teeth and disintegration of the bone. *(b)* Similar effects on a palate (SK 57). *(c)* Disintegration of the bone structure shown by a mandibular ramus (SK 1514).

Makapansgat. As I attempted to lift the skull out of the water it simply collapsed in my hands, and I was left with a white malleable sludge that a few moments before had constituted a perfectly recognizable baboon skull. The teeth had split up into sharp-edged segments, and the entire structure of the bone had disintegrated.

It is not difficult to imagine the effects of overburden pressure on specimens in this condition. Resultant fossils are formless sheets of bone whose anatomical features have largely disappeared. They are not uncommon in parts of the Swartkrans Member 1 deposit, and I have also encountered them in Sterkfontein Member 4.

8 Summary
Bone Accumulations in Southern African Caves: A Search for Interpretative Criteria

The aim of this investigation, which I have discussed in the last five chapters, was to provide insights into the interpretation of bone accumulations in caves—to decide how the bones found their way to these places and to reconstruct something of the behavior of the animals involved in their accumulation. Certain guidelines have emerged.

Separation of an Accumulation into Micro- and Macrovertebrate Components

The difference between the bones classified in these two categories is clearly one of size—body size of the animals that contributed their remains. It is conceivable that abundant bones of microvertebrates, such as insectivores, small rodents, birds, and reptiles should enter caves in a variety of ways; yet experience of contemporary southern African situations has shown that, almost invariably, such bones represent the prey of owls. In comparison with other agents responsible for collecting bones of this size, owls are supremely significant. For this reason it is perhaps justifiable, when speaking of the microvertebrate component of a bone accumulation, to imply the owl-collected component.

The Owl Species Involved

The various species of southern African owls likely to contribute to bone accumulations in contemporary southern African caves are discussed in chapter 6. They are the barn owl and three species of eagle owl; presumably the antecedents of these species behaved like their modern counterparts, though I have no good evidence for this assumption except that fossilized food remains very similar to modern equivalents have been found.

Recognition of Owl Species on the Basis of Their Food Remains

Some indications have emerged whereby the species of owl responsible for a particular bone accumulation may be recognized. No fewer than eighteen published studies on the composition of barn-owl pellet collections from seventy-seven localities in southern Africa are available (table 68). A wide range of animals is eaten, but prey with body weights of 20–40 g appears to be preferred (table 70; fig. 134).

What little evidence there is suggests that it would not be easy to distinguish food remains of barn owls from those of spotted eagle owls if the birds had been hunting over the same range. Prey selection by the two species in the vicinity of Swartkrans has proved remarkably similar (table 66; fig. 125).

Although superficially similar to the spotted eagle owl, the Cape eagle owl has far more powerful feet and takes much larger prey. As is documented in table 67 and figure 128, food remains studied in Zimbabwe came principally from hares, dassies, cane rats, and other mammals such as springhare, mongoose, genet, and juvenile civet. Strictly speaking, such prey would fall into the macro- rather than the microvertebrate component. In East Africa, *B. capensis* showed a strong preference for nesting and roosting close to water and for feeding on freshwater crabs, whose remains accumulated around such sites. The Uitkomst roost in the Transvaal was similarly placed close to water, and freshwater crabs featured in the diet of the Cape eagle owls there as well.

Comparatively large mammalian prey also characterizes the food remains of giant eagle owls, although these birds have not yet been observed to roost *inside* caves. It appears that hedgehogs are much favored prey of this species, and their remains have been found in all collections of giant eagle owl food debris studied to date. I would go so far as to infer that abundant hedgehog remains in a cave are suggestive of *B. lacteus* or *B. capensis* involvement.

Although barn owls roost in a variety of situations, it seems that places in caves selected as long-term roosts and nest sites have some consistent characteristics (fig. 142). The site is usually on a ledge beneath an overhanging roof in subdued light. This being so, large concentrations of barn-owl-like food remains, fossilized in a cave breccia, suggest that the cave originally had ledges beneath an overhanging roof and that that particular part of the cave was within the twilight zone.

Evaluation of the Macrovertebrate Component

This part of a bone accumulation typically consists of remains from larger animals that found their way to the cave in a variety of ways. Elisabeth Vrba (1976*b*) has proposed a valuable scheme for classifying agents responsible for accumulating bones in caves of the Sterkfontein valley type. According to the scheme, biological bone-transport agents may be divided into two groups: *autopod* and *allopod*. Autopod agents involve transport on the animals' own feet; allopod agents, on the feet of others.

Autopod Transport Mechanisms

Animals come to the cave to seek shelter and die there, or they fall accidentally into a precipitous cave entrance and kill themselves. Their articulated, or disarticulated, skeletons may become buried and fossilized, but characteristically the skeletal parts may not have been affected by carnivore action. In the Sterkfontein valley context, such remains are rare and insignificant.

Allopod Transport Agents

With these agents, bones are transported to the cave or its vicinity through the activity of other animals; there are two distinct patterns involving *primary agents*—those predators that bring their own kills for consumption in the cave, and *secondary agents*—those animals that transport bones of animals they did not kill themselves—the scavengers and collectors. Primary and secondary bone-transport agents are likely to be represented mainly by meat-eaters, though vegetarian porcupines are undoubtedly the most important "collectors" of bones in southern African caves. Within the meat-eater category, separation between *predators* and *scavengers* delineates the *primary* from the *secondary* agents. Regrettably, such a delineation proves fuzzy in some cases, since carnivores are invariably opportunists. On one day a hyena or a hominid is a hunter—a primary predator—but on the next it scavenges and may then function as a secondary bone-accumulating agent.

Separating remains derived from primary predation from those of scavenging is not likely to be easy. Vrba (1975, 1976*a,b*) has pointed to the two criteria that, in the bovid context, are very significant. Primary predators tend to prefer prey within a restricted weight bracket (see fig. 97); second, juvenile prey tends to be consumed more completely than are adults of the same species; for this reason, juvenile representation in scavenged samples characteristically tends to be low. The application of these criteria to the various Sterkfontein valley site units will be considered in subsequent chapters.

The first chapters of this book have been devoted to a consideration of the various agents, primary and secondary, responsible for bone accumulations in African caves. I will now summarize some of the criteria whereby the agents may be recognized.

Survival and Disappearance of Skeletal Parts

As was discussed in chapter 2, the individual components of a vertebrate skeleton vary a great deal in their ability to withstand destructive treatment of whatever nature. It is now possible to sort skeletal parts, at least those of bovids, into groups according to their robusticity and ability to resist destruction. In limb bones, ability to withstand carnivore action is closely related to the compactness of the bony part and may be reflected in the bone's specific gravity; it is also affected by the time at which the epiphysis of a particular extremity fuses to its shaft (table 7; fig. 16).

Another result of the investigations described in chapter 2 is that each part of a bovid skeleton may be allocated a *potential-survival rating*, which will indicate how likely

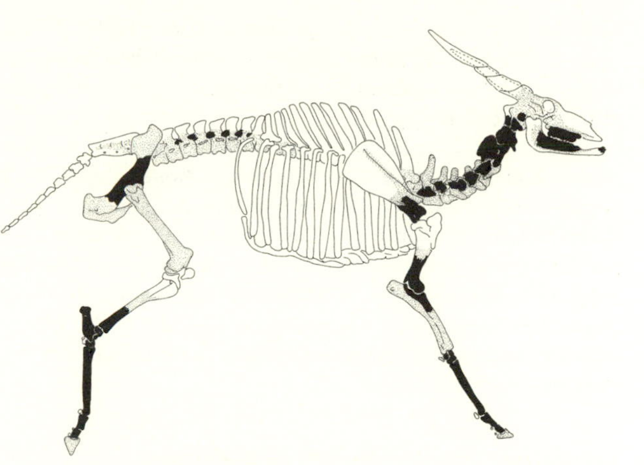

Fig. 148. Diagram of a bovid skeleton in which the various parts have been allocated "potential survival ratings." Parts shown in black have high ratings; stippled elements have intermediate ratings; and unshaded bones have low survival potential.

it is to survive a particular destructive regime. As is shown in figure 148, each part of an antelope skeleton may be placed in one of three categories, having a high, medium, or low potential survival rating. A bone assemblage composed essentially of elements with high survival ratings clearly has been subjected to a good deal of destructive treatment. On the other hand, the skeleton of an animal that has been preserved without carnivore intervention will also retain fragile elements with low survival ratings.

It is not only individual skeletal parts that vary in robusticity and survival potential. As was indicated in chapter 2, skeletons of primates, and probably also of carnivores, are far more susceptible to destruction than are antelope skeletons derived from animals of equivalent live body weight.

Recognizing Collections Made by Porcupines

As was discussed in chapter 5, African porcupines are extremely important collectors of bones in caves. Being vegetarians, porcupines do not have the same interest in bones as carnivores do, yet they collect large numbers of bones and other objects in their lairs and gnaw them at their leisure. As was shown in figure 112, porcupine incisors in the upper and lower jaws occlude against one another, ensuring that sharp chisellike edges are constantly maintained. Nevertheless, unless the porcupines have hard objects to gnaw, the incisors appear to grow unmanageably long (as they frequently do in zoo animals), preventing proper occlusion and sharpening of the edges. Gnawing bones may also provide the porcupines with mineral salts they require. It is unlikely that calcium is the objective of bone-gnawing, since porcupines in the Kalahari, for example, collect and gnaw large numbers of bones despite the highly calcareous nature of the soil and country rock. It is more likely that they are looking for phosphorus, which is generally the motive for bone-chewing by ungulates. It is known that high calcium content in the soil can reduce the availability of phosphorus to plants, and thus to herbivores. The phosphorus deficiency may then be made up by osteophagia—chewing bones (Sutcliffe 1973*b*).

Whatever motive porcupines have for collecting and gnawing bones, that they do these things cannot be disputed. Although porcupine collections have a number of consistent characteristics, as outlined in chapter 5, the surest indicator of porcupine involvement in a bone accumulation is the presence of gnawmarks, caused by the animals' incisors (fig. 113).

Porcupines frequently collect more bones than they gnaw, and the percentage of gnawed bones depends, it seems, on availability. When suitably defatted and weathered bones are abundant in the vicinity of a lair, the percentage of gnawed bones in a porcupine collection may be as low as 20%. When bones are scarce, 100% of them may be gnawed. Gnawing on 60–70% of the pieces may be regarded as typical of bones from lairs in normal circumstances.

Recognizing Primitive Human Food Remains

The Nature of Primitive Human Diets

As I emphasized in chapter 3, the animal remains that concern us here represent only part of the primitive human diet; the more stable, and frequently larger, component was of plant origin. At this time, thirteen analyses of bone accumulations from southern African caves, representing human food remains, are available (table 33; fig. 48). These show that hunter-gatherers are opportunists, making use of whatever food resources happen to be at hand and often participating in a seasonal round to exploit them. The same was probably true for early hominids, and I seriously doubt that primitive human food remains could be positively identified as such on the basis of the kinds of animals that had been hunted. One point that has emerged from the analyses is that remains of *carnivores* are not abundant in the human food refuse. In fact, when *carnivore-ungulate* ratios are calculated for the thirteen accumulations, the mean value proves to be ten. By contrast, bone accumulations built up by certain carnivores themselves—as, for example, brown hyenas—tend to have carnivore-ungulate ratios far above this.

Association with Artifacts

Human food remains are frequently associated with artifacts, and the relative abundance of each kind of trace will determine how a particular site is classified (fig. 47). By the scheme outlined in chapter 3, caves containing human food refuse would usually be classified "camp or occupation sites." In such situations, increased numbers of bone fragments in a particular level are typically associated with more abundant stone artifacts (figs. 26 and 50).

Association with Traces of Fire

Indisputable traces of fire associated with bones in a cave generally point to human involvement. Natural fires caused by lightning in caves have been recorded, but they must surely be rare.

Apart from the presence of ash, charring of bones may be taken as an indicator of fire. As shown in figure 51, light charring blackens the bone; more prolonged incineration causes the bone to revert to a chalky color and consistency.

Bone Damage: Depressed Fractures

Dart (1949*a,b*) described a large series of baboon and australopithecine craniums that showed depressed fractures of their vaults. He attributed the fractures to purposeful hominid bludgeoning with clubs made of antelope humeri.

If it could be demonstrated beyond doubt that the injuries had been inflicted before the skulls were buried in the cave earth, then I would have no hesitation in agreeing that deliberate hominid activity was indicated. But it is extremely difficult to separate the effects of localized pressure within the enclosing matrix from prefossilization injury. Stones or other hard objects in the matrix can cause depressed fractures most suggestive of hominid violence (Brain 1972). Regrettably, most skulls available at present showing this particular kind of damage have been freed from their matrix, and so the controversy concerning them cannot be resolved (chap. 7).

Bone Damage: Shaft Fractures

As was discussed in chapter 1, a good deal has been written about the "crack and twist" technique for breaking antelope limb bones. Typically the technique results in spiral fractures of the bone shafts. If spiral fractures resulted *only* from the cracking and twisting of bones, a uniquely human procedure, then the presence of spirally fractured shafts would be a good indicator of human involvement. But, as is shown in figure 149, spirally fractured shafts can result not only from human bone-breaking, but also from the bone-cracking of spotted hyenas, brown hyenas, and leopards. In his paper "Some Operational Aspects of Human and Animal Bone Alteration," Bonnichsen (n.d.) has shown that a glass tube or cylinder will frequently break with a spiral fracture when struck a vertical blow. This observation implies that a tube without internal structure of any kind in its walls may produce spirally fractured pieces as a result of impact. Long-bone shafts, on the other hand, are not devoid of internal structure. Tappen (1969) has shown that

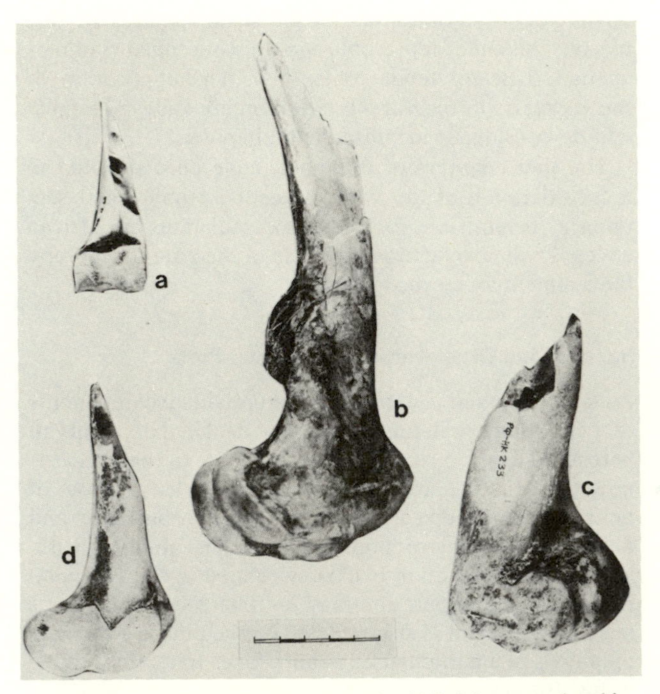

Fig. 149. Examples of spiral fractures on bovid limb bones produced by different agents: *(a)* human action: Namib Hottentots; *(b)* spotted hyena: Kruger National Park; *(c)* brown hyena: Kalahari National Park; *(d)* leopard: Kruger National Park.

superficial cracks that develop on weathered bones have orientations identical with those that may be produced by the artificial split-line technique. Such cracks and lines follow the main structural orientations in bones (Tappen 1970) and, when bones are subjected to impact, the fractures also tend to preferentially follow such orientations. The weathered giraffe humerus in figure 150, shows superficial cracking that follows the structural orientation of the bone in a spiral fashion. When such a humeral shaft fractures, through either stone impact or carnivore tooth pressure, the fracture line is very likely to follow the spiral path dictated by the structural orientation of the bone.

The appearance of spiral fractures on humeral and other limb bone shafts therefore tends to reflect more of the internal structure of the bone than the nature of the trauma the shaft has suffered.

Bone Flakes: Human-Made or Hyena-Made?

The shattering of long-bone shafts through either stone impact or carnivore tooth pressure typically produces bone flakes, as defined in chapter 1. For interpreting bone accumulations, it would be useful if bone flakes produced by the two agencies could be separated. As figure 151 demonstrates, flakes of the two kinds frequently both show impact points, with negative flake-scars on their undersides where the hammerstone blow, or tooth pressure, was applied. I am not confident in my ability to separate human-made from hyena-made bone flakes by shape or by structure.

It has not proved easy to make a large collection of bone flakes from hyena feeding sites or dens. This is not because spotted and brown hyenas do not produce abundant flakes, but because such flakes are rapidly gulped down by the hyenas or carried away by vultures that feed them to their young (Mundy and Ledger 1976; Plug 1978). To date I have been able to examine only 220 bone flakes produced by spotted hyenas and 34 made by brown hyenas in natural circumstances. The lengths of these flakes are given in tables 73 and 74. The impression given by these small samples of bone flakes resulting from hyena action is that many pieces are longer than those typically produced by late human bone-breakage. Much larger hyena-produced samples are required to allow comparison with the extensive human-activity samples described in chapter 3. For what the exercise is worth, length distributions of the 220 flakes produced by spotted hyenas are compared with those of the vastly more numerous flakes from Pomongwe Cave (fig. 152). Whereas most of the flakes from the human refuse have lengths of less than 5 cm, those resulting from hyena feeding have a fairly even length spread between 2 and 12 cm. These results suggest that the variability in hyena-made flake lengths is greater than that in the humanly produced flakes. It will be interesting to see if larger samples collected in the future will support this indication.

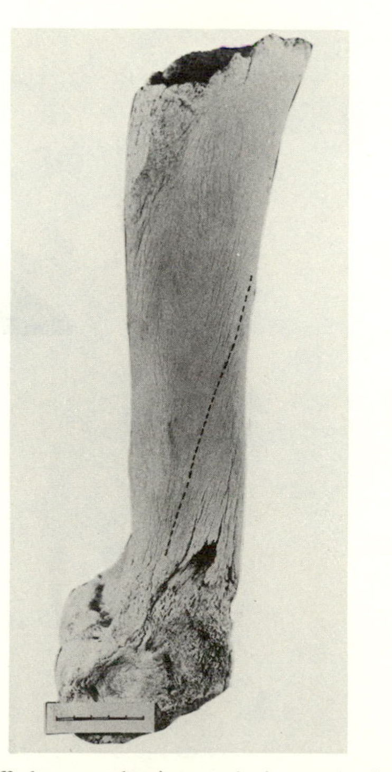

Fig. 150. A giraffe humerus showing weathering cracks that follow the main structural orientation in the bone in a spiral manner. Fracture of the shaft of such a bone is likely to follow a similar spiral course.

Problems of Wear and Polish on Bones

It is tempting to interpret wear and polish on bones as resulting from deliberate human use. Yet, as was explained in chapter 2, such an interpretation may be incorrect. Wear, both localized and general, as well as polish, can result from the abrasion of sand regularly disturbed by the feet of animals. The positive identification of wear or polish resulting from deliberate human use would depend on microscopic examination of scratches or other marks and the exclusion of natural agencies.

Recognizing Hyena Food Remains

The teeth of hyenas are specialized for the destruction of bones, and although the damage they cause may be basically similar to that inflicted by other carnivores, the degree of the damage is greater.

As was outlined in chapter 4, hyena action on bones falls into two clear categories: premolar cracking of manageable pieces and incisor/canine gnawing of larger, uncrackable pieces. The results of the two types of chewing are easily recognizable, as is shown in figures 61–64.

Specific Attributes of Brown Hyena Food Remains

It is probably not possible to separate food remains of spotted and brown hyenas on the basis of damage suffered by individual specimens. However, as was described in chapter 4, brown hyenas actively hunt large numbers of small mammals, particularly other carnivores, and feed the bodies to their cubs. The craniums of these small carnivores frequently remain intact. Bone assemblages in caves with high carnivore-ungulate ratios could very easily result from brown hyena cub rearing.

The Possible Role of Extinct Hyaenids

As was emphasized in chapter 4, the extinct hunting hyenas of the genera *Hyaenictis* and *Euryboas* are well represented among the fossils from the Transvaal caves. They may well have reared their young in these caves. It is likely that they were social, cursorial hunters, like wild dogs, and that they preyed on antelopes of all size classes. They may have contributed very significantly to the fossil bone accumulations.

Recognizing Leopard Food Remains

Evidence presented in chapter 4 makes it clear that contemporary leopards use caves both as breeding lairs and as feeding retreats. Food remains tend to accumulate in both types of site.

As was indicated in table 48 and figure 97, leopards prey on a very wide spectrum of animals but show a strong preference in all study areas for antelopes in size class II. Observations on kills alone, however, are inclined to provide a picture strongly biased toward larger prey (table 50; fig. 98). Analysis of scats provides data on the numerous smaller prey items not normally observed as kills.

Leopards are prone to develop strong preferences for certain types of prey and to seek these preferred animals with considerable perseverence. A preference for primates, either baboons or people, has been shown to develop in various areas throughout the leopard's geographic range.

Leopard Damage to Bones

Some information has come to light on characteristic damage caused by leopards to the skeletons of their prey. In dassies this is highly characteristic (fig. 93). Damage to impala-sized antelopes is equally diagnostic (figs. 103 and

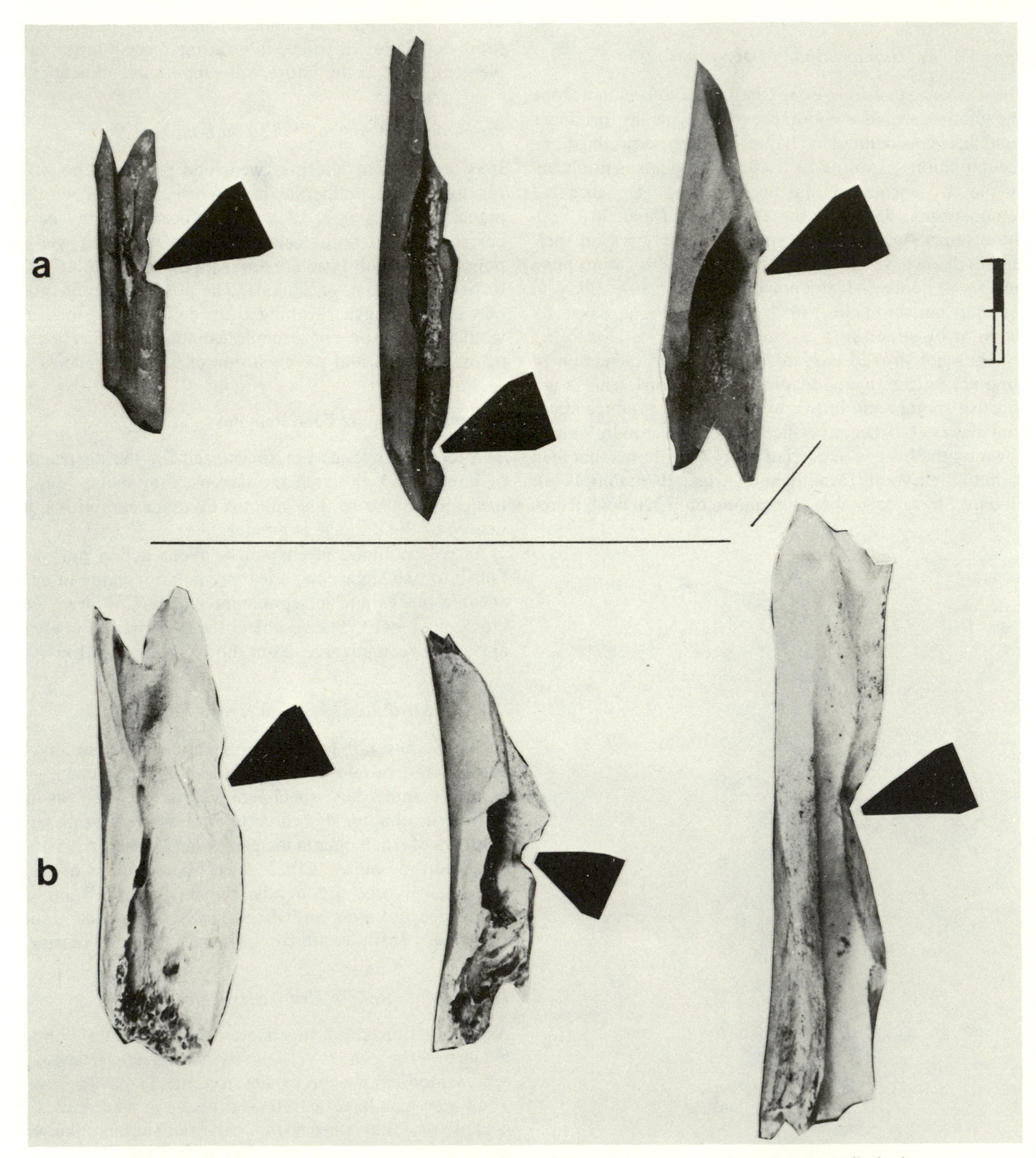

Fig. 151. Examples of bone flakes produced from bovid limb bones by *(a)* a stone chopper on an anvil; *(b)* the premolars of spotted hyenas in the Kruger National Park. Arrows indicate the points at which impact or pressure was applied to cause the fractures.

104), leading to the "eaten-out" appearance of the carcass in which only the head and lower limb segments survive. With primate prey the whole body, except the head, is prone to disappear.

The Possible Role of Extinct Cats

As is discussed in chapters 4 and 9, several extinct cats, particularly the false sabertooths, *Dinofelis*, and the true sabertooths, *Homotherium*, *Machairodus*, and *Megantereon*, appear to have frequented the Transvaal caves and may have contributed to bone accumulations there. *Dinofelis* is thought to have hunted and behaved as a very large leopard may have done, while the true saber-toothed cats would have been well suited to killing very large prey, but quite unsuited to utilizing their bones. True sabertooth food remains, therefore, should be characterized by minimal bone damage.

Matching Observed Damage with Specific Carnivore Action

It should be possible to match the kind of damage observed on bones in an assemblage with that characteristically done by a particular carnivore. Figure 153 shows three humeri whose proximal ends have been removed by carnivores. In each case the distal end of the bone proved to be beyond the chewing ability of the specific carnivore's dental battery, unless unnatural effort were expended. In example *c*, a spotted hyena has gnawed off the proximal end of a giraffe humerus but has abandoned the distal end; in *b* a leopard has done the same to an impala humerus, and example *a* shows a dassie humerus—one of many from the Pomongwe Cave—in which the proximal end was apparently chewed away by a Stone Age hunter, while the distal end was simply discarded into the ash of his fire. In fact, dassie remains prove to be excellent indicators of the carnivores that left them. As was described in chapter 4, leopards and cheetahs characteristically ingest almost the whole body, shearing off the back of the skull and mandible in an unmistakable manner (figs. 93 and 95). Black eagles typically damage the dassie craniums with their sharp beaks, creating equally unmistakable openings (fig. 109), and human dassie-eaters leave a scatter of unchewable skeletal parts (table 15) that conform to a predictable and consistent pattern.

The Incidence of Carnivore Tooth Marks on Bones

In collections of bones accumulated by carnivores it is usually possible to find certain pieces bearing clear evidence of the meat-eater's chewing action. Damage that characteristically results when a particular species of carnivore, such as a leopard, feeds on a specific prey animal, such as a dassie, has already been discussed. Typical results of hyena bone-cracking and bone-gnawing have also been mentioned.

The recognition of carnivore tooth marks on bones is a rather subjective procedure, and not all investigators would agree on the criteria to be considered. For this discussion I will restrict my observations to "tooth marks" in the form of *(a)* punctate marks and gouges caused by pointed teeth and *(b)* ragged-edge damage when a bone has been "worried" by the teeth of a carnivore and an irregular margin has resulted (see figs. 80 and 93).

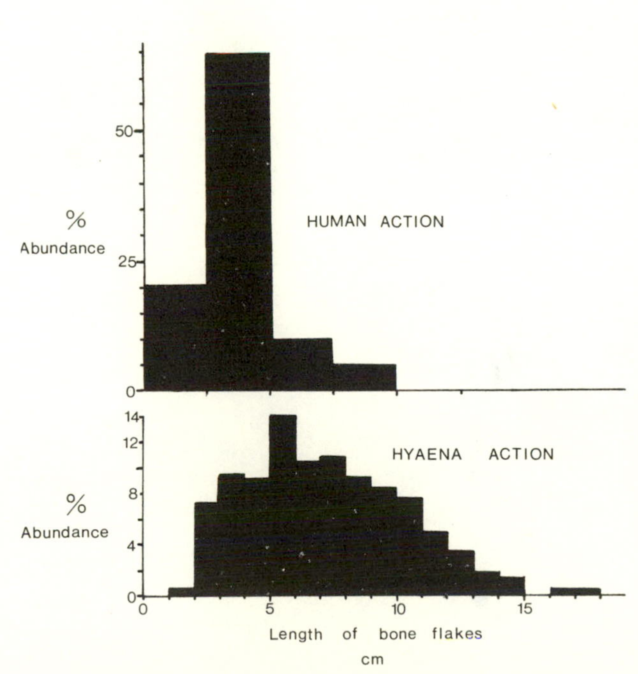

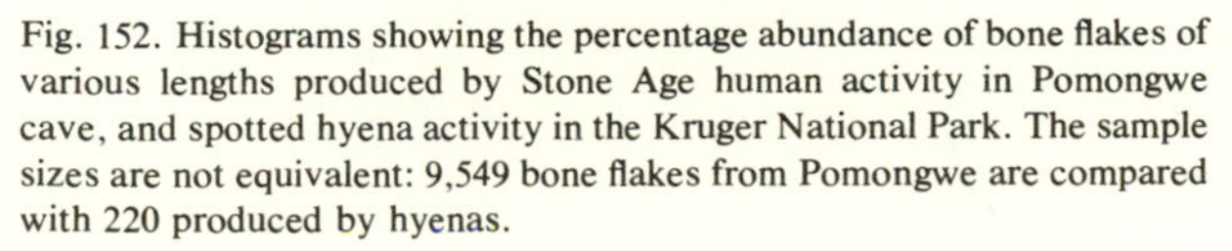

Fig. 152. Histograms showing the percentage abundance of bone flakes of various lengths produced by Stone Age human activity in Pomongwe cave, and spotted hyena activity in the Kruger National Park. The sample sizes are not equivalent: 9,549 bone flakes from Pomongwe are compared with 220 produced by hyenas.

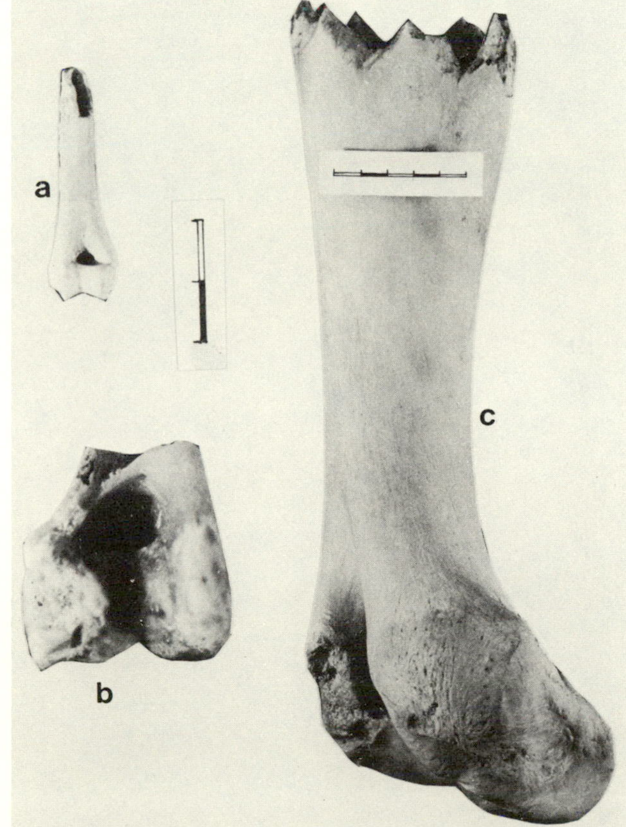

Fig. 153. The damage observed on bones may be matched to that characteristically done by particular carnivores. The three humeri shown here all had their proximal ends removed by carnivore action, and the scale of damage may be equated with the chewing ability of the particular carnivore concerned: *(a)* a dassie humerus damaged by human teeth; *(b)* an impala humerus chewed by a leopard; *(c)* a giraffe humerus gnawed by spotted hyenas.

It would be useful to know the incidence of bones showing these recognizable tooth marks in accumulations where carnivores are known to have been involved. Further studies are urgently needed, but some information may be derived from those presented in this book. Bones from leopard lairs will be considered first. Of the 176 pieces collected in the Suswa leopard feeding cave, 50, or 28.4%, showed clear tooth marks (table 42); 64 bones thought to represent leopard food remains were collected from the Quartzberg cave, and of these 15, or 23.4%, bore tooth marks (table 45). Percentages of tooth-marked bones from the two leopard breeding lairs were much lower, as is perhaps to be expected since leopards rarely bring food back to the places where cubs are being reared. In fact, most of the bones in the Portsmut and Hakos River breeding lairs (tables 43 and 44) were probably introduced by agents other than leopards. The Portsmut lair produced 192 bone pieces, of which 7, or 3.6%, showed tooth marks. Of the 333 bones from the Hakos River lair, only 7, or 2.1%, of the total were so marked.

The collection of 235 bones from brown hyena breeding dens in the Kalahari National Park (table 41) was found to contain 96 tooth-marked bones, implying that 40.8% of the total showed recognizable evidence of carnivore action.

A conclusion to be drawn from these observations is that the frequency of tooth-marked bones in a bone accumulation of known carnivore origin is likely to be surprisingly low. Fewer than 30% of bones left by leopards in feeding lairs may show tooth marks, and fewer than half of the bones transported, fed upon, and discarded by brown hyenas at their dens may bear such traces.

Part 2
Fossil Assemblages from the Sterkfontein Valley Caves: Analysis and Interpretation

9 The Fossil Animals

The remains of animals preserved in the Sterkfontein valley caves fall naturally into two groups according to how they found their way to the fossilization sites. One group of typically larger animals, referred to here as the "macrovertebrate component," contributed their skeletal remains in a variety of ways; the second, consisting of characteristically small animals termed the "microvertebrate component," almost certainly formed the prey of owls that roosted in the caves. These remains became fossilized from pellets that owls regurgitated on the cave floors. The composition of the macro- and microvertebrate components will be considered separately.

The Macrovertebrate Component

The fossils by which each species of animal is represented at the various cave sites and in the different site units will be considered in subsequent chapters devoted to specific localities. The purpose of this chapter is to provide some information about the animals whose remains are represented by the fossils. Some of the species have living representatives and are therefore familiar; others are extinct and, to a greater or lesser extent, unusual or bizarre.

As table 75 shows, the animals are mammals, birds, reptiles, or mollusks. Of the 67 genera represented, 24 are extinct, and 54 of the 91 species involved have no living representatives. The order of presentation in the table and text does not follow a phylogenetic scheme but reflects the taphonomic significance of the taxa involved. The provenance quoted refers to the Sterkfontein valley site units only.

Order Primates

Family Hominidae
Australopithecus africanus Dart
Provenance: Sterkfontein Member 4
A definition:

> A species of the genus *Australopithecus* characterised by the following features: more gracile, lighter construction of the cranium; calvaria hafted to facial skeleton at a high level, giving a distinct though not marked forehead and a high supraorbital height index; ectocranial superstructures and pneumatisation not as marked as in other species; sagittal crest commonly absent though probably present in some individuals; nuchal crest not present, but slight to moderate occipital torus commonly present; bony face of moderate height and varying from moderately flat and orthognathous to markedly prognathous; nasal region slightly elevated from facial plane; ramus of mandible of moderate height and sloping somewhat backward; jaws moderate in size with lesser development of zygomatic arch, lateral pterygoid plate, temporal crest and temporal fossa; palate of more or less even depth, shelving steeply in front of the incisive foramen; premolars and molars of moderate size and not so markedly expanded buccolingually; M^3 smaller than M^2 in mesiodistal diameters, but equal in buccolingual diameters; mandibular canine larger than in other species, and hence more in harmony with the postcanine teeth; degree of molarisation of lower first deciduous molar less complete; cingulum remnans or derivatives present on all maxillary molars, weak on buccal surfaces, pronounced on lingual, representing an earlier or more primitive stage in the trend towards reduction of the cingulum; sockets of anterior teeth arranged in a moderate to marked curve. [Tobias 1968*a*, pp. 293–94]

The first gracile australopithecine specimen to be described was the child skull from Taung that, in his *Nature* paper of 7 February 1925, R. A. Dart designated *A. africanus* in commemoration "first, of the extreme southern and unexpected horizon of its discovery, and secondly, of the continent in which so many new and important discoveries connected with the early history of man have recently been made, thus vindicating the Darwinian claim that Africa would prove to be the cradle of mankind." In his now classic paper, Dart continued (1925*a*, pp. 198–99):

> It will appear to many a remarkable fact that an ultrasimian and prehuman stock should be discovered, in the first place in this extreme southern point of Africa, and, secondly, in Bechuanaland, for one does not associate with the present climatic conditions obtaining on the eastern fringe of the Kalahari desert an environment favourable to higher primate life.... In anticipating the discovery of the true links between the apes and man in tropical countries, there has been a tendency to overlook the fact that, in the luxuriant forests of the tropical belts, Nature was supplying with profligate and lavish hand an easy and sluggish solution, by adaptive specialisation, of the problem of existence in creatures so well equipped mentally as living anthropoids are. For the production of man a different apprenticeship was needed to sharpen the

wits and quicken the higher manifestations of intellect—a more open veldt country where competition was keener between swiftness and stealth and where adroitness of thinking and movement played a preponderating role in the preservation of the species. Darwin has said, "no country in the world abounds in a greater degree with dangerous beasts than Southern Africa" and, in my opinion, Southern Africa, by providing a vast open country with occasional wooded

belts and a relative scarcity of water, together with a fierce and bitter mammalian competition, furnished a laboratory such as was essential to this penultimate phase of human evolution.

In Southern Africa, where climatic conditions appear to have fluctuated little since Cretaceous times and where ample dolomitic formations have provided innumerable refuges during life, and burial-places after death, for our troglodytic forefathers, we may confidently anticipate many complementary discoveries concerning this period in our evolution.

And, in fact, many complementary discoveries were indeed waiting to be made. The next gracile australopithecine specimen came to light eleven years later on 17 August 1936 in the Sterkfontein lime quarry, later known as the Type Site. It consisted of an endocranial cast and part of the skull, TM 1511 (fig. 163), of an adult that Broom tentatively assigned to the same genus as the Taung specimen but that appeared to belong to a new species (Broom 1936, p. 488). After the further discovery of a lower molar, Broom (1937*a*) proposed the name *Australopithecus transvaalensis,* but he created the new genus *Plesianthropus* after the discovery of the mandibular symphysis of a child, TM 1516, which appeared to differ in structure from that of the Taung fossil (Broom 1938*b*). Particulars of the first postcranial skeletal element of *P. transvaalensis* were published in the same year (Broom 1938*c*); the specimen consisted of a left distal femur piece designated TM 1513. Details of an os magnum followed (Broom 1941*a*).

Descriptions of all specimens available at the end of the war were published in two Transvaal Museum memoirs (Broom and Schepers 1946; Broom, Robinson, and Schepers 1950), and renewed Sterkfontein excavations in 1947 led to the finding of Skull 5, "Mrs. Ples," on 18 April (fig. 155). This specimen proved to be the most complete adult australopithecine cranium known. It was pictured in the *Illustrated London News* of 17 May 1947, and the first description appeared in an issue of *Nature* of the same date (Broom 1947). Further anatomical descriptions were published immediately thereafter (Broom and Robinson 1947*a*). A remarkably complete pelvis with articulated vertebral column, found on 1 August 1947 (Broom and Robinson 1947*b*), provided the first information about this region of the skeleton (fig. 165).

After paleontological reconnaissance work in the Makapansgat valley led by P. V. Tobias, the first gracile australopithecine specimen from the Makapansgat Limeworks was found by J. W. Kitching and his brothers in September 1947. It consisted of the occipital region of an adult and was followed the next year by A. R. Hughes's discovery of an adolescent mandible (Dart 1948*a*). Pelvic and other bones subsequently came to light.

The Makapansgat australopithecine was described by Dart (1948*b,c*) as *A. prometheus,* the specific name prompted by Dart's suspicion that blackened bones associated with the hominid remains had been deliberately burned. Dart also concluded that the Makapansgat australopithecines had been responsible for building up the remarkable bone accumulation in the Limeworks gray breccia, or Member 3 of the Makapansgat Formation (Brock, McFadden, and Partridge 1977), as well as for the "osteodontokeratic culture" (Dart 1957*b*). This issue will be touched on again in chapter 13.

Two hominid fossils found at Garusi in Tanzania in 1939 by Kohl-Larsen were described by Weinert (1950, 1951) as *Meganthropus africanus.* Robinson (1953*a*) reexamined the specimens and concluded that they had close affinities with *Plesianthropus* from Sterkfontein. They therefore represented the first indications of a gracile australopithecine outside South Africa.

The following year Robinson (1954) discussed the genera and species of the Australopithecinae in detail and concluded that all known gracile australopithecine specimens should be referred to a single species *Australopithecus africanus,* which would contain two subspecies: *A. a. africanus,* for the Taung fossil, and *A. a. transvaalensis,* for the material from Sterkfontein, Makapansgat, and Garusi. This scheme would involve sinking the genus *Plesianthropus,* the taxon *Meganthropus africanus,* and the species *prometheus.* Subsequently, Dart (1964*a*) agreed that the Makapansgat australopithecine should not be separated specifically from the Sterkfontein one. In the interim he had described remains from higher levels in the Makapansgat sequence: from Member 4 (Dart 1959*d*) and Member 5 (Dart 1955).

Renewed fieldwork at the Sterkfontein site (Tobias and Hughes 1969) has recovered further gracile australopithecine remains, including the fine cranium designated StW 12/13/17 and mandible StW 14 (Tobias 1973*b*). Farther north on the continent new finds attributed to *A.* cf. *africanus* are steadily accumulating.They include the mandible piece from Lothagam Hill in northern Kenya dating to the Pliocene but with clear gracile australopithecine affinities (Patterson, Behrensmeyer, and Sill 1970; Tobias 1975; Smart 1976); the distal humeral fragment from Kanapoi in the same geographic area (Patterson and Howells 1967); part of a right temporal bone from the Chemeron Beds west of Lake Baringo (Martyn and Tobias 1967); some of the specimens from Olduvai Gorge, such as OH 13 and OH 24 and the cranium KMN–ER

1813 from the Koobi Fora Formation (R. E. F. Leakey 1976*a,b*).

Recently some exceptionally fine specimens, apparently representing an early form of gracile australopithecine, have been found in Hadar, Ethiopia. Much of a skeleton (AL 288–1) came to light on 24 November 1974 (Johanson and Taieb 1976), and the following year remarkably complete skeletons of several juvenile and adult individuals were discovered in close approximation—perhaps the remains of an australopithecine band buried by a flash flood (Johanson 1976).

To accommodate these finds, as well as others made by M. D. Leakey at Laetoli, a new taxon, *Australopithecus afarensis,* has been proposed (Johanson, White, and Coppens 1978; Johanson and White 1979). It may well prove to be ancestral to *A. africanus.*

A new development concerning the antiquity of the original Taung skull is of some interest. Two independent studies by Partridge (1973) and Butzer (1974*b*) have suggested that the type specimen of *A. africanus* is likely to be far younger than previously thought. Partridge's geomorphological analysis indicates an age of less than one million years, and Butzer (1974*b*, p. 382) suggests that "Taung is contemporary with or even younger than Swartkrans and Kromdraai, rather than broadly coeval with Sterkfontein and Makapansgat." The implications of this new age assessment have been considered by Tobias (1973*a*). He points out that the time gap between the Taung skull and the next youngest gracile australopithecine specimen may be as much as two million years and that during this gap robust rather than gracile australopithecine populations appear to have existed. Tobias goes so far as to suggest that the Taung child might in fact represent a juvenile robust australopithecine rather than an *A. africanus* of the kind known from Sterkfontein and elsewhere. The kind of taxonomic muddle that would result if the type specimen of *A. africanus* proved to represent the robust rather than the gracile lineage has been considered by Olson (1974).

In a short but significant paper, Washburn and Patterson (1951) made the following statement: "the great evolutionary importance of the 'man-apes' is that they show the *initial* division between ape and man to be a locomotor adaptation." The skeletal remains of *Australopithecus* showed that, contrary to expectation, these hominids walked erect before the major expansion of the brain characteristic of *Homo* took place. As with other animals, various parts of the body evolved at strikingly different rates.

This "mosaic nature" of human evolution has been considered in detail by McHenry (1975*b*). He points out that the idea that bipedalism might precede the expansion of the brain goes back to Lamarck, who expressed his ideas on the subject in 1809; thereafter the concept was further developed by both Haeckel and Darwin before falling into temporary disrepute among paleoanthropologists. McHenry makes the point that the australopithecine pelvis was basically human in structure, though certain changes were to come about in *Homo,* where fetuses with enlarged brains necessitated the evolution of widened birth canals.

Attempts have been made by McHenry (1975*a*) to estimate body weight of *A. africanus* by determining the cross-sectional area of vertebral centra. A figure of 27.6 ± 10.5 kg (60.8 ± 23.0 lb) for the specimen STS 14 (which includes the pelvic bones referred to above) has been obtained, though, as McHenry (1975*b*) points out, this is probably a low value for the population as a whole. On the basis of a larger vertebra (STS 73), McHenry's estimate is 43.0 kg (95 lb).

Various estimates have been made of the cranial capacity of South African gracile australopithecines. Robinson (1966) gives a figure of 430 cm^3 for the mean cranial capacity of six specimens of *A. africanus,* and Holloway (1970*b*) arrives at a figure of 442 cm^3. Holloway's (1970*a*) estimate of the cranial capacity of the Taung specimen was 405 cm^3, which would have led to an estimated adult volume of 440 cm^3—appreciably less than Dart's original estimate of 525 cm^3. Tobias's (1971) estimates are appreciably higher than Holloway's with a mean for six South African gracile australopithecines of 494 cm^3. His estimate of the population range, based on the sample mean ± 3 SD, gives values of 370–618 cm^3.

Australopithecus robustus (Broom)
Provenance: Swartkrans Member 1 and Kromdraai B
A definition:

> A species of the genus *Australopithecus* characterised by the following features: more robust, heavier construction of the cranium; calvaria hafted to facial skeleton at a low level, giving a low or absent forehead and a low supra-orbital height index; well-developed ectocranial superstructures and degree of pneumatisation (more marked than in *A. africanus,* though not as pronounced as in *A. boisei*); moderate to marked supra-orbital torus with no "twist" between the medial and lateral components; sagittal crest normally present; small nuchal crest commonly present; bony face of low to moderate height, and flat or orthognathous; nose set in a central facial hollow; ramus of mandible very high and vertical; jaws large and robust with strong development of zygomatic arch, lateral pterygoid plate, temporal crest and temporal fossa; palate deeper posteriorly than anteriorly; shelving gradually from the molar region forwards; premolars and molars of very large size; M^3 commonly larger than M^2 in both buccolingual and mesiodistal diameters; mandibular canine absolutely and relatively small and hence not in harmony with the postcanine teeth; degree of molarisation of lower first deciduous molar more complete; cingulum remnants only weakly represented on lingual face and absent on buccal face of maxillary molars, representing a more advanced stage in reduction of the cingulum; sockets of anterior teeth arranged in a low to moderate curve. [Tobias 1968*a*, pp. 294–95]

The first specimen of a robust australopithecine came into Broom's hands on 8 June 1938; the circumstances surrounding this discovery are discussed in chapter 12. The specimen Broom (1938*b*) designated as the type of *Paranthropus robustus* (fig. 201) consisted of the left side of a partial cranium and a nearly complete right mandibular ramus. Broom estimated that the cranium had a capacity of 600 cm^3 and observed that the face was remarkably flat and much shorter than in a gorilla. Although the second premolar proved to be half as large again as in *Plesianthropus,* Broom found that the canines and incisors were relatively small.

Later the same year Broom (1938*c*) described the first of the postcranial parts associated with the type skull. They consisted of a right distal humerus, part of a right proximal ulna, and a toe phalanx. These were followed by descriptions of bones from a left hand (Broom 1942) and part of a right talus from the same block (Broom 1943).

Renewed fieldwork at Kromdraai in 1941 led to the discovery of a young *Paranthropus* child mandible that Broom (1941*b*) described as differing from that of *Australopithecus* from Taung in that the second incisors and canines were much smaller in *Paranthropus,* while the first and second milk molars also differed in shape and arrangement of cusps. After a restoration of the Kromdraai skull, Broom wrote (1939*b,* p. 328):

> In my opinion the Kromdraai skull differs from the Sterkfontein in the unusually advanced position of the cheeks. They are so advanced that if a ruler is placed on the two cheeks, the rest of the face is behind the ruler. Also there is a remarkable degree of flattening of the lower part of the face above the incisors and canines.

Later he continued (1943, p. 690):

> *Paranthropus,* in having a brain of about 650 cc. and a very lightly built body of probably not more than 80 lb. or 90 lb., and in being most probably bipedal, with delicate slender hands, must have been an animal much more like a human being than either the chimpanzee or gorilla. Like the other known Australopithecines, it was certainly not a forest-living Primate, but a being who lived among the rocks and on the plains and hills as the baboons do today.

Fieldwork started by Broom and Robinson at Swartkrans in November 1948 resulted in their finding a wealth of robust australopithecine remains. Part of a mandible, now known as SK 6 (fig. 192), was designated the type of a new species, *Paranthropus crassidens,* by Broom (1949). Clearly related to the Kromdraai australopithecine, the new form appeared to have more massive teeth. In the same year Broom and Robinson (1949*b*) reported the discovery of a thumb metacarpal of *P. crassidens,* and the abundant remains from Swartkrans, including a pelvic bone, SK 50, available at the time of Broom's death, were described in a Transvaal Museum memoir (Broom and Robinson 1952).

In his account of the genera and species of the australopithecines, Robinson (1954) listed the main differences, as he saw them, between *Paranthropus* and *Australopithecus.* He remarked on the structure of the deciduous molars and of the anterior part of the nasal cavity floor; he observed that the *Paranthropus* P^3 tended to have multiple roots more commonly than was the case in *Australopithecus;* that the canines tended to be smaller than in *Australopithecus;* and that the skull shape also differed. Robinson proposed that a single species of robust australopithecine should be recognized in South Africa, *P. robustus* with two subspecies, *P. r. robustus* from Kromdraai, and *P. r. crassidens* from Swartkrans.

Robinson (1963) developed his concept that structural differences in the craniums of *Paranthropus* and *Australopithecus* resulted primarily from differences in the masticatory apparatus of the two forms. He pointed out that proportions in the anterior-posterior parts of the *Australopithecus* dentition followed a human pattern while those in *Paranthropus* were strongly aberrant. Anterior teeth of *Paranthropus* appeared to be much reduced in size relative to the postcanine dentition, which was specialized for grinding tough, fibrous food. Robinson explained the differences between the two forms in terms of different dietary specializations—he saw *Paranthropus* as a vegetarian, continuing the conservative tradition of the early hominid stock, and *Australopithecus* as an omnivore, taking meat as well as vegetable food. Robinson visualized two critical points or thresholds where adaptive shifts in the evolution of hominids occurred: the first was the change from quadrupedal to bipedal locomotion; the second was the inclusion of meat in the diet. He saw the second threshold as resulting from increasing aridity of the environment in which *Australopithecus* lived—from the need to supplement dwindling plant food sources with animal protein. Robinson (1963, p. 413) visualized *Paranthropus* as representing the basic stock from which *Australopithecus* and *Homo* arose and as retaining certain apelike characters, such as a relatively long ischium that suggested (Robinson 1972) that *Paranthropus* may have been better adapted to tree-climbing than was *Australopithecus* or *Homo.*

Tobias (1967) could not agree that the anatomical differences between robust and gracile australopithecines were of great significance in terms of diet or ecological separation. Formerly, L. S. B. Leakey, Tobias, and Napier (1964) had proposed that *Australopithecus* and *Paranthropus* should be regarded as subgenera of *Australopithecus;* subsequently Tobias (1967) proposed that

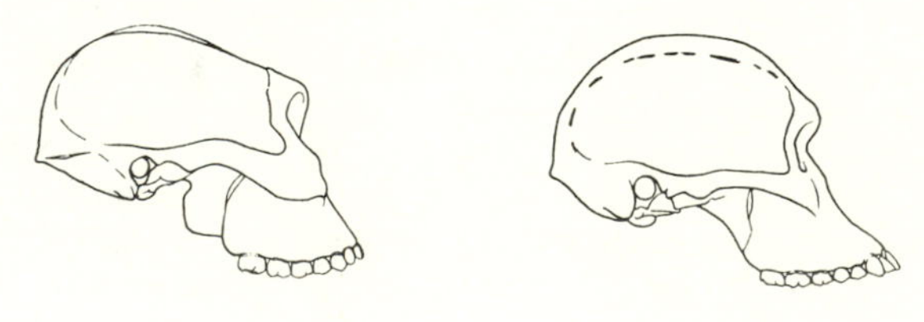

these two subgenera be sunk and that the two South African forms be designated *A. africanus* Dart and *A. robustus* (Broom). This terminology appears to have been followed by most anthropologists and paleontologists since. However, in his recent extensive review of the hominidae, Howell (1978) retained the taxa *A. crassidens* for robust australopithecines from Swartkrans Member 1, *A. robustus* for specimens from Kromdraai B, and *A. boisei* from a variety of East African localities.

Likewise, it is generally agreed that *A. robustus* was a larger animal than *A. africanus*. McHenry's (1975*a*) recent estimate of the body weight of a robust australopithecine, SK 3981, based on vertebral cross-sectional area, gave a figure of 36.1 ± 10.7 kg (79.3 ± 23.5 lb), but this is regarded as a minimum adult value. On the other hand, McHenry regards Robinson's estimate (1972) of the robust australopithecine adult weight range of 150–200 lb as excessive. Despite the larger body size, the cranial capacities of robust australopithecines may not have been greater than those of their gracile counterparts. Holloway (1970*b*, 1972) deduced capacities of 530 cm^3 for both the Swartkrans endocast, SK 1585, and Olduvai Hominid 5 *(Zinjanthropus)*—figures within the sample range for gracile australopithecines (Tobias 1971).

Apart from the two original localities of Kromdraai and Swartkrans, robust australopithecines are now known from a variety of sites. These include Olduvai Bed 1, source of the skull of *Australopithecus boisei,* a cranium that has been treated to more detailed description (Tobias 1967) than any other; Peninj, near Lake Natron (L. S. B. Leakey and M. D. Leakey 1964); East Rudolf (R. E. F. Leakey 1973, 1974, 1976*b*); the Omo valley (Howell and Coppens 1976); and the central Afar (Johanson and Taieb 1976). In addition, Robinson (1953*a*) has claimed that "*Meganthropus palaeojavanicus*" from Java should be regarded as a robust australopithecine. If he is right, these creatures must have ranged widely through Africa and the Far East, a concept that has been challenged by several authors (e.g., Le Gros Clark 1978; Tobias and von Koenigswald 1964; R. E. F. Leakey and Walker 1976) who believe there is no incontestable evidence for the former presence of *Australopithecus* outside of Africa.

Homo sp.

Provenance: Sterkfontein Member 5, Swartkrans Members 1 and 2

A definition of the genus *Homo* Linnaeus:

> A genus of the Hominidae with the following characters: the structure of the pelvic girdle and of the hind-limb skeleton is adapted to habitual erect posture and bipedal gait; the fore-limb is shorter than the hind-limb; the pollex is well developed and fully opposable and the hand is capable not only of a power grip but of, at the least, a simple and usually well developed precision grip; the cranial capacity is very variable but is, on the average, larger than the range of capacities of members of the genus *Australopithecus,* although the lower part of the range of capacities in the genus *Homo* overlaps with the upper part of the range in *Australopithecus;* the capacity is (on the average) large relative to body-size and ranges from about 600 c.c. in earlier forms to more than 1,600 c.c.; the muscular ridges on the cranium range from very strongly marked to virtually imperceptible, but the temporal crests or lines never reach the midline; the frontal region of the cranium is without undue post-orbital constriction (such as is common in members of the genus *Australopithecus*); the supra-orbital region of the frontal bone is very variable, ranging from a massive and very salient supra-orbital torus to a complete lack of any supra-orbital projection and a smooth brow region; the facial skeleton varies from moderately prognathous to orthognathous, but it is not concave (or dished) as is common in members of the Australopithecinae; the anterior symphyseal contour varies from a marked retreat to a forward slope, while the bony chin may be entirely lacking, or may vary from a slight to a very strongly developed mental trigone; the dental arcade is evenly rounded with no diastema in most members of the genus; the first lower premolar is clearly bicuspid with a variably developed lingual cusp; the molar teeth are variable in size, but in general are small relative to the size of these teeth in the genus *Australopithecus;* the size of the last upper molar is highly variable, but it is generally smaller than the second upper molar and commonly also smaller than the first upper molar; the lower third molar is sometimes appreciably larger than the second; in relation to the position seen in the Hominoidea as a whole, the canines are small, with little or no overlapping after the initial stages of wear, but when compared with those of members of the genus *Australopithecus,* the incisors and canines are not very small relative to the molars and premolars; the teeth in general, and particularly the molars and premolars, are not enlarged bucco-lingually as they are in the genus *Australopithecus;* the first deciduous lower molar shows a variable degree of molarization. [Leakey, Tobias, and Napier 1964, pp. 5–6]

Excavations conducted by Robinson at the Sterkfontein Extension Site (Member 5) during 1957 and 1958 produced isolated hominid teeth and a fragmentary juvenile maxilla in direct association with stone artifacts. Robinson and Mason (1957) and Robinson (1958, 1962) concluded that these remains represented *Australopithecus* similar in structure to the more abundant material from the Sterkfontein Type Site (Member 4). On the other hand, Robinson attributed the artifacts to a more advanced hominine, similar to *Telanthropus* (*Homo* sp.) from Swartkrans, that moved into the area in the time span separating Sterkfontein Members 4 and 5 and there-

after displaced *Australopithecus*. Tobias (1965) was inclined to regard some of the teeth as belonging to *Homo*, perhaps *H. habilis*, a form presumably responsible for the stone culture found in association.

New evidence on this controversial matter came to light on 9 August 1976 when Hughes discovered a remarkably complete hominid cranium, StW 53, weathering from the side of a solution pocket in Sterkfontein Member 5. In the opinion of Hughes and Tobias (1977), the new specimen should be referred to *Homo habilis* or to a form closely related; this hominine was presumably the Sterkfontein toolmaker.

On 29 April 1949 a mandible was discovered in the outer cave at Swartkrans that appeared strikingly different from the remains already found there. It was described by Broom and Robinson (1949*a*) as belonging to *Telanthropus capensis*, a form "somewhat allied to Heidelberg man and intermediate between one of the ape-men and true man." The mandible, now designated SK 15 (fig. 186), was closely associated with parts of two premolars (SK 43, 18*a*) and a proximal radius piece (SK 18*b*), all apparently from the same individual.

A right mandibular corpus of an old adult, SK 45, was found in September 1949 and described by Broom and Robinson (1950) in a paper entitled "Man Contemporaneous with the Swartkrans Ape-Man." At about the same time a maxillary fragment (SK 80) was discovered, and this, together with the mandible pieces, was referred to *Telanthropus capensis* by Robinson (1953*b*) in his paper "*Telanthropus* and Its Phylogenetic Significance." The genus *Telanthropus* was subsequently sunk by Robinson (1961), and the Swartkrans specimens were transferred to *Homo erectus*.

In July 1969 R. J. Clarke assembled hominid cranial fragments collectively designated SK 846*b* and 847, which had previously been referred to *Paranthropus robustus*. After assembly it was found that there was a perfect join across the left side between the posterior palatal fragment of SK 847 and the maxillary fragment SK 80, previously classified as *Homo*. The composite cranium SK 846*b*/847/80 (fig. 191) was reallocated to *Homo* sp. in a paper "More Evidence of an Advanced Hominid at Swartkrans," by Clarke, Howell, and Brain (1970).

With the separation of Swartkrans faunal remains by their stratigraphic provenance (Brain 1976*a*), the composite cranium SK 846*b*/847/80 was allocated to Member 1, while the mandible SK 15 and associated remains (SK 18) were referred to Member 2. In the interim, Clarke (1977) has concluded that remains of two more *Homo* individuals are present in the hominid fossil sample from Member 1. These are represented by a flattened juvenile cranium, SK 27, formerly thought to be an aberrant *Paranthropus*, and two upper premolars, SK 2635.

On the basis of the Swartkrans evidence there can be no reasonable doubt that early representatives of the genus *Homo* coexisted with robust australopithecine populations in the Sterkfontein valley. This was pointed out for the first time by Broom and Robinson (1950). Since then, corroborative evidence of such coexistence has been forthcoming from a number of East African localities, including Olduvai Bed I (L. S. B. Leakey, Tobias, and Napier 1964), Member G of the Shungura Formation (Boaz and Howell 1977), the upper member of the Koobi Fora Formation (R. E. F. Leakey and Walker 1976), and Hadar, Ethiopia (Johanson and Taieb 1976).

In conclusion, there is no universal consensus among anthropologists on which fossil forms should be included in the genus *Homo*. While most authorities separate the gracile australopithecines generically from *Homo*, Robinson (1972) does not; instead, he proposes the taxon *Homo africanus*, which he separates from the genus *Paranthropus*.

Family Cercopithecidae

The fossil baboons and monkeys from the Sterkfontein valley caves were initially classified by Freedman (1957, 1970) as follows:

Family: Cercopithecidae Gray, 1821
  Subfamily: Cercopithecinae Blanford, 1888
    Genera: *Parapapio* Jones, 1937
      Papio Erxleben, 1777
      Simopithecus Andrews, 1916
      Dinopithecus Broom, 1936
      Gorgopithecus Broom and Robinson, 1949
  Subfamily: Uncertain
    Genus: *Cercopithecoides* Mollett, 1947

More recently Eisenhart (1974) has reviewed the systematic classification of the Cercopithecoidea and drawn up a scheme based on the writings of Delson (1973), Jolly (1972), Kuhn (1967), Maier (1970*a*), Napier and Napier (1967), and Simons (1972). In this scheme the genus *Cercopithecoides* is placed in the subfamily Colobinae Blyth, 1875, while the genus *Simopithecus* is reduced in status to that of a subgenus of *Theropithecus* I. Geoffroy, 1843.

The Genus *Parapapio*

The generic name *Parapapio* was first used by Jones (1937) in describing a collection of fossil baboon skulls from Sterkfontein. All the specimens in this collection, except one, were thought to belong to a single new species, *P. broomi;* the exception was later found to be a specimen of *Cercopithecoides williamsi*. As more material from Sterkfontein became available, Broom (1940) concluded that there were in fact three species of *Parapapio* present. They differed in size—the largest species he named *P. whitei* and the smallest *P. jonesi*, and he retained the name *P. broomi* for baboons with intermediate dimensions. Representatives of these three species are now known from other southern African cave sites of Taung, Makapansgat, Swartkrans, Bolt's Farm, and Kromdraai A, and the genus also appears to be represented at various East African fossil sites (Howell, Fichter, and Eck 1969; Patterson 1968; Simons and Delson 1978).

It does seem remarkable that three closely related species of *Parapapio*, differing only in size, should have lived synchronously in the immediate vicinity of Sterkfontein and Makapansgat and presumably elsewhere as well. Yet this seems to have been the case, and there is

no evidence at present that the specimens from Member 4 at Sterkfontein are separated in time or that the three species form a chronocline. They occur in approximately the same relative abundance at Sterkfontein and Makapansgat, with *P. broomi* specimens being the most numerous, followed by *P. jonesi*, then *P. whitei*. This study indicates that the collection studied from Member 4 at Sterkfontein contained remains from a minimum of 91 *P. broomi* individuals, 37 *P. jonesi*, and 10 *P. whitei* (fig. 167).

A recent reexamination in statistical terms of the *Parapapio* material from Sterkfontein by Freedman and Stenhouse (1972) has confirmed the validity of the three species apparently occurring there.

Morphologically the dentitions of *Parapapio* and *Papio* are probably indistinguishable. The main anatomical difference between the genera lies in the profile of the muzzle when viewed from the side. In *Parapapio* the nasals and frontals form a straight line or slightly concave curve from the glabella to the posterior margin of the nasal aperture, whereas in *Papio* the profile is sharply concave immediately anterior to the glabella.

In his recent detailed study of the skull of *Papio ursinus*, Jones (1978) concluded that a distinction between

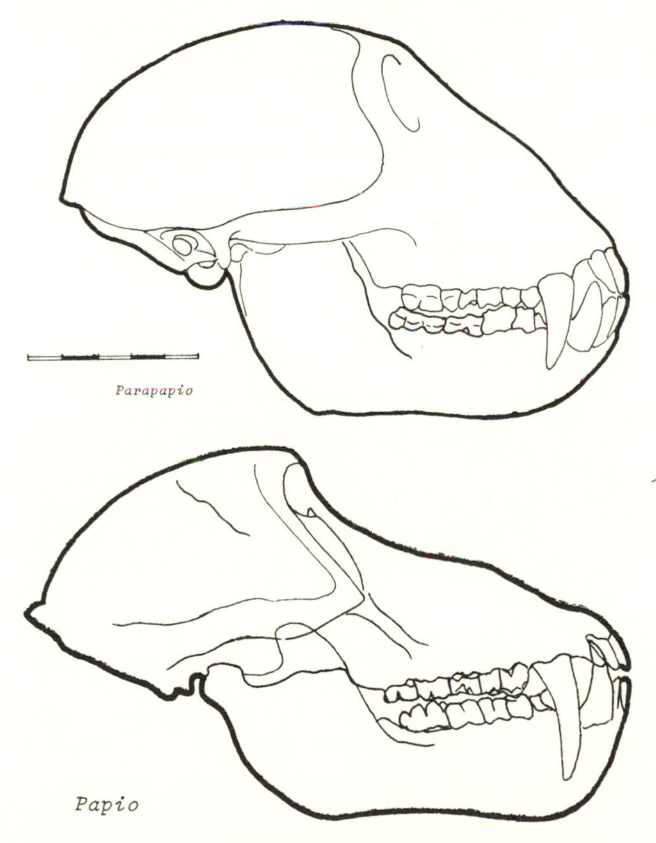

Papio and *Parapapio* based on muzzle profile may not be as reliable as was previously thought. In a single troop of baboons from Zimbabwe, Jones found individuals with both the typical *Papio* and the typical *Parapapio* muzzle profiles.

Parapapio jonesi Broom.
Provenance: Sterkfontein Member 4, Swartkrans Member 1 and Kromdraai A.

Parapapio broomi Jones
Provenance: Sterkfontein Member 4

The Genus *Papio* Erxleben

Living baboons from Africa and fossil forms from both Africa and India are included in this genus, the type species of which is *Papio papio* (Desmarest, 1820). Characteristic of the genus is the long, doglike snout, particularly well developed in males. Viewed from the side, the profile of the muzzle is strikingly different from that of *Parapapio* (see illustration), in that there is a marked drop in the interorbital region and then a fairly gradual slope to the alveolar margin of the incisors. Sexual dimorphism is marked, males being appreciably larger than females and possessing much more robust canines.

Five species of living *Papio—P. anubis, P. cynocephalus, P. hamadryas, P. papio, and P. ursinus*—are currently recognized by many primatologists. These appear to be allopatric species with contact at the margins of their ranges and even a certain amount of hybridization there. Several authors have therefore questioned the status of the five species, and Buettner-Janusch (1966) has proposed that a single polytypic species exists throughout the African range, *Papio cynocephalus*, which incorporates the other four forms, previously considered specifically distinct. For the purpose of this discussion, the only baboon currently occurring in southern Africa, *P. ursinus*, is still regarded as a separate and valid species.

Papio ursinus (Kerr, 1792). Chacma baboon
Provenance: Swartkrans Member 2 (cf.)

The well-known living Chacma baboon of southern Africa, widely distributed through a great diversity of

habitats—in fact, the success of the species can probably be attributed to its adaptability and environmental tolerance. Although several species of baboons coexisted in the Sterkfontein valley previously, *P. ursinus* is the only living baboon in the subcontinent today.

Papio robinsoni Freedman
Provenance: Swartkrans Member 1, Kromdraai A and B

A species of extinct baboon, very similar to *P. ursinus* and almost certainly ancestral to it, originally described by Freedman (1957) on the basis of a large sample from Swartkrans. The main difference between *P. robinsoni* and *Papio ursinus* lies in the shape and structure of the muzzle: that of *P. robinsoni* is flatter, with the maxillae overlapping the nasals to a greater extent than in the living chacma baboon.

According to Freedman (1970) the taxon is also known from Cooper's site, Swartkrans II, Gladysvale, Bolt's Farm, and Skurveberg.

Papio angusticeps (Broom)
Provenance: Kromdraai A and B

The species was originally described by Broom (1940) as *Parapapio angusticeps* on the basis of a female skull from Kromdraai A. The specimen was damaged in the interorbital area, and Broom assumed that the muzzle profile would have been that of a typical *Parapapio*. However, when more complete material became available, Freedman (1957) was able to conclude that the species *angusticeps* definitely belonged in the genus *Papio*, characterized by the steep dip in muzzle profile in the interorbital region.

The skull of *P. angusticeps* is similar to that of *P. ursinus* in almost all respects except size. An adult male of *P. angusticeps* appears to have been typically smaller than a female of *P. ursinus*.

The species has also been recorded from Cooper's site and Minnaar's Cave.

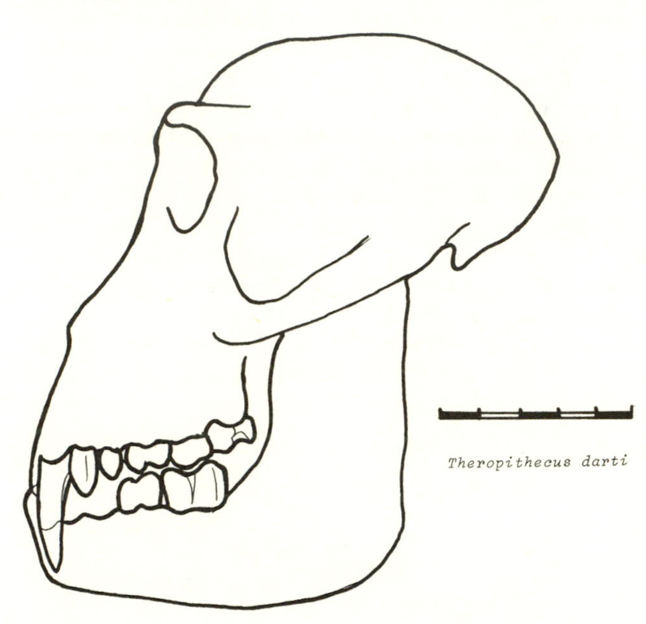

The Genus *Theropithecus* Geoffroy

These are large baboons with high, flat faces that form a striking contrast to the dog-faced baboons of the genus *Papio*. The cheek teeth show an unmistakeable pattern of high cusps, separated by deep clefts, foveae, and fossae. Lingual cusps of upper molars and buccal cusps of lower ones are connected by high ridges of enamel.

The first specimen from a South African cave deposit was described by Broom and Jensen (1946) from Makapansgat Limeworks as *Papio darti* but was subsequently transferred to the genus *Gorgopithecus* by Kitching (1953). In his 1957 study, Freedman placed all the Makapansgat specimens into a single species, *Simopithecus darti*, and also described a second, larger form from Swartkrans as *S. danieli*.

Jolly (1966, 1970, 1972) argues that *Simopithecus* should be regarded as a subgenus of *Theropithecus*, the genus to which the living gelada of Ethiopia belongs. This scheme is followed here. However, Maier (1970*a,b*) favors retention of the genus *Simopithecus* for the fossil forms and places *Theropithecus* on a related lineage that deviated from *Simopithecus* in the course of the Pliocene.

Some of the cranial adaptations found in *Theropithecus* appear to mirror those of the robust australopithecines. There is a reduction in the size of the anterior teeth relative to that of the molars, while the ascending ramus has become very upright. According to Simons and Delson (1978) the forelimb was long but the phalanges were short and stout. These features reflect the terrestrial life-style of *Theropithecus* in open grasslands where resistant grass seeds formed the bulk of the diet.

Theropithecus (Simopithecus) danieli (Freedman)
Provenance: Swartkrans Member 1

A typical theropithecine baboon, known only from Swartkrans, in which sexual dimorphism appears to have

been marked. Adults were large, with females about the size of an adult male chacma baboon.

The Genus *Dinopithecus* Broom

Very large extinct baboons with marked sexual dimorphism, robust skulls, and long muzzles. The teeth are large and broad, and the dental arcade is horseshoe-shaped, somewhat narrowed posteriorly. Maier (1973) has suggested that *Dinopithecus* may have relationships with *Gorgopithecus* and *Papio*.

The genus was originally described by Broom (1937*b*) on the basis of a male mandible from Skurveberg. Freedman (1957) subsequently assigned a large number of specimens from Swartkrans to the same taxon.

Some teeth from Omo were described by Arambourg (1947) as *D. brumpti*, but according to Simons and Delson (1978) most of these can be referred to *Theropithecus*. The latter authors consider, however, that some undescribed specimens from Leba in Angola may represent *Dinopithecus*.

Dinopithecus ingens Broom
Provenance: Swartkrans Member 1

In appearance, *D. ingens* was probably like an exceptionally large chacma baboon, with adult males weighing more than 45 kg (100 lb).

The Genus *Gorgopithecus* Broom and Robinson

The type specimen from the Kromdraai Faunal Site was described originally by Broom (1940) as *Parapapio major* and consisted of two considerably worn upper molars. Later, when more complete material became available from Kromdraai, Broom and Robinson (1949*c*) created the genus *Gorgopithecus* to accommodate it. A second species was subsequently described by Kitching (1953) from Makapansgat Limeworks as *G. wellsi*, but this was sunk into *Simopithecus darti* by Freedman (1957). Thus *Gorgopithecus* remains a monotypic genus.

On the grounds of size in particular, Delson (1975) suggested placing *Gorgopithecus* as a subgenus of *Dinopithecus*, but this was not upheld in a subsequent review (Simons and Delson 1978).

Gorgopithecus major (Broom)
Provenance: Kromdraai A

An exceptionally well preserved male skull from Kromdraai A (fig. 207) shows *G. major* to be a large baboon with a high, narrow, and relatively short muzzle. Other specimens indicate that, unlike *Dinopithecus, Gorgopithecus* had very slight sexual dimorphism. Adults were probably intermediate in size between *Dinopithecus* and *Papio ursinus*.

The Genus *Cercopithecoides* Mollett

It is remarkable that, although all living monkeys in southern Africa belong to the genus *Cercopithecus*, all fossil forms from the australopithecine caves are colobines, having no direct phylogenetic link with the living representatives.

The genus *Cercopithecoides* was first described by Mollett (1947) on the basis of a single male cranium and mandible from Makapansgat Limeworks, attributed to *C. williamsi*. Freedman (1957) subsequently described a second, larger species from Swartkrans that he called *C. molletti*. However, three years later he sank *C. molletti* into *C. williamsi*, which is currently the only recognized species.

Cercopithecoides monkeys were fairly large—certainly heavier than a living vervet monkey—with exceptionally short muzzles and big brains. The large circular orbits were widely spaced and lay beneath pronounced supraorbital tori. The teeth were characteristically colobine, molars showing high cusps separated by deep foveae. According to Simons and Delson (1978), *Cercopithecoides* was most closely related to *Paracolobus* on dental and facial morphology.

Living colobine monkeys are typically arboreal animals restricted to evergreen forest areas. It is not known what the habitat requirements of *Cercopithecoides* monkeys were—wooded country was doubtless required, but perhaps not as thickly forested as is preferred by living representatives such as *Colobus*. In fact, it seems that the *Cercopithecoides* monkeys were using a niche later to be taken over by southern African vervet and samango monkeys of the genus *Cercopithecus*.

Cercopithecoides williamsi Mollett (fig. 167*d*)
Provenance: Sterkfontein Member 4, Swartkrans Member 2, and Kromdraai B

At one time this species was apparently common and widespread. In southern Africa, remains have also been found at Taung, Makapansgat Limeworks, Cooper's site, Swartkrans II, Sterkfontein Graveyard, and Bolt's Farm, while in East Africa very similar remains have been recovered from the Omo Group deposits and the Awash valley (Eck 1976) and the East Rudolf succession (M. G. Leakey 1976).

Order Carnivora

Family Felidae

Remains of cats from the Sterkfontein valley caves come from two familiar species, the leopard and lion, one extinct and poorly known creature with characteristics both of the leopard and the cheetah, and from a range of extinct carnivores showing sabertooth adaptations to a greater or lesser extent. The latter are classified in two subfamilies: the *Felinae* or true cats and the *Machairodontinae* or sabertoothed cats.

Subfamily Felinae
Panthera pardus Linnaeus, 1758. Leopard
Provenance: Sterkfontein Member 4, Swartkrans members 1 and 2, and channel fill, Kromdraai B

On the basis of the fossils from Member 1 at Swartkrans, Ewer described a new subspecies of the leopard, *P. pardus incurva*, which differed from the living southern African leopards on the following points: "angular process of the mandible scooped out rather after the manner of a hyaena; anterior cusp of P^3 more reduced; M^1 larger and milk carnassial with a distinct talonid; probably a trifle smaller than the living leopard" (Ewer 1956*a*, p. 84).

Remains from the site units other than Swartkrans Member 1 are too incomplete to allow allocation to a particular subspecies—all that can be said is that they came from leopards. It appears that leopards were part of the Sterkfontein valley fauna throughout the time spanned by the various cave deposits.

Panthera leo (Linnaeus, 1758). Lion
Provenance: Sterkfontein Member 5, Swartkrans Member 2, and Kromdraai A

The isolated teeth from Kromdraai were described by Ewer (1956*a*) as coming from a lionlike felid but being larger and heavier than the corresponding teeth in a living lion. She tentatively equated them with the large cat previously described by Broom (1948*a*) as *Felis shawi* from Bolt's Farm. Hendey has reexamined the specimens and concluded that they can best be assigned to *P. leo* (pers. comm.).

Ewer (1956*a*) pointed out that, although the Swartkrans fossils belonged to a lionlike cat, the P_4 is exceptionally broad in proportion to its length and also shows unusual features. She concluded: "From the characters of P_4 it seems clear that specimen SK 359 is not identical with the living South African Lion, but whether the difference should be accorded specific or subspecific recognition cannot be decided until further material becomes available" (Ewer 1956*a*).

The specimens from Member 5 at Sterkfontein consist of two isolated teeth and a right distal humerus. These

have been examined by Hendey and assigned (pers. comm.) to cf. *Panthera leo*.

Clearly the taxonomic status of the fossil lions from the Sterkfontein valley caves is somewhat uncertain. What evidence we have suggests that the Swartkrans and Kromdraai lions were more robust than their modern counterparts.

Felis crassidens Broom. The "crassidens cat"
Provenance: Kromdraai A

This remarkable cat was described by Broom (1948*a*) on the basis of two maxillary pieces and one mandibular piece. The teeth have features in common with both the leopard and the cheetah but are typical of neither; they clearly came from a short-faced feline of leopardlike proportions. No further remains of this tantalizing creature have been found in southern Africa thus far; the closest to it is perhaps *F. obscura* Hendey, from "E" Quarry at Langebaanweg. A single specimen is known that Hendey (1974*a*) considered to come from a cat slightly smaller than a leopard with a P^3 reminiscent of a cheetah *Acinonyx jubatus* (Schreber). However, the fact that the Langebaanweg specimen lacked a P^2 indicates that its affinities do not lie with *Acinonyx* or *Panthera*. Hendey points to the similarity between *F. obscura* and cats of the genus *Sivafelis* Pilgrim, known as fossils from India and China.

Remains from East Africa are also coming to light. Ewer (1965) tentatively assigned a mandible from Olduvai Bed II to *F. crassidens*, and three postcranial bones from Olduvai have been referred to *Panthera crassidens* (Broom) by G. Petter (1973). More recently, two postcranial bones from the Koobi Fora Formation of the East Rudolf succession have been described under *P.* cf. *crassidens* by M. G. Leakey (1976). From the Shungura Formation in the Omo valley Howell and G. Petter (1976) have also assigned a mandible and postcranial pieces to *P. crassidens*. They tentatively equate this taxon with *Sivafelis potens* Pilgrim from the Siwaliks of India.

Saber-toothed Cats: False and True

The sabertooth adaptation—the tendency to greatly increase the size of the upper canines as lethal killing weapons, has been experimented with by various carnivores over a long span of geological time. Saber-toothed forms existed in four subfamilies of the Felidae: in the Felinae, Hoplophoneinae, Nimravinae, and Machairodontinae. The only living cat that shows a tendency toward exceptional enlargement of the canines is the clouded leopard of Asia, *Neofelis nebulosa*. In this case, however, there is enlargement of both upper *and* lower canines—a situation not typical of the extinct sabertoothed cats, where the upper canine is greatly enlarged at the expense of the lower.

Ewer has discussed the characteristic sabertooth adaptation (1954*a*, 1973), pointing out that the killing method used by a normal feline is different from that of a sabertooth, for example, a machairodont. Typical anatomical differences between a feline and a machairodont are shown in the illustration. Although sabertooths are now regrettably extinct, they were the first successful

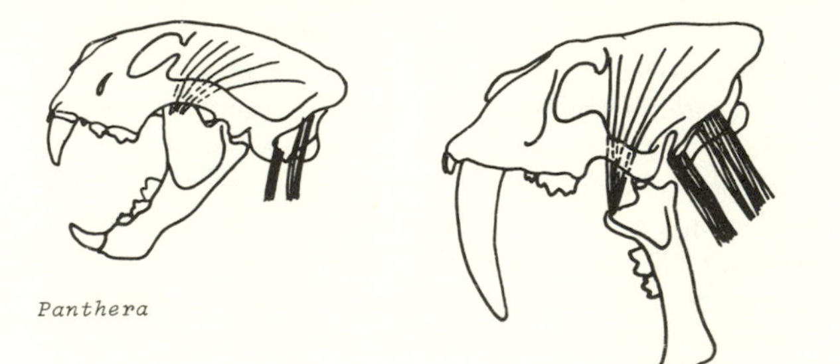

large felid predators, dominating the scene from the Oligocene to the Pliocene.

A normal cat kills its prey with its upper *and* lower canines, forcefully driven in by the closing of the jaws, through the action of the large temporalis muscles that run down on each side of the skull, from the upper part of the braincase to the coronoid processes of the lower jaw. Where the coronoid processes are fairly high, as in typical felines, considerable force can be transmitted even when the jaws are wide open. A low coronoid results in a less favorable mechanical advantage and a weaker bite. Having killed its prey, the cat uses its bladelike carnassial teeth, P^4 above and M_1 below, in a scissor action to slice the meat. Finally, manageable bones may be cracked by the premolars, anterior to the carnassials.

The anatomy of a saber-toothed skull makes it abundantly clear that these carnivores killed with their upper canines only, driving them into the prey not by the action of the temporalis muscles closing the jaws, but by a downward thrust of the head, powered by the neck muscles. These neck muscles, the cleidomastoids and sternomastoids, were attached to the greatly enlarged mastoid process of the skull, which lay farther below the joint of the skull with the neck than in normal cats. The changed position improved the mechanical advantage of the system, adding power to the downward jab of the head.

Other structural changes allowed the lower jaw to be swung down, out of the way of the upper canines. The coronoid processes were reduced in height and breadth, allowing the carnassial teeth to be brought closer to the hinges of the jaw and thus adding power to the scissor action. In fact, considerable force could still be exerted by the temporalis muscles, but now only when the jaws were almost closed. As meat-shearing mechanisms, the sabertooth carnassials were even more efficient than those of true cats. Ewer has described the situation as follows (1954*a*, p. 34):

> In cats, the upper carnassial consists of three lobes lying along the length of the jaw and an inner lobe projecting on to the palate almost at right angles to the main axis of the tooth (see illustration). The two posterior lobes (2, 3) are narrow and form the cutting blade, working against the edge of the lower carnassial while the stouter anterior (1) and (4) lobes form part of the crushing mechanism. In sabre-tooths the whole length of the upper carnassial becomes incorporated in the cutting blade. This invoves the reduction and eventual loss of the inner lobe, so that the whole tooth

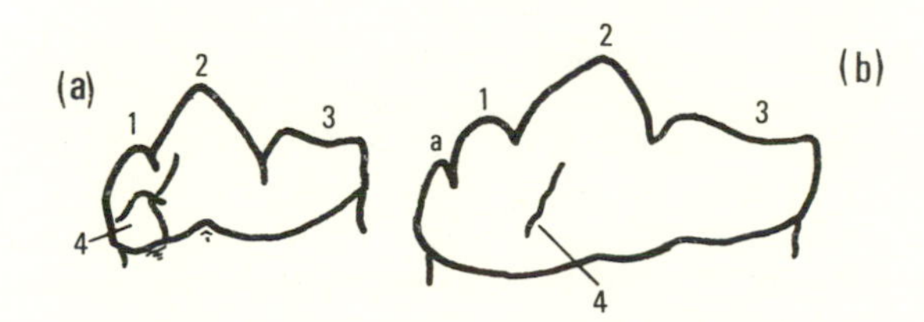

> forms a single blade, and in advanced species the tooth is further lengthened by the development of an extra anterior lobe (a) in line with the rest of the blade (see Figures). In these forms the very long upper carnassial usually cuts against two teeth in the lower jaw; but the bone-crushing teeth in front of the carnassials are so reduced as to be almost functionless, and clearly the sabre-tooths cannot have been able to cope with bones of any size. This is hardly surprising as the enlarged upper canines would surely have made the manipulation of bones very difficult, and might have been in danger of getting broken themselves in the process. The large canines must also have been rather in the way when getting started on a meal, and the enlargement of the incisor teeth helps to get over this difficulty, and would also facilitate the process of grooming.

Among the machairodonts, two main adaptive types had become dominant by the Pliocene, and both are represented in the fossils from the Sterkfontein valley caves. One, typified by the genera *Homotherium* and

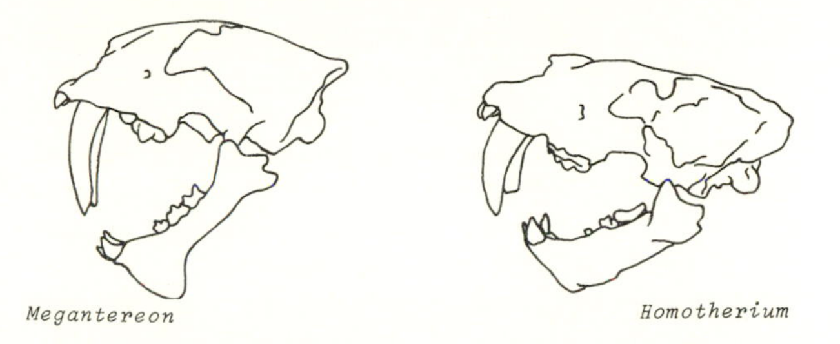

Machairodus, had upper canines in the form of wide, laterally compressed blades, not particularly elongated but with edges serrated like a bread knife. Upper carnassials tend to be very specialized, with the inner lobe reduced or absent and an accessory cusp adding to the length of the blade. The other adaptive type, characterized by the genus *Megantereon,* had very elongated but much less flattened, smooth-edged canines; the carnassials were less specialized, having small inner lobes and no accessory cusps.

It appears that the two kinds of canine would have been used in slightly different ways. The serrated canines of the *Machairodus-Homotherium* group would have been most effective in slicing, while the sharp, smooth-edged sabers of *Megantereon* were stabbing daggers. Ewer (1973) has suggested that *Megantereon* specialized in killing large, heavy prey with a tough protective hide, against which the stabbing daggers would have been effective. On the other hand, the bladelike slicing canines of *Machairodus* or *Homotherium* would have been better suited for use on thinner-skinned prey: they would have been driven in and then, with a backward pull, used to rip open the victim's throat.

In addition to these two machairodont adaptive types, a group of true cats with sabertooth tendencies must be considered. These are placed in the genus *Dinofelis* and show characters intermediate between those of normal felines and machairodonts; the upper canines were not excessively enlarged, nor were the lowers much reduced. The carnassials and other teeth do not show the extreme specializations found in the machairodonts. Cats of the genus *Dinofelis* are referred to as "false sabertooths."

It is interesting to speculate on possible reasons for the extinction of the sabertooths. We do not know exactly when the extinctions occurred in southern Africa. The most recent machairodont known from the subcontinent comes from Elandsfontein. It is a *Megantereon* and may be associated with the "Final Acheulean" stage of the Early Stone Age (Hendey 1974*a*). It may be as much as half a million years old. Elsewhere in the world, sabertooths survived much longer; for instance, the well-known American genus *Smilodon* persisted until the

Holocene. Ewer (1973) has pointed out that the machairodonts were smaller brained and less agile than the felines that became progressively more dominant after the end of the Pliocene. She suggests that they may not have been able to compete with the felines in hunting the modern, swift ungulate fauna that was evolving at that time. Another factor was almost certainly the rise of human intelligence and technology. Sabertooths, like all large carnivores, must surely have posed a threat to early man. One may be sure that human hunters would have taken steps to minimize this threat.

Dinofelis, though a feline, was rather specialized and heavy bodied. It went the way of machairodonts, probably for the same reasons.

There are some indications that the sabertooths suffered from bone diseases that may have weakened them and perhaps contributed to their demise. A complete right radius (SE 680) attributed to cf. *Megantereon* sp. from Sterkfontein Member 5 shows pathology of the distal end, perhaps attributable to osteoarthritis. Among machairodont postcranial material from Langebaanweg, Hendey (1974*a*) found similar evidence of pathology.

A femur and tibia, apparently belonging to the same individual *Megantereon* sp. from Elandsfontein in the Cape, showed osteitis of the distal femur and proximal tibia. Four bones, possibly from a single animal referred to *?Machairodus* sp., show the following effect (Hendey 1974*a,* p. 184):

> Right tibia. Osteitis, mainly near the proximal end. Osteoarthritis at the proximal end, with severe eburnation of the lateral articular facet and lipping of the bone.
>
> Left and right calcanei. Osteitis on lateral surfaces anterior to distal extremity.
>
> Metapodial. Osteitis on dorsal surface. Osteoarthritis (eburnation) of distal articular facet.

In addition, the following effects were noted on bones from other *?Machairodus* sp. individuals:

> Proximal end of left ulna and distal end of left humerus. Severe osteitis on lateral and medial sides of ulna and a less extensive but similar condition on the humerus, particularly the arch enclosing the entepicondylar foramen.
>
> Distal end of right tibia. Severe osteitis on the shaft.
>
> Proximal end of left humerus. Extensive osteitis.
>
> Left radius lacking distal epiphysis (immature). Severe osteitis of the shaft.

The term "osteoarthritis" is used for any condition affecting the joints; "osteitis" refers to any other bone inflammation (see Brothwell 1963). Although the postcranial bones of sabertooths are not common in the southern African deposits, the observed incidence of pathology on them is remarkably high. Could it be that sabertooths were susceptible to chronic disorders of the bones and joints? It is tempting to speculate that they suffered from a calcium deficiency as a result of their inability to crunch up any but the most delicate of bones. This may have been a price they paid for the remarkable dental specializations that improved their hunting performance in other ways.

In an interwoven ecological situation, the extinction of one component of the fauna cannot occur without repercussions elsewhere in the system. In this regard, the fortunes of hyenas appear to have been linked with those of the sabertooths. It is obvious that sabertooths were not effective crackers of bones and that the remains of their prey would have provided a regular source of food for any scavengers able to utilize them. In fact, Ewer (1967) has argued that the rise and decline of the hyaenids was regulated by their relationship with the machairodonts. Not only did the sabertooths provide the food, but the nature of this food made it expedient for the hyenas to perfect a series of bone-crushing adaptations.

Toward the end of the sabertooth's period of dominance, hyenas showed a diversity of form unknown today. During the accumulation of Member 1, at Swartkrans, for instance, six recognizably different hyena taxa had their remains preserved as fossils. Today only two different hyenas, the spotted, *Crocuta crocuta,* and the brown, *Hyaena brunnea,* occur in southern Africa—they represent the remnants of a formerly richer hyaenid fauna, a residue that proves viable in the current, postsabertooth era.

The Genus *Dinofelis* Zdansky

The genus *Dinofelis* contains a group of cats that show a range of sabertooth tendencies but have not specialized as far as the true machairodont sabertooths have. They are thus retained in the subfamily Felinae and are referred to as false sabertooths.

The first southern African representative was described initially by Broom (1937*b*) as *Megantereon barlowi* on the basis of a damaged skull and canine from Sterkfontein Type Site. It was named after G. W. Barlow, manager of the Sterkfontein Limeworks, who had also discovered the first australopithecine specimen there.

Ewer (1955*c*) reexamined Broom's specimens and concluded that they did not rightly belong in the genus *Megantereon.* She found that the canine was not a typical slender, curved blade as a *Megantereon* saber should be, but was an almost straight-edged dagger; likewise, the skull profile was not a typical machairodont one. She therefore transferred the specimens to *Therailurus barlowi* (Broom).

In the same paper Ewer described a new species, *Therailurus piveteaui,* on the basis of a reasonably complete skull and two mandible fragments from Kromdraai A. She pointed out that the specimens showed features intermediate between those of the true cats (Felinae) and the true sabertooths (Machairodontinae). One species had previously been described in the genus *Therailurus: T. diastemata* Astre, 1929, and Ewer found close similarities between this and the Kromdraai cat, although the sabertooth specializations had been carried further in *T. piveteaui* than in *T. diastemata.*

Meanwhile, some saber-toothed cat remains were discovered at the Makapansgat Limeworks by J. W. Kitching and his two brothers. These consisted of a well-preserved mandible piece and articulated maxillary and mandible fragments from a second individual; they were described by Toerien (1955) as belonging to a new species of the genus *Machairodus: M. darti.* The following year Ewer (1956*d*) reexamined the Makapansgat material and showed that *M. darti* should, in fact, be equated with *Therailurus barlowi* from Sterkfontein.

Hemmer (1965) reviewed the nomenclature and distribution of the genus *Dinofelis* Zdansky, 1924, and concluded that the known species of *Therailurus* rightly belonged in the genus *Dinofelis,* which would thus contain four species:

D. abeli Zdansky, 1924 (formerly *Machairodus horribilis*), from the Pontian stage of the Pliocene and the early Pleistocene.

D. diastemata (Astre), 1929, from the Astian stage of the Upper Pliocene of France.

D. barlowi (Broom), 1937, from the Sterkfontein Type Site, Makapansgat Limeworks, and Bolt's Farm.

D. piveteaui (Ewer), 1955, from Kromdraai A.

He concluded that, of these four, *D. diastemata* was the most primitive in the specialization of its teeth while *D. piveteaui* was the most advanced. He suggested that *D. diastemata–D. barlowi–D. piveteaui* formed an evolutionary sequence, while *D. abeli* occupied a more isolated position. This phylogenetic concept has been supported by Hendey (1974*a*).

Some exceptionally fine fossils of *Dinofelis barlowi* were recovered from Bolt's Farm in the course of the University of California's excavations there in 1947–48. A team under the direction of Camp and Peabody found

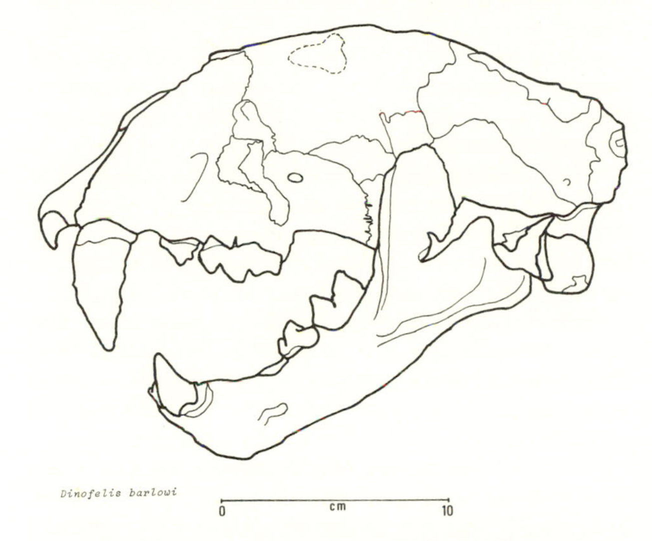

remains of three *Dinofelis* individuals in Pit 23 of the Bolt's Farm complex. These, together with the other fossils collected on the expedition, were taken back to the University of California in Berkeley, where they were studied by H. B. S. Cooke in 1957–58. Cooke (n.d.) concluded that the three skulls, together with their postcranial skeletons, belonged to an adult male, adult female, and young male of *D. barlowi*. The presence of three well-preserved *Dinofelis* skeletons, fossilized in close proximity to one another, is remarkable and suggests some unusual circumstances at the time of death. Cooke has pointed out that the remains of the three cats occurred in association with those of at least eight baboons but that bovid or equine remains, characteristic of the other Bolt's Farm localities, were absent. He suggests that Pit 23 represented some kind of natural trap into which the baboons had fallen. They were followed by the three *Dinofelis* cats, perhaps members of a single family, who then found themselves unable to escape from the trap. The presence of both baboon and cat coprolites supports the suggestion that the animals were alive for a while in the fossilization site before dying.

The large male *Dinofelis* skull from Pit 23 is now housed in the Transvaal Museum, through the courtesy of the University of California and H. B. S. Cooke.

Both cranial and postcranial remains of *Dinofelis diastemata* have been found at Langebaanweg by Hendey. They come from the Quartzose Sand Member and Pelletal Phosphorite Member in E Quarry (Hendey 1974*a*). The postcranial remains are of particular interest because they provide some insight into the build of the animal. For instance, the five caudal vertebrae are about the size of

adult cheetah vertebrae in diameter but are less than half the length. The tail of *Dinofelis* apparently was appreciably shorter than that of an adult cheetah or leopard. Concerning body build, Hendey wrote (1974*a*, p. 176):

> The Langebaanweg *Dinofelis* was evidently a heavily built animal, with the fore- and hindfeet, and perhaps the limbs in general, being more equally proportioned than in either the leopard or cheetah, and possessing a relatively short tail. Its locomotion is likely to have been ambulatory. The indications are thus that it had paralleled the developments in the Machairodontinae in its postcranial skeleton as well as the skull.

More recently, remains of *D. barlowi* and *D.* cf. *piveteaui* have been reported from the Lower and Ileret members respectively of the Koobi Fora Formation at Lake Turkana (M. G. Leakey 1976). There is also a report of *Dinofelis* sp. from the Usno and Shungura Formations in the Omo valley (Howell and Petter 1976). The indications are that the genus had a wide range in both space and time.

The True Sabertooths: Subfamily Machairodontinae

As mentioned earlier, the true sabertooths from the Sterkfontein valley caves fall into two natural groups: the *Machairodus-Homotherium* group with crenulated sabers and highly specialized carnassials, and the *Megantereon* group with smooth-edged sabers and less specialized cheek teeth. In his discussion of European sabertooths, Kurtén (1968) equates the *Machairodus-Homotherium* group with the tribe Homotheriini and places *Megantereon* in the tribe Smilodontini.

The genus *Machairodus* is not represented among the fossils from the cave sites specifically studied here but has been reported from Bolt's Farm, where Broom (1939*a*) described the crown of an upper canine and an isolated upper carnassial (which may or may not be associated) as *M. transvaalensis*. Further remains from the Transvaal have not come to light in the interim, but some have been described from the Quartzose Sand Member in E Quarry at Langebaanweg by Hendey (1974*a*). He points out that if the carnassial from Bolt's Farm is, in fact, referrable to *M. transvaalensis*, then the Langebaanweg specimens must represent a different species, but in the absence of more complete material he refers them to *Machairodus* sp.

The Genus *Homotherium*

The genus is particularly well represented as *H. sainzelli,* the greater scimitar cat, among fossils from the Villafranchian and early Middle Pleistocene of Europe. It is

also well known from North America, China, and elsewhere. Some of its anatomical characteristics have already been discussed.

In her paper on machairodont remains from the Transvaal caves, Ewer (1955*c*) described two lower carnassial teeth (KA 66) from Kromdraai A. She pointed out that the specimens were remarkably similar to *Epimachairodus crenatidens* (Fabrini) of Europe and less so to *Machairodus ultimus* Teilhard, from China. She designated the Kromdraai specimens *Epimachairodus?* sp. More recently they have been reexamined by Hendey, who concluded that they can best be accommodated in *Homotherium* sp. (pers. comm.).

No other traces of *Homotherium* have been found in the Sterkfontein valley caves, but a fine skull with articulated mandible was discovered by Kitching in the banded travertines close to the so-called ancient entrance of the Makapansgat Limeworks. The specimen was described by Collings (1972) as *Megantereon problematicus,* but Hendey (1974*a*) pointed out that it fits best in the genus *Homotherium* and refers to it as *Homotherium* sp. indet. (Hendey 1974*a*, p. 159). More recently Collings et al. (1976) have reassigned the Makapansgat specimen to *H.* cf. *nestianus.*

A very spectacular though incomplete left upper canine has been found in the Pelletal Phosphorite Member of E Quarry at Langebaanweg and referred by Hendey (1974*a*) to cf. *Homotherium* sp. Remains have also been reported from various horizons throughout the Koobi Fora and Koobi Algi formations near Lake Turkana (M. G. Leakey 1976) and from the Usno and Shungura formations in the Omo valley (Howell and Petter 1976).

The Genus *Megantereon*

The genus is well known on the basis of the European species *M. megantereon* Croizet and Joubert, a Villafranchian guide fossil, referred to as the Dirk tooth. Other

species have been described from the Siwalik deposits of India, and from China and elsewhere.

In 1937 Broom described part of a felid left mandible from Schurveberg as *"Felis whitei,"* stressing that he was uncertain about the most appropriate genus for the specimen. Later (1939*a*, 1948*a*) he suggested that the fossil should probably be referred to *Megantereon,* where it has remained pending the discovery of more complete specimens.

A second specimen, consisting of the right mandibular ramus with P_4 and M_1, was described by Broom (1948*a*) from the Sterkfontein Type Site as *Megantereon gracile.* It is similar in many respects to *M. whitei* but differs in the proportions of the carnassial.

The first reasonably complete specimen of a *Megantereon* came from Kromdraai A and was described by Ewer (1955*c*) as *M. eurynodon*—a new species, larger than *M. whitei* but of about the same dimensions as *M. gracile* and *M. megantereon.* The upper canines were long, recurved, much flattened blades with sharp edges, devoid of crenulations.

A single specimen from Swartkrans, consisting of a left mandible fragment with P_{3-4}, was tentatively assigned to *Megantereon* by Ewer (1955*c*). This assignation was subsequently confirmed by Hendey (1974*b*).

Fossils from the Ileret Member of the Koobi Fora Formation have recently been referred to *M. eurynodon* (M. G. Leakey 1976), and representatives of the genus are also known from the Usno and Shungura formations in the Omo valley (Howell and Petter 1976) and from Bed 1 at Olduvai, site FLK NNI (M. G. Leakey 1976). The genus *Megantereon* has also been reported from Elandsfontein (Hendey 1974*a*)—a record that apparently represents the most recent appearance of a sabertooth in southern Africa.

Postcranial remains are not common at the local sites, but what indications we have support Kurtén's view (1968, p. 75) that the European *Megantereon* typically had very powerful neck muscles, used in the stabbing movement, "short but massive front legs and relatively feeble hind quarters. The animal was obviously not a fast runner but relied on its tusks and the great strength of its front paws, so the prey may be visualised as a relatively large, slow-moving animal. In the Villafranchian fauna the rhinoceros may be a possibility, or perhaps young mastodonts or elephants."

As mentioned earlier, two fossil forms, originally described as species of *Megantereon,* have come to rest in other genera. *M. barlowi,* described by Broom in 1937, is now regarded as *Dinofelis barlowi* (Broom), while *M. problematicus* Collings, 1972, has become *Homotherium* cf. *nestianus.*

"Nimravidae Indet."

When Ewer (1955*c*) discussed the fossil machairodonts from the Transvaal caves she considered that a single specimen from Swartkrans may have represented a nimravid. The subfamily *Nimravinae* contains saber-toothed felids, typically Oligocene in age; the discovery of a possible representative at Swartkrans was therefore unexpected.

The specimen, SK 336, consists of a left mandible fragment and M_1. In his reassessment of the Swartkrans carnivores, Hendey (1974*b*) concluded that the fossil actually came from a hunting hyena, *Euryboas nitidula* type A (primitive), so that what was regarded as the most problematical identification among the Swartkrans carnivores has been resolved.

Family Hyaenidae

Hyaenids had their origin in the Miocene but did not become abundant until the Pliocene, when their main radiation occurred. They appear to have arisen from viverrid ancestors, perhaps not unlike the modern civets, and then to have undergone dental adaptations that made the efficient crushing of bones possible. Although hyenas have very effective carnassials for slicing meat, their anterior premolars have become extremely robust to serve as bone crushers. As already mentioned, the hyena radiation was probably made possible by the activities of saber-toothed cats that were unable to eat the bones of their prey themselves. A valuable food source thus became available to any scavenger able to use it. During the Pliocene, the hyenas appear to have made full use of this bony resource provided by the sabertooths.

Fossil hyenas from the Sterkfontein valley caves are currently classified in four genera: *Hyaenictis, Euryboas, Hyaena,* and *Crocuta*. The first two are extinct and appear to have been secondarily predacious carnivores, while the latter two have living representatives almost identical to their fossil antecedents.

The Hunting Hyenas: *Hyaenictis* and *Euryboas*
The Genus *Hyaenictis*

This genus was created when Gaudry described an immature hyaenid mandible from Pikermi and named it *H. graeca*. It was characterized by a low, flat mandible profile, not deepened below M_1, and by the retention of M_2. The genus is now known to have had a mid-Pliocene time range, occurring in Eurasia and Africa.

During the early phase of work at Swartkrans, a hyaenid cranium and mandible (SK 314) was found and considered by Broom and Robinson (1952) to belong in the genus *Hyaenictis*. No specific name was suggested. Ewer (1955*a*) reexamined the specimen and concluded that, although M_2 was indeed present, other characters diagnostic of the genus *Hyaenictis* were not. She therefore described SK 314 as a new species of the genus *Leecyaena* Young and Liu, 1948, naming it *L. forfex* Ewer. Later, Ewer (1967) expressed the opinion that *L. forfex* was not closely related to the only other known species of *Leecyaena, L. lycyaenoides* from the late Pliocene of China. Hendey (1974*b*) tended to agree with her and, because he did not think the two formed parts of the same lineage, he concluded that the Swartkrans fossil should be named *Hyaenictis forfex*. There is, however, still some doubt about its generic identity.

From Langebaanweg, Hendey (1974*a*) described hyaenid remains rather similar to the Swartkrans specimen as *Hyaenictis preforfex*, a form he regarded as ancestral to *H. forfex*. In addition to cranial remains, excellent postcranial specimens were obtained at Langebaanweg that showed that the limbs of *preforfex* were long and slender, indicating a cursorial habit. Hendey remarks (1974*a*):

> This is of interest because of the suggestion (Thenius, 1966) that the long-legged "hunting hyaena" *Euryboas*, was descended from *Hyaenictis graeca*. Pliocene *Hyaenictis* may thus have been a group of long-limbed forms which evolved in two directions, one branch becoming increasingly cursorial and actively predacious *(Euryboas)*, and the other paralleling *Hyaena* and *Crocuta (H. bosei, H. forfex, H. preforfex)*.

Hyaenictis forfex (Ewer, 1955)
Provenance: Swartkrans Member 1

This animal had a head about the size of a brown hyena *(H. brunnea)* but had longer limbs and a more slender body. Dentally it is characterized by anterior cheek teeth resembling those of an advanced *Hyaena*, a relatively long M_1, and the presence of a small M_2.

The Genus *Euryboas*

The genus is best known by the European hunting hyena, *E. lunensis* Del Campana, found as fossils in various sites of Villafranchian age. Although definitely a hyena, *Euryboas* does not show dental specializations for bone crushing carried to an extreme degree. Its teeth are those of a predator, while its limb bones are almost as long and slender as those of a cheetah. One may assume that *Euryboas* was a fast cursorial hunter, running down its prey after the fashion of a cheetah. Kurtén (1968) considers the prey of the European hunting hyena to have been largely the Bourbon gazelle, *Gazella borbonica* Depéret, or the chamois antelope, *Procamptoceras brivatense* Schaub.

The southern African representatives of the genus are known only from Sterkfontein and Swartkrans and were initially placed in the genus *Lycyaena*. The first specimen from Sterkfontein came to light in a rather unusual manner. The Abbé Breuil was in Johannesburg in January 1945 and happened to visit an art gallery owned by H. K. Silberberg. There he saw some fossils Silberberg had picked up at Sterkfontein a few years earlier, and among them was the muzzle of an interesting-looking carnivore. The abbé asked for the specimen and took it to Broom in Pretoria, who immediately recognized that its affinities lay with the Pliocene genus *Lycyaena*, known from Europe and Asia. He described it (Broom 1948*a*) as *Lycyaena silberbergi*.

In her account of the fossil carnivores from the Transvaal caves, Ewer (1955*b*), assigned two further specimens from Sterkfontein and one from Swartkrans to *L. silberbergi* and amplified Broom's description of the species. She also described a new species of *Lycyaena*, named *L. nitidula*, on the basis of a series of specimens from Swartkrans and a single one from Sterkfontein. When reviewing the fossil hyaenids of Africa, Ewer (1967) expressed the opinion that *L. nitidula* should be regarded as a subspecies of *L. silberbergi*. More recently, Hendey (1974*b*) has proposed that the *Lycyaena* specimens from Sterkfontein and Swartkrans should be referred to the genus *Euryboas*. This proposal appears to have been generally accepted. Hendey's reassessment of the specimens has not yet appeared, but he is inclined (pers. comm.) to place all the specimens from Sterkfontein in *E. silberbergi* and all those from Swartkrans in *E. nitidula,* this taxon, however, being split into a type A category, showing some primitive characteristics, and a type B group with more advanced attributes.

Two species of hunting hyena are thus currently recognized in the Sterkfontein valley assemblages:

Euryboas silberbergi (Broom)
Provenance: Sterkfontein Member 2 or Member 3 (see below) and Member 4

The type specimen was picked up underground, in what is now called the Daylight Cave. It probably came from

Member 2 or Member 3 of the Sterkfontein Formation while subsequent specimens were found in Member 4.

Euryboas nitidula (Ewer)
Provenance: Swartkrans Member 1. Both primitive and advanced forms (types A and B) appear to occur.

E. nitidula may be separated from *E. silberbergi* by the overlapping and oblique setting of the premolars and by the arrangement of accessory cusps. A tentative identification of *Euryboas* has recently been made by Howell and Petter (1976) from Member F of the Shungura Formation, Omo Group, of southern Ethiopia.

The Genus *Hyaena*

The genus has two living representatives: *H. hyaena* Linné, the striped hyena of tropical and north Africa, Asia, and formerly of Europe, and *H. brunnea* Thunberg, the brown hyena, now restricted to southern Africa.

The first fossil *Hyaena* specimens from southern Africa were described by Toerien (1952) from the Makapansgat Limeworks and named *H. makapani*. Ewer (1967) did not regard the differences between this species and *H. hyaena* as sufficient to warrant specific separation, and she renamed the Makapansgat material *H. hyaena makapani*.

Hendey (1974*a*) has traced a phylogeny for the genus *Hyaena* from Pliocene ancestors *Ictitherium* or *Palhyaena*. He visualizes the late Pliocene *H. abronia* from Langebaanweg and the mid-Pliocene *H. pyrenaica* from Europe as the earliest recorded representatives of the *H. hyaena* and *H. brunnea* lineages respectively. He writes Hendey 1974*a*, p. 146):

> The *H. hyaena* lineage, which includes *H. h. makapani* from the Transvaal, was a conservative one which underwent comparatively little change from the late Pliocene onwards. The differentiation of this lineage might well have taken place in Africa. An ancestor of *H. brunnea* probably entered Africa from the north during the Pliocene. It is first recorded in South Africa during the Makapanian (Swartkrans) and it apparently replaced *H. hyaena,* which is last recorded earlier in the age (Makapansgat). *H. bellax,* another Makapanian species, is probably an off-shoot of the *H. brunnea* lineage.

The two species involved in the Sterkfontein Valley caves are *H. brunnea* and *H. bellax*.

Hyaena brunnea Thunberg, 1820. Brown hyena
Provenance: Swartkrans members 1 and 2 and Kromdraai B (tentative identification)

The remains from Swartkrans Member 1 have been described by Ewer (1955*a*) as a new subspecies *H. brunnea*

dispar, distinguished from the extant race by "the slightly more primitive character of the upper premolars, the relatively short P_3 and the narrowness of the anterior palatal foramen" (Ewer 1955*a,* p. 824).

Hyaena bellax Ewer, 1954
Provenance: Kromdraai A

This large extinct hyena was described on the basis of a beautiful cranium with articulated mandible, KA 55 (fig. 208). In appearance the animal must have been similar to the brown hyena, to which it was certainly related. Ewer (1954*b*) states that the species shows a combination of primitive and advanced characters, the upper molars relatively as large as those of *H. hyaena* but the carnassial teeth more advanced than those of *H. brunnea*.

More recently Ewer (1967) has suggested that *H. bellax* may have evolved from *Hyaenictis (Lycyaena) forfex,* but this view does not appear to have the support of Hendey (1974*a*).

H. bellax is very probably also represented among fossils from Baard's Quarry, Langebaanweg (Hendey 1974*a*).

The Genus *Crocuta*

The spotted hyena, *Crocuta crocuta* (Erxleben, 1777), is a familiar member of the genus Kurtén (1956) has suggested arose in India, south of the Himalayan range, then spread into Europe and Africa during the Middle Pleistocene; the possibility of an earlier migration to Africa is considered (Kurtén 1957*a,b*). Later Ewer (1967) made the countersuggestion that *Crocuta* may have arisen in Africa, thence spreading to Europe and Asia.

That *Crocuta* was present in Africa before the Middle Pleistocene is well established. Howell and Petter (1976) record it from Member G of the Shungura Formation; M. G. Leakey (1976) records it from the Lower, Upper, and Ileret members of the Koobi Fora Formation, and it is also known from Laetolil (Dietrich 1942), Bed 1 at Olduvai (Ewer 1965), and elsewhere.

Howell and Petter have expressed the opinion that *Crocuta* did, in fact, migrate into Africa as Kurtén suggests, but considerably earlier than he has visualized.

The following forms have been described from cave deposits in the Sterkfontein valley at various times:

Crocuta ultra ultra Ewer, 1954
Based on several cranial and postcranial specimens from Kromdraai A, distinguished inter alia by a short snout, compressed incisor region, and the fact that the frontals and premaxillae do not make contact.

C. ultra latidens Ewer, 1954
Based on three specimens from the Clyde Trading Company site in the Sterkfontein valley and differing from *C. u. ultra* particularly in the structure of the cusps.

C. spelaea capensis Broom, 1939
Based on a cranium from Kromdraai that both Broom (1939*a*) and Ewer (1954*b*) considered to resemble the European *C. spelaea* more closely than it does the living *C. crocuta*. There has been a good deal of controversy over whether *C. spelaea* should be separated from *C. crocuta*. The matter remains unresolved. *C.* cf. *spelaea* has also been recorded from the Sterkfontein Type Site on the basis of a left maxilla (Ewer 1955*a*).

C. venustula Ewer, 1955
Based on two mandible pieces from Member 1 at Swartkrans—a small *Crocuta* with a slender, shallow mandible.

C. crocuta angella Ewer, 1955
Based on several specimens from Member 1 at Swartkrans and differing from the living spotted hyena in certain minor features of the dentition.

In her review of African fossil hyaenids, Ewer (1967) relegated the previous species *C. ultra* and *C. venustula* to subspecies of *C. crocuta*. Subsequently Hendey remarked (1974*b*): "There has probably been more confusion over the taxonomy of *Crocuta* than that of any other carnivore in Africa and there is still a need for a comprehensive review of the *Crocuta* material now available."

He agreed with Ewer that *C. venustula* should not be regarded as a separate species and placed all the Swartkrans material in the taxon *C. crocuta*. Likewise, all other *Crocuta* material from the cave deposits is referred to *C. crocuta*, pending a comprehensive revision of the genus (Hendey, pers. comm.).

Family Hyaenidae
Subfamily Protelinae
Genus *Proteles*

Until recently, the genus contained a single living species, *P. cristatus* Sparrman, 1783, the aardwolf. Although unquestionably a hyaenid, the aardwolf is small and has undergone very marked degeneration of the post-canine teeth in response to its diet of termites and other small invertebrates. Premolars and molars have been reduced to ineffectual pegs. Ewer (1973) has suggested that predacious Pliocene hyenas of the genus *Lycyaena* may have given rise to *Proteles*. She argues that they probably failed in competition both with the specialized scavengers such as *Crocuta* and with the advanced felines, but that they found their salvation in turning to a diet of insects.

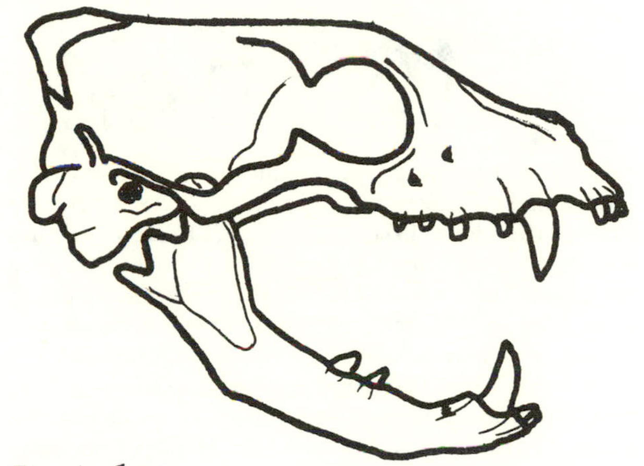

Proteles

An interesting fossil ancestor of the aardwolf has come to light from Member 1 at Swartkrans and is probably also represented at the Kromdraai australopithecine site (Hendey 1973, 1974*b*). Described as *Proteles transvaalensis* Hendey (1974*b*), it proves to be bigger and less degenerate dentally than is the living *P. cristatus*. The remains suggest that some of the dental degeneration seen in the living aardwolf has taken place during the Pleistocene.

Sparse remains of the extant *P. cristatus* are found in Swartkrans Member 2. These do not appear to differ from equivalent parts of the living form. Recent fossils have also been described from the Black Earth Cave at Taung by Gingerich (1974).

Family Canidae

In striking contrast to the felids, the canids have remained generalized feeders, with dentitions capable of coping with either meat or vegetable food. The molars have not been reduced or lost as they have in cats, and they provide a grinding battery in addition to the slicing carnassials. Canid muzzles are thus typically longer than felid muzzles when animals of equivalent body weight are compared.

The smallest members of the family in southern Africa are foxes, which differ from jackals and larger canids in that they do not have frontal sinuses. According to Ewer (1956*b*), the function of the frontal sinuses is to increase the attachment area for the anterior fibres of the temporalis muscles without unduly increasing skull weight. Foxes, with their relatively weak jaws, have not found it necessary to increase the temporalis attachment areas. Canids are well represented among fossils from the Sterkfontein valley caves, specimens having been allocated to four genera: *Otocyon,* the bat-eared foxes; *Vulpes,* the foxes; *Canis,* the jackals; and *Lycaon,* the hunting dogs.

Genus *Otocyon* Müller, 1836. Bat-eared foxes

The bat-eared fox, *O. megalotis* Desmarest, is a familiar carnivore of the more arid regions of southern Africa

and has an extensive range farther north, feeding on a variety of small prey including insects, arachnids, and rodents (Smithers 1971*a*). Hendey (1974*b*, p. 31) has pointed out that "*Otocyon* dentitions are an exception to the general rule among mammals in that the more advanced the species, the greater the number of molars and the greater the complexity of their occlusal surfaces."

Two specimens that may be referred to *Otocyon* have been found in Member 2 at Swartkrans, an incomplete palate and an isolated lower molar. Hendey concluded that these were dentally more primitive than *O. megalotis* and referred them to *O. recki,* a species known from Bed 1 at Olduvai. He pointed out that the teeth of the Swartkrans specimens were not quite as primitive in character as those from Olduvai.

Genus *Vulpes.* Foxes

The Cape fox or "silver jackal," *V. chama* (A. Smith, 1833), is the only southern African representative of the genus and occurs in the more arid areas of the region. Fossil remains of the species have been found at the sites of Elandsfontein, Swartklip, and Sea Harvest in the western Cape (Hendey 1974*a*), while much earlier fossils referrable to *Vulpes* are known from Langebaanweg (Hendey 1976). Broom (1937*c*) described an edentulous mandible from Taung as *V. pattisoni* Broom. This may probably be referred to *V. chama.*

A fossil form intermediate in size between *V. chama* and the European fox, *V. vulpes* Linné, was described by Broom on the basis of a mandible from Kromdraai A. He named it *V. pulcher* Broom. More recently two specimens from Member 1 at Swartkrans have been referred to the species (Ewer 1956*b;* and Hendey 1974*b*). *V. pulcher* was clearly a fox slightly more robust than the living Cape fox. It is known only from Kromdraai A and Swartkrans Member 1.

Genus *Canis.* Jackals

Two species are currently found in southern Africa, the black-backed jackal, *C. mesomelas* Schreber, and the side-striped form, *C. adustus* Sundevall. Fossil remains, particularly of *C. mesomelas,* occur abundantly at a variety of sites.

Three species of jackal have been described or identified among the fossils from the Sterkfontein valley caves. In his early descriptions, Broom (1937*c*, 1939*a*, 1948*a*) used the generic name *Thos* Oken, but Ewer (1956*b*) has argued that *Thos* should be regarded as a subgenus of *Canis.* The three species are as follows:

C. terblanchei (Broom). Terblanche's jackal

A jackal about the size of the living *C. adustus* but differing from it in certain minor aspects of its cranial

anatomy and dentition. The type skull was described by Broom (1948*a*) as coming from Kromdraai A, but Ewer (1956*b*) pointed out that the matrix was unlike that of other KA specimens, and she suggested that the type had, in fact, come from elsewhere. Two specimens referred to *C. terblanchei* by Ewer (1956*b*) had been found at the Cooper's site, but the provenance of the type remains uncertain. Apart from these, the only other specimen known is a left maxillary piece from Member 5 at Sterkfontein. It is tentatively identified as *C.* cf. *terblanchei.*

C. brevirostris Ewer, 1956. Ewer's short-faced jackal

The species is known on the basis of a single cranium from Member 4 at Sterkfontein. According to Ewer's

(1956*b*) description, it was a small, short-faced jackal with well-developed upper canines, extremely shortened premolar series, and rather rectangular upper molars. The dentition suggests that *C. brevirostris* was a less specialized feeder than are the living jackals.

C. mesomelas Schreber. Black-backed jackal
Provenance: Sterkfontein Member 4, Swartkrans members 1 and 2 and channel fill, Kromdraai A, and probably Kromdraai B

Remains of this familiar jackal have been found at all the main fossil-bearing localities in the Sterkfontein valley.

Broom (1937*b*) described the mandible of an immature jackal as representing a new species, *Thos antiquus*

Broom. He stated that it came from "the same cave as that in which the skull of *Australopithecus transvaalensis* was found"; however, the catalog entry records the specimen as coming from Minnaar's Cave, and there is consequently some uncertainty. Ewer (1956*b*) prepared and examined the specimen further and concluded that it should be placed in the taxon *C. mesomelas*. At the same time she examined all the available specimens from the Sterkfontein valley caves and decided that, as a group, the fossils were sufficiently different from living black-backed jackals to warrant creating a new subspecies, *C. mesomelas pappos* Ewer. The points distinguishing the fossil subspecies were summarized as follows (Ewer 1956*b*, p. 113): "differing from the living *C. mesomelas* in having the lower premolars longer in comparison with the length of the carnassial; nasals frequently not extending behind the posterior end of the maxilla."

More recently Hendey (1974*b*) expressed the opinion that *C. mesomelas pappos* constitutes a heterogeneous and unsatisfactory taxon, and the subspecies name is therefore not used in this work.

Genus *Lycaon*. Hunting dogs

The hunting or wild dog, *L. pictus* Temminck, is a familiar African predator, hunting in packs and relying on its speed and endurance to run down its medium-sized ungulate prey. A European species was formerly designated *L. lycaonoides* Kretzoi but has now been transferred to the genus *Xenocyon* by Schütt (1974). The form appeared briefly during the early Middle Pleistocene of Europe but appears to have failed in competition with the Dhole, *Cuon* sp. (Kurtén 1968). The only other species in the genus is *L. atrox,* although the Olduvai form *Canis africanus* should possibly also be included (Q. B. Hendey, pers. comm.).

Lycaon atrox (Broom). Kromdraai hunting dog
Provenance: Swartkrans Member 2 (possibly) and Kromdraai A

In 1939 Broom mentioned the lower carnassial of a large canid from Kromdraai A and subsequently described this tooth (1948*a;* Broom and Schepers 1946) together with an isolated upper molar, referring them to *Canis atrox* Broom. In the course of her reexamination of the carnivore remains from Kromdraai, Ewer (1956*b*) concluded that *C. atrox* had affinities with *C. africanus* from Bed II at Olduvai, though she concluded that they were specifically distinct.

More recently, Hendey (1974*a*) has shown that *Lycaon* remains from Elandsfontein in the western Cape exhibit characters intermediate between those of *C. atrox* and *L. pictus*. He concluded that hunting dogs of the genus *Lycaon* arose from a wolflike *Canis* ancestor during late Pliocene or early Pleistocene times and that *C. atrox* was an early member of the lineage that led to the living *L. pictus*. On these grounds he placed *C. atrox* in the genus *Lycaon*.

Regrettably, we know nothing of the skeleton of the Kromdraai hunting dog, *L. atrox*. We may speculate that,

as a primitive member of the *Lycaon* lineage, its cursorial adaptations may have been less well developed than they are in the living hunting dogs. Its limbs may have been somewhat shorter and less slender.

Apart from the two Kromdraai teeth just discussed, the only other hunting dog fossil known from the Sterkfontein valley caves is an isolated upper canine from Member 2 at Swartkrans. This has been referred to *Lycaon* sp. indet. (Hendey 1974*b*).

Family Mustelidae

Three subfamilies are represented in the living southern African fauna: Mustelinae, which includes polecats (*Ictonyx* Kaup, 1835) and striped weasels (*Poecilogale* Thomas 1883); Mellivorinae, the honey badgers (*Mellivora* Starr, 1780); and Lutrinae, which includes otters of the genera *Aonyx* Lesson, 1827, and *Lutra* Brisson, 1762. These carnivores present a wide spectrum of bodily form, size, and adaptation, the polecats and weasels being small, active, and highly efficient predators, while the honey badger and others are much larger animals in which the crushing function of the dentition is more developed to suit their particular diets.

Mustelid remains are not common among fossils from the Sterkfontein valley caves. In his report on fossil mammals collected by the University of California expedition in 1947–48, H. B. S. Cooke (n.d.) mentioned the presence of the otter, *Aonyx* cf. *capensis* in Pit 3 of Bolt's Farm and described a new species of polecat from Pit 10 of the same site. He suggested that it represented an offshoot from the stock that gave rise to both *Ictonyx* and

Poecilogale, but that it was less specialized than any of the living species of these genera.

The only mustelid from the sites that particularly concern us here is the following:

Mellivora cf. *sivalensis* Falconer and Cautley, 1868. Honey badger
Provenance: Swartkrans Member 2

Two isolated teeth and a mandible piece have been described by Hendey (1974*b*). He found that these teeth

were less primitive than those of *M. punjabiensis* Lydekker, 1884, provisionally identified from E Quarry at Langebaanweg (Hendey 1974*a*) but not as advanced as teeth of the living honey badger *M. capensis* (Schreber, 1776). He therefore referred the Swartkrans specimen to *M.* cf. *sivalensis,* a species from the Pinjor stage of the Siwaliks in India that shows the same set of characters.

The Swartkrans honey badger was probably very similar to the living one in appearance and habits.

Family Viverridae

Carnivores with a very long history, representing the basic stock from which both the hyenas and the cats arose, the viverrids are an old-world family that retains many primitive features. They tend to be mustelinelike in their adaptation. Two subfamilies occur in southern Africa: the Viverrinae, comprising the civets and genets, and the Herpestinae, the mongooses. Skulls of the latter can be separated from those of the former by the presence of well-developed postorbital processes.

Subfamily Viverrinae
Genus *Viverra* Linnaeus, 1758
cf. *Viverra* sp. Civet
Provenance: Kromdraai B

The distal end of a right humerus from Kromdraai B proves to be morphologically indistinguishable from the equivalent part of the living civet, *V. civetta* Schreber or from that of the Pliocene form, *V. leakeyi* Petter, found at Langebaanweg and in East Africa (Hendey 1973). Until more complete specimens come to light, little more can be said than that remains of a civet are represented in the Kromdraai B assemblage. They have not been recorded in the other Sterkfontein valley caves.

Subfamily *Hespestinae.* Mongooses

Fossil mongooses from the sites of Sterkfontein, Swartkrans, and Kromdraai are currently allocated to three genera: *Herpestes, Cynictis,* and *Crossarchus.*

Genus *Herpestes* Illiger, 1811
Herpestes mesotes Ewer, 1956
Provenance: Kromdraai A

Based on a single skull and mandible, this species closely resembles the living large gray mongoose, *H.*

ichneumon Grill, 1858, which has an extremely wide African distribution. The fossil differs from the living form in having more robust cheek teeth, a less reduced M^2, and a less prolonged palate behind the last molars (Ewer 1956*c*). In her description, Ewer pointed out that *H. mesotes* had close affinities with both *H. ichneumon* and the water mongoose *Atilax paludinosus* (G. Cuvier). She considered that the fossil form had characters that might be expected in an ancestor of either of the living animals, though she was inclined to place it low down on the lineage leading to *Atilax.* If this is so, and Hendey (1974*a*) is inclined to believe it is, then the species *mesotes* should be placed in the genus *Atilax* rather than *Herpestes.* Further material is required to settle the question. Fossils of *H.* sp. aff. *ichneumon* Linné have been recorded from Bed 1 at Olduvai (G. Petter 1973), while *Herpestes* sp. is known from Member C of the Shungura Formation, Omo valley (Howell and Petter 1976). More recent fossils of *H. ichneumon* come from the Elandsfontein, Swartklip, and Sea Harvest sites in the western Cape (Hendey 1974*a*).

Herpestes cf. *sanguineus* Rüppel, 1836. Slender mongoose
Provenance: Swartkrans Member 2 and possibly Kromdraai B

Part of a mandible has been described by Hendey (1974*b*). He points out that it could be referred either to *H.* cf. *sanguineus,* the common slender mongoose, or to *H. pulverulentus* Wagner, the Cape gray mongoose. However, since these two forms are apparently very

closely related, the Swartkrans fossil could be ancestral to either.

Two specimens from Kromdraai B have also been referred to *Herpestes* sp. by Hendey (1973), who expressed the opinion that they might come either from "a large variety of a species such as *H. sanguineus* or a small variety of a species such as *H. ichneumon*."

Cynictis penicillata (G. Cuvier). Yellow mongoose
Provenance: Swartkrans Member 2

The familiar and widely distributed yellow mongoose is characterized by a bushy tail with a white tip.

Ewer (1956*c*) described a mandible from Swartkrans, pointing out that it differed from mandibles of the living

C. penicillata in the wide spacing of the cheek teeth and the relatively small size of M_1. Although she felt that the fossil was subspecifically distinct, she declined to name the new subspecies in the absence of more complete specimens. She has, however, expressed the opinion that the Swartkrans subspecies could be the same as *C. penicillata brachyodon* Ewer that she described from the Makapansgat Limeworks on the basis of a reasonably complete cranium.

Crossarchus transvaalensis Broom, 1937
Provenance: Kromdraai A

Described originally on the basis of a right mandibular ramus from Bolt's Farm. Broom (1939*a*) subsequently assigned a fossil palate from Kromdraai A to the same taxon, albeit somewhat tentatively. Remains have not been recorded from other sites, and the relationship of *C. transvaalensis* is somewhat obscure. The genus is represented by three species of living long-nosed mongooses elsewhere in Africa: *C. ansorgei* Thomas from Angola; *C. obscurus* F. Cuvier and *C. alexandri* Thomas and Wroughton from equatorial forest regions, which two species may be conspecific (Coetzee 1971). These are diurnal mongooses, hunting in small parties. They are mainly terrestrial, though they may take refuge in trees.

Order Artiodactyla

Artiodactyls whose fossils have been preserved in the Sterkfontein valley caves are classified in three families: Bovidae, the antelopes; Suidae, the pigs; and Giraffidae, the giraffes.

Family Bovidae

Fossils from the Sterkfontein valley caves may be classified, following Ansell (1971), in four subfamilies that, in turn, are divided into nine tribes as follows:

Subfamily Bovinae	
Tribe Bovini	Buffalo
Tribe Tragelaphini	Eland, kudu, bushbuck, nyala
Subfamily Hippotraginae	
Tribe Alcelaphini	Wildebeest, hartebeest, blesbok
Tribe Hippotragini	Sable, roan
Tribe Reduncini	Waterbuck, reedbuck
Tribe Peleini	Rhebuck
Subfamily Antilopinae	
Tribe Antilopini	Springbok, gazelle
Tribe Neotragini	Steenbok, klipspringer
Subfamily Caprinae	
Tribe Ovibovini	"Makapania"

Subfamily Bovinae

Ansell (1971) includes three tribes in this subfamily: Bovini, Tragelaphini, and Boselaphini, the last without living African representatives but present in India in the form of the nilgai, *Boselaphus tragocamelus* (Pallas), and the four-horned antelope *Tetracerus quadricornis* (De Blainville). Recently Gentry (1974) has described a boselaphine from Pliocene deposits at Langebaanweg as *Mesembriportax acrae* Gentry, the first representative of the tribe to have been found in southern Africa.

Tribe Bovini

Three genera are included in Ansell's (1971) African list: *Bos* Linnaeus, the wild ox; *Bubalus* H. Smith, the water buffalo; *Syncerus* Hodgson, the African buffalo.

Syncerus sp. Buffalo
Provenance: Swartkrans members 1 and 2 and Kromdraai A

The taxon is best represented at Swartkrans, where 13 cranial pieces come from at least 3 individuals; a single juvenile is recorded from Sterkfontein, and 4 teeth have been found at Kromdraai.

Vrba (1976*a*) has pointed out that the dentitions from the Sterkfontein valley sites have clear affinities with that of the extant African buffalo *S. caffer* (Sparrman) but

differ in the following respects: the fossil teeth are consistently larger; they have simpler enamel patterns on their occlusal surfaces; and the paraconid and metaconid of the P_4 remain unfused.

The possibility that the fossils should be referred to the genus of the extinct long-horned buffalo, *Pelorovis* Reck, has been thoroughly discussed by Vrba. They could, for instance, have affinities to the widespread *P. antiquus* (Duvernoy), which is known from many southern African localities and which appears to have become extinct about 10,000 years ago (Klein 1974*a*); other possible relatives are *P. oldowayensis* Reck from Olduvai and a third unnamed species known from Elandsfontein (Hendey 1974*a*).

For several reasons Vrba discounts a close relationship between the Swartkrans fossils and *Pelorovis*. Instead she points to similarities between these specimens and fossils from Bed II at Olduvai that appear to be on the *Syncerus* lineage. The Olduvai species remains to be named.

It may be concluded that the bovine remains from Sterkfontein, Swartkrans, and Kromdraai appear to have come from buffalo somewhat more robust than, but probably close to, the ancestry of the living African buffalo *S. caffer* (Sparrman), which shows a great deal of variability in its wide African range and which appears to comprise at least two subspecies, *S. caffer caffer* in southern Africa and *S. c. nanus* in forests of west and central Africa.

Tribe Tragelaphini

Opinions on the classification of this group of antelope vary considerably, but Ansell (1971) places all living representatives in two genera: *Tragelaphus* De Blainville (bongo, mountain nyala, sitatunga, nyala, bushbuck, and kudu) and *Taurotragus* Wagner, the eland. All are medium to large antelope with spiral horns that may or may not be present in the females; they are browsers preferring a bushy or wooded environment.

Tragelaphus cf. *scriptus* Pallas. Bushbuck
Provenance: Swartkrans Member 2 and Kromdraai A

Some rather fragmentary dentitions have been tentatively referred to this taxon by Vrba (1976*a*). In their study of fossil Bovidae from Makapansgat, Wells and Cooke (1956) described what they considered a new species of duiker, *Cephalophus pricei,* perhaps related to

the red duiker, *C. natalensis* A. Smith. Gentry (1976) has reclassified the Makapansgat dentitions (though not the horn-core) as *Tragelaphus pricei* (Wells and Cooke), suggesting that they have strong affinities with the bushbuck, *T. scriptus*. Gentry also identifies *T.* cf. *pricei* from Member C of the Shungura Formation in the Omo valley, pointing out that this animal differed from the living bushbuck in having horns more vertically inserted and less anteroposteriorly compressed.

Apart from the evidence of these early remains and some much later ones from Klasies River Mouth caves (Klein 1976*a*), the fossil history of the bushbuck is obscure.

Tragelaphus cf. *strepsiceros* (Pallas). Greater kudu
Provenance: Swartkrans members 1 and 2 and Kromdraai A

Vrba (1976*a*) has shown that the remains—a single horn-core piece and various adult and juvenile

dentitions—appear consistently larger than equivalent parts of the living greater kudu. A similarly large form has been described from Makapansgat by Wells and Cooke (1956).

Another large form of kudu was described by L. S. B. Leakey (1965) as *Strepsiceros grandis* from upper Bed II at Olduvai. Vrba (1976*a*) states that the Sterkfontein valley remains are probably similar to these Olduvai fossils, which Gentry (1970*a*) reclassifies as *Tragelaphus strepsiceros grandis* (Leakey).

Fossils that appear to be at least subspecifically different from the living kudu are known from Melkbos and Elandsfontein (Hendey 1968), while others, modern in appearance, are known from many sites such as the Klasies River Mouth caves (Klein 1976*a*) and Cave of Hearths (H. B. S. Cooke 1962).

The living greater kudu has an extremely wide African range and, according to Ansell (1971), four subspecies are currently recognized.

Tragelaphus sp. aff. *angasi* Gray. Nyala
Provenance: Sterkfontein Member 4 and Swartkrans Member 2

A few juvenile dentitions are tentatively referred to this taxon by Vrba (1976*a*), although they are somewhat larger than living counterparts. Fossils that also appear to come from rather large nyalas have been described from Makapansgat by Wells and Cooke (1956), who compared them to the mountain nyala, *T. buxtoni* Lydekker, a species endemic to the Ethiopian highlands.

T. angasi has a limited distribution in southeastern Africa and is typically restricted to thick riverine cover. It is

unusual in that the male is not only larger than the female but strikingly different in appearance.

Taurotragus cf. *oryx* (Pallas). Eland
Provenance: Sterkfontein Member 5, Swartkrans Member 2, and Kromdraai A

The only remains consist of isolated teeth that Vrba (1976*a*) considered morphologically similar to those of

living eland. Similar fossils have been reported from Makapansgat Limeworks (Wells and Cooke 1956), Melkbos (Hendey 1968), Elandsfontein, Swartklip, and Saldanha (Hendey 1974*a*), and many other southern African sites.

A more primitive species of eland, *T. arkelli,* was described by L. S. B. Leakey (1965) from Bed IV at Olduvai but is not known from South Africa.

Ansell (1971) recognizes three subspecies of the living eland, which has a wide, but frequently discontinuous, distribution, mainly south of the equator.

Subfamily Hippotraginae
Tribe Alcelaphini

The tribe contains four living genera confined to Africa: *Connochaetes* Lichtenstein (wildebeest), *Alcelaphus* De Blainville (hartebeest), *Beatragus* Heller (Hunter's hartebeest), and *Damaliscus* Sclater and Thomas (blesbok). Gentry (1978) suggests that *Aepyceros*, the impala, should also be included in the Alcelaphini. Three additional and extinct genera are represented among the Sterkfontein valley fossils: *Parmularius* Hopwood, *Rabaticeras* Ennouchi, and *Megalotragus* van Hoepen. All members of the tribe are medium to large antelopes, and both sexes carry horns. The frontals and horn pedicles are characterized by well-developed, smooth-walled sinuses, while the braincase is steeply angled to the axis of the face, which tends to be long and narrow. The molars are hypsodont and premolar rows are short.

Genus *Connochaetes* Lichtenstein

Two living species are currently placed in this genus: the brindled gnu or blue wildebeest *C. taurinus* (Bur-

chell), with a fairly wide distribution in southern and eastern Africa, and the white-tailed gnu or black wildebeest, *C. gnou* (Zimmerman), which formerly ranged over the open central plateau of South Africa but is now much more restricted. Fossil remains of *Connochaetes* are known from a number of sites in the Orange Free State, from Elandsfontein and elsewhere. The genus is also abundantly represented in assemblages from east African sites, and it or a related genus is present in north African sites.

Cranial remains, mainly dentitions, of wildebeestlike alcelaphines are abundant in the collections from the Sterkfontein valley caves. In her study of fossil Bovidae from Sterkfontein and Kromdraai, Vrba (1976*a*) records the taxon cf. *Connochaetes* sp. from the site units of Sterkfontein members 4 and 5, Swartkrans members 1 and 2, and Kromdraai A. *Connochaetes* sp. is recorded from Kromdraai B and from Swartkrans channel fill. *C.* cf. *taurinus* is recorded from dump D16 at Sterkfontein. Vrba expresses the opinion that all these remains probably came from animals on, or close to, the *C. taurinus* lineage, though some of the teeth from Kromdraai A show unusual morphology suggestive of a specialization away from the lineage.

Dentitions from Member 1 at Swartkrans showed clear differences from those of the living *C. taurinus.* To convert the Swartkrans dental morphology into that of the blue wildebeest as known today, Vrba (1976*a*, p. 10) speculated that the simple tooth shape of the fossils would become more complicated and specialized; the anterior part of the toothrow would assume greater functional significance, despite the gradual loss of P_2; a widening of P_3, P_4, and M_1, together with further molarization of the premolars, would occur, and increasing hypsodonty of the premolars would result in deepening of the mandibular ramus below them.

It is clear that ancestors of the blue wildebeest, known in the area today, were a well represented component of the fauna from the Sterkfontein valley sites.

Genus *Damaliscus* Sclater and Thomas

According to Ansell (1971), the genus contains three living species: *D. dorcas* (Pallas, 1766), the blesbok and bontebok; *D. lunatus* (Burchell, 1823), the tsessebe, topi, and so forth; and *D. hunteri* (P. L. Sclater, 1889), the

Hunter's hartebeest. The last species, *D. hunteri*, has an extremely restricted range between the Tana River, Kenya, and the Juba River, Somalia; it is considered by some mammalogists to fit best in a different genus, *Beatragus* Heller, and in this account of the fossil animals *Beatragus* is maintained as a genus separate from *Damaliscus*.

Small alcelaphine antelope are well represented in the Sterkfontein valley fossil assemblages. Vrba (1976*a*, p. 24) has tabulated differences in the dentitions that allowed her to divide the collections between three taxa: *Damaliscus* sp. 1, or *Parmularius* sp.; *Damaliscus* sp. 2 (*?D. niro*); and *D. dorcas*. Subsequently her further studies have indicated (Vrba 1978) that all the *Damaliscus* sp. 1 or *Parmularius* sp. material belongs to the genus *Parmularius*. A new species, *Parmularius parvus*, was described on the basis of the Kromdraai A small alcelaphine specimens. In addition, a specimen from Swartkrans Member 1, SK 14104, is now regarded as *Parmularius angusticornis* (Schwarz) (Vrba, pers. comm.).

Damaliscus sp. 1 or *Parmularius* sp.
Provenance: This is the dominant antelope species at Kromdraai A and is also represented at Swartkrans

Representatives of the genus *Parmularius* are known from several levels in the Olduvai sequence (L. S. B.

Leakey 1965), from the Shungura Formation of the Omo Group (Gentry 1976), and from the Koobi Fora Formation of the East Rudolf succession (Harris 1976). Recently Vrba (1977) has described a new species of *Parmularius*—the first from South Africa—from members 2, 3, and possibly 4 of the Makapansgat Formation. Previously, Vrba (1976*a*) had expressed the opinion that *P. parvus* from Kromdraai A may be closely related to *P. rugosus* L. S. B. Leakey, as known from Beds III–IV at Olduvai.

Parmularius sp.

Material from members 4 and 5 at Sterkfontein may belong not to *P. parvus*, but to a different species.

Damaliscus sp. 2 (*?niro*)
Provenance: Swartkrans Member 2, abundant (57 specimens from a minimum of 9 individuals); Sterkfontein Member 4, a single specimen that may have come from elsewhere; Sterkfontein Member 5 (7 specimens from a minimum of 4 individuals)

Vrba (1976*a*) created this taxon to accommodate small alcelaphine dentitions similar to those of *D. dorcas* but somewhat larger and with less complex enamel patterns on the occlusal surfaces of the teeth. A single horn-core piece from Member 2 at Swartkrans is strongly reminiscent of *D. niro* (Hopwood).

The species *D. niro* was first described by Hopwood (1936) on the basis of a right horn-core found by L. S. B. Leakey in Bed IV at Olduvai and sent to the British Museum in the early 1930s. Hopwood described it as *Hippotragus niro*, pointing out the similarity of the horn-core to that of the sable antelope, *H. niger*. He derived the trivial name *niro* from a Masai word meaning brown, the color of the holotype.

L. S. B. Leakey (1965) pointed out that specimens of *H. niro* had similarities to the genus *Damaliscus*. Gentry (1965) reexamined the material, including a new specimen from the Peninj Beds, west of Lake Natron and concluded that *Damaliscus* was in fact the most appropriate genus. The taxon thus became *D. niro* (Hopwood). It represents an alcelaphine antelope of medium size with horns reminiscent of those of a sable but differing by prominent widely spaced transverse ridges on their anterior surfaces. The horn-cores are typically shorter, however, and the frontals are less sharply raised between the horn bases.

D. niro is now known from several sites in the Orange Free State, including Cornelia and Florisbad (H. B. S. Cooke 1974), Driefontein near Cradock, Wonderwerk Cave, and Cave of Hearths (H. B. S. Cooke 1962; Wells 1970).

Damaliscus cf. *dorcas* (Pallas, 1766). Blesbok or bontebok
Provenance: Sterkfontein members 5 and 6, Swartkrans Member 2 and channel fill

Remains indistinguishable from equivalent parts of modern blesbok are fairly abundant in the later Sterkfontein valley site units. Vrba states (1976*a*, p. 24):

D. dorcas dentitions can be distinguished from all other extant or extinct alcelaphine species by the unique combination of

1. being among the smallest alcelaphine dentitions known,
2. having long premolar rows with P_2 and P^2 more often present than absent and P_3 and P^3 well developed, and
3. having a complicated molar occlusal surface enamel configuration.

Extinct alcelaphine species almost always seem to have a distinctly less complicated dental enamel configuration.

Extant representatives of the species are divided between two subspecies, according to Ansell (1971): *D. d. dorcas* (Pallas, 1766), the bontebok, restricted to the southwestern Cape Province, *D. d. phillipsi* Harper, 1939, the blesbok, with a much wider distribution in open habitats of the interior.

Rabaticerus porrocornutus (Vrba, 1971)
Provenance: Swartkrans Member 1

The type specimen was described by Vrba (1971) as *Damaliscus porrocornutus* on the basis of a frontlet with

horn-cores, SK 3211*a–b*, from Swartkrans Member 1. The specific name is based on the Latin adverb *porro* meaning "forward" and adjective *cornutus*, "horned"; it refers to the tendency of the horn-cores to sweep farther forward, relative to the face, than they do in related antelopes (Vrba 1971, p. 60). A year after the initial description appeared, the parameters of the type specimen were subjected to multivariate analysis (Laubscher, Steffens, and Vrba 1972), and the conclusion was reached that the type was unlikely to belong correctly in either of the genera *Damaliscus* or *Alcelaphus*. Subsequently, Vrba (1976*a*) reclassified it as *Rabaticeras porrocornutus*.

The genus *Rabaticeras* was first created by Ennouchi (1953) when he described a fossil skull from Rabat as *R. arambourgi*. he placed it in the Ovicaprinae, but it soon became apparent that this was in fact an alcelaphine.

The Swartkrans type specimen differed from *R. arambourgi* in a number of respects, notably that the horn-cores are oriented less forward with respect to the skull as a whole, and they also show more pronounced mediolateral compression (Vrba 1976*a*).

Rabaticeras remains are also known from Olduvai Bed II and the III/IV junction and from Elandsfontein. It has become clear that representatives of the genus were probably related to the ancestors of the living hartebeest species *Alcelaphus lichtensteini* (Peters) and *A. buselaphus* (Pallas).

Cf. *Beatragus* sp.
Provenance: Member 2 at Swartkrans

As mentioned earlier, the genus has a single living representative, Hunter's hartebeest, classified by Ansell

(1971) as *Damaliscus (Beatragus) hunteri* (P. L. Sclater) and restricted to a small area of Kenya and Somalia. It appears, however, that alcelaphines of the genus *Beatragus* were formerly more widespread. L. S. B. Leakey (1965) described *B. antiquus* from Olduvai, where it had been recorded from beds I and II as well as from high up in the Shungura Formation of the Omo valley (Gentry 1976). *B. antiquus* differed from living species in several respects, particularly its larger size and more upright horn insertions.

Remains of *Beatragus* occur at Elandsfontein, and Vrba (1976*a*) quotes Gentry's opinion that *B. antiquus* may have been replaced geographically by a southern species whose remains are found at Swartkrans and Elandsfontein.

Cf. *Megalotragus* sp. Giant hartebeest
Provenance: Sterkfontein Member 4, Swartkrans members 1 and 2, and Kromdraai A

The genus *Megalotragus* was created by van Hoepen (1932*b*) when he described a pair of horn-cores from Cornelia as *M. eucornutus*. The earliest description of a giant hartebeest from southern Africa was that of Broom (1909*a*) when he named the taxon *Bubalis priscus* on the basis of a cranial fragment, with part of the left horn-core, found on the bank of the Modder River between Kimberley and Bloemfontein.

Subsequently, a variety of giant hartebeest specimens were described from the Orange Free State under the

genera *Pelorocerus* van Hoepen, *Megalotragus* van Hoepen, and *Lunatoceras* Hoffman. As H. B. S. Cooke (1974) pointed out, the genus *Megalotragus* has priority over both *Pelorocerus* and *Lunatoceras;* furthermore, Cooke considered that the type specimen of Broom's *Bubalis priscus* was inadequate and hence a *nomen vanum*.

Gentry (1976, 1978), on the other hand, considers *priscus* the valid type species of *Megalotragus*. According to evidence accumulated by Klein (1974*a*) from sites in the southern Cape, *M. priscus* made its last appearance there in terminal Pleistocene times, 12,000–10,000 years B.P. It appears to have been one of a series of mammals that became extinct about that time.

M. priscus was appreciably larger than the blue wildebeest but less massive than the eland. Farther north a second species, *M. kattwinkeli* (Schwarz), is known from Olduvai Bed II and the Shungura Formation of the Omo valley (Gentry 1976).

Tribe Hippotragini

A group of medium to large antelopes with horns present in both sexes. The horns are either straight, scimitar-like, or twisted and generally have hollow pedicles to their cores. The molars are moderately hypsodont with characteristic basal pillars. According to Ansell (1971), living hippotragines fall into three genera: *Addax* Laurillard, containing the addax found in north African deserts; *Hippotragus* Sundevall, the extinct blue buck, the sable, and the roan; and *Oryx* De Blainville, with two species, the scimitar oryx and the gemsbok.

Fossil remains from the Sterkfontein valley caves appear to be restricted to the genus *Hippotragus*.

Hippotragus cf. *equinus* (Desmarest). Roan antelope
Provenance: Sterkfontein Member 4 (but see below) and Kromdraai A

The material is very sparse, consisting of two upper molars from Sterkfontein and a single lower molar from Kromdraai. Previously H. B. S. Cooke (1947) described a mandible from the "upper quarry" at Sterkfontein as *Hippotragoides broomi* Cooke, but both Mohr (1967) and Vrba (1976*a*) agree that the specimen belongs in *H. equinus*.

Vrba (1976*a*) speculates whether Cooke's specimen as well as the two Sterkfontein molars might not have come from a deposit younger than Member 4.

H. B. S. Cooke (n.d.) has described further material among the University of California collection from Gladysvale as *Hippotragus broomi* (Cooke), concluding that it showed characters similar to those of his "*Hippotragoides*" from Sterkfontein.

All that can be said, then, is that remains of roan antelopes come doubtfully from Sterkfontein Member 4 and certainly from Kromdraai A.

Hippotragus cf. *niger* (Harris). Sable antelope
Provenance: Swartkrans Member 2

The sample from Swartkrans came from a minimum of six juveniles and three adults. Morphologically the fossils

cannot be separated from the extant sable antelope, which has a wide distribution in the southern savanna zone of Africa.

?Hippotragini
Provenance: Sterkfontein Member 4 and, very tentatively, Member 5, Swartkrans Member 1, and Kromdraai A

This tentative taxon is best represented in Sterkfontein Member 4 and by single specimens only from the other site units. Vrba (1976*a*) expresses the opinion that most of the specimens probably belong to a single species that is likely to include material from the Makapansgat Limeworks, designated *Taurotragus* cf. *oryx* by Wells and Cooke (1956). She gives reasons for believing that the Sterkfontein valley teeth could well be hippotragine rather than tragelaphine and concludes that they most closely resemble those of the extinct form *Hippotragus gigas* L. S. B. Leakey, although the possibility of other tribal affinities cannot be ruled out (Vrba 1976*a*, p. 46).

Hippotragus gigas was described by L. S. B. Leakey (1965) as "a *Hippotragus* of gigantic proportions, with characters recalling both *H. equinus* and *H. niger*." The holotype comes from Bed II and the paratype from the top of Bed I at Olduvai. A single tooth from the Omo valley could belong to *H. gigas* (Gentry 1976), and the taxon is also recorded in the Koobi Fora Formation (Harris 1976) and from Elandsfontein and possibly Florisbad.

Tribe Reduncini

A group of medium to large antelopes in which only males carry horns; these may be bowed, lyrate, or hooked. Basal pillars are present in both upper and lower molars, and the latter often show goat folds.

There are two genera of living reduncines, *Redunca* H. Smith, which contains three species of reedbuck, and *Kobus* A. Smith, encompassing the waterbuck, lechwes, kob, and puku. Together these animals have a very wide savanna-zone distribution in Africa but are seldom found far from water.

Two fossil representatives of the tribe are found in the Sterkfontein valley assemblages.

Redunca cf. *arundinum* (Boddaert). Reedbuck
Provenance: Sterkfontein Member 4, Swartkrans Member 1, and Kromdraai A

The material is sparse, consisting of single juvenile dentitions from Sterkfontein and Swartkrans and remains

of one adult from Kromdraai A. Vrba (1976*a*) observes that the fossils are morphologically indistinguishable from the extant reedbuck *R. arundinum,* but the possibility also exists that they had affinities with an extinct form *R. darti* Wells and Cooke. In the course of their study of Makapansgat fossil Bovidae, Wells and Cooke (1956) found that the most commonly occurring species in the assemblage was a reedbuck about the size of *R. arundinum,* "but possessing more massive horn cores with a distinctly sigmoid profile" (Wells and Cooke 1956, p. 17). The P^2 was found to be larger than in *R. arundinum,* and the cheek teeth were relatively and absolutely broader. In these respects *R. darti* was reminiscent of the puku, *Kobus vardoni* (Livingstone), and appeared in some respects to be intermediate in form between the puku and the reedbuck.

Dr. Vrba's suspicion that *R. darti* could be represented in the Sterkfontein valley assemblages is strengthened by the fact that a *Redunca* frontlet, found at Sterkfontein about 1935 and preserved in the Anatomical Museum of Witwatersrand University, has been identified by Wells (1969*a*) as belonging to *R. darti.*

Cf. *Kobus ellipsiprymnus* (Ogilby). Waterbuck
Provenance: Swartkrans Member 2

Vrba (1976*a,* p. 27) notes that her identification of the two Swartkrans specimens is extremely tentative.

Waterbuck have a very wide African distribution, and the species is divided by Ansell (1971) into an *ellipsiprymnus* and a *defassa* group. The former contains four subspecies, the latter nine.

Wells (1967) notes that *K. ellipsiprymnus* is known only from late fossil contexts in southern Africa. However, in 1913 Broom described a frontlet and a right horn-core from Haagenstad (now known as Florisbad) as *Kobus venterae,* which he thought to be intermediate in form between the waterbuck and the lechwe, *K. leche*. Similar remains have since been found at various other Orange Free State sites, but there does not seem to be justification for separating *K. venterae* from *K. leche* (Vrba, pers. comm.).

A second species of *Kobus* has been described by H. B. S. Cooke (1949*a*) on the basis of an isolated P^4 from the Vaal River Gravels in the Riverview Estates area. The tooth is described as being similar in form to that of *K. ellipsiprymnus* but twice as large. No further remains of a giant waterbuck appear to have come to light.

According to Gentry (1976), the ancestor of *K. ellipsiprymnus* is *K. sigmoidalis* Arambourg, a form originally described from the Omo valley. Remains are abundant in the Shungura Formation, and Gentry believes that the transition from *K. sigmoidalis* to *K. ellipsiprymnus* is detectable in Member G. Waterbuck remains are extremely numerous in the East Rudolf bovid fauna, but the separation of *K. sigmoidalis* from *K. ellipsiprymnus* does not seem to be as readily definable in the Koobi Fora as it is in the Shungura Formation (Harris 1976).

Tribe Peleini

The tribe contains a single representative, *Pelea capreolus* (Forster), the gray rhebuck or vaal ribbok, confined to South Africa and southeastern Botswana. It is a rather long-necked, graceful antelope weighing 23–27 kg (50–60 lb), with straight, vertical horns carried only by the males. Its preferred habitat is grassy hills and mountains, where it lives in small groups, each with a dominant male. It is exclusively a grazer.

There is some uncertainty about whether *Pelea* should be placed in the subfamily Hippotraginae. Gentry (1970*a*) has suggested that it might possibly be placed in the Caprinae, with affinities to the goats or the chamois, *Rupicapra rupicapra* (Linnaeus).

Pelea cf. *capreolus* (Forster, 1790)
Provenance: Members 1 and 2 at Swartkrans and Kromdraai A

The most complete remains are found in Member 2 at

Swartkrans and include one almost intact skull of a subadult, SK 2735*a–e*. According to Vrba (1976*a*), this shows some interesting differences from extant *Pelea:* the teeth appear to be larger, and the angle between braincase and face is perhaps less acute. Possible subspecific variation is suggested. It is perhaps surprising that the earliest fossils, from Swartkrans Member 1, are indistinguishable in size from the living form, whereas those from the later Member 2 and from Kromdraai A are larger.

Remains of *P. capreolus* are known from the Cave of Hearths (H. B. S. Cooke 1962), Klasies River Mouth caves (Klein 1976*a*), and Wilton Shelter (Brain, this volume).

Subfamily Antilopinae

In his review of African artiodactyls, Ansell (1971) classifies living representatives of the Antilopinae as follows:

Tribe: Antilopini

Genera: *Gazella* De Blainville, gazelles; *Antidorcas* Sundevall, springbok; *Litocranius* Kohl, gerenuk.

Tribe: Ammodorcadini (Or Ammodorcini)

Genus *Ammodorcas* Thomas, dibatag.

Tribe: Neotragini

Genera: *Oreotragus* A. Smith, klipspringer; *Madoqua* Ogilby, dik-diks; *Dorcatragus* Noack, Beira antelope; *Ourebia* Laurillard, oribi; *Raphicerus* H. Smith, steenbok and grysbok; *Neotragus* H. Smith, suni, and so forth.

Tribe Antilopini

Genus *Antidorcas* Sundevall, 1847

The genus contains a single living representative, *A. marsupialis* (Zimmerman), the springbok, of which three subspecies are recognized by Ansell (1971): *A. m. marsupialis* (Zimmerman), widely distributed in South Africa, *A. m. hofmeyri* Thomas, from southern South-West Africa and Botswana, and *A. m. angolensis,* from northern South-West Africa and Angola.

Antidorcas australis Hendey and Hendey and/or *A. marsupialis* (Zimmerman)

Provenance: Swartkrans Member 2 and channel fill

In 1968 Q. B. Hendey and H. Hendey described a new subspecies of springbok as *A. marsupialis australis* on the basis of cranial remains from the Swartklip site on the False Bay coast near Cape Town. The new form differed from the extant springbok particularly in the shape and size of the horn-cores: these were intermediate in size between those of the male and female of the modern form and, although lyrate in form, showed no inward curve toward the tips. The teeth of the two forms appeared morphologically indistinguishable, though those of the fossil tended to be rather narrow (Hendey and Hendey 1968, p. 56). Although the Swartklip springbok was regarded as a subspecific geographic variant of *A. marsupialis,* the authors pointed out that, if *A. m. australis* should prove to have been more widely distributed in the past, specific status would be warranted.

When describing *Antidorcas* fossil material from Swartkrans, Vrba (1973) allocated several dentitions and horn-cores to *A. australis,* pointing out that in view of the temporal and geographic separation of Swartkrans from the western Cape sites, it seemed highly likely that *australis* should be regarded as a species separate from *marsupialis*. More recently it has become apparent that the Swartkrans specimens referred to here come from the Member 2 accumulation phase that, when calcified channel fills are included, certainly spanned a considerable period of time. Vrba (1976*a*) now has queried the conclusion that *A. australis* is definitely present at Swartkrans and suggests that the *A. australis*like specimens may represent pre*marsupialis* evolutionary stage.

Vrba (1973) provided a diagram of possible phylogenetic relationship in the genus *Antidorcas,* suggesting that in Pliocene times the *Antidorcas* lineage separated from the *Gazella;* thereafter the *A. bondi* stream separated from that of *A. recki,* and at a subsequent time *A. recki* gave rise to both *A. australis* and *A. marsupialis.*

Antidorcas bondi (Cooke and Wells, 1951). Bond's springbok

Provenance: tentatively recorded from Sterkfontein Member 4, Kromdraai A and B; positively from Sterkfontein Member 6, Swartkrans Member 2 and channel fill

In 1942, when describing faunal remains from the Vlakkraal thermal springs in the Orange Free State, Wells and Cooke noted that they were unable to assign three very hypsodont lower teeth and a corresponding upper tooth to any known bovid species. The teeth were similar to those of the impala but transversely narrower; they were listed as *Antilope* gen. et sp. indet.

Some years later Cooke and Wells (1951) described fossilized remains from alluvial terraces flanking a tributary of the Umgusa River on Chelmer Farm, 13 mi west-northwest of Bulawayo. Bond and Summers (1951) showed that the fossils came from the top of the older alluvium that rests on Karroo (Forest) sandstone. The assemblage included several teeth similar to those ob-

tained from Vlakkraal and were named *"Gazella" bondi* Cooke and Wells.

In her study of *Antidorcas* fossils from Swartkrans, Vrba (1973) noted that in Gentry's opinion *"Gazella" bondi* should become *Antidorcas bondi* (Cooke and Wells); Vrba concurred with this suggestion and provided a detailed expanded diagnosis of the species on the basis of abundant material from Swartkrans. She showed that the teeth were consistently very tall and that the "hypsodonty index" for M_2 was higher at all wear stages than in other species of *Antidorcas*—that is, *recki, australis,* or *marsupialis*. The horn-cores of the Swartkrans *A. bondi* specimens were found to be less mediolaterally compressed than in other *Antidorcas* material, and they typically diverged slightly basally but converged toward the tips, giving the horns the appearance of parentheses when viewed from the front.

Remains of *A. bondi* are now known from many southern African sites, and the species appears to have been common during later Pleistocene times. Its precise extinction point has not been established.

Antidorcas recki (Schwarz)
Provenance: positively identified from Kromdraai A, tentatively from Sterkfontein members 4 and 5, Swartkrans Member 1, and Kromdraai B

This springbok was appreciably smaller than the living *A. marsupialis* and had proportionally shorter legs. Its horn-cores were bent sharply backward toward their tips and showed more mediolateral compression than does the living form.

The type was originally described by Schwarz as *Adenota recki* on the basis of a skull and right horn-core from Olduvai, but the specimen was destroyed in Munich during the Second World War. More recent material has been referred to *Phenacotragus recki* (Schwarz) but is now regarded as fitting better in the genus *Antidorcas* (see Vrba 1976*a*).

When describing fossil mammals from the Vaal River gravels, H. B. S. Cooke (1949*a*) created the name *Gazella wellsi* Cooke for two mandibular fragments and an isolated left M_3 from Power's site near Kimberley. He remarked that the teeth were remarkably similar to those of the Mongolian *Gazella subgutterosa*.

Gentry (1966) pointed out that the most common gazelline species at Olduvai appeared to be conspecific with *G. wellsi,* particularly on the basis of excellent material from Pit 3 at Bolt's Farm (Cooke, n.d.). More recently Gentry (1978) considered that all the *G. wellsi* material should be reclassified as *Antidorcas recki*.

A remarkable collection of *A. recki* remains was excavated at site SHK II of Olduvai Gorge by L. S. B. Leakey in 1935. These remains appear to have come from a herd of nine or ten individuals, all but two of which were juvenile or young (Gentry 1966, p. 78). The herd may have been driven into a swamp by hominid hunters, since the bones were preserved in clay about 90 m from a living floor in Bed II.

With the recording of *A. recki* in the Koobi Fora Formation (Harris 1976*a*) and possibly in the Shungura Formation (Gentry 1976), it appears that this probable ancestor of the living springbok had a wide African distribution.

Gazella sp.
Provenance: Sterkfontein Member 4, Kromdraai B, and possibly Swartkrans Member 1

No living species of *Gazella* occur in southern Africa today, although ten species are recognized by Gentry (1971*a*) from farther north. Antilopines of the genus *Gazella* De Blainville differ from those of *Antidorcas* particularly in the structure of their frontal bones. In *Antidorcas* the frontal sinus is particularly well developed, causing inflation of the frontal bones and a hollowing of the horn pedicles. For this reason the frontal bones between the horn pedicles are raised above the level of the orbital rims and the openings of the supraorbital canals are sunk into bony tubes. None of these developments may be observed in *Gazella*.

The first fossil assigned to the genus *Gazella* from southern Africa was found at Cornelia in the Orange Free State and described as *G. helmoedi* by van Hoepen (1932*b*). It consisted of a long slender horn-core with a gentle spiral twist; it was perhaps as much as 40 cm long in undamaged form. Initially van Hoepen thought the horn-core was from the left side, but more recently H. B. S. Cooke (1974) has argued that it is from the right and that the basal hollowing is more suggestive of *Antidorcas* than *Gazella,* if, in fact, the specimen is from an antilopine. Alternatively, it could as well be of alcelaphine affinity, as suggested by Vrba (1973), who, in fact, placed it in the Alcelaphini.

From the Limeworks Cave at Makapansgat, Wells and Cooke (1956) described two new species of gazelline antelope: *Gazella gracilior* and *Phenacotragus vanhoepeni*. The former was reminiscent of the red-fronted gazelle, *G. rufifrons,* from west and north Africa but with more delicate horn-cores and a relatively larger M_1. The second species, *P. vanhoepeni,* has subsequently been removed from the genus *Phenacotragus* by Wells (1969*b*) and placed in *Gazella* on account of the structure of its frontal bones and sinuses. Wells and Cooke (1956) pointed out that *G. vanhoepeni* resembled *Antidorcas recki* but had horn-cores that were comparatively massive and laterally compressed, rising vertically from the frontals.

Vrba (1976*a*) has expressed the opinion that the *Gazella* sp. specimens from Sterkfontein, Swartkrans, and Kromdraai could have affinities to *G. vanhoepeni*. The species has apparently not been recognized at any other sites.

Tribe Neotragini

This is a rather heterogeneous group of small antelopes with short spikelike horns and large preorbital fossae. Six genera of living neotragines are recognized, as follows:
Oreotragus A. Smith; klipspringer
Madoqua Ogilby; dik-dik
Dorcatragus Noack; Beira antelope
Ourebia Laurillard; oribi
Raphicerus H. Smith; steenbok and grysbok
Neotragus H. Smith; suni, and so forth

Oreotragus cf. *oreotragus* (Zimmerman). Klipspringer
Provenance: Swartkrans Member 2

This is a small compact antelope adapted for life in rocky habitats, particularly in the structure of the hooves,

which have the consistency of hard rubber and are truncated and blunt, allowing the animal to gain a foothold in the most precarious places. The coat is thick and bristly and appears to constitute a cushion against bumps, to which klipspringers are prone (Dorst and Dandelot 1972).

A single living species is recognized with a very wide distribution in southern and eastern Africa. In all but a few races, the females are hornless. The Swartkrans remains do not appear to differ from the living form.

Oreotragus cf. *major* Wells
Provenance: Sterkfontein Member 5, Swartkrans members 1 and 2

According to the available evidence, this antelope was similar to the living klipspringer but was 15–20% larger in cranial dimensions (Wells and Cooke 1956). It was originally described by Wells (1951) on the basis of a reasonably complete skull from a breccia deposit on a farm adjacent to Makapansgat. No other fossils have been described from this locality, and its age relationships are unknown. Further remains have been described from the Makapansgat Limeworks (Wells and Cooke 1956).

Raphicerus cf. *campestris* (Thunberg). Steenbok
Provenance: Swartkrans Member 2; *Raphicerus* sp.: Swartkrans Member 2, Kromdraai A

This is a familiar small antelope with a wide southern

and more restricted east African distribution. Its habitat tolerance is considerable, ranging from arid plains to open woodland, where it browses on low bushes and shrubs. Steenbok occur either alone or in pairs; females are frequently larger than males but lack horns.

Gentry (1978) is of the opinion that one of the paratypes of *Cephalophus pricei* described by Wells and Cooke (1956) from Makapansgat should be referred to *Raphicerus*.

Fossils of *Raphicerus* are known from Langebaanweg, Elandsfontein, and many other sites (Hendey 1974*a*). Klein (1976*c*) has reviewed the fossil history of the genus in the Cape biotic zone and has suggested that individuals were larger during cold climatic intervals.

Ourebia cf. *ourebi* (Zimmerman). Oribi
Provenance: Swartkrans Member 2

This is a graceful small antelope of the open plains, always close to water, and almost entirely a grazer. According to Ansell (1971), a single species split into thirteen subspecies occurs over an extremely wide African range. The short black tail, prominently displayed, and the "rocking-horse" gait, interspersed with high leaps, are characteristic. The fossil history of the genus appears to be unknown.

Subfamily Caprinae
Tribe Ovibovini

There are two living representatives: the muskox, *Ovibos moschatus* (Zimmerman) of high latitudes in North America, and the Chinese takin, *Budorcas taxicolor* Hodgson.

Gentry (1971*b*) has pointed out that the Ovibovini are well represented as fossils, almost all of which are Eurasian. The only African forms thus far described are specimens of the genus *Makapania* Wells and Cooke from the Transvaal and a tentatively identified horn-core from the Omo valley (Gentry 1970*b*).

Makapania broomi Wells and Cooke
Provenance: *M.* cf. *broomi:* Sterkfontein members 4 and 5; cf. *Makapania* sp.: Swartkrans Member 1

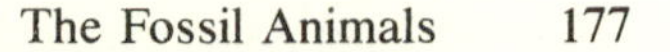

Wells and Cooke (1956) created this taxon for some remarkable fossils from Makapansgat Limeworks that they considered to belong in the tribe Alcelaphini. The most striking feature of these antelope was their horns, which were directed strongly laterally.

In a subsequent reexamination of the Makapansgat material, Gentry (1970*b*) concluded that *Makapania* was in fact an ovibovine. Strong affinities to the French Villafranchian ovibovine *Megalovis latifrons* Schaub were suggested by the structure of the basioccipitals as well as "the wide insertions of the horn cores just behind the orbits, their transverse emergence, the gentle ascent towards the tips, absence of transverse ridges on the horn cores, the short braincase well angled on the facial axis, the extent of the projection of the dorsal parts of the orbital rims, the small localised preorbital fossae, fairly hypsodont teeth" (Gentry 1970*b*, p. 64).

Makapania therefore appears to have been an antelope similar in appearance to the muskox, with laterally projecting horns. Vrba (1976*a*) equates the dentitions from Member 4 at Sterkfontein, and the single tooth from Member 5, tentatively with *M. broomi*. The teeth from Member 1 at Swartkrans, however, are appreciably smaller, though in most respects *Makapania*like. Vrba concludes that the Swartkrans material could represent either a smaller, and later, species of *Makapania* or an aberrant medium-sized alcelaphine.

Family Suidae

Three species of indigenous living pig are currently recognized in sub-Saharan Africa (Ansell 1971): *Potamochoerus porcus* (Linnaeus), the bush pig or red river hog; *Phacochoerus aethiopicus* (Pallas), the warthog; and *Hylochoerus meinertzhageni* Thomas, the giant forest hog. The wild boar, *Sus scrofa* Linnaeus, still occurs in Tunisia, although it has disappeared from most of its former north African range.

Three extinct Suid taxa are currently recognized in the Sterkfontein valley assemblages under consideration.

Phacochoerus modestus (van Hoepen and van Hoepen)
Provenance: Swartkrans Member 2, Kromdraai A and B

This warthog was clearly very similar in appearance to the living form and apparently close to the ancestry of it. The species *P. antiquus* was originally described by Broom (1948*a*) on the basis of an immature skull first thought to have come from Sterkfontein but, in fact, almost certainly from Kromdraai A. Ewer (1956*e*, p. 528) redescribed the specimen, placing it in a new subgenus *Potamochoerops*. She considered it to differ from the living form by the following characters:

> auditory bulla conical and pointed; frontals not concave above the orbits, at least in the immature animal; jugal not forming a well defined post orbital process; side walls of the snout above the canines slightly concave; canines more forwardly directed than in *P. africanus;* cheek teeth very similar to those of the living species, but differing in the following points, P_4 with posterior accessory cusp single, first and second molars distinctly bilobed and with the talon or talonid containing only a single cusp. Root formation on the anterior columns of M_3 starts before the posterior columns come into wear.

In her consideration of adaptive features in the skulls of African pigs, Ewer (1958*b*) pointed out that features in the skull of the bushpig, *Potamochoerus koiropotamus*, were related to this animal's digging habit, while those of the warthog skull were related to a grass-eating habit. She concluded that the adaptations in the skull of *P. antiquus* that allowed for the grinding action of the cheek teeth were far less well developed than in the living warthog.

An exceptionally complete skull with articulated atlas and axis vertebrae was found by the University of California expedition at Pit 3 of Bolt's Farm. It has been described in detail by Cooke (n.d.), who states that its features confirm the conclusions reached by Ewer. Cooke and Wilkinson (1978) use the name *P. modestus* in preference to *P. antiquus*.

Metridiochoerus andrewsi Hopwood
Provenance: Swartkrans Member 1 and Kromdraai A

Shaw (1938) reported the discovery of three third molars from Sterkfontein, presumably from Member 4. They are not included in this analysis.

These large extinct pigs had molars that are morphologically similar to those of *Phacochoerus* but are strikingly different in anterior skull structure. In particular, the canines were comparatively straight, reminiscent of small elephant tusks; the uppers projected laterally from the sides of the muzzle.

The specific name *meadowsi* was created by Broom (1928) when he named an isolated third molar from diamond gravels on the Vaal River near Kimberley as *Notochoerus meadowsi;* a few years earlier he had created the genus (Broom 1925) for an exceptionally large third molar, also from the Vaal River gravels, that he named *N. capensis*.

The genus *Tapinochoerus* was introduced by van

Hoepen and van Hoepen (1932) when they described *T. modestus* from the Cornelia Beds. They expressed the opinion that *Notochoerus meadowsi* should be placed in *Tapinochoerus,* an opinion supported by H. B. S. Cooke (1949*b*) and subsequent authors.

Cooke and Maglio (1972) have expressed the opinion that two forms found at Olduvai, *Afrochoerus nicoli* and *T. meadowsi,* could be regarded as a single species showing considerable sexual dimorphism. The almost straight, elephantlike tusks of *A. nicoli* would find their extreme development in the males. More recently Cooke (1976) abandoned this possibility, sinking *Tapinochoerus* into *Metridiochoerus*.and *Afrochoerus* into *Stylochoerus*.

In their recent review, White and Harris (1977) agree that *Tapinochoerus* should be regarded as a synonym of *Metridiochoerus*. This policy is supported by Cooke and Wilkinson (1978).

Cf. *Metridiochoerus* sp.
Provenance: Sterkfontein Member 4

Family Giraffidae

There are two living representatives of the family: the giraffe, *Giraffa camelopardalis* (Linnaeus), and the okapi, *Okapia johnstoni* Sclater, placed in the subfamilies Giraffinae and Palaeotraginae respectively. The third subfamily, the Sivatheriinae, contains a variety of large fossil forms, including the one recorded from Swartkrans. Giraffids are known from the Sterkfontein valley caves by this single specimen, though Broom (1948*a*) refers to a mandible of "a very large Giraffid," without saying from which site it came. I have not been able to relocate his specimen, though it may be the one described by Cooke and Wells (1947) from the Makapan valley.

Genus *Sivatherium* Falconer and Cautley, 1835

Harris (1976*b*, p. 317) has provided the following diagnosis of this genus, based on an earlier one by Colbert (1935):

> A gigantic Pleistocene giraffid, with four ossicones in the male, an anterior pair arising from the frontals and a posterior pair situated on the parietals. As in the other gigantic sivatheriines there are deep pits in the temporal fossae for the temporal muscles, and on the supraoccipital for the neck muscles. The face is very short, the nasals being retracted and strongly curved. The teeth are large with rugose enamel. Body and limbs heavy, limbs not elongated.

Several forms of sivathere have been described from South Africa, including *Griquatherium cingulatum,* which Haughton (1922) named on the basis of an upper molar from the Vaal River gravels, *G. haughtoni* H. B. S. Cooke, 1949, and *Orangiatherium vanrhyni* von Hoepen, 1932. Singer and Boné (1960) synonymized some of the earlier names, recognizing in southern Africa *Sivatherium cingulatum* as well as three subspecies of *S. olduvaiense*.

In his description of the only sivathere specimen currently known from the Sterkfontein valley caves, a single tooth from Swartkrans, Churcher (1974) concluded that all southern African material known at that time could be classified as *S. maurusium*. This view was supported by Harris (1976*c*), who also described a new species, *S. hendeyi,* from the Varswater Formation of Langebaanweg.

Until more material becomes available, the validity of this new species is questioned by Churcher (1978).

Sivatherium maurusium (Pomel). Sivathere
Provenance: Swartkrans Member 2

The single specimen, an isolated and unerupted upper molar, came from a subadult sivathere of a form that,

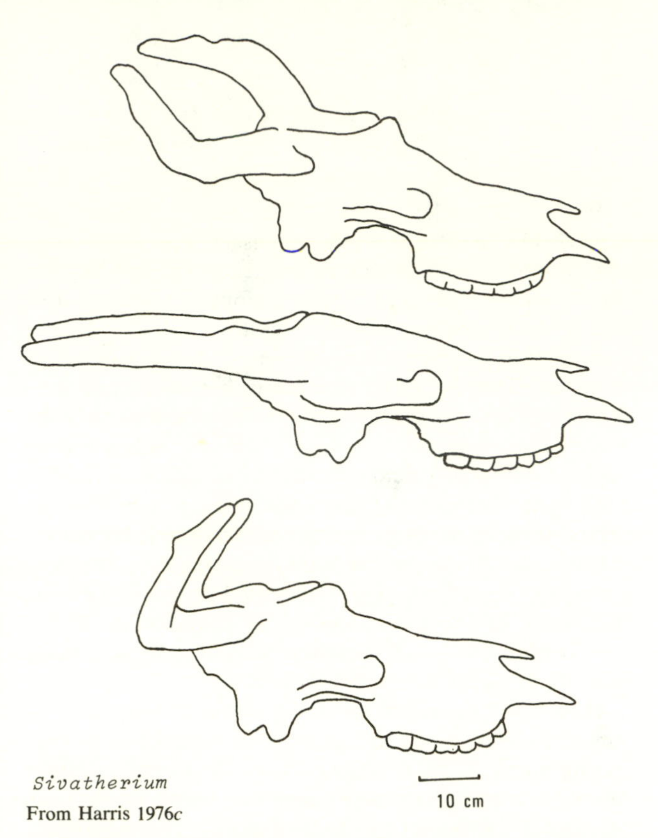

Sivatherium
From Harris 1976*c*

although now extinct, had a wide African range both in space and in time. These were heavy-bodied, short-necked giraffids that must have browsed in the same way that modern giraffes do. According to Harris (1976*c*), the

ossicones of this species varied a good deal in form. He recognized three basic ossicone shapes, as indicated in the illustration.

Order Perissodactyla

Two families of odd-toed ungulates are represented in Africa: Rhinocerotidae, containing two species of rhino, and Equidae, containing one species of ass and four of zebra. No rhino remains have thus far been found in the Sterkfontein valley caves, but equid remains are fairly common.

Family Equidae

All living African equids are placed in the genus *Equus* Linnaeus by Ansell (1971), though the subgenera *Asinus, Dolichohippus,* and *Hippotigris* are retained. Fossil equids to be considered here fall into two genera, *Hipparion* and *Equus*.

The Genus *Hipparion* de Christol, 1832

Horses of the genus *Hipparion* were divided from a *Merychippus* stock in North America and appeared first in Europe 12.5 million years ago (Hooijer 1975). They were generally lightly built animals, about the size of a pony, though some species were somewhat larger, an example being *H. crusafonti* Villalta, the terminal species in Europe that died out during the Villafranchian, though African forms survived much longer.

In *Hipparion* there are three functional toes on each of the four feet, whereas in *Equus* the weight of the horse is carried through the enlarged third metapodials while the second and fourth toes are reduced to insignificant "splint" bones. This means that *Equus* is a single-toed animal, whereas in *Hipparion* the side toes presumably touched the ground during trotting and galloping.

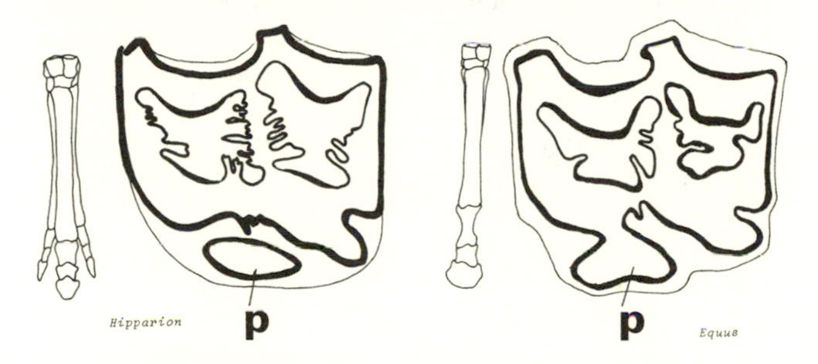

Another clear difference between representatives of the two genera is found in the enamel patterns on the upper cheek teeth. As shown in the illustration, the protocone forms an island in *Hipparion* but not in *Equus*.

The following hipparionid taxa have been described from South Africa:

Hipparion steytleri van Hoepen, 1930, from Uitzoek, near Cornelia, based on four isolated teeth.

Eurygnathohippus cornelianus van Hoepen, 1930, from Cornelia, based on a mandibular symphysis with incisors, now shown to have belonged to a *Hipparion* (L. S. B. Leakey 1965).

Stylohipparion hipkini van Hoepen, 1932, from Uitzoek, near Cornelia, based on an isolated right M_2.

Notohipparion namaquense Haughton, 1932, from a locality 49 mi east of Springbok, Namaqualand, and based on a series of mandibular teeth from a single individual.

Hipparion (Hipparion) albertense baardi Boné and Singer (1965) from Langebaanweg, based on a number of specimens.

Boné and Singer (1965) have proposed two subgenera in which all African forms may be accommodated:

Hipparion (Hipparion) de Christol

Hipparion (Stylohipparion) van Hoepen

They place the Langebaanweg material in the former subgenus, but all the other southern African fossils in the latter. Singer and Boné (1966) were inclined to place all the known *Hippariaon (Stylohipparion)* material in the single species *libycum* Pomel, but this has not been supported by Churcher (1970), who classified the Swartkrans and Kromdraai material as *H. (Stylohipparion) steytleri* van Hoepen.

The taxonomic arrangement of Boné and Singer has not gained wide support. Hooijer (1975) has pointed out that *Stylohipparion* is a synonym of *Eurygnathohippus* and that *H. albertense* is a *nomen vanum*. He regards the form *H. albertense baardi* as a full species *H. baardi*. A full description of the late Pliocene Equidae from Langebaanweg is provided by Hooijer (1976), and he also records *H.* cf. *namaquense* from this site.

In a recent review, Churcher and Richardson (1978) classify all the advanced hipparions from Africa in a single taxon *Hipparion lybicum*. As synonyms from southern Africa they list *Hipparion steytleri, Eurygnathohippus cornelianus,* and *Stylohipparion hipkini*. This arrangement is followed here.

Hipparion lybicum Pomel. Three-toed horse

Provenance: Swartkrans Member 1 and Kromdraai A

Little is known of this animal apart from its teeth, but it appears to have been a lightly built equid, appreciably

smaller than the familiar Burchell's zebra. It has three toes on each foot, but whether it was striped or plain is something we may never know.

There is no certainty about when *Hipparion* died out in South Africa, although the most recent evidence of the animal is found at Cornelia. The fossils are found particularly in Bed b of Member 1 (Butzer 1974*a*), a horizon overlying that in which artifacts are concentrated. According to J. D. Clark (1974), the Cornelia artifacts show similarities to those of Upper Acheulean affinities at Broken Hill in Zambia.

Equus (Dolichohippus) capensis Broom. Giant Cape horse

Provenance: Sterkfontein Member 4, Swartkrans Member 1, and Kromdraai A

This was the first species of fossil horse to be described in South Africa. For some time Broom had suggested that a large extinct equid may have existed in the Cape, but in 1909 (Broom 1909*b*) he was able for the first time to de-

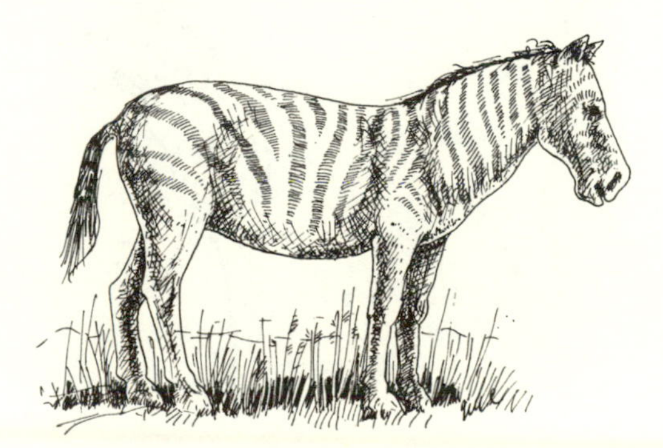

scribe its dentition. The fossil was found in a slab of surface limestone cast up on the beach at Yzerplaatz in Table Bay. It consisted of much of the left mandibular ramus of a horse considerably larger than *Equus caballus* Linnaeus. Further remains from various parts of South Africa allowed Broom (1913*b*, 1928 and Broom and Le Riche 1937) to describe the animal in greater detail. On the evidence of the limb bones, Broom (1913*b*) concluded that the proportions of *E. capensis* differed considerably from those of *E. caballus*. *E. capensis* appears to have been more powerfully built, but did not stand so high, probably no more than fourteen hands. It appears to have had a relatively more massive head.

In his consideration of the nomenclature of the South African fossil equids, Wells (1959) regards Broom's type of *E. capensis* as indeterminable and the name as a *nomen vanum*. He considered that the correct name for the extinct Cape horse should be *E. helmei* Dreyer. In his description of the remains from Sterkfontein, Swartkrans, and Kromdraai, Churcher (1970) does not agree that the type of *E. capensis* is indeterminable or that the name should be discarded. He applies it to the Sterkfontein valley material and suggests that the name *E. capensis* should include *E. helmei* Dreyer, *E. cawoodi* Broom, *E. kuhni* Broom, and *E. zietsmani* Broom as well as some of the teeth referred to *E. harrisi* Broom and *E. plicatus* (van Hoepen). Churcher and Richardson (1978) retain the name *E. capensis* but place the species in the subgenus *Dolichohippus*.

Klein (1974*a*) has shown that the disappearance of *E. capensis* from the southern Cape occurred synchronously with that of other forms such as the giant buffalo, *Pelorovis antiquus*, the giant hartebeest, *Megalotragus priscus*, and at least five other forms. These extinctions seem to coincide with terminal Pleistocene environmental changes dated at 12,000–10,000 years B.P., and Klein suggests that the extinctions may have been caused both by these changes and by more effective human hunting techniques.

It is not known whether the Cape horse survived after 10,000 years B.P. in other parts of southern Africa, though its remains are coming to light at many sites scattered over the subcontinent. It seems quite likely that artists responsible for rock paintings or engravings could have seen the horse alive, and it may be possible, from such artistic records, to establish whether *E. capensis* had zebra stripes. At present this is a matter for speculation.

Equus (Hippotigris) quagga Gmelin. Quagga
Provenance: Swartkrans Member 2 and a tentative identification at Kromdraai A

The quagga was familiar to early travelers in the southern African interior, occurring abundantly in the southern and eastern parts of the great Karroo, with a possible extension into Great Namaqualand (Ansell 1971). It does not appear to have occurred in any numbers north of the Vaal River, so that the recorded presence of the species in the Sterkfontein valley caves represents an extension of the historically observed range.

It appears that quaggas may have survived in the Orange Free State as late as 1878, and we know that the last individuals in European zoos died in 1872 (London), 1875 (Berlin), and 1883 (Amsterdam). A surprising number of mounted specimens are still in existence in museums as listed and described by Rau (1974).

Although the cranial dimensions of *E. quagga* are very similar to those of *E. burchelli*, there are apparently consistent differences in the enamel patterns on the occlusal surfaces of the cheek teeth (H. B. S. Cooke 1943, 1950; Churcher 1970). Remains of *E. quagga* have been recorded at the Wonderwerk Cave near Kuruman (H. B. S. Cooke 1941) and from near Kroonstad in the Orange Free State (Shapiro 1943).

Churcher and Richardson (1978) have cast doubt on Churcher's earlier identification of Swartkrans specimens as *E. quagga*. They state that "it seems best to regard the presence of *E. quagga* at so early a date and so great a distance from its known range with skepticism."

Equus (Hippotigris) burchelli (Gray). Burchell's zebra
Provenance: Sterkfontein members 5 and 6, Swartkrans channel fill, and Kromdraai A

This familiar animal has a wide distribution in southern and eastern Africa. Ansell (1971) recognizes six subspecies and notes that *E. b. burchelli* (Gray, 1824), which formerly occurred in the Orange Free State, northeastern Cape, southern Botswana, and southeastern Transvaal, is now extinct.

H. B. S. Cooke (1950) expresses the opinion that three forms described by van Hoepen as *E. platyconus*, *E. simplicissimus*, and *Kraterohippus elongatus* can be regarded as belonging to *E. burchelli*. The same is true of *E. lylei* Dreyer from Florisbad. Remains of Burchell's zebra are also known from the Vaal River gravels and a variety of later Pleistocene sites.

Order Proboscidea

Family Elephantidae: Elephants

Only one species of living African elephant is currently recognized, *Loxodonta africana* (Blumenbach), although a large number of fossil elephantid genera and species have been described from Africa, creating great confusion. Recently, however, Cooke and Maglio (1972), Maglio (1970, 1973) and Coppens, Maglio, Madden, and

Beden (1978) have furnished a phylogenetic scheme providing comparative simplicity and clarity.

The family Elephantidae is thought to contain twenty-five valid species grouped in two subfamilies, the Stegotetrabelodontinae and the Elephantinae, both of which had their origin in Miocene gomphotheres. It is thought that the Elephantinae arose from the Stegotetrabelodontinae during the Pliocene, though neither of the two described species of *Stegotetrabelodon* appear to have been the direct ancestors.

Within the Elephantinae, three lineages are thought to have arisen about five million years ago; these are represented by the genera *Loxodonta, Elephas,* and *Mammuthus*. They came from *Primelephas,* in which the lower incisors had been greatly reduced, in contrast to the situation in *Stegotetrabelodon,* where lower tusks were typically present.

Fossil representatives of each of the three genera *Loxodonta, Elephas,* and *Mammuthus* are found in South Africa, but only two specimens are known from the cave breccias. These have been identified as *Loxodonta atlantica* (Pomel) from Pit 7, Bolt's Farm by H. B. S. Cooke (1963 and n.d.); as *E. ekorensis* by Maglio (1973); and as *Elephas* cf. *recki* (Dietrich) by Maglio (pers. comm.).

Elephas cf. *recki* (Dietrich)
Provenance: Sterkfontein Member 4, known by a single deciduous lower molar

According to Cooke and Maglio (1972), the *Elephas* lineage, which arose from a *Primelephas* stock 4 to 5

million years ago, showed progressive evolution until its African extinction point in the late Pleistocene. The earliest recognizable form in the lineage is *E. ekorensis,* known from Kanapoi, Ekora, and Kubi Algi. Here the cranial specializations separating *Elephas* from *Loxodonta* were already apparent: a reduction of the tusks and premaxillae, compression of the skull in the facial plane, and an expansion of the parietals and occipitals.

In East Africa, and presumably in South Africa as well, *E. ekorensis* evolved into *E. recki,* which can then be traced through four progressive stages (Maglio 1970) in which the dentition increased its shearing efficiency and the skull continued the trends that began in *E. ekorensis*. It appears that *E. recki* stage 4 gave rise to *E. iolensis,* a species synonymous with *E. transvaalensis* (formerly *Palaeoloxodon*) from the Younger Gravels of the Vaal River (H. B. S. Cooke 1960).

The single deciduous tooth from Sterkfontein has been examined by Maglio, who assigned it tentatively to *E. recki,* stage 1. In their 1972 paper, Cooke and Maglio list the specimen under *E. ekorensis,* but whichever name is used there can be no doubt that the tooth came from an *Elephas* at an early stage of the lineage's progressive development. An age of 3 to 4 million years is indicated by Cooke and Maglio's (1972) calibration diagram.

The Sterkfontein elephant was a juvenile of a form which superficially resembled the living African elephant, except perhaps for smaller tusks.

Order Hyracoidea

Family Procaviidae: Dassies

Three genera of living hyraxes or dassies are recognized by Bothma (1971): *Procavia* Storr, *Heterohyrax* Gray, and *Dendrohyrax* Gray. They are all restricted to Africa with the exception of *P. syriacus* (Schreber), the Syrian dassie, which extends into Israel, Syria, and southeastern Arabia. This was apparently the "coney" of the Bible.

All living dassies are small, compact animals weighing no more than 4.5 kg, with short necks, ears, and legs. The

toes are flattened except for the second, which carries a long curved claw. The feet are well adapted for climbing, with the soles kept damp through glandular secretions. A large gland is also present on the back, surrounded by hair different in color from the rest. Fossils from the Transvaal caves represent two species of *Procavia,* discussed below, and an exceptionally large form, *Gigantohyrax maguirei* Kitching, described from Makapansgat Limeworks (Kitching 1965) and also known from deposits in the Omo valley. Meyer (1978) suggested that both *Procavia* and *Gigantohyrax* probably came from a *Prohyrax*like ancestor that in turn had been derived from a form similar to *Saghatherium*.

Procavia antiqua Broom
Provenance; Sterkfontein Member 4, Swartkrans members 1 and 2, and Kromdraai A

The species was originally described by Broom (1934) on the basis of a number of broken skulls apparently associated with the *Australopithecus* cranium from Taung. Although they resembled *P. capensis,* the living form, in many respects, Broom found the fossil dentitions to be more brachyodont, and he proposed a new subgenus, *P. (Prohyrax) antiqua,* suggesting that this dassie could have been ancestral to both *P. capensis* and *P. arborea* (now *Dendrohyrax arboreus* A. Smith). In a subsequent description of further specimens from Taung, Broom (1948*a*) dropped the subgeneric designation and described, in the same paper, a new species of dassie from Sterkfontein that he named *P. robertsi*. This again was a brachyodont form similar in size to *P. capensis,* but no direct comparison was made between it and *P. antiqua*.

Churcher (1956) synonymized *P. robertsi* with *P. antiqua* and provided a revised definition of the taxon. He found the skull of *P. antiqua* to be slightly smaller and lighter than that of *P. capensis,* with less hypsodont teeth and a generally more primitive, unspecialized aspect.

P. antiqua was clearly a common and widespread dassie in the northern Cape–Transvaal area, and probably elsewhere. Churcher suggests that it was a member of the lineage leading to the modern *Procavia* species.

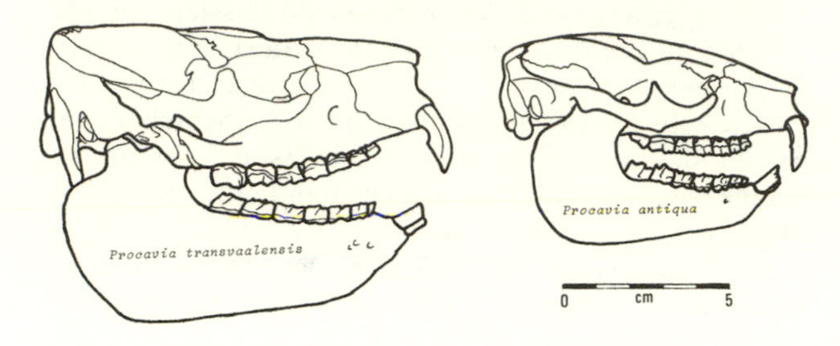

Procavia transvaalensis Shaw
Provenance: Sterkfontein Member 4, Swartkrans members 1 and 2, and Kromdraai A

This extinct dassie was about one and a half times as large as the living *P. capensis,* as indicated by cranial dimensions, and apparently was common in the Transvaal during the Pleistocene, coexisting with *P. antiqua*.

The species was described first in February 1937 by Shaw on the basis of a partial skull and mandible from Cooper's site, situated between Sterkfontein and Kromdraai. In October 1936 Broom (1937*c*) had announced a new, large *Procavia* from Gladysvale that he named *P. obermeyerae*. However, the description was not published until March 1937, so that when *P. transvaalensis* and *P. obermeyerae* were found to be synonymous by Churcher (1956), the former name had priority.

At the time of his description, Broom (1937*c*) pointed out that the new *Procavia* was the largest true dassie known. This remained true until the description of a very much more massive hyracoid, *Gigantohyrax maguirei* from Makapansgat, by Kitching in 1965. Remains of *P. transvaalensis* are now known from all the major Transvaal cave breccias.

Procavia capensis (Pallas). Cape dassie
Provenance: Sterkfontein Member 6 and Swartkrans channel fill

This is the familiar dassie of southern Africa, widely distributed wherever suitable rocky habitat occurs. They live in colonies of up to sixty individuals and are very largely diurnal and entirely vegetarian. They tend to concentrate their droppings in caves and rock shelters, where large accumulations may build up. The crystalline urine, known as "hyracium," has been much prized in the past as a remedy for various ailments including convulsions, epilepsy, and women's diseases.

Order Lagomorpha

Family Leporidae: Rabbits and hares

Rabbits and hares were at one time classified as a suborder of the Rodentia, but they may be distinguished from rodents in various ways, including the presence of two upper incisors on either side, while rodents have only one. F. Petter (1971*a*) recognizes five genera of extant African lagomorphs, three of which, *Lepus* Linnaeus, *Pronolagus* Lyon, and *Bunolagus* Thomas, occur in southern Africa.

No lagomorph taxon has as yet been described from the australopithecine-bearing breccias, though H. B. S. Cooke (1963) provisionally lists the two extant forms *Pronolagus randensis* from Makapansgat limeworks and *Lepus capensis* from Taung, Sterkfontein, and Bolt's Farm.

In the present study unidentified lagomorphs are recorded from Sterkfontein Member 5, Swartkrans Member 2 and channel fill, and Kromdraai A and B. In all probability the two forms mentioned by Cooke are involved: *Pronolagus randensis* Jameson, the red rock hare, a short-eared form occurring in suitable rocky habitats in the Transvaal, Rhodesia, Botswana, and South-West Africa. They frequently live in caves, and for many years a pair has lived and bred in the twilight zone of the Lower Cave at Swartkrans. *Lepus capensis* Linnaeus, the Cape hare, described from the Cape Province but, according to F. Petter (1971*a*) a species that ranges throughout Africa, Europe, and Asia. It has appreciably longer ears than *Pronolagus* and apparently does not make use of caves.

Order Rodentia

Family Hystricidae: Porcupines

Two genera of living African porcupines are currently recognized (Missone 1971): *Atherurus* Cuvier, the brush-tailed porcupines, and *Hystrix* Linnaeus, the common and crested porcupines. No representatives of *Atherurus* occur in southern Africa. Two species of *Hystrix* are known: *H. africaeaustralis* Peters, the common porcupine of southern Africa, and *H. cristata* Linnaeus, the crested porcupine found in the northern half of the continent.

Apart from the forms found in the australopithecine caves breccias of the Transvaal, only one species of fossil porcupine has been described from elsewhere in Africa: *H. astasobae* Bate from Abu Hugar on the Blue Nile in the Sudan. It is of Middle or Upper Pleistocene age and, according to Maguire (1976), has affinities with *H. cristata*.

The Transvaal fossils have been equated with the extant *H. africaeaustralis* and with two extinct forms, *H. makapanensis* Greenwood and *Xenohystrix crassidens* Greenwood.

Hystrix africaeaustralis Peters. Common porcupine
Provenance: Sterkfontein members 4, 5, and 6, Swartkrans members 1 and 2, and Kromdraai A

The common porcupine is extremely important as a collector of bones in caves, as was discussed in chapter 5. It appears to have existed unchanged for at least three million years, and its remains are commonly preserved in Stone Age sites, including Elandsfontein and various

other localities in the southwestern Cape. Fossils from the australopithecine sites have been studied in detail by Maguire (1976).

Some of the material from Makapansgat Limeworks was described by Bakr (1959) as *H. (Hystrix) greenwoodi* sp. nov. This taxon, however, is not considered to be valid by Maguire (1976), who equates it with *H. africaeaustralis*.

Hystrix makapanensis Greenwood
Provenance: Tentative identifications from Swartkrans Member 1 and Kromdraai A

The species was originally described as *H. major* from the Makapansgat Limeworks by Greenwood (1955) on the basis of a left mandibular fragment. It had clear affinities with *H. africaeaustralis* but was about one-third larger. Subsequently it was found that the specific name *major* was preoccupied, having been used by Gervais in 1859 for a specimen from Ratoneau Island near Marseilles in France. Greenwood (1958) therefore renamed her Makapansgat taxon *H. makapanensis*.

Maguire (1976) points out that *H. makapanensis* is the least common of the three Makapansgat fossil porcupine forms. It is intermediate in size between *H. africaeaustralis* and *Xenohystrix crassidens*, but in other respects was clearly very like the living porcupines still found in the area.

Both *H. makapanensis* and *X. crassidens* have been recorded from the Omo group sediments by Coppens and Howell (1976), suggesting a wide African distribution in early Pleistocene times.

Class Aves

Order Struthioformes
Family Struthionidae
Struthio sp. Ostrich
Provenance: Swartkrans Member 2 and Kromdraai B

Ostriches are represented by some pieces of bone and eggshell, but these remains are too fragmentary to yield information on whether the fossil animals were different from contemporary ostriches.

Class Reptilia

Order Crocodilia
Cf. *Crocodylus niloticus*. Nile crocodile
Provenance: Kromdraai B

This tentative identification is based on a single tooth, and it would be unwise to deduce much from it.

Order Squamata
Family Cordyliidae
Cf. *Cordylus giganteus*. Girdled lizard
Provenance: Kromdraai B

The distribution of this lizard is restricted today to a small area of the highveld, and the animal is a clear indicator of open grassland conditions. The tentative identification from Kromdraai is based on caudal vertebrae and dermal plates; very little can be deduced from these remains about the nature of the fossil *Cordylus*.

Family Varanidae
Varanus cf. *niloticus*. Monitor lizard
Provenance: Sterkfontein Member 5

The identification is based on a single vertebra, similar to that of an extant monitor lizard such as still occurs in the area today.

Order Chelonia
Family Testudinidae

Remains of unidentified tortoises have been recovered from Sterkfontein Member 5, Swartkrans Member 2 and channel fill, and Kromdraai A and B

Generally, nothing more than a few pieces of carapace or plastron are available, on the basis of which little of taxonomic value can be said.

Phylum Mollusca

Class Gastropoda
Order Pulmonata
Cf. *Achatina* sp. Land snail
Provenance: Swartkrans members 1 and 2 and Kromdraai A

Some well-preserved shells of these large land snails, still common in parts of the Transvaal today, have been found. They await detailed study.

The Microvertebrate Component

Order Insectivora

Family Macroscelididae: Elephant shrews

Elephant shrews are restricted to Africa and, according to Corbet (1971), are represented by fifteen species in the following four genera: *Rhynchocyon* Peters, *Petrodromus* Peters, *Macroscelides* A. Smith, and *Elephantulus* Thomas and Schwann. They are typically small insectivores, although the forest elephant shrews *Rhynchocyon* and *Petrodromus* may reach a head and body length of almost 30 cm. All have long, mobile snouts and strongly developed hind limbs. Most species are diurnal but fall prey to owls hunting in the late afternoon.

Genus *Elephantulus* Thomas and Schwann, 1906

On the basis of remains from Skurveberg, Broom (1937*c*) created a new genus and species, *Elephantomys langi,* for elephant shrews having "a molariform second upper premolar, no third lower molar and no large development of air cells in the ear region of the skull." In these respects *E. langi* differed from all other known forms except *Elephantulus intufi* Smith, which Broom proposed should also be placed in his new genus *Elephantomys.* Subsequently Broom (1948*a*) transferred the species *langi* to *Elephantulus,* suggesting that two subgenera, *Elephantulus* and *Elephantomys,* were probably warranted. In the same paper, Broom described a new species, *Elephantulus antiquus,* on the basis of a maxillary specimen from Bolt's Farm.

E. langi has been recorded from Sterkfontein (Member 4 or Member 5) by Broom and Schepers (1946), de Graaff (1960*a*), and T. N. Pocock (1969); from Swartkrans (almost certainly Member 1) by Davis (unpublished) and by H. B. S. Cooke (1963), who lists it for Kromdraai A. It is likely to be present in the Kromdraai B microfaunal sample currently being studied by Davis and Pocock.

In their revision of the elephant shrews, Corbet and Hanks (1968) point out that the specific name *langi* is preoccupied and therefore propose the new name *E. broomi* for this taxon. Butler and Greenwood (1976) conclude that *E. broomi* occurred at both Makapansgat Limeworks and Bed 1 of Olduvai, and that it was related to the living species *E. intufi* and *E. rupestris.*

Remains attributed to *E.* cf. *brachyrhynchus,* the living short-snouted elephant shrew, have been recorded from Swartkrans Member 1 and Kromdraai B (Davis, unpublished).

Mylomygale spiersi Broom

Broom (1948*a*) described the left mandible of a remarkable macroscelid from a cave about 1.6 km north of the australopithecine locality at Taung. Its hypsodont teeth appeared unusually rodentlike and, concerning the specimen, Broom wrote (1948*a,* p. 8): "There is not, I think, any reasonable doubt that this fossil mammal is a greatly specialised Menotyphlan or elephant shrew of which the nearest living ally is *Petrodromus.* We may assume as probable that some early Pliocene elephant shrew allied to *Petrodromus* changed its habits and became vegetarian and developed rodentlike molars. This Menotyphlan is one of the most interesting fossil mammals discovered in recent years." An isolated premolar from Sterkfontein (presumably Member 4 or Member 5) has been tentatively assigned to *M. spiersi* by de Graaff (1960*a*).

Family Chrysochloridae: Golden moles

According to Meester (1971, p. 1), this family, which is restricted to Africa, contains seven valid genera of golden moles that are "small to medium-sized, blind burrowing animals, with colour varying from pale fawn-grey to black in different species, always with bright metallic sheen. Adaptations for subterranean existence include a fusiform body, with no external eyes or ears, well-developed forelegs and claws, no visible tail, close-set fur and a muzzle terminating in a thick, leathery pad."

Although golden moles spend most of their time underground, they are occasionally taken by owls, suggesting that they must surface at night. Fossils are not abundant, although remains of *Chrysochloris* sp. occur in fair numbers in the Quartzose Sand Member and Pelletal Phosphorite Member of the Varswater Formation at Langebaanweg (Hendey 1974*a*). Apart from the two taxa mentioned below, the only other fossil form from southern Africa is *Chrysotricha hamiltoni*, described by de Graaff (1957) from the Makapansgat Limeworks.

Proamblysomus antiquus Broom, 1941

The type specimen described by Broom (1941*c*) came from Bolt's Farm and resembled *Amblysomus*, although it differed in having a wider temporal region and large tympanic bullae, unlike those of living golden moles. A specimen from Kromdraai A has been tentatively assigned to this taxon (Broom 1948*a*).

Chlorotalpa spelea Broom

According to Meester (1971), the genus *Chlorotalpa* is closely related to *Amblysomus* and contains five extant

species. *C. spelea* was described (Broom 1941*c*) on the basis of a single well-preserved skull from the Sterkfontein Type Site (probably Member 4); the skull appeared longer and narrower than that of known living species. *C. spelea* has also been identified from Swartkrans Member 1 (Davis, unpublished; H. B. S. Cooke 1963).

Family Soricidae: Shrews

According to Meester (1963), six genera of living African shrews are regarded as valid; these are *Crocidura* Wagler, *Suncus* Hemprich and Ehrenberg, *Myosorex* Gray, *Sylvisorex* Thomas, *Paracrocidura* De Balzac, and *Scutisorex* Thomas. Only the first three of these are represented as fossils in the australopithecine-bearing caves, together with a species of *Diplomesodon* mentioned below.

The genera *Crocidura* and *Suncus* are thought to have evolved directly from an archaic group, represented by *Sylvisorex*, with primitive features in the skull and teeth. In addition, some members of the genus *Suncus* may have evolved into *Crocidura* by the loss of P^3, meaning that *Crocidura*, as now defined, may have had a polyphyletic origin.

Myosorex, on the other hand, appears to represent an entirely different evolutionary line perhaps related to the Soricinae, or red-toothed shrews (Meester 1963, pp. 11–12).

Genus *Crocidura* Wagler

Broom (1948*a*) described *C. taungensis* from Hrdlicka's Cave at Taung, pointing to the resemblance

Crocidura

between this species and the extant *C. bicolor*. In the same paper Broom remarked that another species of *Crocidura* occurs at Sterkfontein, and possibly still another at Kromdraai, though these had not been satisfactorily identified. In his study of South African fossil shrews, Meester (1955) was not able to substantiate the suggestion that *Crocidura* was represented at either Sterkfontein or Kromdraai, though H. B. S. Cooke (1963) lists *C.* cf. *bicolor* from both sites. It is not clear on what material his listing is based.

Genus *Suncus* Ehrenberg

Three extant species are currently recognized in southern Africa: *S. lixus* (Thomas, 1898), *S. varilla* (Thomas, 1895), and *S. infinitesimus* (Heller, 1912). The last two of these are represented as fossils in the Sterkfontein valley caves.

Suncus varilla (Thomas, 1895)

Meester and Meyer (1972) record this species from Sterkfontein members 4 and 5, Swartkrans Member 1, Bolt's Farm, and Witkrans near Buxton.

Suncus infinitesimus (Heller, 1912)

The same authors have identified this species from Sterkfontein members 4 and 5, Gladysvale, and Makapansgat.

Genus *Myosorex* Gray

Until very recently only two species of living *Myosorex* were known from southern Africa: *M. varius* (Smuts) and *M. cafer* (Sundevall). A third species has now been discovered at Knysna, and its description is awaited (Meester and Dippenaar 1978).

Two attempts have been made to explain the distribution of the original *Myosorex* species in terms of past climatic changes (Meester 1958; Brain and Meester 1964). Both species are endemic and are separated by a considerable distance from their east and central African allies. They both prefer moist, densely vegetated and often mountainous habitats. It has been suggested that southern Africa was colonized twice by *Myosorex* during periods of high rainfall, with the migrations proceeding from the Inyanga area of Rhodesia, along the eastern escarpment to the Transvaal, and thence farther. During times of drier climate, it is suggested that the Limpopo valley formed an effective barrier. The first wave led to the establishment of *M. varius,* by way of its ancestral form *M. robinsoni.* The second wave resulted in colonization by *M. cafer.*

Myosorex robinsoni Meester, 1955

The type specimen came from Swartkrans Member 1 (Meester 1955), but material is also known from Sterkfontein (probably Member 4), Gladysvale, and Makapansgat.

Genus *Diplomesodon* Brandt, 1853

The genus is not represented in the contemporary African shrew fauna, but a new species, *D. fossorius,* was described by Repenning (1965) on the basis of fossils from the Makapansgat Limeworks. The fossils were found in a block of breccia prepared at the University of Colorado Museum and are thought to be most closely related to the living *D. pulchellum* of south-central Russia.

Representatives of this genus have not yet been found among the Sterkfontein valley microvertebrate assemblages, but their discovery might well be expected.

Family Erinaceidae: Hedgehogs

No remains of hedgehogs are known from the cave sites of Sterkfontein, Swartkrans, or Kromdraai, but a single skull from Bolt's Farm was described by Broom (1937*b*) as *Atelerix major* and subsequently figured by him (Broom 1948*a,* p. 9). The specimen was subsequently transferred to the taxon *Erinaceus broomi* by Butler and Greenwood (1973).

Order Chiroptera

Occasional remains of bats occur among the microvertebrate remains, but they may well have resulted from natural deaths of cave residents rather than representing the prey of owls. No detailed descriptions have been published, but the following taxa have been recorded: *Rhinolophus* cf. *darlingi* and *R.* cf. *augur,* together with postcranial remains of Vespertilionidae from dump 8 at Sterkfontein (T. N. Pocock 1969); *Myotis* sp. from Bolt's Farm (Broom 1948*a*); *Rhinolophis* cf. *capensis* from Rodent Cave, Makapansgat (de Graaff 1960*a*); *R.* cf. *geoffroyi* and *Miniopterus* cf. *schreibersii* from the Cave of Hearths, Makapansgat (de Graaff 1960*a*); *Rhinolophus* cf. *capensis* from Taung, Makapansgat, and Bolt's Farm (H. B. S. Cooke 1963); and cf. *Myotis* sp. from Bolt's Farm (H. B. S. Cooke 1963).

Order Rodentia

Suborder Hystricomorpha

Family Bathyergidae. Mole rats

The mole rats are characterized by very large incisors, minute eyes and ears, an insignificant tail, and short legs. They are fossorial animals, feeding on roots and bulbs that they find by burrowing tunnels beneath the surface, throwing up heaps of soil at intervals.

According to de Graaff (1971), the following five genera of extant mole rats are considered valid: *Heterocephalus* Rüppell, *Heliophobius* Peters, *Bathyergus* Illiger, *Cryptomys* Gray, and *Georychus* Illiger. The number of cheek teeth is variable, from 4/4 in the southern African genera *Cryptomys, Bathyergus,* and *Georychus* to 3/3 or 2/2 in *Heterocephalus* and 6/6 in *Heliophobius.*

Representatives of two extinct bathyergid genera have been reported from southern Africa: *Bathyergoides neotertiarius* Stromer, from a Miocene horizon in South-West Africa, and three species of *Gypsorhychus*—*G. darti* and *G. minor* from Taung (Broom 1934, 1948*b*) and *G. makapani* from Makapansgat (Broom 1948*b*).

All the bathyergid remains from the Sterkfontein valley caves have been referred to the genus *Cryptomys.*

Genus *Cryptomys* Gray

On the basis of some cranial material from Skurveberg, Broom (1937*c*) described a new species, *C. robertsi,* that

differed from living species in the shape of the posterior part of the mandible. He observed a distinct ridge running forward on the outer side of the mandible from the condyle.

C. robertsi has been recorded from Sterkfontein, probably Member 4 (de Graaff 1960*a*) and from Sterkfontein, Swartkrans, Kromdraai A, and Bolt's Farm by H. B. S. Cooke (1963). *Cryptomys* sp. is recorded from Kromdraai B (Davis 1959). T. N. Pocock (1969) lists *C.* cf. *holosericeus* from dump 8 at Sterkfontein, but de Graaff (1971) regards the species *holosericeus* as a synonym of *hottentotus,* the common living form in the area today.

Suborder Myomorpha

Family Muridae

Subfamily Murinae

The subfamily contains a large number of genera, and the taxonomy of many is confused. The following taxa have been identified by various authors from the Sterkfontein valley caves under consideration:

From Sterkfontein, probably Member 4, by de Graaff (1960*a*): *Dasymys* cf. *incomptus, ?Arvicanthus* sp., *Pelomys fallax, Rhabdomys* cf. *pumilio, Aethomys* cf. *namaquensis, Mastomys natalensis,* and *Leggada minutoides.*

From Sterkfontein, dump 8, by T. N. Pocock (1969): *Dasymys* sp., *Rhabdomys minor, Aethomys* cf. *chrysophilus, Praomys (Mastomys)* cf. *natalensis, Praomys* sp., *Mus.* cf. *musculus.*

From Swartkrans Member 1, by Davis (1955): *Mus* cf. *minutoides, Mus* cf. *triton, Dasymys ?bolti,* and *Lemniscomys* sp.

From Kromdraai B, by Davis ((1959): *Rattus* sp., *Mus* sp., *Rhabdomys* sp., and *Dasymys* sp.

Davis (1962) has expressed the opinion that *Dasymys bolti,* a species named by Broom but not formally described, was directly ancestral to *D. incomtus,* the extant swamp rat.

Subfamily Otomyinae

Rats of this family are characterized by their laminate molars, the first upper molar with three laminae, the second with two, and the third with a varying number from three to nine. In the lower jaw the first molar has from four to seven lamellae, while the second and third lower molars typically have only two lamellae each (Roberts 1951, p. 418).

Only two extant genera are recognized by Misonne (1971): *Parotomys* Thomas and *Otomys* Cuvier. Both are restricted to Africa south of the Sahara; the former contains the Karroo rats, characterized by inflated bullae and other adaptations to arid environments; the latter contains the vlei rats, which are frequently found in marshy habitats. Individuals of both genera are diurnal but regularly fall prey to owls, so that their remains are common in modern and fossil microfaunal accumulations. Davis (1955) is inclined to regard *Myotomys* as a subgenus of *Parotomys,* whereas T. N. Pocock (1976) argues that it should be retained as a valid genus.

In the past, many mammalogists have been inclined to classify the Otomyinae as a subgenus of the Muridae, but the subfamily was transferred to the Cricetidae by Misonne (1969), and this procedure has been followed in the Smithsonian Institution's identification manual of African mammals (Misonne 1971).

Recently T. N. Pocock (1976) has described a remarkable rodent from the Pliocene deposits of E Quarry, Langebaanweg, which appears to form a link in tooth structure between the Otomyinae and the Murinae. He has named it *Euryotomys pelomyoides* and described it as a rat or mouse with a typically murine cusp pattern but differing from all other known murines in possessing an enlarged and complex third upper molar. On the basis of this link fossil, Pocock speculated that the otomyines have evolved from the Murinae within the last 4 to 5 million years, the ancestral murine being of the *Pelomys* type.

Two fossil species of the Otomyinae have been described: *Palaeotomys gracilis* by Broom (1937*c*) from Skurveberg, and *Prototomys campbelli* Broom 1946 from Taung. The latter is known by the type specimen only, but the former represents one of the commonest species in the microfaunal concentrations from the Transvaal caves. In his description of the type, Broom expressed the opinion that *P. gracilis* was probably near to the common ancestor of *Otomys* and *Myotomys.*

Davis (1962) has speculated that *Otomys (Palaeotomys) gracilis/O. (Myotomys) sloggetti* form a "chronospecies" spread over the Pleistocene. *O. sloggetti* does not occur in the Sterkfontein area today—two sibling species, *O. irroratus* and *O. angoniensis,* take its place. Davis suggests that *gracilis* may be closer to the latter species than to the former.

P. gracilis has been recorded from Sterkfontein (probably Member 4) by de Graaff (1960*a*); from Sterkfontein dump 8 by T. N. Pocock (1969); from Swartkrans Member 1 by Davis (1955); and from Kromdraai B by Davis (1959).

Family Cricetidae

Members of this large family of rodents can be separated from members of the family Muridae only by the structure of the teeth. The Muridae are distinguished by the presence of an anterior inner cusp on the first lamella of M^1. In Cricetidae this cusp is absent (Coetzee 1972).

Representatives of the following three subfamilies are to be found among the microfauna from the Sterkfontein valley caves: Cricetinae, Gerbillinae, and Dendromurinae.

Subfamily Cricetinae

A single extant genus and species, *Mystromys albicaudatus* (Smith), occurs in Africa, though the subfamily is well represented in the northern hemisphere. The genus *Mystromys* Wagner is of particular interest in the microfauna from the Sterkfontein valley caves, since individuals belonging to it tend to be extremely abundant as fossils, outnumbering those of murine affinities that are strongly dominant today.

M. albicaudatus, the white-tailed rat or South African hamster, is a savanna species of highveld and montane grasslands, not known north of latitude 25°S but widely distributed in the Transvaal, Orange Free State, Lesotho, and the eastern Cape, with a relict population in the southwestern Cape (Davis 1974).

Three extinct species have been named. *M. antiquus* came from Taung; its teeth were figured by Broom (1948*a*), though a description was never furnished. *M. hausleitneri* was described by Broom (1937*c*) as *M. hauslichtneri,* though the spelling was subsequently corrected (Broom 1948*a*). The type specimens came from Skurveberg and differed in skull structure from *M. albicaudatus,* particularly in the shorter snout and broader palate. Broom (1948*a*) also described a new subspecies *M. hausleitneri barlowi* from a fissure filling at Sterkfontein, apparently younger than the main deposit. Davis (1955) has been inclined to synonymize *M. antiquus* and *M. hausleitneri barlowi* with *M. hausleitneri.* From the

Makapansgat Limeworks, Lavocat (1956) described a new small form as *M. darti*.

M. hausleitneri has been recorded from Sterkfontein (probably Member 4) by de Graaff (1960*a*) and from Sterkfontein dump 8 by T. N. Pocock (1969); from Swartkrans Member 1 by Davis (1955), and from Kromdraai B (Davis 1959).

In his recent review of African fossil rodents, Lavocat (1978) remarks on the abundance of *Mystromys* in the older southern African cave deposits. He believes that it was in the early Pleistocene that the prominence of *Mystromys* declined in favor of the murines.

Subfamily Gerbillinae

The gerbils form a group of nocturnal rodents typically adapted to arid or semiarid open environments. They have well-developed hind limbs, large ears and eyes, and extremely well-developed bullae; their molars are characterized by transverse oval sections of enamel, each surrounding a cement center not in rows of cusps as in the Murinae, nor in layers of transverse enamel as in the Otomyinae. Three genera of living gerbils are currently recognized in southern Africa (Petter 1971*b*): *Tatera* Lataste, *Desmodillus* Thomas and Schwann, and *Gerbillurus* Shortridge. The first two of these have fossil representations in the fauna of the Sterkfontein valley caves.

Genus *Tatera* Lataste

According to Davis (1971), all species of *Tatera* fall into one of two groups: the *afra* group or the *robusta* group. *T. afra* (Gray) is confined to sandy areas of the southwestern Cape, while *T. robusta* (Cretzschmar), the fringe-tailed gerbil, extends from west Africa across the sub-Saharan zone to Ethiopia, Somalia, and Tanzania. Included in the *afra* group are the two southern African species *T. brantsii* (A. Smith), the highveld gerbil, and *T. inclusa* Thomas and Wroughton, the Gorongosa gerbil. The only southern African representative of the *robusta* group is *T. leucogaster* (Peters), the bushveld gerbil.

Tatera

*T.*cf. *brantsii* has been recorded from Sterkfontein (presumably Member 4) by de Graaff (1960*a*) and *Tatera* sp. from Sterkfontein dump 8 (T. N. Pocock 1969), and Kromdraai B (Davis (1959). In an unpublished manuscript, Davis (1955) has proposed a new species from Swartkrans Member 1: *T. robinsoni,* which, when compared with the two living species in the Sterkfontein valley today, comes closer to *T. brantsii* than to *T. leucogaster,* although it is appreciably smaller.

Genus *Desmodillus* Thomas and Schwann

A single species is currently recognized: *D. auricularis* (A. Smith), the Namaqua gerbil, characterized by enormous development of the tympanic bullae and distributed widely through the drier western regions of southern Africa.

Davis (1955) records *D. auricularis* from Taung and Kromdraai B.

Subfamily Dendromurinae

The dendromurines form a group of small mice characterized by grooved incisors and well-developed "masseter knobs" on their maxillae (see Coetzee 1972). Living members in southern Africa are placed in three genera, *Dendromys, Steatomys,* and *Malacothrix,* each of which has representatives among the fossils from the Sterkfontein valley caves.

Genus *Dendromys* A. Smith

The climbing mice are slender animals with prehensile tails and fairly large, rounded ears. They spend much of their time above the ground, either among the tops of long grasses, where they feed on seeds, or in shrubs or trees. Their nests are hollow balls of shredded grass with holes at each end (Roberts 1951).

Misonne (1971) recognizes five species of living *Dendromys,* but only one fossil form has been named from southern Africa: *A. antiquus,* figured and mentioned by Broom and Schepers (1946) but never formally described. The specimen came from Taung, but Davis (1955) has suggested that the taxon may be present in Swartkrans Member 1. He also records *Dendromys* sp. from Kromdraai B (Davis 1959), while de Graaff lists *C.* cf. *mesomelas* from Sterkfontein (probably Member 4) and T. N. Pocock (1969) has identified *D.* cf. *melanotis* among microvertebrates from Sterkfontein dump 8.

Genus *Steatomys* Peters. Fat mice

The following three species are regarded as valid in southern Africa: *S. minutus* Thomas and Wroughton, *S. pratensis* Peters, and *S. krebsii* Peters (Meester, Davis, and Coetzee 1964). The common name "fat mouse" comes from the fact that these rodents tend to lay down appreciable layers of fat and then to estivate for part of the year in their grass-lined underground chambers. The preferred habitat appears to be alluvial floodplains or sandy areas close to water.

Steatomys sp. has been recorded from Kromdraai B (Davis 1959) and from Sterkfontein dump 8 (T. N. Pocock 1969), while *S.* cf. *pratensis* is known from Swartkrans member 1 (Davis 1955).

Genus *Malacothrix* Wagner. Large-eared mice

Only one species, *M. typica* (Smith), is thought to be valid by Misonne (1971), with a wide distribution in the central and western areas of southern Africa. In Botswana, Smithers (1971*a*) found the preferred habitat to be hard ground with short grass around the fringes of pans. Deep inclined burrows are constructed, and the mice have been found to wander far at night in search of food.

De Graaff (1960*a*) has proposed a new fossil form from Makapansgat, which he designated *?Malacothrix makapani* on the basis of an isolated left M^1 that appeared appreciably smaller than that of *M. typica*.

*M.*cf. *typica* has been recorded from Swartkrans Member 1 (Davis 1955), and from Kromdraai, presumably A (H. B. S. Cooke 1963); *Malacothrix* sp. was listed from Sterkfontein dump 8 (T. N. Pocock 1969).

Family Muscardinidae. Dormice

Dormice, widely distributed through Africa, Asia, and Europe were characterized by bushy tails and soft fur. They are nocturnal omnivores, taking large numbers of insects and other small prey. The occlusal surfaces of their incisors are V-shaped when viewed from the front; their cheek teeth are very small compared with those of most other rodents of equivalent size.

Misonne (1971) recognizes only two genera in the family: *Eliomys* Wagner and *Graphiurus* Smuts, the former restricted to North Africa.

The only published record of a fossil dormouse for Sterkfontein valley caves is that of T. N. Pocock (1969), who listed *Graphiurus* sp. from dump 8 at Sterkfontein.

Class Amphibia

Remains of frogs occasionally occur among the microvertebrate remains, but these have not been studied in detail. Bones from two species of frogs have been recorded from dump 8 at Sterkfontein, but no further particulars are available (T. N. Pocock 1969).

Class Reptilia

Occasional remains of lizards are present in the collections, but these have yet to be studied in detail.

Class Aves

The microvertebrate collection from Kromdraai B contains fairly abundant remains of birds, which have been studied by T. N. Pocock (1970). Largely on the basis of postcranial elements, he identified the following forms: *Tyto* sp. (barn owl), *Francolinus* sp. (francolin), *Coturnix* (quail), *Turnix* (button quail), *Crex* sp. (crake), *Apus* sp. (swift), Scolopacidae (two species of sandpiper and ruff), *Agapornis* spp. (larger and smaller species of lovebird), *Sturnidae* (starlings), cf. *Anthus* sp. (pipit), cf. *Estrilda* sp. (waxbill), cf. *Cisticola* sp. (grass warbler), cf. *?Passer* sp. (sparrow), cf. *Ploceus* sp. (weaver), and Hirundinidae (swallows of three different sizes).

Pocock interpreted the bird remains as follows (1970, p. 4):

> Thus thanks to the Pleistocene barn owl sallying forth to prey on the microfauna and returning to deposit its pellets below the cave-ledge roost, we can form a fair picture of the smaller avifauna at the time of the Kromdraai ape-man, *"Paranthropus" robustus*. Quail, button quail, pipits and perhaps also cisticolas were quite common, confirming the grassveld character of the environment. Flocks of starlings were preyed upon as they roosted in the streamside or marsh vegetation. Swallows may have fallen victim in the same way, especially if they were non-breeding migrants, or they may have built their mud nests, perhaps colonially, in the very cave frequented by the owl. Swifts doubtless also used the cave, nesting in crevices, or perhaps commandeering some of the swallow nests. Non-breeding migrants probably did occur, crakes, sandpipers and ruffs from the steppes and tundra of a glaciated Europe.

An interesting conclusion to emerge from the Kromdraai analysis is that, though small members of the parrot family were apparently common, seed-eating passerines such as sparrows, weavers, and queleas were not. Pocock speculated that, during Kromdraai B accumulation times, the niche today filled by abundant seed-eating passerines was occupied by lovebirds and similar psittacine birds. Such a concept supports the contention that weavers of the family Ploceidae are a modern group that may have entered Africa, perhaps from Asia, in relatively recent times and have undergone extensive radiation here. In doing so they came into competition with the resident parrots, displaced them, and reduced their population numbers.

Bird remains from the other cave sites await study. They are likely to yield valuable information not only about avian evolution, but also about avifaunal changes during the Pleistocene.

10 Sterkfontein

A Brief History of Activity

During the 1890s, prospectors and mining companies showed interest in the area north of Krugersdorp, extending to the Bloubank River valley. This interest centered particularly around the outcrops of Black Reef quartzite which, in the vicinity of Kromdraai, proved to be gold-bearing, though ultimately not profitable. Prospectors consequently turned their attention to deposits of pure white travertine that occurred in ancient caves dissolved in the dolomitic country rock. Such caves frequently contained bone-bearing sediments, and it was almost certainly through lime mining that the first fossils came to light. On 1 February 1895 a well-known geologist, David Draper, who was to become the first secretary of the Geological Society of South Africa later that year, sent a sample of fossiliferous breccia to the British Museum, accompanied by the following information:

> Mass of rock containing a number of fragments of bones. This is from a cave on the farm Kromdraai situated about 16 miles west of Johannesburg. There is a bed of stalagmite with masses of rock in which the bone is abundant. It is really a cave deposit containing bone breccia or fragments of bone. I send you the specimen for examination as I should like to know whether the rock is similar to that found in European caves. I am informed that a portion of a human skull and the claw of a lion were found in the cave but I have not seen the specimens nor have I yet inspected the cave though shortly hope to do so.

Three months later on 8 April 1895, Draper contributed to the discussion after the inaugural address by the first president of the newly formed Geological Society of South Africa, H. Exton. He stated that "he had a short time ago visited the Kromdraai caves, or, rather, the formations that once had been caves, and he found them most interesting from a geological point of view. There was much to discover there. He, therefore, suggested that they make a beginning by making up a party to go and explore this interesting geological ground" (Draper 1896, p. 11).

In my opinion there is a strong possibility that the fossiliferous cave deposit from which Draper's sample came, and which he and other geologists visited in 1895, was the surface outcrop on the Sterkfontein hill. This was the spot being worked for lime at the time by G. Martinaglia, a pioneer prospector of the Witwatersrand goldfield. C. van Riet Lowe (1947) has documented this early phase of activity at Sterkfontein and quoted extracts from a letter written to him by G. Martinaglia, son of the original miner. It runs as follows:

> My father was prospecting for lime and one day when I took him some refreshments to the quarry where he was working, he pointed enthusiastically to a dark hole on the side of the quarry and told me in Afrikaans that he had just blasted open a "wondergat." This made a great impression on my boyish mind and I still remember well the words spoken on that occasion by my father. There are only a few people alive to-day who saw these caves in the days of their original splendour before they were destroyed later by commercial exploitation.

The date of this first discovery of the underground caves has been established as shortly after the Jameson Raid, which ended on 2 January 1896 (van Riet Lowe 1947; Malan 1959), but it is obvious that the fossiliferous breccia outcropping on the surface was known at least a year earlier than this. The question whether the surface outcrop was, in fact, linked to the Sterkfontein caves has been considered by B. D. Malan as follows (1960, p. 110):

> In considering the date of discovery of the Sterkfontein Caves, the question arose whether Draper's mention of the Kromdraai Caves on 8th April, 1895, referred to the nearby Sterkfontein Caves, it being quite possible that the name Kromdraai might have denoted a large area. On the strength of recollections of pioneers that the Sterkfontein Caves were discovered after the Jameson Raid [after 2 January 1896] and Draper's reference to "the foundering that had once been caves," it was concluded that Draper's early reference could not have been to the Sterkfontein Caves. This conclusion is now specifically confirmed in Draper's letter to the British Museum where he clearly states that the material came from the *farm* Kromdraai, whereas the Sterkfontein caves lie on the adjacent farm Swartkrans. There can thus now be hardly any doubt that the Sterkfontein Caves were not known when Draper reported on breccias from Kromdraai early in 1895.

Malan is certainly right that the underground Sterkfontein caves were not known when Draper reported on the fossiliferous surface outcrops, but I believe the latter are nevertheless more likely to have been those on the

Sterkfontein hill, being worked by Martinaglia, than any other, such as those that exist at Kromdraai.

This view has, I find, already been expressed by Broom, who stated: "What are today called the Sterkfontein Caves had been known for many years. They are really on part of the original farm Zwartkrans, and adjoining Kromdraai they have been referred to as the Kromdraai Caves; but it seems better to keep the more modern term Sterkfontein Caves for those where the Sterkfontein skull was found, and to reserve the name Kromdraai for the farm where the Kromdraai skull was discovered" (Broom and Schepers 1946, p. 46).

The Kromdraai B and A deposits both formed extremely insignificant surface outcrops before Broom's paleontological exploitation of them in 1938 and 1947 respectively. Draper's description of "a bed of stalagmite with masses of rock in which bone is abundant" is to my mind likely to refer to one of the Sterkfontein hill exposures, perhaps that of the "West Pit" where microfaunal remains are specially abundant. The microfaunal list compiled by de Graaff (1961) from Draper's sample in the British Museum shows a suite of species that, regrettably, could well have been found in any of the Sterkfontein valley site units known at present. Nor are the macrofaunal remains any more diagnostic. Oakley wrote of them (1960, p. 110), "A number of fossils have been extracted from the breccia by the acetic acid technique in our laboratory, including remains of baboon, a small carnivore, a large horse, porcupine, rodents, a small lizard and birds."

An interesting comment on the fauna of the "Kromdraai caves" by M. E. Frames (1898) was elicited by a paper presented to the Geological Society of South Africa by August Prister (1898). Prister presented evidence that he regarded as indicating that the Pretoria and Witwatersrand areas had been subjected to glaciation in Quaternary times. As was to be expected, such an allegation evoked a good deal of comment, most of it unfavorable to Prister's theory. Frames's comment (1898, pp. 94–95) ran as follows:

> A review of the past and present flora and fauna, as an argument bearing on the theory of a glacial period in quaternary times in South Africa, may not be without interest. . . .
>
> To deal thoroughly with this aspect of the question, it will be necessary to consider the animal remains found in the Kromdraai Caves in the Dolomite near Krugersdorp.
>
> Amongst these are those of the horse species, antelopes, monkeys, porcupines, rats, bats, etc., and the presence of the two first in the cave would lead one to infer that they had been dragged there by beasts of prey. The cave is now filled with bones, cave earth and stalagmites to a depth of about 15 feet, and the whole mass is consolidated into hard rock. The latter feature naturally points to the antiquity of the cave and its contents. The roof has, in some cases, been eroded, so that portions of the bone bed are to-day exposed on the surface. Therefore, assuming that the glacial period was prior to the filling of the caves, surely the erosion which removed the roof would have proved sufficient to obliterate all traces of glaciation from the perishable sandstones and quartzites of the Magaliesberg?
>
> Again, if the ice age occurred after the caves were filled, then the same denudation argument would apply, with no reasonable explanation accounting for the remains of living types in the caves, and the absence of arctic types to fill the gap marking the intensely cold period. We cannot believe in the extinction of these animals and their sudden appearance again after the glaciers had melted.

This was the first of a long series of subsequent attempts to interpret the fossil fauna of the Sterkfontein valley caves in climatic terms.

Whatever the precise location of the original fossiliferous breccia outcrop described by the early geologists, there is no doubt about that of the Stekfontein *underground* caves, discovered first by Martinaglia. The quarry in which the "wondergat" first appeared was a short distance southwest of what subsequently became known as the Type Site excavation. The discovery of the specatular subterranean caverns at Sterkfontein aroused interest both locally and overseas. Malan (1959) quoted a report that appeared in the little-known journal *English Mechanic and World of Science* of 27 August 1897:

> Some wonderful caves were discovered recently at a place called Sterkfontein. . . . Limestone has been quarried for some months in a small kopje, and after an explosion after some blasting operations, a cavity of great depth was left . . . a party descended and it was found that they had driven through one of the largest of a series of magnificent caves. The spectacle was one of great beauty, the light carried by the explorers being reflected from thousands of stalactites. . . . The caverns, which have not yet been one-half explored, seem to run in tiers down to a depth of 150 feet.

Republication of this report in the French journal *Cosmos* prompted a Marist brother in Johannesburg (whose identity has not been established) to organize an outing to the cave with seven of his colleagues and thereafter to prepare an article entitled "Les grottes de Sterkfontein" for publication in the same journal during 1898. The author of the article was greatly impressed by the beauty of the formations in the caverns and corridors and remarked on the fossil jaw of a large antelope that was "probably the prey of a carnivore."

The remarkable beauty of the formations in the Sterkfontein caves had already been perceived by geologists in the Transvaal. In his presidential address to the Geological Society of South Africa on 22 February 1898, H. Exton (1899) referred to the unusual branching crystals of aragonite found on the stalactites. He expressed the opinion that, for such aragonite crystals to have formed, the caves must at one time have been filled with hot water.

At a previous meeting of the Society, held on July 12th 1897, Draper (1898, p. 63) stated that steps had been taken to protect the caves for the benefit of the public: "with this object in view he had approached the owners, and they had most willingly acceded to his request to place the caves under the protection of their manager, who would have instructions to prevent any wanton destruction of their interior.

Regrettably, such protection was not as long-lived as might have been hoped, and irreparable damage was caused to the subterranean formations between 1918 and 1920. At that time the caves belonged to E. P. Binet, who was leasing them to a Mr. Nolan. Upon expiration of the lease, the owner was not in favor of renewal, a decision that so annoyed Nolan that he deliberately dynamited what he could of the formations.

Little further interest appears to have been shown in the Sterkfontein caves till after the discovery of the Taung skull in 1924. At that time lime mining was being undertaken in the Sterkfontein area by the Glencairn Lime Company, the operation being managed by G. W. Barlow. When the Taung discovery was announced in 1925, a collection of Sterkfontein fossils was sent to R. A. Dart, who, however, did not investigate the matter further because he thought the deposit was probably relatively recent (Dart and Craig 1959).

During 1931 and 1932 J. H. S. Gear, later to become director of the South African Institute for Medical Research, was describing fossil baboons from Taung and spent a day in the underground caves at Sterkfontein in search of further fossils. He gave particulars in a letter to P. V. Tobias, written in 1972 and quoted by Tobias (1973*b*). A collection of fossils was, in fact, made and taken back to the Institute for Medical Research. Unfortunately it has never been located. However Dr. Gear's discussions at the Anatomy Department lunch table in those days may have led to further Sterkfontein investigations. R. Trevor Jones collected and studied a series of fossil baboon skulls from Sterkfontein in 1935, and the following year two other students of Dart's, G. W. H. Schepers and H. Le Riche, obtained further baboon material at Sterkfontein and showed it to Robert Broom.

Broom had joined the staff of the Transvaal Museum in August 1934 with the intention of furthering his paleontological studies, which he had previously undertaken in his spare time. In the hope of finding remains of an adult *Australopithecus,* Broom started work at old limestone quarries at Skurveberg, Hennops River, and Gladysvale. When Le Riche and Schepers showed him fossil baboon skulls from Sterkfontein, Broom arranged to visit the site the following Sunday. He was met there by the quarry manager, G. W. Barlow, who showed him various specimens on the table in the rondavel that served as a tearoom. The specimens were sold as souvenirs to interested visitors. In the past, Barlow had worked at Taung, and when Broom asked him whether he had ever seen anything like the Taung child skull at Sterkfontein, he replied that he rather thought he had. He promised to keep a sharp lookout.

On the occasion of Broom's third visit to Sterkfontein on Monday, 17 August 1936, Barlow handed him an endocranial cast and asked, "Is this what you're after?" Broom replied "Yes, that's what I'm after." Concerning the specimen Broom wrote as follows (1950, pp. 44–45):

> It was clearly the anterior two-thirds of the brain-cast of an anthropoid ape or ape-man, and in perfect condition. It had been blasted out that morning. I hunted for some hours for further remains, but without success. But I found the natural cast of the top of the skull in the side of the quarry, and had this cut out. So we returned to Pretoria with the fine brain-cast and the cast of the top of the skull. Next day I was back again and I had with me Dr. Herbert Lang, Mr. FitzSimons of the Museum, Mr. White, my assistant, and three Museum native boys. After a long hunt I discovered the base of the skull on which the brain-case had rested, with all the blocks that had been attached to it. I also found parts of the parietals and portions of the frontal. When, after some weeks of work, we had the remains of the skull pretty well cleared of matrix, it was found that we had practically all the base, the front of the brow, and most of the bones surrounding the eyes, and both upper jaws, which had been badly displaced; but the whole face had been considerably crushed. . . . To have started to look for an adult skull of *Australopithecus,* and to have found an adult of at least an allied form in about three months was a record of which we felt there was no reason to be ashamed. And to have gone to Sterkfontein and found what we wanted within nine days was even better.

In fact a strangely prophetic remark, made by R. M. Cooper, owner of the caves at that time, had been fulfilled. In a guidebook to places of interest around Johannesburg, Cooper wrote in 1935, "Come to Sterkfontein and find the missing link." Broom went and found it.

Regular visits to the Sterkfontein quarry provided Broom with further australopithecine and faunal remains. Then in 1937 Barlow stopped work in the Type Site locality and turned his attention to a spot lower down, close to the subterranean cave's exit. Here further australopithecine remains came to light before a fall in the price of lime brought mining at Sterkfontein to a close shortly before the war in 1939. The site then lay dormant till 1947.

After the war Gen. J. C. Smuts, prime minister of the Union of South Africa at the time, asked Broom to continue fieldwork in search of more "missing links" and promised financial support from the government. So on 1 April 1947 Broom's field team, which had been joined by J. T. Robinson, commenced work in the Type Site, close to the spot where the first adult *Australopithecus* had been found nine years before. Important fossils were discovered within the first week, and then on 18 April the most complete adult australopithecine skull, that of "Mrs. Ples," was blasted out. Broom described the incident thus (1950, p. 64):

> just about two weeks from the time we started we blasted a large piece of what looked very unpromising breccia. It was only a yard below where the type skull had been found. When the smoke of the blast blew away, we found that a beautiful skull had been broken in two. The outer part of the rock had the top of the skull and all the lower half was exposed in the wall. As the top of the skull had been split off we could see into the brain-cavity, which was lined with small lime crystals. I have seen many interesting sights in my long life, but this was the most thrilling in my experience.

The Type Site excavations of Broom and Robinson came to an end in January 1949 when the last of the working party was transferred to Swartkrans on the other side of the valley.

During 1956 I was completing a study, "The Transvaal Ape-Man-Bearing Cave Deposits," in which the Sterkfontein site featured prominently (Brain 1958). Under the subheading "Cultural Material from Sterkfontein" I found I had nothing to write and thereupon resolved to spend the next day, 7 May 1956, checking once again that no stone artifacts were to be found in the breccia. To my considerable surprise, breccia exposed in a shallow prospectors' pit, the "West Pit" a few meters west of the Type Site, contained unquestionable artifacts fashioned from quartzite and diabase. J. T. Robinson, head of the Transvaal Museum's paleontology department, in which I worked, was on a visit to the United States and I therefore sent him a telegram announcing the unexpected discovery.

An excavation of the loose overburden between the West Pit and the Type Site undertaken by R. J. Mason and myself at Robinson's request confirmed that the West Pit breccia was in fact continuous with that of the upper levels of the Type Site. Thereafter, during 1957 and 1958, Robinson undertook two seasons of excavation in the West Pit breccia. More than two hundred stone artifacts, several bone tools, and the fossils discussed in this chapter as coming from Member 5 came to light.

For many years the rondavel that had served as the Sterkfontein tearoom was a landmark on the site. Its walls were decorated with newspaper clippings, many of them yellow with age, telling of the discoveries of Sterkfontein

Fig. 154. 18 August 1936. Broom indicates the spot in the Sterkfontein type site quarry where the first australopithecine fossil was found. Also in the picture are G. W. Barlow and, behind him, Saul Sithole, who is still on the Transvaal Museum staff. The photograph was taken by Herbert Lang.

Fig. 155. Sterkfontein type site, 18 April 1947. Broom and John T. Robinson with the newly discovered skull of "Mrs. Ples," STS 5. Photograph the *Star* (Johannesburg).

Fig. 156. Sterkfontein extension site, 26 March 1958. John T. Robinson at the West Pit excavation. With him are B. D. Malan *(left)*, J. D. Clark, and Revil Mason.

fossils. Then in August 1956 the building burned down in a spectacular manner. On that day I was working in the West Pit area when, toward noon, a thunderstorm approached. I drove the Land Rover in under the eaves of the rondavel and waited for the storm to pass, while the African staff sat down to their midday meal in a room adjoining the rondavel. A bolt of lightning struck the thatched roof of the rondavel, showering the Land Rover with pieces of burning grass. In great haste I moved the vehicle and then entered the building to find that the men, including Daniel Mobelhe and Absalom Lobelo, who had worked at Sterkfontein since 1936 and 1948, had been knocked from their stools by the bolt and were lying motionless on the floor. Fortunately they slowly recovered and stumbled from the smoke-filled room.

During 1958 the Stegmann family, owners of the Sterkfontein farm, donated a 20-morgen area surrounding the caves to the University of the Witwatersrand. It became the Isaac Edwin Stegmann Nature Reserve. The blackened remnants of the rondavel and adjoining rooms were demolished to make way for the new restaurant building erected by the university. The building now includes the Robert Broom Museum, opened on 1 December 1966 by Broom's son, Norman Broom. Tourist facilities to the underground caves have been taken care of by Basil Cooper, and electric lighting and recorded tape commentaries have now been installed underground.

The 30th of November 1966 saw the one hundredth anniversary of the birth of Robert Broom, and as part of the celebrations to mark this event, plans were laid in July 1966 by P. V. Tobias and A. R. Hughes for a new, long-term program of work at the site. There were four original objectives (Tobias and Hughes 1969):

1. A vegetation survey of the 20-morgen Sterkfontein site.
2. A contour survey of the same area.
3. An archaeological excavation of the overburden, which would expose the underlying breccia and dolomitic country rock.
4. Systematic excavation of selected parts of the in situ breccia.

Since 1966, work has progressed steadily at Sterkfon-

Fig. 157. Phillip V. Tobias attempts to reposition the cranium of "Mrs. Ples" in the Sterkfontein Type Site, August 1967. Looking on are H. B. S. Cooke, R. A. Dart, and A. R. Hughes.

tein in accordance with the planned objectives. The vegetation survey was undertaken by A. O. D. Mogg and has resulted in a booklet on the indigenous plants of the area (Mogg 1975); an extremely detailed survey of the area has been made by I. B. Watt, and a grid network has been erected; extensive excavation of overburden has revealed that the breccia exposure is a good deal more extensive than had been anticipated, particularly to the west of the West Pit, while limited excavation of in situ breccia has been undertaken. Renewed geological investigations by T. C. Partridge are discussed in the next section. The area of paleontological interest on the crown of the Sterkfontein hill has been surrounded by a security fence, and a large work building has been erected.

The first hominid specimen found during the new fieldwork came to light on 10 June 1968. It consisted of the crown of an isolated left upper molar and was followed by other molars on 25 March and 3 July 1969. More recently a spectacular gracile australopithecine cranium, StW 12/13/17, was found in situ in the Type Site quarry, as well as an adult mandible, StW 14, and adolescent right maxilla, StW 18*a–c*.

But by far the most important discovery to be made at Sterkfontein in recent years occurred on 9 August 1976, precisely forty years after Broom paid his first visit to the site. It took the form of a remarkably complete hominid

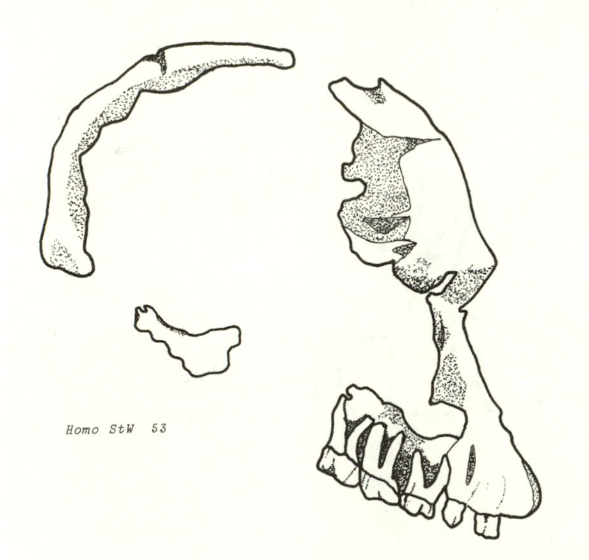

cranium found by A. R. Hughes in a decalcified pocket of deroofed breccia about 2 m from the spot where stone tools were originally found in the Sterkfontein breccia. Fortunately, parts of the skull were still in place, weathering from the wall of the solution pocket and indicating that the fossil was preserved in Member 5 breccia. The specimen, StW 53, consists of most of the face and palate, nine upper teeth, a large part of the vault, and some of the braincase and mandible. Its discovery was reported in the *South African Journal of Science* of August 1976, and a description of it appeared in *Nature* of 27 January 1977 (Hughes and Tobias 1977). According to Tobias's assessment, the new skull shows features that distinguish it clearly from *Australopithecus africanus,* whose remains are found in such abundance in the underlying Member 4 breccia. These features have affinities with the early species of *Homo,* perhaps *H. habilis,* described originally from Olduvai Gorge. One is tempted to think the skull may have belonged to one of the makers of the stone artifacts preserved alongside it in Member 5 of the Sterkfontein Formation.

Some Notes on the Site

As early as 1938 H. B. S. Cooke (1938) provided a plan and section through the breccia-filled cavern at Sterkfontein. He visualized the cave filling as consisting predominantly of stony, bone-poor breccia, but with a layer of finer-grained, richly fossiliferous breccia above it. Bedding in the entire mass was inclined, sloping down from southwest to northeast, which suggested to Cooke that the cavern had probably been not an occupation site, but rather a receptacle into which the sediment had slumped from a rock shelter at a higher level.

Cooke realized, too, that the underground cave system at Sterkfontein took the form of "two lines of galleries developed along two main fissure directions running approximately north from their entrances and following more or less the general dip of the dolomite" (H. B. S. Cooke, 1938, p. 204).

In the interim since 1938 various workers have contributed to an understanding of the form of the fossil and underground caverns, as well as to the nature of their deposits.

As a result of his "Extension Site" excavations of 1957–58, Robinson (1962) concluded that the breccia in the fossil cavern could not be regarded as a single conformable deposit as previous writers had suggested (e.g., H. B. S. Cooke 1938; Brain 1958), but that three superimposed sediments were involved, separated by significant unconformities. He termed the three deposits the Lower, Middle, and Upper breccias (see figs. 158 and 159) and suggested that the last two deposits filled spaces beneath the original cave roof created by periodic slumping of the Lower Breccia into progressively enlarging subterranean chambers.

Subdivision of the Sterkfontein deposit on stratigraphic criteria has been taken further by Partridge (1975), who formally designated six members within the Sterkfontein Formation. He points out that the form of both the fossil and underground caves has been strongly controlled by three sets of fractures, the most prominent being parallel and at right angles to the strike of the dolomitic country rock, with a subordinate set aligned approximately northeast-southwest at an angle of about 40 degrees to the strike. In the vicinity of the Type Site, fractures belonging to the strike and oblique sets intersect, enclosing a parallelogram-shaped area bounding most of the fossil cavern. According to Partridge, the east-west fracture planes run almost vertically through the dolomite and determine the walls of the fossil cavern, while the irregular floor of the cavern may be seen in the "Daylight Cave" (now renamed the Silberberg Grotto by P. V. Tobias), almost 20 m below the present surface of the breccia.

It is in fact the Daylight Cave, situated below the southern wall of the fossil cave, that has provided Partridge with the basis for his stratigraphic column. He is of the opinion that the breccia previously designated Lower, Middle, and Upper breccias above it by Robinson (1962), resulting in a far longer profile than had ever been anticipated.

Partridge has defined and described his six members as follows:

Member 1

Visible above the dolomite floor in the Daylight Cave, it consists of 0.5–2.0 m of white, crystalline to cryptocrystalline calcite, frequently contaminated with fine sediment. It has been divided into a lower Bed A, generally uncontaminated and sterile of bone, and Bed B, darker in color and including angular dolomite and chert pieces as well as some bones.

Member 2

0.5–2.0 m of reddish brown sandy silt resting conformably on Member 1 and variably calcified. Bone is abundant, particularly in the lower levels, and it is from here that the type specimen of the long-legged hunting hyena, *Euryboas silberbergi* (Broom), originally came (see chap. 9). Further carnivore and primate fossils are visible where Member 2 breccia is exposed in the wall of the Daylight Cave.

Member 3

Material assigned to this member forms most of the vertical profile visible in the walls of the Daylight Cave. Par-

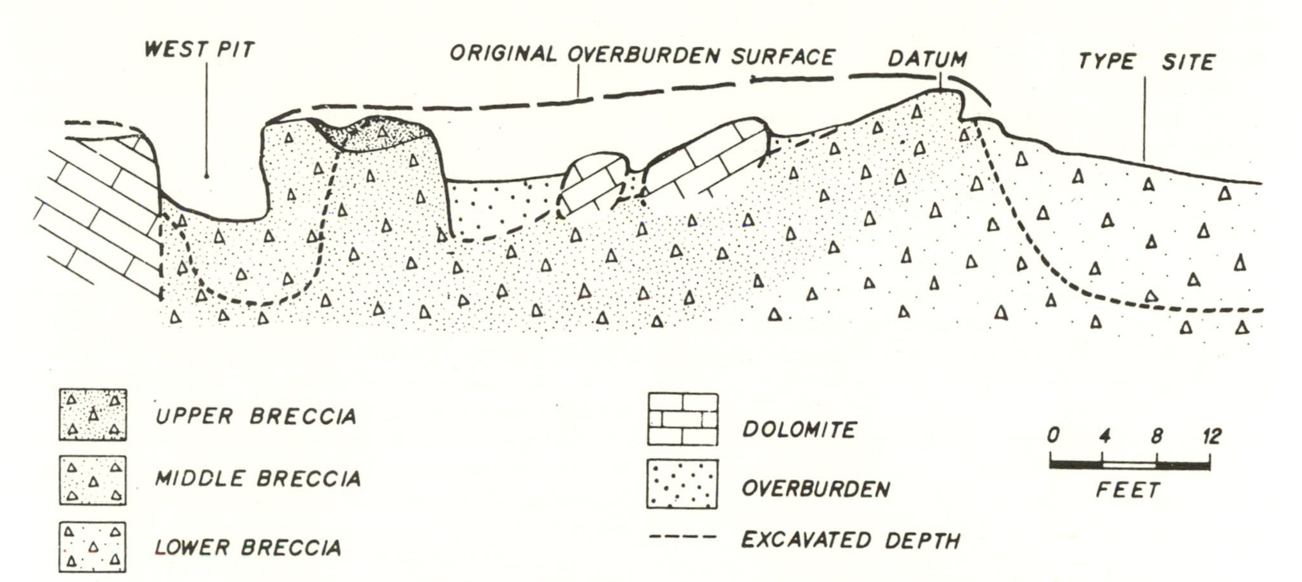

Fig. 158. Stratigraphy of the upper part of the Sterkfontein deposit, as visualized by Robinson (1962). Current terminology for Lower, Middle, and Upper Breccia is Member 4, Member 5, and Member 6.

Fig. 159. T. C. Partridge demonstrates the unconformity between members 4 and 5 in the west wall of the Sterkfontein type site, May 1977.

tridge considers it to consist of 10 m of pale reddish brown, well-calcified sediment, ending perhaps with a layer of steeply dipping travertine that has been observed in the "Middle Pit" of the Type Site and that appears to separate members 3 and 4.

Member 4

Breccia assigned to this member occurs particularly in the Type Site excavation and the Lower Site, close to the exit of the underground cave. It is particularly important from the point of view of the bone accumulation discussed in this chapter. Partridge has subdivided Member 4 breccia into four beds:

Bed A. 2–3 m of brownish red, calcified sandy silt that overlies the inclined travertine at the top of Member 3. Some bone is present.

Bed B. This is essentially a debris cone of fallen roof blocks that had its apex near the eastern end of the southern walls of the Type Site excavation. The blocks of rock typically form a mosaic cemented with calcite and including pockets of calcified reddish sediment that probably filtered down between the rocks after their collapse from the cave roof. The current floor of the Type Site excavation is largely formed of this material; it is not rich in bone, and its sterility led to the termination of the Sterkfontein excavation in 1949.

Bed C. A deposit of dark reddish brown, well-calcified silty sand resting unconformably on the eroded surface of Bed B. Its sediment appears to have been derived from the surface via a cave entrance considerably enlarged by the roof falls that contributed to Bed B. Together with Bed D, this unit constitutes Robinson's Lower Breccia and was the source of many of the australopithecine and other fossils recovered during the Broom era.

Bed D. This unit is separated from Bed C by a marked erosional unconformity and consists of 0.5–2 m of pale brownish red, heavily calcified silty sand. It is represented in both the Type Site and the Lower Sites and was the source of STS 5, the well-known cranium of "Mrs. Ples," as well as the more recently discovered skull designated StW 12/13/17.

Bed E. An insignificant deposit of laminated calcite lenses, up to 75 cm in thickness that concluded the Member 4 depositional phase, frequently just below the dolomite roof.

Member 5

This is the horizon in which the Sterkfontein stone artifacts are concentrated; it represents the Middle Breccia, in Robinson's 1962 terminology. According to Partridge's definition it is a mass of calcified sediment that rests unconformably upon a weathered surface of the underlying Member 4. Faunal evidence (Vrba 1975, 1976*a*) suggests that the time interval between the end of the accumulation phase of Member 4 and the beginning of that of Member 5 may be considerable—perhaps as much as half a million or a million years. Partridge agrees with Robinson in assuming that the available space in the fossil cavern was filled with Member 4 sediment; it was only through subsidence of this filling into a lower-lying void that space was once again created which the sediment represented by Member 5 penetrated.

Partridge considers that this sediment is now represented by 1–5 m of brownish red, well-calcified silty sand containing chert and dolomite inclusions. Bone and stone artifacts are well represented.

Member 6

A rather insignificant deposit, representing the Upper Breccia of Robinson (1962). It overlies Member 5 breccia unconformably in the West Pit area and consists of darker, poorly calcified breccia filling spaces beneath the dolomite roof probably created by minor subsidence of the breccia mass. It is probably considerably younger than Member 5 and contains some bone, as reflected in the analysis presented in this chapter.

In addition to sedimentological analyses of samples from the breccia profile, Partridge and Talma (n.d.) undertook stable isotope measurements of carbon and oxygen in carbonates from the deposits. With certain precautions, these measurements may be used as a guide to the temperature of the groundwater at the time of carbonate deposition. They concluded that deposition temperatures remained constant within 2°C during most of the sediment accumulation period, although an increase of 3°C near the base of Member 4 and 4°C in the middle of Member 5 is suggested; the evidence for these increases is somewhat suspect owing to the high ^{13}C concentration in the samples.

Despite many and varied attempts at absolute dating of the Sterkfontein deposit and its fossils, no conclusive results have been obtained. On the basis of cyclic nickpoint migration in the Bloubank valley, and valley flank recession, Partridge (1973) arrived at a date of 3.26 million years for the first opening of the Sterkfontein fossil cave.

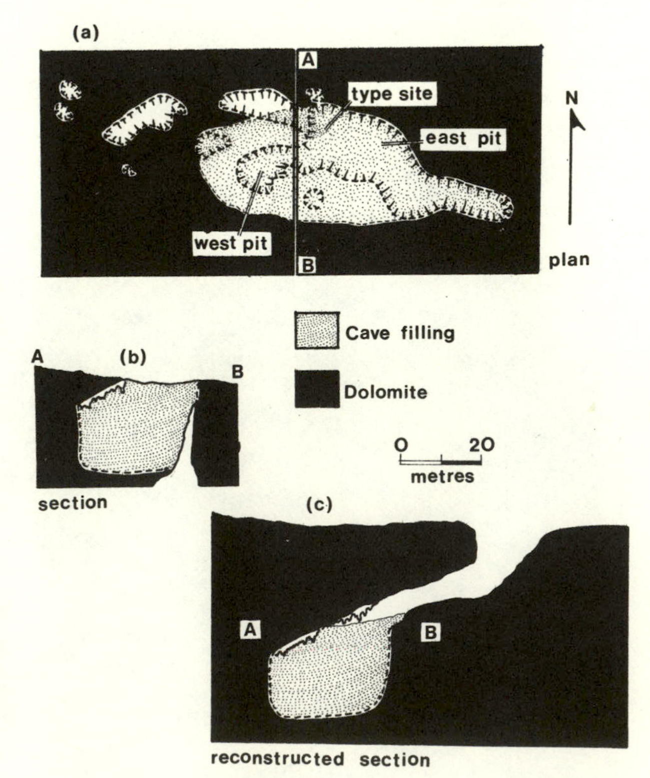

Fig. 160. Outline plan *(a)* and section *(b)* of the Sterkfontein fossil cave, together with the suggested form of a reconstructed section at the time of accumulation of Member 5 sediment.

Although this is an interesting attempt, not everyone would agree that this kind of geomorphological method can be used in precise age estimation.

Turning to the underground cave at Sterkfontein: a remarkably detailed study has been undertaken by Wilkinson (1973). He has shown that the cave system consists of three large chambers: Tourist Cave, Lincoln's Cave, and Fault Cave (see fig. 161), linked by a series of fracture-determined passages. The system underlies the much older fossil cave system that has already been extensively deroofed by prolonged erosion of the Sterkfontein hilltop.

The Sterkfontein Bone Accumulations

The analyses presented here were done on collections in the Transvaal Museum that resulted from two periods of excavation:

a. work in the Type Site between 1936 and 1948; all bones are regarded as coming from Member 4;

b. excavations carried out by J. T. Robinson in the "extension site" during 1957 and 1958. The sample is readily divisible into two components, on the basis of the enclosing matrix, with origin in either Member 5 or Member 6.

The bone accumulations are further divided into macro- and microvertebrate components, the former having presumably entered the cave in a variety of ways, while the latter was almost certainly derived from pellets regurgitated by owls that roosted in the cave.

Remains from Member 4: Macrovertebrate Component

The collection consists of 1,895 fossils from a minimum of 348 individuals, as detailed in table 76. Particulars of 384 bones of bovid origin, assigned to size classes but not specifically identified, are given in table 77. The various identified animals are depicted in figure 162.

I shall now present data on the fossil material by which the various animal taxa are represented.

Class Mammalia
Order Primates
Family Hominidae
Australopithecus africanus

Many scientists have made their own assessments of the Sterkfontein hominid collection, from the point of view of number of individuals involved and the ages at death of these individuals. There have, however, been two specific studies on the subject—that of Tobias (1968*b*, 1974) and that of Mann (1968, 1975). In his consideration of the age of death among the australopithecines, Tobias (1968*b*) made use of the stages of development, eruption, and attrition of teeth, although he pointed out that such features must be applied "with the utmost circumspection." In this study he defined five age categories as follows:

Early childhood—from birth up to the time of eruption of the first permanent teeth.

Later childhood—up to the eruption of all teeth except for the third molars or wisdom teeth, roughly ending with puberty.

Adolescence—from puberty to young adulthood, marked by attrition on the premolars and second molars, the approach to complete development of the third molars, their emergence and their approach to the occlusal plane as well as ongoing fusion of the three components of the hip-bone.

Young adulthood—marked by the complete eruption of the third molars and the early beginnings of signs of attrition on them.

Adulthood—attrition on the fully-erupted third molars is moderate or marked, suggesting that these teeth have been in use for some time.

Tobias was able to assign 47 specimens from Sterkfontein to the five age categories listed above, as follows: early childhood, 4; later childhood, 5; adolescence, 5; young adulthood, 2; adulthood, 30.

In his paleodemographic study of the South African australopithecines, Mann made use of similar age indicators but assumed that tooth eruption times were comparable to those of modern man, namely, that the first permanent molar erupts at six years, the second at twelve, and the third at eighteen. Following this basic assumption, it was possible for Mann to estimate ages at death, expressed in years, for many of the fossils.

In the following listing of the *Australopithecus africanus* material available, Mann's age estimate for each specimen is appended. In cases where no estimate in years is attempted, other categories are used: "infant," before M1 erupts; "juvenile," before M2 erupts; "adolescent," before M3 erupts; "adult, 1," M3 erupted, but little or no wear; "adult, 2," stage 2 wear; "mature adult" or "adult, 3," dentine pits exposed on M3. "Mature" and "immature" are more general divisions employed when the condition of the fossil precluded a more definite statement (Mann 1975, p. 150).

Australopithecus africanus
Material: Craniums in varying degrees of completeness

TM 1511 + STS 60. The holotype of *Plesianthropus transvaalensis* Broom, 1936 (TM 1511), consisting of part of a cranium including left and right maxillae and facial and occipital pieces (fig. 163). The following teeth are present: left P^3, P^4, M^1, M^2, and isolated M^3; right P^4, M^1, and M^2. TM 60 is an associated natural endocranial cast. Age estimate, 22 ± 2 years.

STS 5, probably associated with STS 6. STS 5 is a virtually complete and undistorted cranium, known as "Mrs. Ples," without teeth (fig. 164). Isolated M_1, and M_2 (STS 6) were found close by and may be associated. Age estimate, ± 35 years.

STS 17, part of an adult ♀ cranium with part of the

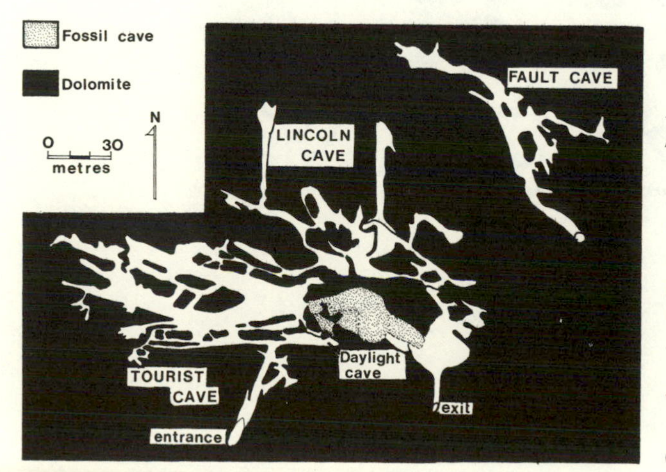

Fig. 161. Plan of the Sterkfontein underground cave based on a survey conducted by Justin Wilkinson (n.d.).

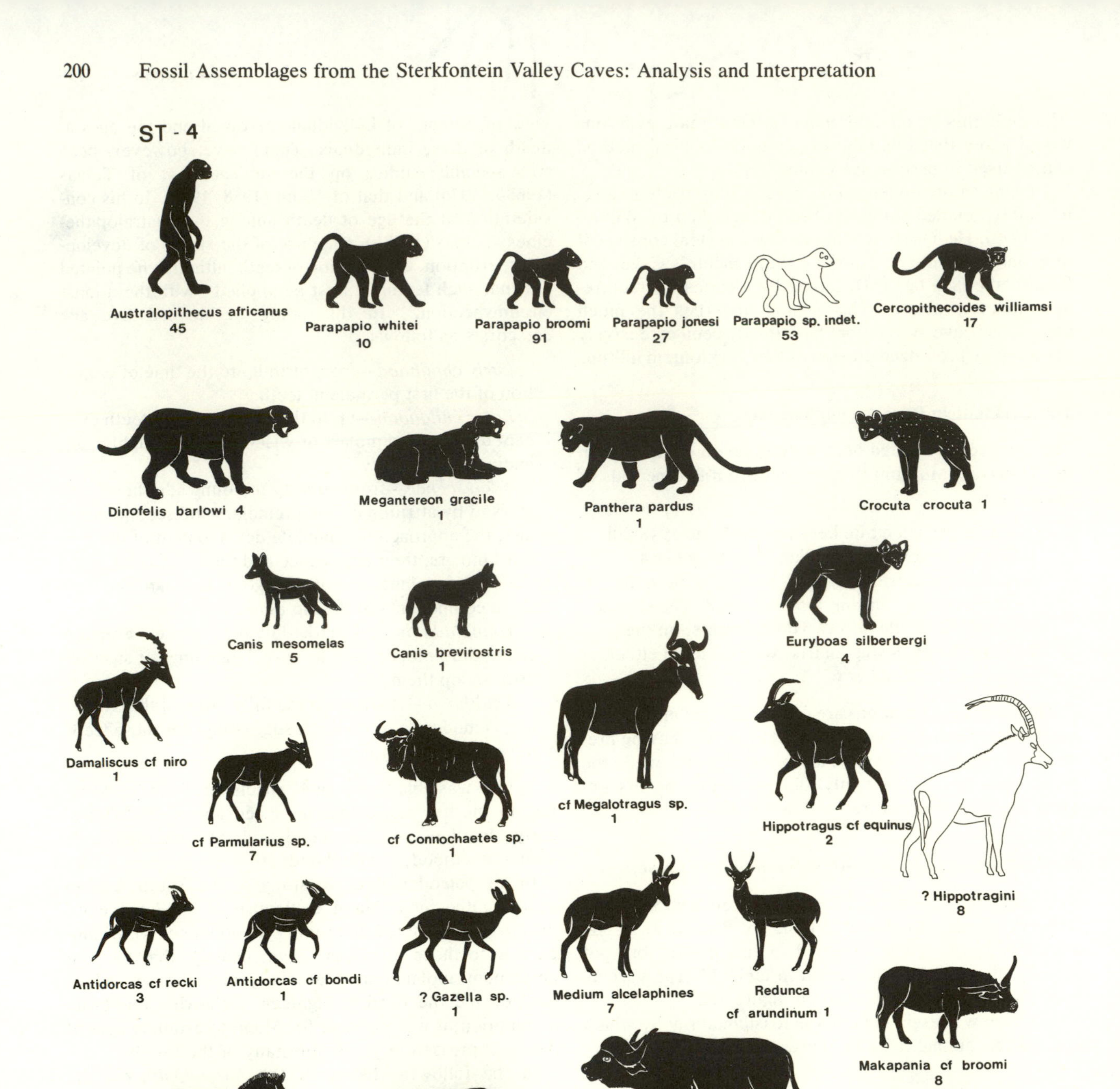

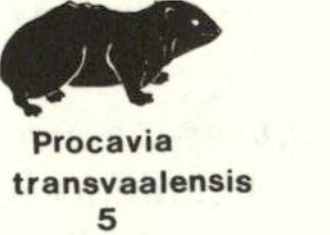

Fig. 162. Animals represented by the fossils in the macrovertebrate component of the Sterkfontein Member 4 sample. Minimum numbers of individuals are indicated.

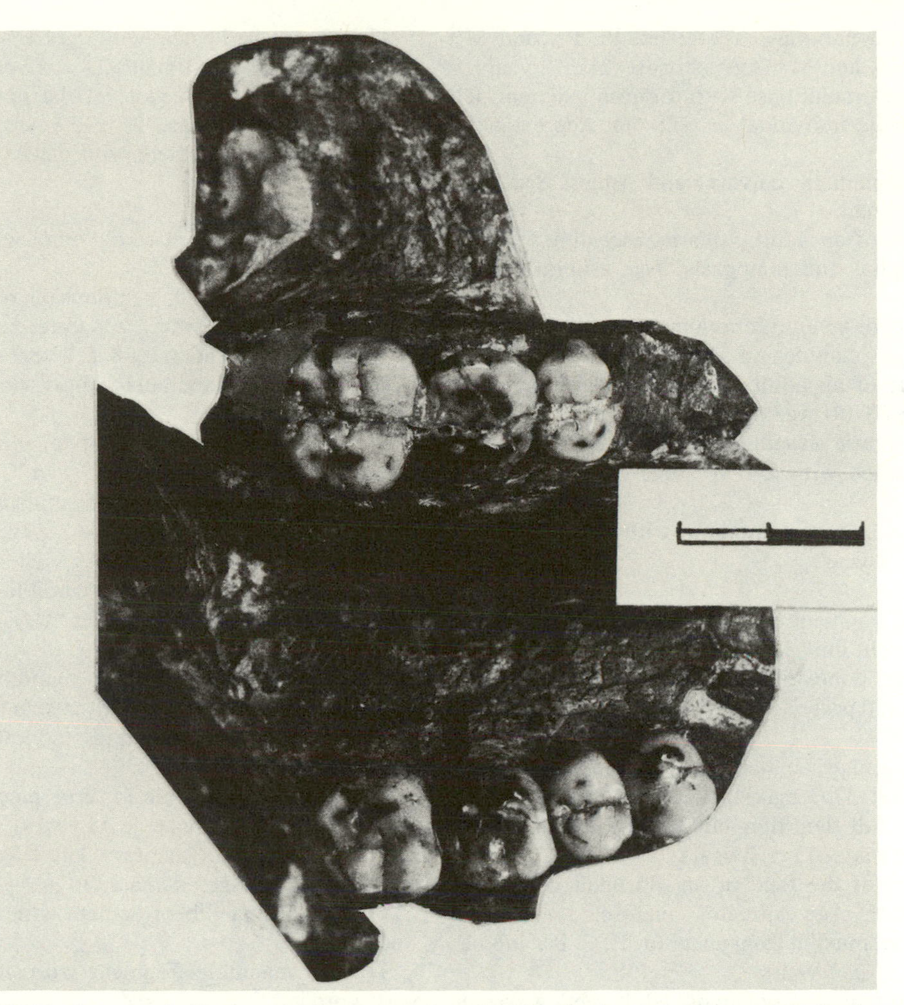

Fig. 163. The first australopithecine specimen to come from Sterkfontein—the type specimen of *Plesianthropus transvaalensis*, TM 1511.

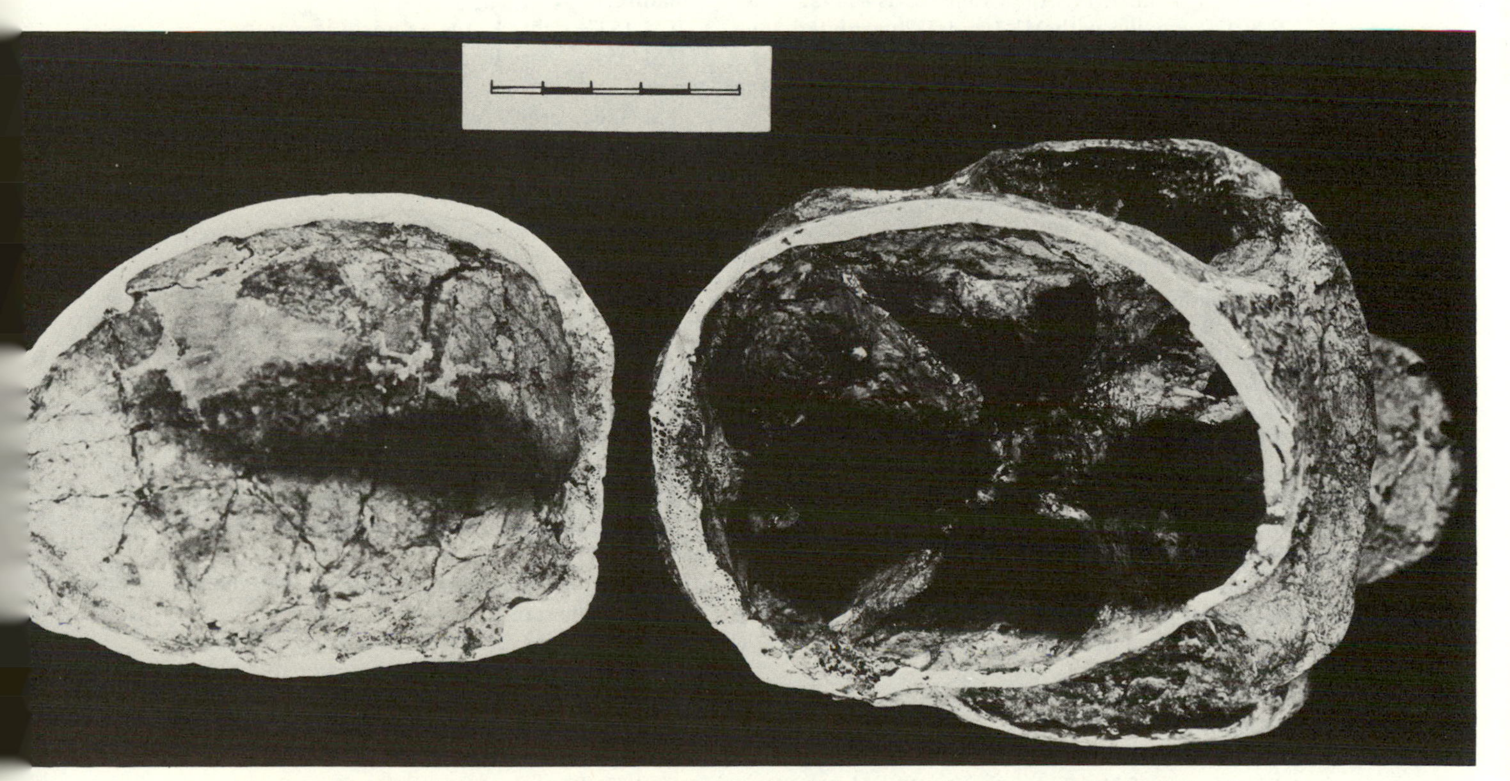

Fig. 164. The well-known cranium of "Mrs. Ples," STS 5. Although the endocranial space was empty, the skull remained undistorted because of its protected fossilization site, close to the back wall of the cave and beneath a stable area of dolomite roof.

face, palate, and parietals. Teeth: left P^3, P^4, and M^1; right P^4, M^1, M^2, and M^3. Age estimate, 21 ± 2 years.

STS 19, adult cranial base with foramen magnum; left M^3 (perhaps same individual as STS 58). Age estimate, 26 ± 3 years.

STS 20, fragmentary calvaria and natural endocast. Age not determined.

STS 25, part of an adult calvaria, including frontals, parietals, occipital and temporals. Age estimate, "mature."

STS 26, parts of an adult cranium (two pieces), including left M^3. Age estimate, "mature."

STS 58, part of an adult calvaria, perhaps the same individual as STS 19. Age estimate, "mature."

STS 67, calvaria fragment with external auditory meatus, associated with left M^3. Age estimate, "adult, 3."

STS 71, right side of a cranium, with P^{3-4}, M^{1-3}. Age estimate, 33 ± 2 years.

Maxillae

STS 2, facial fragment of a juvenile with left dm^{1-2} and unerupted $\underline{C}$; right dm^{1-2}. Age estimate, 4 ± 1 years.

STS 13, facial bones of an old adult with most of the right dentition and part of the left (specimen missing). Age not determined.

STS 52, face and associated mandible of an adolescent, probably female: *52a*, maxilla with full dentition, *52b*, mandible with full dentition but lacking left ascending ramus. Age estimate, 17 ± 2 years.

STS 63, part of the face of an old adult with some fragmentary teeth. Age estimate, "mature."

TM 1512, right maxilla from an adult, I^2, $\underline{C}$, P^3, and M^1. Age estimate, 21 ± 2 years.

TM 1514, left maxilla from an old adult, teeth worn and broken, only M^3 in reasonable condition. Age estimate, 40 ± 5 years.

TM 1535, $\underline{C}$ from 1514. Age estimate, as above.

STS 8, left maxilla with M^{1-2} and M^3 unerupted. Age estimate, 16 ± 2 years.

STS 10, left maxilla with part of M^2 and M^3. Age estimate, 39 ± 1 years.

STS 12, maxilla with left P^{3-4}, M^{1-2}, and part of M^3; right P^4, M^{1-2}. Age estimate, 21 ± 2 years.

STS 27, maxillary piece with M^{2-3} (specimen missing). Age not determined.

STS 32, maxillary fragment with right M^2 and part of M^3. Age estimate, 24 ± 2 years.

STS 35, maxilla with left P^{3-4}, M^{1-2}. Age estimate, 26 ± 3 years.

STS 57, maxillary fragment with left M^1 and unerupted crowns of P^3 and P^4. Age estimate, 6½ ± 1 years.

STS 61, right maxillary piece with P^{3-4}, M^1, and part of M^2. Age estimate, 26 ± 3 years.

STS 66, part of a palate with fragmentary teeth. Age not determined.

STS 69, maxillary fragment from a juvenile individual. Age estimate, "mature."

STS 70, part of a juvenile maxilla with some deciduous tooth fragments and unerupted anterior permanent teeth. Age estimate, "immature."

STS 3009, right maxilla of an old adult with roots of P^4, M^{1-3}. Age estimate, "mature."

STS 1, crushed palate of an adult with left P^4, M^{1-2}; right P^4 and part of M^1. Age estimate, 13 ± 1 years.

STS 29, part of an adult palate with right M^2 and fragments of M^1 and M^3. Age estimate, 32 ± 2 years.

STS 42, crushed palate with right P^{3-4}, and parts of the left toothrow. Age estimate, 33 ± 2 years.

STS 53, palate with most of the postcanine dentition bilaterally. Age estimate, 26 ± 3 years.

STS 64, palate fragment with pieces of teeth. Age not determined.

Mandibles

TM 1515, mandible of an old adult with left P_{3-4}, M_{1-2}. Age estimate, "mature."

TM 1516, + STS 50, symphyseal region of mandible with left and right $\bar{C}$, dm_1, separate, associated with left $\bar{C}$. (STS 50). Age estimate, 8 ± 1 years.

TM 1522, right mandibular ramus with roots of M_3. Age estimate, "mature."

STS 7, crushed mandible of an old adult male with complete dentition except for right $\bar{C}$, lost in life; left ascending ramus; part of right scapula and right proximal humerus associated (STS *7a–b*). Age estimate, 35–40 years.

STS 18, part of a juvenile mandible, right dm_{1-2}, M_1, and M_2 (unerupted); left dm_2 and M_1. Age estimate, 8 ± 1 years.

STS 24, incomplete juvenile mandible with left di_{1-2}, $d\bar{c}$, dm_1; right $d\bar{c}$, dm_{1-2}, and M_1; also unerupted crowns of I_{1-2} and P_3 and parts of right di_2 and left I_1. Age estimate, 6 ± 1 years.

STS 36, adult mandible in three pieces, left I_1, $\bar{C}$–M_3; right M_{2-3}. Age estimate, ± 35 years.

STS 38, very fragmentary left mandible piece with pieces of M_{1-3}. Age estimate, 30 ± 2 years.

STS 41, left mandible fragment with M_3. Age estimate, "adult, 2."

STS 62, mandible fragment with alveoli of incisors, very worn left $\bar{C}$ and roots of P_3. Age estimate, "mature."

Isolated Maxillary Teeth

TM 1524, roots of a very worn M^1. Age estimate, "mature."

TM 1527, right $\underline{C}$. Age estimate, "mature."

TM 1532, very worn left M^3. Age estimate, 33 ± 2 years.

TM 1561, crown of right M^3. Age estimate, 23 ± 1 years.

STS 11, left M^1 and M^2 (specimens missing). Age not determined.

STS 16, very worn upper molar. Age estimate, "mature."

STS 21, crown of upper right molar (M^1 or M^2). Age estimate, 24 ± 2 years.

STS 22, left M^1 and M^2. Age estimate, 17 ± 2 years.

STS 23, part of right M^2. Age estimate, 13 ± 1 years.

STS 28 + STS 37, right $M^{2-3}$and left M^1 (STS 28); left M^{2-3} (STS 37). Age estimate, 19 ± 1 years.

STS 30, right P^4 and M^2. Age estimate, 23 ± 2 years.

STS 31, part of M^3. Age estimate, 25 ± 2 years.

STS 33, lingual half of a worn upper premolar. Age estimate, "mature."

STS 39, considerably worn left P^4 and M^1. Age estimate, "mature."

STS 44, left M^2 and M^3. Age estimate, "adult, 2."

STS 45, part of an upper molar, probably right M^2. Age not determined.

STS 46, most of the crown of right M^3. Age estimate, 32 ± 2 years.

STS 47, P^3. Age estimate, "mature."

STS 48, left $\underline{C}$. Age estimate, "mature."

STS 54, right M^3. Age estimate, 38 ± 2 years.

STS 55, P^3 *(a);* crown of left M_3 *(b)*. Age estimate, 16 ± 1 years.

STS 56, left M^1, M^2, and part of dm^2. Age estimate, 9 ± 2 years.

STS 72, part of crown of right M^3. Age estimate, 26 ± 3 years.

Isolated Mandibular Teeth

TM 1518, right M_3. Age estimate, 24 ± 2 years.

TM 1519, right M_3, very worn. Age estimate, ± 35 years.

TM 1520, left M_3, very worn. Age estimate, 30 ± 3 years.

TM 1523, left P_4. Age estimate, 16 ± 2 years.

TM 1525, roots of a lower premolar. Age not determined.

TM 1528, right $\bar{C}$. Age estimate, "mature."

TM 1533, part of M_3 (specimen missing). Age not determined.

STS 3, right $\bar{C}$, well worn. Age estimate, "mature."

STS 4, left M_2 (matches STW—H1). Age estimate, 25 ± 3 years.

STS 9, right M_1 with unformed roots. Age estimate, 6 ± 1 years.

STS 40, lingual part of a lower canine. Age estimate, "immature."

STS 49, buccal part of crown of P_3. Age estimate, "immature."

STS 51, right P_3 and unerupted crown of right $\bar{C}$. Age estimate, 10 ± 1 years.

STS 59, crown of left M_3. Age estimate, 16 ± 1 years.

STS 1534, 1535, 1374, worn incisors that may or may not be australopithecine. Age not determined.

Postcranial Pieces

TM 1513, distal end of left femur. Age estimate, "mature."

TM 1526, carpal bone: right os magnum. Age not determined.

STS 14, parts of an articulated vertebral column, ribs, pelvis, and hind limb: *a–f,* 6th–1st lumbar vertebrae; *g–i,* 12th–9th thoracic vertebrae; *k–n,* four thoracic vertebrae; *o–p,* vertebral bodies; *q,* sacrum; *r,* left innominate; *s,* right innominate; *t,* right proximal femur; *u,* shaft piece; *w–y,* rib fragments; *z,* vertebral and pelvic fragments (fig. 165). Age estimate, "mature."

STS 34, distal end of right femur. Age estimate, "mature."

STS 65, right pelvic bone with most of ilium and parts of pubis; proximal end of left femur *(a)*. Age estimate, "mature" (some doubt whether this specimen is from a hominid).

STS 68, proximal end of left radius (perhaps cercopithecoid). Age estimate, "mature."

STS 73, vertebral centra. Age not determined.

With fragmentary remains of this nature it is difficult to be certain of the minimum number of individuals that contributed to the sample. After careful consideration of all possible associations of the pieces available, it is estimated that at least 45 individuals are involved.

Table 78 indicates Mann's allocation of 47 specimens from Sterkfontein Member 4 to age categories. If those specimens below twenty years of estimated age at death are regarded as *subadult,* then seventeen out of 47 speci-

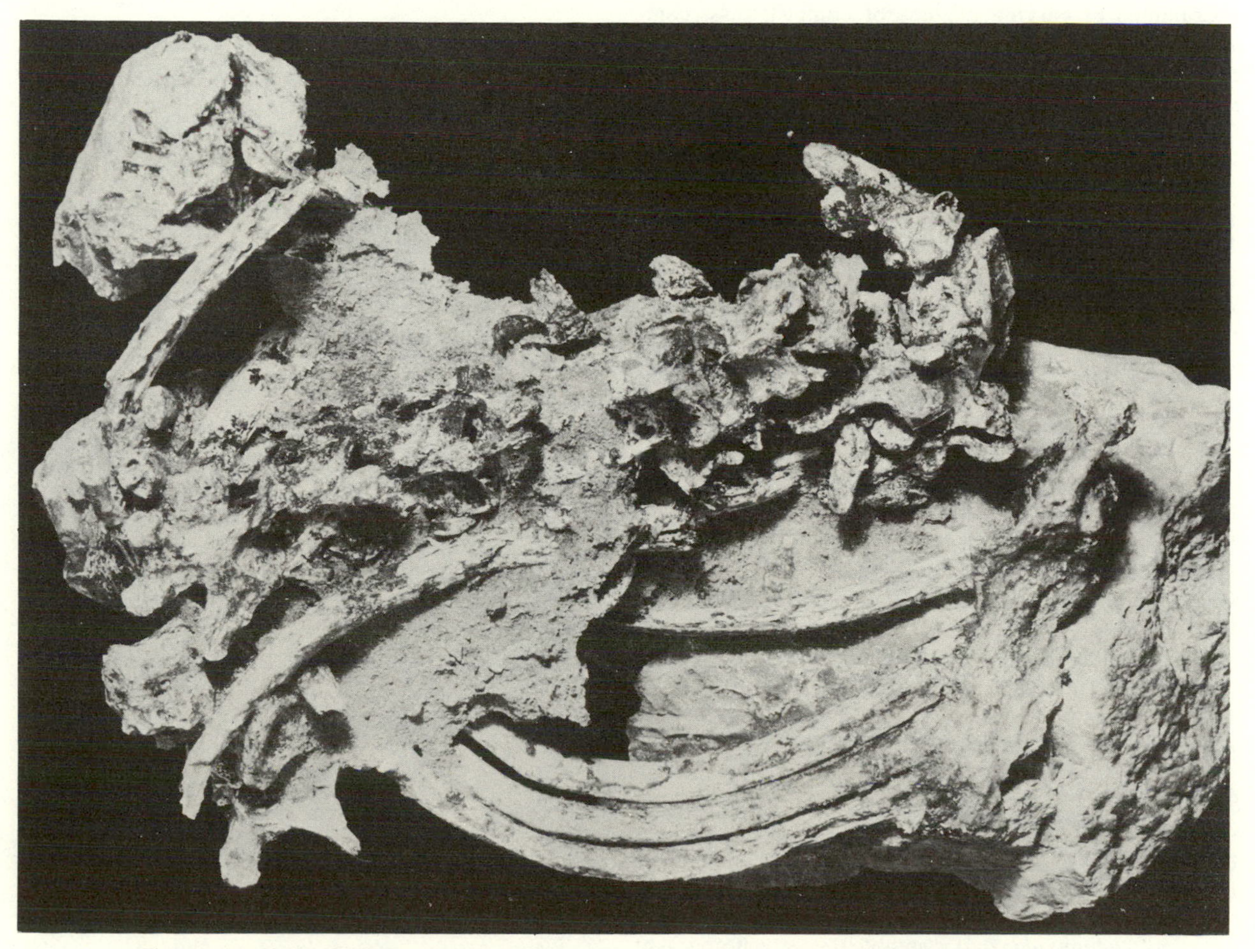

Fig. 165. An articulated segment of vertebral column, pelvis, and other bones of *Australopithecus africanus* (STS 14) from Sterkfontein Member 4. The specimen is shown in course of preparation. Photograph by J. T. Robinson.

mens, or 36% of the ageable sample, was subadult at death, as is shown in figure 166.

Family Cercopithecidae
Parapapio jonesi (fig. 167*a*)
Material:
37 pieces from a minimum of 27 individuals.
Type specimen: an almost complete ♀ cranium with left $\underline{C}$–M^3 and right P^3–M^3.

Craniums in varying degrees of completeness: crushed cranium, ♀, probably complete before being shattered, P^3–M^3 bilaterally, STS 547; base of braincase and posterior part of palate, left P^4–M^3, right M^{2-3}; left side of juvenile ♀ cranium, $\underline{C}$ erupting, M^1, M^{2-3} erupting, STS 333; maxillary pieces, STS 250, 284, 287, 367, 368*a*, 372*a–e*, 390, 441, 456, 458; mandible pieces, STS 276, 302, 306, 307, 313, 317, 329, 334, 340, 348, 355, 381*a–e*, 384, 421, 443, 446, 448, 457, 1925*a;* isolated lower teeth, STS 418*a–c*, 485; articulated atlas and axis vertebrae, STS 368*b;* part of a thoracic vertebra, STS 381*e*.

The 27 individuals may be divided into age classes as defined by Freedman (1957). Particulars are given in table 79: 2 individuals are rated as "juvenile," 2 as "immature adult," 7 as "young adult," 12 as "adult," and 4 as "old adult."

Parapapio broomi (fig. 167*b*)
Material:
100 pieces from a minimum of 91 individuals.
Type specimens: partial cranium ♂, complete endocast, snout, roots of right M^3, STS 564; ♀ mandible with $\bar{C}$–M_3, STS 562.

Craniums in varying degrees of completeness: complete cranium, ♀, left P^4, M^{2-3}, right M^{2-3}, STS 262; complete ♀ cranium except for left side of braincase, STS 254*a*; complete ♀ cranium, flattened laterally, STS 393; almost complete ♀ cranium, with M^{1-3} bilaterally, STS 397; crushed and weathered cranium, adult ♀, left M^{1-3}, STS 535; much of a flattened ♂ cranium, STS 531; right side of ♀ cranium with M^3, STS 396; parts of a shattered ♀ cranium, left P^{3-4}, M^{1-2}, STS 539; part of an immature ♀ cranium, STS 251. Palates, STS 253, 260, 272, 297, 354, 378*a–d*, 388*a–b;* left maxillary pieces, STS 264, 266, 267, 274, 301, 322, 380*a–c*, 385*a–e*, 398*a–c*, 415*a*, 544; right maxillary pieces, STS 277, 325, 379*a,b*, 383*b*, 420*a*, 530; mandible pieces with both rami, STS 258, 261 + 423, 283, 296 + 351, 299, 323, 331, 335, 337, 360, 362, 363, 371, 374*b*, 390*a,b*, 416*a,b*, 426, 533 + 534, 3035; left mandible pieces, STS 255, 256, 268, 271, 278, 285, 286, 312, 314, 326, 328, 346, 353, 386*a*, 409, 414*a,b*, 463, 466, 557; right mandible pieces, STS 270, 280, 289, 298, 309, 311, 338, 339, 356, 369*a,b*, 382*a*, 383*a*, 411*a,b* + 425, 434, 469, 542, 558; isolated teeth, STS 406, 410*a–b*, 419, 437, 438, 445, 484, 491, 511.

A separation of the 91 individuals into sex and age classes is given in table 80; 3 individuals are classed as juvenile, 10 as immature adult, 20 as young adult, 39 as adult, and 19 as old adult.

Parapapio whitei (fig. 167*c*)
Material:
14 pieces from a minimum of 10 individuals.
Type specimen: mandible of a young adult ♀, left I_1–P_4; right I_1–M_3, STS 563.

Articulated maxillae and mandible pieces: left and right maxilla and mandible pieces associated with axis vertebra, STS 263 + 370; right maxilla, P^4–M^1, mandible, left

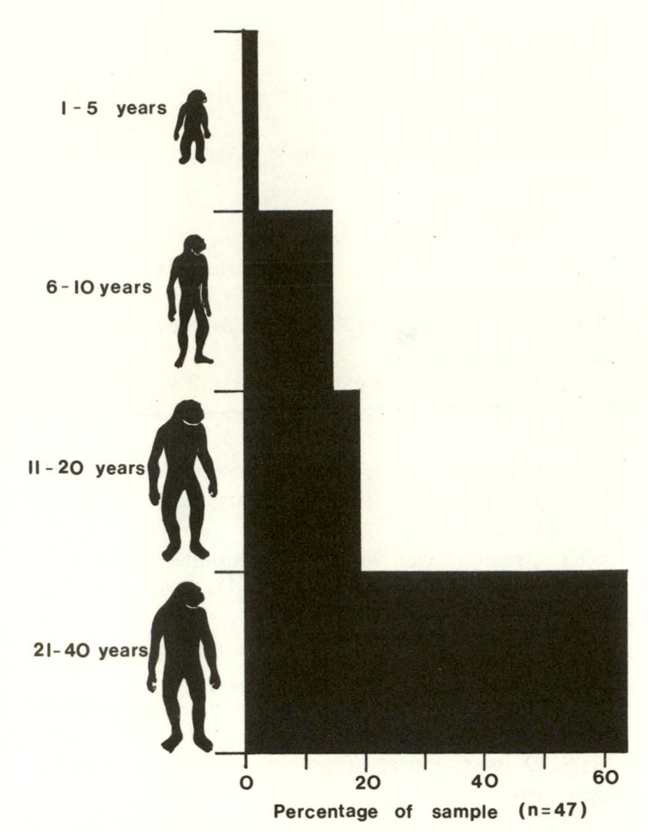

Fig. 166. The allocation of 47 specimens of *Australopithecus africanus* from Sterkfontein Member 4 to age-at-death categories, according to Mann (1975).

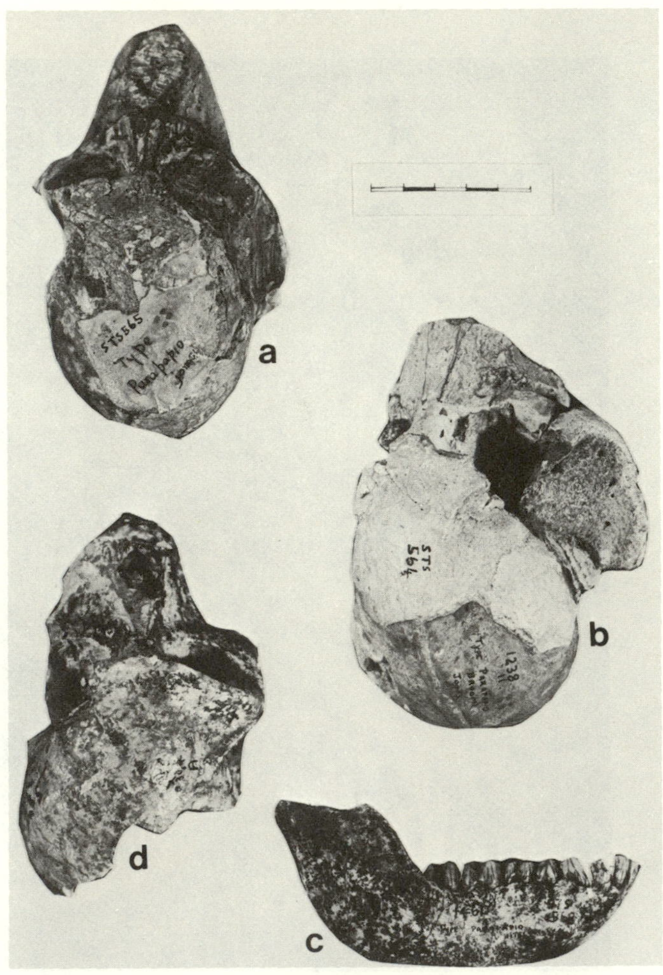

Fig. 167. Examples of cercopithecoid craniums from Sterkfontein Member 4: *(a) Parapapio jonesi*, STS 565; *(b) P. broomi*, STS 564; *(c) P. whitei*, STS 568; and *(d) Cercopithecoides williamsi*, STS 394*a*.

I_1–P_3, right I_1–M_2, STS 389*a,b;* left maxilla, P^4–M^2, mandible with right P_4 and left M_3, STS 548*a–c;* palate, ♀, left P^4, M^2, right P^3–M^3, STS 336; left maxillae, STS 259, 303, 424; right maxilla, STS 462; mandible with both tooth rows, STS 359; left mandible piece, STS 352; right mandible piece, STS 342, 467.

A separation of the 10 individuals into sex and age classes is given in table 81. There are no juveniles in the sample; 1 individual was classed as immature adult, 2 as young adult, 5 as adult, and 2 as old adult.

Parapapio sp. indet.

A total of 201 cranial and 5 postcranial pieces has been assigned to *Parapapio,* but the specimens are not complete enough to be specifically identified. However, 101 pieces could be placed, on the basis of tooth eruption and wear, in age classes.

Material:

Age-determined individuals

101 pieces from a minimum of 38 individuals; 3 individuals were rated as "young juvenile," 22 as "juvenile," 4 as "immature adult," 7 as "young adult," 1 as "adult," and 1 as "old adult."

Craniums in varying degrees of completeness, STS 364, 376, 540, 559*a–c,* 3048, 3057, 3076; craniums or maxillae with articulated mandibles, STS 373*a,* 387*a,b,* 395, 546, 1046, 3067, 3073; palates, STS 305, 319, 324, 345, 1189, 2119; left maxillary pieces, STS 308, 315, 375*a–c,* 431, 488, 3047, 3052; right maxillary pieces, STS 273, 294, 318, 320*b,* 343, 429, 432, 433, 439, 454, 455, 459, 464, 545, 1592, 1593, 1594, 2117, 2215; mandibles with left and right toothrows, STS 281, 351, 468, 1238, 1534, 3037; left mandible pieces, STS 269, 304, 310, 316, 320, 349, 413*a,* 436, 440, 442, 453, 461, 1753, 3053; right mandible pieces, STS 257, 293, 327, 341, 358, 399*a,* 403*a,* 413*b,* 427, 430, 444, 447, 449, 450, 451, 560, 3070, 3078; isolated teeth, STS 291, 402*b,* 407*a,* 412, 460, 465, 486, 504, 509, 510, 520, 522, 524, 527, 528, 529, 3045.

Specimens on which age determinations are not possible

105 pieces from a minimum of about 15 individuals

Craniums in varying degrees of completeness, STS 536, 537, 1026, 1052, 3063; maxillary pieces, STS 428, 3055, 3066, 3072; mandible pieces, STS 275, 470, 543, 1544, 1817, 2067, 2510, 3043, 3054, 3056, 3061, 3062; isolated teeth, STS 254*b,* 292, 386, 391*a–e,* 400*a–c,* 401*a,b,* 402*a,* 403*c,* 404*a,b,* 405*a–c,* 407*b,* 408*a,b,* 417*a,b,* 422, 452, 471, 472, 473, 474, 475, 476, 477, 478, 479, 480, 482, 487, 489, 490, 492, 493, 494, 495, 496, 497, 498, 499, 500, 501, 502, 503*b,* 505, 506, 507, 508, 512, 513, 514, 515, 517, 519, 521, 525, 526, 556, 1184, 1704, 1760, 1895, 2116, 2123, 2348, 2477, 3036, 3038, 3039, 3040, 3041, 3042, 3044, 3046, 3049, 3050, 3051, 3064, 3065, 3068, 3071; thoracic vertebra pieces, STS 3059, 3060; lumbar vertebra piece, STS 415*b;* proximal radius piece, STS 375*e;* proximal femur piece, STS 378*d.*

Cercopithecoides williamsi (fig. 167*d*)

Material:

21 pieces from a minimum of 17 individuals.

Craniums: most of a cranium with full dentition, ♀, articulated mandible, atlas, and axis, STS 394*a–c;* most of calvaria without snout, but mandible with P_4–M_3 bilaterally, STS 541; partial braincase and snout, right M^{1-3}, STS 392; left side of braincase and snout, left M_{2-3}, STS 361; base of calvaria, right M^{2-3}, STS 252; palates: juvenile palate with articulated mandible, STS 290 + 357 + 435; anterior part of ♂ palate, STS 350; right maxillae, STS 295, 347; mandible pieces with both rami, STS 300, 366; left mandible pieces, STS 279, 288, 344, 532; right mandible piece, STS 282; isolated teeth, STS 516, 518, 523.

A separation of the 17 individuals into sex and age classes is given in table 82.

Two individuals were classed as juveniles, 1 as immature adult, 3 as young adult, 10 as adult, and 1 as old adult.

Cercopithecoid indet.

It was found that 413 cranial and 46 postcranial pieces were clearly of cercopithecoid origin but could not be identified to generic or specific level. It is highly likely that most of these specimens came from individuals that have already been listed in the *Parapapio* and *Cercopithecoides* categories. At a rough estimate, the specimens come from a minimum of 100 individuals.

Material:

Cranial pieces with articulated mandibles, STS 423, 477, 1013, 1038, 1047, 1178, 1485, 1784, 1816, 2333; craniums in varying degrees of completeness, STS 421, 422, 430, 476, 494, 943, 945, 1002, 1003, 1024, 1028, 1030, 1031, 1039, 1048, 1053, 1058, 1193, 1238, 1260, 1275, 1449, 1456, 1529, 1635, 1777, 2008, 2013, 2020, 2034, 2118, 2145, 2190, 2244, 2352, 2572, 3091, 3099, 3100, 3103, 3105, 3109, 3111; calvaria and endocast pieces, STS 217, 400, 405, 406, 409, 412, 424, 428, 434, 442, 457, 458, 461, 467, 484, 485, 486, 491, 493, 515, 530, 538, 541, 991, 995, 1000, 1002, 1009, 1011, 1012, 1014, 1017, 1018, 1019 + 1074, 1021, 1023, 1027, 1032, 1034, 1041, 1042, 1043, 1044*a,* 1050, 1054, 1055, 1059, 1063, 1065, 1066, 1067, 1070, 1071, 1072, 1075, 1077, 1078, 1079, 1080, 1081, 1082, 1083, 1088, 1109, 1229, 1230, 1235, 1238, 1253, 1256, 1258, 1267, 1306, 1310, 1334, 1342, 1361, 1442, 1459, 1471, 1474, 1486, 1499, 1506, 1532, 1536, 1547, 1588, 1615, 1639, 1675, 1688, 1740, 1778, 1810, 1841, 1955, 1957, 1960, 1963, 1965, 1969, 1970, 1983, 2009, 2010, 2012, 2015, 2023, 2032, 2035, 2038, 2043, 2045, 2051, 2111, 2143, 2212, 2215, 2220, 2222, 2223, 2224, 2262, 2264, 2377, 2388, 2406, 2450, 3092, 3093, 3094, 3095, 3098, 3102, 3104, 3106, 3108; palates and maxillary pieces, STS 378, 411, 416, 432, 470, 474, 478, 503, 525, 529, 544, 547, 560, 992, 1001, 1005, 1008, 1025, 1037, 1051, 1089, 1188, 1217, 1271, 1305, 1311, 1343, 1455, 1494, 1516, 1541, 1566, 1591, 1640, 1702, 1708, 1770, 1797, 1813, 1837, 1890, 1917, 1929, 1939, 1961, 1966, 1995, 2002, 2015, 2018, 2044, 2071, 2079, 2106, 2113, 2142, 2161, 2205, 2208, 2265, 2303, 2328, 2344, 2350, 2375, 2376, 3097, 3101, 3113, 5394; mandible pieces, STS 398, 407, 413, 427, 436, 444, 446, 481, 508, 558, 947, 1007, 1010, 1036, 1049, 1056, 1085, 1102, 1127, 1195, 1208, 1246, 1314, 1339, 1346, 1372, 1401, 1461, 1519, 1533, 1552, 1568, 1575, 1618, 1713, 1728, 1772, 1788, 1792, 1794, 1801, 1868, 2017, 2021, 2051, 2077, 2087, 2093, 2095, 2103, 2118, 2124, 2139, 2158, 2174, 2224, 2274, 2280, 2295, 2306, 2331, 2365, 2449, 2475, 2513, 2525, 2545, 2558, 2570, 3096, 3107; isolated teeth, STS 126, 377*a–d,* 441, 452, 487, 505, 528, 532, 540, 549, 564, 790, 803, 1022, 1093, 1116, 1160, 1169, 1344, 1387, 1411, 1428, 1472*b,* 1479, 1530, 1601, 1661, 1669, 1683, 1684, 1716, 1722, 1762, 1768, 1780, 1785, 1829, 1836, 1839, 1899, 1927, 1934, 1937, 1954, 1992, 2013, 2016, 2017, 2041, 2105, 2130, 2133, 2144, 2146, 2147, 2163, 2168, 2193, 2196, 2206, 2207, 2209, 2218, 2225, 2231, 2277, 2279, 2292, 2297, 2416, 2433, 2438, 2444, 2488, 2489, 2508, 2512, 2531, 2543, 2548, 3086; cervical vertebra, STS 551; pelvis pieces, STS 1044*b,* 2146; proximal humerus pieces, STS 1458, 2185, 2219; humerus shaft pieces, STS 2308, 1444; distal humerus pieces, STS

1264, 1504, 2074, 2198, 2201, 2380; articulated distal humerus and proximal radius, STS 2198; proximal radius piece, STS 534; distal radius piece, STS 1764; proximal femur pieces, STS 443, 1069, 1089, 1092, 1094, 1469, 1733, 1978, 1992, 2050, 2069, 2259, 2357, 2521, 3109; femur shaft pieces, STS 1846, 2183; distal femur pieces, STS 549, 1614, 1905, 2150, 2188; proximal tibia pieces, STS 1860, 2230, 2562; distal tibia pieces, STS 1204, 1663, 2474, 2563.

Order Carnivora
Family Felidae
Panthera pardus
Material:
A single specimen from 1 individual.
Isolated left P^4, STS 134.

Dinofelis barlowi (fig. 168)
Material:
5 pieces from a minimum of 4 individuals.
The type specimen of *Megantereon (Therailurus) barlowi,* consisting of much of the cranium without the left maxilla and mandible, TM 1541; also isolated left $\underline{C}$, without root, TM 1542; crushed cranium and mandible, STS 131; isolated left I_3, TM 1566; left $\underline{C}$, STS 132; left P^4, TM 1579.

Megantereon gracile
Material:
A single specimen from 1 individual.
The type specimen consisting of a right mandibular fragment with P_4 and M_1, STS 1558.

Family Hyaenidae
Euryboas silberbergi (fig. 169)
Material:
5 pieces from a minimum of 4 individuals.
The type specimen of *Lycyaena silberbergi,* consisting of parts of a crushed cranium and mandible, STS 130; left maxilla with P^{3-4}, STS 127; left maxilla with P^3, STS 135; associated with crowns of P_{2-4}, STS 133; left and right mandible fragments, STS 126. The type may be from Member 2 or 3.

Crocuta crocuta
Material:
A single specimen from 1 individual.
Left maxilla with P^{1-4}, STS 128, listed by Ewer as 218.

Hyaenid indet.
Material:
5 pieces from an estimated minimum of 3 individuals.
Palate fragment, STS 129; mandible pieces, STS 863, 867; isolated tooth fragments, STS 880, unnumbered.

Canis brevirostris
Material:
The type specimen consisting of braincase and left maxilla with $\underline{C}$, P^{1-4}, M^{1-2}; left mandible piece with P_4, M_{1-2}, STS 137.

Canis mesomelas
Material:
9 pieces from an estimated minimum of 5 individuals.
The type specimen of *Canis antiquus,* consisting of left and right mandibular rami with deciduous and permanent teeth, TM 1582; cranium and partial snout, STS 142; maxillary fragment, STS 141; mandible pieces, STS 136, 138, 139, 2022, 2089; isolated right I^3, STS 140.

Carnivore indet.
Material:
32 cranial and 16 postcranial pieces from an estimated minimum of 8 individuals.

Fig. 168. The type specimen of *Dinofelis barlowi,* TM 1541, from Sterkfontein Member 4.

Maxillary pieces, STS 1591, 2113; mandible pieces, TM 869, STS 1402, 1538, 1616, 1678, 1876, 1959, 2059*d*, 2094, 2315; tooth fragments, TM 791, STS 858, TM 866, STS 870, 945, TM 1092, STS 1505, 1518, 1556, 1557, 1604, 1634, 1644, 1709, 1909, 1964, 2101, 2113, 2204, 3124; axis vertebra, STS 1774; 2 articulated lumbar vertebrae, STS 1244; scapula pieces, STS 1400*b*, 1872; humerus pieces, STS 1090, 1832, 2176; femur pieces, STS 1700, 2059*a,c*, 2111, 2273; tibia pieces, STS 550, 1608; metapodials, STS 798*b*, 2190*b*.

Order Artiodactyla
Family Bovidae
Damaliscus cf. sp. 2
Material:
A single specimen from 1 individual.
Right mandible piece with M_{2-3}, STS 2582.

Damaliscus sp. 1 or *Parmularius* sp.
Material:
26 pieces from a minimum of 7 individuals.
Right maxilla with M^{1-2}, associated with parts of a cranium and base of left horn-core, STS 2368*a–f;* maxillary pieces, STS 1592, 2577; mandible pieces, STS 1319*a*, 1563, 1742, 1800*a,b*, 1980, 2027, 2135, 2208, 2581, 2585*a*, 2586; isolated teeth, STS 1100*a,b*, 1687, 1696, 1743, 1910, 1984, 1999, 2005, 2046, 2562, 2580, 2593*a*.

Medium-sized Alcelaphines
Material:
14 pieces from a minimum of 7 individuals.
Frontals with bases of both horn-cores, cf. *Rabaticeras porrocornutus*, STS 2595*a;* maxillary pieces, STS 1551*a*, 1844, 1852; mandible pieces, STS 714, 1114, 1427, 1445; isolated teeth, STS 1317, 1324, 1333, 1334, 2519, 2563.

Connochaetes sp.
Material:
3 pieces from a minimum of 1 individual.
Left maxilla, P^4–M^3, STS 2597; mandible piece with left M_3 and right M_1, STS 2512*b,d;* isolated right M_2 or M_1, STS 2200.

Cf. *Megalotragus* sp.
Material:
A single specimen from 1 individual.
Left M_{2-3}, STS 1339.

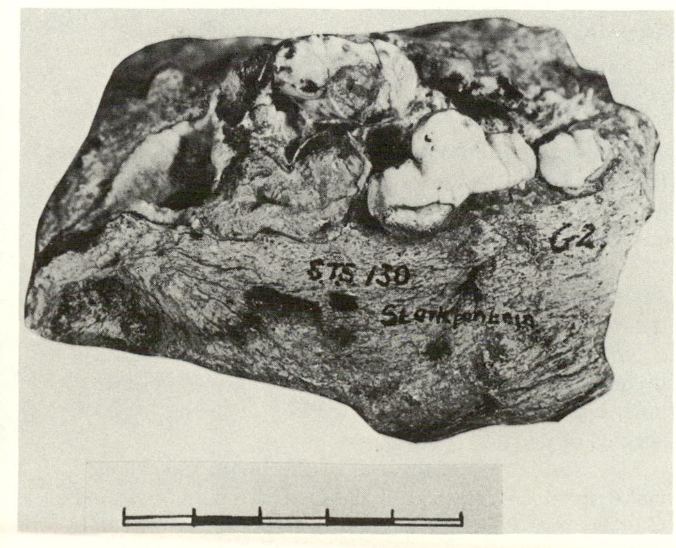

Fig. 169. The type specimen of a hunting hyena, *Euryboas silberbergi*, STS 130, from the lower levels of the Sterkfontein deposit.

Hippotragus cf. *equinus*
Material:
2 pieces from 2 individuals.
Left M^2, STS 1630; left M^2, STS 2599.

?Hippotragini
Material:
23 pieces from a minimum of 8 individuals.
Palate with left P^4–M^3, and right M^{2-3}, associated mandible pieces with left M_{1-2}, and right P_4–M_3, STS 1539*a–c;* maxillary pieces, STS 1137, 1847, 2190*a*, 2336*a;* mandible pieces, STS 1438, 1531, 1589, 1682, 2031, 2055, 22282584; isolated teeth, STS 789, 792, TM 1342, STS 1632, 1866*b*, 1833, 2064, 2145*a*, 2524, 2560.

Redunca cf. *arundinum*
Material:
A single specimen from 1 individual.
Part of a juvenile palate with right dpm^{3-4}.

Antidorcas cf. *recki*
Material:
5 pieces from a minimum of 3 individuals.
Right maxilla with P^3–M^3, STS 1435; right mandible piece with M_{1-2} and M_3 unerupted, STS 1560; right mandible piece with M_3, STS 2369; left mandible piece with M_{2-3}, STS 1944; isolated right M^2, STS 1325*a*.

Antidorcas cf. *bondi*
Material:
A single specimen from 1 individual.
Left mandible piece, P_2–M_2, STS 1125.

?*Gazella* sp.
Material:
2 pieces from 1 individual.
Right mandible piece with M_2, STS 1996; left mandible piece, M_{1-3}, TM 2076.

Syncerus sp.
Material:
A single specimen from 1 individual.
Part of a juvenile mandible, left dpm_4, P_3–M_1, STS 1936*a*.

Tragelaphus sp. aff. *angasi*
Material:
3 pieces from a single individual.
Right maxilla, M^{1-2}, STS 2092; right mandible piece, dpm_4, M_1 unerupted, STS 2493; left mandible piece, dpm_4, M_1 unerupted, STS 1865.

Makapania cf. *broomi* (fig. 170)
Material:
22 pieces from a minimum of 8 individuals.
Maxillary pieces, STS 817, 1573, 1721, 1734, 1756, 2059*b;* mandible pieces, STS 952, 1564*a*, 1879, 1901*a*, 1925, 1938, 2362*a*, 2565, 2588, 2593*a;* isolated teeth, STS 1754*a*, 1824, 1894, 1994, 2121, 2592.

Antelope size class I
Material:
4 cranial and 8 postcranial pieces from a minimum of 2 individuals, as listed in table 77.

Antelope size class II
Material:

81 cranial and 30 postcranial pieces from an estimated minimum of 8 individuals, as listed in table 77.

Antelope size class III
Material:
157 cranial and 89 postcranial pieces from an estimated minimum of 15 individuals, as listed in table 77.

Antelope size class IV
Material:
13 cranial and 2 postcranial pieces from a minimum of 2 individuals, as listed in table 77.

Family Suidae
Metridiochoerus sp.
Material:
2 pieces from 2 individuals.
M^1 (?) in maxillary fragment, STS 2395; left mandible piece with dpm_4 and M_2, STS 3074*a–b*.

Order Perissodactyla
Family Equidae
Equus capensis
Material:
18 pieces from a minimum of 7 individuals.
Left maxilla, P^4–M^2, STS 3000; left maxilla, P^{2-4}, STS 3001; right juvenile maxilla, dpm^{2-4} and part of M^1, STS 3002; parts of 3 maxillary teeth, STS 1571; juvenile mandible pieces, STS 1830*a*, 2102 + 1717, 3006, 3127; adult mandible pieces, STS 1853, 3007; isolated teeth, STS 1888, 1972, 2313, 2316*a–c*, 3003, 3004, 3005, 3126.

Fig. 170. Characteristic teeth of the remarkable bovid *Makapania* cf. *broomi*, well represented in the Sterkfontein Member 4 breccia.

Equus sp.
Material:
2 pieces from 1 individual.
Isolated incisors, STS 1100*b*, 2512*c*.

Order Proboscidea
Family Elephantidae
Elephas recki
Material:
A single specimen from 1 individual.
Isolated dm_1 or dm_2, probably right, STS 1863.

Order Hyracoidea
Family Procaviidae
Procavia antiqua
Material:
17 pieces from a minimum of 8 individuals.
Type specimen of *P. robertsi*, partial cranium and articulated mandible, TM 1197, STS 105; almost complete cranium, STS unnumbered; palates, STS 109, 2140; mandible pieces, STS 102, 103, 104, 107, 108, 140, 1201, 1350*b*, 1829, 2320; isolated incisor, STS 1878.

Procavia transvaalensis
Material:
9 pieces from a minimum of 5 individuals.
Calvaria pieces, STS 2319, 2361; maxillary piece, STS 1535; mandible pieces, STS 101, TM 1212, STS 2117, 2461; isolated incisor, STS 2085.

Order Rodentia
Family Hystricidae
Hystrix africaeaustralis
Material:
6 pieces from a minimum of 5 individuals.
Left maxilla with P^4, M^{1-3}, STS 2078; left premaxillary fragment with part of I^1, STS 2061; parts of left and right mandibular rami with dpm_4, M_{1-2} bilaterally, STS 957*a–e;* left mandible piece with parts of I_1 and P_4, STS 2095; isolated right M^3, STS 1622; isolated incisor, STS 2540.

The Major Features of the Bone Accumulation

The Composition of the Fauna Represented by the Fossils

The information provided in table 76 and depicted in figure 162 may be further summarized. This is done in Table 83 and figure 171. Of the 348 individual animals that contributed bones to the accumulation, no fewer than 243, or 69.8%, were primates. Of these primates, 45 were australopithecines, while the remainder were represented by three species of *Parapapio* and one of *Cercopithecoides*. Of the individual animals, 8.1% were carnivores, represented by a leopard, four *Dinofelis* individuals, one *Megantereon,* two species of hyena, and two species of jackal.

The artiodactyl component of the fauna was found to make up 14.7% of the total and came from 49 bovid individuals, together with two pigs. Dividing the bovid remains into size classes, we obtain the following results:

class 1, 2 unidentified individuals; class II, 14 individuals, including 6 juveniles; class III, 11 individuals including 4 juveniles; and class IV, 18 individuals including 7 juveniles.

The extinct Cape horse, *Equus capensis,* made up 2% of the preserved fauna, and 13 individual dassies of two species also contributed. Remains of 1 juvenile elephant and 5 porcupines are present.

The Representation of Skeletal Parts

There is little doubt, I think, that the fossil sample from Member 4 has been strongly and artificially biased in favor of cranial and readily identifiable specimens. This is a result of the rather haphazard collecting procedure at the cave between 1936 and 1947. During much of this period the site was being mined for lime; it was also frequented by innumerable casual visitors, many of whom carried away samples of bone-bearing breccia as souvenirs. In addition, Broom understandably concentrated on those fossils that were potentially capable of yielding specific identifications. For these reasons I suspect that, although the fossil sample may provide a reasonably accurate reflection of faunal composition in the Member 4 bone accumulation, it cannot be expected to do so for body-part representation. Many postcranial fossils presumably were simply not collected at all by Broom's team and may subsequently have been picked up on the dumps by souvenir-hunters.

Despite this probable bias, the paucity of primate postcranial fossils is striking. To go with the cranial remains of at least 45 australopithecines, there are only 7 catalog entries of postcranial bones. One of these, STS 14 (fig. 165), represented articulated bones from a reasonably complete torso, but the others represent nothing more than scattered skeletal fragments. The same is true for the baboon and monkey remains. Here 769 cranial specimens (some of them, admittedly, isolated teeth) are associated with only 55 postcranial pieces. Despite the artificial sampling bias, it is difficult to avoid the conclusion that most of the primate postcranial skeletons had disappeared before fossilization. The position in respect to antelope remains is only slightly better: here 49 individual bovids, identified largely on 360 cranial remains, are represented by only 129 postcranial bones. The 7 equids, 1 elephant, 13 dassies, and 5 porcupines have all been identified by

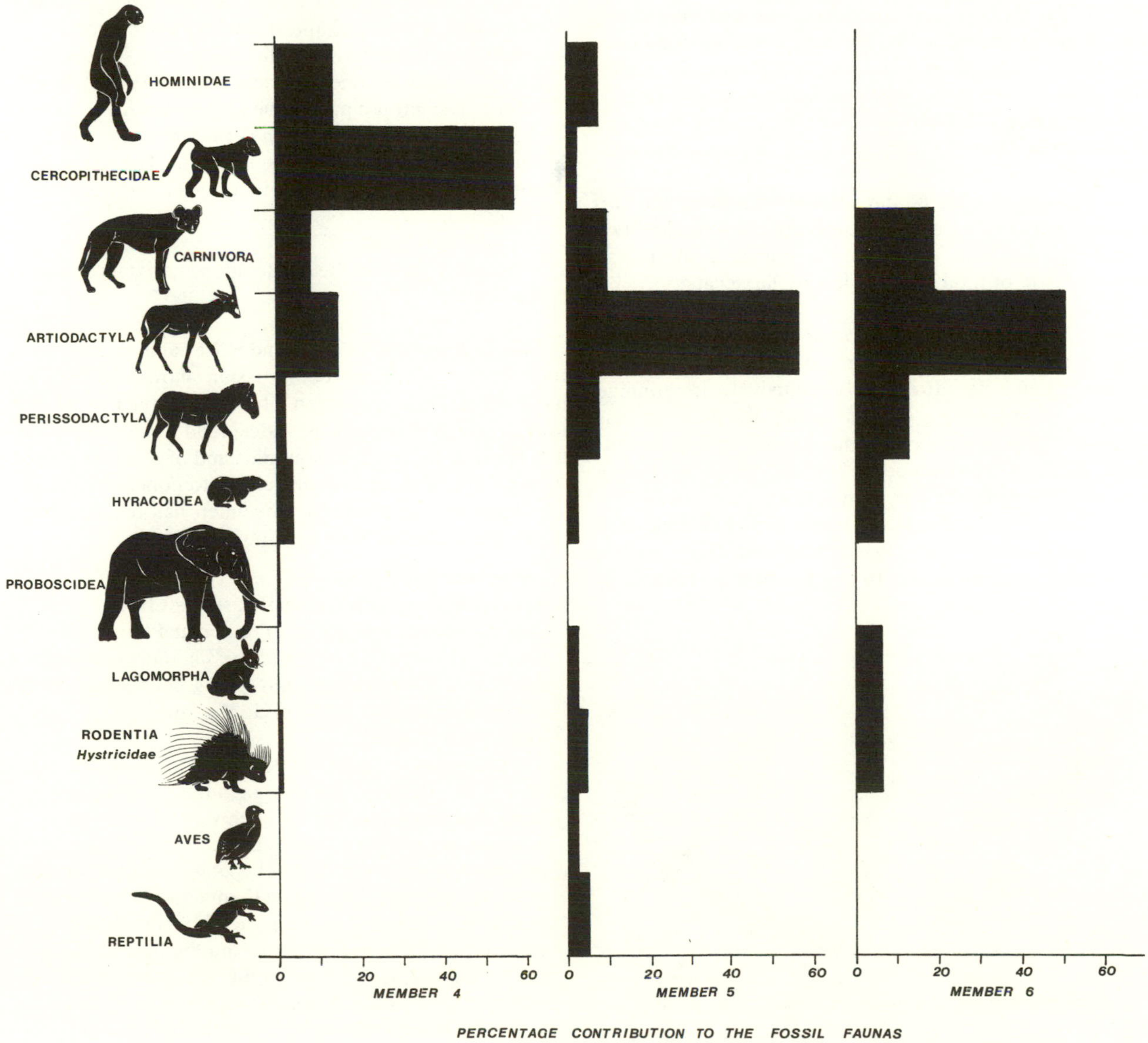

HOMINIDAE
CERCOPITHECIDAE
CARNIVORA
ARTIODACTYLA
PERISSODACTYLA
HYRACOIDEA
PROBOSCIDEA
LAGOMORPHA
RODENTIA
Hystricidae
AVES
REPTILIA
0 20 40 60
MEMBER 4
0 20 40 60
MEMBER 5
0 20 40 60
MEMBER 6
PERCENTAGE CONTRIBUTION TO THE FOSSIL FAUNAS

Fig. 171. Percentage contributions made by various groups of animals to the samples of macrovertebrate fossil fauna from Sterkfontein Members 4, 5, and 6.

cranial elements only, though some of their other bones may well be among the 330 indeterminate fragments, too incomplete to allow taxonomic placing.

Observed Damage to the Bones

During the analysis of this and other fossil assemblages from the Sterkfontein valley site units, each specimen was examined for the presence of diagnostic damage. However, the recognition of such damage has been seriously complicated for Member 4 fossils by postdepositional events. Almost all the bones have been distorted or crushed by compression in the breccia, and most of them also suffered extensive damage during the mining process. Broken edges are seldom those that existed when the bone was buried, and determining what damage occurred before fossilization is extremely difficult.

Details of observed damage are given in table 84 and may be summarized as follows:

Porcupine Gnawing. Gnaw marks were observed on one specimen only—a juvenile *Parapapio* mandible.

Carnivore-Inflicted Damage. Two *Australopithecus* specimens, a juvenile mandible and a palate, show ragged-edge damage that could have resulted from carnivore chewing. In both cases the bone is in poor condition, and such a diagnosis is tentative. Two *Parapapio* mandibles show evidence of carnivore damage—one, STS 351, shown in figure 172, bears two punctures on the inner side of the corpus caused—unquestionably, I think—by carnivore teeth.

As detailed in table 84, an antelope mandible and two distal humeri show traces of carnivore chewing, and two dassie cranial pieces have been damaged in a way characteristic of food remains left by large cats.

Traces of Artificial Bone Alteration. In the course of this study, I observed nothing on any of the bones that, in my opinion, could be attributed conclusively to hominid feeding or butchering. In the past, R. A. Dart has suggested that a number of the Sterkfontein craniums show evidence of purposeful hominid violence. I shall now give some details and comments.

In his paper "The Predatory Implemental Technique of *Australopithecus,*" Dart (1949*a*) proposed that 3 australopithecine specimens and 18 cercopithecoid skulls from Sterkfontein showed traces of hominid battering or feeding activity. In a separate paper (Brain 1972) I have commented on the evidence for interpersonal violence among australopithecines, but I will reiterate my comments relevant to the three Sterkfontein hominid fossils. Endocranial casts of three skulls are involved, previously designated as *Plesianthropus transvaalensis* types 1, 2, and 3. They are now classified as *A. africanus;* type 1 is numbered STS 60, while the other two are unnumbered and were housed in the Anatomy Department of Witwatersrand University for many years, although they are now at the Transvaal Museum. Dart (1949*a*, p. 38) described the specimens and their damage as follows:

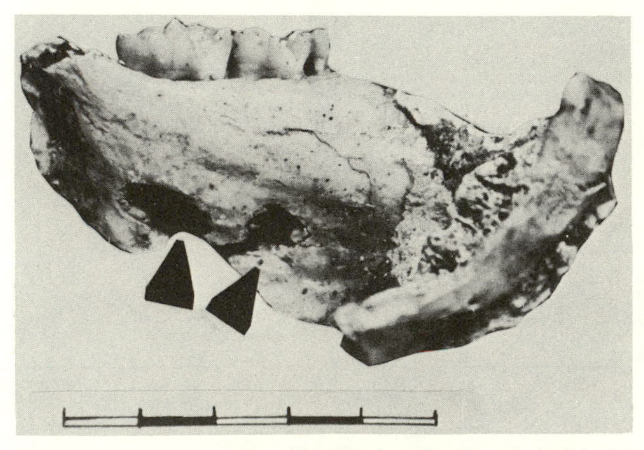

Fig. 172. *Parapapio* mandible, STS 351, from Sterkfontein Member 4, showing carnivore damage in the form of two punctate marks on the inner side of the ramus.

> *Plesianthropus transvaalensis* type 1. A fragmented skull (without mandible or much of face) and a fronto-parietotemporal endocranial cast lacking the right parietal and occipital regions. The right maxilla, premaxilla and molar are fairly complete and little crushed; the left maxilla is better preserved but less complete. The base and the vault are sufficiently preserved to yield accurate reconstruction of the general cranial dimensions.
>
> OBSERVED DAMAGE
>
> The general volume and shape of the endocranial cast have been altered by compression. The squashing of the cast is maximal anteriorly and the temporal height is estimated to be reduced by at least 1 cm by the left temporal bone overriding the parietal at their suture. The left temporal lobe tip was absent from the cast. Crushing and distortion is evident also in the chiasmatic and inferior frontal regions. The left temporal bone was so damaged that it was removed piecemeal.
>
> SUGGESTED CAUSE
>
> A lateral blow on the left temporo-parietal region of the skull.

The skull from which the endocast came is extremely fragmentary, while the compression shown by the cast itself reflects the distortion of the calvaria in which it formed. Such distortion is widespread among Sterkfontein fossil skulls, both of hominids and of other animals; I do not believe there is good reason to conclude that the distortion of this particular specimen has resulted from deliberate violence. Dart continues (1949*a*, p. 38):

> *Plesianthropus transvaalensis* type 2. Fronto-parietal temporal endocranial cast. Left fronto-parietal region deficient and occipital region and base entirely absent. The frontal portion of the sagittal sutural line has been deflected more than 15 degrees to the right of the parietal portion of the sagittal suture.
>
> OBSERVED DAMAGE
>
> The bregmatic angles of both parietal bones are the seat of two depressed fractures in the right parietal bone that hinge together on each side of an irregular line; the left margin of the fractured area follows roughly the sagittal suture. The lateral margins of the area are depressed; the right 12 mm and the left 4 mm below the remainder of the parietal bones. Running across these broken fragments and the frontal bones run radiating fracture lines to the left and right lateral orbital margins both of which were fractured transversely. Owing to the great distortion of the right Sylvian notch region, the temporal bone overrode the parietal for a centimeter.

> Suggested Cause
> A vertical blow just behind and to the right of the bregma with a double-headed object.

Apart from the depressed area in the right parietal bone (fig. 173*a*), Schepers (in Broom and Schepers 1946) showed that the skull in which the endocast formed was distorted in such a way that the anterior part of the mid-sagittal plane is deflected 15 degrees relative to its posterior part (fig. 174).

This kind of distortion is frequently seen in skulls preserved in cave breccia (chap. 7) and undoubtedly occurs after the skull has been enclosed in the surrounding and supporting matrix. Examination of the endocast shows that the skull had been resting upside-down during the fossilization process. Fine-grained sediment entered, presumably through the foramen magnum, but did not fill the endocranial space completely, perhaps because the calvaria collapsed after a certain time.

The roof of the calvaria undoubtedly has suffered a depressed fracture as Dart has claimed. It has occurred in that part of the skull that would have been in contact with the cave floor when it came to rest before becoming buried and fossilized. Unfortunately we have no record of the nature of the breccia that surrounded the skull or of the floor on which it lay. Was there, for instance, a stone in the matrix corresponding to the depressed area? We will never know, and Dart's claim of antemortem injury resulting from violence can be neither substantiated nor refuted.

Dart's description of the third specimen from Sterkfontein (1949*a*, p. 38) is as follows (see fig. 173*b*):

> *Plesianthropus transvaalensis* type 3. Parieto-occipital endocranial cast fractured (in blasting) horizontally through the occipital poles and coronally through the parietal region in the vicinity of the vertex.
> Observed Damage
> Fragments of bone in the substance of the cast show that the skull must have been open in some region or other (as in the Taungs specimen). The sagittal and lambdoid sutures have both been sprung. In addition to minor fracture lines, a major line of fracture runs completely across the cast through the parietal regions and skirting the anterior margin of the right parietal area fragment. The postero-medial portion of the left parietal area is depressed below the right parietal area at the sagittal suture and even more deeply below the antero-lateral portion of the bone both anteriorly and laterally.
> Suggested Cause
> A vertical blow slightly to the left of mid-parietal region with a bludgeon.

Like the preceding endocast, the specimen has lost its context relative to the matrix. No associated skull is known, nor is there any information on the nature of the breccia that originally enclosed it. The skull was certainly fractured in the midparietal region, but it is very doubtful that it could ever be proved whether the damage occurred before or after the burial and fossilization of the specimen.

In addition to these australopithecine specimens, Dart selected 18 *Parapapio* skulls from Member 4 at Sterkfontein, described the damage they had suffered and suggested how such damage could have come about. Only one case will be quoted here (Dart 1949*a*, p. 34):

> 585 Tvl. Posterior half of an infantile skull (1 mm thick) filled with breccia: species indeterminable.
> Observed Damage
> Depressed (4 mm) radiating fracture in left lateral parietal region which caused depression and buckling of skull along line of fracture running transversely through the middle of the parietal bones. The main fracture gives the impression of being due to, or exaggerated by, crushing by left thumb, while cranium was being evacuated of its contents through anterior deficiency. The rock filling at the anterior margin of the skull had been broken away before fossilisation.
> Suggested Cause
> Destruction of anterior part of skull with a blow from front and crushing of posterior portion on left side between fingers and thumb during evisceration of cranial contents.

This and other specimens are figured and described in Dart's paper, which the interested reader can consult. In my opinion all the described damage could be attributed to causes other than hominid bludgeoning or feeding action, but on the basis of these particular specimens the question will never be resolved.

Association with Stone Artifacts

No stone artifacts have been described from Member 4 at Sterkfontein, although a single chert flake came to light during the preparation of an *Australopithecus* maxilla, STS 70. The flake had clearly been detached before being enclosed in the breccia, and its edges were fresh and sharp. Although it is tempting to conclude that this flake resulted from purposeful hominid activity, it could have been detached as a result of a roof-fall.

An interpretation of the Member 4 bone assemblage in terms of accumulating agencies will be attempted in chapter 13.

Clues to the Interpretation of the Member 4 Fossil Assemblage

The skeletons involved in this assemblage are very incomplete, and the disappearance of many parts before fossilization needs to be accounted for. We must consider the possibility that the animals came voluntarily to the cave to die, or that they fell accidentally to their deaths into a natural trap. If this had occurred, and if the carcasses had been protected from scavengers inside the cave, reasonably complete skeletons could have been expected. This, however, is not the case, and other possible agencies will therefore be considered.

Possible Porcupine Collecting Involvement

If the presence of gnaw marks on bones is accepted as a criterion for collecting by porcupines, then the fact that only 1 bone out of a total of 1,891 shows results of gnawing seems conclusive. As mentioned earlier, the fossil bones are generally not in good condition, and the gnaw marks could have been missed in some cases. However, as was discussed in chapters 5 and 8, active porcupine involvement in a bone accumulation typically results in 50% or more of the pieces being gnawed. I have no hesitation in saying that the original evidence of gnawing on Member 4 bones appears to have been low and that porcupines played an insignificant part in the collecting.

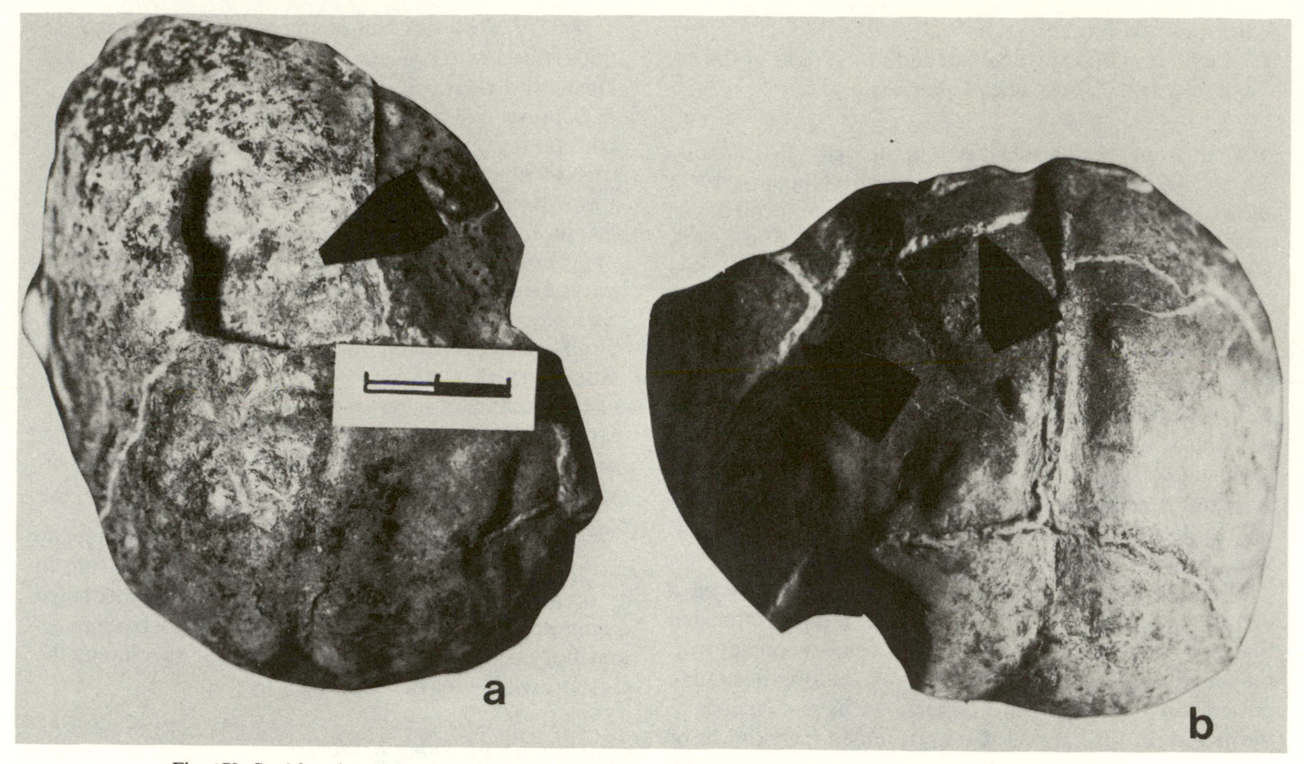

Fig. 173. Sterkfontein specimens on the basis of which claims of interpersonal violence have been made: *(a) Australopithecus* endocast 2 showing its depressed fracture; *(b)* endocast 3 with arrows indicating lines of fracture or displacement.

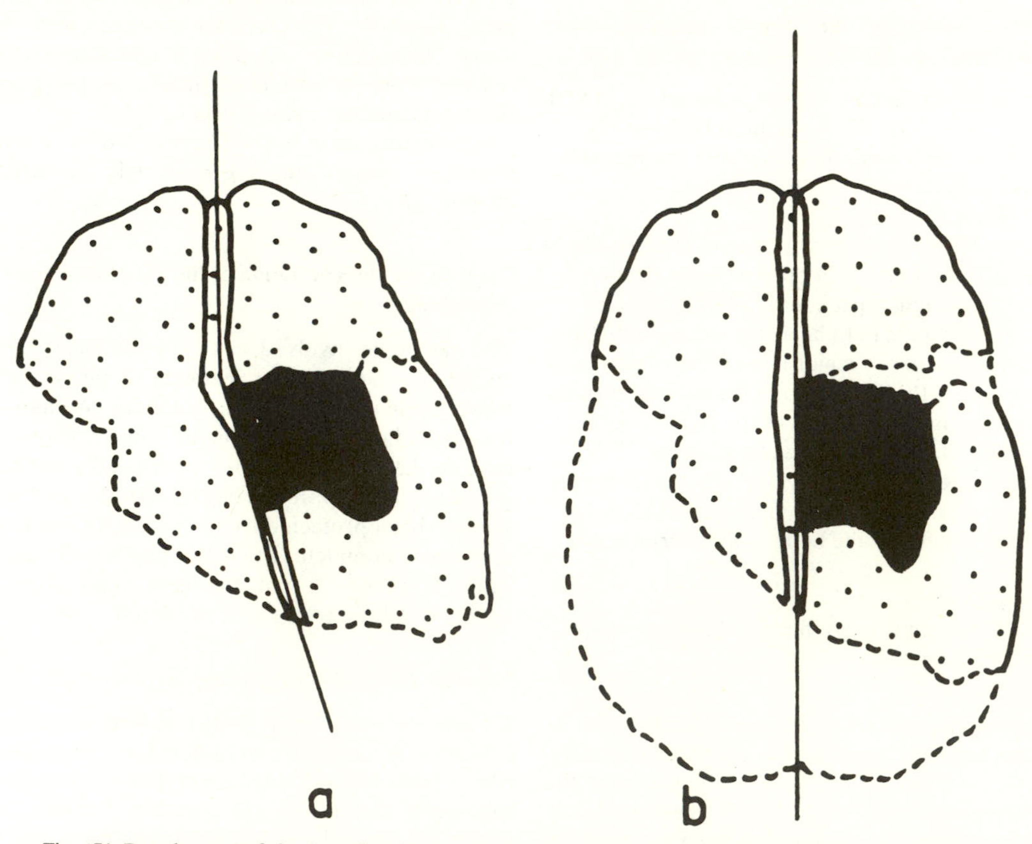

Fig. 174. Dorsal aspect of the *Australopithecus* endocast (Ples. 2) from Sterkfontein Member 4. The black area represents the depressed fracture. *(a)* Distorted form as when found; *(b)* with distortion corrected.

Possible Hominid Involvement

As I mentioned earlier, no stone or bone tools have yet been described from Member 4. Damage on 3 australopithecine craniums and 18 baboon skulls has been interpreted by Dart as having resulted from purposeful hominid bludgeoning or feeding activity, but, for the reasons discussed earlier, I find it impossible to be sure that this damage did not result from compression during fossilization.

On currently available evidence I am inclined to discount hominid involvement in the bone accumulation process.

Possible Carnivore Involvement

The extremely fragmentary nature of the fossil bones, and the fact that tooth marks may be observed on some of them, means that we must consider carnivore involvement in the accumulation. As I mentioned earlier, positive evidence of tooth marks on the bones is not abundant, but this paucity of evidence can probably be attributed, at least in part, to two causes: my conservative approach to the recognition of such evidence, and the poor condition of the fossils involved. Had the fossils been in better condition, I have no doubt that the incidence of recognizably damaged bones would have been higher. Evidence summarized in chapter 8 suggests that, in bone accumulations resulting from leopard feeding only 30% or fewer of the pieces may show undoubted tooth marks. Results of carnivore chewing are often more ambiguous than those of, for instance, porcupine gnawing.

The possibility of carnivore involvement in the Member 4 accumulation is strengthened by the fact that 8% of all the individual animals represented are carnivores and the carnivore/ungulate ratio is 28/58 or 48.3%—a remarkably high figure for a bone accumulation in an African cave.

Carnivores identified among the fossils consist of a leopard, 4 *Dinofelis* cats, a *Megantereon*, 4 hunting hyenas *(Euryboas)*, 1 spotted hyena *(Crocuta)*, and 6 Jackals. With the probable exception of the jackals, all the other carnivores could well have contributed their food remains to the bone accumulation.

The Member 4 fossil assemblage is remarkable chiefly for the great preponderance of primates represented. They account for 69.8% of all the animals identified in the sample. A special explanation has to be sought for their presence, and some possibilities are explored in chapter 13.

In her study of the fossil Bovidae, Vrba (1975, 1976*a*) pointed out that the Member 4 antelopes were mostly of medium to large size, while the strong representation of juveniles suggests the results of primary predation rather than scavenging. To account for the very large bovids represented, she suggested that "the major predators were the false and true saber-tooths, and possibly other, as yet undiscovered cats of sufficient size and/or large-prey adaptation" (Vrba 1976*a*, p. 67).

Although I agree with this proposition, I would not exclude smaller social carnivores, such as the *Euryboas* hunting hyenas may well have been, as likely hunters of large prey as well as likely accumulators of bones in the cave.

It is unfortunate that the artificial collecting bias referred to earlier has prevented us from obtaining a true sample of postcranial bones from Member 4. Had such a sample been available, it might have been possible to identify the carnivores involved with greater certainty. Food remains of the true saber-toothed cats, for instance, would presumably have consisted of relatively undamaged skeletons, whereas those of the hyenas would have suffered a great deal more.

Sterkfontein Member 4: Microvertebrate Component

Since very little microfaunal material definitely attributable to Member 4 is available in the Transvaal Museum collections, I will present no new analysis here. When de Graaff (1960*a*) undertook his preliminary investigation of mammalian microfauna from the Transvaal caves, he identified a number of species from breccia blocks found in the vicinity of the Type Site. Some of these blocks are likely to have been derived from Member 4; others perhaps come from Member 5, where the microvertebrate concentration is greater. The species listed by de Graaff are as follows (for details of these fossil animals, see chap. 9): *Chlorotalpa spelea* Broom, *Elephantulus langi* Broom, *?Mylomygale spiersi* Broom, *Myosorex robinsoni* Meester, *Suncus* Hemprich and Ehrenberg sp., *Mysotromys hausleitneri* Broom, *Tatera* cf. *brantsi* (Smith), *Dasymys* cf. *incomtus* (Sundevall), *?Arvicanthis Lesson* sp., *?Pelomys* cf. *fallax* Thomas and Wroughton, *Rhabdomys* cf. *pumilio* (Sparrman), *Aethomys* cf. *namaquensis* (Smith), *?Mastomys* cf. *natalensis* (Smith), *Leggada* cf. *minutoides* (Smith), *Dendromys* cf. *mesomelas* (Brants), *Palaeotomys gracilis* Broom, and *Cryptomys robertsi* Broom.

Remains from Member 5: Macrovertebrate Component

The collection consists of 1,202 fossils from a minimum of 41 individual animals as detailed in table 85. Particulars of 317 bones of bovid origin, assigned to size classes but not specifically identified, are given in table 86. The various animals identified are depicted pictorially in figure 175.

Data on the fossil material, by which the various animal taxa are represented, will now be presented.

Class Mammalia
Order Primates
Family Hominidae
Cf. *Homo* sp. (or *Australopithecus africanus*)
Material:

SE 255, fragmentary maxilla of a juvenile with part of dm^1, dm^2, and crown of M^1. Age estimate, 6 ± 1 years.

SE 1508, isolated right M^2. Age estimate, adult.

SE 1579, part of the crown of left M^1 or M^2. Age estimate, indeterminate.

SE 1937, isolated left $\bar{C}$. Age estimate, adult.

SE 2396, lingual half of left P^3. Age estimate, adolescent.

SE 2398, tooth fragment (specimen missing). Age estimate, indeterminate.

Family Cercopithecidae
Gen. et sp. indet.
Material:
2 pieces from a single individual.

Maxillary fragment with unerupted molar, SE unnumbered; part of an isolated canine, SE 350.

Order Carnivora
Family Felidae
Cf. *Panthera leo*
Material:
3 pieces from a single individual.
Isolated left I^3, SE 1500*a;* left M_1, SE 1500*b;* right distal humerus, SE 524.

Cf. *Megantereon* sp.
Material:
3 pieces, probably from a single individual.
Complete right radius with pathology of the distal end, SE 670; right calcaneus, SE 1692; left distal tibia, SE 1832.

Family Hyaenidae
Proteles sp.
Material:
A single specimen.
Piece of right mandible, SE 562.

Family Canidae
Canis cf. *terblanchei*
Material:
A single specimen.
Left maxillary piece with canine and alveoli, SE 125.

Carnivore indet.
Material:

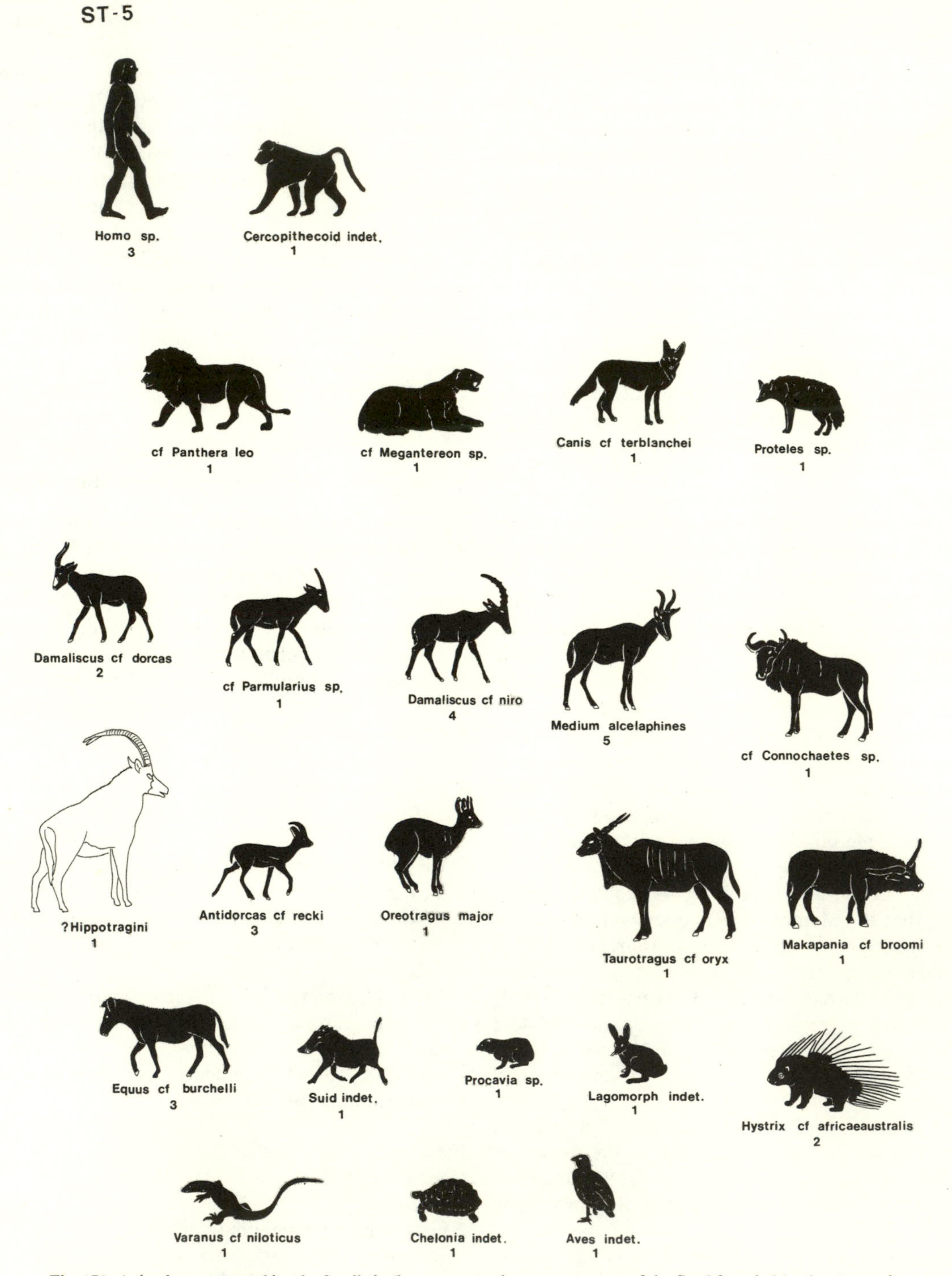

Fig. 175. Animals represented by the fossils in the macrovertebrate component of the Sterkfontein Member 5 sample. Minimum numbers of individuals are indicated.

12 pieces from a minimum of 3 individuals.

Maxillary piece, SE 1154; mandible pieces, SE 56, 1629; tooth fragments, SE 1172, 1477, 1883; humerus pieces, SE 1075, 1627; metapodial pieces, SE 1520*a,b*.

Order Artiodactyla
Family Bovidae
Damaliscus cf. *dorcas*
Material:
4 pieces from a minimum of 2 individuals (1 juvenile, 1 adult).

Left maxilla with dpm$^{2-4}$, SE 1185; left mandible piece, M_{1-3}, SE 1728; isolated right M^3, SE 1218, 1770.

Damaliscus cf. sp. 2
Material:
7 pieces from a minimum of 4 adult individuals.

Left maxilla with P^4, M^{1-3}, SE 794; right maxilla with M^{2-3}, SE 588; left mandible piece, P_4–M_3 and right M_2, SE 1233; left mandible piece, M_{1-3}, SE 1614; left M^2, SE 1334; right M^1, SE 1381; right M_3, SE 1754.

Damaliscus sp. 1, or *Parmularius* sp.
Material:
A single specimen from an adult individual.

Left mandible piece, P_3–M_1, SE 192.

Medium-sized Alcelaphines
Material:
7 pieces from a minimum of 5 individuals (1 juvenile and 4 adults).

Right mandible piece, M_{2-3}, SE 464; right mandible piece M_{2-3}, SE 1763; isolated left dpm$_4$, SE 2133; right M_1, SE 1828; right M_1 and M_2, SE 627; right M_2, SE 1424; right M_2, SE 535.

Cf. *Connochaetes* sp.
Material:
A single specimen from an adult individual.

Isolated right M_2, SE 2601.

?Hippotragini
Material:
A single specimen from an adult individual.

Part of an isolated left M_3, SE 1125.

Antidorcas cf. *recki*
Material:
5 specimens from a minimum of 3 individuals (1 juvenile and 2 adults).

Left mandible piece, dpm$_4$, M_{1-2}, M_3 unerupted, SE 1258; left mandible piece, P_{2-3}, SE 125; left mandible piece, P_4–M_3, SE 535; left mandible piece P_4–M_3, SE 1855; right mandible piece, dpm$_{3-4}$, SE 1313.

Oreotragus major
Material:
3 pieces from a single adult individual.

Crushed palate with left P^2–M^3; right P^3–M^3; parts of both mandibular rami, left P_3–M_3; right M_{2-3}, M 8361*a,b* (ex Bernard Price Institute).

Taurotragus cf. *oryx*
Material:
A single specimen from an adult individual.

Isolated left M_1, SE 196.

Makapania cf. *broomi*
Material:
A single specimen from an adult individual.

Part of a left M^3, SE 1425.

Antelope size class I
Material:
1 cranial and 19 postcranial pieces from an estimated minimum of 3 individuals, as listed in table 86.

Antelope size class II
Material:
29 cranial and 122 postcranial pieces from an estimated minimum of 6 individuals, as listed in table 86.

Antelope size class III
Material:
32 cranial and 108 postcranial pieces from an estimated minimum of 5 individuals, as listed in table 86.

Antelope size class IV
Material:
1 cranial and 5 postcranial pieces from a minimum of 2 individuals, as listed in table 86.

Family Suidae
Suid, gen. et sp. indet.
Material:
2 pieces from a single individual.

Mandible piece with M_2, SE 1069; tooth fragment, SE 1624.

Order Perissodactyla
Family Equidae
Equus cf. *burchelli*
Material:
18 cranial and 5 postcranial pieces from a minimum of 3 individuals (1 juvenile and 2 adults).

Isolated left P^4, SE 2752 (also numbered 1602); isolated tooth fragments, SE 3, 63, 74, 116, 189, 214, 245, 274, 401, 687, 1246, 1382, 1391, 1628 1629, 1855, 2397; proximal femur piece, SE 1440; articulated distal femur and proximal tibia, SE 1605; distal tibia piece, SE 1795; carpal bone, SE 100.

Order Hyracoidea
Family Procaviidae
Procavia sp.
Material:
3 pieces probably from a single individual.

Cranial fragments, SE 744; part of an isolated incisor, SE 1655; distal humerus fragment, SE 723.

Order Lagomorpha
Family Leporidae
Lagomorph, gen. et sp. indet.
Material:
A single specimen.

Parts of a mandible, SE 222.

Order Rodentia
Family Hystricidae
Hystrix cf. *africaeaustralis*
Material:
6 cranial and 1 possible postcranial piece from a minimum of 2 individuals (1 juvenile, 1 adult).

Isolated teeth and tooth fragments, SE 411, 1249, 1255, 1262, 1263, 1269; left calcaneus, may be porcupine, SE 2024.

Class Aves
Bird indet.
Material:
1 cranial and 1 postcranial piece, possibly from 1 individual.
Shattered cranium, SE 1342; proximal ulna piece, SE 2225.

Class Reptilia
Order Squamata
Family Varanidae
Varanus cf. *niloticus*
Material:
A single isolated vertebra, SE 817.

Order Chelonia
Family Testudinidae
Gen. et sp. indet.
Material:
A single carapace piece, SE unnumbered.

Sterkfontein Member 5: The Major Features of the Bone Accumulation

The Composition of the Fauna Represented by the Fossils

The information provided in table 85 and depicted in figure 175 may be further summarized. This is done in the second columns of table 83 and of figure 171. The overall composition of the preserved fauna in Member 5 differs markedly from that of the underlying Member 4. Whereas primates dominated in the Member 4 assemblage, artiodactyls are by far the most important group in Member 5. In fact, while 56% of the individual Member 4 animals were cercopithecoids, the same percentage of Member 5 individuals are bovids. The difference between the two assemblages clearly results from a fundamental change in accumulation pattern between the two members, a matter to be discussed in chapter 13.

The few primate remains present in the Member 5 collection came from 3 hominid individuals and 1 baboon. The hominids, 2 adults and a child, have been variously classified as *Australopithecus* and as *Homo*, and the baboon remains are too fragmentary to allow firm identification.

Carnivores are represented by a single lion, a *Megantereon*like sabertooth, an aardwolf, and a jackal. The 22 bovid individuals may be placed into size classes: class I, represented by 3 adult klipspringers, *Oreotragus major;* class II by 10 individuals, including 2 juveniles, of *Damaliscus, Parmularius,* and *Antidorcas;* class III by 6 individuals, including 1 juvenile, of medium-sized alcelaphines, including *Connochaetes;* class IV by 3 adult individuals that include an eland and a *Makapania*. Among the artiodactyl fossils are remains of a single unidentified pig.

The rest of the animals made a small contribution to the overall faunal picture. Of interest is the presence of monitor lizard *(Varanus)* and tortoise remains—animals that feature consistently in the diet of Stone Age hunter-gatherers of more recent times.

The Representation of Skeletal Parts

Most of the animals are very poorly represented and have been identified on little more than scraps. Yet this is a true reflection of the position in Member 5, since the sample resulted from systematic excavation by Robinson and from careful acetic acid preparation of many of the specimens. The only animals whose remains are reasonably abundant are the bovids (table 86). In size class II, 10 individuals contributed 151 bones, and 6 class III animals are represented by 140 pieces. The skeletal parts table shows that there is a scatter of fossils from all parts of the skeleton, ranging from the head to the foot bones. This does not necessarily imply that the animals were brought back whole to the cave, but rather that a wide range of parts, at any rate, found their way there.

Observed Damage to the Bones

From the point of view of estimating damage to individual bones, the Member 5 sample is much more satisfactory than the Member 4 equivalent. The fossils had been carefully excavated and prepared so that damaged edges and surfaces could be studied with more definitive results. Details are provided in table 87.

Porcupine and Small-Rodent Gnawing. Fifteen bone pieces showed clear evidence of porcupine gnawing, and 17 bore small-rodent tooth marks. I am not able to suggest what species of rat or mouse was responsible for the latter marks, though it appears to have been an animal about the size of a gerbil. An excellent example of a rodent-gnawed bone from Member 5 has been figured by Robinson (1959).

Carnivore-Inflicted Damage. Clear evidence of carnivore chewing was noted on 23 bovid bones (see table 87 for details) as well as on 14 bone flakes and fragments. Probable carnivore damage was seen on 7 more specimens.

Traces of Artificial Bone Alteration. A bone tool, fashioned on a flake and bearing two worn facets (SE 612, fig. 176*a*) came from the Member 5 excavation and was described by Robinson (1959). To my mind there is little doubt that this is a genuine artifact as Robinson has claimed and that the facets were worn through use on skins or other similarly soft surfaces. A second bone flake, SE 1000, has worn edges that may have been artificially caused. A third specimen, SE 1524, is shown in figure 176*b;* it appears to be a piece of a bovid horn-core, three surfaces of which have been worn to a tapered point. It may well have been employed for the same purpose as SE 612.

A specimen showing convincing evidence of hominid action is SE 1729 (fig. 176*c*), the shaft of a small bovid humerus bearing three clear chop marks that were presumably caused by the edge of a stone tool.

Several specimens show oval openings up to 2 cm in diameter in bone shafts. These are rather larger than the punctate marks usually caused by carnivore teeth and may have resulted from hominid use of pointed objects.

Association with Artifacts

The extension site excavation conducted by Robinson during 1957–58 yielded 284 stone artifacts (Robinson 1957, 1962). These were in situ in Member 5 and directly associated with the bones under discussion. They have been described by Mason (1962*b*) and by M. D. Leakey (1970) and consist mainly of core or chopping tools without the flakes, which must have been detached elsewhere.

Clues to the Interpretation of the Member 5 Fossil Assemblage

Possible Porcupine Collecting Involvement

The porcupine-gnawing incidence is 15 bones out a total of 1,202, or 1.3%. By this criterion porcupines may be excluded as significant collectors of the Member 5 bones.

Possible Hominid Involvement

The 1,202 bones in this sample were directly associated with 284 stone artifacts and at least 2 bone tools. This implies a ratio of 4.2 bone pieces to 1 stone artifact. At least 1 bone piece shows indisputable chop marks caused by a sharp-edged tool.

The remarkable density of artifacts in the excavated part of Member 5 strongly suggests that the cave was intensively occupied during this accumulation phase. It would therefore be remarkable if the bone pieces associated with the artifacts did not represent hominid food remains.

Possible Carnivore Involvement

Only 4 of the 41 animals represented in the fossils were carnivores—a lion, a saber-toothed cat, an aardwolf, and a jackal, while the carnivore/ungulate ratio proved to be 4/26, or 15.4. Despite such pointers to minimal carnivore involvement, at least 37 bones show clear carnivore tooth marks, indicating that carnivores were undoubtedly involved in the history of some of the bones.

From her study of the bovid fossils, Vrba (1976*a*) concluded that, on account of the low juvenile percentage and the distribution of antelope individuals through all weight classes, the assemblage was predominantly a *non-primary, scavenged* one. She suggested that, on the evidence of the bovid food remains, the Member 5 hominids were *scavengers* rather than effective hunters. In the light of evidence I have assembled since Elisabeth Vrba wrote, I have no reason to dispute her conclusion, implications of which are discussed further in chapter 13.

Remains from Member 5: Microvertebrate Component

T. N. Pocock (1969) provided a list of microfauna provisionally identified from sieved material recovered from Sterkfontein dump 8. According to the site plan given by Tobias and Hughes (1969), dump 8 was a long, narrow one extending along the northern margin of the West Pit and Extension Locality; it is thus almost certainly derived from Member 5 breccia.

The following taxa were identified by Pocock; minimum numbers of individuals are given in parentheses: *Elephantulus langi* (31); *Myosorex* sp. (9); *Suncus* cf. *orangiae* (6); *Suncus* cf. *gracilis* (4); Chrysochloridae (postcranial material); *Rhinolophus* cf. *darlingi* (2); *R.* cf. *augur* (1); Vespertilionidae (postcranial material); Leporidae (1); *Mystromys hausleitneri* (286); *Tatera* sp. (31); *Dasymys* sp. (2); "Rhabdomys minor"

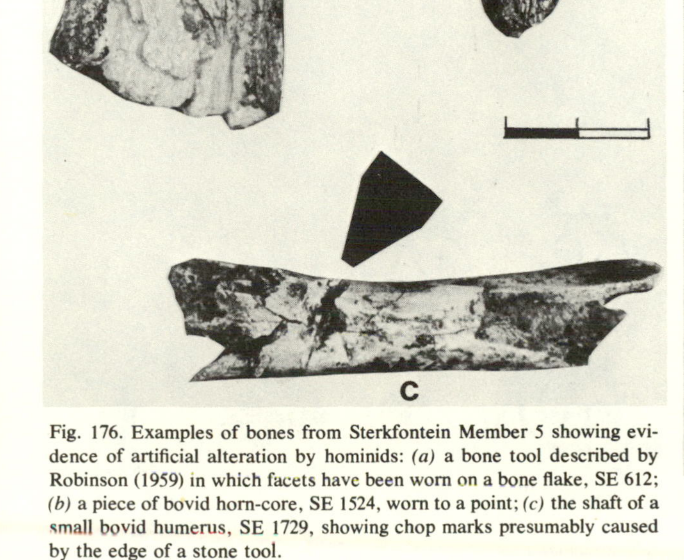

Fig. 176. Examples of bones from Sterkfontein Member 5 showing evidence of artificial alteration by hominids: *(a)* a bone tool described by Robinson (1959) in which facets have been worn on a bone flake, SE 612; *(b)* a piece of bovid horn-core, SE 1524, worn to a point; *(c)* the shaft of a small bovid humerus, SE 1729, showing chop marks presumably caused by the edge of a stone tool.

Fig. 177. A selection of the very numerous stone artifacts from Sterkfontein Member 5. Core and chopping tools predominate, the flakes from which do not appear to have been detached in the cave.

(6); *Aethomys* cf. *chrysophilus* (2); *Praomys (Mastomys)* cf. *natalensis* (8); *Praomys* sp. (6); *Mus.* cf. *musculus* (2); *Dendromys* cf. *melanotis* (36); *Malacothrix* sp. (19); *Steatomys* sp. (6); *Otomys* cf. *gracilis* (43); *Graphiurus* sp. (1); *Cryptomys* cf. *holosericeus* (30); unidentified remains of frogs, lizards, and birds.

In the course of his Extension Site excavations of 1957–58, Robinson recovered samples of breccia rich in microfaunal remains. These came from the upper levels of Member 5 breccia on the western side of the West Pit. Some of this breccia in the Transvaal Museum collection was prepared by the acetic acid method in the course of this study. The sample was found to contain remains from a minimum of 644 individual animals as listed in table 88. The faunal composition of this sample is presented pictorially in figure 178. As is apparent, the fauna is dominated by *Mystromys,* dendromurines, and shrews.

Remains from Member 6: Macrovertebrate Component

The sample discussed here is derived from Robinson's 1957–58 Extension Site excavation. On the basis of the appearance of the enclosing matrix it is fairly easy to separate fossils derived from members 5 and 6. This separation has been made, and the sample consists of 454 fossils from a minimum of 19 individual animals, as listed in table 89. The various identified animals are depicted in figure 179.

Particulars of 66 bones of bovid origin, placed in size classes but not specifically identified, are given in table 90.

Details of the fossil material by which the various animal taxa are represented will now be presented.

Phylum Chordata
Class Mammalia
Order Carnivora
Carnivore indet.
Material:
3 cranial and 6 postcranial pieces from an estimated minimum of 3 individuals.

Mandible piece, SE 2104; tooth fragments, SE 2290, 2329; distal humerus, SE 864; ?tibia shaft, SE 2059; metapodial pieces, SE 1140, 2062, 2095; phalanx, SE 731.

Order Artiodactyla
Family Bovidae
Antidorcas bondi
Material:
3 cranial pieces from a minimum of 2 adult individuals.

Right mandible piece, M_{2-3}, SE 690; isolated left M^2, SE 829, 875.

Damaliscus cf. *dorcas*
Material:
A single specimen, right maxilla, P^2–M^2, SE 1318.

Antelope size class I
Material:
10 postcranial pieces from a minimum of 2 individuals, as listed in table 90.

Antelope size class II
Material:
3 cranial and 34 postcranial pieces from an estimated minimum of 4 individuals, as listed in table 90.

Antelope size class III
Material:
4 cranial and 15 postcranial pieces from a minimum of 4 individuals as listed in table 90.

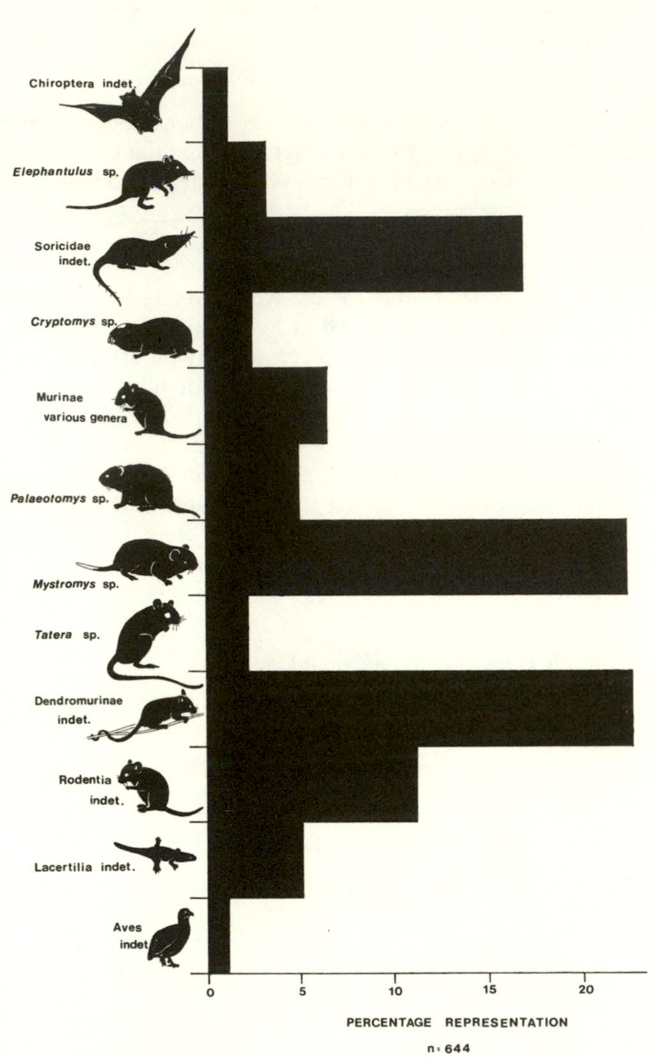

Fig. 178. The percentage representation of animals of various kinds in the microvertebrate component of the Sterkfontein Member 5 fossil sample; 644 individual animals are involved, and the sample is dominated by shrews, dendromurines, and the white-tailed rat, *Mystromys*.

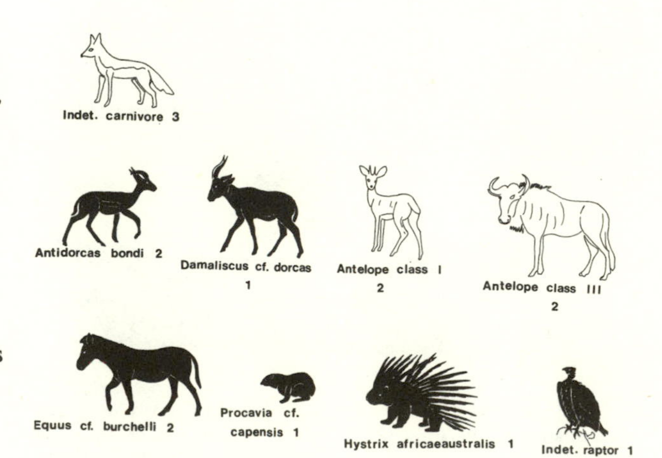

Fig. 179. Animals represented by the fossils in the macrofaunal component of the Sterkfontein Member 6 sample. Minimum numbers of individuals are indicated.

Order Perissodactyla
Family Equidae
Equus cf. *burchelli*
Material:
6 cranial pieces from a minimum of 2 individuals.
Isolated tooth fragments, SE 686, 693, 704, 794, 818, unnumbered.

Order Hyracoidea
Family Procaviidae
Procavia cf. *capensis*
Material:
3 cranial and 2 postcranial pieces from 1 individual.
Left mandible piece, SE 806; isolated teeth, SE 928, 1050; radius, SE 995; pelvis piece, SE 703.

Order Rodentia
Family Hystricidae
Hystrix africaeaustralis
Material:
2 fragments of incisor, probably from a single individual, SE 921, 994.

Class Aves
Order Falconiformes
Family Accipitridae
Bird, large raptor
Material:
2 postcranial pieces, probably from a single individual.
Piece of an ulna, SE 1888; isolated claw, SE 1.

The Major Features of the Bone Accumulation

The Composition of the Fauna Represented by the Fossils

The information provided in table 89 and depicted in figure 179 may be further summarized. This is done in the third columns of table 83 and figure 171. Member 6 is a rather insignificant stratigraphic unit representing the infilling of a low space between the top of Member 5 and the overhanging dolomite roof. Of the 454 bone pieces from it in the present sample, 290 were bone flakes. The rest were found to come from a minimum of 16 individual animals, 8 of which were bovids. Among these, only *Antidorcas bondi* and *Damaliscus* cf. *dorcas* were identifiable.

Representation of Skeletal Parts

A sample of this size cannot be expected to provide much information about skeletal part representation. As table 90 shows, the 8 bovid individuals were represented by 59 postcranial pieces, including parts from the lower limbs and feet. A far larger sample would be required before conclusions could be drawn about butchery practices or carnivore feeding patterns.

Observed Damage to the Bones

As with bones from other site units, each piece from Member 6 was examined for diagnostic damage traces. Results were as follows (table 91):

Porcupine and Small-Rodent Gnawing. Eight bone pieces showed clear evidence of porcupine gnawing, and 4 bore small-rodent tooth marks.

Carnivore-Inflicted Damage. Clear evidence of carnivore chewing was noted on 3 bovid fragments and 5 other fragments. Probable carnivore damage was apparent on 2 other pieces.

Traces of Artificial Bone Alteration. Cut marks were noted on a mandible of *Antidorcas bondi* (SE 690), and 3 bone flakes showed wear and rounding that may have been artificially induced.

Association with Artifacts

No artifacts are known to have been definitely associated with the Member 6 bones.

Clues to the Interpretation of the Member 6 Fossil Assemblage

Possible Porcupine Collecting Involvement

Eight bone pieces out of 454 showed evidence of porcupine-gnawing, a percentage of 1.8. By this criterion porcupines may be excluded as significant contributors to the collection.

Possible Hominid Involvement

No artifacts have yet been found in Member 6, but at least one bovid mandible shows cut marks that were certainly caused by sharp-edged tools. That 290 of the 454 bone pieces were bone flakes is strongly reminiscent of the situation in Stone Age human food remains.

Possible Carnivore Involvement

Parts of about 3 unidentifiable carnivores were found, which means that the carnivore/ungulate ratio is 3/10, or 30.0%. Clear evidence of carnivore chewing was observed on 8 bones.

The sample is too small to allow firm conclusions, but it is obvious that both hominids and carnivores have been involved to some extent in the taphonomic process. Perhaps further excavation of Member 6 breccia will elucidate the situation.

Remains from Member 6: Microvertebrate Component

No microvertebrate remains from Member 6 were available at the time of the study.

11 Swartkrans

A Brief History of Activity

During 1948 the University of California's African expedition was searching for australopithecine fossils in the Transvaal. In September of that year the expedition's leader, Wendell Phillips, approached Broom and offered financial help for exploiting a new fossil locality. At the time Broom and J. T. Robinson were excavating in the Sterkfontein Type Site, but they agreed to move part of their work force across the valley to Swartkrans, where the presence of bone-bearing breccia had been known for some time. The quarrying of both sites was placed under the supervision of Mr. van der Nest.

The Swartkrans locality was immediately productive, and within the first week the mandible of what appeared to be a new species of robust australopithecine, *Paranthropus crassidens* Broom, had come to light (fig. 193). During the subsequent months many valuable finds were made, including the jaw of an early form of true man, found by J. T. Robinson on 29 April 1949. This was named *Telanthropus capensis* Broom and Robinson but subsequently was transferred to *Homo* (fig. 187).

By March 1949 most of the California expedition members were back in the United States, and no further funds were forthcoming from this source. The Swartkrans excavation was nevertheless continued throughout 1949, at

Fig. 180. Searching for fossils among blocks of breccia in 1948. Broom may be seen in the background. Photograph by J. T. Robinson.

Transvaal Museum expense, until financial problems brought the work to an end in November. By this time a substantial seam of pure travertine (fig. 181) had been exposed along the north wall of the excavation. This attracted the attention of a local lime miner named Fourie, who started blasting at Swartkrans in December 1949, to the dismay of the Transvaal Museum team, which was powerless to stop it. The mining continued throughout 1950 and 1951 and resulted in the finding of some spectacular specimens, such as the well-known cranium SK 48 (fig. 144) and the beautiful mandible SK 23 (fig. 194).

Robert Broom died in Pretoria on 6 April 1951, but Robinson was able to recommence paleontological work at Swartkrans with financial assistance from the Nuffield Foundation. This continued into 1953 and yielded many valuable specimens. Thereafter the site was abandoned for twelve years.

Active research was conducted at the Transvaal Museum between 1953 and 1957 on various aspects of the fauna and geology of Swartkrans and other Sterkfontein valley caves. This resulted in many papers by various specialists, as well as two museum memoirs—on the dentition of the Australopithecinae (Robinson 1956) and on the geology of the cave deposits (Brain 1958).

In April 1965 paleontological work was resumed at Swartkrans, with generous support from the Wenner-Gren Foundation for Anthropological Research in New York and, subsequently, from the South African Council for Scientific and Industrial Research (Brain 1967*a*). The immediate aim of this work was to restore order after the chaotic mining episode of 1950–51, but it soon became apparent that 1951 was not the first time the site had been mined. An extensive dump on the hillside below the cave contained thousands of tons of rock blasted from the site before paleontological work started there in 1948. This was unexpected, since the site had shown little obvious evidence of such activity. It was only when attention was given to clearing the cave itself of rubble that the extent of the first mining operations became apparent. Not only had considerable quantities of rock been blasted out, but the quarry had been back-filled with rubble before the miners left it.

Local residents were asked about the identity of the first miner, and firsthand information was provided by M. Bolt, owner of Bolt's Farm adjoining Swartkrans. She was able to tell us of a prospector named Knowlan who had been active in the Swartkrans area during the early 1930s and who had also worked at Taung. An aged African was then found who had actually been employed by Knowlan during the Swartkrans quarrying, probably about 1932. He remembered the back-filling after the blasting, which enabled the miners to lay a cocopan track across the main outer cave area when they turned their attention to the deep shaft at the eastern end of the site.

Fig. 181. The very large stalagmitic boss that originally separated the Outer Cave from the Inner Cave at Swartkrans. A hominid cranium is visible in the breccia adjacent to the stalagmite.

The project of removing and sorting the miner's rubble has been extremely tedious, involving seven years of labor (Brain 1973). It has resulted in the finding of many fossils, none of them, regrettably, in situ; more significantly, however, it has clarified the structure of the cave and the stratigraphic relationships of its filling.

A comprehensive study of the fossil bovid remains from Swartkrans was undertaken between 1969 and 1973 by E. S. Vrba (1975, 1976*a*). This contributed greatly to a proper understanding of the age relationships between the breccias in the Outer Cave. Previously it had been thought (Brain 1958) that most of the Outer Cave filling was formed by breccia of a single time unit. Gradually, however, it became apparent that at least two deposits were involved, differing greatly in age and faunal composition (Brain 1976*a*). These were formally named members 1 and 2 of the Swartkrans Formation by K. W. Butzer (1976), who had been undertaking detailed stratigraphic studies at the site for some years.

The University of Witwatersrand acquired the Swartkrans farm and cave site in 1968. Through the generosity of the Bernard Price Institute of the University, it has been possible for Transvaal Museum operations to continue at Swartkrans with gratifying results.

Although the first evidence for co-existence of an early form of *Homo* with robust australopithecines at Swartkrans came to light in 1950 (Broom and Robinson 1950), further significant information was forthcoming twenty years later (Clarke, Howell, and Brain 1970). After R. J. Clarke's assembly of a composite cranium (fig. 192) it was possible to confirm that the two hominids had coexisted during Member 1 times.

Current work at Swartkrans aims to define the extent of the breccia mass in the Outer Cave and to elucidate stratigraphic relationships within this. Carefully controlled excavation is for the first time providing specimens whose provenance is accurately known. Swartkrans has suffered from the disadvantage that, although numerous hominid specimens have come from the site, very few of them can be accurately positioned within the cave and its stratigraphy.

Since the manuscript of this book was completed, new excavations have shown that the eastern end of the Outer Cave is extensive, containing a major sediment accumulation that entered through a shaft above the southeastern cave wall. The main Member 1 sediment mass entered through an opening above the north wall, and the relationship of the two bodies is now being studied.

Some Notes on the Site

The Swartkrans cave complex is on the southeastern aspect of the Swartkrans hill at an altitude of 1,480 m. A trigonometric beacon on the crown of the hill stands at 1,497 m, and the level of the Bloubank streambed below the cave is 1,454 m. The hill lies in a zone where intense faulting has occurred; many of the outcrops are of a shatterrock consisting of angular fragments of chert, produced by the faulting, recemented in a dolomitic matrix. Making use of a large-scale aerial photograph, the nature of the rock outcrops are plotted, particularly whether they consisted mainly of undisturbed dolomite, undisturbed chert, or shatter rock. It was found that wide bands of fault-shattered rock traverse the hill, but the cave system has developed in a thick dolomitic band, comparatively free of chert bands. The aerial photograph showed two lineaments in the country rock intersecting at approximately right angles, and the cave system has developed close to the intersection of these two lines. Two natural shafts—the "eastern shaft" within the Swartkrans cave system itself and the "southwestern shaft" about 80 m distant—are both aligned with the lineaments.

The Swartkrans cave system consists of what have become known as the Outer, Inner, and Lower caves. The Outer Cave has no roof and is elongated east-west; it stretches for about 45 m as shown in figure 184. Its northern and western walls are clearly delimited by the outcropping dolomitic country rock, but for most of its length the southern wall is undefined. Here the dolomite wall is not apparent on the surface, and further excavation is being undertaken to establish its position.

The Outer Cave is separated from the Inner and Lower caves by the "floor block"—a vast piece of dolomite that separated from the roof at the western end of the original solution cavern and divided the space there into three separate volumes. The floor block hinged downward from the northern wall of the cavern so that the gap between its upper surface and that of the remaining roof is greatest on the southern side—a vertical measurement of about 3 m. The block was extensively cracked by its fall, and the cracks have been filled with white travertine. In the area separating the Inner Cave from the Outer Cave, a very substantial boss of travertine was built up (fig. 181) before the first direct opening of the cave to the surface. This all but sealed the connection between the two caves.

The Inner Cave retains its dolomitic roof, which is as much as 10 m thick. Downward, the Inner Cave leads beneath the floor block to the Lower Cave, typically low

Fig. 182. Looking down on the contents of the Outer Cave at Swartkrans in 1969. Miners' rubble is being cleared from the outcrops of fossiliferous breccia.

roofed at the western end but opening toward the east into a less confined chamber, hung with stalactites. Still farther to the east the Lower Cave is connected with the surface by an irregular vertical shaft about 9 m deep.

In a cave system like the one at Swartkrans, where the vertical depth of a chamber frequently exceeds its width, drawing an adequate plan is difficult. In the Outer and Inner caves, a series of five horizontal lines was painted around the cave walls, using a surveyor's level. On each line a large number of points were surveyed, and plots of each line were used to build up composite plans and sections.

Fig. 183. Excavations carried out during 1978 that aim to define the southeastern limits of the Swartkrans deposit. In the picture are George Moenda, foreman at the site, and Bella, "the spirit of Swartkrans."

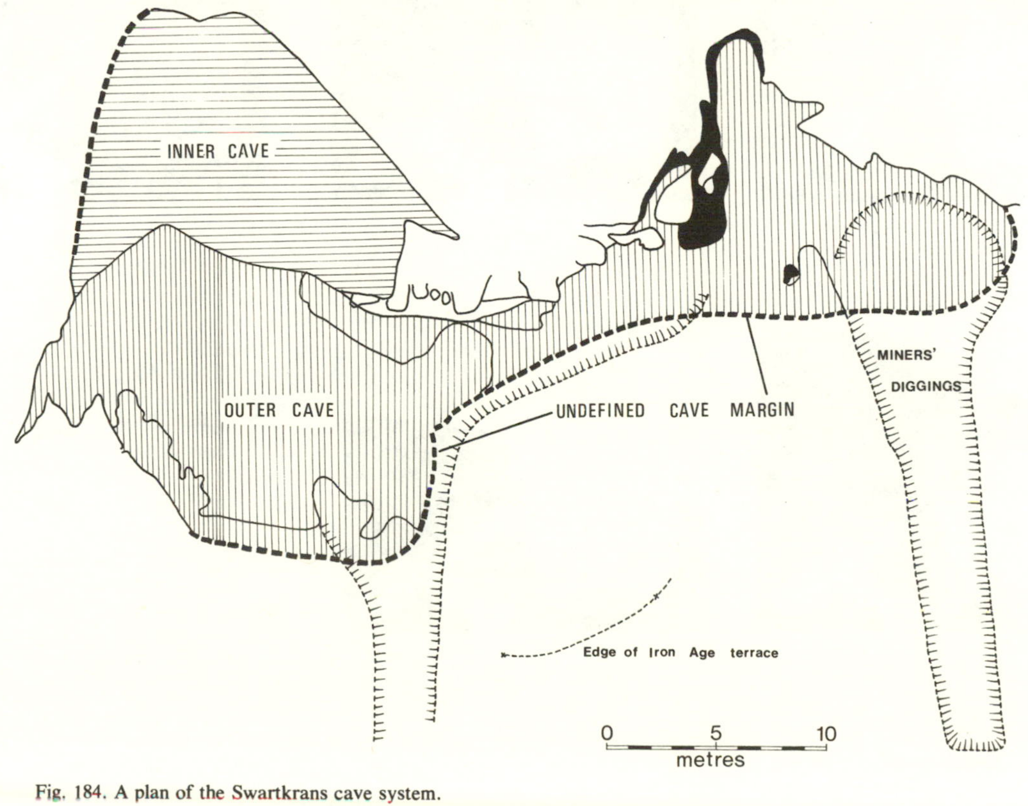

Fig. 184. A plan of the Swartkrans cave system.

Suggested Stages in the Formation of the Cave System

In reading the discussion that follows, refer to the vertical sections in figure 185. These are reconstructed and somewhat idealized sections through the western part of the Outer, Inner, and Lower caves, running southeast to northwest. They do not take into account the newly discovered talus cone filling the eastern end of the Outer Cave and penetrating beneath the mass of Member 1 in the western end. This deposit is now being excavated, and results of the current study will certainly influence the scheme presented here.

Stage 1 (fig. 185)

An irregularly shaped cavern has been dissolved in the dolomite below the level of standing water.

Stage 2

The level of standing water has dropped through incision of the Bloubank River valley in the vicinity of the cave. A large piece of dolomite, destined to become the "floor block," has become detached, hingeing down from the north wall.

Stage 3

Indirect connection with the surface has led to ventilation of the cave; travertines, in the form of stalactites and a massive boss of stalagmite over the floor block, are developing. Planes of weakness in the dolomite above the

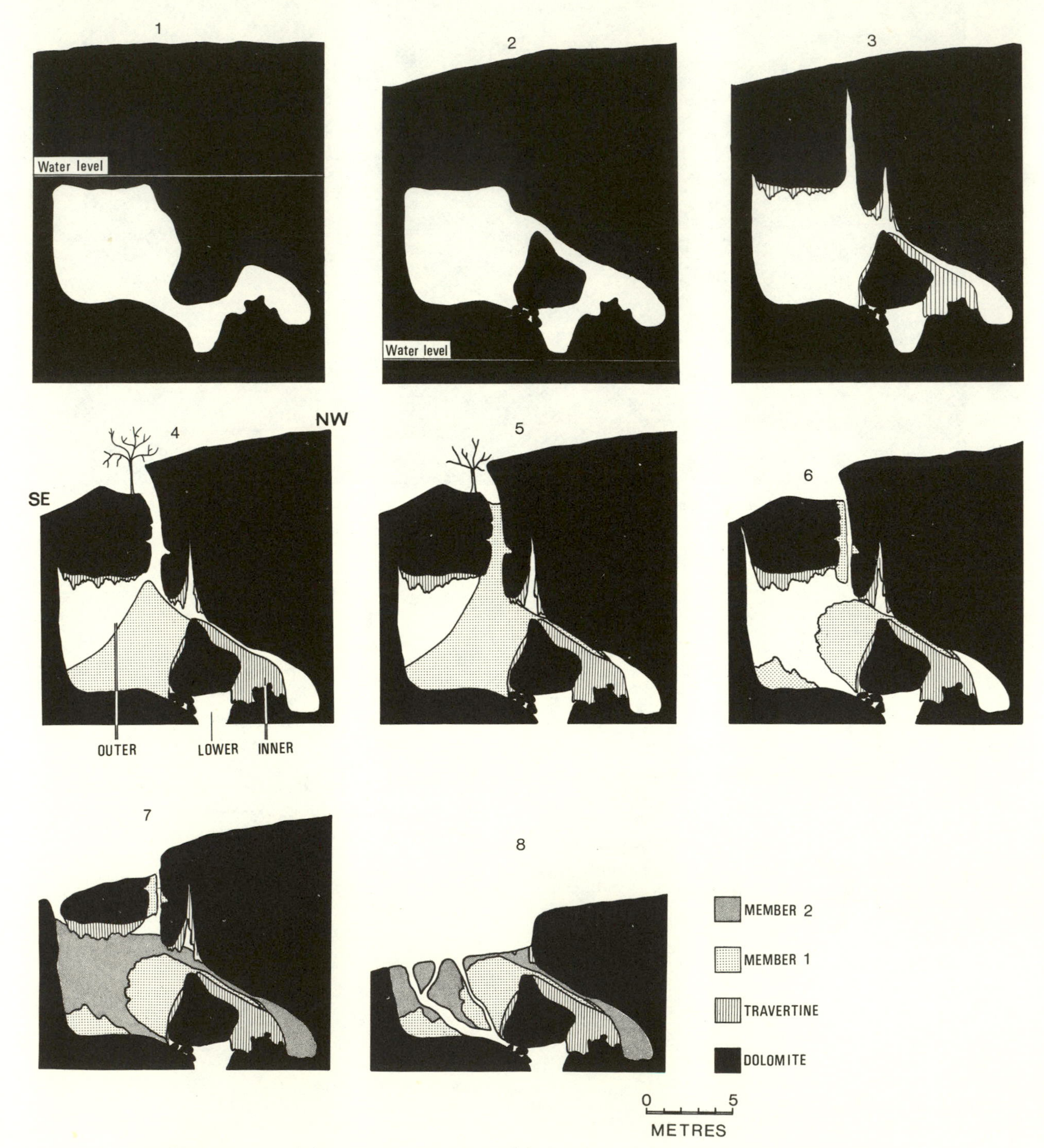

185. Reconstructed sections through the western end of the Swartkrans cave system showing suggested stages of development.

Outer Cave are being enlarged by solution, leading to the development of avens.

Stage 4

A vertical shaft has developed, linking the Outer Cave with the surface. It appears to have had suitable ledges around it that served as owl roosts. The Outer Cave has started to fill up with bone-bearing sediment destined to become Member 1 of the Swartkrans Formation.

Stage 5

Accumulation of Member 1 sediment has completely choked the shaft.

Stage 6

Over a long period, perhaps as much as a million years, the calcified Member 1 sediment was severely eroded by water passing through the cave. Progressive denudation of the hillside above the cave continued.

Stage 7

Another shaft has once again linked the Outer Cave with the surface, this time somewhere close to the cave's southern margin. Through the shaft came sediment destined to become Member 2 breccia; it infiltrated the eroded surface of Member 1 and gradually filled the remaining space in the Outer and Inner caves. The Lower Cave remained empty.

Stage 8

As time passed, the Outer Cave's roof was eroded away, exposing the cave filling on the hillside. A system of irregular channels developed through both breccia members, serving as stormwater drains from the surface to the Lower Cave and thence to the subterranean reservoirs in the dolomite. These channels were partly filled with later sediments.

Stratigraphy

In the course of an early study of Swartkrans stratigraphy (Brain 1958), researchers appreciated that breccias of two types and ages existed in the cave. These were termed the Outer Cave pink breccia and the Inner Cave stratified brown breccia. The former, being the source of the abundant australopithecine fossils, was obviously the older, and the two were separated by a clear unconformity.

More recently, on the basis of paleontological and geological evidence (Vrba 1975, 1976*a*, Brain 1976*a*), it became clear that a considerable volume of the Outer Cave filling consisted of breccia younger than the australopithecine-bearing matrix. The more recent breccia was browner than the older pink equivalent but generally lacked the layering of the "stratified brown breccia" previously described from the Inner Cave. Where contact between the two Outer Cave breccias could be observed, it was invariably unconformable and often intricate. Fortunately it has proved possible to separate the two breccias fairly easily in the hand specimen.

On the basis of his detailed sedimentological study, Butzer (1976) has formally designated the older and younger deposits as Member 1 and Member 2 of the Swartkrans Formation. Member 1 includes the basal travertine and a breccia mass that was the source of the "Member 1" fossil assemblage described in this book.

At the time of Butzer's analysis, the extensive filling in the eastern end of the Outer Cave had not been uncovered or recognized. The unexpected deposit will be described elsewhere.

Concerning Member 2, Butzer (1976) noted that the properties of the sediment differed somewhat between the Outer Cave and the Inner Cave, but he did not propose any stratigraphic subdivisions. He concluded that the Member 2 material in the Outer Cave suggested a cavern with direct access to a doline in process of active enlargement, while the Inner Cave facies of Member 2 represented a "filtered," sorted variant of the Outer Cave material, lacking both the coarsest fraction (trapped in the Outer Cave) and much of the clay (flushed out through the Lower Cave).

As I mentioned earlier, the Outer Cave filling has been traversed by a series of irregular solution channels. These make their way down from the original surface of the exposed breccia to the Lower Cave. In places the channels are completely or partially filled with sediment in various degrees of calcification, and these fillings are likely to vary in age. A volume of about 2 m^3 of unconsolidated sediment, undisturbed by lime-mining operations, has been excavated. This yielded a large number of bone fragments, details of which are given in the section on "Channel Fill," but no artifacts. The positions and inclinations of the channel system through the western end of the Outer Cave filling are shown in figure 186.

Stratigraphic Relationships of the *Homo* Remains

During the first year of paleontological work at Swartkrans, 1948–49, abundant remains of robust australopithecines were found. These were interspersed with sporadic finds of a more advanced hominine, originally designated *Telanthropus capensis* Broom and Robinson but subsequently transferred to *Homo* sp. (for particulars, refer to the section on *Homo* sp. in chap. 9).

The original "*Telanthropus*" find, made on 29 April 1949 (fig. 187), was preserved in a browner matrix that differed in appearance from the main mass of australopithecine-bearing breccia. Concerning the mandible SK 15, together with two associated teeth and proximal radius, Broom and Robinson wrote (1949*a*, p. 322):

> Though this [the mandible, SK 15] was discovered in the same cave as the large ape-man, it is clearly of considerably later date. In the main bone breccia of the cave deposit there has been a pocket excavated and refilled by a darker type of matrix. The pocket was of very limited extent, being only about 4 ft by 3 ft, and about 2 ft in thickness. The deposit was remarkably barren, there being no other bones in it except the human jaw and a few remains of very small mammals. We are thus at present unable to give an age to the deposit except to say that it must be considerably younger than the main deposit. If the main deposit is Upper Pliocene, not improbably the pocket may be Lower Pleistocene.

When Robinson (1953*b*) wrote his paper "*Telanthropus* and Its Phylogenetic Significance," he described the second specimen (maxillary fragment SK 80) from the main Outer Cave breccia and inclined to the view that all the "*Telanthropus*" specimens, including the controversial SK 15 mandible, were contemporaneous with the australopithecine fossils. Meanwhile, K. P. Oakley undertook a flourine test on bone from the SK 15 pocket, and compared the results with those on bones from the undoubted australopithecine breccia. No significant difference in flourine content was found between the two samples,

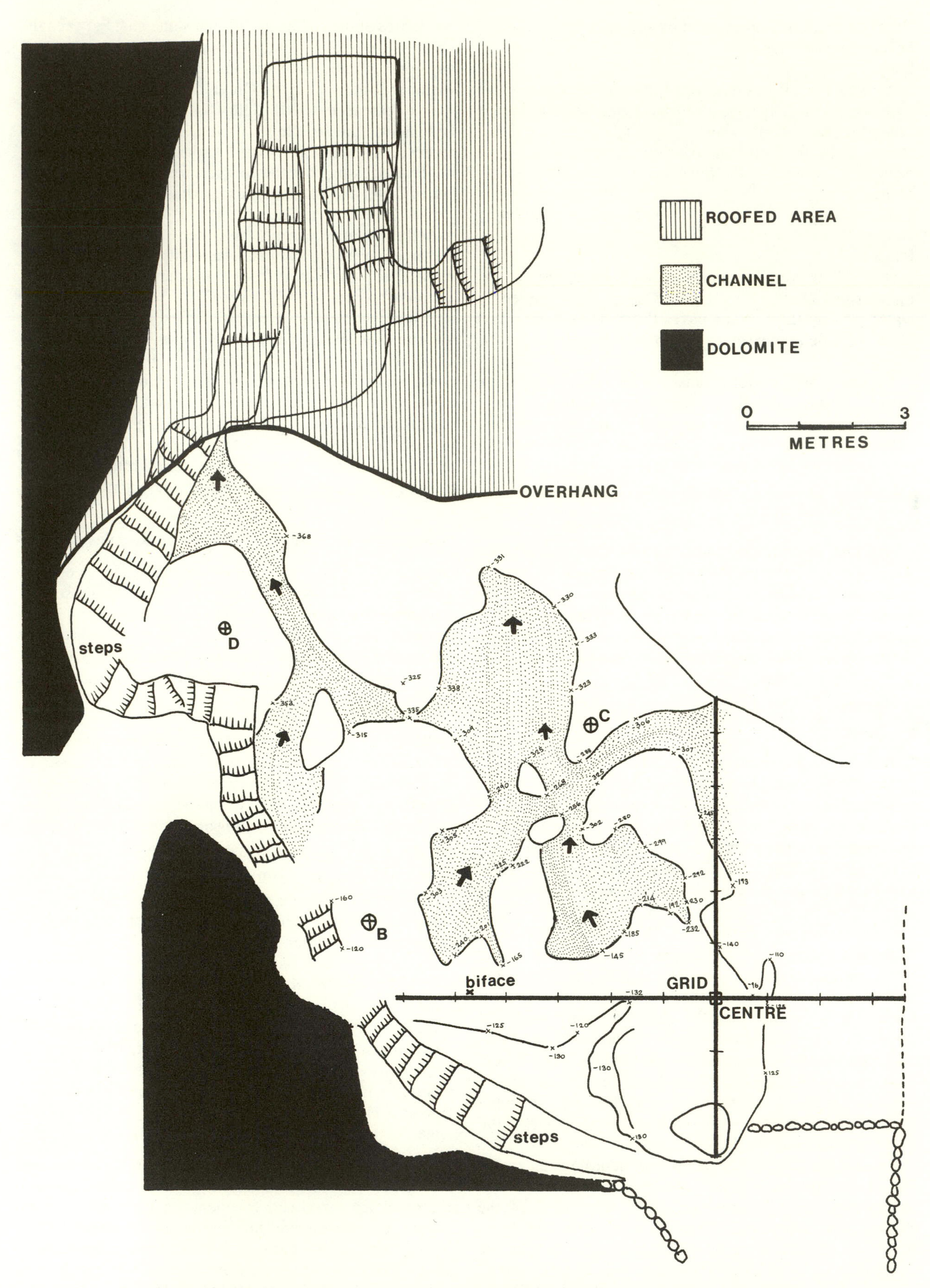

Fig. 186. Plan of the western end of the Outer Cave at Swartkrans showing the positions of the main channel systems.

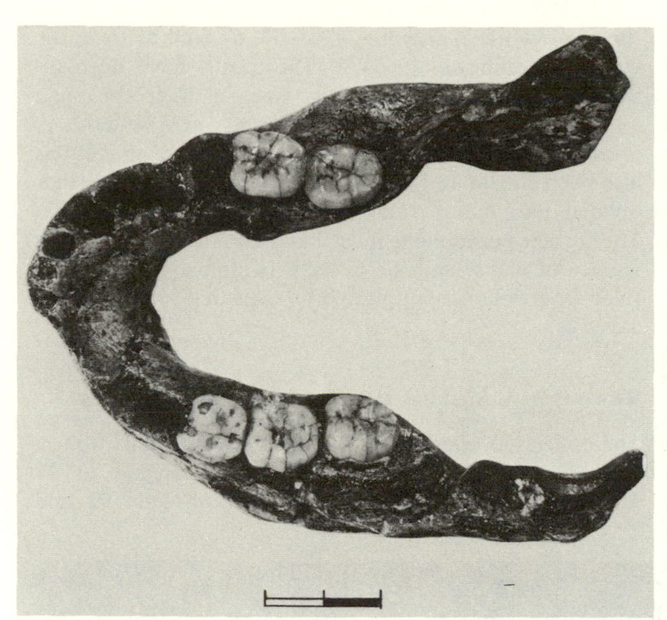

Fig. 187. The *Telanthropus* mandible, SK 15, from Member 2 breccia in the Outer Cave.

suggesting that any large age difference was unlikely provided the flourine method could reliably be applied in calcified sediments, which has unfortunately not yet been established. Oakley (1954*a,b*) was of the opinion that the "*Telanthropus*" and robust australopithecine remains from Swartkrans were probably contemporaneous.

Another piece of evidence that contributed to Robinson's (1953*b*) suggestion that the SK 15 pocket was of the same age as the main deposit came from sediment studies undertaken at the time (Brain 1958). Grading of the SK 15 pocket sediment was found to be broadly similar to that of the adjoining breccia mass, although the carbonate content was found to be abnormally low.

When the significance of the calcified channel fills in the Swartkrans Outer Cave was first appreciated in 1974 (Brain 1977), I speculated that the SK 15 pocket could well have represented part of such a filled channel system. Shortly thereafter, however, I reached the conclusion (Brain 1976*a*) that the Outer Cave did in fact contain a large volume of brown Member 2 breccia that, in the contact zone, had infiltrated the eroded surface of the Member 1 mass.

It is unfortunate that the "*Telanthropus*" pocket no longer exists and cannot be repositioned in the cave stratigraphy as we now understand it. My current opinion, however, is that the SK 15 mandible and associated fragments should be allocated to Member 2, while the other remains of *Homo* sp.—that is, the mandible piece SK 45, the composite cranium SK 846*b*/847/80, the juvenile cranium SK 27, and upper premolar SK 2635—are undoubtedly derived from the later component of Member 1.

Age Relationships of the Swartkrans Members

Repeated attempts to obtain absolute dates for the Swartkrans deposits and fossils have so far been fruitless. Fairly recently a series of samples was collected for paleomagnetic evaluation, but the results proved ambiguous (Brock, McFadden, and Partridge 1977). An attempt by Partridge (1973) to date the first opening of each of the australopithecine caves by measurements of cyclic nickpoint migration and valley flank regression gave a figure of 2.57 million years for the first opening of the Swartkrans cavern. Although this figure is interesting, there is a good deal of disagreement among specialists on whether a geomorphological method of this kind is reliable.

On the basis of her study of fossil bovids, Vrba (1975, 1976*a*) suggested that remains from Member 1 (the later component of Member 1 as we now know) could be referred to the Swartkrans faunal span with an age indication of between 1 and 2 million years. On bovid faunal grounds Vrba also divided the Member 2 assemblage into older and younger groups termed "earlier and later SKB" and spanning, in all probability, part of the past half million years. In the present study I make no attempt to separate the Member 2 fossil assemblage into older and younger components. I would have no confidence in doing so, since very few of the fossils have come from controlled excavations, and stratigraphic information is therefore lacking. The stratigraphic relation of the Outer Cave breccias is exceedingly complex owing to periodic erosion of earlier breccias *within* the cave, followed by infilling of the irregularly eroded surface by later sediment. Examples of the result of this process may be seen in figure 189. In *a*, a flattened australopithecine skull is shown that had much of its braincase removed by erosion after being embedded in Member 1 sediment. The space was then filled with browner Member 2 material. In *b* the intricate relationship between Member 1 and Member 2 breccias can be seen.

Evidence of Stone Culture at Swartkrans

Site clearing during the late 1960s resulted in the discovery of a series of stone artifacts embedded in breccia blocks that had been blasted from the site by lime miners. Although the stratigraphic origin of none of these was known in detail, it was certain that they all came from the Outer Cave, which at that time was thought to contain a single breccia mass. The available sample of 30 artifacts was described by M. D. Leakey (1970) on the assumption that they all came from breccia of robust australopithecine age. As discussed above, the initial assumption that all Outer Cave breccia was of one depositional phase and age proved incorrect (Brain 1976*a*). Both Member 1 and Member 2 were, in fact, well represented in the Outer Cave, and it was therefore necessary to try to assign the artifacts to the members in which they originally lay. This was a difficult and unsatisfactory task, since in some cases very little or no adherent breccia had been left on the tools.

From the characteristics of the matrix, and considering the artifacts described in detail by Leakey, we may conclude that only one tool in this collection is definitely from Member 1 breccia (fig. 190*a*). This was classified as a "heavy-duty scraper" and described as follows: "a quartzite flake, 84 mm long and 88 mm wide, struck from a cobblestone. Approximately 50 percent of the circumference has been steeply but irregularly trimmed. The edge is chipped and blunted by use on the dorsal aspect and there is also a crushed notch, 12 mm wide and approximately 4mm deep." The specimen shows a certain amount of general abrasion as if it had been transported in the streambed some distance before it entered the cave.

Six other specimens in the original collection could have originated in Member 1 breccia, though the quantity of adhering matrix is too small to allow certain allocation. These include a second "heavy-duty scraper," SK 3960, a "bifacial side chopper," SK 3967, and three unnamed tools, SK 3946, 3961, and 7878. All the remaining tools, including the well-made diabase cleaver SK 3962 (fig. 190*b*), appear to have come from Member 2 breccia.

The position regarding the occurrence of stone culture at Swartkrans has clearly been most unsatisfactory, though it is improving now as a result of carefully controlled excavation.

The Swartkrans Bone Accumulations

The analysis presented here has been done on the fossil collections in the Transvaal Museum that resulted from the early work at the site, 1948–53, as well as from the current operations, 1965–75. The fossils have been divided into two groups, on the basis of their enclosing matrix, as to their origin in Member 1 or Member 2. A third group from an excavated channel fill (see above) has also been considered. The bone accumulations are further divided into macro- and microvertebrate components. The former component must have entered the cave in a variety of ways, and the latter was almost certainly derived from pellets regurgitated by owls that roosted in the cave.

Remains from Member 1: Macrovertebrate Component

The collection consists of 2,381 individual fossils from a minimum of 339 individual animals belonging to 41 identified taxa. Details are provided in table 92 and

Fig. 188. A face in the Outer Cave filling showing a solution channel traversing the older Member 1 breccia. The channel is partly filled with younger Member 2 sediment.

particulars of 644 bovid bones, assigned to size classes but not specifically identified, are given in table 93. The various identified animals are depicted in figure 191.

Details of the fossil material by which the various animal taxa are represented will now be presented.

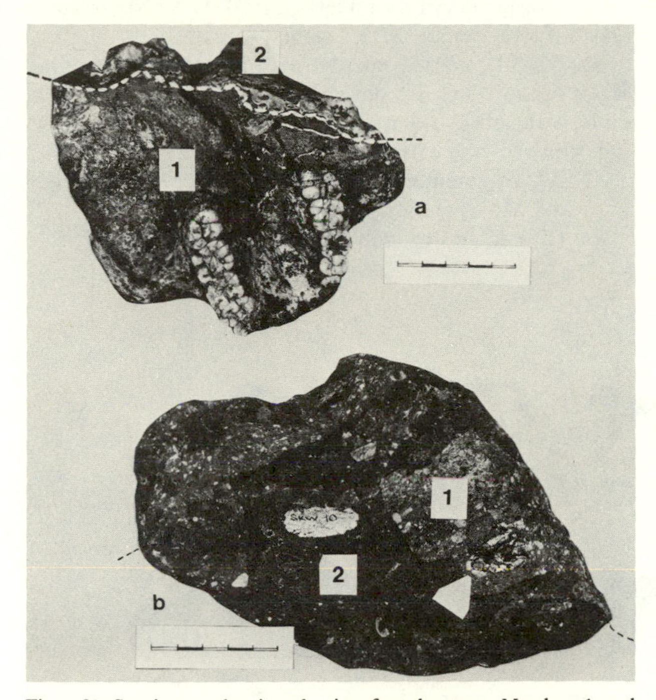

Fig. 189. Specimens showing the interface between Member 1 and Member 2 breccias. *(a)* A flattened australopithecine cranium, SK 79, which had been embedded in Member 1 sediment. Part of the specimen was then eroded away and the space filled with browner Member 2 material. *(b)* A piece of breccia showing the intricate form of the Member 1/Member 2 interface. The surface of the Member 1 breccia was eroded before infilling with Member 2 sediment occurred. A hominid tooth, SKW 10, is preserved in the older breccia and is indicated by the arrow.

Fig. 190. *(a)* The only artifact positively known to have come from Member 1 breccia: a quartzite flake, SK 3963. *(b)* A diabase cleaver and *(c)* biface from Member 2.

Class Mammalia
Order Primates
Family Hominidae
Homo sp.
Material:

SK 80 + 846*b* + 847 + 45 (fig. 192), maxillary piece from an old individual with heavily worn left I^2 and part of P^3; 2 isolated teeth (SK 80); part of left temporal with partially preserved auditory meatus, most of mastoid process, and the petrous portion, SK 846*b;* cranial pieces with left M^3, SK 847; right mandible pieces with M_{1-2} and alveolus M_3.

SK 27, flattened juvenile cranium with left I^2, dm^2, P^4, unerupted, M^1; right M^1, unerupted $\underline{C}$, P^3, and M^2 removed; SK 2635, isolated left P^{3-4}, showing strong lingual wear.

Australopithecus robustus
Material:
Craniums in varying degrees of completeness

SK 46, cranium from an adult, probably female, severely crushed; left half of braincase, part of face, palate, and maxilla with left P^4, M^{1-3}; right P^{3-4}, M^{1-3}. Age estimate, 34 ± 3 years.

SK 47, partial cranium from an adolescent probably female; most of base and palate well preserved, with left M^{1-2}, right P^4 (unerupted), M^{1-2} and M^3 erupting. Age estimate, 13 ± 2 years.

SK 48 + SKW 7 (fig. 144), reasonably complete cranium from an adult, probably female; palate with left P^4 (roots), M^{1-3}; right $\underline{C}$, P^3, P^4 (roots), M^{1-3} (roots). SKW 7 consists of crowns of right P^4, M^{1-2}. Age estimate, 20 ± 1 years.

SK 49, badly crushed cranium of an adult, with parieto-occipital piece separate; left and right P^{3-4}, M^{1-3}. Age estimate, 19 ± 2 years.

SK 52, cranium of an adolescent, probably male, with lower part of the face, part of the right side and cranial base; left I^2, P^{3-4}, M^1; right I^2, P^{3-4}, M^1, and unerupted M^3. Age estimate, 16 ± 1 years.

SK 54, part of a juvenile calotte with two depressed fractures near lambda (fig. 197*a*). Age estimate, immature.

SK 79, anterior part of an adult cranium with much of the face and palate; left P^{3-4}, M^{1-3}; right P^{3-4}, M^{1-3} (fig. 189*a*). Age estimate, 32 ± 2 years.

SK 83, most of a badly damaged adult cranium; right I^{1-2} (roots), M^3; left $\underline{C}$, P^{3-4}, M^{1-3}. Age estimate, 32 ± 2 years.

SK 821, part of a face, including left P^3 somewhat worn. Age estimate, 26 ± 2 years.

SK848, fragment of the auditory region of a cranium, showing the extermal auditory aperture and glenoid fossa. Age estimate, mature.

SK 859, most of the occiput and parts of the parietals of a young juvenile. Age estimate, immature.

SK 878, small piece of calotte with tooth fragments. Age estimate, mature.

SK 1585, approximately right half of an endocranial cast. Age estimate, indeterminate.

SK 1590, fragments of palate with right I^2, $\underline{C}$, P^{3-4}, M^1; crushed and distorted mandible in very poor condition; femoral head and parts of an innominate. Age estimate, 27 ± 2 years.

SK 14003, part of a crushed skull with left M^1, right M^1 (part), M^{2-3}. Age estimate, 32 ± 2 years.

SK 826*a* + 877 +SKW 31 + SK 843 + 846*a*, maxillary fragment with very worn and broken left P^{3-4}, M^{1-2}, (826*a*); part of a maxilla with right P^4, M^{1-2} (SK 877); part of left M^2 in maxillary fragment (SKW 31); left corpus of mandible with slightly worn M_{1-2} and M_3 unerupted (SK 843); isolated right M_1 (SK 846*a*). Age estimate, 14 ± 2 years.

Maxillae

SK 11 + 10, part of an adult face with left P^{3-4}, M^{1-3}, right P^{3-4}, M^{1-2} (SK 11); left mandible piece with parts of M_{2-3} (SK 10). Age estimates, 27 ± 2–31 ± 2 years.

SK 12*a,b,* associated maxilla and mandible of an elderly male; maxilla with left I^2 (root), P^3 (roots), P^4, and M^1 (roots); right P^{3-4}, M^{1-2}; mandible with left and right P_3 (roots), P_4, M_{1-3}. Age estimate, 33 ± 3 years.

SK 13–14, part of an adolescent face with left P^{3-4}, and M^1, and M^3 (unerupted); right P^{3-4}, M^{1-2}, M^3 (unerupted) (SK 13); isolated left M^2 with pronounced dental caries (SK 14). Age estimate, adolescent.

SK 21–21*a*, maxilla with left $\underline{C}$, P^{3-4}, M^{1-2}; M^3 separate (21*a*). Age estimate, 37 ± 2 years.

SK 55*a,b,* juvenile maxilla with left I^{1-2}, $\underline{C}$, P^3, dm^2, M^{1-2}; right I^{1-2}, $\underline{C}$, P^3, dm^2 (55*a*); part of associated mandible with left P_3, dm_2, M_{1-2}; right dm_2, M_{1-2}, and M_3 (unerupted). Age estimate 13 ± 2 years.

SK 57, fragmentary palate with left P^{3-4}, M^{1-3}; right P^{3-4}, and M^1. Age estimate, 25 ± 2 years.

SK 65 + 67 + 74*c*, adult maxilla with left I^{1-2}, $\underline{C}$, P^{3-4}

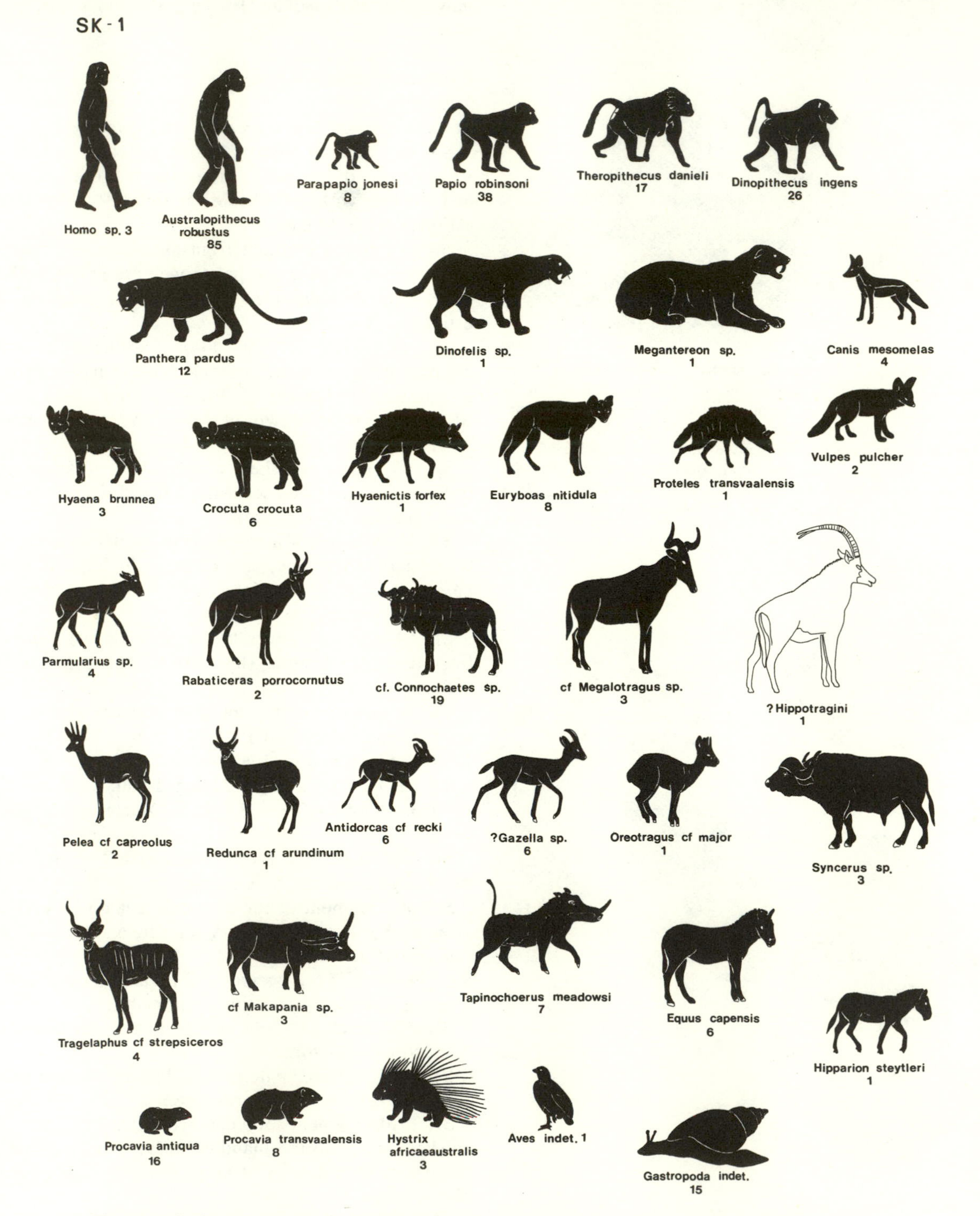

Fig. 191. Animals represented by the fossils in the macrovertebrate component of the Swartkrans Member 1 sample. Minimum numbers of individuals are indicated.

(SK 65); isolated right $\underline{C}$ (SK 65*a*); isolated right I^1 (SK 67); isolated right P^4 (SK 74*c*). Age estimate, 26 ± 2 years.

SK 66, juvenile maxilla with roots of right $d\underline{c}$, dm^{1-2}, incompletely developed crowns of I^{1-2}. Age estimate, 4½ ± 1 years.

SK 831*a*, adult maxilla with left M^{2-3} (not related to SK 831). Age estimate 28 ± 2 years.

SK 838*a* + 102, part of a juvenile maxilla with right dm^2, M^1 (SK 838*a*); crown of left M^1 probably associated (SK 102). Age estimate 8 ± 2 years.

SK 839 + 852, part of a juvenile palate with damaged left and right $d\underline{i}$, dm^2, and M^1; probably associated with juvenile mandible in poor condition, left dm_{1-2} and right dm_1 (SK 852). Age estimate, 5 ± 1 years.

SK 845, part of an adult face, crushed, with left I^2, $\underline{C}$, P^{3-4}, M^{1-2}; right P^{3-4}, M^1. Age estimate, adult.

SK 881 + 882, part of a maxilla with left P^{3-4}, M^1; probably associated with part of a left upper molar (SK 882). Age estimate, 26 ± 2 years.

SK 1512, part of a palate with tooth roots and premolar fragment. Age estimate, mature.

SK 1592, part of a palate with broken right P^4, M^{1-3}. Age estimate, 32 ± 3 years.

SK 1595, part of a juvenile maxilla with left dm^2 and pieces of erupting teeth. Age estimate, 4 ± 1 years.

SKW 11, maxilla, in situ against west wall of cave; unprepared.

SKW 12, part of a maxilla with left P^{3-4}, M^{1-3}. Age not estimated.

SKW 33, part of a maxilla with right M^{1-2} (previously cataloged as SK 14129*a*). Age estimate, 14 ± 2 years.

Mandibles

SK 6 + 100, the holotype of *Paranthropus crassidens* (fig. 193), part of a mandible with left P_{3-4}, M_{1-2}, and unerupted M_3; right P_4 (abnormal), M_{1-2} and M_3 (unerupted) (SK 6); isolated right P_3 (SK 100). Age estimate, 15 ± 1 years.

SK 23, mandible of adult, probably female, with full dentition in excellent condition (fig. 194). Age estimate, 19 ± 2 years.

SK 25 + 832, part of a juvenile mandible with left P_4 (erupting), M_{1-2}; right P_4, M_{1-2} (SK 25); probably associated with crown of left M^1 (SK 832). Age estimate, 9 ± 3 years.

SK 34, almost complete mandible of a large adult, probably male, in two halves, left P_4, M_{1-3}; right I_{1-2}, $\bar{C}$, P_{3-4}, M_{1-3}. Age estimate, 23 ± 3 years.

SK 37, part of a mandible with left M_2 and small piece of M_1. Age estimate, 12 ± 1 years.

SK 61, part of a juvenile mandible with left and right di_{1-2}, $d\bar{c}$, dm_{1-2} and right M_1 erupting. Age estimate, 6 ± 2 years.

SK 62, part of a juvenile mandible with left di_2, $d\bar{c}$, dm_{1-2}, and M_1 unerupted; right I_1, $d\bar{c}$, dm_{1-2}. Age estimate, 6 ± 1 years.

SK 63 + 89*a* + 90 + 91, almost complete juvenile mandible with left and right dc, dm_{1-2}, M_1, and M_2 (unerupted, right unerupted crowns of I_1 and $\bar{C}$ (SK 63); unworn crowns of left and right M^1 (SK 89*a*); left dm^2 (SK 90); right dm^1 (SK 91). Age estimate, 7 ± 1 years.

SK 64, part of a juvenile or infant mandible with right dm_{1-2}. Age estimate, 2½ ± ½ years.

SK 74*a*, part of an adult mandible, showing possible mental eminence, with left P_{3-4}, M_{1-2}; right I_2, P_{3-4}, M_{1-2}. Age estimate, 28 ± 3 years.

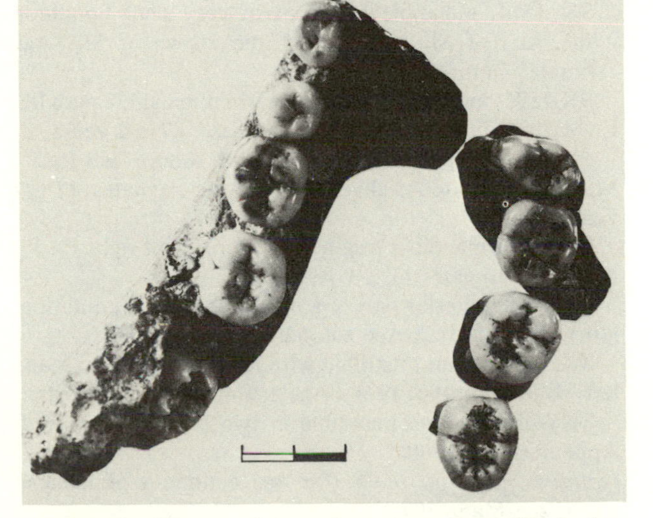

Fig. 193. The holotype of *Paranthropus crassidens*, SK 6, from Swartkrans Member 1.

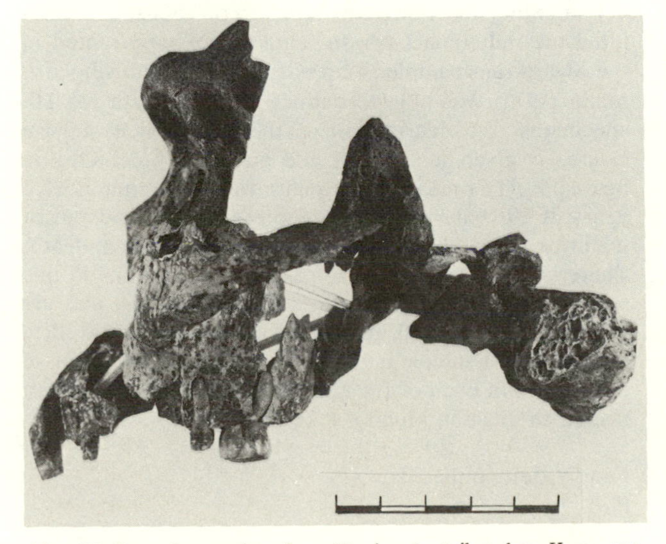

Fig. 192. Part of a cranium from Member 1 attributed to *Homo* sp.: SK 80 + 846*b* + 847 + 45.

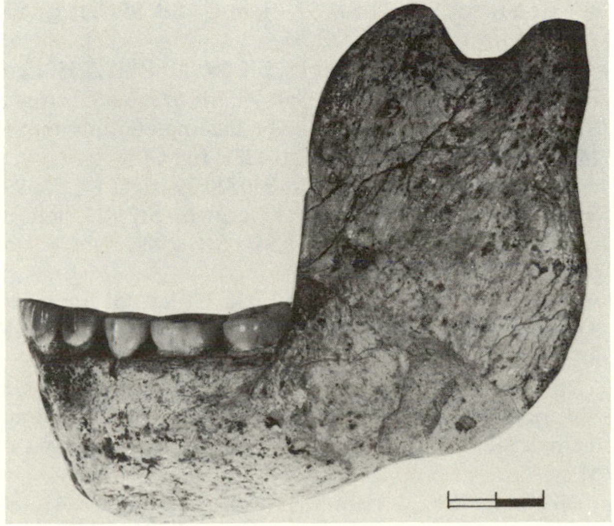

Fig. 194. A remarkably complete and well-preserved mandible of *Australopithecus robustus*, SK 23, from Member 1.

SK 81, part of a mandible in poor condition with left P_{3-4}, M_{1-3}; right I_2, $\bar{C}$, P_{3-4}, M_{1-3}. Age estimate, 26 ± 1 years.

SK 96, part of a juvenile mandible with roots of left dm_1, unerupted left $\bar{C}$ and P_3. Age estimate, 6 ± 1 years.

SK 438, part of a juvenile mandible with left unerupted dm_2. Age estimate, 2 ± ½ years.

SK 841*a*, mandible fragment with left dm_2 and part of M_1 (not related to 841*b*). Age estimate, 2 ± ½ years.

SK 842 + 869, part of a juvenile mandible with roots of dm_1 (SK 869); left dm_2 (SK 842). Age estimate, 2½ ± ½ years.

SK 844, part of an adult mandible in poor condition, left M_{2-3}. Age estimate, 25 ± 2 years.

SK 862, part of a mandible with roots of right P_4 and M_1, M_{2-3}. Age estimate, adult.

SK 876, part of a crushed mandible in poor condition, left $\bar{C}$ root, P_{3-4}, M_{1-3}; right I_2, $\bar{C}$–M_3. Age estimate, 28 ± 4 years.

SK 858 + 861 + 883, part of an adult mandible with left I_{1-2}, $\bar{C}$, P_{3-4}, M_{1-2}; right I_{1-2}, $\bar{C}$, P_{3-4}, M_{1-3} (SK 858); mandible fragment with parts of M_{1-2} (SK 861); part of right M_3 and angle of mandible. Age estimate, 19 ± 2 years.

SK 1514, mandible fragment in very poor condition with roots of M_1 and part of the crown of M_2. Age estimate, mature.

SK 1586, most of a poorly preserved mandible with left I_1, M_{1-3}; right I_{1-2}, M_{1-3}. Age estimate, 27 ± 2 years.

SK 1587*a,b*, part of a mandible with roots of left P_3, P_4, M_{1-2} *(a);* isolated right M_2 *(b)*. Age estimate, 17 ± 1 years.

SK 1588, mandible fragment with roots of right P_3, P_4, M_1. Age estimate, 13 ± 1 years.

SK 1648, anterior part of a mandible in poor condition, left M_1; right M_{1-2}. Age estimate, 29 ± 2 years.

SK 3978, infant mandible with left and right dm_{1-2} and left M_1 in crypt (fig. 197*b*). Age estimate, 2½ ± ½ years.

SKW 5*a,b*, adult mandible in two parts, unprepared. Appears to be adult.

Isolated maxillary teeth (for age estimates of isolated teeth, see Mann 1975)

Incisors: SK 68, left I^1; SK 69 and 73, left I^1 and right I^1; SK 2, right I^1; SK 40, SK 3, and SK 4, left I^1, right I^1, and right $\underline{C}$; SK 70, left I^2; SK 71, right I^2.

Canines: SK 38, right $\underline{C}$; SK 85*a* and SK 93, right $\underline{C}$ and left $\underline{C}$; SK 86, left $\underline{C}$; SK 92, right $\underline{C}$; SK 95, left $\underline{C}$; SK 1596, $\underline{C}$.

3d premolars: SK 24, left P^3; SK 28, left P^3; SK 33, right P^3; SK 44, right P^3; SK 101, left P^3; SK 822, left P^3 crown; SK 823, right P^3; SK 867, P^3, incomplete crown; SK 14001, left P^3; SKW 32 (SK 14128), right P^3.

4th premolars: SK 32, right P^4; SK 39, right P^4; SK 99, left P^4 crown; SK 824, left P^4 crown; SK 825, left P^4 crown; SK 856, right P^4 and M^1; SK 1589, right P^4 unerupted crown.

1st molars: SK 17, right M^1; SK 35, left M^1 fragment; SK 829, left M^1; SK 833, left M^1; SK 849, right M^1; SK 872, left M^1; SK 1591 and SK 16, left M^1 and M^2.

2d molars: SK 42, right M^2; SK 98, left M^2 crown; SK 834, right M^2; SK 837, right M^2; SK 868, right M^{2-3} fragments; SKW 33 (SK 14129*a*), right M^2 and fragment of M^1.

3d molars: SK 31, right M^3; SK 36, right M^3; SK 41, left M^3; SK 105, left M^3 crown; SK 835, left M^3; SK 836, left M^3; SK 870, left M^3; SK 1524, left M^3, unerupted crown; SK 3975, left M^3; SK 3977, right M^3.

Isolated mandibular teeth

Incisors: SK 74*b*, right I_1.

Canines: SK 29, right $\bar{C}$; SK 87, right $\bar{C}$ crown; SK 94, right $\bar{C}$; SK 884, left $\bar{C}$ and fragment of mastoid.

3d premolars: SK 30, left P_3; SK 72, left P_3; SK 831, right P_3, not related to 831*a;* SK 850, right P_3; SK 857, right P_3; SK 1593, right P_3.

4th premolars: SK 7, right P_4; SK 9, left P_4; SK 88, left P_4; SK 826*b*, left P_4, not related to 826*a;* SK 827, left P_4; SK 830, left P_4.

1st molars: SK 20, left M_1; SK 104, right M_1, unerupted crown; SK 828, left M_1; SK 838*b*, left M_1, not related to 838*a;* SK 864, probably M_1, severely damaged; SK 1594, right M_1, buccal half; SK 3974, right M_1 crown.

2d molars: SK 1, left M_2, complete; SK 5, left M_2; SK 3976, left M_2, complete.

3d molars: SK 19, right M_3, probably; SK 22, right M_3; SK 75, right M_3, unerupted crown; SK 840, left M_3; SK 841*b*, left M_3; SK 851, right M_3; SK 855, right M_3; SK 871, left M_3, unerupted crown; SK 880, left M_3, associated with two tooth fragments; SK 885, left M_3.

Isolated tooth fragments

SK 863, 865, 866, 873, 874, 875, 879, 14000, 14080; SKW 6, 8*a,b* 10, 14, 15, 30.

Postcranial bones

SK 50, pelvis bone from adult, probably male; most of right innominate, pubis and iliac crest missing. Age estimate, adult.

SK 3155*b*, right ilium with complete acetabulum. Age estimate, mature.

SK 82, proximal end of right femur from a mature adult. Age estimate, mature.

SK 97, proximal end of right femur from a mature adult. Age estimate, mature.

SK 860, distal end of right humerus (considered by some to be cercopithecoid). Age estimate, mature.

SK 853, isolated lumbar vertebra. Age estimate, immature.

SK 854, axis vertebra. Age estimate, mature.

SK 3981*a,b*, last thoracic vertebra *(a);* last lumbar vertebra, *(b)*. Age estimate, adult.

SK 84, 1st left metacarpal. Age estimate, mature.

SK 85, part of 4th left metacarpal. Age estimate, mature.

SKW 27 (SK 14147), adult 5th left metacarpal. Age not estimated.

Estimating the minimum number of robust australopithecine individuals whose remains are represented in the Swartkrans sample is not easy, and figures will vary. Mann (1975) was able to deduce ages at death for 113 specimens; the distribution of these specimens in age classes is given in table 94 and graphically depicted in figure 195. The mean age at death for the sample is 17.2 years; if 20 years is taken as representing the advent of maturity, then 68% of the sample was immature at death. There is no question that some of the specimens in this sample of 112 came from the same individuals, and my estimate of the actual number of individuals involved in the Member 1 sample is 87. However, the proportions of individuals in each of the age classes was probably very similar to those in Mann's sample of 113.

Family Cercopithecidae

Parapapio jonesi

Material:

29 cranial pieces from a minimum of 8 individuals.

♂ palate with I^1–$\underline{C}$ bilaterally, SK 588*a;* crushed calvaria and face with right M^3, SK 2127; braincase with part of left M^3, SK 556; maxilla with left I^1–M^2 and right I^1, SK 573*b;* left maxillae, SK 543 and unnumbered; right maxillae, SK 442, 462, 14151; ♀ mandible with full dentition, SK 573*a;* ♀ mandible with P_3–M_3 bilaterally, SK 414; mandible with left dm_{1-2}, M_{1-2}, and right dc, dm_2, and part of M_1, SK 418; left mandible pieces, SK 433, 437; isolated associated teeth, left $\bar{C}$, P_4, M_1, and right P_4, SK 588*b;* isolated teeth, SK 468, 493, 501, 502, 505, 511, 512, 533, 601*b*, 612, 635, 1835, 14164.

The 8 individuals have been placed in age classes (Freedman and Brain 1977) as indicated in table 95. The sample contains 2 juveniles only.

Papio robinsoni
Material:
121 cranial pieces from a minimum of 38 individuals.

The type specimen, consisting of the snout and parts of the orbits of a young adult ♂, with left $\underline{C}$, P^4–M^3, and right P^4–M^3, SK 555; almost complete ♀ skull, SK 558; left half of a skull with P^3–M^3, SK 557; complete muzzle and partial braincase, SK 560; complete muzzle with P^3–M^3 bilaterally, SK 562; anterior part of muzzle and palate, SK 565; palate with M^{1-3} bilaterally, SK 566; flattened cranium, SK 590; shattered skull, SK 602; left side of cranium, SK 2202; parts of a fragmentary cranium, SK 14163; maxillary pieces, SK 436, 439, 447, 449, 456, 458, 466, 469, 476, 497, 536, 537, 538, 540, 544, 549, 552, 571*b*, 608, 614, 629, 631, 2151, 2164, 2171, 2180; mandible pieces, SK 406, 407, 408, 409, 410, 416, 417, 419, 420, 421, 423, 425, 427, 429, 430, 431, 435, 445, 446, 453, 457, 459, 460, 463, 464, 550, 568, 570, 571*a*, 572, 596, 609, 610, 615, 617, 622, 2161, 2175, 2319, 2694, 2924, 3211*b*, 14083, 14152, 14156; isolated teeth, SK 465, 469, 475, 477, 478, 482, 483, 484, 485, 486, 499, 500, 508, 509, 516, 518, 520, 522, 529, 530, 531, 534, 582, 587, 595, 598, 611, 613, 619, 620, 621, 623, 625, 626, 2146, 2187, 14153.

The various individuals have been placed in age classes (Freedman and Brain 1977) as indicated in table 96. The sample contains 4 juveniles.

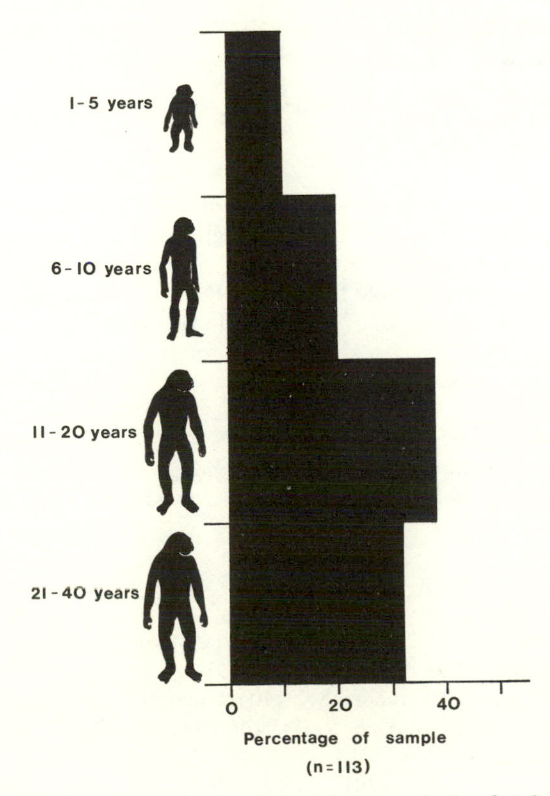

Fig. 195. Histograms showing estimated ages-at-death of 113 australopithecine specimens from Swartkrans Member 1. Data from Mann (1975).

Theropithecus danieli
Material:
31 cranial pieces from a minimum of 17 adult individuals.

The type specimen, consisting of a ♀ muzzle with P^3–M^3 bilaterally, left mandible piece with $\bar{C}$–M_3, and right mandible piece with P_3–M_3, SK 563 + 402 + 405; part of a shattered cranium with left I^1–M^3, right I^1, $\underline{C}$–M^3, SK 561; maxillary pieces, SK 448, 461, 464, 567, 575*b,c,* 593, 597, 1607, 2148, 2181, 2193; mandible pieces, SK 403, 411, 426, 432, 491, 530, 569, 575*a*, 2177; isolated teeth, SK 479, 495, 507, 521, 581, 2158, 2172, 3529, 14162.

The various individuals have been placed in age classes (Freedman and Brain 1977) as indicated in table 97. No juveniles are present in the sample.

Dinopithecus ingens
Material:
57 cranial pieces from a minimum of 26 individuals.

Complete skull, ♀, with deciduous and permanent dentition, SK 554; ♀ skull with P^4 and M^3 bilaterally, SK 600; crushed ♀ skull with P^4–M^2 bilaterally, SK 603; right half of a skull with P^4–M^3, SK 553; crushed skull without teeth, SK 599; maxillary pieces, SK 440, 441, 443, 542*a*, 545, 546, 548, 571*b*, 574, 576*a*, 578*a,b*, 604*a*, 630, 1518, 14004; mandible pieces, SK 401, 404, 413, 415, 422, 424, 428, 455, 470, 589, 2407; isolated teeth, SK 473, 474, 487, 489, 492, 498, 503, 510, 513, 514, 526, 527, 528, 532, 577, 585, 586, 618, 628*a,b*, 636, 1583, 2163, 2169, 2625, 14157, 14159.

The various individuals have been placed in age classes (Freedman and Brain 1977) as indicated in table 98. There are 4 juveniles in the sample.

Cercopithecoid indet.
Material:
134 cranial and 31 postcranial pieces from an estimated minimum of 28 individuals.

Calvaria pieces, SK 559, 634, 1584*a,b*, 1599, 2105, 2120, 2123, 2124, 2130, 2131, 2132, 2134, 2136, 2137, 2138, 2139, 2140, 2141, 2174, 2176, 2183, 2189, 2190, 2195, 2199, 2201, 2203, 2204, 2205, 2206, 2208, 2210, 2211, 2213, 2216, 2775, 2976, 3241, 3256, 3260, 3266, 3280, 3282, 3290, 3450, 3863, 3885, 5582, 7144, 7198, 7284, 14025, 14160, 14161; maxillary pieces, SK 601, 2170, 2173, 2196, 2352*b*, 2758, 2842, 3522, 7032, 7462, 9345, 14154; mandible pieces, SK 541, 594, 2125, 2128, 2143, 2152, 2159, 2160, 2165, 2179, 2182, 2188, 2221, 2222, 2427, 2710, 2858, 2891, 3187, 3258, 3272, 4019, 7333; isolated teeth and tooth fragments, SK 467, 471, 472, 480, 481, 490, 494, 504, 506, 523, 525, 584, 627, 1861, 2014, 2144, 2145, 2149, 2154, 2162, 2168, 2192, 2194, 2276, 2467, 2524, 2692, 2789, 2890, 2912, 3189, 3309, 3430, 3502, 3516, 3602, 3678, 3895, 6996, 7094, 7274, 14158; pelvis piece, SK 3434; proximal radius, SK 1815, 1816, 1867, 1870; radius shaft piece, SK 2813; proximal ulna, SK 591; proximal humerus, SK 2601, 14029; distal humeri, SK 1506, 2 unnumbered; proximal femur, SK 607, 1513, 1823, 2595, 2810, 2836, 8065, 14015, 14024; distal femur, SK 605, 1500, 1817, 3254, 5236, 2 unnumbered; proximal tibia, SK unnumbered; phalanx, SK 6759.

Order Carnivora
Family Felidae
Panthera pardus
Material:
29 cranial and 3 postcranial pieces from a minimum of 12 individuals, of which 1 was juvenile and 1 immature.

The type specimen of *P. pardus incurva,* consisting of a cranium and mandible in three parts, SK 349; parts of crushed craniums, SK 353, 354, 1932; almost complete braincase, without muzzle, SK 14010; left zygoma and cranial base, SK 14185; palates, SK 351, 352; left maxillae, SK 355, unnumbered; right maxilla with P^4, SK 5960; three parts of a mandible with left I_{2-3}, $\bar{C}$, P_{3-4}, M_1, and right P_{3-4}, SK 1866 + 343*a,b;* left mandibular pieces, SK 339, 341, 342, 344, 2797, 6892; right mandible pieces, SK 340, 345, 350, 357, 1806, 7132; isolated teeth, SK 347, 356, 358, 7088; distal humeri, SK 1639, 1899; distal radius, SK 3060; metapodial fragment, SK 2605.

Dinofelis sp.
Material:
2 cranial and 2 postcranial pieces, probably from 1 individual.

Right mandible fragment with P_4 and M_1, SK 335; isolated right $\bar{C}$, SK 372; metatarsal pieces, SK 1848, 1860.

Megantereon sp.
Material:
A single cranial piece.

Right mandible fragment with P_{3-4}, SK 337.

Family Hyaenidae
Hyaena brunnea dispar
Material:
7 cranial pieces from a minimum of 3 individuals—2 adults and 1 immature.

The type specimen consisting of a right maxilla with P^{3-4} and M^1, SK 326; left maxilla with P^4 erupting, SK 331; right maxilla with P^{2-4}, SK 332; left mandible fragment with M_1, SK 329; isolated right I^3, SK 330; right P^2, SK 333; left P^4, SK 327.

Crocuta crocuta venustula
Material:
3 cranial pieces from 2 adult individuals.

The type specimen consisting of 3 mandible pieces with right P_{2-4} and M_1, SK 317; left mandible fragment with P_2–M_1, SK 318.

Crocuta crocuta ultra
Material:
9 cranial pieces from a minimum of 4 adult individuals.

The former type specimen of *C. crocuta angella,* consisting of much of a cranium without mandible, extensively damaged on left side, SK 319; parts of a cranium and mandible, 5 pieces, SK 320; right maxilla with P^{3-4}, SK 323; left mandible fragment with P_2–M_1, SK 322 + 325; right mandible fragment with P_{2-4} amd M_1, SK 321; isolated P^3, SK 1807; left P^4, SK 324; right P^4, SK 2452.

Hyaenictis forfex
Material:
Parts of a single specimen.

The former type of *Leecyaena forfex,* consisting of a complete cranium with full dentition, muzzle crushed dorsally, SK 314; associated with left and right mandibular rami, SK 315, 316.

Euryboas nitidula, type B, advanced
Material:
9 cranial pieces from a minimum of 5 individuals.

The former type of *Lycyaena nitidula,* consisting of a right mandibular ramus with P_{3-4}, SK 301; left maxillary pieces, SK 307, 312; right mandible piece with $\bar{C}$, P_{2-4}, SK 304; isolated left P^2, SK 310, 379; right P^2, SK 311; left P^3, SK 309, 313.

Euryboas nitidula, type A, primitive
Material:
5 cranial pieces from a minimum of 2 individuals.

Right maxillary piece with P^{2-4}, SK 305; right maxillary piece with P^{3-4}, SK 306; left mandible piece with $\bar{C}$, P_{2-4}, SK 302; left mandible fragment with M_1, SK 336; right mandible piece with P_2–M_1, SK 14005.

Euryboas nitidula, type uncertain
Material:
4 cranial pieces, possibly from a single individual.

Isolated right $\underline{C}$, SK 1853; left $\bar{C}$, SK 1831; left P^4, SK 308; left P_2, SK 303.

Hyaenid indet.
Material:
16 cranial pieces from an estimated minimum of 4 individuals.

Parts of 2 crushed skulls, SK 2732, 14082; mandible pieces, SK 1840, 2300, unnumbered; isolated teeth and tooth fragments, SK 1818, 1827, 1830, 1844, 1856, 1857, 1858, 1872, 1873, 2695, 2922.

Proteles transvaalensis
Material:
2 cranial pieces, probably from a single individual.

Left maxillary fragment, SK 1851; mandible fragment with roots of 2 postcanine teeth, SK 3173.

Family Canidae
Canis mesomelas pappos
Material:
9 cranial pieces from a minimum of 4 adult individuals.

Complete cranium, SK 375; right maxillary fragment with M^1, SK 370; left mandible pieces, SK 361, 362, 14107; right mandible fragment, SK 363; isolated right $\underline{C}$, SK 1868; left P^4, SK unnumbered; M^1, SK 3284.

Vulpes pulcher
Material:
2 cranial pieces from 2 individuals.

Left mandible fragment with $\bar{C}$–M_2, SK 376; left mandible piece with alveoli of premolars and part of M_1, SK 12707.

Carnivore indet.
Material:
28 cranial and 28 postcranial pieces from an estimated minimum of 15 individuals.

Maxillary fragment, SK 14042; mandible pieces, SK 1608, 1805, 2769, 3366, 5927; isolated teeth, SK 535, 1874, 2119, 2166, 2236, 2353, 2395, 2415*b,* 2550, 2675, 2708, 2721, 2743, 2911, 2913, 3661, 3695, 3798, 6944, 7082, 14130; axis vertebra, SK unnumbered; scapula piece, SK

6643; proximal humerus, SK 1894; proximal radius, SK 3337, 10535; proximal femur, SK 606, 1802, 3269, 8014; distal tibia, SK 1879, two unnumbered; metapodial pieces, SK 1812, 1822, 1832, 1854, 1876 + 1859, 2270, 2815, 3265, 3690, 5063, 8025, 12017; phalanges, SK 1569, 1834, unnumbered.

Order Artiodactyla
Family Bovidae
Damaliscus sp. 1 or *Pamularius* sp.
Material:
8 cranial pieces from a minimum of 1 juvenile and 3 adult individuals.

Part of a palate, left M^{1-2} and other tooth fragments, SK 3832; left mandible piece, M_{1-3}, SK 3127; left mandible piece, M_{2-3}, SK 2957; left mandible fragment, M_3, SK 2064; mandible piece, dpm_4, M_{1-2}, SK 10500; right mandible piece, P_4–M_1, SK 1999; right mandible piece, P_4, M_{1-2}, SK 2000; right mandible piece, M_{2-3}, SK 3135.

Rabaticeras porrocornutus
Material:
3 cranial pieces from 2 adult individuals.

The type specimen consisting of left and right frontal pieces with incomplete horn-cores, SK 3211*a,b;* almost complete braincase with both orbits and horn-core bases, SK 14104 + 2620 + 2865.

Medium-sized alcelaphines, including *Rabaticeras* dentitions
Material:
97 cranial pieces from a minimum of 4 juveniles and 18 adults.

Left and right mandible pieces associated with horn-core pieces, SK 3213*a–d;* left juvenile maxilla, SK 2274; right juvenile maxillae, SK 1633, 3115; left adult maxillae, SK 2032, 2076, 2314, 2662, 2950, 2989, 3056, 3111, 3153, 12193; right adult maxillae, SK 1616, 1624, 2092, 2107, 2114, 2116, 2239, 2257, 2286, 2318, 2326, 2336, 2510, 2987, 3013, 3053, 3087, 3108, 3118, 3126, 3142, 3207, 14117; left juvenile mandible pieces, SK 2082, 14214; right juvenile mandible pieces, SK 2964; left adult mandible pieces, SK 1613*e*, 1656*a*, 2478, 2492, 2523, 2974, 2983, 2992, 3040, 3067, 3146; right adult mandible pieces, SK 1623, 1961, 2083, 2287, 2316, 2529, 2971, 2991, 2985, 2996, 3002, 3004, 3043, 3046, 3089, 3125, 3141, 3151; isolated deciduous teeth, SK 3498, 12633; isolated permanent teeth, SK 1991, 2006, 2048, 2049, 2068, 2232, 2269, 2278, 2296, 2302, 2364, 2406, 2426, 2438, 2448, 2457, 2526, 2527, 3050, 3081, 5941, 14056, 14124.

Cf. *Connochaetes* sp.
Material:
70 cranial and 4 postcranial pieces from a minimum of 6 juvenile and 13 adult individuals.

Cranial piece associated with a vertebral fragment, left astragalus, calcaneus, and left naviculocuboid attached to piece of proximal metatarsal, SK 3812*a–f;* fragmentary palate with left and right M^{2-3}, SK 2224; left and right mandible pieces with left M_{1-3}, right M_{2-3}, SK 3100; left juvenile maxillae, SK 3229, 14207; right juvenile maxillae, SK 2066, 5991; left adult maxillae, SK 1634, 2966, 3066, 3097, 3128, 5946; right adult maxillae, SK 2061, 2097, 2225, 2261, 2591*c*, 2686, 2982, 3080, 14120; left juvenile mandible pieces, SK 1618, 3090, 5203, 5906 + 4479; right juvenile mandible pieces, SK 2973, 7315; left adult mandible pieces, SK 1630, 2065, 2354, 3091, 3104, 3131, 3134, 3156, 7216; right adult mandible pieces, SK 2069, 2352*a*, 2358, 2697, 2986, 3010, 3045, 3052, 3061, 3068, 3105, 3137, 6073; isolated deciduous teeth, SK 4089, 5185; isolated permanent teeth, SK 1652, 2025, 2054, 2109, 2110, 2284, 2379, 2482, 2483, 2498, 2667, 2749, 3008, 3018, 3041, 3047, 3102, 4244.

Cf. *Megalotragus* sp.
Material:
8 cranial pieces from 3 adult individuals.

Left maxillary fragment with P^4–M^3, SK 3031; left mandible pieces, SK 1944, 2063, 2081, 3099; right mandible pieces, SK 2118, 3132, 14113.

?Hippotragini
Material:
2 cranial pieces from a single juvenile individual.

Right maxilla with dpm^4, M^{1-3}, SK 3107 + 3139.

Redunca cf. *arundinum*
Material:
A single cranial piece from a juvenile.

Right maxilla with dpm^{3-4}, SK 3533.

Pelea cf. *capreolus*
Material:
2 cranial pieces from a juvenile and an adult individual.

Left maxilla with dpm^4, M^{1-2}, and erupting premolars, SK 2682; right mandible piece with M_3, SK 2378.

Antidorcas cf. *recki*
Material:
12 cranial pieces from a minimum of 4 juveniles and 2 adult individuals.

Fragmentary palate, left M^3 and right M^{1-3}, SK 3033; right juvenile mandible, dpm_4, M_{1-2}, SK 3094; left maxillary fragment, SK 3169; right maxilla, M^{1-3}, SK 2567; mandible with M_{1-2} bilaterally, SK 2310; left mandible piece, SK 2545; right mandible pieces, SK 1931, 2235, 3095, 3501; isolated right dpm_4, SK 2079; left M_1, SK 3759.

?Gazella sp.
Material:
8 cranial pieces from a minimum of 1 juvenile and 5 adult individuals.

Incomplete snout and palate, left P^2–M^2, found in contact with hominid innominate, (SK 3155*b*), SK 3155*a;* palate with left and right P^2–M^3, SK 14063; complete palate, right P^3–M^3; left P^4–M^3, SK 3012; left maxillae, SK 2495, 2972; right maxilla, SK 3261; right mandible piece, SK 10440; isolated left M^1, SK 14068.

Antilopine or *Neotragine* indet.
Material:
3 cranial pieces from 3 adult individuals.

Right mandible piece, P_2–M_3, SK 3019 + 2509; right mandible piece, M_{2-3}, SK 2665; right mandible piece with M_3, SK 3025.

Oreotragus cf. *major*
Material:
A single cranial specimen.

Left mandible piece, P_{3-4}, M_1, SK 14059.

Syncerus sp.
Material:
13 cranial pieces from 1 juvenile and 2 adult individuals.

Maxilla fragment, right M^2, SK 2077; left juvenile mandible piece, dpm_{2-4}, M_{1-2}, SK 3064; left adult mandible pieces, SK 2491, 3130; right adult mandible pieces, SK 1972, 2968; isolated teeth, SK 1983, 2028, 2480, 2517, 3034, 3059, 3074.

Tragelaphus cf. *strepsiceros*
Material:
12 cranial pieces from 1 juvenile and 3 adult individuals.

Part of a right horn-core, SK 3171; right juvenile maxilla, dpm^{3-4}, SK 2271; left adult maxillae, SK 2281, 2681; right adult maxillae, SK 2095, 3000, 14112?; right mandible piece, P_4, M_{1-3}, SK 3110; isolated teeth, SK 2304, 2541, 2576, 3023.

Cf. *Makapania* sp.
Material:
10 cranial pieces from a minimum of 3 adult individuals.

Left maxillary piece with P^4, M^{1-3}, SK 3150; right maxillary pieces, SK 2759, 3005, 3065; left mandible fragment, SK 2965; right mandible pieces, SK 1627, 3113; isolated teeth, SK 2373, 2693, 2849*a*.

Antelope size class I
Material:
25 cranial and 11 postcranial pieces from an estimated minimum of 5 individuals, as listed in table 93.

Antelope size class II
Material:
117 cranial and 113 postcranial pieces from an estimated minimum of 20 individuals, as listed in table 93.

Antelope size class III
Material:
199 cranial and 162 postcranial pieces from an estimated minimum of 25 individuals, as listed in table 93.

Antelope size class IV
Material:
15 cranial and 2 postcranial pieces from an estimated minimum of 2 individuals as listed in table 93.

Family Suidae
Metridiochoerus andrewsi
Material:
8 cranial pieces from a minimum of 2 juvenile and 5 adult individuals.

Anterior part of juvenile mandible with left di_1, dc, dpm_{3-4}, SK 381; juvenile mandible piece with left dpm_3, right dpm_{3-4}, and part of M_1, SK 380; maxillary fragment and part of ?right M^3, SK 393 + 391; cranial fragment with parts of left and right M^3, SK 394; isolated left dpm^3, SK 14240; isolated left M_3, SK 387; right M_3, SK 388; ?left M^3, SK 392; fragment of M^3, SK 2088, 2380; part of M_3, SK 390; right M_2, SK 389.

Suid indet.
Material:
9 cranial pieces from an estimated minimum of 3 individuals.

Isolated tooth fragments, SK 384, 2456, 2677, 2811, 3445, 4743, 6018, 9423, 14148.

Order Perissodactyla
Family Equidae
Equus capensis
Material:
9 cranial pieces from a minimum of 1 juvenile and 5 adult individuals.

Palate with parts of P^2–M^3 bilaterally, SK 3983; left mandible fragment with M_1, SK 2584; left mandible piece with fragmentary molars, SK 1619; right mandible piece, P_{2-3}, SK 1626; isolated teeth, SK 2104, 3159, 3311, 3993, 3996.

Hipparion libycum
Material:
3 isolated teeth from a single old individual.

Left P^3, P^4, and right P^4, SK 3278 + 3982 + 2307.

Equid indet.
Material:
14 cranial and 3 postcranial pieces from an estimated minimum of 4 individuals.

Mandible fragment, SK 1945; isolated teeth and tooth fragments, 2018, 2070, 2481, unnumbered, 2539, 3250, 3530, 3534, 3830, 7146, 7357, 9095, 12029; distal femur, SK 1896; lateral metapodial, SK unnumbered; navicular, SK 6691.

Order Hyracoidea
Family Procaviidae
Procavia antiqua
Material:
35 cranial pieces from a minimum of 16 individuals.

Craniums in varying degrees of completeness, SK 143, 145, 172, 2794, 6012, 6072, 14142, 14239; maxillary pieces, SK 135, 163, 164, 178, 204, 1993, 3698, 14140*b;* mandible pieces, SK 128, 136, 138, 140, 161, 162, 182, 205, 1924, 3220, 3285, 3449, 4076, 5969, 14090, 14134, 14140*a;* isolated teeth, SK 2493, 3367.

Procavia transvaalensis
Material:
20 cranial pieces and 1 postcranial piece from a minimum of 8 individuals.

Craniums with articulated mandibles, SK 184, 196; craniums in varying degrees of completeness, SK 188, 207, 209, 211, 2768; maxillary piece, SK 208; mandible pieces, SK 112*a*, 113, 117, 126, 201*a*, 2935, 4216, 14135; isolated teeth, SK 3223, 12709, 14141, 14144; left distal humerus, SK 5255.

Hyracoid indet.
Material:
6 cranial pieces from a minimum of 2 individuals.

Partial cranium, SK 3244; maxillary pieces, SK 2126, 3510; mandible pieces, SK 2889, 2934; part of an isolated incisor, SK 3153.

Order Rodentia
Family Hystricidae
Hystrix africaeaustralis
Material:
6 cranial pieces from a minimum of 3 adult individuals.

Incomplete palate with left M^{2-3} and right M^3, SK 3082; right M^2 and M^3, SK 2466; isolated left I^1, SK 4315; ?right I_1, SK 2875; left M^1, SK 14236.

Hystrix ?makapanensis
Material:
An isolated left P_4, SK 14237, has been tentatively referred.

Class Aves
Bird indet.
Material:
A complete cranium from an as yet unidentified bird, SK 1001.

Phylum Mollusca
Class Gastropoda
Order Pulmonata
Land snail, cf. *Achatina* sp.
Material:
15 shells in varying degrees of completeness.
SK 2405, 2627, 2723, 2817, 3186, 3333, 3343, 3353, 3608, 4814, 4958, 7476, 7488, 8090, 9192.

The Major Features of the Bone Accumulation

The Composition of the Fauna Represented by the Fossils

The information provided in table 92 and depicted in figure 191 may be further summarized. This is done in the first columns of table 99 and figure 196. Of the 339 animals that contributed bones to the assemblage, 179, or 52.9%, were primates. These were almost equally divided between hominids and the four species of baboons. Next in numerical importance were 82 bovid individuals covering a wide range of body sizes. In size class I, 5 adults are represented; in class II, 20 individuals including 8 juveniles are found; in class III, there are 50 individuals with 11 juveniles, and class IV is represented by 7 individuals, among which are 2 juveniles.

The carnivore component in the Member 1 fauna is important, with 36 individuals making up 10.6% of the total. Cats are represented by 12 individual leopards, a

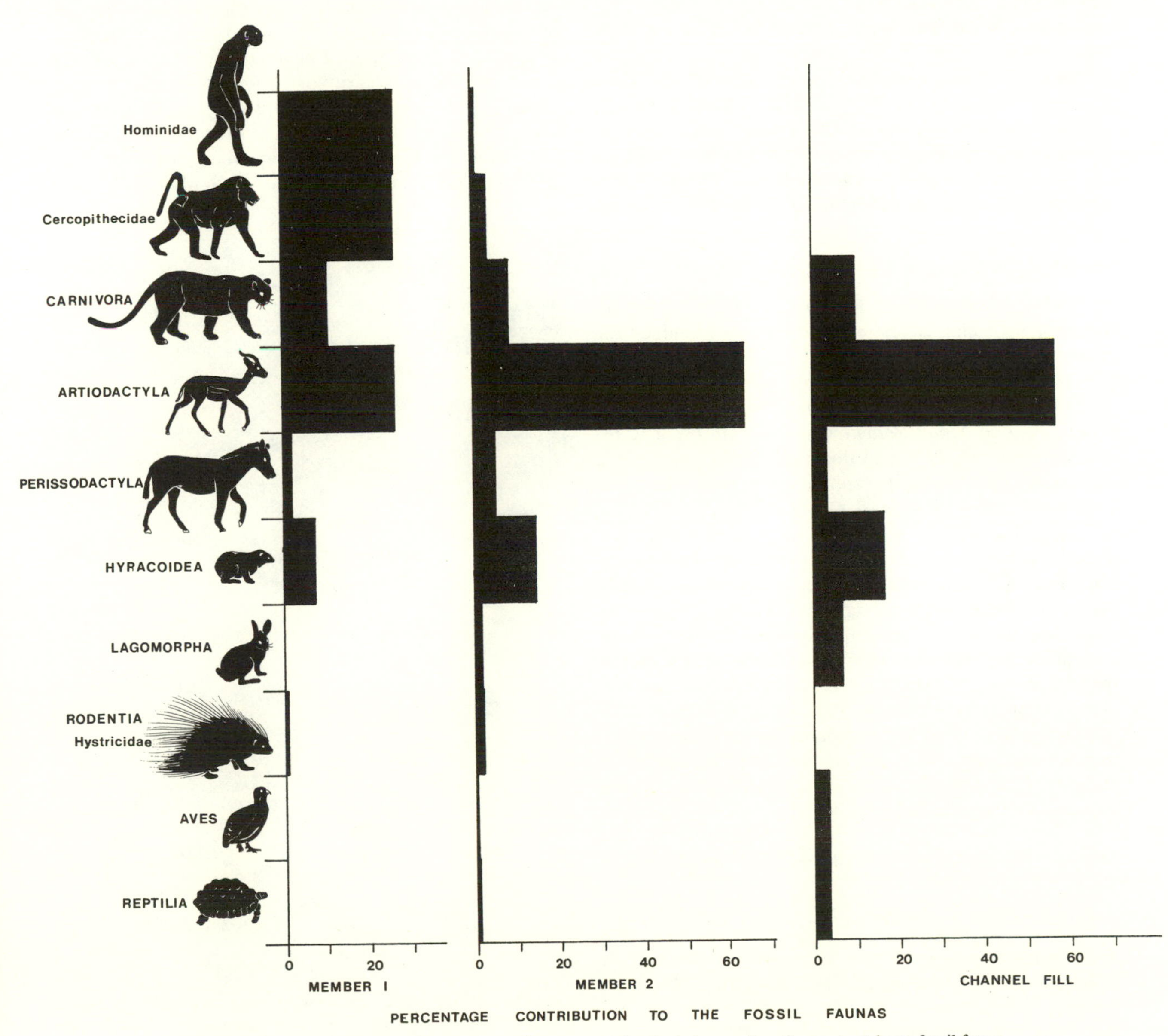

Fig. 196. Percentage contributions made by various groups of animals to samples of macrovertebrate fossil fauna from Swartkrans members 1 and 2 and from channel fill.

single *Dinofelis* and a *Megantereon*. Remains of 3 brown and 2 spotted hyena individuals have been found, together with those of 8 long-legged hunting hyenas. Parts of 1 aardwolf, 4 jackals, and 2 foxes are known.

Other faunal elements are 7 equids, 24 dassies, and 3 porcupines.

The Representation of Skeletal Parts

The Member 1 sample has undoubtedly been artificially biased in favor of identifiable cranial fossils, but this bias could not account, I think, for the almost total absence of primate postcranial parts. Study of the assemblage strongly suggests that the bodies of the hominids and baboons had largely disappeared before fossilization of the skulls, or parts thereof, took place. In the case of *Australopithecus robustus,* for instance, only 11 postcranial bones are associated with 218 cranial pieces, although the latter do include isolated teeth. The situation among the baboon fossils is even more extreme, where only 31 postcranial pieces accompany 372 cranial pieces.

Most of the carnivores, too, are known by cranial specimens only, and this is also true of the dassie component of the assemblage.

For the bovids, skeletal representation is certainly more comprehensive. In the various size classes, relative numbers of cranial/postcranial pieces are as follows: class I, 25/11; class II, 117/113; class III, 199/162; class IV, 15/12. Details of the parts actually preserved may be found in table 93, but the numbers are too small to let us discern a pattern. In size classes II and III parts from most of the body are represented, albeit in small numbers.

Observed Damage to the Bones

As with the assemblages from other site units in the Sterkfontein valley caves, each piece was examined for diagnostic damage marks. Details of such damage are given in table 100 and may be summarized as follows:

Porcupine and Small-Rodent Gnawing Marks. The incidence is extremely low in this assemblage, there being one specimen showing porcupine gnaw marks and one with small-rodent gnaw marks.

Carnivore-Inflicted Damage. Clear carnivore damage was observed on 3 australopithecine specimens: two punctate marks on the juvenile calvaria SK 54 (fig. 197*a*) to be discussed further; ragged-edge damage on the juvenile mandible SK 3978 (fig. 197*b*); and tooth marks on the innominate SK 3155*b*. Positive carnivore damage was also seen on 10 bovid fossils.

Damage that could very probably be attributed to carnivore action was observed on the *Homo* mandible SK 45, on 14 australopithecine specimens, and on 1 cercopithecoid calvaria (table 100).

Traces of Artificial Bone Alteration. None was observed.

Association with Stone Artifacts

As I mentioned earlier, only one stone artifact may be said with positive assurance to have come from Member 1 breccia. It is somewhat abraded, as if water-worn, and thus may not be in true association with the bone assemblage.

Clues to the Interpretation of the Member 1 Fossil Assemblage

Possible Porcupine Collecting Involvement

Only 1 bone, out of 2,366, shows porcupine gnaw marks, a percentage of less than 0.1. By this criterion porcupines may be excluded as significant collectors of the Member 1 bones.

Possible Hominid Involvement

As I mentioned earlier, only one artifact has definitely come from Member 1 breccia thus far, and this shows signs of stream abrasion before incorporation in the deposit. No bone damage clearly attributable to hominids has been observed. On the current evidence, therefore, hominid involvement in the accumulation process is not indicated, though further excavation of Member 1 breccia may change this conclusion.

Possible Carnivore Involvement

Fossils from Swartkrans Member 1 are generally in better condition than those from Sterkfontein Member 4, though many have also suffered badly from rough handling during mining or excavation. I have no doubt that my list of specimens showing carnivore tooth marks and other damage would have been a good deal longer if all the fossils had been removed from the breccia with the care that is essential for proper taphonomic assessment. It is significant that 18 of 29 specimens from Member 1 that I

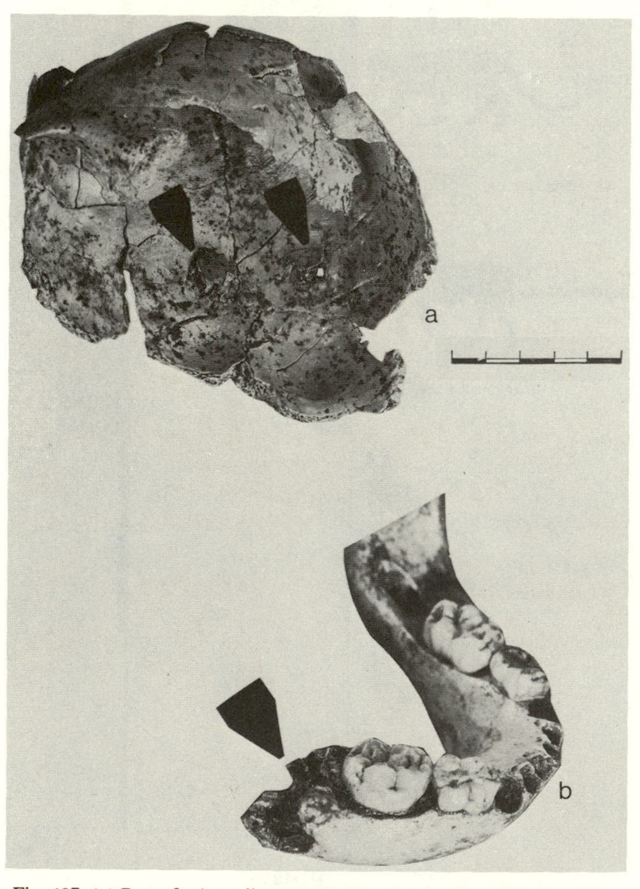

Fig. 197. *(a)* Part of a juvenile australopithecine cranium, SK 54, with a punctate mark in each of the parietal bones. *(b)* A juvenile australopithecine mandible, SK 3978, showing ragged-edge damage to its ramus.

have listed as bearing traces of carnivore damage (table 100) are from hominids. It has not been possible to process the majority of Swartkrans fossils with the care that has been devoted to the hominid finds, and so details of damage most of the bones may have suffered are still obscured by enclosing breccia.

In some cases the absence of skeletal parts is as indicative of carnivore action as is the presence of tooth marks. Member 1 hyracoid or dassie remains come from a minimum of 26 animals. This estimate is based on 61 cranial specimens that *are associated with only one postcranial bone*. In chapter 4 a description is given of the characteristic way leopards consume dassies. The entire body is typically eaten and only skull pieces remain, as with the Swartkrans hyracoid fossils.

A remarkable range of carnivores is represented in the Member 1 fossil assemblage, as was discussed earlier. The carnivore/ungulate ratio is 39/96, or 37.5%, a high figure for a bone accumulation in an African cave.

An explanation has to be sought for the presence of very numerous hominid and cercopithecoid remains: this is further explored in chapter 13.

After her study of the fossil bovids, Elisabeth Vrba (1975, 1976*a*) pointed out that, though many of the antelopes represented in Member 1 could have been killed by leopards, about half of them were too large to be regarded as typical leopard prey. Moreover, the proportion of juveniles did not rise as the body size of the species increased. For this reason Vrba suggested that false and true saber-toothed cats could also have contributed significantly to the accumulation.

We should not lose sight of the fact that, in addition to the felids known from Member 1, hyenas are also well represented. Remains of 3 brown and 2 spotted hyenas have been found together with those of at least 9 hunting hyenas of the genera *Hyaenictis* and *Euryboas*. It would be remarkable indeed if these had not contributed significantly to the bone accumulation. I suggested earlier that the hunting hyenas may well have been social carnivores that used the caves as breeding lairs. With cooperative effort they would have been able to hunt the largest bovids and could easily have brought parts of their prey back to the cave for their cubs.

In conclusion I have no hesitation in confirming my earlier suggestion (e.g., Brain 1970) that the Member 1 bone accumulation resulted very largely from carnivore feeding. In my opinion, both cats and hyenas are likely to have been involved, as is further discussed in chapter 13.

Remains from Member 1: Microvertebrate Component

Acetic acid preparation of australopithecine and other fossils during the early phases of Swartkrans paleontological work produced a by-product of microfaunal bones. In about 1960 the available collection, which had been sorted and provisionally identified by J. T. Robinson, was submitted to D. H. S. Davis, who drew up a detailed report (1955) that, regrettably, was never published. The manuscript, referred to here, was found in the Transvaal Museum files. The following taxa were recorded:

Order Insectivora
Family Macroscelididae
Elephantulus (Nasilio) cf. *brachyrhynchus* (2 individuals)
E. (Elephantomys) langi (4 individuals)

Family Chrysochloridae
Amblysomus (Chlorotalpa) spelea (number not recorded)

Order Lagomorpha
Lepus sp. nov. (?) (1 individual)

Order Rodentia
Family Bathyergidae
Cryptomys robertsi (2 individuals)

Family Muridae
Subfamily Murinae
Mus cf. *minutoides* (16 individuals)
M. cf. *triton* (1 individual)
Dasymys ?bolti (1 individual)
Lemniscomys sp. nov. (about 5 individuals)

Subfamily Otomyinae
Otomys (Palaeotomys) gracilis (9 individuals)

Family Cricetidae
Subfamily Dendromurinae
Steatomys cf. *pratensis* (2 individuals)
Dendromys ?antiquus (5 individuals)
Malacothrix cf. *typica* (2 individuals)

Subfamily Cricetinae
Mystromys hausleitneri (24 individuals)

Subfamily Gerbillinae
Tatera robinsoni sp. nov. (3 individuals)

Some elements of the microvertebrate assemblage, such as the soricids, reptiles, and birds, were not included in the analysis, and it is no longer possible to be sure of the faunal composition of the sample as a whole. For this reason two new samples were prepared from Swartkrans Member 1 breccia, one from the "rodent breccia" that outcrops around the entrance to the "eastern shaft" in the Outer Cave, the other from the "orange breccia," designated Member 1C by Butzer (1976). It is likely that the "rodent breccia" represents a very early phase in the accumulation of the Member 1 material and that Member 1C represents the terminal phase. The two samples therefore appear to provide information about the fauna in the vicinity of Swartkrans at each end of the Member 1 time span. On the other hand, the original sample studied by Davis would have been intermediate in age between the other two.

The samples from the "rodent breccia" and from Member 1C were found to contain remains from a minimum of 271 and 256 individual animals respectively. The faunal composition of the two samples is given in table 101. The composition is remarkably similar in each case, the only important difference being that dendromurines are more abundant in the "rodent breccia" sample than in the Member 1C sample; the reverse is true for remains of lizards. Further taxonomic study of the samples is required, although a broad picture of the faunal composition has emerged. This is shown in figure 198.

Remains from Member 2: Macrovertebrate Component

Evidence is accumulating that Member 2 is a rather heterogeneous stratigraphic entity, embracing breccias and calcified channel fills of several ages. Collectively these deposits are all younger than Member 1 and, for the time being, the fossils they have yielded are considered as a unit. Fieldwork is under way that should clarify the relationships of the deposits in the Outer Cave and provide information on the valid subdivisions of Member 2.

At present the Member 2 fossil sample consists of 5,894 specimens from a minimum of 258 individual animals, details of which are provided in table 102. The various identified animals are depicted in figure 199. The sample was found to include 3,811 specimens of bovid origin, mainly from the postcranial skeleton, that could not be specifically identified. They were placed in bovid size classes (see chap. 1 for definition of the classes); particulars are provided in tables 103 and 104. The remains detailed in table 104 consist of 2,860 specimens from antelopes classed as II*a*, or animals with body weights low in the class II range. All appear to have come from the same species, and there is little doubt that the taxon involved is *Antidorcas bondi,* whose identified cranial remains are similarly abundant. The significance of this strong representation of *A. bondi,* together with implications of the observed damage to bones from Member 2, will be considered further.

Details of the fossil material representing the various animal taxa will now be presented.

Phylum Chordata
Class Mammalia
Order Primates
Family Hominidae
Homo sp. (fig. 187)

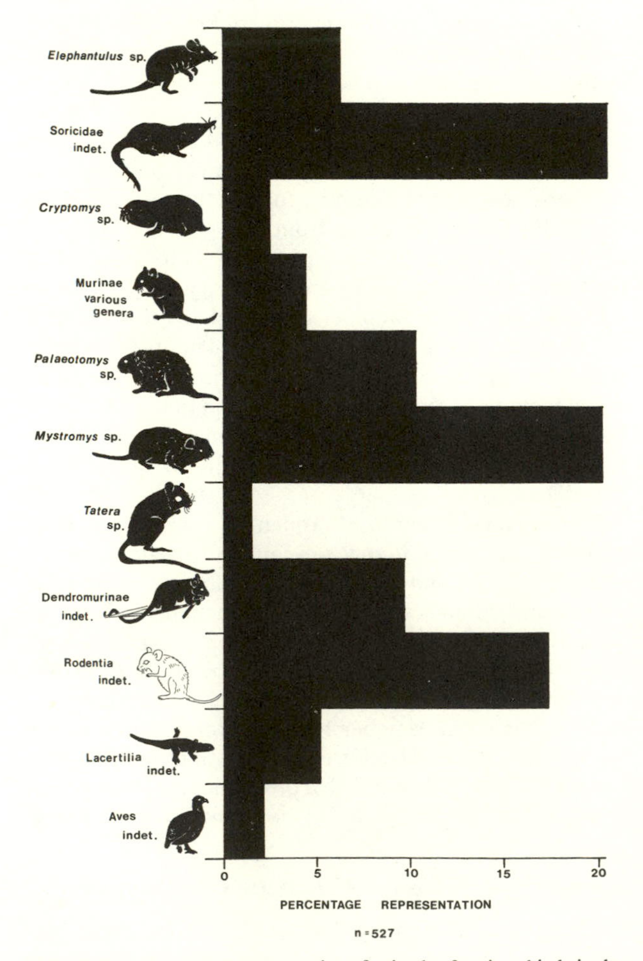

Fig. 198. The percentage representation of animals of various kinds in the microvertebrate component of the Swartkrans Member 1 sample. A total of 527 animals is involved.

Material:
An adult mandible, SK 15, with left M_{1-3} and right M_{2-3}; associated with it, part of right P_4, SK 43; left P_3, SK 18*a;* and the proximal end of a radius, SK 18*b*.

Family Cercopithecidae
Papio sp.
Material:
Articulated parts of a ♂ maxilla and mandible, SK 547; ♂ left maxilla, SK 444; ♂ left mandible piece, SK 434; ♀ palatal piece with broken teeth, SK 592; right mandible piece, SK 14100; right mandible piece from an immature individual, SK 616.

Cercopithecoides williamsi
Material:
♂ palate with right $\underline{C}$–M^3 and left P^4–M^3 (former type specimen of *Cercopithecoides molletti,* SK 551); maxillary fragment, SK 2135; 3 mandible pieces, SK 412, 579, 624.

Cercopithecoid indet.
Material:
Calvaria fragment, SK 3145; maxillary pieces, SK 3446, 4974, and 14150; mandible pieces, SK 2121, 2150, 4012, 5165; tooth fragments, SK 2185, 14202, 24567; left femur, SK 591*a;* humerus, SK 591*c;* distal tibia, SK 3211*a;* and phalanx, SK 4145.

Order Carnivora
Family Felidae
Panthera aff. *leo*
Material:
Left mandible piece, juvenile with $d\bar{c}$ and dpm_4; left and right mandible pieces associated with 2 thoracic vertebrae, SK 359; isolated teeth, SK 360, 1865, 1869, 11818; metapodial pieces, SKW 463, SK 3133, 6684, 6747.

Panthera pardus
Material:
Left mandible piece with part of $\bar{C}$ and P_3, SK 346; right mandible piece with part of $\bar{C}$, P_{3-4}, and M_1, SK 348; left distal humerus, SK 1810.

Felid indet.
Material:
Metatarsal, SK 6669; phalanx, SK 6736.

Family Hyaenidae
Hyaena brunnea
Material:
Almost complete cranium, SK unnumbered; right maxillary piece with P^3 and part of P^4; left mandible piece with P_{2-4}, SK 328.

Proteles cristatus
Material:
Right maxillary piece with $\underline{C}$, SKW 122; left mandible piece with 2 postcanine teeth, SK 11400.

Hyaenid indet.
Material:
Isolated premolar, SK 14189; proximal end of a metapodial, SK 10365.

Family Canidae
Canis mesomelas
Material:
Almost complete cranium, SK 371; maxillary pieces, SK 367, 368, 2771, unnumbered, 3512, 10388, 10529, 10828; mandible pieces, SK 1637, 2266, 2756, 4035, 6902, 7275, 10658, 10997, 11378, 11426; isolated teeth, SK 3719, 10543, 11312, 11484, 11815, 11823, 11927; distal humeri, SK 195, 6378, 14017; distal radius, 10591; left tibia, SK 10760; metapodial pieces, SK 10948, 11173, 11498, 11597.

Cf. *Lycaon* sp.
Material:
Isolated right $\underline{C}$, SK 1843.

Otocyon recki
Material:
Maxillary piece, SK 5972; isolated right M_1, SK unnumbered.

Family Mustelidae
Mellivora aff. *sivalensis*
Material:
Right mandible piece, SK 6918; isolated right M_1, SK 1829; isolated left $\bar{C}$, SK 1226.

Family Viverridae
Cf. *Herpestes sanguineus*

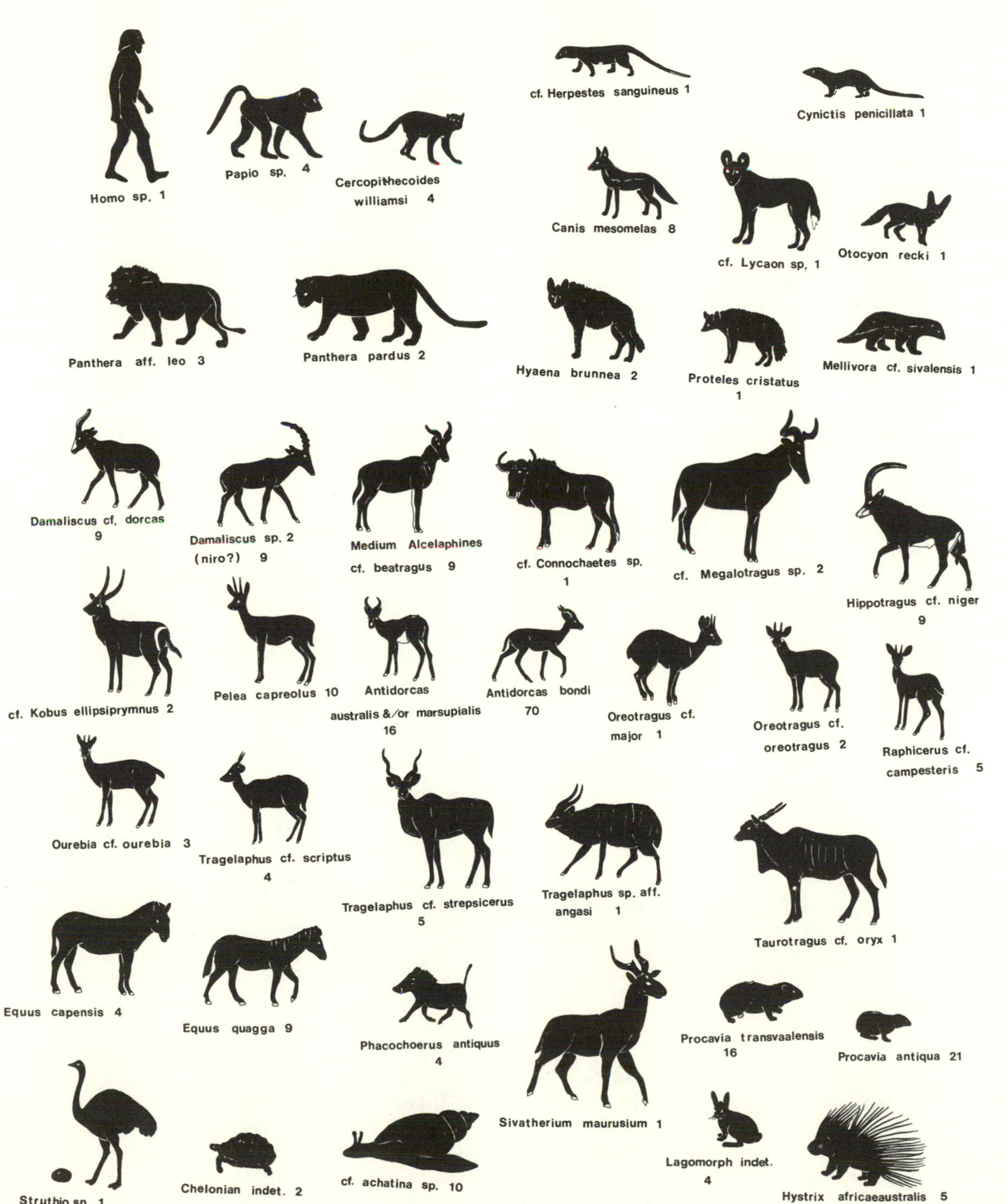

Fig. 199. Animals represented by fossils in the macrovertebrate component of the Swartkrans member 2 sample. The minimum number of individuals is indicated in each case.

Material:
Right mandible piece with $\bar{C}$, P_2, P_4, and M_1, SK 11184.

Cynictis penicillata
Material:
Right mandibular ramus, type specimen, SK 377.

Carnivore indet.
Material:
Calvaria fragment, 1; mandibular pieces, 4; isolated tooth fragments, 8; postcranial pieces, 43.

Order Artiodactyla
Family Bovidae
Damaliscus cf. *dorcas*
Material:
Maxillary pieces, SK 2116, 3123, 5996, 10941, 12003; mandible pieces, SK 6037, 10421, 10867, 11238, 11777, 11889, 11939; isolated teeth, SK 4015, 4056, 4574, 5397, 9341, 9897, 10653, 11271, 14111.

Damaliscus sp. 2 *(?niro)*
Material:
Horn-core piece, SK 2862; maxillary pieces, SK 1520, 2516, 3129, 3148, 4036, 4044, 5954, 10521, 10797, 11244, 11404, 11504, 12485*b;* mandible pieces, SK 4016, 4219, 5123, 5180, 5920, 5979, 7050, 7335, 7716, 11003, 11390, 11827, 11851, 14054; isolated teeth, SK 1971, 2003, 2017, 2242, 2540, 3306, 4065, 4075, 4572, 5023, 5172, 5208, 5942, 5999, 6000, 6014, 6029, 6032, 7791, 8007, 10841, 10906, 11117, 11178, 11391, 11477, 12145, 14122, 14205.

Medium-sized alcelaphines (including cf. *Beatragus* sp.)
Material:
Maxillary pieces, SK 1523, 2010, 6064, 10917, 11124, 12000; mandible pieces, SK 2978, 2993, 3003, 3083, 3143, 4013, 4020, 6057, 12201, 14048, 14211, 14212; isolated teeth, SK 2341, 3251, 5900, 5949, 5951, 5967, 5977, 5978, 6008, 6090, 11199.

Beatragus sp.
Material:
Frontal piece with base of a left horn-core, SK 14183.

Cf. *Connochaetes* sp.
Material:
Isolated left M_1 or M_2, SK 6004; isolated left M_3, SK 6059.

Cf. *Megalotragus* sp.
Material:
Left maxillary piece with M^2 and M^3, SK 14218; right mandible piece with M_3, SK 1953; isolated right M_3, SK 3249.

Hippotragus cf. *niger*
Material:
Maxillary pieces, SK 2663, 5926, 5993, 14046; mandible pieces, SK 1428, 1977, 1992, 2072, 2285, 2548, 2954, 3032, 6001, 6002, 6005, 6124, 14214, 14242, 14243; isolated teeth, SK 1947, 1980, 2355, 5909, 8010, 11641.

Cf. *Kobus ellipsiprymnus*
Material:
Left mandible piece with P_4, M_1, and M_2, SK 2960; isolated left M^2, SK 11297.

Pelea cf. *capreolus*
Material:
Parts of a complete but disintegrated skull, SK 2735*a,b;* an almost complete palate, SK 2990; maxillary pieces, SK 2923, 4030, 4040, 6087, 14049, 14116; mandible pieces, SK 1429, 1995, 2246, 2273, 2311, 2455, 2468, 2981, 3015, 3035, 3042, 3085, 6047, 10694, 11221, 14055, 14241; isolated teeth, SK 2090, 2308, 3124, 4029, 4087, 9911, 10741, 12531.

Antidorcas australis and/or *marsupialis*
Material:
Horn-core pieces, SK 1428, 3011, 3071, 7281, 7436, 9524, 10597, 14216; maxillary pieces, SK 1960, 2115, 2730, 3055*a*, 4022, 5418, 5427, 5995, 6051, 11287; mandible pieces, SK 2027, 2253, 2293, 2362, 2381, 2479, 2535, 2664, 2685, 2702, 2953, 2956, 2961, 2979, 3037, 3057, 3075, 3116, 3138, 3838, 4006, 4043, 4081, 4305, 5154, 5175, 5958, 5982, 9201, 11724, 12051, 12056, 12125, 14169; isolated teeth, SK 2547, 3941, 4021, 4039, 4054, 4068, 11073, 11683, 14064, 14070, 14123.

Antidorcas bondi
Material:
A partial ♀ braincase with right orbit and both horn-cores, mandible pieces, 4 thoracic vertebrae, 2 rib pieces, parts of both scapulae, proximal metapodial and naviculo-cuboid, SK 14126*a–p;* right ♂ frontal with complete horn-core, right maxilla with M^2 and M^3, SK 3152; calvaria pieces, SK 2948, 3084, 3152, 5488; frontals and horn-cores, SK 1223, 2640, 2641, 2647, 2722, 2781, 2880, 2946, 2949, 3001, 3192, 3253, 3571, 3773, 4010, 4534, 5115, 5277, 6924, 6959, 6973, 6987, 7010, 7115, 7254, 8544, 9221, 9321, 9393, 9658, 10402, 10634, 10702, 10943, 11070, 11105, 11433, 11748, 11862 + 11596, 12689, 14125*a,* 14167, 14176, 14215, 14221; palates, SK 3014, 3103; left maxillae, SK 1640, 1930, 2046, 2115, 2328, 2531, 3048, 3112, 4059, 5938, 7079, 7426, 10670, 10724, 11031, 11036, 12671, 14047; right maxillae, SK 1958, 2366, 2384, 2439, 2984, 3117, 3122, 3147, 4240, 5910, 5914, 5975, 5976, 5992, 7694, 7703, 10350, 11557, 12578; left mandible pieces, SK 1921, 1957, 1965, 1987, 2020, 2096, 2113, 2306, 2375, 2574, 2578, 2952, 2958, 2962, 2963, 2977, 2998, 3030, 3049, 3054, 3073, 3092, 4032, 4041, 4042, 4046, 4051, 4445, 4497, 4570, 5130, 5143, 5354, 5704, 5901, 5922, 5929, 5944, 5959, 5962, 5974, 5984, 5988, 6021, 6043, 6052, 6075, 6080, 6081, 6088, 6093, 6116, 7435, 7521, 7698, 9385, 10038, 10417, 10489, 10577, 10622, 10663, 11084, 11099, 11272, 11389, 11933, 12067, 12135, 12623, 12677, 14050, 14051, 14064, 14225, 14226, 14227; right mandible pieces, SK 1628, 1920, 2030, 2085, 2250, 2291, 2315, 2321, 2338, 2351, 2409, 2490, 2518, 2532, 2568, 2705, 2970, 2999, 3006, 3009, 3029, 3062, 3076, 3079, 3140, 3144, 3841, 4011, 4024, 4049, 4074, 4086, 5059, 5155, 5204, 5880, 5899, 5905, 5907, 5908, 5934, 5945, 5956, 5986, 5987, 6023, 6045, 6076, 11412, 11986, 12273, 12472, 12526; isolated teeth, SK 2051, 2067, 2264, 2277, 2289, 2292, 2367, 2372, 2387, 2393, 2399, 2414, 2417, 2465, 2474, 2486, 2506, 2530, 2553, 2720, 3248, 3931, 4023, 4025, 4061, 4062, 4063, 4064, 4071, 4080, 4285, 4626, 4633, 5057, 5404, 5731, 5882, 5890, 5902, 5956, 5990, 6038, 6044, 6084, 6095, 6101, 6106, 6108, 6109, 6117, 6118, 7559, 7920, 10278, 10520, 10555, 10601, 10611, 10804, 11068, 11122, 11167, 11168, 11345, 11506, 11514, 11561, 11609, 11637, 11801, 11899, 11946, 12324, 12501, 12596, 12628, 12630, 12669, 12753, 14066, 14069, 14072.

Oreotragus cf. *major*
Material:
An isolated right horn-core, SK 14243.

Oreotragus cf. *oreotragus*
Material:
Right mandible pieces, SK 1631, 4052; left mandible piece, SK 11405.

Raphicerus cf. *campestris*
Material:
Almost complete skull, SK 1515; maxillary piece, SK 12363; mandible pieces, SK 2024, 2108, 2719, 4287, 5930, 14060.

Cf. *Raphicerus* sp.
Material:
Left mandible piece, SK 2040; horn-core pieces, SK 7880, 14170.

Ourebia cf. *ourebi*
Material:
Maxillary pieces, SK 5892, 5893; mandible pieces, SK 6995, 14168; isolated tooth, SK 4060.

Tragelaphus cf. *scriptus*
Material:
Mandible pieces, SK 2329, 3114, 4261, 14052, 14205.

Tragelaphus cf. *strepsiceros*
Material:
Maxillary piece, SK 3098; mandible pieces, SK 1941, 2500, 3086, 6860; isolated teeth, SK 1989, 2230, 5888, 5923, 10848, 14012.

Tragelaphus sp. aff. *angasi*
Material:
Mandible piece, SK 2980; isolated tooth, SK 4028.

Taurotragus cf. *oryx*
Material:
Isolated left M_3, SK 14171.

Antelope size class I
Material:
239 pieces, as listed in table 103.

Antelope size class II*a*
Material:
2,860 pieces, as listed in table 104.

Antelope size class II*b*
Material:
458 pieces, as listed in table 103.

Antelope size class III
Material:
245 pieces, as listed in table 103.

Antelope size class IV
Material:
9 pieces, as listed in table 103.

Family Suidae
Phacochoerus modestus
Material:
Palate with left and right M^2 and M^3, SK 4005; mandible with right $\bar{C}$, P_4, M_{1-2}, left M_2 and M_3 associated with left and right M^2, SK 382, 385, 386; left and right maxillary pieces, SK 6030, 14131*a–c;* right maxilla and right mandible piece, SK 2359, 5989.

Suid indet.
Material:
Mandible piece, SK 8604; canine fragment, SK 3825*b*.

Family Giraffidae
Sivatherium maurusium
Material:
Isolated and unerupted molar, probably left M^2, SK 14045.

Order Perissodactyla
Family Equidae
Equus quagga
Material:
Parts of a palate and mandible with full dentition, SK 3997*a–z;* palate with associated mandible, SK 3998*a,b;* mandible pieces, SK 2243, 3989, 3994, 3999; isolated teeth, SK 2339, 2424, 2736, 3166, 3986, 3987, 3988, 3991, 3992, 3995; two carpal bones associated with left metacarpals, 1st and 2d phalanges, SK 4000*a–f*.

Equus capensis
Material:
Mandible piece, SK 1942; isolated teeth, SK 1632, 2626, 3160, 3164, 3984, 3990, 14133; two distal metapodial pieces, 2d phalanx and sesamoid, SK 4002*a–d;* left astragalus, SK 4001.

Equid indet.
Material:
Tooth pieces, SK 1837, 2047, 2052, 2317, 2537, 3190, 4403, 5000, 5119, 5120, 5134, 5167, 5275, 5287, 5375, 8014, 10380, 10735, 12088, 12445, 14222; carpal, SK 4721; metapodial pieces, SK 5070, 9469; astragali, SK 4924, 6257; 2d phalanx, SK 7423.

Order Hyracoidea
Family Procaviidae
Procavia cf. *antiqua*
Material:
Calvariae, SK 139, 144, 147, 148, 149, 174, 1615, 1959, 6111, 14081; maxillary pieces, SK 123, 129, 130, 134, 150, 154, 159, 166, 169, 173, 176, 215, 1620, 4094, 4628, 4663, 5940, 6017, 6899, 9295, 10458, 11110, 11183, 11507, 11565, 12525, 12675; mandible pieces, SK 114, 118, 125, 127, 131, 132, 133, 137, 142, 150, 158, 167, 170, 171, 175, 180, 181, 185, 202, 210, 1614, 1621, 2212, 2536, 2746, 2951, 3026, 3027, 4017, 4204, 4647, 5948, 6015, 6023, 6085, 6102, 6110, 7205, 7338, 7342, 7501, 10017, 10534, 10539, 11169, 11353, 11518, 11602, 11809, 11930, 12680, 12702, 14143, 14146; isolated teeth, SK 168, 190, 3324, 3361, 3425, 3454, 3476, 6853, 10295, 11204, 14145, 14165; distal humeri, SK 4150, 9802, 11306, 11308, 11411, 11687, 11955, 14136.

Procavia transvaalensis
Material:
Complete cranium with mandible, SK 216 + 111*a,b;* maxilla with associated mandible, SK 183*a,b;* calvaria, SK 112*b,* 115, 157, 199, 200, 212, 217, 2197, 7061; maxil-

lary pieces, SK 193, 194*a,b*, 206, 3479, 6031, 7420; mandible pieces, SK 122, 151, 152, 153, 155, 172, 187, 191, 213, 214, 2106, 3474; isolated teeth, SK 2899, 3243; associated postcranial bones: right scapula, SK 192; right distal humerus, SK 198; right radius and ulna and carpal bones, SK 197; left distal radius and ulna and carpals, SK 201.

Procavia sp.
Material:
Calvaria pieces, SK 3301, 3327, 3730, 3732, 3736; maxillary pieces, SK 3196, 3716, 7093, 7262, 7529, 9411; mandible pieces, SK 2508, 3304, 4260, 4568, 7135, 7651, 9555, 10376, 14126, 14200; isolated teeth, SK 3775, 3796, 5886, 7627; humerus shaft, SK 10117; distal humerus, SK 4409.

Order Lagomorpha
Family Leporidae
Lagomorph gen. et sp. indet.
Material:
Craniums with associated mandibles, SK 3345, 14097; calvaria pieces, SK 1617, 3299 + 3734, 3326; maxillary piece, SK 3485; mandible pieces, SK 2895, 3226, 3354 + 3394 + 3604, 3391, 3407, 3818, 11071.

Order Rodentia
Family Hystricidae
Hystrix africaeaustralis
Material:
Maxillary piece, SK 2835; mandible pieces, SK 3063, 6945, 14235; isolated teeth, SK 2040, 2045, 7155, 14238.

Class Aves
Order Struthioformes
Family Struthionidae
Struthio sp.
Material:
Indet. bone pieces, SK 7728, 8633; eggshell pieces, SK 7171, 14126.

Class Reptilia
Order Chelonia
Family Testudinidae
Chelonian indet.
Material:
Humerus pieces, SK 9866, 11947; carapace pieces, SK 4678, 4775, 7739, 9835, 12342, 12408.

Phylum Mollusca
Class Gastropoda
Order Pulmonata
Land snail, cf. *Achatina* sp.
Material:
Shells in varying degrees of completeness, SK 2609, 2660, 2887, 3185, 6946, 7628, 8126, 8213, 9264, 11537.

Indeterminate fragments
A total of 723 pieces constitutes this category.

Bone flakes
Material:
330 pieces with lengths as listed in table 102.

Coprolites
Material:
Parts of two carnivore coprolites, SK 7542, 7667.

Swartkrans Member 2: The Major Features of the Bone Accumulation

The Composition of the Fauna Represented by the Fossils

The information provided in table 102 and depicted in figure 198 may be further summarized. This is done in the second columns of table 99 and figure 196. The faunal composition of this assemblage forms a striking contrast to that from the older Member 1 at Swartkrans. Australopithecines have disappeared; a single *Homo* is represented together with only 8 individual cercopithecoids. The collection is dominated by no fewer than 160 individual antelopes, 118 of which fall into size class II. The class has been further subdivided into II*a* and II*b*, the former containing the smaller antelopes, particularly *Antidorcas bondi*, of which at least 70 individuals are represented.

In size class I, 12 individuals of klipspringer, steenbok, and oribi are found, while 27 individuals of class III antelopes are represented. Of these, 17 are juveniles. Remains of 3 adult individuals of larger, class IV bovids have also been found.

Fossils of at least 37 individuals of two species of dassies have been recovered, together with parts of at least 21 carnivores. These include 3 lions, 2 leopards, 2 brown hyenas, 1 aardwolf, 8 jackals, and various smaller animals as depicted in figure 198.

The Representation of Skeletal Parts

Particulars of the various skeletal parts making up the assemblage are given in the various tables. A most striking and valuable feature of the accumulation is the collection of 2,860 bones of class II*a* antelope, which without doubt may be associated with the 70 *Antidorcas bondi* individuals listed on the basis of cranial fossils. These bones come, perhaps entirely, from the younger component of Member 2, which occurs as an irregular sheet of lightly calcified breccia in the central area of the Outer Cave (see above). The sample is large enough to provide useful insights into those parts of the springbok skeletons that survived and those which disappeared before fossilization.

On the basis of the listing of skeletal parts provided in table 104, it has been possible to work out percentage survival figures for many of these parts, assuming that a minimum of 70 individual springbok contributed to the sample. Results are given in table 105 and are shown pictorially in figure 200. Taking certain cranial parts as representing 100% survival, a very considerable variation in representation occurs from one area of the skeleton to the next. The vertebral column and rib cage have suffered particularly severly, together with vulnerable segments of the limb bones such as the proximal humerus and proximal tibia. Many of the lower limb segments are particularly well represented. The significance of the survival pattern shown by the *A. bondi* skeletons will be considered shortly.

Observed Damage to the Bones

Details of specific damage observed on bone pieces from Member 2 are given in table 106 and may be summarized as follows:

Porcupine and Small-Rodent Gnaw Marks. The typical gnaw marks of porcupines were observed on 12 pieces, and 163 specimens bore marks caused by the incisors of small rodents.

Carnivore-Inflicted Damage. As listed in table 106, no fewer than 291 bones, out of a total of 5,884, showed clear evidence of carnivore damage, and another 123 bore less positive traces. The great majority of the damaged pieces were bovid skeletal parts, and, from the interpretative point of view, bones from class II*a* are of particular interest here. As detailed in the table, carnivore damage was observed on 17 vertebrae, 6 scapulae, 9 pelvis pieces, 55 humeri, 15 radii, 46 femurs, 19 tibiae, 3 calcanei, and 39 metapodials from bovid class II*a*.

Traces of Artificial Bone Alteration. Cut marks, certainly produced by a sharp-edged artifact, have been observed on the mandible of an oribi and on 2 other bones. Undoubted chop marks were also found on a limb bone shaft.

Association with Artifacts. The rather unsatisfactory situation regarding stone artifacts from Member 2 has already been discussed. A single biface has been found in situ in the younger component of Member 2 (fig. 188), and 24 other tools were embedded in Member 2 matrix, though it is not known from what part of the deposit they came. Until carefully controlled excavation is undertaken it will not be possible to say whether tools occur as a scatter throughout the depth of the Member or whether certain layers reflect more intensive human occupation.

Clues to the Interpretation of the Member 2 Fossil Assemblage

Possible Porcupine Collecting Involvement

Porcupine gnaw marks were observed on 12 bones out of 5,884 specimens, giving a gnawed-bone percentage of 0.2%. By this criterion, porcupine involvement in the collecting process may be discounted. A larger number of bones, 163, showed traces of small-rodent gnawing. Since many of the bone pieces weighed more than the estimated liveweight of the rodents that gnawed them, it is likely that the gnawing was done where the bones were found.

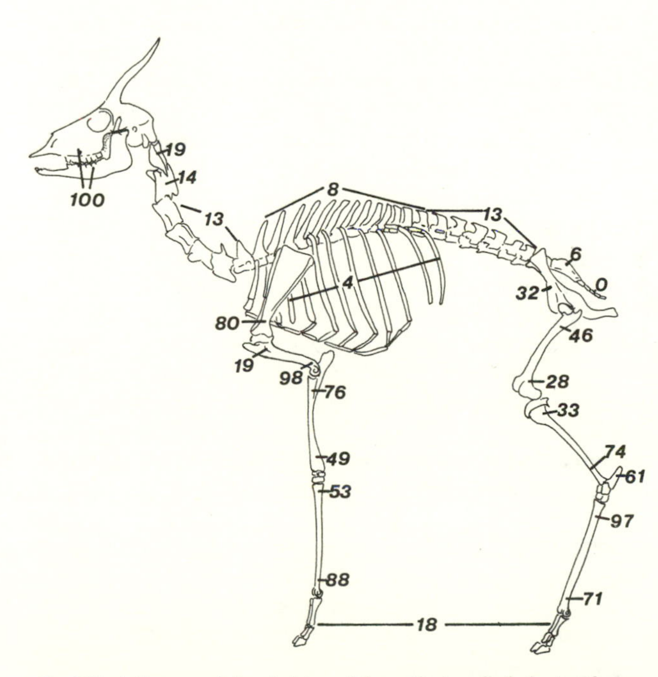

Fig. 200. A diagram of the skeleton of the extinct springbok, *Antidorcas bondi*, of which remains from 70 individuals have been recovered from Swartkrans Member 2 breccia. The percentage survival of various skeletal parts in the sample is indicated. Further details are given in table 105.

Possible Hominid Involvement

The Acheulean tools occasionally found in Member 2 clearly indicate that the cave was intermittently occupied by humans during the accumulation of this deposit. This is borne out by the fact that a few of the bone pieces show unmistakable cut and chop marks. As I mentioned earlier, Member 2 is not a homogeneous stratigraphic unit, and the fossils under consideration here come from older *and* younger breccia components. It is not impossible that the tools were concentrated in only one part of the profile, but this cannot be established without careful excavation.

Possible Carnivore Involvement

Of the 258 individual animals identified from Member 2, 21 are carnivores, giving a carnivore/ungulate ratio of 21/179, or 11.7%. The presence of remains from 3 lions is somewhat unexpected, since these cats normally have little to do with caves. Carnivores represented in the member that might well have contributed to the accumulation process are leopards and brown hyenas; the other smaller species are, to my mind, more likely to have formed part of the prey.

From her study of the fossil bovids, Elisabeth Vrba (1975, 1976*a*) pointed out that the Member 5 bovid assemblage was characterized by animals of low liveweight, there being 130 individuals in classes I and II out of a total of 160. Moreover, of the remaining 30 larger bovids, no fewer than 17 were juveniles. These facts led Vrba to suggest that the collecting agents were either carnivores of fairly small prey-adaptation, hominid hunters, or both.

Let us first consider who, or what, could most likely have been responsible for the *Antidorcas bondi* component of the fossil fauna. At least 70 individuals of this small springbok are involved, of which 17 are juveniles. The evidence I have presented indicates that, although parts of the skulls and lower legs are very well represented, the vertebral columns and rib cages have in many cases disappeared. Evidence of carnivore damage to the various skeletal parts is widespread and convincing. I have little hesitation in suggesting that the springbok were killed by carnivores and consumed within the catchment of the Swartkrans cave entrance as it was in Member 2 times. The pattern of damage and part-survival is very similar to what I described in chapter 4 where leopards were observed to feed on impala. To my mind, the obvious choice for the killer of most of the Member 2 springbok was the leopard.

As with Member 1 fossils, the hyracoid or dassie remains from Member 2 are very suggestive of leopard involvement. A minimum of 47 animals are involved, represented by 160 cranial pieces, *with which are associated only 14 postcranial bones*. The presence of skull pieces, with an absence of postcranial bones, could well have resulted from the kind of felid feeding described in chapter 4.

Despite this conclusion, the evidence is clear that human hunters were also involved in building up the Member 2 bone assemblage, and it would be very surprising if the cave had not been used as a brown hyena breeding lair as well. Numerous remains of small carnivores could very well have come from prey that brown hyenas fed to their young in the cave.

Remains from Member 2: Microvertebrate Component

Microfaunal remains are surprisingly rare in Member 2 and, although a large volume of breccia has been dissolved in acetic acid, the yield of small bones has not been sufficient to permit a faunal analysis.

Remains from the Channel Fill: Macrovertebrate Component

As I mentioned earlier, the sample of bones to be discussed here came from approximately 2 m^3 of unconsolidated sediment excavated from a solution channel passing through both Member 1 and Member 2 in the Outer Cave. The position of the excavated area is indicated on the plan (fig. 186).

In all, 837 bone pieces were recovered (table 107), including an australopithecine maxillary fragment that had obviously weathered out from the wall of the channel. It is likely that some of the other bones in the sample were likewise derived from older breccia in the walls. The animals represented by the fossils are depicted in figure 201.

Among the bone pieces were 178 specimens of bovid origin, largely postcranial, that were not specifically identified but were simply assigned to antelope size classes (as defined in chap. 1). Particulars are given in table 108.

The material by which each identified taxon is represented is now presented.

Phylum Chordata
Class Mammalia
Order Primates
Family Hominidae
Australopithecus robustus
Material:
A single cranial piece, weathered from Member 1 breccia of the channel wall.

Left maxilla with P^{3-4}, M^{1-2} and part of M^3, SKW 12.

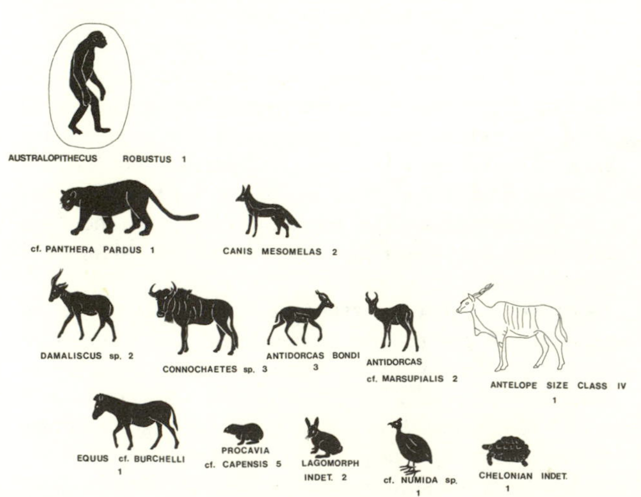

Fig. 201. Animals represented by the fossils in the macrovertebrate component of the Swartkrans channel fill sample. Minimum numbers of individuals are indicated in each case. The single *Australopithecus* individual is represented by a maxillary fossil derived from Member 1 breccia forming the wall of the channel.

Order Carnivora
Family Felidae
Cf. *Panthera pardus*
Material:
1 cranial and 1 postcranial piece, probably from a single individual.

Isolated carnassial, SKW 2669; distal radius, SKW 2859.

Family Canidae
Canis mesomelas
Material:
3 cranial pieces from a minimum of 2 individuals.

Piece of maxilla with articulated mandibular fragment, SKW 2616; maxillary fragment, SKW 2463; mandible fragment, SKW 3054.

Carnivore indet.
Material:
12 postcranial pieces from an estimated minimum of 3 individuals.

Distal humerus, SKW 2809; ulna shaft, SKW 2866; proximal femur, SKW 2838; distal tibia, SKW 2825, 2873; metapodial pieces, SKW 2841, 2670; pelvis fragments, SKW 2436, 2456; calcaneus, SKW 2647; phalanx, SKW 2399.

Order Artiodactyla
Family Bovidae
Medium alcelaphine, including *Damaliscus* sp. and *Connochaetes* sp.
Material:
17 cranial pieces from a minimum of 5 individuals.

Isolated left dpm_3, SKW 2638; M^1, SKW 2466; left M_3, SKW 2462, 2484; right M_3, SKW 2479; tooth fragments, SKW 2464, 2471, 2472, 2478, 2483, 2490, 2645, 2652, 2695, 3059, 3095, 3181.

Antidorcas bondi
Material:
6 cranial pieces from a minimum of 1 juvenile and 2 adults.

Left mandible piece with dpm_3, and M_1, SKW 2465; isolated right dpm_3, SKW 2668; left mandible piece, P_2 (roots), P_{3-4}, M_{1-3}, SKW 2455; right mandible piece with M_3, SKW 2494; isolated right M_2, SKW 2486, 2605.

Antidorcas cf. *marsupialis*
Material:
5 cranial pieces from a minimum of 2 adults.

Horn-core pieces, SKW 2495, 2628; isolated left M_2, SKW 2467, 2631; M^2, SKW 2603.

Antelope size class I
Material:
4 cranial and 37 postcranial pieces from an estimated 3 individuals, as listed in table 108.

Antelope size class II
Material:
33 cranial and 73 postcranial pieces from an estimated minimum of 10 individuals, as listed in table 108.

Antelope size class III
Material:
6 cranial and 24 postcranial pieces from an estimated minimum of 3 individuals, as listed in table 108.

Antelope size class IV
Material:
A single postcranial piece, detailed in table 108.

Order Perissodactyla
Family Equidae
Equus cf. *burchelli*
Material:
4 cranial and 3 postcranial pieces, probably from 1 individual.
Tooth fragments, SKW 2459, 2468, 2498, 2602; metapodial pieces, SKW 2360, 2867; phalanx, SKW 2819.

Order Hyracoidea
Family Procaviidae
Procavia cf. *capensis*
Material:
11 cranial and 16 postcranial pieces from a minimum of 5 individuals.
Maxillary fragments, SKW 2449, 2476, 2492, 2609, 3128; mandible pieces, SKW 2606, 2623, 2640, 2652; isolated teeth, SKW 2644, 2655; distal humeri, SKW 2678, 2820, 2822, 2823, 2826, 2827, 2834, 2835, 2842, 3005; humerus shaft, SKW 2831; proximal radius, SKW 2879; femur shaft, SKW 2836; distal tibia, SKW 2843; calcaneus, SKW 2549; phalanx, SKW 2833.

Order Lagomorpha
Family Leporidae
Gen. et sp. indet. Hare
Material:
3 cranial pieces from 2 individuals.
Maxillary pieces, SKW 2618, 2649; molar fragment, SKW 3198.

Class Aves
Order Galliformes
Family Numididae
Cf. *Numida* sp. Guinea fowl
Material:
5 postcranial pieces, probably from 1 individual.
Proximal ulna, SKW 2846; tibiotarsus piece, SKW 2806; femur shaft piece, SKW 3004; tarsometatarsus pieces, SKW 2839, 2899.

Class Reptilia
Order Chelonia
Family Testudinidae
Gen. et. sp. indet. Tortoise
Material:
3 carapace pieces, probably from 1 individual.
SKW 2493, 2619, 3166.

Indeterminate fragments
193 pieces.

Bone flakes
375 pieces as listed in table 107.

Swartkrans Channel Fill: The Major Features of the Bone Accumulation

The Composition of the Fauna Represented by the Fossils

The information provided in table 107 and depicted in figure 201 may be further summarized. This is done in the third columns of table 99 and figure 196. This small sample, derived from about 30 animals, is dominated by bones from 17 individual antelopes, 3 of which were of size class I, 10 from class II, 3 from class III, and 1 from class IV. Second in number to the bovids were 5 dassies, followed by 3 carnivores and 2 hares.

Representation of Skeletal Parts

Particulars of the bovid skeletal parts are given in table 108. The sample is too small to reveal much of a feeding or butchery pattern, although in the first three size classes a fair scatter of parts throughout the skeletons does appear to be preserved, from skull to feet.

Observed Damage to the Bones

Details of specific damage observed on bone pieces from the channel fill are given in table 109 and may be summarized as follows:

Porcupine and Small-Rodent Gnaw Marks. Porcupine gnawing was observed on 1 specimen; small-rodent gnaw marks appeared on 2 bones.

Carnivore-Inflicted Damage. Clear traces of this kind were found on 4 bones, less positive traces appeared on 2 others.

Traces of Artificial Bone Alteration. A probable chop mark was found on a bovid horn-core, and one bone flake has worn edges, suggestive of human agency.

Association with Artifacts. No tools were found directly associated with these bones, though flakes have been observed in other similarly loose channel fills of the Swartkrans cave system.

Clues to the Interpretation of the Channel Fill Bone Assemblage

The assemblage must be regarded as one of mixed provenance, since some specimens, such as the australopithecine maxilla SKW 12, have weathered out from older breccias forming the channel wall.

Absence of any porcupine-gnawed pieces but one excludes porcupine collecting as an important agency in this case. The bones show indications of both human and carnivore involvement, and the presence of leopard remains suggests that these cats may have used the ramifying channels as a lair. On the other hand, that 375 of the 837 bone pieces proved to be bone flakes points to possible human food remains. As with bones from most of the other Sterkfontein valley site units, the assemblage suggests involvement of more than one accumulating agency.

Remains from the Channel Fill: Microvertebrate Remains

No microfaunal remains were present in the excavated sediment.

12 Kromdraai

A Brief History of Activity

It was at Kromdraai that the first robust australopithecine known to science was discovered. Circumstances surrounding that discovery are best told in the words of Robert Broom (Broom and Schepers 1946, p. 3):

> One Wednesday in June, 1938, I visited Sterkfontein, when the caretaker, Mr. G. W. Barlow, showed me a palate with a molar tooth that had been picked up. I immediately purchased it from him; but he did not seem inclined to tell me how he had obtained it, or where it had been picked up. It was manifestly the palate of an anthropoid allied to the Sterkfontein skull, but apparently different. Some teeth had been freshly broken off, and there were other evidences of fresh fractures. The matrix was different from that in the Sterkfontein caves, and I felt sure it had come from elsewhere. I returned to Sterkfontein on Saturday, when I showed the palate to the Kafirs, who worked in the quarry; but none had seen the specimen before. Barlow was away. Determined to get to the bottom of the mystery I was again at Sterkfontein on the Tuesday following, when I insisted on Barlow telling me where the specimen had come from. He told me that it was a school boy, Gert Terblanche, who had picked it up at Kromdraai, two miles east of Sterkfontein. Gert, who was about 15 years of age, acted as a guide to visitors to the caves on Sundays. Of course I immediately set off in pursuit of Gert. I found he lived on a small farm about two miles away, and I went to his home.
>
> Gert was at school, another two miles away; but I saw his mother and sister; and his sister took me to the top of an adjacent hill where the specimen had been broken out of a weathered outcrop of bone breccia. I picked up two or three nice fragments, and a couple of teeth; but the sister told me that Gert had four beautiful teeth with him at school; and she was sure he had some other nice pieces hidden away somewhere.
>
> I went off to the school. About a mile of the way was so rocky that it was impossible to go by car. It was playtime, about 12.30 pm., when I arrived. I saw the principal, and told him what I had come about. Gert was found, and drew from his trouser pocket four of the most wonderful teeth ever seen in the world's history. Two of these fitted on to the palate which I had in my pocket. The other two were teeth of the other side.
>
> Gert told me that there were some more bits on the hillside hidden away. As the school only broke up at 2 o'clock I suggested to the principal that I would give a lecture to the teachers and children, on how caves were formed, and how animal bones got into them. He was delighted. So I lectured to four teachers and about 120 children for over an hour. As it was now almost 2 o'clock the principal dismissed the classes; and Gert came away with me, and took me to the place on the hill where he had carefully hidden away a nice lower jaw. I gathered every fragment.
>
> When I returned to Pretoria, and cleaned up and joined the fragments, I found I had most of the palate, much of the left side of the face, almost the whole of the left zygomatic arch, and the left side of the base of the skull, a considerable portion of the parietal, and the greater part of the right mandible, with most of the teeth. Almost the whole of the right side of the skull had been weathered away with the back of the occiput and the top of the brain case; but otherwise the skull was well preserved, and quite uncrushed.
>
> By later sieving the ground I was fortunate in finding a number more teeth; and we now have nearly every fragment that was there when Gert broke the skull out of the matrix with a hammer. Believing this Kromdraai skull to be very different from the Sterkfontein form I have called it *Paranthropus robustus*.

After the original Kromdraai discovery in 1938, the tempo of paleontological work in the Sterkfontein valley slowed for a while. A fall in the price of lime brought quarrying at Sterkfontein to a halt in 1939, and soon thereafter World War II broke out. Nevertheless, in February 1941 Broom sent a small work party to Kromdraai for a short period to investigate breccia adjacent to the spot where the type skull had been found. This work resulted in the discovery of a juvenile mandible with excellent deciduous dentition. The work was stopped when the underlying stony breccia appeared to be sterile.

Paleontological work at Kromdraai A, or the Faunal Site, appears to have been restricted to the period January to April 1947, before Broom resumed his work at Sterkfontein. A good deal of breccia was blasted from the site, and promising blocks were taken back to the Transvaal Museum, where they were processed over a number of years. The resulting fossil collection is described in this chapter.

In an attempt to obtain further information about the extent and nature of the Kromdraai B deposit, I carried out fieldwork there from March 1955 until May 1956. Very little solid breccia was removed, but decalcified breccia along the north wall of the cave was excavated to a depth of 4.9 m, or 16 ft (fig. 203). The excavation revealed that the deposit was considerably more extensive than we had thought, and that the stratigraphy was inclined from east to west. Bones recovered from the decalcified breccia, including robust australopithecine teeth and part of a pelvis, are included in the analysis to be presented here.

A new excavation of the Kromdraai B deposit (fig. 204) was initiated by E. S. Vrba of the Transvaal Museum toward the end of 1976. Cores have been drilled at selected spots along the length of the deposit and have revealed a considerable depth and an interesting succession of breccias. Samples have been taken for paleomagnetic orientation, and the results are awaited with interest. At this time Vrba and her team are preparing the first australopithecine specimen (fig. 205) to come

Fig. 203. Excavations conducted at Kromdraai B by the author in 1955, the aim being to define the northern margin of the cave deposit.

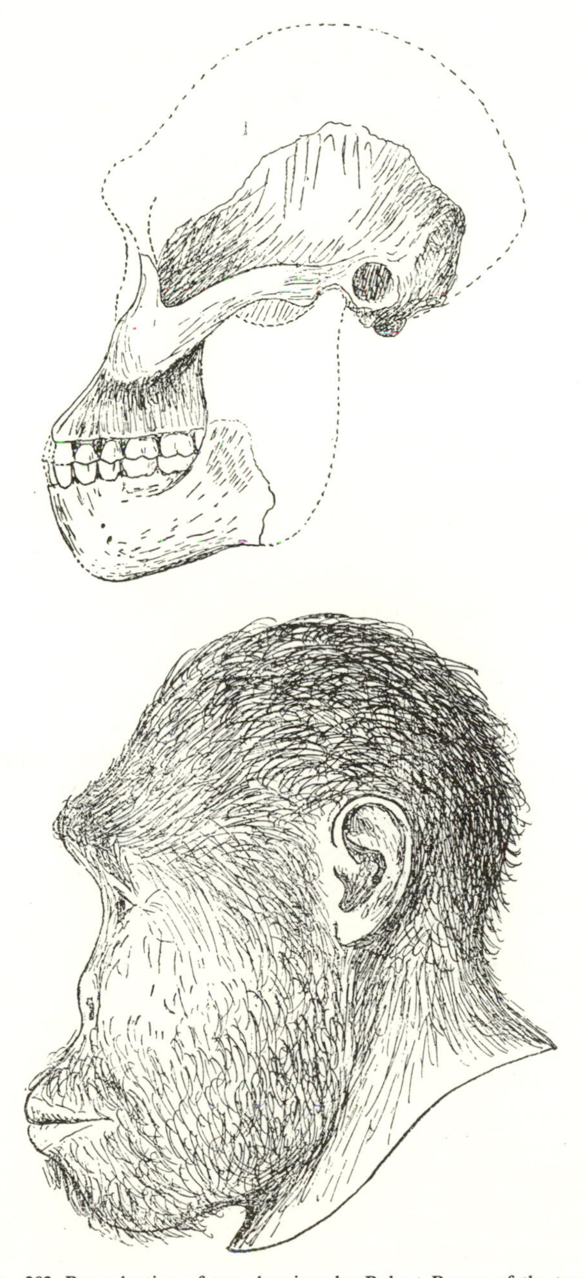

Fig. 202. Reproduction of two drawings by Robert Broom of the type specimen of *Paranthropus robustus*, fossil and reconstruction. From Broom 1950.

Fig. 204. Renewed excavations at Kromdraai B in 1977, under the direction of Elisabeth Vrba.

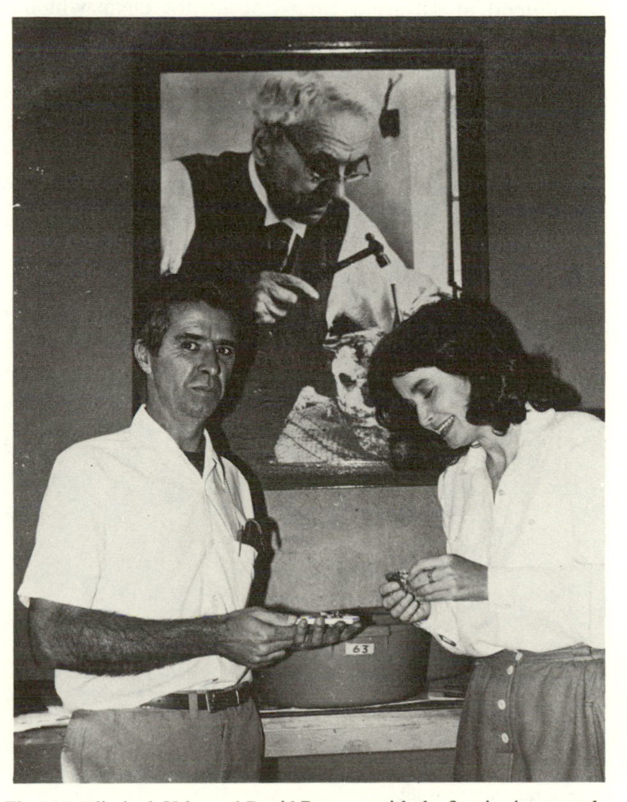

Fig. 205. Elisabeth Vrba and David Panagos with the first in situ australopithecine specimen to come from their new excavations at Kromdraai B.

from the new excavation. It is an especially significant fossil in that it may be tied precisely to the rather complex stratigraphy of the site, which is now starting to unfold.

Some Notes on the Kromdraai Sites

The Kromdraai cave sites lie clustered on the south side of the Bloubank River, 1,750 m east-northeast of Sterkfontein. The Kromdraai A and B deposits both consist of calcified fillings of narrow solution galleries, parallel to one another, trending east-west and no more than 17 m apart (see fig. 206). The dolomitic country rock appears reasonably undisturbed, dipping in a generally northerly direction at angles of 11°–25° in the immediate vicinity of the fossil sites. It appears that at Kromdraai, as at Sterkfontein, the country rock has developed a rectangular joint pattern trending north-south and east-west. The two main fossiliferous deposits at Kromdraai fill cavities that resulted from enlargement of east-west lineaments, while others such as Kromdraai C (KC in fig. 206) are aligned north-south. Kromdraai A and B caves have both lost their dolomite roofs through erosion, and the vertical extent of the KA deposit is not known. Inclination of the stratigraphy, downward toward the west in the Kromdraai B deposit strongly suggests that the original opening to the cave was at a higher level and at the eastern end. It is not impossible that the Kromdraai C deposit is an extension of the KB accumulation to the east. Vrba's current excavation will clarify this.

The alignment and proximity of the KA and KB caves suggest that they may have formed two parallel galleries of a single cavern system. In the past it has been customary to associate the fauna from KA with the hominids from KB and to assume contemporaneity. Faunal studies now suggest that the deposits were almost certainly not contemporaneous (e.g., Freedman and Brain 1972; Hendey 1973; Brain 1975*b*). It therefore appears that the fillings of the two parallel and adjacent galleries were introduced at different times. It is not clear which is definitely the older.

The Kromdraai A Bone Accumulation: Macrovertebrate Component

The collection consists of 1,847 fossils from a minimum of 194 individuals, as detailed in table 110. Particulars of 702 bones of bovid origin, assigned to size classes but not specifically identified, are given in table 111. The various identified animals are depicted in figure 207. Data on the fossil material by which the various animal taxa are represented will now be presented.

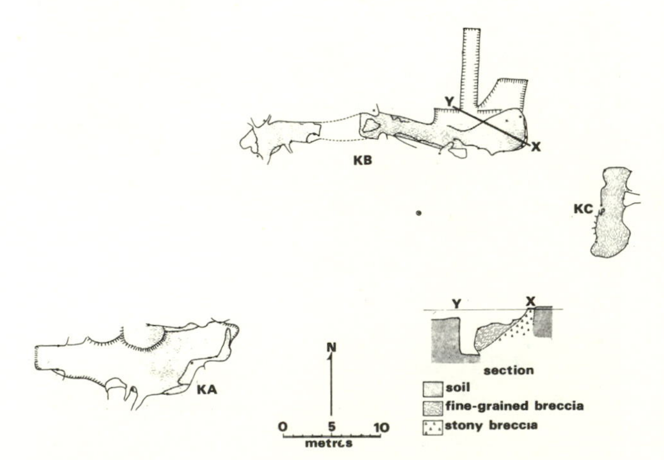

Fig. 206. Plan of the Kromdraai A, B, and C deposits, based on a level survey by the author in 1973. A section through the eastern end of Kromdraai A is also shown.

Class Mammalia
Order Primates
Family Cercopithecidae
Parapapio jonesi
Material:
Right maxilla with M^{1-3}, KA 175; right mandible piece with M_{2-3}, KA 176.

Papio robinsoni
Material:
Left maxillary piece with M^2 and M^3, KA 160.

Papio angusticeps
Material:
Type specimen, an almost complete ♀ cranium, without mandible, KA 194; snout with dm^1, dm^2, and M^1 bilaterally, KA 155; snout with parts of both toothrows, KA 156; snout with left P^4–M^3, and right M^3, KA 157; maxillary and mandible pieces, associated with parts of a distal humerus, KA 168*a–d;* right maxilla with P^3–M^1, associated with parts of an articulated hand, KA 167*a–f;* right mandible piece with P_3–M_3, associated with 2 isolated teeth, distal humerus, proximal tibia, and part of a phalanx, KA 166*a–g;* left side of a cranium with dm^1, and M^1, KA 151; right maxillary piece, KA 161; mandible pieces, KA 163, 165, 179.

Gorgopithecus major (fig. 208)
Material:
Type specimen, left M^2 and M^3, KA 193; complete cranium, without mandible, KA 192; maxillary pieces, KA 153, 154, 169, 524, 605; mandible pieces, KA 150, 152, 173, 198, 1148; isolated teeth, KA 170, 180, 182, 183, 620.

Cercopithecoid indet.
Material:
Calvaria pieces, KA 976, 1689, 2454; maxillary pieces, KA 159, 944, 1142, 1421, 1434, 2333; mandible pieces, KA 158, 164, 167, 171, 172, 177, 613, 1240, 1254, 1489, 1538, 2327, 2331, 2373; isolated teeth and tooth fragments, KA 181, 184, 185, 186, 187, 188, 189, 708, 1171, 1231, 1327, 1439, 1562, 1627, 2332, 2337, 2340, 2341, 2345, 2348, 2349, 2361, 2362, 2367, 2375, 2384; thoracic vertebrae, KA 745, 2379,; lumbar vertebra, KA 507; proximal femur, KA 1658, 1798; distal femur, KA 190, 191, 699.

Family Felidae
Panthera leo
Material:
Left maxillary piece with P^4, KA 67; isolated left P^3, KA 68; right P_3, KA 69; right M_1, KA 70.

Felis crassidens
Material:
The type specimen, which consists of right maxillary piece with P^3 and P^4, KA 87; also left maxilla and right mandible piece referred to by Broom (1948), but now missing.

Felis sp.
Material:
Isolated left P^4, KA 1544.

Homotherium sp.
Material:
Isolated left and right M_1, KA 66*a,b*.

Megantereon eurynodon
Material:
The type specimen, which consists of most of the cranium with articulated mandible, severely decayed; articulated atlas, axis, and parts of other cervical vertebrae; head of left scapula and proximal humerus; and various other fragments, KA 64.

Dinofelis piveteaui (fig. 105)
Material:
The type specimen, which consists of a complete cranium, KA 61; right mandible piece with P_4, KA 62; left mandible piece with M_1, KA 63.

Family Hyaenidae
Crocuta crocuta
Material:
The type specimen of *C. crocuta ultra,* which consists of

Fig. 207. Animals represented by the fossils in the macrofaunal component of the Kromdraai A sample. Minimum numbers of individuals in each case are indicated.

many separate parts: snout and right zygoma, posterior part of braincase, left and right mandible pieces, various isolated teeth, thoracic vertebra and 7 vertebral disks, 8 rib pieces, proximal tibia, 10 metapodial pieces, 3 carpal bones, 22 phalanges, KA 58; the type specimen of *C. spelaea capensis,* an almost complete cranium, KA 56; left mandible piece with $\bar{C}$, P_2, and P_3, associated with 2 skull pieces, KA 57*a–c;* isolated right M_1, KA 59; isolated P^3, KA 60.

Hyaena bellax (fig. 209)
Material:
An almost complete cranium with articulated mandible, the type specimen, KA 55.

Hyaenid indet.
Material:
Right maxillary fragment with P^{3-4}, KA 538; part of a braincase, KA 1664; isolated teeth, KA 861, 954, 985, 1429, 1437, 1525.

Family Canidae
Lycaon sp.
Material:
Former type specimens of *Canis atrox,* isolated right M_1, KA 1288; left M^1, KA 1556.

Canis mesomelas
Material:
The type specimen of *C. mesomelas pappos,* which consists of left and right mandibular rami, isolated right I^2, and right M_1, KA 73*a,b;* complete cranium and half-mandible, KA 71; crushed skull, KA 88; maxillary pieces, KA 77, 85, 691, 1118, 1297; mandible pieces, KA 74, 75, 766, 986, 1456, 2086; isolated teeth, KA 76, 78, 79, 948; left scapula head, KA 72.

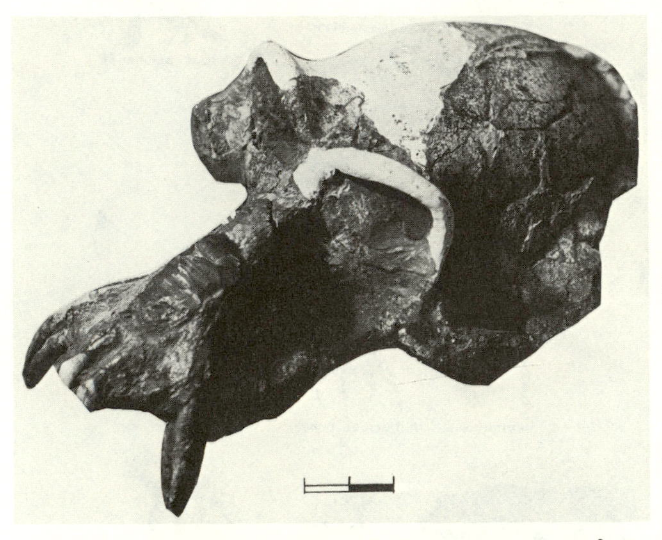

Fig. 208. A remarkably complete cranium of *Gorgopithecus major* from Kromdraai A.

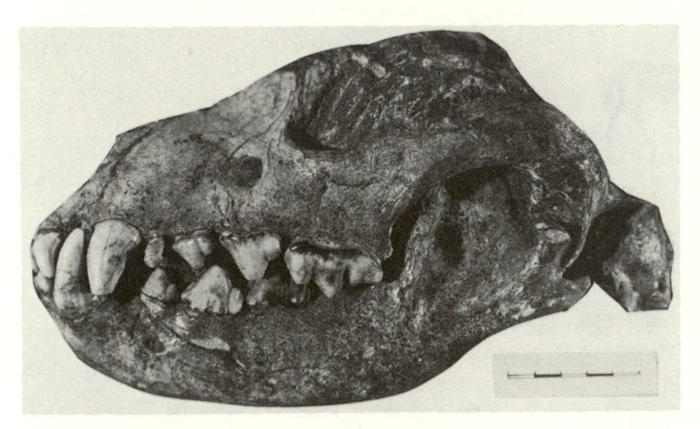

Fig. 209. The type specimen of *Hyaena bellax,* an extinct species related to the brown hyena, from Kromdraai A.

Canis terblanchei
Material:
The type specimen, consisting of most of a crushed cranium with mandible, KA 1290.

Vulpes pulcher
Material:
The type specimen, consisting of the left mandibular ramus with $\bar{C}$–M_3, KA 1289.

Family Viverridae
Herpestes mesotes
Material:
The type specimen, consisting of a cranium and mandible, 3 cervical vertebrae, left humerus, right proximal humerus, distal radius, and ulna fragments, 4 metapodials and phalanges, KA 86.

?Crossarchus transvaalensis
Material:
A snout with both toothrows, KA 1569.

Viverrid indet.
Material:
A cranium and endocast, KA 1488; right mandible piece with P_4 and part of M_1, KA 1508.

Carnivore indet.
Material:
Calvaria pieces, KA 681, 1677; maxillary pieces, KA 676, 1580, 1600, 2360; mandible pieces, KA 1444, 1570; atlas vertebra, KA 1164; scapula piece, KA 1723; ribs, KA 741, 1226; humerus, KA 642; distal radius, KA 1145; femur pieces, KA 547, 637, 643, 849, 1208, 1696; astragalus, KA 756; metapodial pieces, KA 629, 634, 795, 920, 943, 1037, 1054, 1126, 1148, 1773, 1788, 1815, 1842, 1875; phalanges, KA 553, 556, 593, 693, 932, 956, 996, 1612, 1647.

Carnivore coprolites
Material:
2 of jackal size, KA 1420; hyena size, KA 732, 1490, 1796.

Order Artiodactyla
Family Bovidae
Damaliscus sp. 1 or *Parmularius*
Material:
Complete but shattered cranium with articulated mandible, atlas, and axis, KA 1592 + 1716; right side of cranium with articulated mandible, atlas, and axis, KA 731; upper part of cranium, snout, and horn-core bases, right P^4 and M^3, KA 1601*a–c;* cranial piece and almost complete left mandible, KA 931*a,b;* maxillary and mandible pieces, KA 709*a–g;* base of right horn-core, KA 540; maxillary pieces, KA 514, 525, 526, 564, 587, 635, 646, 697, 731, 750*a,* 762, 877, 897, 960, 970, 991*a–d,* 994, 1127, 1619*a,* 1750, 2511*b,* 2512*a;* mandible pieces, KA 541, 566, 569, 576, 578, 646*a,* 700, 725, 728, 749, 751, 758, 770, 776, 833, 855, 858, 867, 913, 929, 935, 951, 969, 1004, 1010, 1101, 1102, 1122*b,* 1134, 1136, 1153, 1183, 1198, 1204, 1246, 1296, 1484, 1516, 1553, 1587, 1635, 1653, 1691, 1668,

1687*a*, 1739, 1747, 1774, 1775*a*, 1784, 1803, 1827*a*, 2353, 2450, 2511, 2608*a*, 2611, 2613; isolated teeth, KA 516, 565, 583, 585, 587, 606, 608, 611, 632, 656, 659, 665, 668, 673, 692*b*, 703, 718, 743, 780, 822, 832, 836, 838, 841, 862, 868, 872, 898, 916, 918, 920, 922, 926, 947, 1041, 1095, 1096, 1097*b*, 1104, 1108, 1113, 1115, 1117, 1140, 1168, 1170*a*, 1173, 1174, 1186, 1195, 1211, 1244*a*, 1258*a*, 1272, 1306, 1344, 1460, 1631, 1636, 1642, 1646, 1660, 1681, 1686, 1688, 1693, 1701, 1711, 1737, 1745, 1751, 1752, 1776, 1800, 1824, 2481, 2483, 2612.

Medium-sized alcelaphines
Material:
Maxillary pieces, KA 794*a*, 924, 1067, 1273, 1781, 2410, 2500; mandible pieces, KA 542, 680*a*, 1156*a,b*, 2453, 2497, 2514; isolated teeth, KA 532, 650, 771, 825, 906, 992, 1022, 1151, 1177, 1225, 1235, 1291, 1309, 1520, 1671, 2424, 2457, 2460, 2461, 2466, 2609.

Cf. *Connochaetes* sp.
Material:
Maxillary pieces, KA 1584*a*, 2513; mandible pieces, KA 740, 827, 883, 981, 1069, 1147, 1278; isolated teeth, KA 615, 782*a,b*, 865, 1066, 1293, 2409, 2449, 2482.

Cf. *Megalotragus* sp.
Material:
Left mandible piece with M_{1-2}, KA 1371; isolated left M_2, KA 1292.

Hippotragus cf. *equinus*
Material:
Left mandible piece with M_2, KA 2491.

?Hippotragini
Material:
Parts of an unprepared and unnumbered skull.

Redunca cf. *arundinum*
Material:
Right mandible piece with P_4, KA 1349.

Pelea cf. *capreolus*
Material:
Maxillary pieces, KA 527, 1164; mandible pieces, KA 903, 1149, 1285, 1766*b,c*.

Antidorcas recki
Material:
Almost complete skull, KA 1779; maxillary and mandible pieces, KA 603*a–f;* frontal pieces with horn-core bases, KA 1567, 1577; maxillary pieces, KA 765, 881*a*, 901, 925, 964*c*, 1046, 1111, 1213*a*, 1310, 1517*b*, 2501, 2610; mandible pieces, KA 506, 769, 821, 842, 964*a,b*, 1002, 1093, 1114, 1123*a–e*, 1205, 1517*a*, 1632, 1867, 2474; isolated teeth, KA 520, 679, 864, 881*b*, 1119, 1278*a,b*, 1352*a*, 1453*d*, 1639, 1817, 2512*b*.

Antidorcas bondi
Material:
Maxillary pieces, KA 1157, 2465; mandible pieces, KA 537, 648, 1162, 1163, 1676, 2172, 2464, 2508; isolated teeth, KA 999, 2472.

Cf. *Raphicerus* sp.
Material:
Right mandible piece with P_2–M_3, KA 1152.

Syncerus sp.
Material:
Isolated teeth, right M^2, KA 1630; right M^3, KA 752; left M^2, KA 1451; right M_2, KA 1268.

Tragelaphus cf. *scriptus*
Material:
Right mandible piece, P_2–M_1, KA 2498.

Tragelaphus cf. *strepsiceros*
Material:
Left maxilla, dpm^4, M^{1-3}, KA 644*a;* right mandible, dpm_4, KA 856; isolated teeth, KA 1269, 1298, 1616, 2451*b*, 2459, 2541*a–c*.

Taurotragus cf. *oryx*
Material:
Mandible piece, left M_3, KA 1303; isolated right M^2, KA 2469.

Antelope size class I
Material:
7 cranial and 56 postcranial pieces, as listed in table 111.

Antelope size class II
Material:
214 cranial and 299 postcranial pieces, as listed in table 111.

Antelope size class III
Material:
49 cranial and 73 postcranial pieces, as listed in table 111.

Antelope size class IV
Material:
2 cranial and 2 postcranial pieces, as listed in table 111.

Family Suidae
Phacochoerus modestus
Material:
The type specimen, consisting of a juvenile skull and mandible lacking the braincase and occiput; dpm^{2-4} and M^1 in wear, M^2 unerupted, KA 89.

Metridiochoerus andrewsi
Material:
The former type specimen of *Notochoerus meadowsi*, consisting of cranium with articulated mandible, complete except for posterior part of calvaria.

Order Perissodactyla
Family Equidae
Hipparion libycum
Material:
Isolated right M_2.

Equus burchelli
Material:
Isolated teeth: right M^1, KA 1213; left M^1, KA 1909; right M^3, KA 1130, 1906; left M_2, KA 1908.

Equus capensis
Material:
Maxillary pieces, KA 712, 1179, 1898; mandible pieces, KA 685, 1906*a–c*, TM 1250; isolated teeth, KA 534*a–c*, 575, 610*a–c*, 623, 624*a–d*, 729 + 1351, 735, 787*a,b*, 817*a–g*, 824, 860*a*, 869, 892, 899*a–e*, 910*a,b*, 927*a*,

939*a–f,* 966, 972*a–e,* 1086*a–e,* 1091, 1102*a–c,* 1162, 1624, 1648, 1684, 1703, 1713, 1772, 1879, 1895*a–j,* 1896, 1897, 1899, 1900, 1901*a,b,* 1909, 1910, 1902*a–d,* 1903, 1904*a,* 1905*a–c,* 1911*a, b,* TM 1123, 1187, 2445, 2507; femur pieces, KA 508, 557, 1008, 1112; phalanges, KA 796, 1505.

Equid indet.
Material:
Maxillary pieces, KA 1217; isolated teeth and tooth fragments, KA 641, 1044, 1129, 1158, 1192, 1234, 1377, 1652, 1596, 1708, 1743, 1789, 1846, 2475, 2478, 2489; proximal femur pieces, KA 557, 1008, 1112; distal femur piece, KA 508.

Order Hyracoidea
Family Procaviidae
Procavia antiqua
Material:
Most of a skull with 8 articulated vertebrae, KA 807; partial cranium, KA 3; calvaria pieces, KA 36, 1488; palates and maxillae, KA 11, 12, 15, 16, 21, 22, 28, 31, 34, 39, 54; mandible pieces, KA 6, 7, 8, 14, 17, 18, 19, 24, 27, 33, 44, 49, 1724; isolated incisor, KA 13.

Procavia transvaalensis
Material:
Complete cranium without mandible, KA 48; calvaria piece, KA 1173; maxillary pieces, KA 23, 25, 35, 53; mandible pieces, KA 1, 2, 4, 5, 9, 10, 26, 40, 43; isolated teeth, KA 29, 37, 42, 47; distal humerus, KA 50.

Hyracoid indet.
Material:
Almost complete cranium, KA 51; mandible piece, KA 1822; isolated incisors, KA 1216.

Order Lagomorpha
Family Leporidae
Lagomorph indet.
Material:
Complete cranium, KA 588; cranial piece, KA 1560; mandible pieces, KA 1233, 1316.

Order Rodentia
Family Hystricidae
Hystrix africaeaustralis
Material:
Right juvenile mandible with $\bar{I}$, dpm_4, M_{1-2}, KA 1432; left mandible piece, P_4–M_3, KA 757; isolated teeth, KA 674, 690, 1546, 1607.

Hystrix cf. *makapanensis*
Material:
Isolated right M^3, KA 1912.

Class Aves
Cf. *Columba* sp.
Material:
Part of an eggshell, KA 1167.

Class Reptilia
Order Chelonia
Family Testudinidae
Cf. *Testudo*
Material:
Carapace piece, KA 1198*a*.

Phylum Mollusca
Class Gastropoda
Order Pulmonata
Cf. *Achatina* sp.
Material:
Land snail shells, KA 736, 1265, 1458, 1529, 1810, 5991.

The Major Features of the Bone Accumulation

The Composition of the Fauna Represented by the Fossils

The information provided in table 110 and depicted in figure 207 may be further summarized. This is done in the first columns of table 112 and figure 210. In terms of numbers of individual animals, the fauna represented by the fossils is dominated by bovids (48%), followed by equids (15%), carnivores (12%), and baboons (11%).

The antelope remains came from animals of widely varying sizes: in size class I, 5 adult steenbok are represented; in class II, parts of 56 individuals are involved, of which 22 were juveniles; in class III, 23 individuals, including 10 juveniles, are present; and in the largest size group, IV, 7 individuals including 3 juveniles, are represented.

More individual equids are represented at Kromdraai A than in any other site unit; at least 23 individuals of the large extinct Cape horse, *Equus capensis,* are involved, of which 3 were juveniles. The 5 Burchell's zebras included 2 colts, and the single *Hipparion* was mature.

The Kromdraai A assemblage comprises a remarkable variety of carnivores, including a lion, 2 *Felis* species, 2

Fig. 210. Percentage contributions made by various groups of animals to the samples of macrovertebrate fossil fauna from Kromdraai A and B.

saber-toothed cats and a false sabertooth, 2 species of hyena, 4 different canids, and 2 mongooses.

Among the baboons, 1 species of *Parapapio* is involved and 2 of *Papio,* as well as 8 individuals of the remarkable animal *Gorgopithecus major* (fig. 208). Other animals include 6 individuals of the large dassie, *Procavia transvaalensis,* and 10 of the smaller form *P. antiqua.*

The Representation of Skeletal Parts

As will be seen in the listings under individual taxa, most identifications are based on cranial remains, and there may have been some artificial bias toward such parts in the collecting process. Like the Sterkfontein Member 4 and Swartkrans Member 1 assemblages, the Kromdraai A sample is likely to give a truer reflection of species present than of skeletal part representation. For this reason I am not inclined to attempt much in the way of deductions from the presence or absence of skeletal parts.

It is worth noting, however, in the bovid size class II, that the 56 individuals are represented by 299 postcranial pieces (table 111) that cover virtually the whole range of skeletal parts. Furthermore, these parts are not typically the most resistant ones, there being more proximal humeri, for instance, than distal ones (17 : 12). In the tibial fossils, 10 proximal ends are found in association with 4 distal ends. We may conclude that destructive action on the bones was not intensive, resulting in survival of resistant elements only, but rather unintensive, permitting preservation of parts with low survival ratings. Rapid burial by incoming sediment is also indicated.

Observed Damage to the Bones

Like the fossils from other site units, Kromdraai A bones were individually examined for characteristic damage marks. Results are presented in table 113 and may be summarized as follows:

Porcupine and Small-Rodent Gnaw Marks. Only 5 fossils in the assemblage showed evidence of porcupine gnawing, and 2 bore traces of small-rodent tooth marks.

Carnivore-Inflicted Damage. Clear evidence of carnivore damage was observed on 19 fossils, and damage probably attributable to carnivores was seen on 12 more specimens. Undoubted carnivore damage was seen on a *Gorgopithecus* mandible, on a wide range of antelope bones, and on craniums of both species of dassie.

Traces of Artificial Bone Alteration. None was observed.

Association with Artifacts. No stone artifacts have yet been found in the breccia from which the fossil sample came.

Clues to the Interpretation of the Kromdraai A Fossil Assemblage

Possible Porcupine Collecting Involvement

Only 5 out of 1,841 bone pieces were found to be porcupine-gnawed—a gnawed-bone incidence of 0.3%. One may confidently conclude that porcupine involvement in the collection of the bones was minimal. The few pieces gnawed by small rodents were probably worked upon where they lay in the cave, before being covered by sediment.

Possible Hominid Involvement

We have no evidence at present, in the form of either artifacts or artificial alteration of bone, to suggest hominid involvement in the bone accumulation.

A Possible Death Trap?

The Kromdraai A assemblage is unusual among the other Sterkfontein valley site units in that articulated skeletal parts (see fig. 211) are comparatively common. This could indicate that the Kromdraai A cave opening served as a natural death trap into which the animals fell. If this were so, one would expect the skeletons to be more complete than they are. Furthermore, that some of the bones show clear signs of carnivore chewing suggests that the site was more probably a carnivore lair than a death trap.

Kromdraai A: A Carnivore Lair?

It would be difficult, I think, to avoid the conclusion that the site had served as a carnivore feeding retreat, a breeding lair, or both. It is always tempting to simplify what must obviously have been a complex situation and to make a dogmatic statement on the course of events. I will try to resist this temptation. The carnivores known from the site consist of at least 5 species of felid, 2 species of hyena, 4 different canids, and 2 mongooses. The mongooses and similar canids may be excluded as significant accumulation agents; they are more likely to have formed part of the prey of the larger predators. The evidence suggests that some of the animals were killed and eaten in the cave by felids, perhaps sabertooths, that did comparatively little damage to the skeletons, while other, more fragmentary remains were left by hyenas. The felids may have been responsible for the articulated and relatively undamaged bovid specimens that characterize the Kromdraai A assemblage, together with the typically damaged dassie craniums, while the sparse and fragmented remains of equids could well represent parts brought back and discarded by hyenas.

In the course of her study of the fossil bovids, Elisabeth Vrba (1975, 1976*a*) pointed out that in size class II, 38% of the *Damaliscus* or *Parmularius* individuals (now recognized as *Parmularius* by Vrba 1978) were juveniles, suggesting that they had been consumed by primary predators within the cave. Interestingly enough, Vrba found

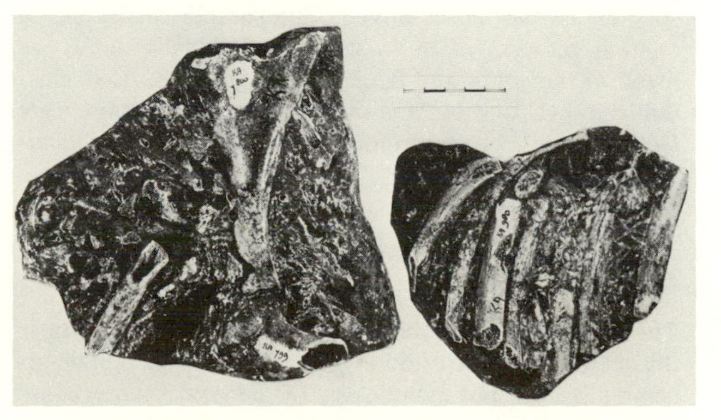

Fig. 211. Many of the fossils from Kromdraai A consist of articulated skeletal parts. Two examples are shown here.

that, as the bovid liveweight rose in the Kromdraai A assemblage, the proportion of juvenile individuals represented did not increase. It appeared that, whatever predator had been involved, it was powerful enough to cope with adult individuals of very large prey species. Suitable predators known from Kromdraai would have been *Homotherium, Megantereon,* and *Dinofelis,* and *Felis crassidens* may also have been involved in hunting the smaller prey, though we know nothing of this cat's habits. That 11% of the animals represented at the site were baboons suggests that the predators involved took baboons as a significant part of their prey spectrum.

As I have mentioned elsewhere in this work, the active hunting of large prey does not necessarily mean that single powerful predators must have been involved. Social carnivores, hunting in packs or clans, may certainly overpower large prey animals. Among the Kromdraai carnivores, both spotted hyenas, *Crocuta crocuta,* and a close relative of the brown hyena, *Hyaena bellax,* are represented. Both may well have hunted cooperatively and have brought back parts of their kills or their scavenged meals to the Kromdraai cave lair.

The carnivore/ungulate ratio (as defined in chap. 3) proves to be 25/123, or 20.3%. This is not as high as might be expected in an assemblage from a brown hyena feeding lair (see chap. 4), but it is higher than is typical in Stone Age human food remains. The figure is consistent with a mixed assemblage collected by both felid and hyenid occupants of the cave.

The Kromdraai A Bone Accumulation: Microvertebrate Component

As I mentioned in chapter 10, some breccia samples containing microvertebrate remains from the "Kromdraai Caves" were sent by D. Draper to the British Museum in 1895. The samples were rediscovered in the museum collections by L. S. B. Leakey in 1958, as reported by Oakley (1960). Preparation of the breccia blocks in acetic acid allowed de Graaff (1961) to list the animal species represented, and it now appears likely that the "Kromdraai Caves" referred to by Draper were, in fact, the surface exposures on the Sterkfontein hill. The samples studied by de Graaff came, in all probability, from Member 5 of the Sterkfontein Formation.

The breccia from Kromdraai A contains a fair scatter of microvertebrate bones, though these do not seem to be as concentrated as in parts of Sterkfontein Member 5 and Swartkrans Member 1. A large sample of Kromdraai A breccia has recently been prepared in acetic acid, and I have provisionally identified the microvertebrate bones, though further taxonomic work is needed before a full species list can be compiled. The sample was found to contain remains of at least 273 individual animals, as listed in table 114 and depicted in figure 212. It is clear that the owls that roosted in the Kromdraai A cave were feeding largely on dendromurines and on rats of the genus *Mystromys.*

The Kromdraai B Bone Accumulation: Macrovertebrate Component

The bones that form the sample analyzed here are those that came from the site between 1938 and 1956. The assemblage does not include any of the newer material resulting from Vrba's current excavation. The fieldwork of 1955–56 concentrated mainly on decalcified breccia along the northern wall of the deposit, which was explored to a depth of 4.9 m (16 ft). Microvertebrate bones were found in very large numbers, and these are still being studied by D. H. S. Davis and T. N. Pocock. Larger bones were not as abundant, but almost 5,000 were found, and these have been grouped according to their depth in the excavation:
Layer 1: surface–1.8 m (0–6 ft)
Layer 2: 1.8–3.7 m (6–12 ft)
Layer 3: 3.7–4.9 m (12–16 ft)

The stratification of the breccia is fairly steeply inclined from east to west, which means that the horizontal delineation indicated above has little meaning. Fortunately the new excavations being conducted by Vrba are already providing in situ fossils whose stratigraphic relationships are precisely known.

The sample being considered here consists of 4,985 specimens from a minimum of 94 individual animals, as listed in table 115 and depicted in figure 213. Particulars of 197 bones of bovid origin, assigned to size classes but not specifically identified, are given in table 116. Included in the fossil sample were 2,887 bone flakes; particulars of their lengths are given in table 117.

Data on the fossil material by which the various animal taxa are represented will now be presented:

Phylum Chordata
Class Mammalia
Order Primates
Family Hominidae
Australopithecus robustus

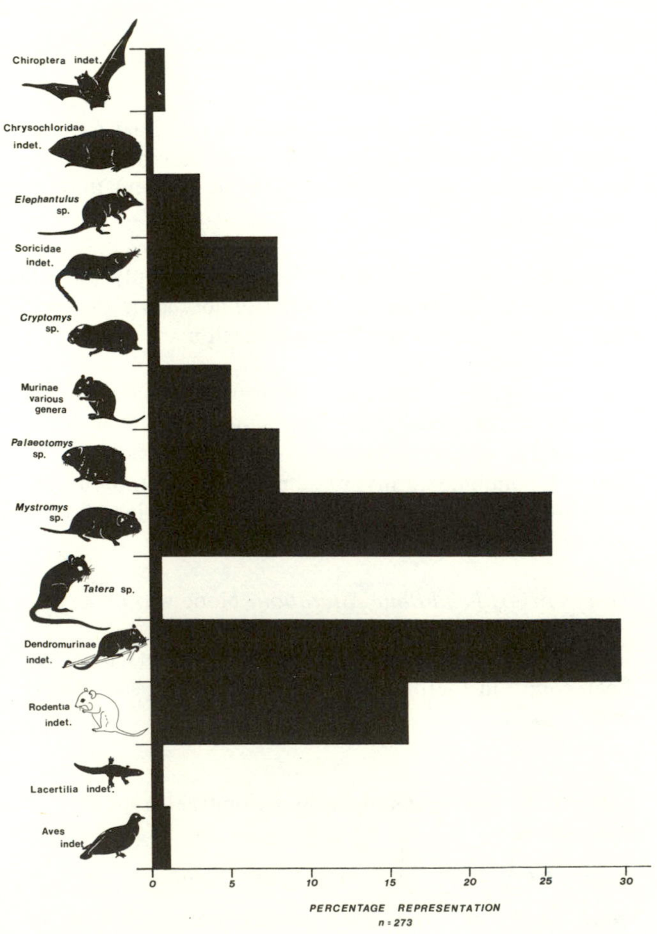

Fig. 212. Percentage representation of animals of various kinds in the microvertebrate component of the Kromdraai A sample. A total of 273 individual animals is involved, and the sample is dominated by dendromurines and the white-tailed rat, *Mystromys.*

TM 1517, holotype of *Paranthropus robustus*, consisting of left side of cranium including parietals, temporals, face, and palate; left maxilla with P^{3-4}, M^{1-2}; isolated crowns of right P^{3-4}, M^{1-3}; mandible piece with right P^{3-4}, M^{1-3}, and cast of right $\bar{C}$; isolated left P^{3-4}; proximal end of right ulna, distal end of right humerus; left metacarpal II; right talus; proximal phalanx (manus); proximal phalanx and distal phalanx (pedis) (fig. 214). Age estimate, 19–20 years.

TM 1603, left M^3 crown from same individual as 1517.

TM 1536, mandible of a juvenile with right I_1, di_2, $d\bar{c}$, dm_{1-2}, M_1; left $d\bar{c}$, dm_1. Age estimate, 2½ ± ½ years.

TM 1600, mandible fragments with left P_3, M_{2-3}. Age estimate 19 ± 1 years.

TM 1601, isolated teeth, left dm_1, crowns of $\bar{C}$, P_{3-4}, and left M^1. Age estimate, 2 ± 1 years.

TM 1602, right maxillary piece with roots of M^{1-3} and portion of palate. Age estimate, mature.

TM 1604, isolated left dm_2. Age estimate, 6 ± 1 years.

TM 1605, ilium, including part of acetabulum, but without iliac crest. Age estimate, mature.

Summarizing the results of his age estimations, Mann (1975) concluded that 2 individuals fall into the 1–5 year age class, 2 into the 6–10 year class, and 2 into the 16–20 year class, the latter being regarded as mature.

Family Cercopithecidae
Papio robinsoni

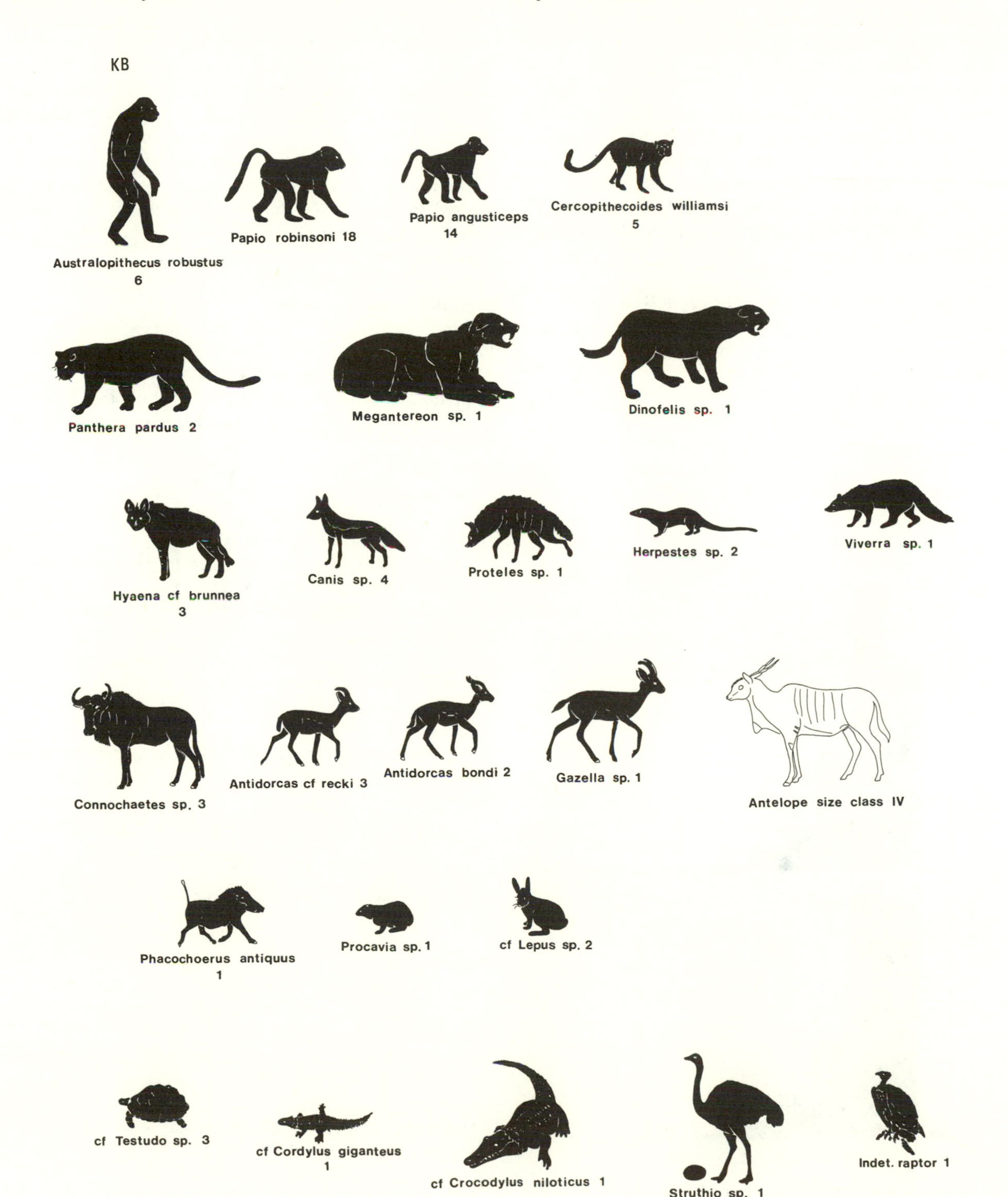

Fig. 213. Animals represented by fossils in the macrovertebrate component of the Kromdraai B sample. Minimum numbers of individuals are indicated in each case.

Material:
Layer 1: Juvenile specimens

Part of a juvenile maxilla with right dm^{1-2}, KB 687; fragments of maxilla and mandible with deciduous and unerupted permanent teeth, pieces numbered individually KB 688–703; isolated right di_1, KB 243, 270; right di_2, KB unnumbered; right $d\bar{c}$, KB 244, 269, 274, 282, 289; left $d\bar{c}$, KB 252, 259; $d\bar{c}$ fragment, KB 271; left dm_1 KB 272, 285, 286; right dm_1, KB 234, 249, 260, 273, 279; left dm_2, KB 261; right dm_2, KB 251, 281; left di^1, KB 226; right di^1, KB 239, 283; left di^2, KB 267, unnumbered; left and right $d\underline{c}$, each with two roots, a very odd occurrence for a primate, KB 233, 224; left $d\underline{c}$, KB 277, 278; right $d\underline{c}$, KB 228, 255, 258; left dm^1, KB 241, 734, 755, and unnumbered; right dm^1, KB 245, 265, 284; dm^1, side uncertain, KB 280; left dm^2, KB 738, 759; right dm^2, KB 747, 756, 767.

Layer 1: Adult specimens

Mandible fragment with parts of M_{2-3}, KB 21; anterior part of a mandible with P_3 bilaterally, KB 683; left I^1, KB 728; right I^1, KB 268; I^1, side uncertain, KB 736; left I_2, KB 237; right I^2, KB 238, 248, unnumbered; right $\underline{C}$, KB 230; right P^3, KB 749; right P^4, KB 246, 735; right M_1, KB 242, 742, 750; left M^2, KB 764; right M^2, KB 739, 748, 762, 763; left M_2, KB 682, 732; right M_2, KB 685; left M^3, KB 231, 765.

Layer 2

Isolated left $d\bar{c}$, KB 3390; right $d\bar{c}$, KB 3428; right I_1, KB 3357; right $\bar{C}$, KB 2919, 3346; right P_3, KB 3416; right M_1, KB 3360.

Layer 3

Isolated right di^1, KB 3225; right di_2, KB 3117; right dm_1, KB 3233; left dm_2, KB 3115; left I^1, KB 3112; left $\underline{C}$, KB 3223; molar crown, KB 3222.

The individuals may be allocated to age classes as indicated in table 118. Of the 18 individuals, 11 proved to be juveniles. For fuller particulars see Freedman and Brain (1972).

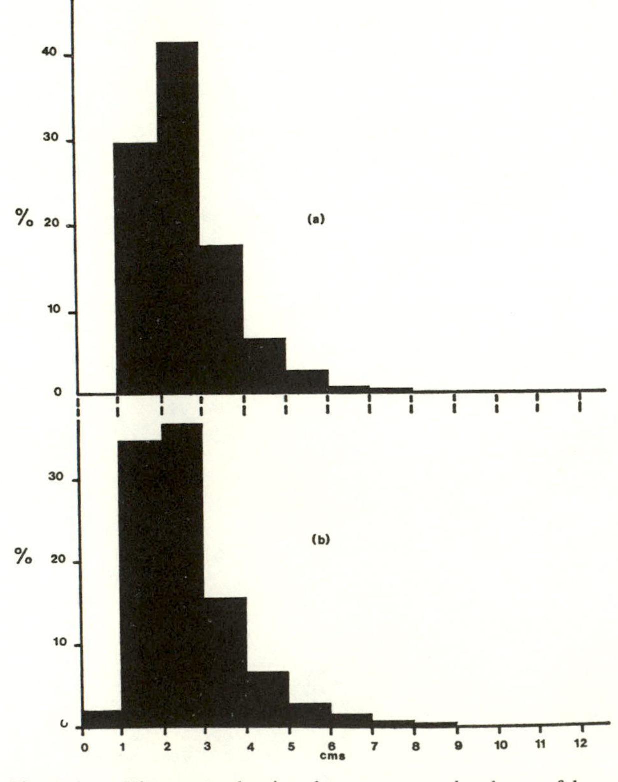

Fig. 214. *(a)* Histograms showing the percentage abundance of bone flakes of various lengths from Kromdraai B; the analysis is based on 2,887 pieces. *(b)* A similar analysis of all 4,985 fragments of bone in the sample from the deposit.

Papio angusticeps
Material:
Layer 1

Part of juvenile mandible, unerupted mesial incisors, KB 684; isolated left dm^1, KB 236, 254; right dm^1, KB 253; mandible with left and right $\bar{C}$–M_1, KB 680; mandible piece, right M_{2-3}, KB 686; mandible piece with right M_{2-3}, KB 197; mandible piece, ♀, left P_3–M_2, KB 196; isolated left I^1, KB 229, 2365; left P^3, KB 746; right P_3, KB 288; left P^4, KB 740; left M^1, KB 250, 251; right M^1, KB 751; right M_3, KB 257.

Layer 2

Isolated left $d\bar{c}$, KB 3427; left M^1, KB 3363.

Layer 3

Maxillary piece with left P^{3-4}, and ♂ canine socket, KB 3118; maxillary fragment with left M^3, KB 3120; isolated right dm_1, KB 3123; right ♂ $\underline{C}$, KB 3107; right P^4, KB 3227; right M^3, KB 3109; left M_3, KB 3114.

The individuals may be allocated to age classes as indicated in table 118. Of the 14 individuals, 3 proved to be juveniles. Fuller particulars may be had from Freedman and Brain (1972).

Cercopithecoides williamsi
Material:
Layer 1

Almost complete calvaria, without snout, KB 195; isolated left M^3, KB 225; right M^3, KB 256; unerupted upper molars, KB 745, 754; unerupted lower molars, KB 730, 758.

Layer 3

Left mandible fragment with dm_{1-2}, socket for M_1, KB 3108 + 2230; isolated right M_3, KB 3124; isolated upper molars, KB 3113, 3119, 3121, 3229.

The individuals may be allocated to age classes as indicated in table 118. Of the 5 individuals, 2 proved to be juvenile. Fuller particulars may be had from Freedman and Brain (1972).

Cercopithecoid sp. indet.
Material:
Layer 1

71 cranial and 194 postcranial pieces, as listed in table 119.

Layer 2

12 cranial and 37 postcranial pieces, as listed in table 119.

Layer 3

19 cranial and 34 postcranial pieces, as listed in table 119.

Order Carnivora
Family Felidae
Panthera pardus
Material:
6 postcranial pieces from a minimum of 2 adult individuals.

Layer 1

Distal radius, KB 2885; proximal ulna, KB 2901; navicular, KB 2903.

Layer 3

Left astragalus, KB 3259; phalanges, 3249, 3252.

?Megantereon sp.

Material:
9 postcranial pieces, probably from 1 individual.
Layer 2
Right astragalus, KB 2942; right calcaneus, KB 2946; metapodial pieces, KB 2932, 2933, 2935, 2937, 2939, 2948; 2d phalanges, KB 2934, 2938, 2940.

?Dinofelis sp.
Material:
A single postcranial piece.
Layer 3
Distal end of a metapodial, KB 3248.

Family Hyaenidae
Hyaena cf. *brunnea*
Material:
1 cranial and 2 postcranial pieces, probably from 3 individuals.
Layer 1
Isolated M_1, KB 295.
Layer 2
Distal metapodial, KB 2936.
Layer 3
1st phalanx, KB 3250.

Proteles sp.
Material:
A single cranial specimen.
Layer 2
Anterior part of right mandibular ramus with alveoli, KB 2945.

Family Canidae
Canis sp.
Material:
7 cranial and 2 postcranial pieces from a minimum of 4 individuals.
Layer 1
Right maxillary fragment with incisors and canine socket, KB 292.
Layer 2
Part of M_1, KB 2947; isolated canine, KB 3320; fragment of proximal metapodial, KB 2930; 2d phalanx, KB 2931.
Layer 3
Maxillary fragment, KB 3255; isolated incisor, KB 3258; isolated canines, KB 3246, 3253.

Family Viverridae
Herpestes sp.
Material:
2 cranial pieces from 2 individuals.
Layer 1
Isolated left P^4, KB 290.
Layer 2
Mandible fragment, KB 2944.

Viverra sp.
Material:
A single postcranial piece.
Layer 3
Distal humerus, KB 3258.

Carnivore indet.
Material:
16 cranial and 26 postcranial pieces, as listed in table 120.

Order Artiodactyla

Family Bovidae
Connochaetes sp.
Material:
6 cranial pieces from a minimum of 2 juveniles and 1 adult individual.
Layer 1
Left mandibular ramus with dm_{1-2}, right dm_3, KB 382, 383, 388; fragment of left horn-core, KB 376.
Layer 2
Isolated right dpm_3, KB 3009; M_2, KB 3006 + 3008.
Layer 3
Isolated left dpm^3, KB 3226; right dpm^4, KB 3178; horn-core fragment, KB 3187.

Antidorcas cf. *recki*
Material:
6 cranial pieces from a minimum of 2 or 3 individuals, including 1 juvenile.
Layer 1
Horn-core fragments, KB 79, 377, 381.
Layer 2
Isolated lower molar, KB 2999.
Layer 3
Isolated right M_3, KB 3224; fragment of horn-core base, KB 3190.

Cf. *Antidorcas bondi*
Material:
4 cranial pieces from a minimum of 2 adult individuals.
Layer 1
Horn-core pieces, KB 372 + 373, 375.
Layer 3
Horn-core piece, KB 3191.

Gazella sp.
Material:
1 cranial piece from 1 individual.
Layer 1
Frontal piece with left horn-core, KB 380.

Bovid incertae sedis
Material:
3 pieces from 1 adult individual.
Layer 3
Horn-core fragments, KB 3188 + 3193 + 3195.

Antelope size class I
Material:
11 cranial and 26 postcranial pieces from a minimum of 6 individuals, as listed in table 116.

Antelope size class II
Material:
26 cranial and 72 postcranial pieces from an estimated minimum of 10 individuals, as listed in table 116.

Antelope size class III
Material:
15 cranial and 45 postcranial pieces from an estimated minimum of 8 individuals, as listed in table 116.

Antelope size class IV
Material:
2 postcranial pieces from 1 individual, as listed in table 116.

Family Suidae
Phacochoerus modestus

Material:
1 cranial piece from a juvenile individual.
Layer 3
Isolated right M^2, unerupted, KB 3276.

Order Hyracoidea
Family Procaviidae
Procavia sp.
Material:
A single postcranial piece.
Layer 1
Distal humerus fragment, KB 7.

Order Lagomorpha
Family Leporidae
Cf. *Lepus* sp.
Material:
2 cranial and 4 postcranial pieces from a minimum of 2 individuals.
Layer 1
Isolated molar, KB 1476; atlas vertebra, KB 706; proximal humerus, KB 1819; distal humerus, KB 562; metapodial, KB 705.
Layer 3
Mandible fragment, KB 3254.

Class Aves
Struthio sp.
Material:
A single piece of eggshell, KB 1082, from Layer 1.

Bird indet. Large raptor
Material:
2 postcranial pieces from a single individual.
Layer 2
2 terminal phalanges, KB 3031, 3366.

Class Reptilia
Order Squamata
Cf. *Cordylus giganteus*
Material:
8 postcranial pieces, probably from 1 individual.
Layer 1
6 pointed scale plates, probably from the base of the tail, KB 2617; caudal vertebrae, KB 704, 707.

Order Crocodilia
Cf. *Crocodylus niloticus*
Material:
A single cranial piece.
Layer 3
Isolated tooth, KB 3235.

Order Chelonia
Cf. *Testudo* sp.
Material:
7 postcranial pieces from a minimum of 3 individuals.
Layer 1
Proximal femur, KB 10; limb bone pieces, KB 263, 583, 2732; plastron piece, KB 142; carapace piece, KB 903.
Layer 3
Coracoid, KB 3237.

Coprolites
Material:
1 of mongoose size, KB 39; 3 of jackal size, KB 37, 38, 40.

Indeterminate fragments
Material:
838 pieces from Layer 1; 212 pieces from Layer 2; 224 pieces from Layer 3; giving a total of 1,274 fragments.

Bone flakes
Material:
2,887 pieces as detailed in table 117.

Kromdraai B: The Major Features of the Bone Accumulation

The Composition of the Fauna Represented by the Fossils

The information provided in table 115 and depicted in figure 213 may be further summarized. This is done in the second columns of table 112 and figure 210. Whereas the fauna from Kromdraai A was dominated by bovids, that from Kromdraai B is similarly dominated by primates—australopithecines, two species of baboon, and a colobid monkey. Bovids are well represented by 6 individuals in class I, 10 in class II, 8 in class III, and 1 in class IV. Carnivores are represented by 15 individuals, and a crocodile has been identified on the basis of a single tooth.

The Representation of Skeletal Parts

The most characteristic feature of the sample from the decalcified breccia is the extreme fragmentation of the bones, making detailed skeletal part studies very difficult. What information I have been able to extract is presented in the various tables. It is worth noting that the australopithecine individual now regarded as the type of *A. robustus* was represented not only by the skull, but also by fragments of an arm, hands, and feet. The cercopithecoids were represented by 210 postcranial pieces (table 119) from most parts of the skeleton, and the same is true of the bovid postcranial remains (fig. 208). Unfortunately, the samples are too small and the fragmentation too great to allow worthwhile conclusions on the reasons for the presence or absence of skeletal parts in the sample.

Observed Damage to the Bones

The bones are generally so fragmented that it is difficult to be sure of the damage they may originally have suffered. What effects have been observed are detailed in table 121. A single piece was found to bear traces of porcupine-gnawing, and probable carnivore tooth marks may be seen on the australopithecine pelvis piece TM 1605 and on 3 other bones.

In connection with the *Australopithecus robustus* type skull, TM 1517*a*, a claim has been made for deliberate hominid violence. After initial description by Broom, the specimen was entrusted to G. W. H. Schepers for a study of its endocranial contours. It was his task to remove the matrix that filled the left parietal region of the braincase, and in this connection he wrote (Broom and Schepers 1946, p. 174):

> While excavating this matrix a large flint-like rock was found embedded in it. The parietal bone had been driven into the endocranial cavity ahead of this rock. It could not be preserved as it was necessary to undercut the rock by destroying the bone, otherwise it would have been well-nigh impossible to remove it.
>
> The presence of this rock is evidence suggestive of the claims that the Homunculi represented by the

australopithecoid and Plesianthropoid fossils, were skilled enough to employ missiles or weapons for defensive, offensive and predatory purposes.

In his paper on the predatory implemental technique of *Australopithecus,* Dart (1949*a*) provides a photograph of the internal aspect of the Kromdraai specimen with the stone to which Schepers refers still in place. From this it can be established that the stone was oval in outline, about 3.5 cm long and 3 cm across. Its thickness cannot be judged but appears to have been between 1 and 2 cm. The breccia matrix from which the specimen came is very rich in chert blocks and fragments derived from the dolomite country rock by normal weathering. These bear a superficial resemblance to flint, and it is extremely likely, though it cannot be proved, that the stone Schepers observed was a piece of chert. On the basis of the estimated size of the piece, its weight would have been about 75 g. If a stone of this size were to have penetrated the parietal bone of a living australopithecine, it would have had to be hurled at a very considerable velocity.

Since only a piece from the left side of the cranium has been found, it is impossible to establish whether the calvaria was more complete when it came to rest in the cave. One can argue that the entire right side of the cranium was missing before fossilization and that the stone, together with the associated matrix, simply filled in the hollow skull fragment, the stone causing damage to the parietal during the consolidation of the breccia.

In my opinion, Schepers's claim for deliberately inflicted injury to the Kromdraai skull is based on an uncritical appraisal of the evidence. A natural explanation for the observed damage can readily be found.

As is detailed in table 121, 70 bone pieces showed rounding and wear as if they had been water-worn. With them were a number of pebbles of similar appearance. As is apparent from the table, the great majority of worn bone pieces were between 1 and 3 cm in length. I am at a loss to explain the wear on these bones unless it was caused by a stream flowing through the cave. Perhaps similarly worn bones will be found in situ in the excavation being conducted by Vrba; if so, it should be possible to decide what mechanism caused the wear.

Association with Artifacts

Throughout the excavation of decalcified breccia at Kromdraai B, I examined all available stones for signs of artificial fracture. The result was that 1 unquestionable flake of chert was found (figured in Brain 1958, p. 98), together with 3 less convincing chert flakes and a broken pebble of quartzite. No artifacts have yet been found in the solid breccia.

Clues to the Interpretation of the Kromdraai B Fossil Assemblage

Possible Porcupine Collecting Involvement

Since only one bone out of a total of 4,981 shows evidence of porcupine gnawing, the involvement of porcupines as collecting agents can be discounted.

Possible Hominid Involvement

A single convincing flake has been found in decalcified breccia, but no artifacts are yet known from in situ breccia. On grounds of associated artifacts, hominid involvement is thus very dubious. In previous writing (Brain 1975*b*) I have suggested that the extreme fragmentation of the bone pieces (see fig. 214), and the fact that 57.9% of all pieces were bone flakes, indicated that hominids had broken up the bones with stone hammers. In fact, the condition of the Kromdraai B bones is very reminiscent of that to be seen in Later Stone Age human food remains as described in chapter 3.

Since originally writing on the subject, however, I have had serious reservations about the reasonableness of my conclusion. Could the fragmentation, for instance, have been a result of the breccia decalcification process? If they had been broken by hammerstones, why are not more artifacts found in the breccia? These questions cannot be answered at present, though I suspect they will be as the current Kromdraai excavation progresses.

Possible Carnivore Involvement

That remains of two leopards, of probable *Megantereon* and *Dinofelis* individuals, and of three brown hyenas are found in the assemblage strongly suggests involvement by these animals in the bone accumulation. Moreover the extremely high carnivore/ungulate ratio of 15/26, or 57.7%, adds support to the suggestion.

Conclusion

Vrba's new excavation will confirm whether or not the observed fragmentation of the bones was a result of decalcification. If it were, then the indications of hominid involvement would diminish considerably. In the interim I suggest that predation by cats and hyenas could best explain the origin of the fossil assemblage.

The Kromdraai B Bone Accumulation: Microvertebrate Component

The excavation of 1955–56 produced a wealth of microfaunal bones which were submitted to D. H. S. Davis for study. Preliminary results on his analysis of rodent remains from the upper levels of the excavation have appeared (Davis 1959, 1962) and may be tabulated as follows (Davis 1959, p. 151).

	Number	*%*
Cricetinae *(Mystromys)*	135	42
Elephant shrews *(Elephantulus)*	117	36
Otomyinae *(Otomys)*	44	13½
Bathyergidae *(Cryptomys)*	11	3½
Murinae *(Rattus, Mus, Rhabdomys, Dasymys)*	6	2
Gerbillinae *(Tatera)*	6	2
Shrews *(Crocidura, Suncus, Myosorex)*	2	½
Bats	2	½
Dendromurinae *(Steatomys, Dendromys)*	known to occur	

Davis remarked on the exceptionally high proportions of *Mystromys* and *Elephantulus* in the fossil sample—far higher than those he had observed in modern owl pellet accumulations from the Sterkfontein valley.

More recently T. N. Pocock (1970) has published the results of his analysis of bird remains from the Kromdraai B microfaunal accumulation. His faunal list and the interpretation based on it has already been discussed in the section of chapter 9 dealing with the class Aves.

13 A Note on Taung and Makapansgat

The project described in this book is concerned with fossil assemblages from the Sterkfontein valley caves; yet I wish to make a few passing comments on the assemblages from Taung and Makapansgat, the two other southern African sites that have yielded australopithecines, since these have a bearing on the questions at issue.

Taung

The original *Australopithecus* cranium, together with other faunal remains, came to light as a result of travertine quarrying at Buxton-Norlim, which by 1924 had destroyed the actual site where the hominid skull had lain. A few years after his announcement of *Australopithecus,* R. A. Dart described the bone accumulation as follows (1929, p. 648):

> Examination of the bone deposit at Taungs shows that it contains the remains of thousands of bone fragments. It is a cavern lair or kitchen-midden heap of a carnivorous beast. It is not a water-borne deposit and the Taungs remains could not have been washed into the cavern from the surface. The bones are chiefly those of small animals like baboons, bok, tortoises, rodents, bats and birds. Egg shells and crab shells have also been found. The fauna is one which is not characteristic of the lair of a leopard, hyena or other large carnivore, but is comparable with the cave deposits formed by primitive man. The deposit was, therefore, formed by primitive man or by Australopithecus, an advanced ape with human carnivorous habits. As no human remains have been found there, and as no Australopithecus remains have been found elsewhere in known Pleistocene deposits, I am of the opinion that the deposit was formed by the Taungs sub-man himself.

Since Dart wrote this account, we have learned a good deal about the nature of bone accumulations in caves, and I have little doubt that this Taung assemblage included a microvertebrate owl-collected component as well as a macrovertebrate component that is likely to have found its way to the cave in a variety of ways. The association of eggshells, crab carapaces, and a diversity of small animal remains is very suggestive of the possibility that the cave had been used as an eagle owl breeding place. As was described in chapter 6, the Cape eagle owl, *Bubo capensis,* has the habit of nesting in caves, frequently close to water, and of accumulating crab carapaces and other food remains around its nest site. Eggshells in such an assemblage are then, as likely as not, derived from the hatched eggs of the owls themselves. Remains of baboons and bucks will obviously represent a different accumulation pattern, to be discussed in a moment.

During the late 1940s the California Africa expedition team made a very thorough study of the Buxton-Norlim travertines (Peabody 1954) and excavated fossiliferous cave fillings adjacent to the spot where the *Australopithecus* skull had been found. Using all methods at his disposal, Peabody tried to reconstruct the stratigraphy of the original cave, concluding that there had been four deposits, each resulting from a different accumulation phase. From oldest to youngest, these were:

1. A fossiliferous, red sandy limestone, representing a dry climatic phase.
2. A much purer calcareous deposit, enclosing patches of fossiliferous sandy limestone and indicating a wet accumulation period.
3. A sterile deposit of red sand.
4. An unconsolidated black earth with Middle Stone Age remains.

Peabody concluded that the australopithecine skull had been preserved in the second, wet-phase deposit.

More recently Butzer (1974*b*) has reinvestigated the Buxton-Norlim travertines and the sediments enclosing the fossils. He has confirmed that the australopithecine specimen did, in fact, come from the wet-phase deposits, whereas most of the faunal remains, including the numerous baboon skulls, had been preserved in the older, dry-phase sediments. Thus, unfortunately, the taphonomic interpretation of the fossil assemblage as a whole has been greatly complicated.

Most of the numerous baboon skulls which came from Taung in the early days have lost their context relative to one another and to the strata in which they lay. In his paper "The Predatory Implemental Technique of *Australopithecus*," Dart (1949*a*) lists 21 baboon craniums from Taung, many of which he considered to show antemortem damage indicative of deliberate hominid bludgeoning. The depressed fracture in the right frontal bone of specimen 5365 (South African Museum catalog; the specimen was made available to me by courtesy of Q. B. Hendey) is shown in figure 215. About it Dart wrote (1949*a*, p. 27):

> 5365 ct. Skull and posterior (circumorbital) region of a juvenile (calvaria 3 mm thickness) face of in-

Fig. 215. A baboon skull, South African Museum catalog number 5365, from Taung, showing a depressed fracture in the right frontal bone.

determinate sex: no teeth present. Probably *Parapapio africanus,* Gear. Nature of the damage: A depressed (8 mm depth) fracture (30 × 27 mm) of right frontal bone severing frontal bone along the mid-line into the right orbit and along the right fronto-parietal suture, shattering the medial halves of the orbital roofs and partially crushing down the left orbit and muzzle. The frontal bone is distorted and cracked on both sides, but especially on the right side where the fractured portion is hinged along its parietal and sphenoidal sutural margin. This hinged region and the V-shaped island of bone left standing above the obvious depression of the cranium shows that the implement used to smash it was double-headed. The parietal bones also exhibit splits radiating from the site of the principal fracture. Remainder of braincase intact: no evidence of opening the skull.

Estimated cause of the damage: A direct downward blow upon the right frontal bone and muzzle delivered from the front with a V-shaped double-headed object having vertical internal borders or sharp margins (e.g., ungulate humerus) and measuring approximately 30 mm between the two heads (e.g., epicondylar ridges).

The damage to this and several of the other baboon skulls from Taung is very interesting, and it is not surprising that Dart's imagination was stirred. The skull, 5365, described above, had been completely filled with a stone-free sandy matrix, but, in addition to the depressed fracture, the entire calvaria shows signs of compressional stress. We will unfortunately never know whether a hard object, such as another fossil, lay in contact with the depressed area during fossilization and, in my opinion, will never be certain whether the damage resulted from an antemortem injury or from postdepositional compression.

Although all the baboon skulls recovered early from the site are now preserved as isolated specimens, it is fortunate that one large block of fossiliferous breccia, or at least a cast of it, is still in existence. The block was excavated by the 1948 California expedition in Pit 5*b* and is numbered 56831 in the catalog of the Museum of Palaeontology of the University of California at Berkeley. The block, which has been mentioned and figured by Freedman (1965), is shown in figure 216. It contains a number of relatively undamaged baboon skulls surrounded by postcranial skeletal pieces that show indisputable evidence of carnivore damage. I would not have the slightest hesitation in concluding that these fossils represent food remains of a carnivore, very probably

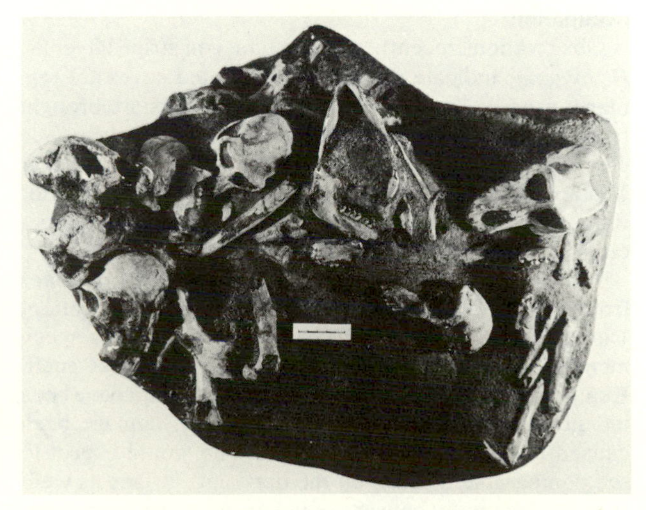

Fig. 216. A block of breccia from Pit 5*b* at Taung, showing numerous baboon skulls and carnivore-damaged postcranial bones.

a leopard, that was preying on baboons and using the cave as a feeding retreat.

That the baboon skulls came, in all probability, from a different accumulation phase from that of the australopithecine skull means that taphonomic deductions valid for the baboon fossils are not relevant to the hominid fossil. Unless a new cave deposit in the Taung travertine is located and excavated with due attention to taphonomic detail, the origin of the bone accumulations there will remain a matter of conjecture.

Makapansgat

I am aware that taphonomic studies of the Makapansgat fossil assemblage are being conducted by Alun Hughes, Judy Maguire, and James Kitching, and for this reason I have not done any research of my own there. I have simply made use of the accounts published by Dart (e.g., 1957*a*) in drawing certain deductions.

As I demonstrated in chapter 2, the remarkable disproportions in bovid skeletal parts that Dart observed for the first time in his gray breccia fossil sample need not be interpreted as the result of artificial hominid selection. The assemblage is clearly composed of the more resistant skeletal elements—those with high survival ratings as outlined in chapter 8—which are typically left when skeletons are subjected to a good deal of destruction that might include carnivore action.

At the time of Dart's original study, it seemed reasonable to assume that hyenas did not accumulate significant numbers of bones in cave lairs. As was discussed in chapter 4, however, current evidence indicates that hyenas *do* accumulate bones at their breeding lairs, and, if these lairs are in caves that have served as bone-preservation sites for thousands of years, the ultimate accumulations could be very considerable indeed. Some of the specimens Dart described, such as the "skull bowls," for instance, are in my opinion unquestionably the product of hyena gnawing.

Most of the gray breccia fossils have been extracted from blocks of rock blasted from the site by lime miners. The few pockets of bone-rich gray breccia I have seen still in situ at Makapansgat had accumulated in what were then low recesses beneath the dolomite roof, where this stepped down considerably around the edges of the major caverns. Such places were almost certainly dark or dimly lit and were too low to have been suitable for hominid occupation.

Observations recently made in Israel on striped hyenas, *H. hyaena,* indicate that dark, low-roofed caves are regularly used as breeding dens and that bones are brought back to such places in very considerable numbers (see chap. 4). *H. hyaena* is known to occur as fossils at Makapansgat and could well have been responsible for introducing many of the bones preserved in the gray breccia.

Although I cannot agree that the vast majority of bones from the gray breccia show signs of hominid modification, Dart and Kitching (fig. 217) have shown me some specimens bearing signs that could be interpreted as such. Examples are bovid metapodials (fig. 218*a*) that have been indented on one surface only. Had such damage been caused by the teeth of a carnivore, one would expect to see comparable damage on the opposing surface as well. A second category consists of bones that have been thrust into the cavities of other bones (fig. 218*b*). Again, hominid involvement could be invoked, although in a densely packed mass of 100,000 bone pieces, or more, the chance of certain bones' entering the cavities of others must be considered. If hominids had been involved, it would be interesting to know from which part of the very extensive cavern system such artificially modified pieces came. Had they perhaps been left close to the cave entrance and subsequently transported to the inner recesses of the cavern by resident porcupines or hyenas?

Over the years, a picture has developed in my mind of how the Limeworks cave may have looked when the bones were accumulating there. I visualize an extensive amphitheater that had resulted from a collapse of part of the cavern system's roof, while from this amphitheater openings to the cavern system we know today led downward. I visualize, too, a permanent water hole in the amphitheater, perhaps at the point where the Makapansgat stream descended into the subterranean chambers. Finally, I visualize large numbers of animals regularly visiting the water hole and some of them being killed there by carnivores that perhaps included hominids. Their bones would lie about in abundance within the catchment area of the cavern's mouth. Some would be modified by australopithecines, all would be worked over by scavengers, and large numbers would be transported to the inner recesses of the cavern by breeding hyenas and resident porcupines. While lying in the much disturbed sand around the fringes of the water hole, some of the bones

Fig. 217. James Kitching demonstrates bones from the Makapansgat gray breccia that show signs possibly attributable to hominid activity

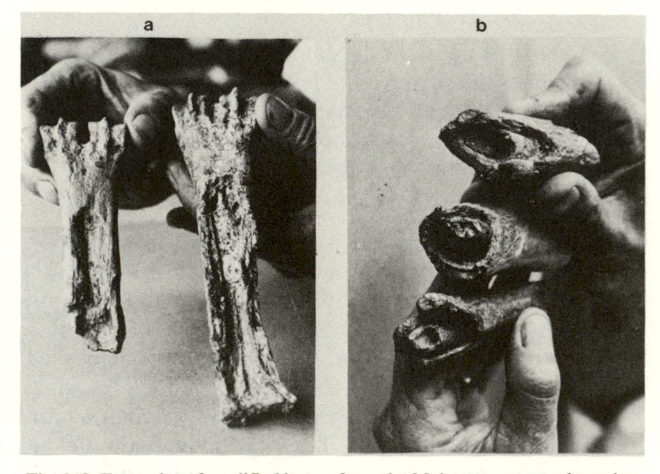

Fig. 218. Examples of modified bones from the Makapansgat gray breccia: *(a)* bovid metapodials indented on one surface only: *(b)* shafts of limb bones into which other bones have been thrust.

would acquire the wear and polish so characteristic of certain specimens in the gray breccia assemblage (see chap. 2 for a discussion on this "pseudotool" production process).

Like all other fossil assemblages in caves, the Makapansgat bones could be taphonomically interpreted with assurance only if they were excavated with due regard to subtle detail. I have no doubt that if an in situ deposit of bone-rich gray breccia could be stripped of its overburden and if the individual fossils could then be chipped out as they lay in a carefully controlled grid system, it would be possible to assess with confidence the accumulation pattern that originally operated. Such a task would be difficult, but it would be highly rewarding in the interpretation of a situation that has excited the imagination of paleontologists for years.

14 Who Were the Hunters and Who the Hunted?

Who Killed the Australopithecines? A Lineup of Potential Carnivore Culprits

On the basis of evidence presented in earlier chapters, I have been led to the conclusion that the australopithecine remains found in the three Sterkfontein valley caves were not from animals that went voluntarily to the sites to die. They came, rather, from hominids that had been taken to the caves by carnivores and eaten there, the fossils representing discarded food remains. I believe the same is true of the cercopithecoid remains that are abundantly associated with those of the australopithecines. If this interpretation is true, we have to account for the deaths of the following primate individuals represented in the three site units:

Sterkfontein Member 4: australopithecines, 47 individuals (17 subadult); cercopithecoids, 198 individuals (± 29 subadult).

Swartkrans Member 1: australopithecines, 87 individuals (58 subadult); cercopithecoids, 89 individuals (27 subadult).

Kromdraai B: australopithecines, 6 individuals (6 subadult); cercopithecoids, 37 individuals (17 subadult).

(In calculating these figures, "subadult" australopithecines are regarded as those with an estimated age-at-death of twenty years and less, although this age barrier may be regarded by some as too high; "subadult" cercopithecoids are taken to include those in the "juvenile" and "immature adult" age classes.)

What carnivores could have killed and partly eaten these primates? The possibility always exits that some unknown predator, whose remains are not represented at the sites, could have been involved. I admit this possibility but realize that further discussion about it would be fruitless. Of the carnivores known by remains in the relevant site units, felids and larger hyaenids must be considered as potential culprits; I am inclined to regard the aardwolf (an aberrant smaller hyaenid), the jackals, and the mongooses as most likely prey of the larger carnivores.

As figure 219 shows, the lineup of potential primate-killers represented in each of the three relevant site units is remarkably similar. In each case *Panthera pardus*, *Dinofelis* sp., and *Megantereon* sp. are represented, though identification of the latter two in Kromdraai B are somewhat tentative. In each site unit, hyaenid remains also have been found, those of "normal" hyenas—the brown and/or spotted—at all three sites, and those of "hunting" hyenas at Sterkfontein and Swartkrans. None of these carnivores can be positively excluded as potential killers of australopithecines or cercopithecoids, and the possible role of each will be considered further.

The Swartkrans Leopard Hypothesis

Some years ago I speculated (Brain 1970) on how the abundant australopithecine remains may have found their way into the Swartkrans cave. It was clear to me then that the very fragmentary hominid fossils probably represented carnivore food remains, and since leopards were well represented in the Swartkrans assemblage it seemed reasonable to assume that they had been involved. I had made the observation that when leopards are harried by spotted hyenas (whose remains were also known from Swartkrans) the leopards are obliged to take their prey into a tree or other inaccessible place. If they fail to do this, they are likely to lose their meal to the hyenas, which are generally dominant in a competitive feeding situation. In woodland areas the leopards simply take their prey up into the nearest tree, but on the open highveld, which appears to have been largely a grassland in Swartkrans Member 1 times as well, trees were certainly less abundant. In the dolomitic areas of the Transvaal highveld, large trees of the genera *Celtis* and *Ficus* are typically associated with the shaftlike openings of caves. In a generally treeless habitat, the cave entrances provide shelter from frost and fire to saplings that would not readily survive on the exposed hillsides. So the very trees available for leopard prey storage were those that overhung the shafts leading downward to the underground fossilization sites. I therefore speculated that, if in Swartkrans Member 1 times leopards were preying on australopithecines and baboons, they may well have fed on them in succeeding generations of trees that overhung the Swartkrans shaft. Food remains falling from the trees, fortuitously passing the waiting hyenas and gravitating into the subterranean cavern, ended up as the fossils by which the Swartkrans australopithecines are now known. This situation is stylistically depicted in figure 220.

The leopard hypothesis apparently was supported by the nature of the damage observed on the cranium of an australopithecine child, SK 54 from Swartkrans, which has been described and figured in chapter 11 (see fig. 197*a*). It consists of two holes, almost certainly caused by the canines of a carnivore, in the parietal bones. As is shown in figure 221, the spacing of these holes, 33 mm

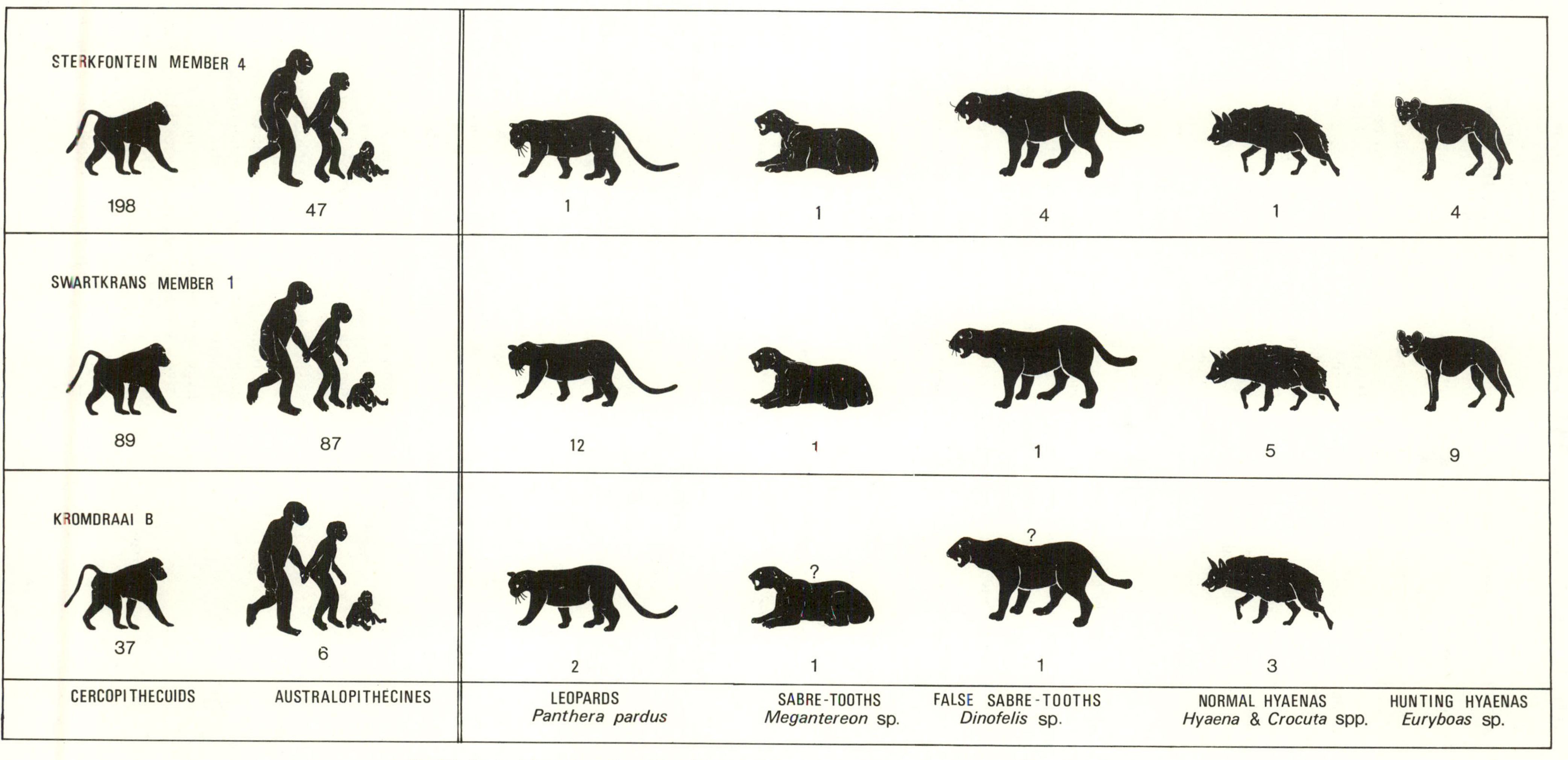

Fig. 219. On the left, primates from three site units whose deaths must be accounted for; on the right, a lineup of potential carnivore culprits. As will be apparent, these are remarkably similar in each case.

apart, precisely matches that of the lower canines of a leopard, SK 349, from the same deposit at Swartkrans. As described earlier (Brain 1970), the damage could have been caused in the manner portrayed in figure 222, where a leopard picked up by the head the child it had killed and dragged it to a secluded feeding place.

If the damage was inflicted by the canines of an adult carnivore, then the spacing suggests that the killer was a leopard rather than any of the other known carnivores (see Brain 1970 for tables and canine-spacing graph). The possibility cannot be excluded, however, that a juvenile of a larger carnivore was responsible.

Although this leopard hypothesis seemed plausible enough when it was formulated, and still does in certain respects, I have had reservations about its validity on two counts.

Reservation I

From Elisabeth Vrba's 1975 study of fossil bovids from the Sterkfontein valley caves, it became apparent that many of the fossils came from animals too large to have formed typical leopard prey. The relevant evidence, presented in detail in chapters 10, 11, and 12, is graphically depicted in figure 223. Class III bovids outnumber their class II equivalents in both Sterkfontein Member 4 and Swartkrans Member 1, and they form an important part of the assemblage at Kromdraai B as well. On the graph, this situation is contrasted with that in four samples of observed leopard prey, details of which are provided in chapter 4. Whereas the leopards in the four study areas concentrated almost exclusively on antelopes in the class II weight category, the three Sterkfontein valley assemblages under review here contained bones from significant numbers of heavier antelopes. Apart from these bovid remains, some of the fossils were also derived from other large animals, such as the extinct Cape horse, *Equus capensis,* adult specimens of which were certainly too bulky to have been transported to the cave by leopards.

Fig. 220. A stylized impression of the Swartkrans leopard hypothesis. An australopithecine is being consumed in a *Celtis* tree overhanging the entrance to a subterranean dolomite cave, while hyenas of several species wait hopefully below.

To explain the presence of remains of class III and IV bovids at Sterkfontein and Swartkrans, Vrba (1975) suggested that, in addition to leopards, false and true saber-toothed cats had been involved in the bone accumulation process. By extrapolation, one would be inclined to believe that these larger cats may also have participated in the primate-predation process.

Now that it is possible to view the fossil assemblages as a whole, I am convinced that contributions will have been made not only by the cats, but also by the hyenas. The fossil assemblage from Swartkrans Member 1 contains a hint that two separate predation patterns may have been involved in its accumulation. The first pattern is indicated by the bovid remains in which, of the 82 individuals represented, 25 came from the weight classes I and II, while 57 were derived from heavier antelopes. Of these larger animals, 12 were juveniles, which could conceivably have been caught by leopards, but 45 were adults that must have fallen prey to carnivores adapted to larger prey.

Turning to the primate component of the fossil fauna, a different pattern is discernible. Here five species are involved that clearly varied a good deal in liveweight, as is indicated by these tentative estimates: *Parapapio jonesi,* 25–40 lb, or 11.3–18.1 kg; *Papio robinsoni,* 35–80 lb, or 15.9–36.3 kg; *Theropithecus danieli* and *Dinopithecus ingens,* 60–100 lb, or 27.2–45.4 kg; *Australopithecus robustus,* 80–150 lb, or 36.3–68.0 kg. There appears to be a correlation between the estimated liveweight of these primates and the percentage of subadult individuals represented in each taxon. Figures are as follows: *Parapapio jonesi,* 8 individuals, 2 subadult: subadult percentage 25; *Papio robinsoni,* 38 individuals, 9 subadult: subadult percentage 23.7; *Theropithecus danieli,* 17 individuals, 6 subadult: subadult percentage 35.3; *Dinopithecus ingens,* 26 individuals, 10 subadult: subadult percentage 38.5; *Australopithecus robustus,* 87 individuals, 58 subadult: subadult percentage 66.7 (if the subadult/adult transition is taken as 16 rather than 21 years, then the number of subadult individuals would be 42 and the subadult percentage 48.3).

These data are depicted graphically in figure 224, from which it will be seen that, as the estimated liveweights of the primates increase, so also do the percentages of subadult individuals. A possible interpretation is that the predator involved selected, by choice, primates with liveweights of about 30–100 lb, or 13–45 kg. This pattern is very different from that suggested by the antelope remains, where the majority of prey animals had liveweights well over 200 lb, or 90 kg.

While the primates could well have fallen prey to leopards, the heavier bovids could represent the prey of the larger cats, *Dinofelis* or *Megantereon,* or of socially hunting hyenas, particularly *Euryboas* or *Hyaenictis*. The bovid remains could equally well have been brought back by brown or spotted hyenas, which appear to have frequented the cave.

The conclusion I draw here is that, if the cave had been used by several species of cat as well as several different hyenas, it would be remarkable indeed if the bone accumulation did not include contributions made by each of these carnivores. The indications are that this was, in fact, the case.

Reservation 2

My second reservation about the validity of the Swartkrans leopard hypothesis concerns the preponderance of primates in the prey sample. As figure 196 indicates, more than 50% of all the animals in the Member 1 macrovertebrate assemblages are either hominids or cercopithecoids. This is a remarkable state of affairs, and one is prompted to ask why, if leopards were the main predators of these primates, they were concentrating so intensively on this kind of prey. Studies of contemporary

Fig. 221. Some evidence for the Swartkrans leopard hypothesis: the spacing of holes in the parietal bones of a juvenile australopithecine cranium, SK 54, matches closely the spacing of the lower canines of a leopard, SK 349, from the same deposit.

leopards, described in chapter 4, suggest that primates typically form only a small part of leopard diet, except in particular circumstances such as the baboon sleeping situation on Mount Suswa.

Evidence presented in chapter 11 suggests that leopards had been the main predators of the very numerous small springbok, *Antidorcas bondi,* whose remains dominate the Swartkrans Member 2 assemblage. The leopards appear to have used the cave itself, or trees overhanging its entrance, as a feeding lair, to which they brought back the bodies of their prey. The composition of this assemblage conforms closely to what might be expected for leopards hunting over a grassland area in normal circumstances. Antelopes in size class II were the dominant prey, and baboons constituted only 3% of animals represented.

The question, and crux of my second reservation, therefore is this: Why, if the Swartkrans leopards hunted a "normal" spectrum of prey in Member 2 times, did they concentrate so heavily on primates during the accumulation of Member 1? I believe the answer to this question will elucidate the situation not only for the Swartkrans australopithecines, but also for the primate-dominated assemblages in Sterkfontein Member 4 and Kromdraai B.

There appear to be two possibilities, which I will discuss as hypotheses A and B.

Hypothesis A: A Specialized Predator of Primates? It would be possible to account for the abundance of primate remains at the three caves if these places had served as lairs for specialized predators that hunted baboons and hominids to the virtual exclusion of other prey. No living African carnivores are known to do this, except in special circumstances such as those described in Hypothesis B, below. Nevertheless, we cannot exclude the possibility that some of the extinct carnivores may have been so adapted.

Hominids and baboons must be regarded as dangerous prey for any predator—not because the individual primate is a formidable adversary, but simply because an attack on one individual is likely to precipitate retaliation by the whole band unless the predator operates at night or silently takes a straggler from the fringes of the group.

In the Sterkfontein valley context the potentially specialized predators of primates were the false sabertooths, *Dinofelis,* the true sabertooth, *Megantereon,* and the hunting hyenas of the genera *Euryboas* and *Hyaenictis.* To my mind, *Dinofelis,* would have been well suited for the task: it was clearly a cat that hunted by stealth, it had

Fig. 222. A reconstruction to account for the damage observed on the skull of an australopithecine child, SK 54. The lower canines appear to have perforated the parietals while the uppers gripped the child's face. From Brain 1970.

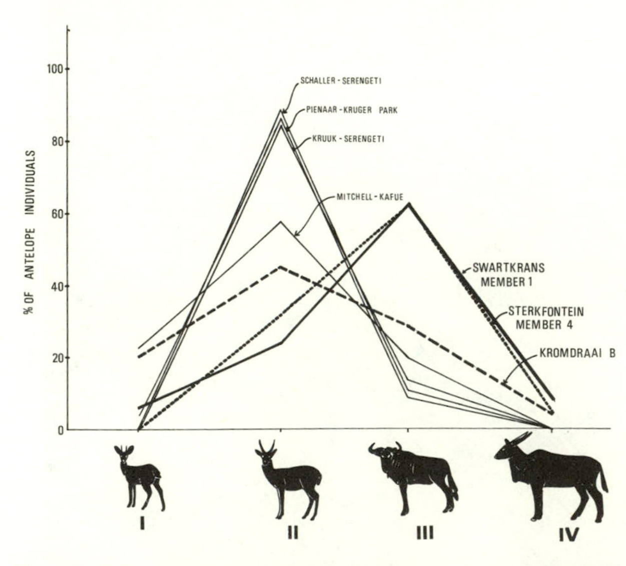

Fig. 223. The percentage representation of bovids of different weight classes in four leopard prey samples, described in the text, and in the fossil assemblages from Sterkfontein Member 4, Swartkrans Member 1, and Kromdraai B.

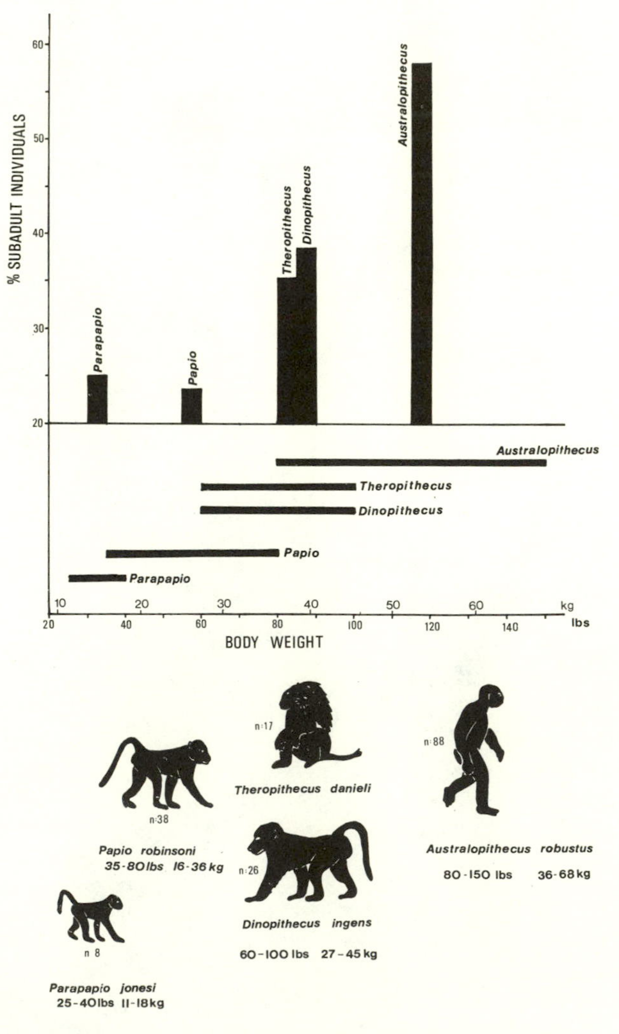

Fig. 224. Histograms showing the percentage abundance of subadult individuals among primates of different adult liveweights, whose fossils are preserved in Swartkrans Member 1.

powerful forequarters to hold the prey in position and long canines to ensure rapid killing. A combination of robust jaws and a well-developed crushing component in the dentition would have allowed *Dinofelis* to eat all parts of a primate skeleton except the skull. The hypothesis that *Dinofelis* was a specialized killer of primates is persuasive.

The true sabertooth, *Megantereon,* was perhaps less suited to the role. Its extremely long canines may have been more effective on prey larger than australopithecines or baboons where there was less danger of damaging the sabers on bones of the prey skeleton. Its postcanine dentition was so highly adapted for slicing meat that it could have caused very little damage to even a primate skeleton. The extensive damage observed in the fossil primate skeletons, particularly the disappearance of postcranial elements, is unlikely to have resulted from *Megantereon* action alone. Superimposition of hyena action upon that of the sabertooths would have to be invoked.

We know regrettably little about the behavior of the extinct hunting hyenas. They appear, from the abundance of their fossils, to have frequented caves, perhaps when rearing their young there. They were cursorial hunters, presumably running down their prey after the manner of wild dogs. It is not impossible that they had acquired the habit of hunting primates and of taking parts of this prey back to their breeding lairs.

In conclusion, it would be unrealistic I think, to ignore the possibility that both *Dinofelis* and the hunting hyenas could have been specialized primate-killers. But the time, regrettably, is long past when they might be observed as they went about their business.

Hypothesis B: A Sleeping-Site Situation? In the first hypothesis I considered the possibility that predators actively went out to hunt primates and brought them back to the cave lairs. The main postulate of the second hypothesis is that the primates actually came to the caves by their own volition and were killed and eaten there by carnivores that were exploiting the situation in an opportunistic manner. Such a situation has already been described in chapter 4 where, on Mount Suswa in Kenya, baboons regularly come to sleep around the edges of sinkholes. Resident leopards kill them there and drag the bodies into subterranean lava tunnels to eat them undisturbed.

The Transvaal highveld is characterized by cold winter nights, and it is known that baboon troops living in this area use caves as sleeping places during the winter months. Over a number of years I have made observations at a dolomite cave 12 km north-northeast of Swartkrans, on the farm Uitkomst, which is used by a baboon troop as a winter sleeping site. The cave opens into the side of a rocky hillside by way of a high, narrow doorway (fig. 225) partly obscured by a large *Ficus* tree. Inside, a steeply inclined talus slope leads down into a large cavern whose form is shown in figure 226. The cavern is dominated by a gigantic block of dolomite that has parted from the roof and come to rest on the floor. The flat top of this block, determined by a dolomitic bedding plane, is where the baboon troop normally sleeps, and an extensive deposit of baboon droppings has accumulated around the base of the block, as indicated on the plan. Lighting in the sleeping area is very subdued—on a bright day it is just possible to make out

the internal contours of the cave. I have seen a troop of about 30 baboons use the cave on nights between May and September; during warmer months they use an open cliff face about 1 km to the northeast. The baboons typically enter the cave shortly before sunset and leave it after sunrise. On one occasion I hid inside the cavern, making my presence known only after the baboons had taken up their sleeping places. Although pandemonium broke out in the cave, the baboons could not be induced to leave the place in the dark.

In view of a baboon's characteristic fear of the dark, some observations by Chris Gow (1973) are of special interest. He described a sinkhole 30 m deep in Tertiary limestones near Bredasdorp in the southeastern Cape that is regularly visited by a baboon troop. At the base of the sinkhole a dark passage extends 130 m to a subterranean stream, and this running water attracts the baboons, which are prepared to brave the darkness to reach it. After they drink, the baboons generally spend the night in the cave.

The current distribution of baboons on the Transvaal highveld is certainly regulated by the availability of suitably sheltered sleeping sites (L. P. Stoltz, pers. comm., and personal observation). Areas devoid of such sites do not support resident troops.

So it was, perhaps, during australopithecine times. I have suggested in chapter 3 that Sterkfontein valley hominids may have participated in a seasonal round that took them into warmer bushveld areas during the coldest months and brought them back onto the highveld in the spring. Although baboons do not do this today, australopithecines may have, in which case they probably used cave sleeping sites only on cold nights in spring and autumn.

Erosion has removed the original entrance areas of the Sterkfontein, Swartkrans, and Kromdraai caves, which must have been intact when the australopithecines inhabited the valley. Yet the abundance of owl-pellet-derived microvertebrate bones in the cave deposits argues in favor of dimly lit entrance areas with suitable owl roosts beneath their overhanging roofs. Such places may well have attracted baboons and hominids as sleeping places, while the deeper recesses of the caves were used as lairs and feeding retreats by the cats who preyed upon the primates.

Fig. 225. The entrance and talus cone in the baboon sleeping cave on the farm Uitkomst, at the edge of the Transvaal highveld.

Conclusions

Some of the Swartkrans evidence suggests that leopards preyed on primates there. The same may well have been true at the caves of Sterkfontein and Kromdraai. If leopards were involved, however, the overwhelming preponderance of primate remains suggests that a sleeping site was being exploited rather than that the leopards were hunting normally in open country.

If the abundance of primate remains in the Sterkfontein valley caves can be attributed to opportunistic predation at sleeping sites, as I think it can, it is probable that more than one species of cat was involved in the predation process. Besides leopards, it is higly likely that *Dinofelis* cats also participated in this process.

The suggestion that *Dinofelis* and the hunting hyenas were active, selective predators of primates cannot be proved or disproved—the matter therefore remains open.

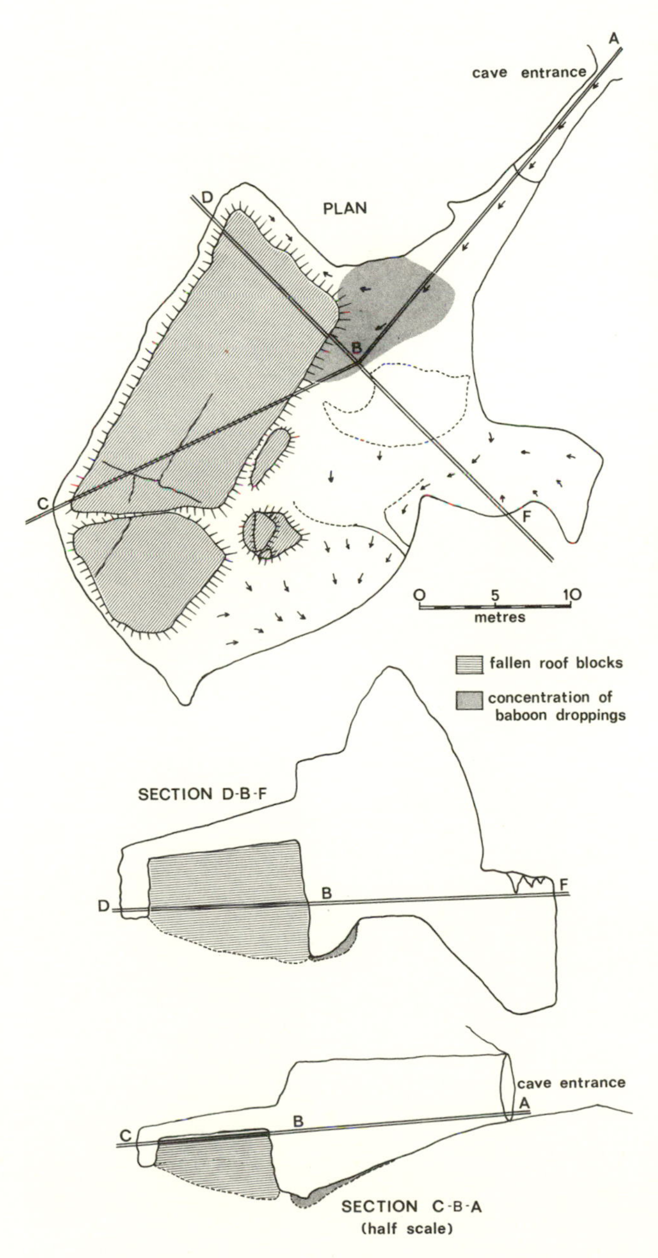

Fig. 226. The Uitkomst baboon sleeping cave in plan and section. The baboon troop spends winter nights on the fallen roof blocks inside the main cavern.

It is to be expected that each of the three bone accumulations has resulted from the independent activity of cats *and* hyenas. From the point of view of taphonomic interpretation, this complicating fact is as unfortunate as it is true.

The Hunted Become the Hunters

At Sterkfontein, the interface between the top of Member 4 and the bottom of Member 5 represents a time interval crucial in the course of human evolution. During this interval the gracile australopithecines disappeared from the Transvaal scene and the first men appeared. In this interval, too, the evolving men mastered a threat to their security that had been posed by the cave cats over countless generations. During Member 4 times the cats apparently controlled the Sterkfontein cave, dragging their australopithecine victims into its dark recesses. By Member 5 days, however, the new men not only had evicted the predators, but had taken up residence in the very chamber where their ancestors had been eaten.

How the people managed this is not recorded, but it could surely have been achieved only through increasing intelligence reflected in developing technology. It is tempting to suggest that the mastery of fire had already been acquired and that this, together with the development of crude weapons, tipped the balance of power in their favor. The tipping of this balance represented, I think, a crucial step in the progressive manipulation of nature that has been so characteristic of the subsequent course of human affairs. It was a step the robust australopithecines apparently failed to take, and their extinction was doubtless hastened by predators they were powerless to control.

Yet the evidence, which flowed first from the researches of Elisabeth Vrba, does not suggest that the first men to live at Sterkfontein were more than amateurs at hunting. The nature of their antelope food remains, preserved in the upper levels of the cave, suggests that they depended heavily on the kills of professional carnivores *before* they progressively developed their own prowess as

hunters. This interesting and significant conclusion is corroborated by the taphonomic studies described in this book.

Postscript

In many ways, the first part of this book is more satisfactory than the second. It explores the potential usefulness of bone accumulations in caves as indicators of past events. For the Sterkfontein valley assemblages used in the study, only a small part of this potential could be realized because the fossils had not been recovered with the regard for subtle detail that is a prerequisite for reliable taphonomic reconstructions. We may hope the guidelines that have emerged in the course of this study will help ensure that fossil assemblages as yet unearthed will not lack the detailed information so vital for their proper evaluation.

Appendix: Tables

Table 1. Separation of the Extant Southern African Bovid Fauna into Size Classes, Based on Liveweights

Species	Weight Range Lb	Weight Range Kg
Antelope class I		
Dikdik, *Madoqua kirki*	10–11	4.5–5
Suni, *Nesotragus moschatus*	10–15	4.5–7
Blue duiker, *Cephalophus monticola*	12–16	6–7
Cape grysbok, *Raphicerus melanotus*	15–20	7–9
Sharpe's grysbok, *Raphicerus sharpei*	15–20	7–9
Red duiker, *Cephalophus natalensis*	20–30	9–14
Klipspringer, *Oreotragus oreotragus*	21–36	10–16
Steenbok, *Raphicerus campestris*	24–33	11–15
Common duiker, *Sylvicapra grimmia*	24–45	11–21
Oribi, *Ourebia ourebi*	30–42	14–19
Antelope class II		
Springbok, *Antidorcas marsupialis*	40–115	18–52
Mountain reedbuck, *Redunca fulvorufula*	50–60	23–27
Gray rhebuck, *Pelea capreolus*	50–60	23–27
Bushbuck, *Tragelaphus scriptus*	50–182	23–83
Blesbok, *Damaliscus dorcas*	70–180	32–81
Impala, *Aepyceros melampus*	80–151	36–69
Reedbuck, *Redunca arundinum*	98–228	45–104
Puku, *Kobus vardoni*	124–84	56–84
Antelope class III		
Lechwe, *Kobus leche*	170–285	77–130
Nyala, *Tragelaphus angasi*	200–250	91–114
Sitatunga, *Tragelaphus spekei*	200–250	91–114
Tsessebe, *Damaliscus lunatus*	258–350	117–58
Red hartebeest, *Alcelaphus bucelaphus*	232–380	106–72
Lichtenstein's hartebeest, *Alcelaphus lichtensteini*	322–450	146–205
Kudu, *Strepsiceros strepsiceros*	330–651	150–296
Black wildebeest, *Connochaetes gnou*	350–400	158–82
Defassa waterbuck, *Kobus defassa*	350–450	158–205
Waterbuck, *Kobus ellipsiprymnus*	350–600	158–272
Gemsbok, *Oryx gazella*	400–524	182–238
Sable, *Hippotragus niger*	450–580	205–64
Blue wildebeest, *Connochaetes taurinus*	450–602	205–74
Roan, *Hippotragus equinus*	491–658	223–99
Antelope class IV		
Buffalo, *Syncerus caffer*	808–1,841	367–837
Eland, *Taurotragus oryx*	870–2,078	396–945

Table 2. Details of the Inhabitants of Eight Villages on the Kuiseb River, South-West Africa, in March 1966

Village	Human Males	Human Females	Dogs	Goats
Ossewater	5	6	8	460
Natab	2	4	7	188
Soutrivier	10	11	3	177
Klipneus	7	4	6	200
Swartbank	3	4	3	125
Itusib	6	6	3	44
Ururas	17	25	6	320
Rooibank	11	12	4	240
Total	61	72	40	1,754

Table 3. List of Goat Bones Collected at Kuiseb River Hottentot Villages

Skeletal Part	Numbers	Skeletal Part	Numbers
Skull	*512*	*Radius and ulna*	*207*
Horns and cores	385	Complete bones	3
Cranial fragments	70	Proximal ends	62
Maxillary fragments	57	Distal ends	19
		Shaft fragments	123
Mandible	*188*		
Complete half-mandibles	38	*Femur*	*115*
Mandible fragments	150	Proximal ends	18
		Distal ends	9
Loose teeth	*15*	Shaft fragments	88
Vertebrae	*115*	*Tibia*	*237*
1st cervical (atlas)	12	Proximal ends	13
2d cervical (axis)	14	Distal ends	72
Other cervical	12	Shaft fragments	152
Thoracic	21		
Lumbar	31	*Metacarpal*	*100*
Sacral	1	Complete bones	8
Caudal	0	Proximal ends	24
Fragments	24	Distal ends	15
		Shaft fragments	53
Ribs	*174*		
		Metatarsal	*101*
Scapula	*59*	Complete bones	9
Head portion	28	Proximal ends	30
Other fragments	31	Distal ends	11
		Shaft pieces	51
Pelvis	*55*		
Acetabular portion	34	*Astragalus, complete*	*16*
Other fragments	21		
		Calcaneus, complete	*14*
Humerus	*196*		
Proximal ends	0	*Phalanges, complete*	*21*
Distal ends	82		
Shaft fragments	114	*Bone flakes*	*248*
		Total	*2,373*

Table 4. Estimated Ages for Kuiseb River Goats Based on Tooth Eruption in Left and Right Half-Mandibles

Age Class	Number of Goats	
	Left Side	Right Side
Under 6 months	1	0
9–12 months	17	23
15–30 months	7	6
More than 30 months	28	35
Total	53	64

Table 5. Parts of Kuiseb Goat Skeletons Arranged in Descending Order of Percentage Survival

Part	Number Found	Original Number	Percentage Survival
Half-mandibles	117	128	91.4
Humerus, distal	82	128	64.0
Tibia, distal	72	128	56.3
Radius and ulna, proximal	65	128	50.8
Metatarsal, proximal	39	128	30.4
Scapula	35	128	27.4
Pelvis, half	34	128	26.6
Metacarpal, proximal	32	128	25.0
Axis vertebrae	14	64	21.9
Atlas vertebrae	12	64	18.8
Metacarpal, distal	23	128	18.0
Radius and ulna, distal	22	128	17.2
Metatarsal, distal	20	128	15.6
Femur, proximal	18	128	14.1
Astragalus	16	128	12.5
Calcaneus	14	128	10.9
Ribs	170	1,664	10.2
Tibia, proximal	13	128	10.1
Lumbar vertebrae	31	384	8.1
Femur, distal	9	128	7.0
Cervical 3–7 vertebrae	12	320	3.8
Phalanges	21	768	2.7
Thoracic vertebrae	21	832	2.5
Sacrum	1	64	1.6
Caudal vertebrae	0	1,224	0
Humerus, proximal	0	128	0

Table 6. Ages at Which Fusion of Epiphysis to Shaft in Limb Bones of Uncastrated Goats Occurs (in months)

Part	Age Range	Median
Humerus		
Proximal	23–48	36
Distal	11–13	12
Radius		
Proximal	4–9	7
Distal	33–48	41
Metacarpal		
Proximal	before birth	
Distal	23–36	30
Femur		
Proximal	23–36	30
Distal	23–48	36
Tibia		
Proximal	23–36	30
Distal	19–24	22
Metatarsal		
Proximal	before birth	
Distal	23–36	30

Source: Data from Noddle (1974).

Table 7. Data on Goat Bones: Percentage Survival of Parts in the Kuiseb R Sample, Specific Gravity of Parts, and Epiphyseal Fusion Time

Part	Percentage Survival	SG	Fusion Time (months)
Humerus			
Proximal	0	.58	36
Distal	64.0	.97	12
Radius			
Proximal	50.8	1.10	7
Distal	17.2	.97	41
Metacarpal			
Proximal	25.0	.98	before birth
Distal	18.0	.91	30
Femur			
Proximal	14.1	.75	30
Distal	7.0	.72	36
Tibia			
Proximal	10.1	.82	30
Distal	56.3	1.17	22
Metatarsal			
Proximal	30.4	.96	before birth
Distal	15.6	.90	30

Table 8. Parts of Fossil Antelopes from Makapansgat Limeworks, Arranged in Descending Order of Percentage Survival

Part	Number Found	Original Number	Percentage Survival
Half-mandibles	369	586	62.9
Humerus, distal	336	586	57.3
Radius and ulna, proximal	279	586	47.6
Metacarpal, distal	161	586	27.4
Metacarpal, proximal	129	586	22.0
Scapula	126	586	21.5
Tibia, distal	119	586	20.3
Radius and ulna, distal	114	586	19.5
Metatarsal, distal	110	586	18.8
Metatarsal, proximal	107	586	18.3
Pelvis, half	107	586	18.3
Calcaneus	75	586	12.8
Tibia, proximal	64	586	10.9
Astragalus	61	586	10.4
Femur, distal	56	586	9.6
Axis vertebrae	25	293	8.5
Atlas vertebrae	20	293	6.8
Humerus, proximal	33	586	5.6
Sacrum	16	293	5.5
Femur, proximal	28	586	4.8
Cervical 3–7 vertebrae	47	1,465	3.2
Lumbar vertebrae	30	1,758	1.7
Phalanges	47	3,516	1.3
Ribs	66	7,618	.9
Thoracic vertebrae	24	3,809	.6
Caudal vertebrae	1	4,688	0

Source: Data from Dart (1957*a*).

Table 9. Overall Analysis of Bone Fragments from Pomongwe Cave, Matopo Hills, Zimbabwe

Level	Skull pieces	Vertebrae	Ribs	Scapula and Pelvis	Limb Bone Pieces	Foot Bones	Miscellaneous Fragments	Tortoise Carapace and Plastron	Ostrich Eggshell Pieces	Land Snail Pieces	Mussel Shell	Long-Bone Flakes	Totals
1	104	134	17	21	176	55	28	112	126	130	0	299	1,202
2	46	1	13	21	140	20	26	64	74	2	0	104	511
3	145	294	59	76	450	82	100	321	37	0	0	393	1,957
4	409	204	150	145	331	204	238	390	66	0	2	3,387	5,526
5	211	146	115	93	217	149	223	189	5	0	1	2,289	3,638
6	189	93	56	58	108	128	164	175	2	0	0	1,731	2,704
7	97	26	14	19	84	25	50	49	8	0	0	661	1,033
8	25	14	4	5	30	16	101	31	28	0	0	257	511
9	51	9	5	15	34	9	38	22	18	3	0	365	569
10	6	3	0	1	11	1	13	4	2	1	0	63	105
Total	1,283	924	433	454	1,581	689	981	1,357	366	136	3	9,549	17,756

Note: Total number of bone fragments: 17,756.

Table 10. Lengths of Bone Flakes from Various Levels in the Pomongwe Cave Deposit, Zimbabwe

Level	Length (in inches)												Total	
	0–1		1–2		2–3		3–4		4–5		5–6			
	N	%	N	%	N	%	N	%	N	%	N	%	N	%
1 + 2	126	31.3	215	53.3	57	14.1	5	1.2					403	99.9
3	47	12.0	260	66.1	74	18.8	9	2.3	3	.8			393	100.0
4	804	23.7	2,281	67.3	268	7.9	32	.9	2	.1			3,387	99.9
5 + 6	738	18.4	2,549	63.4	568	13.9	163	4.1	1	.3	1	.3	4,020	100.1
7 + 8 + 9	207	16.1	798	62.2	249	19.4	26	2.0	3	.3			1,283	100.0
10	4	6.3	52	82.5	5	7.9	1	1.6	1	1.6			63	99.9
Total	1,926		6,155		1,211		236		10		1		9,549	

Table 11. Minimum Numbers of Individual Animals Represented in Pre-Middle Stone Age Levels at Pomongwe Cave, with Their Percentage Contributions to the Diet of the People

Animal	N	Lb Each	Total Weight (lb)	70% of Total Weight: Lb	70% of Total Weight: Kg	Percentage of Total Diet
Antelope class I						
Duiker	2	35	70.0	49.0	22.2	1.89
Steenbok	2	29	58.0	40.6	18.4	1.57
Grysbok	2	18	36.0	25.2	11.4	.97
Klipspringer	2	29	58.0	40.6	18.4	1.57
Antelope class III						
Waterbuck	1	475	475.0	332.5	150.5	12.86
Tsessebe	1	325	325.0	227.5	102.9	8.80
Sable	1	515	515.0	360.5	163.1	13.94
Other mammals						
Dassie	15	9	135.0	94.5	42.8	3.65
Hare	3	5	15.0	10.5	4.8	.41
Warthog	1	140	140.0	98.0	44.3	3.79
Zebra	1	800	800.0	560.0	253.4	21.65
Extinct horse	1	1,000	1,000.0	700.0	316.7	27.06
Porcupine	1	30	30.0	21.0	9.5	.81
Leopard	1					
Springhare	1	7	7.0	4.9	2.2	.19
Cane rat	1	2	2.0	1.4	.6	.05
Reptiles						
Tortoise	4	3	12.0	8.4	3.8	.32
Monitor lizard	2	7	14.0	9.8	4.4	.38
Snake indet.	2	1	2.0	1.4	.6	.05
Birds						
Ostrich egg	—	—	—	—	—	—
Rock pigeon	1	1	1.0	.7	.3	.03
Invertebrates						
Land snail	—	—	—	—	—	—
Total	45	3,431	3,695.0	2,586.5	1,170.3	99.99

Table 12. Miminum Numbers of Individual Animals Represented in Middle Stone Age Levels at Pomongwe Cave, with Their Percentage Contributions to the Diet of the People

Animal	N	Lb Each	Total Weight (lb)	70% of Total Weight: Lb	70% of Total Weight: Kg	Percentage of Total Diet
Antelope class I						
Duiker	5	35	175.0	122.5	55.4	1.29
Steenbok	4	29	116.0	81.2	36.7	.85
Klipspringer	4	29	116.0	81.2	36.7	.85
Grysbok	2	18	36.0	25.2	11.4	.26
Antelope class II						
Reedbuck	4	135	540.0	378.0	171.0	3.97
Bushbuck	1	116	116.0	81.2	36.7	.85
Antelope class III						
Tsessebe	5	325	1,625.0	1,137.5	514.7	11.96
Waterbuck	3	475	1,425.0	997.5	451.4	10.48
Sable	2	515	1,030.0	721.0	326.2	7.59
Wildebeest	1	526	526.0	368.2	166.6	3.87
Kudu	1	491	491.0	343.7	155.5	3.61
Roan	1	575	575.0	402.5	182.1	4.23
Antelope class IV						
Eland	1	1,474	1,474.0	1,031.8	466.9	10.86
Other mammals						
Dassie	11	9	99.0	69.3	31.4	.73
Warthog	4	140	560.0	392.0	177.4	4.12
Zebra	3	800	2,400.0	1,680.0	760.2	17.66
Extinct horse	2	1,000	2,000.0	1,400.0	633.5	14.71
Hare	2	5	10.0	7.0	3.2	.07
Bushpig	1	180	180.0	126.0	57.0	1.32
Porcupine	1	30	30.0	21.0	9.5	.22
Ratel	1	20	20.0	14.0	6.3	.15
Springhare	1	7	7.0	4.9	2.2	.05
Reptiles						
Tortoise	5	3	15.0	10.5	4.8	.11
Monitor Lizard	3	7	21.0	14.7	6.7	.15
Snake indet.	3	1	3.0	2.1	1.0	.02
Plated lizard	1	1	1.0	.07	.3	.01
Birds						
Ostrich egg	—	—	—	—	—	—
Rock pigeon	1	1	1.0	.07	.3	.01
Invertebrates						
River mussel	1	—	—	—	—	—
Total	74	6,947	13,592.0	9,514.4	4,305.1	100.00

Table 13. Minimum Numbers of Individual Animals Represented in Later Stone Age Levels at Pomongwe Cave, with Their Percentage Contributions to the Diet of the People

Animal	N	Lb Each	Total Weight (lb)	70% of Total Weight		Percentage of Total Diet
				Lb	Kg	
Antelope class I						
Steenbok	6	29	174	122	55.2	1.80
Klipspringer	5	29	145	101	45.7	1.49
Duiker	5	35	175	122	55.2	1.80
Grysbok	1	18	18	13	5.9	.19
Antelope class II						
Bushbuck	1	116	116	81	36.7	1.18
Antelope class III						
Sable	4	515	2,060	1,442	652.5	21.26
Wildebeest	4	526	2,104	1,472	666.1	21.70
Tsessebe	2	325	650	455	205.9	6.70
Other mammals						
Dassie	65	9	585	409	185.1	6.03
Hare	13	5	65	46	20.8	.68
Warthog	5	140	700	490	221.7	7.22
Zebra	3	800	2,400	1,680	760.2	24.77
Leopard	2					
Antbear	1	130	130	91	41.2	1.34
Bushpig	1	180	180	126	57.0	1.86
Springhare	1	7	7	5	2.2	.07
Reptiles						
Tortoise	20	3	60	42.0	19.0	.62
Monitor lizard	10	7	70	49.0	22.2	.72
Plated lizard	4	1	4	2.8	1.3	.04
Snake indet.	3	1	3	2.1	1.0	.03
Python	2	10	20	14.0	6.3	.21
Birds						
Rock pigeon	4	1	4	2.8	1.3	.04
Vulture	2	7	14	9.8	4.4	.14
Francolin	1	2	2	1.4	.6	.02
Goose	1	5	5	3.5	1.6	.05
Ostrich egg	—	—	—	—	—	—
Invertebrates						
River mussel	2	—	—	—	—	—
Land snail	—	—	—	—	—	—
Total	168	2,901	9,164	6,784	3,069	99.97

Table 14. Percentage Contributions of Various Animal Groups to the Diet of Stone Age People at Pomongwe Cave

Animal Group	Percentage Representation		
	Pre-MSA	MSA	LSA
Antelope class I	6.0	3.3	5.3
Antelope class II		4.8	1.2
Antelope class III	35.6	41.7	49.7
Antelope class IV		10.9	
Other mammals	57.6	39.0	42.0
All other animals	.8	.3	1.9
Total	100.0	100.0	100.1

Table 15. Skeletal Parts of Dassies, *Procavia capensis* and *Heterohyrax brucei*, from All Levels of the Pomongwe Cave

Minimum number of individuals	
Based on maxillary fragments: 74	
Based on mandible fragments: 76	
Based on distal humeri: 96	
Composite minimum number in all levels: 114	
Total number of bone fragments	*1,192*
Skull	*473*
Maxillary pieces	187
Mandible pieces	199
Isolated teeth	16
Miscellaneous skull pieces	71
Vertebrae	*106*
Atlas	18
Axis	9
Other	79
Scapula, pieces	*38*
Pelvis, pieces	*36*
Ribs, pieces	*32*
Humerus	*207*
Proximal ends	21
Distal ends	186
Ulna	*66*
Complete bones	5
Proximal ends	56
Distal ends	5
Radius	*21*
Complete bones	6
Proximal ends	6
Distal ends	9
Femur	*85*
Proximal ends	38
Distal ends	18
Shaft pieces	29
Tibia	*45*
Complete bones	2
Proximal ends	30
Distal ends	13
Fibula, fragments	*3*
Phalanges, complete	*3*
Pieces of lib bone shaft	*77*
Overall total	1,192

Table 16. Lengths of Bone Flakes from Various Levels at Bushman Rock Shelter, Eastern Transvaal

	Flake Sizes (inches)						
Level	0–1	1–2	2–3	3–4	4–5	5–6	Total
1	211	151	20	—	1	—	383
2	51	47	5	3	—	—	106
Total	*262*	*198*	*25*	*3*	*1*	—	*489*
3	48	73	11	1	1	—	134
4	—	—	—	—	—	—	—
5	1	—	—	—	—	—	1
6	35	35	8	—	—	—	78
7	38	53	12	1	—	—	104
8	46	51	15	1	—	—	113
9	33	94	44	6	1	—	178
10	18	32	4	—	—	—	54
11	19	34	13	3	1	—	70
12	35	66	24	2	—	—	127
13	22	31	19	1	—	—	73
14	39	74	34	6	1	—	154
15	24	59	49	16	—	—	148
16	34	138	90	17	—	—	279
17	26	48	38	6	1	—	119
18	7	24	20	6	1	—	58
19	2	2	—	—	—	—	4
20	6	8	1	2	—	—	17
21	1	3	6	3	—	—	13
22	5	13	3	—	—	—	21
23	12	12	17	2	—	1	44
24	—	6	4	3	—	—	13
25	1	3	2	—	—	—	6
26	6	14	13	—	—	—	33
Total	*458*	*873*	*427*	*76*	*6*	*1*	*1,841*

	Flake Sizes (inches)						
Level	0–1	1–2	2–3	3–4	4–5	5–6	Total
27	9	20	24	4	—	1	58
28	8	43	19	3	—	4	77
29	8	20	4	—	—	—	32
30	1	6	1	—	—	—	8
31	32	22	6	1	—	—	61
32	4	11	2	—	—	—	17
33	—	6	4	3	—	—	13
34	2	6	7	3	—	—	18
35	7	22	10	2	—	—	41
36	1	2	—	—	—	—	3
37	1	6	2	1	—	—	10
38	5	11	13	2	—	—	31
39	—	6	3	—	—	—	9
40	2	2	—	—	—	—	4
41	1	5	2	—	—	—	8
42	—	—	—	—	—	—	—
43	—	1	—	2	—	—	3
Total	*81*	*189*	*97*	*21*	—	*5*	*393*
Overall Total	*801*	*1,260*	*549*	*100*	*7*	*6*	*2,723*

Table 17. Lengths of Bone Flakes from Various Cultural Horizons at Bushman Rock Shelter

Horizon	Flake Sizes (inches) 0–1	1–2	2–3	3–4	4–5	5–6	Total
Bantu							
N	262	198	25	3	1	0	489
%	53.6	40.5	5.1	.6	.2	0	100.0
LSA							
N	458	873	427	76	6	1	1,841
%	24.9	47.4	23.2	4.1	.3	.05	100.0
MSA							
N	81	189	97	21	0	5	393
%	20.6	48.1	24.7	5.3	0	1.3	100.0
Total	801	1,260	549	100	7	6	2,723

Table 18. Occurrence of Various Animal Species at Bushman Rock Shelter and Parts by Which They Are Represented

Animal	Total Number of Individuals	Layer Number (number of individuals in brackets)	Number of Skeletal Parts Used in Identification
Tortoise (sp. indet.)	34	1 (1); 2 (1); 3 (1); 6 (1); 7 (1); 8 (1); 9 (1); 10 (1); 11 (1); 12(1); 13(1); 14 (1); 15 (1); 16 (1); 17 (1); 18 (1); 19 (1); 20 (1); 21 (1); 22 (1); 23 (1); 24 (1); 25 (1); 26 (1); 27 (1); 28 (1); 29 (1); 31 (1); 32 (1); 34 (1); 35 (1); 38 (1); 39 (1); 43 (1)	Carapace: 588 Pectoral and pelvic girdles: 26 Limb bones: 59 Vertebrae: 1
Zebra *(Equus burchelli)*	11	13 (1); 15 (1); 16 (1); 17 (1); 21 (1); 22 (1); 23 (1); 24 (1); 27 (1); 29 (1); 30 (1)	Teeth: 24
Reedbuck *(Redunca arundinum)*	9	8 (1); 9 (1); 10 (1); 16 (1); 17 (1); 18 (2); 21 (1); 24 (1)	Teeth: 15 Mandibles: 2
Warthog *(Phacochoerus aethiopicus)*	8	16 (1); 18 (1); 19 (1); 23 (1); 24 (1); 26 (1); 28 (1); 41 (1)	Teeth: 8
Hartebeest *(cf. Alcelaphus)*	6	11 (1); 12 (1); 17 (1); 18 (1); 26 (1); 43 (1)	Teeth: 8
Wildebeest *(Connochaetes taurinus)*	6	9 (1); 11 (1); 16 (1); 22 (1); 26 (1); 27 (1)	Teeth: 6
Dassie *(Procavia capensis)*	6	9 (1); 22 (1); 23 (3); 41 (1)	Mandibles and maxillae: 6 Limb bones: 1
Monitor lizard *(Varanus* sp.*)*	6	1 (1); 2 (1); 3(1); 15 (1); 21 (1); 23 (1)	Mandibles: 3 Vertebrae: 5
Impala *(Aepyceros melampus)*	4	7 (1); 9 (1); 15 (1); 16 (1)	Teeth: 6
Kudu *(Tragelaphus strepsiceros)*	4	15 (1); 17 (1); 23 (1); 28 (1)	Teeth: 4
Sable *(Hippotragus niger)*	4	15 (1); 16 (1); 17 (1); 27 (1)	Teeth: 5
Duiker *(Sylvicapra grimmia)*	3	6 (1); 7 (1); 18 (1)	Teeth: 4
Steenbok *(Raphicerus campestris)*	3	6 (1); 8 (1); 10 (1)	Teeth: 2 Mandible: 1
Hare *(Lepus* sp.*)*	3	1 (1); 6 (1); 9 (1)	Maxillae: 1 Limb bones: 2
Python *(Python sebae)*	2	26 (1); 27 (1)	Vertebrae: 2
Extinct horse *(Equus capensis)*	1	41 (1)	Teeth: 1
Tsessebe *(Damaliscus lunatus)*	1	9 (1)	Teeth: 1
Small carnivore (sp. indet.)	1	23 (1)	Teeth: 1
Bird (cf. *Bubo* sp.)	1	1 (1)	Limb bone: 1
Mussel *(Unio caffer* [?] *)*	1	11 (1)	Hinge joints: 1
Ostrich eggshell	67 pieces	1 (1); 2 (2); 3 (2); 6 (2); 7 (3); 8 (1); 9 (1); 10 (3); 11 (1); 12 (2); 13 (2); 14 (6); 15 (5); 16 (14); 17 (3); 18 (2); 21 (1); 23 (1); 24 (1) 25 (1); 26 (2); 27 (1); 28 (4); 29 (2); 31 (4)	Fragments of shell: 67
Land snail *(Achatina zebra* [?] *)*	38	1 (2); 2 (1); 3 (1); 5 (1); 6 (1); 7 (1); 8 (1); 9 (5); 10 (1); 11 (1); 12 (1); 13 (1); 14 (1); 15 (1); 22 (1); 23 (1); 25 (1); 26 (1); 27 (1); 28 (1); 29 (1); 30 (1); 31 (2); 32 (1); 33 (1); 34 (1); 35 (1); 36 (1); 37 (1); 38 (1); 39 (1); 40 (1)	Fragments of shell: 490

Table 19. Minimum Numbers of Individual Animals Represented in the Middle Stone Age Layers at Bushman Rock Shelter, with Their Percentage Contributions to the Diet of the People

Animal	N	Lb Each	Total Weight (lb)	70% of Total Weight Lb	70% of Total Weight Kg	Percentage of Total Diet
Antelope class III						
Hartebeest	1	386	386.0	270.2	122.3	6.84
Wildebeest	1	526	526.0	368.2	166.6	9.31
Kudu	1	491	491.0	343.7	155.5	8.69
Sable	1	515	515.0	360.5	163.1	9.12
Other mammals						
Zebra	3	800	2,400.0	1,680.0	760.2	42.50
Warthog	2	140	280.0	196.0	88.7	4.96
Extinct horse	1	1,000	1,000.0	700.0	316.7	17.71
Dassie	1	9	9.0	6.3	2.9	.16
Reptiles						
Tortoise	10	3	30.0	21.0	9.5	.53
Python	1	10	10.0	7.0	3.2	.18
Birds						
Ostrich egg	—	—	—	—	—	—
Invertebrates						
Land snail	—	—	—	—	—	—
Total	22	3,880	5,647	3,952.9	1,788.7	100.00

Table 20. Minimum Numbers of Individual Animals Represented in the Later Stone Age Layers at Bushman Rock Shelter, with Their Percentage Contributions to the Diet of the People

Animal	N	Lb Each	Total Weight (lb)	70% of Total Weight Lb	70% of Total Weight Kg	Percentage of Total Diet
Antelope class I						
Duiker	3	35	105.0	73.5	33.3	.61
Steenbok	3	29	87.0	60.9	27.6	.51
Antelope class II						
Reedbuck	9	135	1,215.0	850.5	384.8	7.09
Impala	4	106	424.0	296.8	134.3	2.47
Antelope class III						
Hartebeest	5	386	1,930.0	1,351.0	611.3	11.26
Wildebeest	5	526	2,630.0	1,841.0	833.0	15.35
Kudu	3	491	1,473.0	1,031.1	466.6	8.60
Sable	3	515	1,545.0	1,081.5	489.4	9.02
Tsessebe	1	325	325.0	227.5	102.9	1.90
Other mammals						
Zebra	8	800	6,400.0	4,480.0	2,027.1	37.35
Warthog	6	140	840.0	588.0	266.1	4.90
Dassie	5	9	45.0	31.5	14.3	.26
Hare	2	5	10.0	7.0	3.2	.06
Small carnivore	1	3	3.0	2.1	.95	.02
Reptiles						
Tortoise	22	3	66.0	46.2	20.9	.39
Varanus	4	7	28.0	19.6	8.9	.16
Python	1	10	10.0	7.0	3.2	.06
Birds						
Ostrich egg	—	—	—	—	—	—
Invertebrates						
River mussel	1	—	—	—	—	—
Land snail	—	—	—	—	—	—
Total	86	3,525	17,136.0	11,995.6	5,427.9	100.01

Table 21. Percentage Representation of Various Animal Groups in the Middle and Later Stone Age Levels at Bushman Rock Shelter

	Percentage Representation	
Animal Group	Middle Stone Age	Later Stone Age
Antelope class I		1.1
Antelope class II		9.6
Antelope class III	34.0	46.1
Antelope class IV		
Other mammals	65.3	42.6
All other animals	.7	.6
Total	100.0	100.0

Table 22. Lengths of Unidentifiable Bone Fragments from Each Stratigraphic Unit in the Wilton Large Rock Shelter

Level or Stratigraphic Unit	Bone Fragment Length (inches)					
	0–1	1–2	2–3	3–4	4–5	Total
1	181	37	4	1	—	223
2A	1,045	65	2	—	—	1,112
2B	1,269	47	1	—	—	1,317
2C	204	14	1	—	—	219
3A	1,876	96	2	—	—	1,974
3B	1,950	107	4	—	—	2,061
3C	2,294	141	2	1	—	2,438
3D	2,100	124	8	—	—	2,232
3D, Hearth 1	193	10	2	—	—	205
3D, Hearth 2	1,026	76	3	2	—	1,107
3D, Hearth 3	1,569	129	5	—	—	1,703
3E	3,919	344	23	1	1	4,288
3E, Hearth 4	92	6	—	—	—	98
3E, Hearth 5	457	21	1	—	—	479
3F	2,869	193	10	1	—	3,073
3G	2,485	226	17	2	—	2,730
3G, Hearth 6	31	—	—	—	—	31
3G, Hearth 7	341	31	—	—	—	372
3H	1,090	106	6	—	—	1,202
3I	521	42	4	—	—	567
3J	180	25	3	—	—	208
4	6,252	—	—	—	—	6,252
Total	31,944	1,840	98	8	1	33,891

Table 23. Occurrence and Percentage Representation of Unidentifiable Bone Fragments in Each Layer at the Wilton Large Rock Shelter

Layer		Bone Fragment Length (inches)					
		0–1	1–2	2–3	3–4	4–5	Total
1	N	181	37	4	1	—	223
	%	81.2	16.5	1.8	0.4	—	99.9
2	N	2,518	126	4	—	—	2,648
	%	95.5	4.7	.2	—	—	100.4
3	N	22,993	1,677	90	7	1	24,768
	%	92.8	6.7	.3	.03	.004	99.8
4	N	6,252	—	—	—	—	6,252
	%	100.0	—	—	—	—	100.0
Total		31,944	1,840	98	8	1	33,891

Table 24. Overall Analysis of Bone Fragments from the Wilton Large Rock Shelter

Skeletal Parts	Levels																					
	1	2A	2B	2C	3A	3B	3C	3D	Hearth 1	Hearth 2	Hearth 3	3E	Hearth 4	Hearth 5	3F	3G	6	7	3H	3I	3J	4	Total
Bovid Remains																2							3
Horn-cores					1														2			4	6
Cranial fragments												1											2
Maxillary fragments	1															1							4
Mandible fragments		1				1			1			7			3	11		3	3	9		93	196
Tooth fragments	1	19		2	7	26	7	1		4		4			1							2	9
Atlas vertebrae					1			1				4			1				1			2	11
Axis vertebrae	1		1					1				4			1	5			2	3	1	7	30
Other cervical vertebrae	1	1	2		1				1	1		7			4	8			1		1	4	43
Thoracic vertebrae	3		3	1	1	1		4		5		6		1	7	3						2	36
Lumbar vertebrae	2	1			1	5	2	5		1		2				5							7
Caudal vertebrae												38	1	1	13	15			6	9		27	203
Indeterminate vertebrae	1	16	12	1	1	4	10	33	1	14		4			1	3			1				12
Scapula								2		1												3	4
Pelvis							1					20		1	23	7		1			1	25	123
Ribs	5	6	1		4	9	4	13		3					1								2
Sternum								1														1	1
Humerus, proximal													1		1					1		2	10
Humerus, distal		1				1		3				1			1						1		4
Radius, proximal		1										2			1				1			1	6
Radius, distal									1			4			3					1		1	21
Ulna, proximal		3	1	1	1	2	2	2				3			2							1	13
Femur, proximal	2		1	1	1	1		1							1								1
Femur, distal																							3
Tibia, proximal							1			2		1		1								2	6
Tibia, distal								1		1		31		3	18	10		2	4	2	1	17	143
Carpals, tarsals, sesamoids	3	8	7	2	12	1	2	17		3		9			2	4		1		3		5	39
Astragalus		3	1		2	1	3	3	2			6		1	1				1				18
Calcaneus		2	2			1	1	3															1
Metacarpal, distal	1																						1
Metatarsal, distal						1						28	2		13	4			3	1	2	10	117
Metapodial fragments	2	15	4	1	4	4	8	9	1	6		60	1	7	14	21		4	8	11		26	267
Phalanges	9	27	21	7	6	8	7	21	3	6		19			11	12		2	1	3		9	114
Terminal phalanges	3	6	8	1	7	7	2	18	4	1													
Mammalian, nonbovid remains																							1
Cranial fragments	1											2				5		1				4	18
Maxillary fragments		1			2	1	2					1							1			3	18
Mandible fragments					2	5				1		2		1		1						5	22
Tooth fragments	1	8	1		2		1															1	1
Atlas vertebrae																							1
Lumbar vertebrae								1				2										1	14
Indeterminate vertebrae				2	2	6		1								2							6

Continued on following page

Table 24. *Continued*

Skeletal Parts	1	2A	2B	2C	3A	3B	3C	3D	Hearth 1	Hearth 2	Hearth 3	3E	Hearth 4	Hearth 5	3F	3G	6	7	3H	3I	3J	4	Total
	Levels																						
Scapula			1		2			1															4
Pelvis	1				3											1							16
Ribs		4	6			2				3													2
Humerus, proximal		1				1						3		3	3							6	20
Humerus, distal	2				3										1							2	3
Radius, proximal														1	1							5	14
Ulna, proximal			1		3	2										1			1			3	7
Femur, proximal					2																		2
Tibia, proximal					1	1																	1
Tibia, distal					1																		1
Carpals and tarsals					1										1			1					6
Astragalus						1		3				3				2		1					11
Calcaneus					3	2																1	4
Metapodial fragments		3										3				3						7	38
Phalanges	3	1	3		6	3	1	6	2														1
Terminal phalanges					1							33	1		24	275		17	94	20		168	905
Microfaunal fragments	40	38	32	16	22	40	29	34		22													
Bird remains																1						3	5
Bird skeletal parts		1									4	11						2					21
Ostrich eggshell fragments		1	3																				
Reptile remains																							
Tortoise, carapace and plastron fragments	22	88	100	10	25	113	253	118	7	196	56	352	3	12	248	230	4	24	111	60	38	516	2,586
Tortoise, other skeletal parts	14	15	9	1	13	8	7	9		6		28		1	12	13		2	10	4	5	58	215
Snake, skeletal parts		10	4	1	18	8	34	5		6		13			6	35		5	2			4	151
Lizard, skeletal parts								1															1
Fish																						1	2
Fish, skeletal parts							1																
Invertebrate remains																					1		3
Marine mollusk fragments					2						15	33		6		23		1	4			9	204
Terrestrial mollusk fragments		2			25	10	20	25		31	2	8			2	6	1	2					118
Freshwater mollusk fragments		40	20		5	17	9	3		3	30	155	10	18	136	101	1	9	27	15	6	11	830
Crab fragments	3	18	23		4	6	133	84		40													
Indeterminate flakes and fragments	223	1,112	1,317	219	1,974	2,061	2,438	2,232	205	1,107	1,703	4,288	98	479	3,073	2,730	31	372	1,202	567	208	6,252	33,891
Miscellaneous skeletal parts	50	149	85	21	142	106	162	251	31	136	35	570	8	27	307	243	5	88	53	48	39	480	3,036
Total	396	1,602	1,669	287	2,314	2,465	3,140	2,913	256	1,600	1,845	5,770	125	558	3,938	3,787	42	538	1,539	757	304	7,784	43,629

Table 25. List of Animal Taxa Identified from the Wilton Remains, with Minimum Numbers Represented in Each Stratigraphic Unit

Animal	Stratigraphic Unit																					Total
	1	2A	2B	2C	3A	3B	3C	3D	Hearth 1	Hearth 2	Hearth 3	3E	Hearth 4	Hearth 5	3F	3G	Hearth 6	Hearth 7	3H	3I	3J	4	(no. of individuals)
Blue duiker					1	1		1								1						1	5
Gray duiker	1																						1
Indeterminate antelope I	1	3	3	1	1		3	1	2	2		4	1	1	3	2		2	2	2	1	2	37
Vaal rhebuck	1																					1	2
Bushbuck (?)						1																	1
Indeterminate antelope II	1	2	1	1	2	1	1	3		1		2		2	2	1			2	2		1	25
Domestic cow (?)	1																						1
Large alcelaphine (cf. wildebeest)																						1	1
Indeterminate antelope III								1				1											2
Common dassie	1	1			1	2						1			1	1			1			3	12
Scrub hare	1	1			1	1	1			1		2			1	1							10
Skunk	1																						1
Small carnivore (indet.)		1										1											2
Chacma baboon			1		1																		2
Bushpig																1						1	2
Small rodents (sp. indet.)	3	3	2	2	5	2	2	2		1		1			1	4		2	3	2		3	38
Insectivores (shrews, sp. indet.)	1	1	1			1		1		1						2							8
Small bird (cf. sparrow)		1																					1
Barn owl																1						1	2
Ostrich egg		1	1								1	1											4
Tortoise (sp. indet.)	2	2	2	1	4	2	2	2	1	2	1	3	1	1	2	2	1		2	1	2	5	41
Puff adder								1															1
Snake (sp. indet.)		1	1	1	1	1	1			1		1			1	1		1	1			1	13
Lizard (sp. indet.)								1															1
Fish (sp. indet.)							1															1	2
Freshwater crab (sp. indet.)	1	1	1		1	1	8	4		1	1	8	1	1	7	6	1	2	2	2	1	2	52
Freshwater snail (sp. indet.)		2	2		1	2	1	1		1	2	1			1	2	1				1		18
Land snail		1			2	2	1	1		1	1	1		1		1		2	1			3	18
Marine mollusk (*Patella* sp.)					1																		1
Total	15	21	15	6	22	17	21	19	3	12	6	27	3	6	19	26	3	9	14	9	5	26	304

Table 26. Inventory of Skeletal Parts on Which Identifications Were Based in the Wilton Remains

Animal	Part	Animal	Part
Blue duiker	1 horn-core, 2 tooth fragments (3A); 2 teeth, 1 distal metatarsal (3B); 1 tooth (3D); 2 horn-cores, 1 tooth (3G); 1 tooth (4)	Tortoise	11 limb bones (1); 15 limb bones (2A); 7 limb bones (2B); 1 limb bone (2C); 9 limb bones (3A); 8 limb bones (3B); 7 limb bones (3C); 8 limb bones (3D); 7 carapace fragments (3D, Hearth 1); 4 limb bones (3D, Hearth 2); 56 carapace fragments (3D, Hearth 3); 22 limb bones (3E); 3 carapace fragments (3E, Hearth 4); 1 limb bone (3E, Hearth 5); 7 limb bones (3F); 12 limb bones (3G); 2 limb bones (3G, Hearth 7); 8 limb bones (3H); 4 limb bones (3I); 3 limb bones (3J); 39 limb bones (4)
Gray duiker	1 tooth (1)		
Vaal rhebuck	1 maxillary fragment (1); 2 teeth (4)		
Bushbuck	1 tooth (3B)		
Domestic cow (?)	1 terminal phalanx (1)		
Large alcelaphine (wildebeest)	3 teeth (4)		
Common dassie	1 distal humerus, 1 occipital (1); 1 maxilla, 7 teeth (2A); 1 maxilla, 1 mandible, 2 distal humeri (3A); 1 maxilla, 4 mandible fragments (3B); 1 maxilla (3E); 2 distal humeri (3F); 3 maxilla, 1 distal humerus (3G); 2 mandible fragments, 1 proximal femur (3H); 4 maxillae, 3 mandibles, 1 tooth, 6 distal humeri, 1 proximal femur, 1 proximal ulna (4)	Puff adder	1 maxilla (3D)
		Snake (indet.)	1 rib (2C); vertebrae: 10(2A); 4(2B); 18(3A); 8(3B); 34(3C); 4(3D); 6(3D, Hearth 2); 13 (3E); 6(3F); 35(3G); 5(3G, Hearth 7); 2(3H); 4(4)
Scrub hare	1 distal humerus (1); 1 proximal humerus (2A); 1 maxilla, 2 teeth, 2 distal humeri (3A); 1 mandible (3B); 1 maxilla, 1 tooth (3C); 1 mandible (3D, Hearth 2); 1 maxilla, 3 distal humeri (3E); 1 distal humerus (3F); 2 maxillae, 1 distal humerus (3G)	Lizard (indet.)	1 mandible (3D)
		Fish (indet.)	1 vertebra (3D); 1 vertebra (4)
		Crab (indet.)	Fragments: 3(1); 4(3A); 6(3B); 1(3G, Hearth 6); Mandibles: 3(2A); 1(2B); 32(3C); 13(3D); 3(3D, Hearth 2); 1(3D, Hearth 3); 29(3E); 2(3E, Hearth 4); 2(3E, Hearth 5); 25 (3F); 22(3G); 6(3G, Hearth 7); 6(3H); 5(3I); 3(3J); 6(4)
Skunk	1 tooth (1)		
Small carnivore	1 tooth (2A); 2 teeth (3E)		
Chacma baboon	1 tooth (2B); 1 tooth (3A)	Freshwater snails (indet.)	Fragments: 37(2A); 20(2B); 5(3A); 17(3B); 9(3C); 3(3D); 3(3D, Hearth 2); 2(3D, Hearth 3); 8(3E); 2(3F); 6(3G); 1(3G, Hearth 6); 1(3J); 9(4)
Bushpig	1 tooth fragment (3G); 4 tooth fragments (4)		
Small rodents	Skull parts: 3(1); 6(2A); 4(2B); 2(2C); 14 (3A); 5(3B); 6(3C); 5(3D); 2(3D, Hearth 2); 1(3E); 1(3F); 18(3G); 2(3G, Hearth 7); 11 (3H); 2(3I); 7(4)		
		Land snails	Fragments: 2(2A); 25(3A); 10(3B); 20(3C); 25(3D); 31(3D, Hearth 2); 15(3D, Hearth 3); 33(3E); 6(3E, Hearth 5); 23(3G); 3(3G, Hearth 7); 4(3H)
Insectivores	Skull parts: 1(1); 1(2A); 1(2B); 1(3B); 1(3D); 1(3D, Hearth 2); 2(3G)		
Bird	1 maxilla (2A)		
Barn owl (?)	1 claw (3G); 3 claws (4)	Marine mollusks	2(3A)
Ostrich eggshell	Fragments: 1(2A); 3(2B); 4(3D, Hearth 3); 11(3E)		

Note: Stratigraphic units are given in parentheses.

Table 27. Minimum Numbers of Individual Animals Represented in the Later Stone Age Layers at Wilton Large Rock Shelter, with Their Contributions to the Diet of the People

Animal	N	Lb Each	Total Weight (lb)	70% of Total Weight: Lb	70% of Total Weight: Kg	Percentage of Total Diet
Antelope class I						
Blue duiker	5	14	70.0	49.0	22.2	1.14
Sp. indet.	36	25	900.0	630.0	285.1	14.70
Antelope class II						
Gray rhebuck	1	55	55.0	38.5	17.4	.90
Bushbuck	1	116	116.0	81.2	36.7	1.89
Sp. indet.	24	118	2,832.0	1,982.4	897.0	46.24
Antelope class III						
Wildebeest	1	526	526.0	368.2	166.6	8.59
Sp. indet.	2	418	836.0	585.2	264.8	13.65
Other mammals						
Rodents	35					
Insectivores	7					
Dassie	11	9	99.0	69.3	31.4	1.62
Hare	9	5	45.0	31.5	14.3	.73
Small carnivore	2	3	6.0	4.2	1.9	.10
Baboon	2	70	140.0	98.0	44.3	2.29
Bushpig	2	180	360.0	252.0	114.0	5.88
Reptiles						
Tortoise	40	3	120.0	84.0	38.0	1.96
Snake indet.	13	1	13.0	9.1	4.1	.21
Puff adder	1	2	2.0	1.4	.6	.03
Birds						
Ostrich egg						
Barn owl	2					
Sparrow	1					
Fish						
Sp. indet.	2	2	4.0	2.8	1.3	.07
Invertebrates						
River crab	51					
River mussel	17					
Land snail	18					
Marine mollusk	1					
Total	284	1,547	6,124.0	4,286.8	1,939.7	100.00

Table 28. Percentage Contributions of Various Animal Groups to the Diet of the Later Stone Age Inhabitants of Wilton Shelter

Animal Group	Percentage Representation
Antelope class I	15.8
Antelope class II	49.0
Antelope class III	22.3
Antelope class IV	—
Other mammals	10.6
All other animals	2.3
Total	100.0

Table 29. Lengths of Bone Flakes from the Later Stone Age Levels at Fackelträger Shelter, South-West Africa

Length (inches)	N	%
0–1	909	56.3
1–2	600	37.8
2–3	65	5.0
3–4	12	.8
4–5	2	.1
Total	1,588	100.0

Table 30. Animal Taxa from the Later Stone Age Layers at Fackelträger Shelter, South-West Africa, and Parts Used in Their Identification

Raphicerus campestris, steenbok: 9 individuals
Maxillary pieces, 3; mandible pieces, 2; calvaria fragment, 1; isolated teeth and tooth fragments, 5; distal tibia fragment, 1; pelvis pieces, 2

Antelope class I, cf. *Raphicerus:* 6 individuals
Mandible pieces, 3; tooth fragments, 3; horn-core pieces, 2; atlas vertebra, 1; proximal femur, 1; astragali, 4; proximal metapodials, 2; distal metapodials, 12; phalanges,7

Antidorcas marsupialis, springbok: 1 individual
Isolated molar 1

Antelope class II, cf. *Antidorcas:* 8 individuals
Mandible fragment, 1; tooth fragments, 9; vertebral fragment, 1; distal humerus, 1; proximal ulna, 1; distal radius, 1; astragalus, 1; sesamoid, 1; distal metapodials, 4; phalanges,2

Antelope class III, cf. *Oryx:* 2 individuals
Tooth fragments, 3; calvaria piece, 1; accessory carpal, 1

Procavia capensis, dassie: 17 adults, 1 juvenile
Maxillary pieces, 9; mandible pieces, 19; isolated molar, 1; isolated incisor pieces, 14; proximal humeri, 2; distal humeri, 16; proximal ulna, 1; pelvis pieces, 3; distal femur, 1; femur shaft, 1; metapodial, 1

Lepus capensis, scrub hare: 10 individuals
Maxilla, 1; mandible pieces, 2; isolated teeth and tooth fragments, 16

Petromus typicus, rock rat: 6 individuals
Maxillary pieces, 2; mandible pieces, 5

Elephantulus sp., elephant shrew: 3 individuals
Mandible pieces, 3

Small carnivore, cf. mongoose: 2 individuals
Mandible fragments, 2

Large carnivore, cf. leopard: 1 individual
Metapodial fragments, 2

Diceros bicornis, black rhino: 1 individual
Tooth fragments, 5

Chelonian indet., tortoise: 17 individuals
Carapace and plastron fragments, 140; vertebrae, 4; fragments of pectoral and pelvic girdle, 8; limb bones, 47; miscellaneous fragments, 9

Varanus cf. *albigularis*, monitor lizard: 10 individuals
Maxillary fragment, 1; mandible pieces, 6; vertebrae, 11

Lizard, indet.: 2 individuals
Maxillary fragment, 1; mandible pieces, 2

Snake, indet.: 2 individuals
Vertebrae, 2

Cf. *Columba guinea*, rock pigeon: 8 individuals
Limb bones, 15; eggshell piece, 1

Birds indet., larger species: 7 individuals
Limb bone pieces, 11; phalanx, 1; eggshell piece, 1

Table 31. Minimum Numbers of Individual Animals Represented in the Later Stone Age Layers at Fackelträger Shelter, South-West Africa, with Their Percentage Contributions to the Diet of the People

Animal	N	Lb Each	Total Weight (lb)	70% of Total Weight Lb	70% of Total Weight Kg	Percentage of Total Diet	Percentage of Diet without Rhino
Antelope class I							
Steenbok	9	29	261.0	182.7	82.7	5.67	10.9
Indet. (cf. steenbok	6	29	174.0	121.8	55.1	3.78	7.2
Antelope class II							
Springbok	1	78	78.0	54.6	24.7	1.69	3.2
Indet. (cf. springbok)	8	78	624.0	436.8	197.6	13.54	25.9
Antelope class III							
Indet. (cf. gemsbok)	2	450	900.0	630.0	285.1	19.54	37.4
Other mammals							
Dassie	18	9	162.0	113.4	51.3	3.52	6.7
Hare	10	5	50.0	35.0	15.8	1.09	2.1
Rock rat	6	1	6.0	4.2	1.9	.13	.3
Elephant shrew	3	1	3.0	2.1	1.0	.07	.1
Mongoose	2	1	2.0	1.4	.6	.04	.1
Large carnivore	1						
Black rhino	1	2,200	2,200.0	1,540.0	696.8	47.75	
Reptiles							
Tortoise	17	3	51.0	35.7	16.2	1.11	2.1
Varanus	10	7	70.0	49.0	22.2	1.52	2.9
Lizard	2	1	2.0	1.4	.06	.04	
Snake	2	1	2.0	1.4	.6	.04	
Birds							
Cf. rock pigeon	8	1	8.0	5.6	2.5	.17	.3
Indet. (larger spp.)	7	2	14.0	9.8	4.4	.30	.6
Total	113	2,896	4,607.0	3,224.9	1,459.1	100.00	99.8

Table 32. Percentage Representation of Various Animal Groups in the Diet of Inhabitants of the Fackelträger Shelter, South-West Africa

Animal Group	With Rhino N	With Rhino %	Without Rhino N	Without Rhino %
Antelope class I	15	9.5	15	18.1
Antelope class II	9	15.2	9	29.1
Antelope class III	2	19.5	2	37.4
Antelope class IV				
Other mammals	41	52.6	40	9.3
All other animals	46	3.2	46	5.9
Total	113	100.0	112	99.8

Table 33. Representation of Various Mammalian Orders in the Human Food Remains from Various Southern African Cave Sites

Mammalian Order	Pomongwe (this study)		Bushman Rock (this study)		Wilton (this study)		Fackelträger (this study)		Nelson Bay (Klein 1974*b*)		Die Kelders 1 (Klein 1975*b*)		Klasies 1 (Klein 1975*b*)	
	N	%	N	%	N	%	N	%	N	%	N	%	N	%
Primates, other than *Homo*	—	—	—	—	2	2.1	—	—	23	3.1	—	—	7	1.1
Carnivora	4	1.6	1	1.5	2	2.1	3	5.2	38	5.1	28	1.1	38	6.1
Artiodactyla	85	35.7	48	70.6	72	75.0	26	44.8	344	46.2	246	10.0	415	66.0
Perissodactyla	10	4.2	11	16.2	—	—	1	1.7	3	.4	8	.3	9	1.4
Hyracoidea	114	47.9	6	8.8	11	11.5	18	31.0	143	19.2	86	3.5	51	8.1
Lagomorpha	18	7.6	2	2.9	9	9.4	10	17.2	3	.4	200	8.1	1	.2
Rodentia (large forms)	6	2.5	—	—					10	1.3	1,866	75.8	30	4.7
Pholidota, Tubulidentata, Proboscidea	1	.5	—	—					3	.4	—	—	3	.5
Pinnipedia	—	—	—	—					165	22.2	26	1.1	65	10.4
Cetacea	—	—	—	—					12	1.6	4	.2	9	1.4
Total	238	100.0	68	100.0	96	100.1	58	99.9	744	99.9	2,464	100.1	628	100.0

Mammalian Order	Klasies, Other Sites (Klein 1976*a*)		Scott's Cave (Klein and Scott 1974)		Andrieskraal 1 (Hendey and Singer 1965)		Elands Bay (Parkington 1976)		De Hangen (Parkington and Poggenpoel 1971)		Redcliff (Klein, n.d.)	
	N	%	N	%	N	%	N	%	N	%	N	%
Primates, other than *Homo*	8	2.1	—	—	1	1.5	3	1.4	1	1.1	8	1.0
Carnivora	13	3.3	3 (12)	5.9	10	14.5	18	8.4	2 (14)	2.3	35	4.5
Artiodactyla	209	53.5	42	82.4	31	44.9	116	54.2	14	15.7	500	63.9
Perissodactyla	14	3.6	—	—			4	1.9	1	1.1	143	18.3
Hyracoidea	47	12.0	2	3.9	9	13.0	25	11.6	64	71.9	71	9.1
Lagomorpha	1	.3	3	5.9	16	23.2	8	3.7	6	6.7	16	2.1
Rodentia (large forms)	6	1.5			2	2.9	35	16.3	1	1.1	7	.9
Pholidota, Tubulidentata, Proboscidea	1	.3					3	1.4			2	.2
Pinnipedia	88	22.5										
Cetacea	4	1.0					2	.9				
Total	391	100.1	51	100.1	69	100.0	214	99.8	89	99.9	782	100.0

Note: Minimum numbers of individuals have been used in all cases except Redcliff, where numbers of identified bones are listed.

Table 34. Percentage Representation of Various Mammalian Orders in Food Remains from Thirteen Southern African Cave Sites Known to Have Been Occupied by Stone Age Peoples

Mammalian Order	Percentage Representation	
	Range	Mean
Primates, excluding *Homo*	0– 3.1	1.0
Carnivora	1.1–14.5	4.7
Artiodactyla	10.0–82.4	51.0
Perissodactyla	0–18.3	3.8
Hyracoidea	3.5–71.9	19.3
Lagomorpha	.2–23.2	6.8
Rodentia, large forms	0–75.8	8.2
Pholidota, Tubulidentata, Proboscidea	0– 1.4	.3
Pinnipedia	0–22.5	4.3
Cetacea	0– 1.6	.4

Table 35. Bone Remains Found in the Vicinity of Six Spotted Hyena Breeding Dens in the Timbavati Area, Eastern Transvaal

Skeletal Part	N	%
Skull	8	2.0
Turbinae	2	.5
Jaw or tooth	20	4.9
Vertebrae	42	10.3
Rib	3	.7
Scapula	13	3.2
Pelvis	12	2.9
Podial	9	2.2
Horn/hoof	20	4.9
Long bones	37	9.0
Articular ends	22	4.5
Shaft pieces		
Ungulate > 84 kg	3	.7
Ungulate < 84 kg	12	2.9
Bone flakes		
Ungualate > 84 kg	43	10.5
Ungulate < 84 kg	136	33.3
Fragments	27	6.6
Total	409	100.0

Source: Data from Bearder (1977).

Table 36. Occurrence of Remains of Various Animal Taxa in Droppings, Regurgitations, and Bone Collections at Breeding Dens of Spotted Hyenas in the Timbavati Area, Eastern Transvaal

Taxon	Scats		Regurgitations		Bones at Dens	
	N	%	N	%	N (pieces)	%
MAMMALIA						
Primates						
Papio ursinus, baboon	4	.5	5	1.9		
Carnivora						
Herpestinae, mongoose	1	.1				
Perissodactyla						
Equus burchelli, zebra	36	4.5	9	3.4	18	14.6
Artiodactyla						
Giraffa camelopardalis, giraffe	304	38.3	43	16.3	20	16.3
Bovid size class I						
Sylvicapra grimmia, duiker	17	2.1	7	2.7		
Raphicerus campestris, steenbok	2	.3	4	1.5		
Bovid size class II						
Aepyceros melampus, impala	292	36.8	159	60.2	59	48.0
Bovid size class III						
Connochaetes taurinus, wildebeest	108	13.6	19	7.2	16	13.0
Kobus ellipsiprymnus, waterbuck	8	1.0	1	.4		
Tragelaphus strepsiceros, kudu			1	.4	10	8.1
Rodentia						
Hystrix africaeaustralis, porcupine	2	.3				
Lagomorpha						
Lepus saxtilis, hare	2	.3	4	1.5		
REPTILIA						
Testudo pardalis, mountain tortoise	1	.1	4	1.5		
Serpentes indet., snake			2	.8		
AVES						
Indet., unidentified bird	4	.5	2	.8		
UNIDENTIFIED TAXA	9	1.1				
Total	794	100.0	264	100.1	123	100.0

Source: Data from Bearder (1977).

Table 37. Food of Spotted Hyenas in the Serengeti and Ngorongoro Areas, as Determined by Direct Observation

	Serengeti				Ngorongoro			
	Carcasses from Kills + Scavenging		Carcasses from Kills Only		Carcasses from Kills + Scavenging		Carcasses from Kills Only	
Prey	N	%	N	%	N	%	N	%
Wildebeest, adult	199	38.8	85	38.4	131	44.1	109	44.6
Wildebeest, calf	64	12.5	33	14.9	74	24.9	69	28.3
Zebra, adult	55	10.7	19	8.6	36	12.2	29	11.1
Zebra, foal	13	2.5	4	1.8	18	6.1	18	7.4
Thomson's gazelle, adult	101	19.7	32	14.5	5	1.7		
Thomson's gazelle, fawn	49	9.6	30	13.6	10	3.4	7	2.9
Grant's gazelle, adult	4	.8	1	.5	5	1.7	1	.4
Grant's gazelle, fawn	3	.6	3	1.4	1	.3	1	.4
Topi, adult	2	.4	1	.5				
Topi, juvenile	3	.6	3	1.4				
Kongoni, adult	1	.2	1	.5				
Waterbuck, adult	1	.2	1	.5	1	.3	1	.4
Eland, adult	2	.4	1	.5				
Buffalo, adult	3	.6			1	.3	1	.4
Impala, adult	1	.2	1	.5				
Warthog	4	.8	1	.5				
Hare	1	.2	1	.5	1	.3	1	.4
Springhare	1	.2	1	.5				
Domestic stock					1	.3		
Porcupine					1	.3	1	.4
Bat-eared fox	1	.2	1	.5	1	.3	1	.4
Golden jackal	2	.4	1	.5				
Lion					1	.3		
Hyena					5	1.7	1	.4
Puff adder					1	.3	1	.4
Ostrich eggs	1	.2	1	.5				
Termites					3	1.0	3	1.2
Afterbirth	2	.4			1	.3		
Total	513	100.2	221	100.6	297	99.8	244	99.9

Source: Data from Kruuk (1972).

Table 38. Animals Found to Have Been Killed by Spotted Hyenas in the Kruger National Park

Prey	N	%
Impala	110	58.8
Waterbuck	24	12.8
Wildebeest	21	11.2
Kudu	19	10.2
Bushbuck	5	2.7
Buffalo	2	1.1
Scaly anteater	2	1.1
Reedbuck	1	.5
Sharpe's grysbok	1	.5
Warthog	1	.5
Zebra	1	.5
Total	187	99.9

Source: Data from Pienaar (1969).

Table 39. Lengths of Muscles Involved in the Closure of *Crocuta* Jaws, with Percentage Changes of Muscle Components at Two Jaw Openings

Muscle Component	Jaws at Rest (length, cm)	Jaws Half Open		Jaws Fully Open	
		Length, cm	% Change	Length, cm	% Change
Temporalis 1	50	64	28.0	82	64.0
Temporalis 2	76	97	27.6	116	52.6
Temporalis 3	80	98	22.5	117	46.3
Masseter 1	60	72	20.0	76	26.7
Masseter 2	57	68	19.3	75	31.6
Masseter 3	70	78	11.4	80	14.3

Table 40. Composition of Bone Accumulations Associated with *Hyaena brunnea*

A. Composition of 39 bone pieces from the vicinity of five *H. brunnea* dens on the farm Tweeputkoppies

Bos taurus, domestic ox: 2 adults, 2 juveniles; 21 pieces
2 skulls, 3 mandibles, 1 scapula, 8 metacarpals, 1 metatarsus, 3 femurs, 3 tibiae
Tragelaphus strepsiceros, kudu: 2 adults; 3 pieces
1 metacarpal, 2 hooves
Aepyceros melampus, impala: 2 adults, 1 juvenile; 8 pieces
1 horn, 2 femurs, 5 metacarpal fragments
Sylvicapra grimmia, duiker: 1 adult; 1 piece
1 femur
Phacochoerus aethiopicus, warthog: 3 adults; 4 pieces
2 vertebrae, 1 metacarpal, 1 hoof
Canis mesomelas, black-backed jackal: 1 adult; 1 piece
1 skull and mandible
Small carnivore, ?mongoose: 1 adult; 1 piece
1 partial cranium

B. Composition of 6 bone pieces from a *H. brunnea* breeding lair on the farm Leeufontein

Bos taurus, domestic ox: 1 adult; 1 piece
Isolated teeth
Phacochoerus aethiopicus, warthog: 1 adult; 1 piece
1 mandible
Equus sp., horse or zebra: 1 adult; 1 piece
Isolated teeth
Papio ursinus, baboon: 1 adult; 1 piece
Unidentified bovid, 2 fragments

Source: Data from Skinner (1976).

Table 41. Analysis of Bones Collected around *H. brunnea* Dens in the Kalahari National Park

Bos taurus, domestic ox: 1 juvenile; 7 pieces
7 pieces from a single cranium
Ovis/Capra sp., sheep or goat: 1 adult; 1 piece
Mandible piece, 1
Connochaetes taurinus, wildebeest: 2 adults; 3 pieces
Maxillary piece, 1; left horn, 1; metacarpal, 1
Alcelaphus buselaphus, hartebeest: 1 adult; 2 pieces
Mandible pieces, 2
Oryx gazella, gemsbok: 2 adults, 3 juveniles; 7 pieces
Maxillary pieces, 2; mandible pieces, 2; horn sheath, 2; hoof sheaths, 2
Antidorcas marsupialis, springbok: 10 adults, 3 juveniles; 41 pieces
Maxillary pieces, 6; mandible pieces, 7; horn-core pieces, 12; calvaria fragment, 1; atlas vertebra, 1; cervical vertebrae, 9; metacarpal, 1; proximal metatarsal, 1; hoof sheaths, 3
Raphicerus campestris, steenbok: 9 adults, 2 juveniles; 15 pieces
Calvaria pieces, some with horn-cores, 10; mandible pieces, 3; horn sheath, 1; terminal phalanx, 1
Sylvicapra grimmia, duiker: 1 adult, 1 juvenile; 3 pieces
Mandible pieces, 3
Felis caracal, lynx: 4 adults, 1 juvenile; 6 pieces
Skulls, 4; maxillary pieces, 2
Canis mesomelas, black-backed jackal: 13 adults, 1 juvenile; 13 pieces
Skulls with articulated mandibles, 3; maxillary pieces, 3; mandible pieces, 4; calvaria pieces, 5
Otocyon megalotis, bat-eared fox: 13 adults; 13 pieces
Calvariae in various degrees of completeness, 12; articulated forefoot, 1
Proteles cristatus, aardwolf: 1 adult; 3 pieces
Calvaria, 1; mandible pieces, 2
Mellivora capensis, ratel: 2 adults; 2 pieces
Skull with articulated mandible, 1; mandible piece, 1
Orycteropus afer, antbear: 1 adult; 2 pieces
Calvaria, 1; innominate piece, 1
Hystrix africaeaustralis, porcupine: 1 adult; 2 pieces
Quills, 2; probably not food remains.
Struthio camelus, ostrich: 2 adults, 1 eggshell; 8 pieces
Tibiotarsal piece, 1; shaft piece, 1; claw sheath, 1; eggshell pieces, 5
Antelope class I: 18 pieces
Cervical vertebrae, 3; thoracic vertebra, 1; lumbar vertebrae, 4; specimen of articulated thoracic and lumbar vertebrae, 1; pelvic pieces, 4; femur, 1; proximal tibia, 1; distal tibia, 2; hoof sheath, 1
Antelope class II: 14 pieces
Mandible piece, 1; atlas, 1; axis, 1; cervical vertebra, 1; thoracic vertebra, 1; rib, 1; scapula pieces, 2; innominate piece, 1; distal humerus, 1; proximal radius, 1; distal radius, 1; distal metacarpal, 1; metatarsal,1
Antelope class III: 24 pieces
Cervical vertebrae, 4; thoracic vertebra, 1; scapula pieces, 2; pelvis pieces, 5; distal humerus, 1; proximal radius, 1; distal radius, 1; distal femur, 1; proximal tibia, 1; distal tibia, 1; proximal metacarpal, 1; metatarsal articulated with tarsals 1; hoof sheaths, 4
Carnivore indet.: 24 pieces
Calvaria piece, 1; mandible piece, 1; atlas, 1; cervical vertebrae, 2; thoracic vertebra, 1; lumbar vertebra, 1; articulated vertebrae, 13; pelvis piece, 1; proximal humerus, 1; tibia, 1; claw sheath, 1
Indeterminate fragments: 6 pieces
Bone flakes: 21 pieces

Table 42. Particulars of Remains from Suswa Locality 36E, the "Suswa Lair"

Papio anubis, anubis baboon
Craniums in varying degrees of completeness
52. Calvaria of a subadult ♂, without mandible, complete except for the following damage: right orbit extensively damaged with tooth marks around it; left brow ridge damaged with a punctate mark; puncture in left parietal; depressed fracture in right occipital; three punctate marks in right auditory region; right zygoma missing and left damaged; two irregular holes in left orbit and temporal.
52 + 54. Cranium of adult ♂, full dentition except for right M^3. Damage to pterygoids; both zygomata missing; two punctures in right temporal, one in left temporal. Mandible complete except for chewing on right coronoid process and right angle of the jaw.
55 + 56. Adult ♂ palatal piece, right $\underline{C}$, M^{2-3}, left M^{2-3}. Almost complete mandible except for damage to coronoid processes; canines and 2 incisors missing.
59 + 60. Adult ♀ calvaria without anterior teeth; pterygoids chewed away; puncture at base of right zygoma and one in midline of palate. Mandible with full dentition except for right I_2; complete except for right ascending ramus, which has been chewed off; tooth marks present.
63 + 64. Cranium of adult ♀. Calvaria without incisors; damage to ventral surface of snout; both zygomata missing; porcupine gnawing on borders of orbits. Mandible without anterior teeth; ascending rami damaged; weathered.
61 + 62. Cranium of adult ♀. Complete calvaria except for anterior teeth; porcupine gnawing above orbits; damage to sides of orbits and snout. Mandible without ascending rami; tooth marks and porcupine gnawing present.
65. Adult ♀ braincase without snout; extensively damaged and showing porcupine gnawing.
57 + 58. Adult ♀ cranium. Calvaria without anterior teeth; zygomata missing; porcupine gnawing on both orbits and right premaxilla. Mandible without anterior teeth; damage to left coronoid process and right ascending ramus; tooth marks present.
41. Juvenile ♂ mandibular fragment. Canines and premolars unerupted; incisors and 1 deciduous molar present. Clear chewing damage to both rami and a punctate mark on lingual side of left ramus.
44. Palate from a juvenile of undetermined sex. Canines unerupted; P^3–M^2 present bilaterally; the calvaria has clearly been chewed away from above.
Maxillary pieces
39. Right maxilla, P^3–M^2.
40. Left maxilla, $\underline{C}$–M^3.
73, 74, 75, 77, 78, and 122. Small maxillary pieces.
Mandible pieces
48. Mandible with M_3 not fully erupted; porcupine gnawing along right lower margin.
49. Adult mandible without anterior teeth or ascending rami. Some possible porcupine gnawing.
50. Old adult mandible without anterior teeth and showing damage to ascending rami.
Various cranial pieces
42. Right parietal with sutured fragments of occipital and frontal. Two holes, apparently caused by canine teeth, spaced 22 mm apart.
43, 47, 76, 92, 93. Miscellaneous cranial pieces.
Isolated teeth
81–86. Isolated canines.
87–91. Isolated postcanine teeth.
Vertebrae
102. Atlas vertebra.
99, 101, 104. damaged thoracic vertebrae.
100. Damaged lumbar vertebra.
103. Damaged sacrum.
106. Damaged caudal vertebra.
80, 82, 105, 107, 119. Undamaged caudal vertebrae.
Scapulae
95, 96, 98. Scapulae showing extensive damage to the blades.
Pelvis pieces
94, 97, 112, 113, 114, 115, 116, 117. Pelvis pieces extensively chewed, some showing punctate tooth marks.
Rib piece
150. A single damaged fragment.
Forelimb
15. Right humerus, complete.
16. Complete right ulna, and 17, complete right radius; both articulate with 15.
18. Complete left humerus, probably same individual as 15.
4, 5, 8, 9, 24, 34, 165. Distal humeri, extensively chewed with shafts often showing spiral fractures.
25, 35, 36. Humerus shafts, extensively chewed at both ends.
19, 166, 167, 168, 169. Radii in varying degrees of completeness, extensively chewed.
26, 30, 170, 171. Pieces of ulna, extensively damaged.
152, 153, 154, 161. Metacarpals, almost undamaged.
155–60, 162. Complete phalanges.
Hind limb
176. Femur, almost complete but showing porcupine gnawing.
10, 46, 70, 172, 173, 174. Proximal femur pieces, extensively damaged and showing tooth marks.
2, 6, 7, 12, 13, 20, 23, 32, 175. Femur shaft pieces showing extensive carnivore damage.
3, 29, 45, 164. Tibia shaft pieces, extensively damaged.
163. Distal tibia piece showing tooth marks.
108. Undamaged astragalus.
109–11. Damaged calcanei.
118. Articulated astragalus and calcaneus, showing tooth marks.
11, 14, 21, 22, 28, 31, 33, 37, 67. Long-bone shaft pieces showing extensive damage.
Panthera pardus, leopard
1. Complete braincase, without snout.
Oreotragus oreotragus, Klipspringer
51. Almost complete calvaria of an adult ♂. No mandible.
38. Pelvis bone showing chewing around the margins and 3 clearly defined tooth holes.
Birds
71, 72. Long-bone fragments from large and small unidentified birds.
Unidentified bone fragments: 13 pieces.
Bone flakes: 22 pieces. 1–2 cm, 5; 2–3 cm, 6; 3–4 cm, 2; 4–5 cm, 1; 5–6 cm, 1; 6–7 cm, 6; 8–9 cm, 1.

Source: Data recorded in Nairobi, June 1970, by C. K. Brain.

Table 43. Remains from the Portsmut Leopard Breeding Lair

Papio ursinus, baboon: 1 adult ♂; 20 pieces.
Almost complete cranium and mandible, 1; left femur shaft, 1; metapodials, 9; phalanges, 8
Panthera pardus, leopard: 1 adult; 50 pieces
Partial cranium, 1; isolated teeth, 9; atlas vertebra, 1; caudal vertebrae, 6; left femur, 1; ulnae, 2; right radius, 1; left astragalus, 1; right calcaneus, 1; metapodials, 8; phalanges, 19
Oreotragus oreotragus, klipspringer: 2 juveniles; 26 pieces
Mandible pieces, 2; isolated teeth, 3; cervical vertebrae, 4; thoracic vertebrae, 3; lumbar vertebra, 1; left scapula, 1; articulated left forefoot, 1; ulna piece, 1; tibia piece, 1; astragalus, 1; metacarpal, 1; tarsal, 1; phalanges,6
Antelope class II: 1 adult, 1 juvenile; 10 pieces
Scapula piece, 1; pelvis piece, 1; femur piece, 1; metapodial pieces, 4; tarsal, 1; phalanges, 2
Antelope class III: 1 adult; 2 pieces
Tibia piece, 1; calcaneus, 1
Equus zebra, mountain zebra: 2 adults; 13 pieces
Right humeri, 2; tibiae, 2; cervical vertebrae, 2; thoracic vertebrae, 3; lumbar vertebrae, 3 articulated; sacrum, 1
Procavia capensis, dassie: 1 adult, 3 juveniles; 15 pieces
Maxillae, 2; mandible pieces, 3; cervical vertebra, 1; scapulae, 2; pelvis pieces, 4; femur shaft, 1; radii, 2
Chelonia indet., tortoise: 1 individual; 11 pieces
Carapace and plastron pieces, 10; limb bone, 1
Indeterminate fragments: 45 pieces
Cultural objects: 17 pieces
Grooved stones, 2; grooved pottery piece, 1; stone pestles, 2; potsherd, 1; hematite pieces, 6; quartz pieces showing flake scars, 5

Table 44. Remains from the Hakos River Leopard Breeding Lair

Papio ursinus, baboon: 1 adult; 1 piece
Metacarpal, 1
Panthera pardus, leopard: 1 adult, 1 juvenile; 26 pieces
Adult remains: humerus, 1; calcaneus, 1; astragalus, 1
Juvenile remains: cervical vertebrae, 2; thoracic vertebrae, 5; lumbar vertebrae, 8; caudal vertebrae, 2; tibia shaft, 1; distal femoral epiphysis, 1; calcaneus, 1; astragalus, 1; innominate pieces, 2
Lycaon pictus, wild dog: 1 adult; 39 pieces
Isolated teeth, 4; atlas vertebra, 1; thoracic vertebrae, 4; caudal vertebra, 1; ribs, 4; innominate, 1; tarsals, 2; metapodials, 7; phalanges, 15
Ovis/Capra sp., sheep or goat: 1 adult; 4 pieces
Scapulae, 2; terminal phalanges, 2
Bos sp., cattle: 1 adult; 7 pieces
Articulated right hind foot, 1; metatarsal, 1; terminal phalanges, 3; hoof sheaths, 2
Oreotragus oreotragus, klipspringer: 1 adult; 4 pieces
Palate, 1; isolated incisor, 1; ulna, 1; tibia, 1
Tragelaphus strepsiceros, kudu: 2 juveniles; 8 pieces
Maxillary pieces, 3; mandible pieces, 2; tooth fragments, 3
Oryx gazella, gemsbok: 1 subadult; 1 piece
Cranium without mandible, 1
Antelope class II: 1 adult; 7 pieces
Thoracic vertebra, 1; left humerus, 1; phalanges, 4; hoof sheath, 1
Antelope class III: 3 adults, 3 juveniles; 96 pieces
Calvaria pieces, 3; mandible pieces, 2; tooth fragments, 4; vertebral pieces: atlas, 2; cervical, 5; thoracic, 4; lumbar, 6; ribs, 3; scapula, 1; pelvis pieces, 2; humeri, 6; radii, 2; ulnae, 2; femur pieces, 6; tibia pieces, 6; metapodial pieces, 3; calcanei, 5; carpals/tarsals, 9; sesamoid, 1; phalanges, 24 pieces
Equus cf. *zebra,* mountain zebra: 1 subadult, 1 juvenile; 6 pieces
Tibia, 1; humerus, 1; sesamoid, 1; hoof pieces, 3
Procavia capensis, dassie: 2 subadults; 4 pieces
Damaged skull and mandible, 1; isolated incisor, 1; ribs, 2
Chelonia indet., tortoise: 1 individual; 6 pieces
Plastron piece, 1; vertebra, 1; girdle fragments, 4
Bone flakes: 13 pieces
Indeterminate fragments: 117

Table 45. Bones Found in the Quartzberg Leopard Lair

A. Probable Porcupine-Collected Component

Gemsbok *(Oryx gazella)*	
Cranial pieces	3
Zebra *(Equus zebra)*	
Cranial pieces	2
Postcranial pieces	3
Ox *(Bos sp.)*	
Cranial piece	1
Baboon *(Papio ursinus)*	
Cranial piece (mandible)	1
Indet. bovid postcranial pieces	106
Indet. bone pieces	20
Bone flakes	11
Total	147

B. Probable Leopard-Collected Component

Klipspringer *(Oreotragus oreotragus)*	
Cranial pieces	6
Postcranial pieces (some articulated)	32
Steenbok *(Raphicerus campestris)*	
Cranial pieces	1
Postcranial pieces	2
Gemsbok *(Oryx gazella)* calf	
Postcranial pieces (articulated)	3
Domestic calf *(Bos sp.)*	
Postcranial piece	1
Zebra *(Equus zebra)* colt	
Postcranial piece	1
Baboon *(Papio ursinus)*	
Cranial piece	1
Postcranial pieces	2
Dassie *(Procavia capensis)*	
Cranial pieces	12
Postcranial pieces	3
Total	64
Overall total	211

Table 46. Minimum Number of Individual Animals Represented by the Bones in the Quartzberg Leopard Lair

A. Probable Porcupine-Collected Component		B. Probable Leopard-Collected Component	
Gemsbok adult	1	Klipspringer male	2
Gemsbok calf	1	Klipspringer female	2
Domestic calf	1	Steenbok male	1
Mountain zebra adult	1	Gemsbok calf	1
Mountain zebra colt	1	Domestic calf	1
Baboon	1	Mountain zebra colt	1
Total	6	Baboon female	1
		Dassie	6
		Total	15

Incidence of porcupine gnawing on bones	No. Gnawed	No. Ungnawed	Total	% Gnawed Bones
Probable porcupine-collected component	118	29	147	80.2
Probable leopard-collected component	15	49	64	23.4

Table 47. Bones from Various Zones in the Quartzberg Leopard Lair

	Presumed Porcupine-Collected Component		Presumed Leopard-Collected Component	
Cave Zone	N	%	N	%
Light zone	5	3.4	23	35.9
Twilight zone	57	38.8	15	23.4
Dark zone	85	57.8	26	40.6
Total	147	100.0	64	99.9

Table 48. Prey of Leopards as Reflected by Observed Kills

Prey Species	Kafue National Park (Mitchell, Shenton, and Uys 1965) N	%	Serengeti (Kruuk and Turner 1967) N	%	Serengeti (Schaller 1972) N	%	Matopos (Smith 1977) N	%	Kruger National Park (Pienaar 1969) N	%
Antelope class I										
Gray duiker	11	11.5					4	10.5	148	2.0
Grysbok	4	4.2							31	.4
Oribi	3	3.1							2	—
Steenbok							1	2.6	66	.9
Klipspringer							1	2.6	35	.5
Antelope class II										
Reedbuck	19	19.8	6	10.9	19	11.6	5	13.2	222	3.0
Mountain reedbuck									5	.1
Puku	15	15.6								
Impala	8	8.3	9	16.3			12	31.6	5,511	73.8
Bushbuck	4	4.2	1	1.8					465	6.2
Thomson's gazelle			15	27.3	104	63.4				
Grant's gazelle			2	3.6	10	6.1				
Antelope class III										
Hartebeest	9	9.4			2	1.2				
Nyala									41	.6
Kudu	3	3.1					1	2.6	208	2.8
Lechwe	3	3.1								
Wildebeest	1	1.0	5	9.0	11	6.7	3	7.9	85	1.1
Topi			1	1.8	3	1.8				
Waterbuck					1	.6	1	2.6	287	3.8
Sable							4	10.5	10	.1
Roan									2	—
Tsessebe							1	2.6	14	.2
Antelope class IV										
Eland							2	5.3	9	.1
Buffalo									4	.1
Other mammals										
Warthog	2	2.1			1	.6			153	2.1
Zebra			4	7.2	2	1.2	1	2.6	75	1.0
Hyrax			1	1.8			1	2.6		
Ant bear									9	.1
Baboon	2	2.1	2	3.6	1	.6			58	.8
Vervet monkey	3	3.1							3	—
Bat-eared fox					2	1.2				
Golden jackal					1	.6				
Black-backed jackal			1	1.8	1	.6				
Leopard									2	—
Cheetah			1	1.8					2	—
Serval	1	1.0			2	1.2				
Genet	1	1.0								
Civet	1	1.0							4	.1
Hare	1	1.0							1	—
Springhare	2	2.1	1	1.8						
Porcupine	1	1.0							5	.1
Cane rat	1	1.0							1	—
Birds										
European stork			2	3.6	4	2.4				
Secretary bird			1	1.8						
Guinea fowl			1	1.8						
Vulture			1	1.8						
Ostrich							1	2.6	6	.1
Reptiles										
Python			1	1.8					1	—
Fish										
Catfish	1	1.0								
Total	96	99.7	55	99.5	164	99.8	38	99.8	7,465	100.0

Table 49. Broad Categories of Leopard Prey as Percentage of Total, Based on Observed Kills

Prey Species	Kafue National Park (Mitchell, Shenton, and Uys 1965)	Serengeti (Kruuk and Turner 1967)	Serengeti (Schaller 1972)	Matopos (Smith 1977)	Kruger National Park (Pienaar 1969)	Mean
Antelope class I	18.8	—	—	15.7	3.8	7.7
Antelope class II	47.9	59.9	81.1	44.8	83.1	63.4
Antelope class III	16.6	10.8	10.3	26.2	8.6	14.5
Antelope class IV	—	—	—	5.3	.2	1.1
Primates	5.2	3.6	.6	—	.8	2.0
Carnivores	3.0	3.6	3.6	—	.1	2.1
Other mammals	7.2	10.8	1.8	5.2	3.3	5.7
Birds, reptiles, fish	1.0	10.8	2.4	2.6	.1	3.4
Total	99.7	99.5	99.8	99.8	100.0	99.9

Table 50. Food Items Represented in Leopard Scats from the Matopos Area, Zimbabwe

Prey Species	Percentage
Dassie	46.1
Klipspringer	10.2
Hare	8.4
Duiker	7.1
Rats and mice	4.8
Sable	4.0
Cane rat	3.2
Bird	2.8
Baboon	1.9
Impala	1.7
Reedbuck	1.7
Springhare	1.3
Steenbok	.8
Wildebeest	.5
Eland	.5
Leopard	.4
Genet	.4
Snake	.3
Bushpig	.3
Porcupine	.6
Hedgehog	.3
Scorpion	.3
Bushbuck	.2
Mongoose	.2
Unidentified	1.8
Total	99.8

Source: Data from Smith (1977).

Table 51. Distances over Which Leopards Have Dragged Their Prey in the Matopos National Park, 1971–74

	Distance Moved (in meters)					
	Wet Season			Dry Season		
Prey Species	N	Mean	Range	N	Mean	Range
Impala	6	170	(50– 300)	3	25	(0– 50)
Steenbok	1	200				
Sable antelope	2	650	(300–1,000)			
Common duiker	1	300		5	125	(0–400)
Wildebeest	1	150		1	50	
Eland				2	200	(0–400)
Reedbuck				4	175	(0–400)
Tsessebe				1	40	
Klipspringer				1	50	

Source: Data from Smith (1977).

Table 52. Food Remains of Black Eagles from below Nests and Perches in the Matopo Hills, Rhodesia, 1975

Procavia capensis	
Adult craniums with articulated mandibles	2
Adult craniums without mandibles	4
Mandibles	6
Isolated upper incisors	3
Total	15
Heterohyrax brucei	
Adult and subadult craniums, without mandibles	23
Adult and subadult mandibles	26
Mandibular rami, left	10
Mandibular rami, right	14
Juvenile cranium, without braincase	1
Calvaria pieces	9
Maxillary piece	1
Isolated upper incisors	3
Total	87
Procavia or *Heterohyrax*	
Articulated right humerus and radius/ulna	1
Articulated right tibia with calcaneus and astragalus	1
Pelvis pieces	2
Left femur, partly digested from regurgitation	1
Total	5
Kinyxis belliana, hinged tortoise	
Plastron pieces	3
Loose scale	1
Total	4
Overall total	111

Table 53. Food Remains of Black Eagles from a Feeding Perch at Wonderbroom, Pretoria, 1969

Procavia capensis, dassie	
Adult craniums with articulated mandibles	7
Adult maxillary piece	1
Adult mandible	1
Subadult craniums with articulated mandibles	4
Subadult cranium without mandible	1
Juvenile maxilla	1
Juvenile mandible pieces	4
Calvaria pieces	2
Isolated upper incisors	4
Vertebrae	8
Sacrum	1
Rib piece	1
Scapula	1
Femur	1
Articulated femur and left innominate	1
Complete pelvises	3
Innominate pieces	3
Total	44
Lagomorph, probably *Lepus* sp.	
Mandible pieces	2
Tibia, right, one with articulated foot	2
Total	4
Bird, francolin size	
Tibiotarsus, right	2
Sacra	2
Humerus	1
Total	5
Testudo pardalis, mountain tortoise	
Carapace pieces	3
Overall total	56

Table 54. Food Remains of Black Eagles from below a Feeding Perch, Portsmut Farm, South-West Africa, 22 June 1969

Procavia capensis, dassie	
Adult craniums with articulated mandibles	4
Adult maxillae	2
Adult maxillary fragments	5
Adult mandible pieces	8
Subadult cranium with articulated mandible	1
Juvenile mandible pieces	3
Pelvis pieces	9
Tibia pieces	3
Limb bones and bone fragments, partly digested, from regurgitations	95
Total	130
Lagomorph, probably *Lepus* sp.	
Distal humerus	1
Overall total	131

Table 55. Species Represented in the Bone Accumulations from the "Human Site," Andrieskraal 1, and the "Porcupine Lair," Andrieskraal 2

	Minimum Number of Individuals	
Taxon	Andrieskraal 1	Andrieskraal 2
MAMMALIA		
Primates		
Papio ursinus, chacma baboon	1	2
Carnivora		
Genetta genetta, small spotted genet	1	
Herpestes pulverulentis, gray mongoose	3	
?Herpestes sp.	2	
Mellivora capensis, honey badger		1
Felis lybica, wild cat	2	
Panthera ?pardus, leopard	1	1
Canis mesomelas, black-backed jackal	1	
Proboscidea		
Loxodonta africana, African elephant		1
Hyracoidea		
Procavia capensis, dassie	9	10
Perissodactyla		
Diceros bicornis, black rhino		1
Equus asinus, domestic donkey		3
Artiodactyla		
Hippopotamus amphibius	1	2
Potamochoerus porcus	1	4
Bos sp., domestic ox		6
Capra sp., domestic goat		6
Tragelaphus scriptus, bushbuck	2	6
Tragelaphus strepsiceros, kudu	1	
Cephalophus monticola, blue duiker	2	2
Sylvicapra grimmia, gray duiker	4	
Raphicerus campestris or *melanotis*, steenbok or grysbok	20	1
Rodentia		
Hystrix africaeaustralis, porcupine	2	5
Indet. small rodents	9	
Lagomorpha		
Lepus sp., hare	16	
REPTILIA		
Tortoise or turtle	23	1
AVES		
Struthio australis, ostrich	1	1
Indet. smaller birds	6	3
Total	108	56

Source: Data from Hendey and Singer (1965).

Table 56. Analysis of Objects from the Nossob Porcupine Lair

Object	Collection 1956	Collection 1968	Collection Total	Object	Collection 1956	Collection 1968	Collection Total
Gemsbok *(Oryx gazella)*				Lion *(Panthera leo)*			
Cranial pieces	7	2	9	Cranial pieces	1		1
Horns	21	7	28	Hunting dog *(Lycaon pictus)*			
Springbok *(Antidorcas marsupialis)*				Cranial pieces	3	3	6
Cranial pieces	12	5	17	Mongoose (cf. *Cynictis* sp.)			
Horns	46	26	72	Cranial pieces	2		2
Wildebeest *(Connochaetes taurinus)*				Skunk *(Ictonyx striatus)*			
Horns	5	3	8	Cranial pieces		1	1
Hartebeest *(Alcelaphus buselaphus)*				Indet. carnivore postcranial pieces	2	5	7
Cranial pieces	2		2	Porcupine *(Hystrix africaeaustralis)*			
Horns	11	6	17	Cranial pieces	1		1
Steenbok *(Raphicerus campestris)*				Tortoise *(Testudo occulifera)*			
Horns	2	4	6	Carapace pieces and scales	56	3	59
Duiker *(Sylvicapra grimmia)*				Ostrich *(Struthio camelus)*			
Cranial pieces	2		2	Postcranial pieces	7	4	11
Goat				Eggshell pieces	7		7
Horns	2	1	3	Large bird (gen. et sp. indet.)			
Cattle				Postcranial piece		1	1
Horns	5		5	Miscellaneous bone fragments	364	31	395
Indet. bovid postcranial pieces	405	243	648	Bone flakes	277	28	305
Equid				Pieces of wood	68	3	71
Cranial pieces	1		1	Metal objects	13	4	17
Postcranial pieces	6	4	10	Total	1,328	384	1,712

Table 57. Minimum Numbers of Individual Animals Represented in the Bone Accumulation from the Nossob Porcupine Lair

Animal	Collection 1956	Collection 1968	Total
Springbok	25	15	40
Gemsbok	11	4	15
Hartebeest	6	3	9
Wildebeest	3	2	5
Steenbok	2	3	5
Duiker	2	0	2
Cattle	3	0	3
Goat	1	1	2
Horse	2	2	4
Hunting dog	1	3	4
Mongoose	2	0	2
Skunk	0	1	1
Lion	1	0	1
Porcupine	1	0	1
Tortoise	6	3	9
Ostrich	1	1	2
Bird	0	1	1
Total	67	39	106

Table 58. Kalahari Park Area Census 1973/74 (De Graaff et al.)

Species	August	November	February	April	Annual Average	Percentage of Total
Springbok	17,560	18,228	24,041	12,894	18,181	40.2
Gemsbok	11,183	7,384	16,073	15,915	12,693	27.9
Hartebeest	14,178	1,408	5,200	5,011	6,449	14.2
Wildebeest	6,378	2,230	4,067	2,522	3,799	8.4
Eland	1,433	611	6,569	1,235	2,462	5.4
Steenbok	1,402	1,073	1,051	1,645	1,293	2.9
Duiker	262	381	710	497	463	1.0
Total	52,396	31,315	57,711	39,719	45,286	100.0

Table 59. Bovid Skeletal Parts, 1968 Collection, Nossob Porcupine Lair

	Antelope Size Class			
Part	I	II	III	Total
Horn pieces	0	28	17	45
Other cranial pieces	0	7	10	17
Vertebrae				
Atlas	0	2	6	8
Axis	0	5	9	14
Cervical, 3–7	0	17	19	36
Thoracic	3	7	6	16
Lumbar	2	20	18	40
Sacral	0	2	2	4
Caudal	0	0	0	0
Rib pieces	0	5	0	5
Scapula pieces	0	3	16	19
Pelvis pieces	3	7	20	30
Humerus				
Complete	0	1	1	2
Proximal	0	1	2	3
Distal	0	2	2	4
Shaft	0	1	0	1
Radius and ulna				
Complete	0	2	3	5
Proximal	0	2	3	5
Distal	0	0	0	0
Shaft	1	0	0	1
Carpal bones	0	0	1	1
Metacarpal pieces	0	1	2	3
Femur				
Complete	0	0	0	0
Proximal	0	1	10	11
Distal	0	0	8	8
Shaft	0	1	2	3
Tibia				
Complete	0	1	4	5
Proximal	1	2	4	7
Distal	0	0	4	4
Shaft	1	0	0	1
Calcaneus	0	1	1	2
Astragalus	1	1	2	4
Metatarsal pieces	0	3	1	4
Metapodial pieces	2	1	7	10
Phalanges	1	8	3	12
Total	15	132	183	330

Table 60. Bovid Skeletal Parts in the Bone Accumulation from the Nossob Porcupine Lair

	Collection		
Part	1956	1968	Total
Cranium			
Complete skulls	0	0	0
Maxillary pieces	11	4	15
Mandible pieces	19	5	24
Isolated teeth	27	0	27
Other pieces	16	9	25
Horn pieces	112	45	157
Vertebrae			
Atlas	17	8	25
Axis	9	14	23
Cervical, 3–7	40	36	76
Thoracic	40	13	53
Lumbar	36	40	76
Sacral	3	4	7
Caudal	1	0	1
Ribs	38	5	43
Scapula pieces	24	18	42
Pelvis pieces	36	30	66
Humerus			
Complete	2	2	4
Proximal	2	3	5
Distal	12	4	16
Shaft	3	1	4
Radius and ulna			
Complete	0	5	5
Proximal	13	5	18
Distal	9	0	9
Shaft	1	1	2
Carpal bones	12	5	17
Metacarpal pieces	7	3	10
Femur			
Complete	1	0	1
Proximal	5	11	16
Distal	6	8	14
Shaft	7	3	10
Tibia			
Complete	3	5	8
Proximal	6	7	13
Distal	9	4	13
Shaft	3	1	4
Calcaneus	16	2	18
Astragalus	15	4	19
Other tarsal bones	9	0	9
Metatarsal pieces	10	4	14
Metapodial pieces	14	10	24
Phalanges	48	12	60
Sesamoids	1	0	1
Total	643	331	974

Table 61. Percentage Survival of Bovid Skeletal Parts (Nossob sample: 81 bovid individuals)

Part	Number Found	Original Number	Survival (%)	Survival of Hottentot Goat Bone Sample (%)
Horn pieces	157	162	96.9	94.7
Pelvis pieces	66	162	40.7	9.0
Atlas vertebrae	25	81	30.9	6.3
Axis vertebrae	23	81	28.4	7.4
Scapula pieces	42	162	25.9	9.2
Cervical vertebrae, 3–7	76	405	18.8	1.2
Maxillae	15	81	18.5	26.3
Metatarsal, proximal	26	162	16.0	10.3
Metatarsal, distal	26	162	16.0	5.3
Lumbar vertebrae	76	486	15.6	2.7
Half-mandibles	24	162	14.8	30.7
Radius and ulna, proximal	23	162	14.2	17.1
Metacarpal, proximal	22	162	13.6	8.4
Metacarpal, distal	22	162	13.6	6.0
Tibia, proximal	21	162	13.0	3.4
Tibia, distal	21	162	13.0	19.0
Humerus, distal	20	162	12.3	21.5
Astragalus	19	162	11.7	4.2
Calcaneus	18	162	11.1	3.7
Femur, proximal	17	162	10.4	4.7
Femur, distal	15	162	9.3	2.4
Sacral vertebrae	7	81	8.6	.5
Radius and ulna, distal	14	162	8.6	5.8
Phalanges	60	972	6.2	.9
Humerus, proximal	9	162	5.5	0
Thoracic vertebrae	53	1,053	5.0	.9
Ribs	43	2,106	2.0	3.4
Caudal vertebrae	1	810	.1	0

Table 62. Weights of Bones from the Nossob Porcupine Lair

Weight (g)	1956 Collection N Ungnawed	N Gnawed	Total	1968 Collection N Ungnawed	N Gnawed	Total	Overall Total
0– 10	388	283	671	31	30	61	732
10– 20	45	100	145	19	34	53	198
20– 30	21	70	91	11	19	30	121
30– 40	16	42	58	4	18	22	80
40– 50	16	29	45	3	15	18	63
50– 60	6	20	26	6	15	21	47
60– 70	6	35	41	5	20	25	66
70– 80	5	15	20	4	10	14	34
80– 90	4	16	20	4	12	16	36
90–100	4	13	17	2	12	14	31
100–150	14	59	73	9	41	50	123
150–200	13	31	44	4	20	24	68
200–250	2	16	18	2	9	11	29
250–300	5	13	18	3	8	11	29
300–350	3	5	8		4	4	12
350–400	5	3	8		2	2	10
400–450	1	6	7		2	2	9
450–500	1	2	3				3
500–550	1	3	4				4
550–600	1	3	4		1	1	5
600–650		3	3		1	1	4
650–700	1	1	2				2
700–750		2	2				2
Total	558	770	1,328	107	273	380	1,708

Table 63. Lengths of Bones from the Nossob Porcupine Lair

Length		1956 Collection			1968 Collection			
(cm)	(inches)	N Ungnawed	N Gnawed	Total	N Ungnawed	N Gnawed	Total	Overall Total
0– 2.5	0– 1	59	14	73	2	3	5	78
2.5– 5.1	1– 2	216	185	401	19	16	35	436
5.1– 7.6	2– 3	113	134	247	15	37	52	299
7.6– 10.2	3– 4	55	121	176	13	47	60	236
10.2– 12.7	4– 5	26	71	97	9	41	50	147
12.7– 15.2	5– 6	21	65	86	15	34	49	135
15.2– 17.8	6– 7	11	34	45	6	20	26	71
17.8– 20.3	7– 8	9	30	39	6	14	20	59
20.3– 22.9	8– 9	13	30	43	3	13	16	59
22.9– 25.4	9–10	7	19	26	8	17	25	51
25.4– 27.9	10–11	3	10	13	4	7	11	24
27.9– 30.5	11–12	6	21	27	2	6	8	35
30.5– 33.0	12–13	4	4	8		8	8	16
33.0– 35.6	13–14	1	8	9	1	7	8	17
35.6– 38.1	14–15	1	2	3		1	1	4
38.1– 40.6	15–16	1	3	4				4
40.6– 43.2	16–17	1	4	5				5
43.2– 45.7	17–18	1	1	2				2
45.7– 48.3	18–19	2		2				2
48.3– 50.8	19–20		2	2	1	1	2	4
50.8– 53.3	20–21							
53.3– 55.9	21–22		1	1				1
55.9– 58.4	22–23		1	1				1
58.4– 61.0	23–24	2	2	4	2		2	6
61.0– 63.5	24–25	1		1	1		1	2
63.5– 66.0	25–26		1	1				1
66.0– 68.6	26–27		1	1		1	1	2
68.6– 71.1	27–28							
71.1– 73.7	28–29							
73.7– 76.2	29–30							
76.2– 78.7	30–31	1		1				1
78.7– 81.3	31–32							
81.3– 83.8	32–33	4		4				4
83.8– 86.4	33–34		1	1				1
86.4– 88.9	34–35							
88.9– 91.4	35–36		1	1				1
91.4– 94.0	36–37		1	1				1
94.0– 96.5	37–38		2	2				2
96.5– 99.1	38–39		1	1				1
99.1–101.6	39–40							
Total		558	770	1,328	107	273	380	1,708

Table 64. Percentage Abundance of Gnawed Bones and Other Objects in Various Weight Classes from the Nossob Porcupine Lair

Weight (g)	N (bones)	% of Total	N (gnawed bones)	% of Total	% Bones Gnawed in This Weight Class
0– 50	1,194	69.9	640	37.5	53.6
50–100	214	12.5	168	9.8	78.5
100–150	123	7.2	100	6.4	81.3
150–200	68	4.0	51	3.0	75.0
200–250	29	1.7	25	1.4	82.8
250–300	29	1.7	21	1.3	72.4
300–350	12	.7	9	.5	75.0
350–400	10	.6	5	.3	50.0
400–450	9	.5	8	.5	88.9
450–500	3	.2	2	.1	66.7
500–550	4	.2	3	.2	75.0
550–600	5	.3	4	.2	80.0
600–650	4	.2	4	.2	100.0
650–700	2	.1	1	.1	50.0
700–750	2	.1	2	.1	100.0
Total	1,708	99.9	1,043	61.1	

Table 65. Percentage Abundance of Gnawed Bones of Various Lengths from the Nossob Porcupine Lair

Length (cm)	Length (inches)	N (bones)	% of Total	N (gnawed bones)	% of Total	% Gnawed Bones in This Size Class
0– 2.5	0– 1	78	4.6	17	1.0	21.8
2.5– 5.1	1– 2	436	25.5	201	11.8	46.1
5.1– 7.6	2– 3	299	17.5	171	10.0	57.2
7.6–10.2	3– 4	236	13.8	168	9.8	71.2
10.2–12.7	4– 5	147	8.6	112	66	76.2
12.7–15.2	5– 6	135	7.9	99	5.8	73.3
15.2–17.8	6– 7	71	4.1	54	3.2	76.1
17.8–20.3	7– 8	59	3.5	44	2.6	74.6
20.3–22.9	8– 9	59	3.5	43	2.5	72.9
22.9–25.4	9–10	51	3.0	36	2.1	70.6
25.4–27.9	10–11	24	1.4	17	1.0	70.8
27.9–30.5	11–12	35	2.1	27	1.6	77.1
30.5–33.0	12–13	16	.9	12	.7	75.0
33.0–35.6	13–14	17	1.0	15	.9	88.2
35.6–38.1	14–15	4	.2	3	.2	75.0
38.1–40.6	15–16	4	.2	3	.2	75.0
40.6 a	16 +	37	2.2	21	1.2	56.8
Total		1,708	100.0	1,043	61.1	

Table 66. Prey Animals Identified from Pellet Collections of *Bubo africanus* and *Tyto alba* at Swartkrans, between 1972 and 1975

Taxon	*Bubo africanus* N[a]	*Bubo africanus* %	*Tyto alba* N	*Tyto alba* %
Order Insectivora				
Family Soricidae	2	5.6	8	5.9
Order Rodentia				
Family Muridae				
Subfamily Murinae				
Praomys natalensis	8	22.2	53	39.0
Mus minutoides	9	25.0	37	27.2
Aethomys chrysophilus	3	8.3	3	2.2
Subfamily Otomyinae				
Otomys irroratus	5	13.9	4	2.9
Family Cricetidae				
Subfamily Gerbillinae				
Tatera sp.			4	2.9
Subfamily Dendromurinae				
Dendromys sp.	4	11.1	5	3.7
Steatomys pratensis	1	2.8	5	3.7
Dendromurine indet.			11	8.1
Class Aves				
Indet.	2	5.6	4	2.9
Class Insecta				
Indet.	2	5.6	2	1.5
Total	36	100.1	136	100.0

[a] Individual animals.

Table 67. Prey of the Cape Eagle Owl, *Bubo capensis mackinderi*, Determined from Analysis of Food Remains Collected in the Matopo Hills, Zimbabwe

Prey Species	Estimated Mass (kg)	N[a]	Total Mass (kg)	Relative % Mass
Pronolagus crassicaudatus, red rock hare	2.0	416	906.0	63.5
Lepus saxatilis, scrub hare	1.6	50	88.5	6.2
Procavia capensis and *Heterohyrax brucei*, dassies	.9	275	247.5	17.3
Thryonomys gregorianus, lesser cane rat	3.6	25	90.0	6.3
T. swinderianus, greater cane rat	4.5	9	40.5	2.8
Elephantulus myurus, elephant shrew	.5	21	10.5	.7
Otomys angoniensis, vlei rat	.2	22	4.2	.3
Other murids, rats and mice	.1	15	.9	.1
Pedetes capensis, springhare	2.7	6	16.2	1.1
Erinaceus frontalis, hedgehog	.3	4	1.2	.1
Paraxerus cepapi, bush squirrel	.2	2	.4	< .1
Herpestes sanguineus, slender mongoose	.5	1	.5	< .1
Paracynictis selousi, juvenile, Selous's mongoose	.5	1	.5	< .1
Genetta sp., genet	2.0	4	8.0	.6
Viverra civetta, juvenile, civet	4.5	1	4.5	.3
Birds		12	7.5	.5
Scorpions		10	.2	< .1
Insects		6	.1	< .1
Lizards		2	.1	< .1
Total		925	1,427.3	100.0

Source: Data from Gargett and Grobler (1976).
[a] Individual animals.

Table 68. Prey Species Most Abundant in Analyzed *Tyto alba* Pellet Collections from Southern Africa

Order Insectivora
- Family Soricidae
  - *Crocidura bicolor*, Luanda, Angola (Dean 1974)
  - *Crocidura* sp., Quibala, Angola (Dean 1974)
- Family Chrysochloridae
  - *Eremitalpa granti namibensis*, Urihauchab Mountains, South-West Africa (Vernon 1972)

Order Rodentia
- Family Muridae
  - Subfamily Murinae
    - *Praomys natalensis*, Onderstepoort, Transvaal (Kolbe 1946); Matope and Lunzu, Malawi (Hanney 1962); Tamara forest, Kingwilliamstown (Skead 1963); Sabie Poort, Nwanedzi, and Hape, Kruger National Park (Coetzee 1963); Pietermaritzburg, Ashburton, Otto's Bluff, and Polly Shorts, Pietermaritzburg area (Vernon 1972); Umfolozi, Natal (Vernon 1972); Naboomspruit, Settler's, Rus de Winter, Van Riebeek Nature Reserve, Bloemhof, Cullinan, and Jack Scott Nature Reserve, Transvaal (Vernon 1972); Warmbaths, Transvaal (Dean 1973*a*); Cuanzo Sul, Huambo, Huila, and Lucala, Angola (Dean 1974); Swartkrans, Bolt's Farm, and Makapansgat Limeworks, Transvaal (Brain, this study)
    - *Mus minutoides*, Kingston, Kimberley area, Cape (Dean 1975)
    - *Aethomys namaquensis*, Valencia Ranch and Gamsberg, South-West Africa (Dean 1975)
  - Subfamily Otomyinae
    - *Otomys irroratus*, Sulekama, Qumbu District, Cape (Crass 1944); Kingwilliamstown, Cape (Bateman 1960); Irene Old Quarry and Irene New Quarry, Transvaal (Brain, this study)
    - *Otomys* sp., Mossel Bay, Cape, and Krugersdorp, Transvaal (Vernon 1972)
- Family Cricetidae
  - Subfamily Gerbillinae
    - *Gerbillurus paeba*, Twee Rivieren, Ky Ky, Monro, Auchterlonie, and Kamfersboom, Kalahari National Park, Cape (Davis 1958); Samevloeiing, Auchterlonie A and B, Kransbrak, and Kij Kij (or Ky Ky), Kalahari National Park (Nel and Nolte 1965); Upington, Cape, Vanzylsrus, Cape, Auob River, Gochas, Stampriet, Koichab Pan, and Tsumis Estate, South-West Africa (Vernon 1972); Mirabib, South-West Africa (Brain 1974*b*); Gochas and Heuningvlei, South-West Africa (Dean 1975)
    - *Desmodillus auricularis*, Craighlockhart, Kalahari National Park, Cape (Nel and Nolte 1965); Colesberg, Cape (Vernon 1972); Windhoek District, South-West Africa (Dean 1975)
    - *Tatera leucogaster*, Mbangari, Machindudzi, Malahla-Panga, Pretorius Kop Rest Camp, and Matupa Cave, Kruger National Park (Coetzee 1963); Namutoni, South-West Africa (Winterbottom 1966)
    - *Tatera* sp., Bolt's Farm, Transvaal (Davis 1959); Batulama, Kalahari National Park, Cape (Nel and Nolte 1965); Marrick, Kimberley District, Cape (Dean 1975)
  - Subfamily Dendromurinae
    - *Steatomys pratensis*, Skukuza koppies, Kruger National Park (Coetzee 1963)
    - *Malacothrix typica*, Springbok, Cape (Vernon 1972)

Class Aves
- Various, mostly passerine birds, Bryanston, Johannesburg (Davis 1959); Bryanston, Johannesburg (de Graaff 1960*b*); Kingwilliamstown, Cape (Skead 1956); Graaff Reinet, Cape and Kub, South-West Africa (Vernon 1972)

Class Reptilia
- Family Gekkonidae, Mirabib, South-West Africa (Vernon 1972)

Table 69. Prey Species in a Sample of *Tyto alba* Pellets from Warmbaths, Transvaal

Species	N[a]	Estimated Mean Weight per Individual (g)
Mammals		
Elephantulus brachyrynchus	1	46.5
Suncus lixus	1	6.0
Crocidura hirta	6	14.1
C. cyanea/silacea/pilosa	5	8.1
C. bicolor	9	5.7
Tatera leucogaster	22	52.0
Otomys angoniensis	17	100.5
Saccostomus campestris	14	41.7
Steatomys pratensis	48	26.0
Mus minutoides	12	5.8
Praomys natalensis	431	38.8
Aethomys chrysophilus	2	81.3
Lemniscomys griselda	5	59.3
Rhabdomys pumilio	1	35.2
Birds		
Streptopelia capicola	1	130.0
Cisticola chiniana	3	13.0
Prinia flavicans	1	9.9
Lanius collaris	1	44.6
Passer domesticus	1	20.5
P. griseus	1	20.8
P. melanurus	1	18.6
Sporopipes squamifrons	4	10.9
Ploceus velatus	17	23.3
Quelea quelea	34	19.4
Euplectes orix	3	23.0
E. afer	1	15.4
Uraeginthus angolensis	4	9.7
Total	646	

Source: Data from Dean (1973*a*).
[a] Individual animals.

Table 70. Percentage Contributions Made to the Diet of Barn Owls by Animals of Different Weights

Weight Range (g)	N[a]	% of Total
0– 20	81	12.5
20– 40	502	77.7
40– 60	43	6.7
60– 80		
80–100	2	.3
100–120	17	2.6
120– 40	1	.2
Total	646	100.0

[a] Individual animals.

Table 71. Faunal Content of Barn Owl Pellet Collections from Several Localities Referred to in the Text

Taxon	Irene Old Quarry		Irene New Quarry		Bolt's Farm		Makapansgat Limeworks	
	N[a]	%	N	%	N	%	N	%
Order Insectivora								
Family Soricidae	6	6.0	3	1.7	10	3.6	19	6.8
Order Rodentia								
Family Muridae								
Subfamily Murinae								
Praomys natalensis	17	17.0	41	23.4	146	52.9	134	47.9
Aethomys chrysophilus					12	4.3	13	4.6
Mus minutoides			1	.6	5	1.8	6	2.1
Rhabdomys pumilio	2	2.0	5	2.9				
Subfamily Otomyinae								
Otomys irroratus/angoniensis	32	32.0	89	50.6	10	3.6	32	11.4
Family Cricetidae								
Subfamily Cricetinae								
Mystromys albicaudatus			1	.6				
Subfamily Gerbillinae								
Tatera sp.	26	26.0	9	5.1	42	15.2	15	5.4
Subfamily Dendromurinae								
Dendromys sp.			1	.6	4	1.4	10	3.6
Steatomys pratensis			8	4.6	23	8.3	20	7.1
Family Bathyergidae								
Cryptomys hottentotus	10	10.0	4	2.3			5	1.8
Rodentia indet.					14	5.1	24	8.6
Class Aves, indet.	7	7.0	12	6.8	10	3.6	2	.7
Class Insecta, indet.			1	.6				
Total	100	100.0	175	99.8	276	99.8	280	100.0

[a] Individual animals.

Table 72. Vegetation Occurring in Circles of 2 Km Radius around the Positions of Barn Owl Roosts at Bolt's Farm, Swartkrans, and Makapansgat Limeworks

	Percentage of Total Area		
Vegetation Type	Bolt's Farm	Swartkrans	Makapansgat
Cultivated fields	8.8	12.1	8.5
Wooded (other than bushveld)	1.6	1.0	
Open grassland	89.6	86.9	
Mixed bushveld			90.5
Evergreen forest			1.0
Total	100.0	100.0	100.0

Table 73. Lengths of Bone Flakes Produced by Spotted Hyenas at Kills in the Kruger National Park and Dens in the Kalahari Gemsbok National Park

Length	Kruger Park		Kalahari Park		Combined Samples	
(cm)	N	%	N	%	N	%
0– 1						
1– 2	1	.6			1	.5
2– 3	15	9.6	1	1.6	16	7.3
3– 4	19	12.1	2	3.2	21	9.5
4– 5	16	10.2	4	6.3	20	9.1
5– 6	25	15.9	6	9.5	31	14.1
6– 7	17	10.8	6	9.5	23	10.5
7– 8	16	10.2	8	12.7	24	10.9
8– 9	12	7.6	8	12.7	20	9.1
9–10	9	5.7	10	15.9	19	8.6
10–11	9	5.7	8	12.7	17	7.7
11–12	6	3.8	5	7.9	11	5.0
12–13	7	4.5	1	1.6	8	3.6
13–14	1	.6	3	4.8	4	1.8
14–15	3	1.9			3	1.4
15–16						
16–17			1	1.6	1	.5
17+	1	.6			1	.5
Total	157	99.8	63	100.0	220	100.1

Table 74. Lengths of Bone Flakes Produced by Brown Hyenas at Dens in the Kalahari National Park

Length (cm)	N	%
0–1		
1–2		
2–3		
3–4	1	2.9
4–5	1	2.9
5–6	1	2.9
6–7	1	2.9
7–8	2	5.9
8–9	2	5.9
9–10	4	11.8
10–11	6	17.7
11–12	2	5.9
12–13	5	14.7
13–14		
14–15	4	11.8
15–16	3	8.8
16–17		
17+	2	5.9
Total	34	100.0

Table 75. Numbers of Genera and Species Represented in the Macrovertebrate Component of the Sterkfontein Valley Caves

Taxon	Genera		Species	
	Total	Extinct	Total	Extinct
Phylum Chordata				
Class Mammalia				
Order Primates				
Family Hominidae	3	3	3	3
Family Cercopithecidae	6	5	9	9
Order Carnivora				
Family Felidae	6	4	9	7
Family Hyaenidae	5	2	8	5
Family Canidae	4	1	6	4
Family Mustelidae	1		1	1
Family Viverridae	4		5	2
Order Artiodactyla				
Family Bovidae	19	5	26	13
Family Suidae	3	2	3	3
Family Giraffidae	1	1	1	1
Order Perissodactyla				
Family Equidae	2	1	4	2
Order Proboscidea				
Family Elephantidae	1		1	1
Order Hyracoidea				
Family Procaviidae	1		3	2
Order Lagomorpha				
Family Leporidae	?2		?2	
Order Rodentia				
Family Hystricidae	1		2	1
Class Aves				
Order Struthioformes				
Family Struthionidae	1		1	
Order Falconiformes				
Family Accipitridae	1		1	
Order Galliformes				
Family Numididae	1		1	
Class Reptilia				
Order Crocodilia				
Family Crocodylidae	1		1	
Order Chelonia				
Family Testudinidae	1		?1	
Order Squamata				
Family Varanidae	1		1	
Family Gerrhosauridae	1		1	
Phylum Mollusca				
Class Gastropoda				
Order Pulmonata	1		?1	
Total	67	24	91	54

Table 76. Remains from Sterkfontein Member 4: Overall Analysis, Taxa Represented, and Minimum Number of Individual Animals Involved

Taxon	Number of Specimens	Minimum Number of Individuals
Phylum Chordata		
Class Mammalia		
Order Primates		
Family Hominidae		
Australopithecus africanus		
Cranial pieces	84	± 45
Postcranial pieces	8	
Family Cercopithecidae		
Parapapio jonesi		
Cranial pieces	35	27
Postcranial pieces	2	
Parapapio broomi		
Cranial pieces	100	91
Parapapio whitei		
Cranial pieces	13	10
Postcranial pieces	1	
Parapapio sp. indet.		
Cranial pieces	201	53
Postcranial pieces	5	
Cercopithecoides williamsi		
Cranial pieces	20	17
Postcranial piece	1	
Cercopithecoid indet.		
Cranial pieces	413	± 100
Postcranial pieces	46	
Order Carnivora		
Family Felidae		
Panthera pardus		
Cranial piece	1	1
Dinofelis barlowi		
Cranial pieces	5	4
Megantereon gracile		
Cranial piece	1	1
Family Hyaenidae		
Euryboas silberbergi		
Cranial pieces	5	4
Crocuta crocuta		
Cranial piece	1	1
Hyaenid indet.		
Cranial pieces	5	± 3
Family Canidae		
Canis brevirostris		
Cranial piece	1	1
Canis mesomelas		
Cranial pieces	9	± 5
Carnivore indet.		
Cranial pieces	32	± 8
Postcranial pieces	16	
Order Artiodactyla		
Family Bovidae		
Damaliscus cf. sp. 2		
Cranial piece	1	1
Damaliscus sp. 1 or *Parmularius* sp.		
Cranial pieces	26	7
Medium-sized alcelaphines		
Cranial pieces	14	7
Cf. *Connochaetes* sp.		
Cranial pieces	3	1
Cf. *Megalotragus* sp.		
Cranial pieces	1	1
Hippotragus cf. *equinus*		
Cranial pieces	2	2
?Hippotragini		
Cranial pieces	23	8
Redunca cf. *arundinum*		
Cranial piece	1	1
Antidorcas cf. *recki*		
Cranial pieces	5	3
Antidorcas cf. *bondi*		
Cranial piece	1	1
?*Gazella* sp.		
Cranial pieces	2	1
Syncerus sp.		
Cranial piece	1	1
Tragelaphus sp. aff. *angasi*		
Cranial pieces	3	1
Makapania cf. *broomi*		
Cranial pieces	22	8
Antelope class I		
Cranial pieces	4	2
Postcranial pieces	8	
Antelope class II		
Cranial pieces	81	± 8
Postcranial pieces	30	
Antelope class III		
Cranial pieces	157	± 15
Postcranial pieces	89	
Antelope class IV		
Cranial pieces	13	2
Postcranial pieces	2	
Family Suidae		
Cf. *Pronotochoerus* sp.		
Cranial pieces	2	2
Order Perissodactyla		
Family Equidae		
Equus capensis		
Cranial pieces	18	7
Equus sp.		
Cranial pieces	2	1
Order Proboscidea		
Family Elephantidae		
Elephas cf. *recki*		
Cranial piece	1	1
Order Hyracoidea		
Family Procaviidae		
Procavia antiqua		
Cranial pieces	17	8
Procavia transvaalensis		
Cranial pieces	9	5
Order Rodentia		
Family Hystricidae		
Hystrix africaeaustralis		
Cranial pieces	6	5
Carnivore coprolite	1	
Insect pupae	3	
Indeterminate fragments	330	
Bone flakes	12	
2–3 cm 1		
3–4 cm 1		
4–5 cm 4		
5–6 cm 4		
6–7 cm 1		
7–8 cm 1		
Total	1,895	470

Table 77. Skeletal Part Analysis of Bovid Remains from Sterkfontein Member 4

Part	Number of Specimens			
	Size Class I	Size Class II	Size Class III	Size Class IV
Calvaria pieces	1	1	1	
Horn-core pieces		3	2	1
Maxillary pieces		9	13	1
Mandible pieces	3	17	33	1
Isolated upper teeth		20	37	5
Isolated lower teeth		10	19	
Tooth fragments		21	52	5
Vertebrae				
Atlas				
Axis				
Cervical, 3–7			8	
Thoracic			4	
Lumbar		4	14	
Sacral				
Caudal			2	
Rib pieces			13	
Scapula pieces			3	
Pelvis pieces				
Humerus				
Proximal pieces			2	
Distal pieces		1	3	
Shaft pieces		1		
Articulated distal humerus/ proximal radius and ulna				
Radius and ulna, proximal pieces			2	
Radius				
Distal pieces		2	2	
Shaft pieces				
Femur				
Proximal pieces			4	
Distal pieces		1	7	1
Shaft pieces				
Patella		1		
Tibia				
Proximal pieces	1	2	4	
Distal pieces				
Shaft pieces				
Metacarpal				
Proximal pieces	2	1		
Distal pieces			3	
Shaft pieces		2		
Metatarsal				
Proximal pieces			2	
Distal pieces		2		
Shaft pieces	1	2		
Metapodial				
Proximal pieces				
Distal pieces	1	4	3	
Shaft pieces			3	
Astragalus	2	1	5	
Calcaneus		1	1	1
Tarsal bones				
Carpal bones			1	
1st phalanges	1	3	2	
2d phalanges		1	1	
Terminal phalanges		1		
Articulated foot bone pieces				
Total	12	111	246	15

Table 78. Allocation to Age Categories of Fossils Representing 47 Individuals of *Australopithecus africanus*

1–5 Years	6–10 Years	11–15 Years	16–20 Years	21–25 Years	26–30 Years	31–35 Years	36–40 Years
STS 2	STS 57	STS 1	STS 52	TM 1511+ STS 60	STS 19	STS 5	TM 1514
	TM 1516	STS 23	STS 8	STS 17	STS 35	STS 71	STS 10
	STS 18		STS 22	TM 1512	STS 61	STS 29	STS 7
	STS 24		STS 28+37	STS 12	STS 53	STS 42	STS 36
	STS 56		STS 55	STS 32	STS 38	TM 1532	STS 54
	STS 9		TM 1523	TM 1561	STS 72	STS 46	TM 1519
	STS 51		STS 59	STS 21	TM 1520		
				STS 30			
				STS 31			
				TM 1518			
				STS 4			
1	7	2	7	11	7	6	6

Source: Following Mann (1975).

Table 79. Allocation to Age and Sex Classes of Fossils Representing 27 Individuals of *Parapapio jonesi*

Age Class	Male	Female	Sex Unknown	Total
Juvenile		333+340		2
		384		
Immature adult		306	457	2
Young adult	421	287+329	332	7
	334	372		
	381+446			
	418+458			
Adult	367	565		12
	485+390	448		
	1925+443	317		
	348+302	313		
		355		
		307		
		456		
		284		
Old adult	250	441		4
		276		
		368		
Total	9	16	2	27

Note: All specimens are from Sterkfontein Member 4 and have numbers prefixed by STS.

Table 80. Remains of *Parapapio broomi* from Sterkfontein Member 4: Separation into Age and Sex Classes

Age Class	Male	Female	Sex Unknown	Total
Juvenile	558		419	3
	283			
Immature adult	270	354		10
	279	411+425		
		383	328	
		277		
		268		
		266		
		251		
Young adult	301	398*a–c*+280	438	20
	414*a,b*	353	544	
		371	388*a,b*	
		390*a,b*	410	
		378	261	
		325	256	
		322+437	382*a*	
		274	289	
		262	286	
Adult	564 (type)	562	271	39
	272	254+383*a*	298	
	267	297	326	
	260	385*a–e*	434	
	346	397+285	406*a–c*	
	253	3035	445	
	311	530+374*b*	491	
	314	535	420	
	323	369*a,b*	511	
	484	338		
	278	557		
	258	426		
	542	409		
		386*a*		
		255		
		299		
		335		
Old adult	533+534	396*a*+362		19
	531+312	264		
	296+351	393		
	416	331		
	360	356		
	339	363	380*a–c*	
	337	415*a*	309	
		466		
		469		
		539		
Total	26	43	22	91

Note: All specimen numbers are prefixed STS.

Table 81. Remains of *Parapapio whitei* from Sterkfontein Member 4: Separation into Age and Sex Classes

Age Class	Male	Female	Sex Unknown	Total
Juvenile	0	0	0	0
Immature adult		263+370		1
Young adult	342	352+467		2
Adult	424+462	259	548	5
		336	303	
Old adult	389			2
	359			
Total	4	4	2	10

Note: All specimen numbers are prefixed STS.

Table 82. Remains of *Cercopithecoides williamsi* from Sterkfontein Member 4: Separation into Age and Sex Classes

Age Class	Male	Female	Sex Unknown	Total
Juvenile	300			2
	290+357+435			
Immature adult			282	1
Young adult			344	
			518+516	3
			523	
Adults	347	394	279	10
	350	532	288	
	366+392		295	
			361	
			252	
Old adult			541	1
Total	5	2	10	17

Note: All specimen numbers are prefixed STS.

Table 83. Fossil Assemblages from Sterkfontein Members 4, 5, and 6: Contributions of Individual Animals of Various Taxa to the Preserved Faunas

	Member 4		Member 5		Member 6	
Taxon	N[a]	%	N	%	N	%
Mammalia						
Primates						
Hominidae	47	13.4	3	7.3		
Cercopithecidae	198	56.6	1	2.4		
Carnivora	28	8.0	4	9.8	3	18.8
Artiodactyla	51	14.6	23	56.1	8	50.0
Perissodactyla	7	2.0	3	7.3	2	12.5
Hyracoidea	13	3.7	1	2.4	1	6.3
Proboscidea	1	.3				
Lagomorpha			1	2.4		
Rodentia	5	1.4	2	4.9	1	6.3
Aves			1	2.4	1	6.3
Reptilia			2	4.9		
Total	350	100.0	41	99.9	16	100.2

[a] Individual animals.

Table 84. Observed Damage to Bones from Sterkfontein Member 4

Porcupine-gnawed piece
Parapapio sp., juvenile mandible, STS 320.

Pieces showing carnivore-inflicted damage
Australopithecus africanus, part of a juvenile mandible, STS 18, showing ragged-edge damage to symphyseal area; palate, STS 53, showing ragged-edge damage around upper margin. Extensive shattering has also occurred, and diagnosis of carnivore-damage on both these specimens is tentative.
Parapapio jonesi, mandible, STS 334, possible carnivore damage to posterior margin of ramus; *Parapapio* sp., mandible, STS 351, two punctate marks on inner side of horizontal ramus.
Tragelaphus sp. aff. *angasi,* mandible, STS 1865, probable carnivore damage.
Procavia antiqua, palate and base of braincase, STS 109, showing typical pattern of leopard damage.
P. transvaalensis, mandible, STS 101, evidence of carnivore chewing.
Antelope class III, distal humerus, STS 1930, extensive tooth marks around distal end; distal humerus, STS 1584, similar damage.

Table 85. Remains from Sterkfontein Member 5: Overall Analysis, Taxa Represented, and Minimum Numbers of Individual Animals Involved

Taxon		Number of Specimens	Minimum Number of Individuals
Phylum Chordata			
Class Mammalia			
Order Primates			
Family Hominidae			
cf. *Homo* sp.			
Cranial pieces		6	3
Family Cercopithecidae			
Cercopithecoid, gen. et sp. indet.			
Cranial pieces		2	1
Order Carnivora			
Family Felidae			
Cf. *Panthera leo*			
Cranial pieces		2	1
Postcranial pieces		1	
Cf. *Megantereon* sp.			
Postcranial pieces		3	1
Family Hyaenidae			
Proteles sp.			
Cranial piece		1	1
Family Canidae			
Canis cf. *terblanchei*			
Cranial piece		1	1
Carnivore indet.			
Cranial pieces		7	3
Postcranial pieces		5	
Order Artiodactyla			
Family Bovidae			
Damaliscus cf. *dorcas*			
Cranial pieces		4	2
Damaliscus cf. sp. 2			
Cranial pieces		7	4
Damaliscus sp. 1 or *Parmularius* sp.			
Cranial piece		1	1
Medium-sized alcelaphines			
Cranial pieces		7	5
Cf. *Connochaetes* sp.			
Cranial piece		1	1
?Hippotragini			
Cranial piece		1	1
Antidorcas cf. *recki*			
Cranial pieces		5	3
Orteotragus major			
Cranial pieces		3	1
Taurotragus cf. *oryx*			
Cranial piece		1	1
Makapania cf. *broomi*			
Cranial piece		1	1
Antelope class I			
Cranial piece		1	
Postcranial pieces		19	±3
Antelope class II			
Cranial pieces		29	
Postcranial pieces		122	±6
Antelope class III			
Cranial pieces		32	
Postcranial pieces		108	±5
Antelope class IV			
Cranial piece		1	1
Postcranial pieces		5	
Family Suidae			
Suid, gen. et sp. indet.			
Cranial pieces		2	1
Order Perissodactyla			
Family Equidae			
Equus cf. *burchelli*			
Cranial pieces		18	3
Postcranial pieces		5	
Order Hyracoidea			
Family Procaviidae			
Procavia sp.			
Cranial pieces		2	1
Postcranial piece		1	
Order Lagomorpha			
Family Leporidae			
Gen. et sp. indet.			
Cranial piece		1	1
Order Rodentia			
Family Hystricidae			
Hystrix cf. *africaeaustralis*			
Cranial pieces		6	2
Postcranial piece		1	
Class Aves			
Gen. et sp. indet.			
Cranial piece		1	1
Postcranial piece		1	
Class Reptilia			
Order Squamata			
Family Varanidae			
Varanus cf. *niloticus*			
Postcranial piece		1	1
Order Chelonia			
Family Testudinidae			
Gen. et sp. indet.			
Carapace piece		1	1
Indet. fragments		580	
Bone flakes			
1–2 cm	12	206	
2–3 cm	57		
3–4 cm	48		
4–5 cm	47		
5–6 cm	14		
6–7 cm	15		
7–8 cm	2		
8+ cm	11		
Total		1,202	58

Table 86. Skeletal Part Analysis of Bovid Remains from Sterkfontein Member 5

	Number of Specimens			
Part	Size Class I	Size Class II	Size Class III	Size Class IV
Calvaria pieces		1	7	
Horn-core pieces		5	6	
Maxillary pieces		1	1	
Mandible pieces	1	3	4	
Isolated upper teeth		6	5	
Isolated lower teeth		6	2	
Tooth fragments		7	7	1
Vertebrae				
Atlas			1	
Axis		2	1	
Cervical, 3–7		3	1	
Thoracic	2	5	6	
Lumbar	2	2	11	
Sacral		1		
Caudal				
Rib pieces		18	25	
Scapula pieces	1	6	4	
Pelvis pieces		3	2	
Humerus				
Proximal pieces				
Distal pieces	3	7	1	
Shaft pieces	3	1	2	
Articulated distal humerus/ proximal radius and ulna				
Radius and ulna, proximal pieces		2	3	1
Radius				
Distal pieces		4	1	
Shaft pieces				
Femur				
Proximal pieces		2	3	
Distal pieces	1	1	3	
Shaft pieces	1	1	1	
Patella		1	2	
Tibia				
Proximal pieces	1	1	4	
Distal pieces		4	1	
Shaft pieces		2		
Metacarpal				
Proximal pieces		1	2	
Distal pieces		2	3	
Shaft pieces			1	
Metatarsal				
Proximal pieces	3	3	1	
Distal pieces	1	1	3	
Shaft pieces		1		
Metapodial				
Proximal pieces				
Distal pieces		9	4	1
Shaft pieces		2		
Astragalus	1	2	3	
Calcaneus		3	1	2
Tarsal bones		5	3	
Carpal bones		3	4	
1st phalanges		15	6	1
2d phalanges		7	1	
Terminal phalanges		2	1	
Articulated foot bone pieces			3	
Total	20	151	140	6

Table 87. Observed Damage to Bones from Sterkfontein Member 5

Porcupine-gnawed pieces
Antelope class I: femur shaft, SE 839
Antelope class II: calcaneus, unnumbered; sacrum, 1885; proximal radius, 2031
Antelope class III: metacarpal, 1458; metatarsal, 127
Bone flakes: 64, 127, 804, 1011, 1458, 2061, unnumbered
Shaft pieces: 1439, unnumbered

Pieces gnawed by small rodents
Antidorcas cf. *recki,* mandible, 535
Antelope class I: scapula, 657
Antelope class II: phalanges, 826, 681; humerus, 1670; femur 1884; metatarsals, 1693, 1401
Antelope class III: rib, 645; phalanx, 745; femur, 745
Bone flakes, 117, 657, 745, 815
Indet. rib, 653; indet. scapula, 657

Pieces showing carnivore-inflicted damage
Clear carnivore damage
Antelope class I: humerus, 1528; humerus, 72 (punctate mark)
Antelope class II: rib, 1535; calcaneus, 1707; scapulae, 835, 1361, 1265; phalanx, 1263; thoracic vertebrae, 629, 742, 1376; lumbar vertebra, 491; axis, 1264; cervical vertebra, 1321; radii, 1336, unnumbered
Antelope class III: cranium, 571*a,b* (punctate marks); thoracic vertebra, 2162; pelvis, 11; ribs, 1510, 625 (punctate mark); femur 521 (punctate mark); metacarpal, 1292
Bone flakes: 653, 804, 1000, 2027; Indet. pieces: 135, 481, 494, 620, 745, 825, 902, 1783, 2009, 2024
Probable carnivore damage
Antelope class II: rib, 1308
Suid: mandible, 1069
Bone flake: 402
Indet. pieces: 435, 1543, 1647, 2164

Pieces showing artificial marks
Antelope class I: humerus shaft, 1729 (clear chop mark)
Antelope class III: horn-core, 2031 (hole through horn-core); horn-core piece, 1524 (worn to a point)
Bone flakes: 612 (worn facets, apparently a bone tool); 1000 (smoothed edges)
Piece showing possible insect boring: shaft piece, 1363

Note: All specimen numbers are prefixed SE.

Table 88. Microfaunal Remains from Sterkfontein Member 5

Taxon	Minimum Number of Individuals	%
Order Chiroptera (indet.)	8	1.2
Order Insectivora		
Family Macroscelididae		
Elephantulus sp.	21	3.3
Family Soricidae		
Soricid indet.	109	16.9
Order Rodentia		
Family Bathyergidae		
Cryptomys sp.	16	2.5
Family Muridae		
Subfamily Murinae, various genera	42	6.5
Subfamily Otomyinae		
Cf. *Palaeotomys gracilis*	32	5.0
Family Cricetidae		
Subfamily Cricetinae		
Mystromys sp.	144	22.3
Subfamily Gerbillinae		
Tatera sp.	14	2.2
Subfamily Dendromurinae		
Dendromurine indet.	146	22.6
Rodentia (Myomorpha) indet.	72	11.2
Class Reptilia		
Order Squamata		
Lacertilia indet.	33	5.1
Class Aves		
Indet.	7	1.1
Total	644	99.9

Table 89. Remains from Sterkfontein Member 6: Overall Analysis, Taxa Represented, and Minimum Numbers of Individual Animals Involved

Taxon	Number of Specimens	Minimum Number of Individuals	Taxon		Number of Specimens	Minimum Number of Individuals
Phylum Chordata			Order Hyracoidea			
Class Mammalia			Family Procaviidae			
Order Carnivora			*Procavia* cf. *capensis*			
Gen. et sp. indet.			Cranial pieces		3	1
Cranial pieces	3	± 3	Postcranial pieces		2	
Postcranial pieces	6		Order Rodentia			
Order Artiodactyla			Family Hystricidae			
Family Bovidae			*Hystrix* cf. *africaeaustralis*			
Antidorcas bondi			Cranial pieces		2	1
Cranial pieces	3	2	Class Aves			
Damaliscus cf. *dorcas*			Order Falconiformes			
Cranial piece	1	1	Family Accipitridae			
Antelope class I			Postcranial pieces		2	1
Postcranial pieces	10	1–2	Indet. fragments		70	
Antelope class II			Bone flakes		290	
Cranial pieces	3	± 4	1–2 cm	17		
Postcranial pieces	34		2–3 cm	112		
Antelope class III			3–4 cm	85		
Cranial pieces	4	± 2	4–5 cm	41		
Postcranial pieces	15		5–6 cm	17		
Order Perissodactyla			6–7 cm	8		
Family Equidae			7–8 cm	7		
Equus cf. *burchelli*			8 + cm	3		
Cranial pieces	6	2	Total		454	19

Table 90. Skeletal Part Analysis of Bovid Remains from Sterkfontein Member 6

Part	Number of Specimens		
	Size Class I	Size Class II	Size Class III
Calvaria pieces			
Horn-core pieces			1
Maxillary pieces			
Mandible pieces		1	
Isolated upper teeth			1
Isolated lower teeth			
Tooth fragments		2	2
Vertebrae			
Atlas			
Axis			
Cervical, 3–7			
Thoracic	1		
Lumbar		1	
Sacral			
Caudal			
Rib pieces	2	11	6
Scapula pieces		1	1
Pelvis pieces		1	1
Humerus			
Proximal pieces			
Distal pieces	3	2	
Shaft pieces			
Articulated distal humerus/ proximal radius and ulna			
Radius and ulna, proximal pieces			
Radius			
Distal pieces			
Shaft pieces			
Femur			
Proximal pieces			1
Distal pieces			
Shaft pieces			
Patella			
Tibia			
Proximal pieces			
Distal pieces			
Shaft pieces		2	1
Metacarpal			
Proximal pieces			
Distal pieces			
Shaft pieces			
Metatarsal			
Proximal pieces		1	2
Distal pieces	1	1	
Shaft pieces		2	
Metapodial			
Proximal pieces			
Distal pieces	1	3	
Shaft pieces		2	
Astragalus			
Calcaneus	1	2	
Tarsal bones			
Carpal bones			
1st phalanges		3	2
2d phalanges	1	2	1
Terminal phalanges			
Articulated foot bones			
Total	10	37	19

Table 91. Observed Damage to Bones from Sterkfontein Member 6

Porcupine-gnawed pieces
Shaft pieces: SE 702, 1526
Bone flakes: 737, 867, 907, 1006, 2 unnumbered pieces

Pieces gnawed by small rodents
Antelope class I: phalanx, 984
Antelope class II: phalanx, 878
Shaft piece, unnumbered, and bone flake, 719

Pieces showing carnivore-inflicted damage
Clear carnivore damage
Antelope class II: pelvis, 697; phalanx, 1890
Antelope class III: rib, 843
Indet. pieces: 913, unnumbered
Bone flakes: 725, 932, 985
Probable carnivore damage
Indet. pelvis piece: 2084
Bone flake, 848

Pieces showing artificial marks
Antidorcas bondi, mandible, 690 (cut marks)
Bone flakes, 1054, 2384, unnumbered (rounded and worn)

Note: All specimen numbers are prefixed SE.

Table 92. Remains from Swartkrans Member 1: Overall Analysis, Taxa Represented, and Minimum Numbers of Individual Animals Involved

Taxon	Number of Specimens	Minimum Number of Individuals
Phylum Chordata		
Class Mammalia		
Order Primates		
Family Hominidae		
Homo sp.		
Cranial pieces	4	3
Australopithecus robustus		
Cranial pieces	218	87
Postcranial pieces	11	
Family Cercopithecidae		
Parapapio jonesi		
Cranial pieces	29	8
Papio robinsoni		
Cranial pieces	121	38
Theropithecus danieli		
Cranial pieces	31	17
Dinopithecus ingens		
Cranial pieces	57	26
Cercopithecoid indet.		
Cranial pieces	134	±28
Postcranial pieces	31	
Order Carnivora		
Family Felidae		
Panthera pardus		
Cranial pieces	29	12
Postcranial pieces	3	
Dinofelis sp.		
Cranial pieces	2	1
Postcranial pieces	2	
Megantereon sp.		
Cranial piece	1	1
Family Hyaenidae		
Hyaena brunnea dispar		
Cranial pieces	7	3
Crocuta crocuta venustula		
Cranial pieces	3	2
Hyaenictis forfex		
Cranial pieces	1	1
Euryboas nitidula (type B, advanced)		
Cranial pieces	9	5
Euryboas nitidula (type A, primitive)		
Cranial pieces	5	2
Euryboas nitidula (type uncertain)		
Cranial pieces	4	1
Proteles transvaalensis		
Cranial pieces	2	1
Hyaenid indet.		
Cranial pieces	16	±4
Family Canidae		
Canis mesomelas pappos		
Cranial pieces	9	4
Vulpes pulcher		
Cranial pieces	2	2
Carnivore indet.		
Cranial pieces	28	±15
Postcranial pieces	28	
Order Artiodactyla		
Family Bovidae		
Damaliscus sp. 1 or *Parmularius* sp.		
Cranial pieces	8	4
Rabaticeras porrocornutus		
Cranial pieces	3	2

Continued on following page

Table 92. *Continued*

Taxon		Number of Specimens	Minimum Number of Individuals
Medium-sized alcelaphines			
Cranial pieces		97	22
Cf. *Connochaetes* sp.			
Cranial pieces		70	19
Postcranial pieces		4	
Cf. *Megalotragus* sp.			
Cranial pieces		8	3
?Hippotragini			
Cranial pieces		2	1
Redunca cf. *arundinum*			
Cranial pieces		1	1
Pelea cf. *capreolus*			
Cranial pieces		2	2
Antidorcas cf. *recki*			
Cranial pieces		12	6
?Gazella sp.			
Cranial pieces		8	6
Antilopine or Neotragine indet.			
Cranial pieces		3	3
Oreotragus cf. *major*			
Cranial pieces		1	1
Syncerus sp.			
Cranial pieces		13	3
Tragelaphus cf. *strepsiceros*			
Cranial pieces		12	4
Cf. *Makapania* sp.			
Cranial pieces		10	3
Antelope class I			
Cranial pieces		25	±5
Postcranial pieces		11	
Antelope class II			
Cranial pieces		117	±20
Postcranial pieces		113	
Antelope class III			
Cranial pieces		199	±25
Postcranial pieces		162	
Antelope class IV			
Cranial pieces		15	2
Postcranial pieces		2	
Family Suidae			
Tapinochoerus meadowsi			
Cranial pieces		8	7
Suid indet.			
Cranial pieces		9	±3
Order Perissodactyla			
Family Equidae			
Equus capensis			
Cranial pieces		9	6
Hipparion steytleri			
Cranial pieces		1	1
Equid indet.			
Cranial pieces		14	±4
Postcranial pieces		3	
Order Hyracoidea			
Family Procaviidae			
Procavia antiqua			
Cranial pieces		35	16
Procavia transvaalensis			
Cranial pieces		20	8
Postcranial pieces		1	
Hyracoid indet.			
Cranial pieces		6	2
Order Rodentia			
Family Hystricidae			
Hystrix africaeaustralis			
Cranial pieces		5	2
Hystrix ?makapanensis			
Cranial pieces		1	1
Class Aves			
Indet.: cranial piece		1	1
Phylum Mollusca			
Class Gastropoda			
Order Pulmonata			
Cf. *Achatina* sp.			
Shells		15	15
Indeterminate fragments		500	
Bone flakes		49	
2–3 cm	7		
3–4 cm	11		
4–5 cm	15		
5–6 cm	10		
6–7 cm	4		
7–8 cm	1		
8+ cm	1		
Total		2,381	463

Note: Corrected minimum number of individuals, 339.

Table 93. Skeletal Part Analysis of Bovid Remains from Swartkrans Member 1

Part	Number of Specimens			
	Size Class I	Size Class II	Size Class III	Size Class IV
Calvaria pieces	2	1	4	1
Horn-core pieces	1	1	6	
Maxillary pieces	2	14	21	2
Mandible pieces	9	42	48	2
Isolated upper teeth	4	23	53	9
Isolated lower teeth	3	18	39	1
Tooth fragments	4	18	28	
Vertebrae				
Atlas		1	1	
Axis			2	
Cervical, 3–7		4	3	
Thoracic		13	1	
Lumbar	1	7	2	
Sacral			2	
Caudal				
Rib pieces		4	1	
Scapula pieces	1	5	5	
Pelvis pieces		7	4	
Humerus				
Proximal pieces		1	9	
Distal pieces	3	1	9	
Shaft pieces	1		3	
Articulated distal humerus/ proximal radius, and ulna			1	
Radius and ulna, proximal pieces		4	1	
Radius				
Distal pieces			1	
Shaft pieces			1	
Femur				
Proximal pieces		2	6	
Distal pieces		5	9	
Shaft pieces		1		
Patella			1	
Tibia				
Proximal pieces		1	1	1
Distal pieces		2	4	
Shaft pieces			1	
Metacarpal				
Proximal pieces		1	2	
Distal pieces			3	
Shaft pieces		3		
Metatarsal				
Proximal pieces		2	2	
Distal pieces		1	1	
Shaft pieces	1	2	3	
Metapodial				
Proximal pieces				1
Distal pieces		5	6	
Shaft pieces		7	5	
Astragalus		4	9	
Calcaneus		3	9	
Tarsal bones		3	4	
Carpal bones	1	1	9	
1st phalanges	1	12	28	
2d phalanges		5	9	
Terminal phalanges	2	5	2	
Articulated foot bone pieces		1	2	
Total	36	230	361	17

Table 94. Allocation of 113 Australopithecine Specimens from Swartkrans Member 1 to Age-at-Death Categories

Age Category (years)	N	%
1– 5	11	10.0
6–10	22	20.0
11–15	21	19.0
16–20	21	19.0
21–25	12	11.0
26–30	14	12.0
31–35	11	10.0
36–40	1	.1
Total	113	101.1

Source: Data from Mann (1975).

Table 95. *Parapapio jonesi* from Swartkrans Member 1: Allocation of Specimens to age Classes

	Maxillae			Mandibles			
Age Class	Male	Female	Sex Unknown	Male	Female	Sex Unknown	Minimum Number of Individuals
Young juvenile							
Juvenile			462			418, 433	2
Immature adult							
Young adult				588*b*			1
Adult	612			1835			1
Old adult	588	573*b*, 543		14164	573*a*, 414		4
Very old adult							

Note: All specimens are prefixed SK.

Table 96. *Papio robinsoni* from Swartkrans Member 1: Allocation of Specimens to Age Classes

	Maxillae			Mandibles			
Age Class	Male	Female	Sex Unknown	Male	Female	Sex Unknown	Minimum Number of Individuals
Young juvenile							
Juvenile			447, 610			453, 446, 420, 463	4
Immature adult	536, 436, 458, 456, ?497			14083, 445, 453			5
Young adult	602, 631, 544, 555 (type)	558, 562, 565, 476		408, 570, 429, 419	407, 409, 410, 457, 540, 2319, 427		11
Adult	14006, 550	557, 549, 2180, 537		417, 435, ?468	321*b*		7
Old adult	590, 629, 14163	571*b*		572, 423, 431, 538	421, 571*a*, 459		7
Very old adult	560	566		406, 430, 2161, ?2171	416, 14156		6

Note: All specimen numbers are prefixed SK.

Table 97. *Theropithecus danieli* Remains from Swartkrans Member 1: Allocation of Specimens to Age Classes

	Maxillae			Mandibles			
Age Class	Male	Female	Sex Unknown	Male	Female	Sex Unknown	Minimum Number of Individuals
Young juvenile							
Juvenile							
Immature adult		564, 593, 448, 507		426, 403			6
Young adult		567, 597, 561		569	411		4
Adult		479, 563 (goes with 405/402)		581	439, 575*a*, 405 + 402		4
Old adult		575*b*,*c*					1
Very old adult		461		491			2

Note: All specimen numbers are prefixed SK.

Table 98. *Dinopithecus ingens* Remains from Swartkrans Member 1: Allocation of Specimens to Age Classes

	Maxillae			Mandibles			
Age Class	Male	Female	Sex Unknown	Male	Female	Sex Unknown	Minimum Number of Individuals
Young juvenile						415, 413	2
Juvenile		554	1518			589	2
Immature adult	447	574, 548, 440, 443		532, 2407			6
Young adult	546	600, 603, 542*a*		404, 428 492, 455			7
Adult	578*a*,*b* 2625	604, 571*b*, 576*a*		474			5
Old adult	545	553, 473		422			3
Very old adult				401			1

Note: All specimen numbers are prefixed SK.

Table 99. Fossil Assemblages from Swartkrans Members 1 and 2 and Channel Fill: Contributions of Individual Animals of Various Taxa to the Preserved Faunas

	Member 1		Member 2		Channel Fill	
Taxon	%	N	%	N	%	N
Mammalia						
Primates						
Hominidae	90	26.6	1	.4		
Cercopithecidae	89	26.3	8	3.1		
Carnivora	36	10.6	21	8.1	3	10.0
Artiodactyla	89	26.3	166	64.3	17	56.7
Perissodactyla	7	2.1	13	5.0	1	3.3
Hyracoidea	24	7.1	37	14.3	5	16.7
Lagomorpha			4	1.5	2	6.7
Rodentia	3	.9	5	1.9		
Aves	1	.3	1	.4	1	3.3
Reptilia			2	.8	1	3.3
Total	339	100.2	258	99.8	30	100.0

Table 100. Observed Damage to Bones from Swartkrans Member 1

Porcupine-gnawed piece
Antelope class III: metapodial, SK 11819

Pieces gnawed by small rodents
Antelope class I: phalanx, 14020

Pieces showing carnivore-inflicted damage
Clear carnivore damage
Australopithecus robustus: juvenile calvaria, 54 (two punctate marks, one in each parietal); juvenile mandible, 3978 (ragged-edge damage to right horizontal ramus); innominate, 3155*b* (tooth marks around several margins)
Antelope class I: humerus, 14074
Antelope class II: mandible, 4209; femurs, 14076*a*, 6294; tibia, 1803; metacarpal, 2646; rib, 3093
Antelope class III: scapula, 4203; calcaneus, 6694
Indet. shaft piece, 8165
Probable carnivore damage
Homo sp.: mandible, 45 (ragged-edge break through horizontal ramus)
Australopithecus robustus: endocranial cast, 1585 (punctate depression in right frontal, probably caused by a tooth); adult mandibles, 74*a*, 1587, 1588 (damage to lower margin of ramus); juvenile mandibles, 61, 62, 64, 438, 869 (damage to horizontal or ascending rami); axis vertebra, 854 (edge damage); adult snout, 11 (damage to right side of maxilla); adolescent snout, 13 (irregular damage all around); innominate, 50 (damage to iliac margin); distal humerus, 860 (damage to shaft)
Cercopithecoid indet.: calvaria, 2132 (three punctate marks)

Note: All specimen numbers are prefixed SK.

Table 101. Swartkrans Member 1: Analyses of Microvertebrate Remains

Taxon	Rodent Breccia N[a]	Rodent Breccia %	Member 1C N[a]	Member 1C %	Mean (%)
Order Insectivora					
Family Macroscelididae					
Elephantulus sp.	12	4.4	21	8.1	6.3
Family Soricidae					
Soricid indet.	50	18.5	57	22.0	20.3
Order Rodentia					
Family Bathyergidae					
Cryptomys sp.	5	1.8	8	3.1	2.5
Family Muridae					
Subfamily Murinae, various genera	12	4.4	11	4.3	4.4
Subfamily Otomyinae					
Cf. *Palaeotomys gracilis*	23	8.5	31	12.0	10.3
Family Cricetidae					
Subfamily Cricetinae					
Mystromys sp.	57	21.0	50	19.3	20.2
Subfamily Gerbillinae					
Tatera sp.	6	2.2	2	.8	1.5
Subfamily Dendromurinae					
Dendromurine indet.	49	18.1	3	1.2	9.7
Rodentia (Myomorpha) indet.	47	17.3	45	17.4	17.4
Class Reptilia					
Order Squamata					
Lacertilia indet.	2	.7	25	9.7	5.2
Class Aves					
Indet.	8	3.0	3	1.2	2.1
Total	271	99.9	256	99.1	99.9

[a] Minimum number of individuals.

Table 102. Remains from Swartkrans Member 2: Overall Analysis, Taxa Represented, and Minimum Numbers of Individual Animals Involved

Taxon	Number of Specimens	Minimum Number of Individuals
Phylum Chordata		
Class Mammalia		
Order Primates		
Family Hominidae		
Homo sp.		
Cranial pieces	1	1
Postcranial pieces	1	
Family Cercopithecidae		
Papio sp.		
Cranial pieces	7	4
Cercopithecoides williamsi		
Cranial pieces	5	4
Cercopithecoid indet.		
Cranial pieces	10	4
Postcranial pieces	4	
Order Carnivora		
Family Felidae		
Panthera aff. *leo*		
Cranial pieces	4	3
Postcranial pieces	6	
Panthera pardus		
Cranial pieces	2	2
Postcranial pieces	1	
Felid indet.		
Postcranial pieces	2	1
Family Hyaenidae		
Hyaena brunnea		
Cranial pieces	3	2
Proteles cristatus		
Cranial pieces	2	1
Hyaenid indet.		
Cranial pieces	1	1
Postcranial pieces	1	
Family Canidae		
Canis mesomelas		
Cranial pieces	27	8
Postcranial pieces	9	
Cf. *Lycaon* sp.		
Cranial piece	1	1
Otocyon recki		
Cranial pieces	2	1
Family Mustelidae		
Mellivora aff. *sivalensis*		
Cranial pieces	3	1
Family Viverridae		
Herpestes sanguineus		
Cranial piece	1	1
Cynictis penicillata		
Cranial piece	1	1
Carnivore indet.		
Cranial pieces	12	8
Postcranial pieces	44	
Order Artiodactyla		
Family Bovidae		
Damaliscus cf. *dorcas*		
Cranial pieces	21	9
Damaliscus sp. 2 *(niro?)*		
Cranial pieces	57	9
Medium-sized alcelaphines (including cf. *Beatragus* sp.)		
Cranial pieces	30	9
Cf. *Connochaetes* sp.		
Cranial pieces	2	1
Cf. *Megalotragus* sp.		
Cranial pieces	3	2
Hippotragus cf. *niger*		
Cranial pieces	25	9
Cf. *Kobus ellipsiprymnus*		
Cranial pieces	2	2
Pelea capreolus		
Cranial pieces	33	10
Antidorcas australis and/or *marsupialis*		
Cranial pieces	64	16
Antidorcas bondi		
Cranial pieces	308	70
Postcranial pieces	10	
Oreotragus cf. *major*		
Cranial piece	1	1
Oreotragus cf. *oreotragus*		
Cranial pieces	3	2

Continued on following page

Table 102. *Continued*

Taxon		Number of Specimens	Minimum Number of Individuals
Raphicerus cf. *campestris*			
Cranial pieces		10	5
Cf. *Raphicerus* sp.			
Cranial piece		1	1
Ourebia cf. *ourebia*			
Cranial pieces		5	3
Tragelaphus cf. *scriptus*			
Cranial pieces		5	4
Tragelaphus cf. *strepsiceros*			
Cranial pieces		11	5
Tragelaphus sp. aff. *angasi*			
Cranial pieces		2	1
Taurotragus cf. *oryx*			
Cranial piece		1	1
Antelope class I			
Cranial pieces		15	
Postcranial pieces		224	4
Antelope class II*a*			
Cranial pieces		152	
Postcranial pieces		2,708	75
Antelope class II*b*			
Cranial pieces		39	
Postcranial pieces		419	±10
Antelope class III			
Cranial pieces		42	
Postcranial pieces		203	7
Antelope class IV			
Cranial pieces		1	
Postcranial pieces		8	2
Family Suidae			
Phacochoerus antiquus			
Cranial pieces		8	4
Suid indet.			
Cranial pieces		2	1
Family Giraffidae			
Sivatherium maurusium			
Cranial piece		1	1
Order Perissodactyla			
Family Equidae			
Equus quagga			
Cranial pieces		16	9
Postcranial pieces		5	
Equus capensis			
Cranial pieces		8	
Postcranial pieces		4	4
Equid indet.			
Cranial pieces		21	
Postcranial pieces		6	3
Order Hyracoidea			
Family Procaviidae			
Procavia cf. *antiqua*			
Cranial pieces		103	21
Postcranial pieces		8	
Procavia transvaalensis			
Cranial pieces		33	
Postcranial pieces		4	16
Hyracoid indet.			
Cranial pieces		24	
Postcranial pieces		2	10
Order Lagomorpha			
Family Leporidae			
Lagomorph, gen. et sp. indet.			
Cranial pieces		13	4
Order Rodentia			
Family Hystricidae			
Hystrix africaeaustralis			
Cranial pieces		8	±5
Class Aves			
Order Struthioformes			
Family Struthionidae			
Struthio sp.			
Postcranial pieces		2	1
Eggshell pieces		2	
Class Reptilia			
Order Chelonia			
Family Testudinidae			
Chelonian indet.			
Postcranial pieces		2	2
Carapace pieces		2	
Phylum Mollusca			
Class Gastropoda			
Order Pulmonata			
Land snail, cf. *Achatina* sp.			
Shells		10	10
Indet. fragments		723	
Bone flakes		330	
0–1 cm	1		
1–2 cm	6		
2–3 cm	38		
3–4 cm	104		
4–5 cm	94		
5–6 cm	45		
6–7 cm	19		
7–8 cm	12		
8+ cm	11		
Total		5,894	392

Note: Corrected minimum number of individuals, 258.

Table 103. Skeletal Part Analysis of Bovid Remains from Swartkrans Member 2

	Number of Specimens			
Part	Size Class I	Size Class II	Size Class III	Size Class IV
Calvaria pieces	0	1	1	0
Horn-core pieces	3	4	1	0
Maxillary pieces	0	1	2	0
Mandible pieces	10	3	14	1
Isolated upper teeth	0	12	11	1
Isolated lower teeth	2	17	7	0
Tooth fragments	0	1	6	0
Vertebrae				
Atlas	3	2	0	0
Axis	2	2	2	0
Cervical, 3–7	7	9	7	0
Thoracic	10	10	6	0
Lumbar	4	24	5	0
Sacral	0	3	1	0
Caudal	0	0	2	0
Rib pieces	16	26	18	0
Scapula pieces	5	10	10	0
Pelvis pieces	6	4	3	1
Humerus				
Proximal pieces	1	2	6	0
Distal pieces	6	7	14	0
Shaft pieces	17	5	2	0
Articulated distal humerus/ proximal radius, and ulna	0	1	0	0
Radius and ulna, proximal pieces	4	21	8	1
Radius				
Distal pieces	2	4	4	0
Shaft pieces	23	15	6	0
Femur				
Proximal pieces	3	8	7	1
Distal pieces	3	4	5	2
Shaft pieces	16	9	1	0
Patella	0	0	2	0
Tibia				
Proximal pieces	6	5	1	0
Distal pieces	4	17	5	0
Shaft pieces	5	8	2	0
Metacarpal				
Proximal pieces	5	16	5	0
Distal pieces	5	18	3	0
Shaft pieces	19	18	1	0
Metatarsal				
Proximal pieces	5	18	7	0
Distal pieces	2	12	1	0
Shaft pieces	23	19	1	0
Metapodial				
Proximal pieces	0	1	0	0
Distal pieces	1	16	15	1
Shaft pieces	3	6	0	0
Astragalus	1	6	5	0
Calcaneus	4	13	11	1
Tarsal bones	1	5	1	0
Carpal bones	0	8	1	0
1st phalanges	8	39	21	0
2d phalanges	2	21	9	1
Terminal phalanges	2	7	3	0
Total	239	458	245	9

Table 104. Skeletal Part Analysis of Bovid Remains (Class II*a*) from Swartkrans Member 2

Part	Number of Specimens	Part	Number of Specimens
Calvaria pieces	7	Femur, distal pieces	39
Horn-core pieces	6	Left 17	
Maxillary pieces	21	Right 16	
Mandible pieces	41	? side 6	
Isolated upper teeth	42	Femur, shaft pieces	139
Isolated lower teeth	35	Left 31	
Vertebrae		Right 27	
Atlas	13	? side 81	
Axis	10	Tibia, proximal pieces	46
Cervical, 3–7	47	Left 20	
Thoracic	68	Right 20	
Lumbar	84	? side 6	
Sacral	4	Tibia, distal pieces	104
Indet. vertebral pieces	5	Left 44	
Rib pieces	66	Right 52	
Scapula pieces	112	? side 8	
Left 41		Tibia, shaft pieces	163
Right 45		Left 15	
? side 26		Right 14	
Pelvis pieces	45	? side 134	
Left 16		Metacarpals, complete	5
Right 18		Left 2	
? side 11		Right 3	
Humerus, proximal pieces	26	Metacarpals, proximal pieces	68
Left 9		Left 32	
Right 9		Right 32	
? side 8		? side 4	
Humerus, distal pieces	136	Metacarpals, distal pieces, ? side	77
Left 71		Metacarpals, shaft pieces, ? side	161
Right 56		Metatarsals, proximal pieces	135
? side 9		Left 53	
Humerus, shaft pieces, ? side	70	Right 61	
Humerus, distal, articulated to radius and ulna,		? side 21	
proximal, left	1	Metatarsals, distal pieces, ? side	58
Radius and ulna, proximal	106	Metatarsals, shaft pieces, ? side	96
Left 35		Metapodial, proximal pieces, ? side	2
Right 62		Metapodial, distal pieces, ? side	82
? side 9		Metapodial, shaft pieces, ? side	7
Radius, distal pieces	69	Astragalus	66
Left 33		Left 40	
Right 30		Right 22	
? side 6		? side 4	
Radius, shaft pieces	115	Calcaneus	85
Left 37		Left 37	
Right 23		Right 46	
? side 55		? side 2	
Femur, proximal pieces	64	Tarsal bones	15
Left 22		Carpal bones	15
Right 24		1st phalanges	250
? side 18		2d phalanges	40
		Terminal phalanges	12
		Sesamoid bones	2
		Total	2,860

Note: Cranial parts represent only those specimens not assigned to specific taxa.

Table 105. Percentage Survival of Skeletal Parts Attributed to *Antidorcas bondi* from Swartkrans Member 2

Part	Number Found	Original Number	% Survival	Part	Number Found	Original Number	% Survival
Atlas vertebrae	13	70	18.6	Distal radius/ulna	69	140	49.3
Axis vertebrae	10	70	14.3	Proximal femur	64	140	45.7
Cervicals, 3–7	47	350	13.4	Distal femur	39	140	27.9
Thoracic vertebrae	68	884	7.6	Proximal tibia	46	140	32.9
Lumbar vertebrae	84	672	12.5	Distal tibia	104	140	74.3
Sacral vertebrae	4	70	5.7	Proximal metacarpal	74	140	52.9
Caudal vertebrae	0	300	0	Distal metacarpal	123	140	87.9
Ribs	66	1,820	3.6	Proximal metatarsal	136	140	97.1
Scapula	112	140	80.0	Distal metatarsal	99	140	70.7
Pelvic bone	45	140	32.1	Astragalus	66	140	47.1
Proximal humerus	26	140	18.6	Calcaneus	85	140	60.7
Distal humerus	137	140	97.9	Phalanges	302	1,680	18.0
Proximal radius/ulna	107	140	76.4				

Note: A minimum of 70 individuals are represented by cranial remains.

Table 106. Observed Damage to Bones from Swartkrans Member 2

Porcupine-gnawed pieces

Antidorcas bondi: mandible, SK 7698; frontal, 2880; horn-core, 14215

Hippotragus cf. *niger:* juvenile mandible, 1992

Antelope class II*a:* mandible, 2034; tibia, 6229; indet. shaft piece, 8232

Antelope class III: horn-core, 9282; humerus, 2623; indet. shaft piece, 9267

Antelope class IV, calcaneus, 14199

Indet. fragment, 3753

Pieces gnawed by small rodents

Hippotragus cf. *niger:* juvenile mandible, 14243

Procavia antiqua: mandible, 202

Large carnivore: metapodial, 463

Antelope class I: pelvis, 9604; radius, 11611; tibia, 7788; metapodials, 4502, 5076, 5189, 5400, 5662, 7100, 7752, 8957; calcaneus, 6754; phalanx, 11672

Antelope class II*a:* mandibles, 3124, 3116; lumbar vertebra, 12738; scapula, 9082; humeri, 4343, 6308, 6339, 6345, 7689, 9351, 9432; radii, 2653, 10788, 10811, 12009, 12087, 12552; femurs, 1901, 2618, 2642, 3629, 5125, 5299, 6281, 8730, 9235, 9284, 10854, 11787, 14037; tibias, 1906, 5027, 5794, 6196, 6521, 7390, 8173, 9305, 9564; calcanei, 4130, 5271, 5276, 6735, 8079, 10922, 11788, 12151, 12497; metapodials, 4594, 4709, 5081, 5245, 5252, 5309, 5394, 5600, 5625, 5635, 5696, 5702, 5710, 5714, 5715, 5763, 5787, 5799, 5840, 5895, 7108, 7695, 7894, 7924, 8077, 8186, 8284, 8648, 9256, 9304, 9320, 9596, 9804, 10550, 10767, 10938, 10990, 11240, 11293, 11719, 11728, 11978, 12001, 12036, 12049, 12058, 12086, 12484, 2 unnumbered; phalanges, 1530, 1576, 4376, 4401, 4493, 5146, 5264, 5391, 5876, 6742, 6746, 6820, 6830, 6836, 10079, 10738, 11593, 11991; indet. shaft pieces, 5257, 7287, 7850, 8728, 12406, 12492

Antelope class II*b:* Radius, 7384; femur, 14027; tibia, 10341; calcaneus, 11058; metapodial, 5825, 9210, 14177; phalanx, 10434, unnumbered

Antelope class III: humeri, 4406, 14031; femur, 1597; phalanx, 6732

Bone flakes: 4182, 5413, 9877, 12708

Indet. pieces: 6016, 7142, 7415, 8289, 8383, 10049, 10470, 12083

Pieces showing carnivore-inflicted damage

Indeterminate pieces

Clear carnivore damage: 4804, 5088, 6850, 6917, 7065, 7589, 9253, 9332, 9377, 9537, 10057, 10430, 10461, 11460, 11776

Probable carnivore damage: 4601, 5761, 6366, 7142, 9168, 10641, 11290, 11972, 12766

Bone flakes

Clear carnivore damage: 5381, 9068, 10208, 10937, 11838, 12124, 12398, unnumbered

Carnivore pieces (sp. indet.)

Clear carnivore damage: 6291, 6785, unnumbered

Antelope class I

Clear carnivore damage: thoracic vertebra, 5364; humerii, 6391, 7797, 10547, 10567, 11350; radii, 10366, 10888; femurs, 7539, 7789, 9770, 11211, unnumbered; tibia, 4223; metacarpals, 8078, 8482, 9974, 11023; metatarsals, 9536, 11024, 11633

Probable carnivore damage: horn-core, 9136; mandible, 11458; femur, 10346; metacarpals, 9905, 11594; metatarsals, 6676, 8914, 9232

Antelope class II*a*

Clear carnivore damage: mandible, 10410; atlas vertebra, 10234; axis vertebrae, 3844, 7604, 10783; other cervical vertebrae, 5451, 9734; thoracic vertebrae, 5062, 5564, 9651, 11572; lumbar vertebrae, 5464, 10164, 10589, 10833, 10988, 11892; sacrum, 11382; scapulae, 9924, 10118, 10716, 11812, 12576, 12580; pelvises, 6604, 6622, 7587, 9313, 9352, 9529, 10884, 10928, 11280; humerii, 2885, 4340, 4343, 4388, 4422, 4774, 4826, 5016, 5296, 5338, 5593, 6247, 6306, 6311, 6313, 6325, 6353, 6358, 6370, 6399, 6404, 6922, 7174, 7546, 7772, 8140, 8475, 9351, 9370, 9552, 9588, 9741, 10034, 10173, 10495, 10508, 10587, 10606, 10747, 10752, 10813, 11103, 11180, 11736, 11765, 12080, 12103, 12374, 12428, 12509, 12634, 14018, 14032, unnumbered; radii and ulnae, 4331, 4359, 5431, 6140, 6234, 6409, 7775, 7891, 9461, 9684, 10015, 10016, 10423, 10846, 10975; femurs, 4116, 4122, 4246, 4278, 4358, 4710, 4855, 5053, 5104, 5117, 5345, 5361, 5675, 6251, 6255, 6263, 6269, 6280, 6290, 7147, 7582, 7698, 8164, 8185, 8204, 8228, 8722, 8857, 9233, 9327, 9391, 9395, 9523, 9549, 9559, 9685, 10540, 10581, 10618, 10854, 10989, 10999, 11217, 12028, 12420, 12488; tibiae, 4658, 5010, 5199, 5319, 6154, 6197, 6249, 6965, 7296, 7351, 7624, 7755, 8040, 8950, 9216, 9375, 9413,

Continued on following page

Table 106. *Continued*

10673, 12202; calcanei, 5212, 6112, 6780; metacarpals, 5216, 5429, 5616, 5757, 5811, 9110, 9788, 10219; metatarsals; 4686, 5215, 5594, 5596, 5598, 5669, 5800, 5847, 5845, 5859, 7723, 8205, 9288, 9441, 10037, 10196, 10571, 10732, 10945, 10978, 11528, 11810, 11881, 11978, 12070, 12111, 12413, unnumbered; metapodials, 5653, 8043, unnumbered; indet. shaft pieces, 1579, 5970, 6985, 7006, 8225, 9826, 9832, 10156, 10247, 10488, 10893, 11325

Probable carnivore damage: mandible, 7724; thoracic vertebra, 11361; lumbar vertebra, 14011; scapulae, 9082, 9430, 9848, 10645, 10886, 11427, 11797, 12594; pelvises, 5089, 6006, 6638, 11466; humeri, 5337, 6317, 6336, 9567, 10222, 10250, 10264, 10757, 11909, 12574, 12587; radii, 6191, 6393, 8794, 9474; femurs, 1913, 4472, 4676, 4725, 5150, 5198, 6646, 7479, 8012, 8522, 8668, 10011, 10387; tibiae, 4398, 4617, 4880, 5171, 7000, 7148, 7849, 9266, 9824, 9868, 11139, 11421, 12070, 12150; astragalus, 11770; calcaneus, 6771; metacarpals, 5807, 5814, 5869, 9518, 9692, 10086, 10781, 12358; metatarsals, 4143, 4500, 4627, 4720, 4903, 5230, 5612, 5745, 5752, 5852, 8099, 8233, 9438, 9486, 9643, 10102, 10475, 11333, 11535, 12104, unnumbered; indet. shaft pieces, 4872, 5242, 5772, 7004, 7287, 7405, 9101, 11150, 12102

Antelope class II*b*

Clear carnivore damage: scapula, 10227; ribs, 5661, 7972; radius, 7687; femurs, 6307, 9315, 10996; calcaneus, 8854; metacarpals, 3368, 7710, 11337; metatarsals, 5648, 5677, 5698, 10642

Probable carnivore damage: pelvis, 12040; radii and ulnae, 6163, 12383; femur, 9611; tibia, 5025; metacarpals, 4567, 9058; metatarsal, 10351; indet. shaft piece, 7074

Antelope class III

Clear carnivore damage: mandibles, 6063, 7476; axis vertebra, 11066; humeri, 2884, 6641, 14127; radius, 10955; calcaneus, 10151; metacarpal, 5857; metatarsal, 5753

Artificial cut marks

Ourebia cf. *ourebi:* mandible, 14168 (two cut marks on lower margin of ramus)

Antelope class II*b:* proximal radius and ulna, 6136 (cuts and scratches)

Indet. fragment: 4729 (cut marks)

Artificial chop marks

Indet. shaft piece: 8289 (chop marks)

Note: All specimen numbers are prefixed SK.

Table 107. Remains from the Swartkrans Channel Fill: Overall Analysis, Taxa Represented, and Minimum Numbers of Individual Animals Involved

Taxon	Number of Specimens	Minimum Number of Individuals
Phylum Chordata		
Class Mammalia		
Order Primates		
Family Hominidae		
Australopithecus robustus		
Cranial piece	1	1
Order Carnivora		
Family Felidae		
Cf. *Panthera pardus*		
Cranial piece	1	1
Postcranial piece	1	
Family Canidae		
Canis mesomelas		
Cranial pieces	3	2
Carnivore indet.		
Postcranial pieces	12	± 3
Order Artiodactyla		
Family Bovidae		
Medium alcelaphine, including *Damaliscus* sp. and *Connochaetes* sp.		
Cranial pieces	17	5
Antidorcas bondi		
Cranial pieces	6	3
Antidorcas cf. *marsupialis*		
Cranial pieces	5	2
Antelope class I		
Cranial pieces	4	
Postcranial pieces	37	± 3
Antelope class II		
Cranial pieces	33	
Postcranial pieces	73	± 10
Antelope class III		
Cranial pieces	6	
Postcranial pieces	24	± 3
Antelope class IV		
Cranial pieces	0	
Postcranial pieces	1	1

Taxon		Number of Specimens	Minimum Number of Individuals
Order Perissodactyla			
Family Equidae			
Equus cf. *burchelli*			
Cranial pieces		4	1
Postcranial pieces		3	
Order Hyracoidea			
Family Procaviidae			
Procavia cf. *capensis*			
Cranial pieces		11	
Postcranial pieces		16	5
Order Lagomorpha			
Family Leporidae			
Gen. et sp. indet.			
Cranial pieces		3	2
Class Aves			
Order Galliformes			
Family Numididae			
Cf. *Numida* sp.			
Postcranial pieces		5	1
Class Reptilia			
Order Chelonia			
Family Testudinidae			
Gen. et sp. indet.			
Carapace pieces		3	1
Indeterminate fragments		193	
Bone flakes		375	
0–1 cm	2		
1–2 cm	29		
2–3 cm	97		
3–4 cm	106		
4–5 cm	60		
5–6 cm	63		
6–7 cm	9		
7–8 cm	4		
8 + cm	5		
Total		837	44

Table 108. Skeletal Part Analysis of Bovid Remains from Swartkrans Channel Fill

Part	Number of Specimens			
	Size Class I	Size Class II	Size Class III	Size Class IV
Calvaria pieces		2		
Horn-core pieces	1	6	1	
Maxillary pieces				
Mandible pieces	1	17	2	
Isolated upper teeth				
Isolated lower teeth	3	5	1	
Tooth fragments		3	2	
Vertebrae				
Atlas		3		
Axis		1	1	
Cervical, 3–7		3		
Thoracic	1	5	2	
Lumbar				
Sacral				
Caudal				
Rib pieces			1	
Scapula pieces		2	1	
Pelvis pieces	1	5	2	
Humerus				
Proximal pieces				
Distal pieces	2	3	2	
Shaft pieces	2		1	
Articulated distal humerus/ proximal radius, and ulna				
Radius and ulna, proximal pieces	4	3	2	
Radius				
Distal pieces		2		
Shaft pieces	1	1		
Femur				
Proximal pieces	1		1	
Distal pieces			2	
Shaft pieces		1		
Patella				
Tibia				
Proximal pieces				
Distal pieces	2	4		
Shaft pieces				
Metacarpal				
Proximal pieces	3	2		
Distal pieces	1			
Shaft pieces	1			
Metatarsal				
Proximal pieces	1	1		
Distal pieces				
Shaft pieces		1		
Metapodial				
Proximal pieces			2	
Distal pieces	4	9	1	
Shaft pieces	4			
Astragalus	1	3		
Calcaneus	1	5	1	1
Tarsal bones		2		
Carpal bones		4	1	
1st phalanges		10	2	
2d phalanges	4	3	1	
Terminal phalanges	2		1	
Articulated foot bone pieces				
Total	41	106	30	1

Table 109. Observed Damage to Bones from Swartkrans Channel Fill

Porcupine-gnawed piece
Shaft piece, SKW 2681

Pieces gnawed by small rodents
Antelope class II, phalanx, 2390; indet. rib, 2424

Pieces showing carnivore-inflicted damage
Clear carnivore damage
Antelope class I: phalanx 2381
Antelope class II: pelvis, 2352; phalanx, 2376; bone flake, 3042
Possible carnivore damage
Bone flakes, 3013, and 3088

Pieces showing artificial marks
Bone flake: 3220 (worn edges)
Antelope class II: horn-core, 2628 (probable chop mark)

Note: All specimen numbers are prefixed SKW.

Table 110. Remains from Kromdraai A: Overall Analysis, Taxa Represented, and Minimum Numbers of Individual Animals Involved

Taxon	Number of Specimens	Minimum Number of Individuals
Phylum Chordata		
Class Mammalia		
Order Primates		
Family Cercopithecidae		
Parapapio jonesi		
Cranial pieces	2	2
Papio robinsoni		
Cranial piece	1	1
Papio angusticeps		
Cranial pieces	12	11
Postcranial pieces	5	
Gorgopithecus major		
Cranial pieces	17	8
Cercopithecoid indet.		
Cranial pieces	50	± 12
Postcranial pieces	8	
Order Carnivora		
Family Felidae		
Panthera leo		
Cranial pieces	4	1
Felis crassidens		
Cranial piece	1	1
Felis sp.		
Cranial piece	1	1
Homotherium sp.		
Cranial piece	1	1
Megantereon eurynodon		
Cranial pieces	1	
Postcranial pieces	6	1
Dinofelis piveteaui		
Cranial pieces	3	1
Family Hyaenidae		
Crocuta crocuta		
Cranial pieces	7	
Postcranial pieces	53	3
Hyaena bellax		
Cranial piece	1	1
Hyaenid indet.		
Cranial pieces	9	3
Family Canidae		
Lycaon sp.		
Cranial pieces	2	1
Canis mesomelas		
Cranial pieces	17	8
Postcranial piece	1	
Canis terblanchei		
Cranial piece	1	1
Vulpes pulcher		
Cranial piece	1	1
Family Viverridae		
Herpestes mesotes		
Cranial pieces	2	
Postcranial pieces	12	1
Crossarchus transvaalensis		
Cranial piece	1	1
Viverrid indet.		
Cranial pieces	2	2

Continued on following page

Table 110. *Continued*

Taxon		Number of Specimens	Minimum Number of Individuals
Carnivore indet.			
Cranial pieces		8	
Postcranial pieces		39	± 10
Carnivore coprolites		5	
Order Artiodactyla			
Family Bovidae			
Damaliscus sp. 1 or *Parmularius*			
Cranial pieces		166	32
Medium-sized alcelaphines			
Cranial pieces		35	12
Cf. *Connochaetes* sp.			
Cranial pieces		17	5
Cf. *Megalotragus* sp.			
Cranial pieces		2	2
Hippotragus cf. *equinus*			
Cranial piece		1	1
?Hippotragini			
Cranial piece		1	1
Redunca cf. *arundinum*			
Cranial piece		1	1
Pelea cf. *capreolus*			
Cranial pieces		6	3
Antidorcas recki			
Cranial pieces		43	13
Postcranial pieces		1	
Antidorcas bondi			
Cranial pieces		12	6
Raphicerus sp.			
Cranial piece		1	1
Syncerus sp.			
Cranial pieces		4	2
Tragelaphus cf. *scriptus*			
Cranial piece		1	1
Tragelaphus cf. *strepsiceros*			
Cranial pieces		8	5
Taurotragus cf. *oryx*			
Cranial pieces		2	2
Antelope class I			
Cranial pieces		7	
Postcranial pieces		56	± 5
Antelope class II			
Cranial pieces		214	± 12
Postcranial pieces		299	
Antelope class III			
Cranial pieces		49	± 8
Postcranial pieces		73	
Antelope class IV			
Cranial pieces		2	1
Postcranial pieces		2	
Family Suidae			
Phacochoerus antiquus			
Cranial piece		1	1
Tapinochoerus meadowsi			
Cranial pieces		2	2
Order Perissodactyla			
Family Equidae			
Hipparion steytleri			
Cranial piece		1	1
Equus burchelli			
Cranial pieces		5	5
Equus capensis			
Cranial pieces		54	23
Postcranial pieces		5	
Equid indet.			
Cranial pieces		17	± 4
Postcranial pieces		5	
Order Hyracoidea			
Family Procaviidae			
Procavia antiqua			
Cranial pieces		29	10
Procavia transvaalensis			
Cranial pieces		19	6
Postcranial piece		1	
Hyracoid indet.			
Cranial pieces		3	2
Order Lagomorpha			
Family Leporidae			
Lagomorph indet.			
Cranial pieces		4	2
Order Rodentia			
Family Hystricidae			
Hystrix africaeaustralis			
Cranial pieces		6	3
Hystrix cf. *makapanensis*			
Cranial piece		1	1
Class Aves			
Cf. *Columba* sp.			
Eggshell piece		1	1
Class Reptilia			
Order Chelonia			
Family Testudinidae			
Cf. *Testudo* sp.			
Carapace piece		1	1
Phylum Mollusca			
Class Gastropoda			
Order Pulmonata			
Cf. *Achatina* sp.			
Shell pieces		6	6
Indet. fragments		396	
Bone flakes		14	
2–3 cm	1		
3–4 cm	1		
4–5 cm	7		
5–6 cm	1		
6–7 cm	2		
7–8 cm	1		
8–9 cm	1		
Total		1,847	253

Note: Corrected minimum number of individuals, 194.

Table 111. Skeletal Part Analysis of Bovid Remains from Kromdraai A

Part	Number of Specimens			
	Size Class I	Size Class II	Size Class III	Size Class IV
Calvaria pieces		4	2	1
Horn-core pieces			2	
Maxillary pieces	1	10		
Mandible pieces		25	4	
Isolated upper teeth	1	62	12	1
Isolated lower teeth	2	53	18	
Tooth fragments	3	60	11	
Vertebrae				
Atlas		6		
Axis		4	2	
Cervical, 3–7	1	3	1	
Thoracic	7	14	2	
Lumbar	5	9		
Sacral		4	2	
Caudal			1	
Rib pieces		16	10	
Scapula pieces		17	1	
Pelvis pieces	1	2	2	
Humerus				
Complete bones		3		
Proximal pieces	2	17	7	
Distal pieces	2	12	1	
Shaft pieces	2	4		
Articulated distal humerus/ proximal radius, and ulna		1		
Radius and ulna, proximal pieces	3	16	3	
Radius				
Distal pieces	2	4	3	
Shaft pieces	4	10	3	
Femur				
Complete bones		1		
Proximal pieces	1	9	1	
Distal pieces		13	5	
Shaft pieces		2		
Tibia				
Proximal pieces	1	10		
Distal pieces	2	4		
Shaft pieces		4	1	
Metacarpal				
Complete bones		1	1	
Proximal pieces	1	10		
Distal pieces	1	4	2	
Shaft pieces	1	11	1	
Metatarsal				
Proximal pieces	6	8		
Distal pieces	1	6	1	
Shaft pieces	3	9		
Metapodial				
Proximal pieces				
Distal pieces	1	6		
Astragalus	1	10	2	
Calcaneus	2	8	4	
Tarsal bones		3		
Carpal bones		3	3	
1st phalanges	2	20	10	
2d phalanges	1	12	3	2
Terminal phalanges		1	1	
Articulated foot bone pieces	3	2		
Total	63	513	122	4

Table 112. Fossil Assemblages from Kromdraai A and B: Contributions of Individual Animals of Various Taxa to the Preserved Faunas

	Kromdraai A		Kromdraai B	
Taxon	N	%	N	%
Mammalia				
Primates				
Hominidae	—	—	6	6.4
Cercopithecidae	22	11.3	37	39.4
Carnivora	25	12.9	15	16.0
Artiodactyla	94	48.5	26	27.7
Perissodactyla	29	15.0	—	—
Hyracoidea	16	8.2	1	1.1
Lagomorpha	2	1.0	2	2.1
Rodentia	4	2.1	—	—
Aves	1	.5	2	2.1
Reptilia	1	.5	5	5.3
Total	194	100.0	94	100.1

Table 113. Observed Damage to Bones from Kromdraai A

Porcupine-gnawed pieces
Procavia transvaalensis: mandible, KA 4
Antelope class III: ulna, 1767
Bone flake: 1305
Indet. shaft pieces: 1312, 1594

Pieces gnawed by small rodents
Procavia antiqua: maxilla, 11
Antelope class II: tibia, unnumbered

Pieces showing carnivore-inflicted damage
Gorgopithecus major: mandible, 198 (punctate tooth marks)
Antidorcas recki: metacarpal, 1123*b* (clear tooth marks)
Damaliscus sp. or *Parmularius* sp.
Clear carnivore damage: mandible, 566, 929; cranium with articulated vertebrae, 731
Probable carnivore damage: mandible pieces, 1101, 1739, 1827*a*
Procavia transvaalensis
Clear carnivore damage: mandible pieces, 1, 5
Probable carnivore damage: partial snout, 53; mandible, 4
Procavia antiqua
Clear carnivore damage: mandible pieces, 8, 17, 19
Probable carnivore damage: mandible pieces, 6, 7
Antelope class I
Clear carnivore damage: humerus, 646*d;* calcaneus, 1619*c*
Antelope class II
Clear carnivore damage: humerus, 724; metacarpals, 1343, 1619; tibiae, 591, 911, 1135
Probable carnivore damage: humeri, 531, 961; femur, 513
Indet. fragments
Clear carnivore damage: 1584*b*
Probable carnivore damage: 535, unnumbered

Note: All specimen numbers are prefixed KA.
Damage apparently caused by insects that bored into the bone was observed on six specimens: 646, 883, 1069, 1204, 1392, 1573.

Table 114. Microvertebrate Remains from Kromdraai A

Taxon	Minimum Number of Individuals	%
Order Chiroptera (indet.)	3	1.1
Order Insectivora		
Family Chrysochloridae (indet.)	1	.4
Family Macroscelidae		
Elephantulus sp.	9	3.3
Family Soricidae (indet.)	22	8.1
Order Rodentia		
Family Bathyergidae		
Cryptomys sp.	2	.7
Family Muridae		
Subfamily Murinae, various genera	14	5.1
Subfamily Otomyinae		
Cf. *Palaeotomys gracilis*	21	8.0
Family Cricetidae		
Subfamily Cricetinae		
Mystromys sp.	69	25.3
Subfamily Gerbillinae		
Tatera sp.	2	.7
Subfamily Dendromurinae (indet.)	81	29.7
Rodentia (Myomorpha) indet.	44	16.1
Class Reptilia		
Order Squamata		
Lacertilia indet.	2	.7
Class Aves (indet.)	3	1.1
Total	273	100.3

Table 115. Remains from Kromdraai B: Overall Analysis, Taxa Represented, and Minimum Numbers of Individual Animals Involved

Taxon	Layer 1		Layer 2		Layer 3		Total	
	Number of Specimens	Minimum Number of Individuals	Number of Specimens	Minimum Number of Individuals	Number of Specimens	Minimum Number of Individuals	Number of Specimens	Minimum Number of Individuals
Phylum Chordata								
Class Mammalia								
Order Primates								
Family Hominidae								
Australopithecus robustus								
Cranial pieces	6	6					6	6
Postcranial pieces	2						2	
Family Cercopithecidae								
Papio robinsoni								
Cranial pieces	75	11	7	4	7	3	89	18
Papio angusticeps								
Cranial pieces	17	9	2	1	7	4	26	14
Cercopithecoides williamsi								
Cranial pieces	7	2			6	3	13	5
Cercopithecoid indet.								
Cranial pieces	71	indet.	12	indet.	19	indet.	102	indet.
Postcranial pieces	194		37		34		265	
Order Carnivora								
Family Felidae								
Panthera pardus								
Cranial pieces								
Postcranial pieces	3	1			3	1	6	2
?*Megantereon* sp.								
Postcranial pieces			9	1			9	1
?*Dinofelis* sp.								
Postcranial pieces					1	1	1	1
Family Hyaenidae								
Hyaena cf. *brunnea*								
Cranial pieces	1	1		1		1	1	3
Postcranial pieces			1		1		2	
Proteles sp.								
Cranial pieces			1	1			1	1
Family Canidae								
Canis sp.								
Cranial pieces	1	1	2	1	4	2	7	4
Postcranial pieces			2				2	
Family Viverridae								
Herpestes sp.								
Cranial pieces	1	1	1	1			2	2
Viverra sp.								
Postcranial piece					1	1	1	1
Carnivore indet.								
Cranial pieces	14	indet.	2	indet.		indet.	16	indet.
Postcranial pieces	24		2				26	
Order Artiodactyla								
Family Bovidae								
Connochaetes sp.								
Cranial pieces	2	1	2	1	1	1	4	3
Antidorcas cf. *recki*								
Cranial pieces	3	1	1	1	2	1	6	2 or 3

Continued on following page

Table 115. *Continued*

Taxon	Layer 1		Layer 2		Layer 3		Total	
	Number of Specimens	Minimum Number of Individuals	Number of Specimens	Minimum Number of Individuals	Number of Specimens	Minimum Number of Individuals	Number of Specimens	Minimum Number of Individuals
Cf. *Antidorcas bondi*								
Cranial pieces	3	1			1	1	4	2
Gazella sp.								
Cranial pieces	1	1					1	1
Incertae sedis								
Cranial pieces					3	1	3	1
Antelope class I								
Cranial pieces	6	2	2	1	3	3	11	6
Postcranial pieces	7		2		7		26	
Antelope class II								
Cranial pieces	20	± 5	4	± 3	2	2	26	10
Postcranial pieces	42		15		15		72	
Antelope class III								
Cranial pieces	2	3	12	3	1	2	15	8
Postcranial pieces	20		15		10		45	
Antelope class IV								
Cranial pieces								
Postcranial pieces					2	1	2	1
Family Suidae								
Phacochoerus antiquus								
Cranial pieces					1	1	1	1
Order Hyracoidea								
Family Procaviidae								
Procavia sp.								
Postcranial pieces	1	1					1	1
Order Lagomorpha								
Family Leporidae								
Cf. *Lepus* sp.								
Cranial pieces	1	1			1	1	2	2
Postcranial pieces	4						4	
Class Aves								
Struthio sp.								
Eggshell piece	1	1					1	1
Bird: indet. large raptor								
Postcranial pieces			2	1			2	1
Class Reptilia								
Order Squamata								
Cf. *Cordylus giganteus*								
Postcranial pieces	8	1					8	1
Order Crocodilia								
Cf. *Crocodylus niloticus*								
Postcranial pieces					1	1	1	1
Order Chelonia								
Cf. *Testudo* sp.								
Postcranial pieces	6	2			1	1	7	3
Coprolites	4	indet.		indet.		indet.	4	indet.
Bone flakes	1,512	indet.	627	indet.	748	indet.	2,887	indet.
Indet. fragments	838	indet.	212	indet.	224	indet.	1,275	indet.
Total	2,897	52	972	20	1,116	32	4,985	104

Table 116. Skeletal Part Analysis of Bovid Remains from Kromdraai B

Part	Number of Specimens										
	Layer 1			Layer 2			Layer 3				
	Class I	Class II	Class III	Class I	Class II	Class III	Class I	Class II	Class III	Class IV	Total
Calvaria pieces			1								1
Horn-core pieces		5	1	1			3	1			11
Maxillary pieces											
Mandible pieces		3									3
Isolated upper teeth											
Isolated lower teeth					2						2
Tooth fragments	6	12		1	2	12		1	1		35
Vertebrae											
Atlas						1					1
Axis											
Cervical, 3–7		2									2
Thoracic	1	6	1			2		1			11
Lumbar	1	1						1			3
Sacral											
Caudal									2		2
Rib pieces		10	9					1	1		21
Scapula pieces					1	1					2
Pelvis pieces											
Humerus											
Proximal pieces											
Distal pieces		2		1			1				4
Shaft pieces	1										
Articulated distal humerus/proximal radius and ulna											1
Radius and ulna, proximal pieces											
Radius						1	1		2		4
Distal pieces			2								
Shaft pieces						1	1		1		5

Continued on following page

Table 116. *Continued*

Part	Number of Specimens										
	Layer 1			Layer 2			Layer 3				
	Class I	Class II	Class III	Class I	Class II	Class III	Class I	Class II	Class III	Class IV	Total
Femur											
Proximal pieces					1						1
Distal pieces			1								1
Shaft pieces											
Patella		1									
Tibia											
Proximal pieces											1
Distal pieces											
Shaft pieces					1	1	2	1			5
Metacarpal											
Proximal pieces											
Distal pieces								1			1
Shaft pieces											
Metatarsal											
Proximal pieces							1	1			2
Distal pieces											
Shaft pieces											
Metapodial											
Proximal pieces											
Distal pieces		2			1	1	1			1	6
Shaft pieces											
Astragalus	1					2	1				4
Calcaneus	1	4			1	1	1	1			9
Tarsal bones		1					1				2
Carpal bones		1	2		1					1	5
1st phalanges		7	2	1	4	4	4	5	1		28
2d phalanges		1			2		1	2	2		8
Terminal phalanges	2	3	2		3		2		1		13
Sesamoid		1	1					1			3
Articulated foot bone pieces											
Total	13	62	22	4	19	27	20	17	11	2	197

Table 117. Lengths of Bone Flakes from Kromdraai B

Length (cm)	Layer 1		Layer 2		Layer 3		Total	
	N	%	N	%	N	%	N	%
0–1	3	.2	3	.5	9	1.3	15	.5
1–2	524	34.7	131	20.9	198	26.5	853	29.6
2–3	652	43.1	268	42.7	278	37.2	1,198	41.5
3–4	205	13.6	141	22.5	157	21.0	503	17.5
4–5	72	4.8	59	9.4	56	7.4	187	6.5
5–6	35	2.4	15	2.4	31	4.1	81	2.8
6–7	9	.6	7	1.1	11	1.5	27	.9
7–8	8	.5	3	.5	4	.6	15	.5
8–9	1				1	.1	2	
9–10					2	.3	2	.2
10–11	2	.1					2	
11–12	1				1	.1	2	
Total	1,512	100.0	627	100.0	748	100.0	2,887	100.0

Table 118. Cercopithecoid Remains from Kromdraai B: Minimum Number of Individuals per Age Class

Age Class	Layer 1	Layer 2	Layer 3	Total Number of Individuals
Papio robinsoni				
Juveniles	7	2	2	11
Immature adults				
Young adults	1	1		2
Adults	2	1	1	4
Old adults	1			1
Total	11	4	3	18
Papio angusticeps				
Juveniles	1	1	1	3
Immature adults	1			1
Young adults	1	1	2	4
Adults	4		1	5
Old adults	1			1
Total	8	2	4	14
Cercopithecoides williamsi				
Juveniles	1		1	2
Immature adults				
Young adults	1		1	2
Adults				
Old adults			1	1
Total	2	0	3	5

Table 119. List of Skeletal Parts from Cercopithecoids from Kromdraai B That Have Not Been Specifically Identified

Part	Layer 1	Layer 2	Layer 3	Total
Fragment of				
Calvaria	40		14	54
Maxilla				
Mandible	1			1
Tooth	30	12	5	47
Vertebrae				
Cervical		7		7
Thoracic	2		1	3
Lumbar	1	2	1	4
Sacral				
Caudal	33	2	6	41
Rib	52			52
Scapula	1		1	2
Pelvis	2	3	1	6
Humerus				
Proximal				
Distal	2		1	3
Shaft			1	1
Radius				
Proximal	4	2		6
Distal	1			1
Shaft	3		1	4
Ulna				
Proximal	4		2	6
Distal				
Shaft				
Femur				
Proximal	6	1	1	8
Distal	2	1	1	4
Shaft	1	1	1	3
Tibia				
Proximal	1			1
Distal	3			3
Shaft				
Metapodial	10	5	3	18
Calcaneus	5	2	2	9
Astragalus	2		1	3
Phalanges	50	11	6	67
Miscellaneous	9		4	13
Total	265	49	53	367

Table 120. Skeletal Parts from Unidentified Carnivores from Kromdraai B

Part	Layer 1	Layer 2	Layer 3	Total
Fragment of				
Tooth	10	2		12
Skull	4			4
Pelvis		1		1
Proximal ulna		1		1
Metapodial	9			9
Phalanges	15			15
Total	38	4		42

Table 121. Observed Damage to Bones from Kromdraai B

Porcupine-gnawed piece
Carnivore mandible piece, KB 2884

Pieces showing carnivore tooth marks
Australopithecus robustus: pelvis piece, TM 1605 (the ragged-edged margins probably represent carnivore damage)
Antelope class II: tibia, 3283; shaft pieces, 2275, unnumbered

Pieces showing rounding and wear
Proteles sp.: mandible, 2945.2
Cercopithecoid: isolated teeth, 247, 996, 1396, unnumbered; phalanx, 1196; caudal vertebrae, 429, 450
Bovid horn-cores, 127, 378; miscellaneous pieces, 20, 137, 243, 403, 592, 647, 872, 942, 1342, 1377, 1459, 1709, 1907, 2016, 2049, 2123, 2224, 2329, 2675, 2765, unnumbered
Bone flakes, 60, 838, 930, 1024, 1056, 1367, 1397, 1491, 1524, 1585, 1601, 1693, 1758, 1867, 1892, 1983, 2116, 2131, 2294, 2414, 2565, 2592, 2653, 2670, 2677, 2690, 2705, 2867, 11 unnumbered pieces

Lengths of bone pieces showing rounding and wear

Length of piece (cm)	N
0–1	1
1–2	28
2–3	26
3–4	8
4–5	6
5–6	1
Total	70

Note: All specimen numbers are prefixed KB unless otherwise noted.

References

Acocks, J. P. H. 1953. Veld types of South Africa. *Botanical Survey S. Afr. Mem.* 28:1–192.

Alexander, A. J. 1956. Bone carrying by a porcupine. *S. Afr. J. Sci.* 52 (11): 257–58.

Alexander, J. E. 1838. *An expedition of discovery into the interior of Africa, through the hitherto undescribed countries of the Great Namaquas, Boschmans and Hill Damaras.* London: Henry Colburn.

Ansell, W. F. H. 1971. Order Artiodactyla. In *The mammals of Africa: An identification manual,* ed. J. Meester and H. W. Setzer, part 15. Washington, D.C.: Smithsonian Institution Press.

Arambourg, C. 1947. Contribution à l'étude géologique et paléontologique du bassin du Lac Rudolphe et de la basse vallée de l'Omo: *Deuxième partie, paléontologie. Mission Sci. Omo 1932–1933, T. Géol. Anth.* 1 (3): 231–562.

Arkell, A. J. 1957. A possibly palaeolithic bone spatula from Egypt. *Proc. Prehist. Soc.* 23:234–36.

Armstrong, A. L. 1936. A bull-roarer of Mousterian age from Pin Hole Cave, Cresswell Crags. *Antiqua. J.* (London) 16:322–23.

Astley Maberly, C. T. 1953. A plea for the leopard. *Afr. Wild Life* 7 (1): 19–27.

Bakr, A. 1959. Ph.D. diss. Biology Dept., Harvard University.

Balestra, F. A. 1962. The man-eating hyaenas of Mlanje. *Afr. Wild Life* 16:25–27.

Banks, R. C. 1965. Some information from barn owl pellets. *Auk* 82:506.

Bateman, J. A. 1960. Owl habits from pellets. *Bull. Zool. Soc. S. Afr.* 1 (3): 20–21.

Bates, H. J. 1971. Baboons. *Afr. Wild Life* 25 (4): 154.

Bearder, S. K. 1977. Feeding habits of spotted hyaenas in a woodland habitat. *E. Afr. Wildl. J.* 15:263–80.

Beaumont, P. B. 1973. Border Cave: A progress report. *S. Afr. J. Sci.* 69:41–46.

Behrensmeyer, A. K. 1975*a*. The taphonomy and paleoecology of Plio-Pleistocene vertebrate assemblages east of Lake Rudolf, Kenya. *Bull. Mus. Comp. Zool., Harv.* 146:473–578.

———. 1975*b*. Taphonomy and paleoecology in the hominid fossil record. *Yearb. Phys. Anthrop.* 19:36–50.

Benson, C. W. 1962. The food of the spotted eagle-owl *Bubo africanus. Ostrich* 33:35.

Benson, C. W., and Irwin, M. P. S. 1967. The distribution and systematics of *Bubo capensis* Smith (Aves). *Arnoldia* (Rhodesia) 3 (19): 1–19.

Bere, R. 1966. *The African elephant.* New York: Golden Press.

Best, A. A., ed. 1973. *Rowland Ward's records of big game.* 15th ed. (Africa). London: Rowland Ward, Ltd.

Bigalke, E. 1973. The exploitation of shellfish by coastal tribesmen of the Transkei. *Ann. Cape Prov. Mus. (Nat. Hist.)* 9 (9): 159–75.

Binford, L. R., and Bertram, J. B. 1977. Bone frequencies—and attritional processes. In *For theory building in archaeology: Essays on faunal remains, aquatic resources, spatial analysis and systematic modeling,* ed. L. R. Binford, pp. 77–153. New York: Academic Press.

Boaz, N. T., and Behrensmeyer, A. K. 1976. Hominid taphonomy: Transport of human skeletal parts in an artificial fluviatile environment. *Am. J. Phys. Anthrop.* 45 (1): 53–60.

Boaz, N. T. and Howell, F. C. 1977. A gracile hominid cranium from upper Member G of the Shungura Formation, Ethiopia. *Am. J. Phys. Anthrop.* 46: 93–108.

Bodenheimer, F. S. 1949. *Problems of vole populations in the Middle East.* Report on the population dynamics of the Levant vole. Jerusalem: Research Council of Israel.

Bond, G., and Summers, R. 1951. The Quarternary succession and archaeology at Chelmer near Bulawayo, Southern Rhodesia. *S. Afr. J. Sci.* 47:200–204.

Boné, E. L., and Singer, R. 1965. *Hipparion* from Langebaanweg, Cape Province, and a revision of the genus in Africa. *Ann. S. Afr. Mus.* 48:273–397.

Bonnichsen, R. N.d. Some operational aspects of human and animal bone alteration. In *Mammalian osteoarchaeology: North America,* ed. B. M. Gilbert. Columbia, Mo.: Missouri Archaeological Society.

Bothma, J. du P. 1971. Order Hyracoidea. In *The mammals of Africa: An identification manual,* ed. J. Meester and H. W. Setzer, part 12. Washington D.C.: Smithsonian Institution Press.

Brain, C.K. 1958. *The Transvaal ape-man-bearing cave deposits.* Transvaal Museum Memoirs, no. 11.

———. 1967*a*. The Transvaal Museum's fossil project at Swartkrans. *S. Afr. J. Sci.* 63:378–84.

———. 1967*b*. Hottentot food remains and their bearing on the interpretation of fossil bone assemblages. *Sci. Papers Namib Desert Res. Station* 32:1–11.

———. 1967*c*. Bone weathering and the problem of bone pseudo-tools. *S. Afr. J. Sci.* 63 (3): 97–99.

———. 1969*a*. New evidence for climatic change during Middle and Late Stone Age times in Rhodesia. *S. Afr.*

Arch. Bull. 24:127–43.

———. 1969*b*. The contribution of Namib Desert Hottentots to an understanding of australopithecine bone accumulations. *Sci. Papers Namib Desert Res. Station* 39:13–22.

———. 1969*c*. Faunal remains from the Bushman Rock Shelter, Eastern Transvaal. *S. Afr. Archaeol. Bull.* 24 (94): 52–55.

———. 1970. New finds at the Swartkrans australopithecine site. *Nature* (London) 225 (5238):1112–19.

———. 1972. An attempt to reconstruct the behaviour of australopithecines: The evidence for interpersonal violence. *Zool. Afr.* 7 (1): 379–401.

———. 1973. Seven years' hard labour at Swartkrans. *Bull. Transv. Mus.* 14:5–6.

———. 1974*a*. Some suggested procedures in the analysis of bone accumulations from southern African Quaternary sites. *Ann. Transv. Mus.* 29 (1): 1–8.

———. 1974*b*. The use of microfaunal remains as habitat indicators in the Namib. *S. Afr. Archaeol. Soc.*, Goodwin ser. no. 2, pp. 55–60.

———. 1974*c*. Human food remains from the Iron Age at Zimbabwe. *S. Afr. J. Sci.* 70:303–9.

———. 1975*a*. An introduction to the South African australopithecine bone accumulations. In *Archaeological studies,* ed. A. T. Clason, pp. 109–19. Amsterdam: North Holland.

———. 1975*b*. An interpretation of the bone assemblage from the Kromdraai australopithecine site, South Africa. In *Palaeoanthropology, morphology and paleoecology,* ed. R. H. Tuttle, pp. 225–43. The Hague: Mouton.

———. 1976*a*. A re-interpretation of the Swartkrans site and its remains. *S. Afr. J. Sci.* 72:141–46.

———. 1976*b*. Some principles in the interpretation of bone accumulations associated with man. In *Human origins: Louis Leakey and the East African evidence,* ed. G. Ll. Isaac and E. R. McCown, pp. 96–116. Menlo Park, Calif.: Staples Press.

———. 1978. Some aspects of the South African australopithecine sites and their bone accumulations. In *African Hominidae of the Plio-Pleistocene,* ed. C. Jolly. London: Duckworth.

Brain, C. K., and Brain, V. 1977. Microfaunal remains from Mirabib: Some evidence of palaeoecological changes in the Namib. *Madoqua* 10 (4): 285–93.

Brain, C. K., and Cooke, C. K. 1967. A preliminary account of the Redcliff Stone Age cave site in Rhodesia. *Bull. S. Afr. Archaeol. Soc.* 21 (84): 171–82.

Brain, C. K., and Meester, J. 1964. Past climatic changes as biological isolating mechanisms in southern Africa. In *Ecological studies in Southern Africa,* ed. D. H. S. Davis, pp. 332–40. Den Haag: Junk Publishers.

Brelsford, V. 1950. Unusual events in animal life—IV. *Afr. Wild Life* 4 (1): 67.

Brock, A.; McFadden, P. L.; and Partridge, T. C. 1977. Preliminary palaeomagnetic results from Makapansgat and Swartkrans. *Nature* (London) 266:249–50.

Brooke, R. K. 1967. On the food of the wood owl. *Ostrich* 38 (1):55.

———. 1973. Notes on the distribution and food of the Cape eagle owl in Rhodesia. *Ostrich* 44:137–39.

Broom, R. 1907. On some new species of *Chrysochloris. Ann. Mag. Nat. Hist.* 19:262–68.

———. 1909*a*. On a large extinct species of *Bubalis. Ann. S. Afr. Mus.* 7:279–80.

———. 1909*b*. On evidence of a large horse recently extinct in South Africa. *Ann. S. Afr. Mus.* 7:281–82.

———. 1913*a*. Man contemporaneous with extinct animals in South Africa. *Ann. S. Afr. Mus.* 12:13–16.

———. 1913*b*. Note on *Equus capensis. Bull. Am. Mus. Nat. Hist.* 32:437–39.

———. 1925. On evidence of a giant pig from the Later Tertiaries of South Africa. *Rec. Albany Mus.* 3:307–8.

———. 1928. On some new mammals from the diamond gravels of the Kimberley district. *Ann. S. Afr. Mus.* 22:439–44.

———. 1931. A new extinct giant pig from the diamond gravels of Windsorton, South Africa. *Rec. Albany Mus.* 4:167–68.

———. 1934. On the fossil remains associated with *Australopithecus africanus. S. Afr. J. Sci.* 31:471–80.

———. 1936. A new fossil anthropoid skull from South Africa. *Nature* (London) 138 (3490): 486–88.

———. 1937*a*. Discovery of a lower molar of *Australopithecus. Nature* (London) 140:681–82.

———. 1937*b*. Notes of a few more new fossil mammals from the caves of the Transvaal. *Ann. Mag. Nat. Hist.* 20 (10): 509–14.

———. 1937*c*. On some new Pleistocene mammals from limestone caves in the Transvaal. *S. Afr. J. Sci.* 33:750–68.

———. 1938*a*. More discoveries of *Australopithecus. Nature* (London) 141:828–29.

———. 1938*b*. The Pleistocene anthropoid apes of South Africa. *Nature* (London) 142 (3591): 377–79.

———. 1938*c*. Further evidence of the structure of the South African Pleistocene anthropoids. *Nature* (London) 142 (3603): 897–99.

———. 1939*a*. A preliminary account of the Pleistocene carnivores of the Transvaal caves. *Ann. Transv. Mus.* 19:331–38.

———. 1939*b*. A restoration of the Kromdraai skull. *Ann. Transv. Mus.* 19 (3): 327–29.

———. 1940. The South African Pleistocene cercopithecoid apes. *Ann. Transv. Mus.* 20:89–100.

———. 1941*a*. Structure of the Sterkfontein ape. *Nature* (London) 147:86.

———. 1941*b*. Mandible of a young *Paranthropus* child. *Nature* (London) 147:607–8.

———. 1941*c*. On two Pleistocene golden moles. *Ann. Transv. Mus.* 20:215–16.

———. 1942. The hand of the ape-man, *Paranthropus robustus. Nature* (London) 149:513–14.

———. 1943. An ankle-bone of the ape-man, *Paranthropus robustus. Nature* (London) 152:689–90.

———. 1947. Discovery of a new skull of the South African ape-man, *Plesianthropus. Nature* (London) 159:672.

———. 1948*a*. Some South African Pliocene and Pleistocene mammals. *Ann. Transv. Mus.* 21:1–38.

———. 1948*b*. The giant rodent mole, *Gypsorhychus. Ann. Transv. Mus.* 21:47–49.

———. 1949. Another new type of fossil ape-man. *Nature* (London) 163:57.

———. 1950. *Finding the missing link.* London: Watts.

Broom, R., and Jensen, J. S. 1946. A new fossil baboon from the caves at Potgietersrust. *Ann. Transv. Mus.* 20:337–40.

Broom, R., and Le Riche, H. 1937. The dentition of *Equus capensis* Broom. *S. Afr. J. Sci.* 33:769–70.

Broom, R., and Robinson, J. T. 1947*a*. Two features of

the *Plesianthropus* skull. *Nature* (London) 159:809–10.

———. 1947*b*. Further remains of the Sterkfontein ape-man, *Plesianthropus*. *Nature* (London) 160:430–31.

———. 1949*a*. A new type of fossil man. *Nature* (London) 164:322.

———. 1949*b*. Thumb of the Swartkrans ape-man. *Nature* (London) 164:841–42.

———. 1949*c*. A new type of fossil baboon, *Gorgopithecus major*. *Proc. Zool. Soc. Lond.* 119:379–86.

———. 1950. Man contemporaneous with the Swartkrans ape-man. *Am. J. Phys. Anthrop.*, n.s., 8:151–56.

———. 1952. *Swartkrans ape-man* Paranthropus crassidens. Transvaal Museum Memoirs, no. 6.

Broom, R.; Robinson, J. T.; and Schepers, G. W. H. 1950. Part 1: *Further evidence of the structure of the Sterkfontein ape-man* Plesianthropus. Part 2: *The brain casts of the recently discovered* Plesianthropus *skulls*. Transvaal Museum Memoirs, no. 4.

Broom, R., and Schepers, G. W. H. 1946. *The South African fossil ape-men: The Australopithecinae*. Transvaal Museum Memoirs, no. 2.

Brothwell, D. R. 1963. *Digging up bones*. London: British Museum (Natural History).

Brown, L. H. 1952. On the biology of the large birds of prey of Embu District, Kenya Colony. *Ibis* 94:577–620.

———. 1955. Supplementary notes on the biology of the large birds of prey of Embu District, Kenya Colony. *Ibis* 97:38–64.

———. 1965. Observations on Verreaux's eagle owl in Kenya. *J. E. Afr. Nat. Hist. Soc.* 25 (2): 101–7.

———. 1966. Observations on some Kenya eagles. *Ibis* 108:531–72.

———. 1970. *African birds of prey*. London: Collins.

Buckland, W. 1822. Account of an assemblage of fossil teeth and bones of elephant, rhinoceros, hippopotamus, bear, tiger, and hyaena and sixteen other animals: Discovered in a cave at Kirkdale, Yorkshire, in the year 1821; with a comparative view of five similar caverns in various parts of England and others on the Continent. *Phil. Trans. Roy. Soc. Lond.* 112:171–237.

———. 1823. *Reliquae diluvianae; or Observations on the organic remains contained in cave fissures, and other diluvial gravel, and on other geological phenomena attesting the action of a universal deluge*. 2d ed. London: Murray.

Buettner-Janusch, J. 1966. A problem in evolutionary systematics: Nomenclature and classification of baboons, genus *Papio*. *Folia Primat.* 4:288–308.

Butler, P. M., and Greenwood, M. 1973. The early Pleistocene hedgehog from Olduvai, Tanzania. *Foss. Vert. Afr.* 3:7–42.

———. 1976. Elephant shrews (Macroscelididae) from Olduvai and Makapansgat. *Foss. Vert. Afr.* 4:1–56.

Butzer, K. W. 1974*a*. Geology of the Cornelia Beds, northeastern Orange Free State. In *The geology, archaeology and fossil mammals of the Cornelia Beds, O.F.S.*, pp. 7–32. Mem. National Mus., Bloemfontein, no. 9.

———. 1974*b*. Paleoecology of South African australopithecines. Taung revisited. *Curr. Anthrop.* 15 (4): 367–82.

———. 1976. Lithostratigraphy of the Swartkrans Formation. *S. Afr. J. Sci.* 72:136–41.

Campbell, A. M. 1959. Apple-scoops. *Nature* (London) 183:1542.

Carnegie, A. J. M. 1961. The stomach contents of a spotted eagle owl *(Bubo africanus)*. *Ostrich* 32:97.

Carter, P. L. 1970. Late Stone Age exploitation patterns in southern Natal. *S. Afr. Archaeol. Bull.* 25 (98):55–58.

Chaplin, R. E. 1971. A study of animal bones from archaeological sites. London and New York: Seminar Press.

Chave, K. E. 1964. Skeletal durability and preservation. In *Approaches to paleoecology*, ed. J. Imbrie and N. Newel, pp. 377–87. New York: John Wiley.

Chitty, D. 1938. A laboratory study of pellet formation in the short-eared owl. *Proc. Zool. Soc. Lond.*, ser. A, 108:2.

Churcher, C. S. 1956. The fossil Hyracoidea of the Transvaal and Taungs deposits. *Ann. Transv. Mus.* 22 (4): 477–501.

———. 1970. The fossil Equidae from the Krugersdorp Caves. *Ann. Transv. Mus.* 26 (6): 145–68.

———. 1974. *Sivatherium maurusium* (Pomel) from the Swartkrans australopithecine site, Transvaal (Mammalia: Giraffidae). *Ann. Transv. Mus.* 29 (6): 65–69.

———. 1978. Giraffidae. In *Evolution of African mammals*, ed. V. J. Maglio and H. B. S. Cooke, pp. 509–35. Cambridge: Harvard University Press.

Churcher, C. S., and Richardson, M. L. 1978. Equidae. In *Evolution of African mammals*, ed. V. J. Maglio and H. B. S. Cooke, pp. 379–422. Cambridge: Harvard University Press.

Clancey, P. A. 1964. *The birds of Natal and Zululand*. Edinburgh: Oliver and Boyd.

Clark, J. D. 1972. Palaeolithic butchery practices. In *Man, settlement and urbanism*, P. J. Ucko, R. Tringham, and G. W. Dimbleby, pp. 149–56. London: Duckworth.

———. 1974. The stone artefacts from Cornelia, O.F.S., South Africa. In *The geology, archaeology and fossil mammals of the Cornelia Beds, O.F.S.*, pp. 33–61. Mem. National Mus., Bloemfontein, no. 9.

Clark, J. D., and Haynes, C. V. 1970. An elephant butchery site at Mwanganda's Village, Karonga, Malawi, and its relevance for palaeolithic archaeology. *World Archaeology* 1 (3): 390–411.

Clark, R. J. 1972. Pellets of the short-eared owl and marsh hawk compared. *J. Wildl. Mgmt.* 36:962–64.

Clark, W. E. Le Gros. 1978. *The fossil evidence for human evolution*. 3d. ed., ed. B. Campbell. Chicago: University of Chicago Press.

Clarke, R. J. 1977. A juvenile cranium and some adult teeth of early *Homo* from Swartkrans, Transvaal. *S. Afr. J. Sci.* 73:46–49.

Clarke, R. J.; Howell, F. C.; and Brain, C. K. 1970. More evidence of an advanced hominid at Swartkrans. *Nature* (London) 225:1217–20.

Clason, A. T. 1972. Some remarks on the use and presentation of archaeozoological data. *Helinium* 12:139–53.

Coetzee, C. G. 1963. The prey of owls in the Kruger National Park as indicated by owl pellets collected during 1960–61. *Koedoe* 6:115–25.

———. 1971. Carnivora (excluding the family Felidae). In *The mammals of Africa: An identification manual*, ed. J. Meester and H. W. Setzer. Washington, D.C.: Smithsonian Institution Press.

———. 1972. The identification of southern African small mammal remains in owl pellets. *Cimbebasia*, ser. A., 2:53–64.

Colbert, E. H. 1935. Siwalik mammals in the American

Museum of Natural History. *Trans. Am. Phil. Soc.*, n.s., 26:1–401.

Collings, G. E. 1972. A new species of machaerodont from Makapansgat. *Palaeont. Afr.* 14:87–92.

Collings, G. E.; Cruickshank, A. R. I.; Maguire, J. M.; and Randall, R. M. 1976. Recent faunal studies at Makapansgat Limeworks, Transvaal, South Africa. *Ann. S. Afr. Mus.* 71:153–65.

Cooke, C. K. 1963. Report on excavations at Pomongwe and Tshangula caves, Matopo Hills, Southern Rhodesia. *S. Afr. Archaeol. Bull.* 18 (71): 73–151.

———. 1966. Reappraisal of the industry hitherto named Proto-Stillbay. *Arnoldia* (Rhodesia) 2 (22): 1–14.

———. 1968. The Early Stone Age in Rhodesia. *Arnoldia* (Rhodesia) 3 (39): 1–12.

———. 1969. A re-examination of the "Middle Stone Age" industries of Rhodesia. *Arnoldia* (Rhodesia) 4 (7): 1–20.

Cooke, H. B. S. 1938. The Sterkfontein bone breccia: A geological note. *S. Afr. J. Sci.* 35:204–8.

———. 1941. A preliminary account of the Wonderwerk Cave, Kuruman District. Section II: The fossil remains. *S. Afr. J. Sci.* 37:300–12.

———. 1943. Cranial and dental characters of the recent South African Equidae. *S. Afr. J. Sci.* 40:254–57.

———. 1947. Some fossil hippotragine antelopes from South Africa. *S. Afr. J. Sci.* 63:226–31.

———. 1949*a*. Fossil mammals of the Vaal River deposits. *Geol. Surv. S. Afr. Mem.* 35 (3): 1–117.

———. 1949*b*. The fossil Suina of South Africa. *Trans. Roy. Soc. S. Afr.* 32:1–44.

———. 1950. A critical revision of the Quaternary Perissodactyla of southern Africa. *Ann. S. Afr. Mus.* 31:393–479.

———. 1960. Further revision of the fossil Elephantidae of southern Africa. *Palaeont. Afr.* 7:46–58.

———. 1962. Notes on the faunal material from the Cave of Hearths and Kalkbank. In *Prehistory of the Transvaal,* ed. R. J. Mason, pp. 447–53. Johannesburg: Witwatersrand University Press.

———. 1963. Pleistocene mammal faunas of Africa, with particular reference to southern Africa. In *African ecology and human evolution,* ed. F. C. Howell and F. Bourliere, pp. 65–116. Viking Fund Publications in Anthropology, no. 36. New York: Wenner-Gren Foundation.

———. 1974. The fossil mammals of Cornelia, O.F.S., South Africa. In *The geology, archaeology and fossil mammals of the Cornelia Beds, O.F.S.,* pp. 63–84. Mem. National Mus., Bloemfontein, no. 9.

———. 1976. Suidae from Plio-Pleistocene strata of the Rudolf basin. In *Earliest man and environments in the Lake Rudolf basin,* ed. Y. Coppens, F. C. Howell, G. Ll. Isaac, and Richard E. F. Leakey, pp. 251–63. Chicago: University of Chicago Press.

———. N.d. South African Pleistocene mammals in the University of California collections. Typescript.

Cooke, H. B. S., and Maglio, V. J. 1972. Plio-Pleistocene stratigraphy in East Africa in relation to proboscidean and suid evolution. In *Calibration of hominoid evolution,* ed. W. W. Bishop, and J. A. Miller, pp. 303–29. Edinburgh: Scottish Academic Press.

Cooke, H. B. S., and Wells, L. H. 1947. Fossil mammals from the Makapan valley, Potgietersrust. III. Giraffidae. *S. Afr. J. Sci.* 43:232–35.

———. 1951. Fossil remains from Chelmer, near Bulawayo, Southern Rhodesia. *S. Afr. J. Sci.* 47:205–9.

Cooke, H. B. S., and Wilkinson, A. F. 1978. Suidae and *Tayassuidae.* In *Evolution of African mammals,* ed. V. J. Maglio and H. B. S. Cooke, pp. 435–82. Cambridge: Harvard University Press.

Coppens, Y., and Howell, F. C. 1976. Mammalian faunas of the Omo group: Distributional and biostratigraphical aspects. In *Earliest man and environments in the Lake Rudolf basin,* ed. Y. Coppens, F. C. Howell, G. Ll. Isaac, and R. E. F. Leakey, pp. 177–92. Chicago: University of Chicago Press.

Coppens, Y.; Maglio, V. J.; Madden, C. T.; and Beden, M. Proboscidea. In *Evolution of African mammals,* ed. V. J. Maglio and H. B. S. Cooke, pp. 336–67. Cambridge: Harvard University Press.

Corbet, G. B. 1971. Family Macroscelididae. In *The mammals of Africa: An identification manual,* ed. J. Meester and H. W. Setzer. Washington, D.C.: Smithsonian Institution Press.

Corbet, G. B., and Hanks, J. 1968. A revision of the elephant shrews, family Macroscelididae. *Bull. Brit. Mus. Nat. Hist. (Zool.)* 16:47–111.

Corbett, J. 1954. *The man-eating leopard of Rudraprayag.* London: Oxford University Press.

Cornwall, I. W. 1956. *Bones for the Archaeologist.* London: Phoenix House.

Coryndon, S. C. 1964. Bone remains in the caves. In *The lava caves of Mount Suswa, Kenya,* ed. P. E. Glover et al. *Studies in Speleology* 1 (1): 60–63.

Crass, R. S. 1944. The birds of Sulenkama, Qumbu District, Cape Province. *Ostrich* 15 (1): 10–20.

Cullen, A. 1969. *Window onto wilderness.* Nairobi: East African Publishing House.

Cummings, J. H.; Duke, G. E.; and Jegers, A. A. 1976. Corrosion of bone by solutions simulating raptor gastric juice. *Raptor Research* 10 (2): 55–57.

Dale, M. M. 1948. New fossil Suidae from the Limeworks Quarry, Makapansgat, Potgietersrust. *S. Afr. J. Sci.* 2:114–16.

Dart, R. A. 1925*a*. *Australopithecus africanus:* The man-ape of South Africa. *Nature* (London) 115 (2884): 195–99.

———. 1925*b*. A note on Makapansgat: A site of early human occupation. *S. Afr. J. Sci.* 22:454.

———. 1929. A note on the Taungs skull. *S. Afr. J. Sci.* 26:648–58.

———. 1948*a*. An adolescent promethean australopithecine mandible from Makapansgat. *S. Afr. J. Sci.* 45:73–75.

———. 1948*b*. The Makapansgat proto-human *Australopithecus prometheus. Am. J. Phys. Anthrop.*, n.s., 6:259–84.

———. 1948*c*. The adolescent mandible of *Australopithecus prometheus. Am. J. Phys. Anthrop.*, n.s., 6:391–412.

———. 1949*a*. The predatory implemental technique of *Australopithecus. Am. J. Phys. Anthrop.*, n.s., 7:1–38.

———. 1949*b*. The bone-bludgeon hunting technique of *Australopithecus. S. Afr. J. Sci.* 2:150–52.

———. 1953*a*. The predatory transition from ape to man. *Int. Anthrop. Ling. Rev.* 1 (4): 201–18.

———. 1953*b*. The proto-human inhabitants of southern africa. In *Africa's place in the human story,* pp. 19–22. Johannesburg: South African Broadcasting Corporation.

———. 1954. The adult female lower jaw from Makapansgat. *Nature* (London) 173:286.

———. 1955. The first australopithecine fragment from

the Makapansgat pebble culture stratum. *Nature* (London) 176:170.

———. 1956*a*. The myth of the bone-accumulating hyena. *Am. Anthrop*. 58 (1): 40–62.

———. 1956*b*. Cultural status of the South African man-apes. *Smithsonian Report*, n. 4240, pp. 317–38.

———. 1957*a*. The Makapansgat australopithecine osteodontokeratic culture. *Proceedings of the Third Pan-African Congress on Prehistory* (Livingstone, 1955). London: Chatto and Windus.

———. 1957*b*. *The osteodontokeratic culture of* Australopithecus prometheus. Transvaal Museum Memoirs, no. 10.

———. 1957*c*. An australopithecine object from Makapansgat. *Nature* (London) 179:693–95.

———. 1958*a*. The minimal bone-breccia content of Makapansgat and the australopithecine predatory habit. *Am. Anthrop*. 60:923–31.

———. 1958*b*. Bone tools and porcupine gnawing. *Am. Anthrop*. 60 (4): 715–24.

———. 1959*a*. *Africa's place in the emergence of civilisation*. Johannesburg: South African Broadcasting Corporation.

———. 1959*b*. The ape-man tool-makers of a million years ago: South African *Australopithecus*—His life, habits and skills. *Illus. Lond. News* 234:798–801.

———. 1959*c*. An australopithecine scoop from Herefordshire. *Nature* (London) 183:844.

———. 1959*d*. The first *Australopithecus* cranium from the pink breccia at Makapansgat. *Am. J. Phys. Anthrop*. 17 (1): 77–82.

———. 1959*e*. Cannon-bone scoops and daggers. *S. Afr. J. Sci*. 55 (3): 79–82.

———. 1959*f*. Osteodontokeratic ripping tools and pulp scoops for teething and edentulous australopithecines. *J. Dent. Ass. S. Afr*. 14 (5): 164–78.

———. 1959*g*. Further light on australopithecine humeral and femoral weapons. *Am. J. Phys. Anthrop*., n.s., 17 (2): 87–94.

———. 1960*a*. *Pithecanthropus* and *Australopithecus*. *Z. Morph. Anthrop*. 50 (3): 261–74.

———. 1960*b*. The bone tool-manufacturing ability of *Australopithecus prometheus*. *Am. Anthrop*. 62 (1): 134–43.

———. 1960*c*. Africa's place in the evolution of man. *Ann. Proc*., 1959–60, Associated Scientific and Technical Societies, South Africa, pp. 21–41.

———. 1960*d*. The place of antelope cannon-bones (or metapodials) in australopithecine economy. *Z. Wiss. Zool*. 68:1–15.

———. 1960*e*. The persistence of some tools and utensils found first in the Makapansgat grey breccia. *S. Afr. J. Sci*. 56 (3): 71–74.

———. 1961*a*. An australopithecine scoop made from a right australopithecine upper arm bone. *Nature* (London) 191:372–73.

———. 1961*b*. Further information about how *Australopithecus* made bone tools and utensils. *S. Afr. J. Sci*. 57 (5): 127–34.

———. 1962*a*. The continuity and originality of australopithecine osteodontokeratic culture. *Actes IV Congrès Panafrican de Prehistoire et de l'étude du Quaternaire*, pp. 27–41.

———. 1962*b*. Substitution of stone tools for bone tools at Makapansgat. *Nature* (London) 196:314–16.

———. 1962*c*. Stalactites as tool material for the australopithecines: A missing cultural link between skeletal and stone tool-making from the Makapansgat stalactitic cavern. *Illus. Lond. News*. 242:1052–55.

———. 1962*d*. Siegfried and *Australopithecus*. *Probe: J. Sci. Stud. Council Univ. Witwatersrand* 1:14–21.

———. 1962*e*. The gradual appraisal of *Australopithecus*. In *Evolution und Hominisation*, ed. G. Kurth, pp. 141–56. Stuttgart: Gustav Fischer Verlag.

———. 1962*f*. From cannon-bone scoops to skull bowls at Makapansgat. *Am. J. Phys. Anthrop*. 20 (3): 287–96.

———. 1964*a*. A brief review of the Makapansgat investigations 1925–1963. *Academia das Ciências de Lisboa, Memorias* 9:1–17.

———. 1964*b*. The Abbé Breuil and the osteodontokeratic culture. *Instituto de Prehistoria y Arqueologia, Monografias* 11:347–70.

———. 1964*c*. The Ecology of the South African man-apes. In *Ecological studies in Southern Africa*, ed. D. H. S. Davis, pp. 49–66. Den Haag: Junk Publishers.

———. 1965*a*. The unavoidable osteodontokeratic culture. In *Dr. D. N. Wadia Commemorative Volume*, pp. 231–54. Lucknow: Mining and Metallurgical Institute of India.

———. 1965*b*. Australopithecine cordage and thongs. In *Homenaje a Juan Comas en su 65 aniversario*, 2:43–61. Mexico.

———. 1965*c*. Pounding as a process and the producer of other artefacts. *S. Afr. Archaeol. Bull*. 20 (79): 141–47.

———. 1965*d*. Tree chopping with an elephant rib. *S. Afr. J. Sci*. 61 (11): 395.

Dart, R. A., and Craig, D. 1959. *Adventures with the missing link*. New York: Harper; London: Hamish Hamilton.

Dart, R. A., and Kitching, J. W. 1958. Bone tools at the Kalbank Middle Stone Age site and the Makapansgat australopithecine locality, Central Transvaal. Part 2. The osteodontokeratic contribution. *S. Afr. Archaeol. Bull*. 13 (51): 94–116.

Davis, D. H. S. 1955. Taxonomic report: Swartkrans fossil microfauna. Unpublished report.

———. 1958. Notes on the small mammals in the Kalahari Gemsbok National Park, with special reference to those preyed upon by barn owls. *Koedoe* 1:184–88.

———. 1959. The barn-owl's contribution to ecology and palaeoecology. *Ostrich*, suppl. 3:144–53.

———. 1962. Distribution patterns of southern African Muridae, with notes on some of their fossil antecedants. *Ann. Cape Prov. Mus*. 2:56–76.

———. 1971. Genera Tatera and Gerbillurus. In *The mammals of Africa: An identification manual*, ed. J. Meester and H. W. Setzer. Washington, D.C.: Smithsonian Institution Press.

———. 1974. The distribution of some small southern African mammals (Mammalia: Insectivora, Rodentia). *Ann. Transv. Mus*. 29:135–84.

Dawkins, W. B. 1874. *Cave hunting*. London.

———. 1877. On the mammalian fauna of the cave of Cresswell Crags. *Q. J. Geol. Soc. Lond*. 33:606–7.

Deacon, H. J. 1969. Melkhoutboom Cave, Alexandria District, Cape Province: A report on the 1967 investigation. *Ann. Cape Prov. Mus*. 6 (13): 141–69.

———. 1970. Plant remains from Melkhoutboom Cave, South Africa. *Proc. Transkei and Ciskei Res. Soc*. 1:13–15.

———. 1972. A review of the post-Pleistocene in South Africa. *S. Afr. Archaeol. Soc*., Goodwin ser. 1:26–45.

———. 1975. Demography, subsistence, and culture

during the Acheulean in southern Africa. In *After the australopithecines,* ed. K. W. Butzer and G. Ll. Isaac, pp. 543–69. The Hague: Mouton.

Deacon, J. 1965. Part 1. Cultural material from the Gamtoos valley shelters (Andrieskraal 1). *S. Afr. Arch. Bull.* 20 (80): 193–200.

———. 1972. Wilton: An assessment after fifty years. *S. Afr. Archaeol. Bull.* 27: 10–48.

Dean, W. R. J. 1973*a*. Analysis of a collection of barn owl *Tyto alba* pellets from Warmbaths, Transvaal. *Zool. Afr.* 8 (1): 75–82.

———. 1973*b*. Age distribution of *Praomys natalensis* prey in *Tyto alba* pellets. *Zool. Afr.* 8 (1): 140.

———. 1974. Analysis of *Tyto alba* pellets from Angola. *Zool. Afr.* 9 (1): 89–90.

———. 1975. *Tyto alba* prey in South West Africa and the northern Cape. *Zool. Afr.* 10:217–19.

Deane, N. N. 1962. The spotted hyaena, *Crocuta c. crocuta*. *Lammergeyer* 2:26–44.

de Graaff, G. 1957. A new chrysochlorid from Makapansgat. *Palaeont. Afr.* 5:21–27.

———. 1960*a*. A preliminary investigation of the mammalian microfauna in Pleistocene deposits of caves in the Transvaal System. *Palaeont. Afr.* 7:59–118.

———. 1960*b*. 'n Ontleding van uilklonte van die Nonnetjiesuil, *Tyto alba*. *Ostrich* 31 (1): 1–5.

———. 1961. On the fossil mammalian microfauna collected at Kromdraai by Draper in 1895. *S. Afr. J. Sci.* 56:259–60.

———. 1971. Family Bathyergidae. In *The mammals of Africa: An identification manual,* ed. J. Meester and H. W. Setzer. Washington, D.C.: Smithsonian Institution Press.

Delson, E. 1973. Fossil colobine makeup of the circum-Mediterranean region and the evolutionary history of the Cercopithecidae (Primates, Mammalia). Ph.D. diss., Columbia University.

———. 1975. Evolutionary history of the Cercopithecidae. *Contrib. Primatol.* 5:167–217.

Dietrich, W. O. 1942. Altestquartäre Säugetiere aus der südlichen Serengeti, Deutsch-Ostafrika. *Paläeontographica* 94:43–133.

Dorst, J., and Dandelot, P. 1972. *A field guide to the larger mammals of Africa*. London: Collins.

Draper, D. 1896. Report of meeting, 8th April, 1895, Geological Society of South Africa. *Trans. Geol. Soc. S. Afr.* 1 (1): 11.

———. 1898. Meeting of the Geological Society of South Africa, 12 July 1897. *Trans. Geol. Soc. S. Afr.* 3:63.

Duke, G. E.; Jegers, A. A.; Loff, G.; and Evanson, O. A. 1975. Gastric digestion in some raptors. *Comp. Biochem. Physiol.* 50A:649–56.

Eck, G. 1976. Cercopithecoidea from the Omo Group deposits. In *Earliest man and environments in the Lake Rudolf basin,* ed. Y. Coppens, F. C. Howell, G. Ll. Isaac, and R. E. F. Leakey, pp. 332–44. Chicago: University of Chicago Press.

Edwards, D. 1974. Survey to determine the adequacy of existing conserved areas in relation to vegetation types: A preliminary report. *Koedoe* 17:1–38.

Efremov, I. A. 1940. Taphonomy: A new branch of palaeontology. *Pan-Am. Geol.* 74:81–93.

Eisenberg, J. F. 1970. A splendid predator does its own thing untroubled by man. *Smithsonian* 1 (6): 48–53.

Eisenberg, J. F., and Lockhart, M. 1972. An ecological reconnaissance of Wilpattu National Park, Ceylon. *Smithsonian Contrib. Zool.,* no. 101, pp. 1–118.

Eisenhart, W. L. 1974. The fossil Cercopithecoids of Makapansgat and Sterkfontein. B.A. thesis, Dept. of Anthropology, Harvard College.

Eloff, F. C. 1964. On the predatory habits of lions and hyaenas. *Koedoe* 7:105–12.

———. 1975. The spotted hyaena *Crocuta crocuta* (Erxleben) in arid regions of southern Africa. *Pub. Univ. Pretoria,* n.s., 97:35–39.

Eloff, J. F. 1969. Bushman Rock Shelter, eastern Transvaal: Excavations, 1967–68. *S. Afr. Archaeol. Bull.* 24 (94): 60.

English Mechanic and the World of Science. 1897. No. 1692 (27 August), pp. 36.

Ennouchi, E. 1953. Un nouveau genre d'Ovicapriné dans un gisement pléistocène de Rabat. *C. R. Séanc. Soc. Géol. Fr.* 1953:126–28.

Estes, R. D. 1967. Predators and scavengers. *Nat. Hist. N.Y.* 76 (2): 20–29; 76 (3): 38–47.

Ewer, R. F. 1954*a*. Sabre-toothed tigers. *New Biology* 17:27–40.

———. 1954*b*. The Hyaenidae of Kromdraai. *Proc. Zool. Soc. Lond.* 124:565–85.

———. 1955*a*. Hyaenidae, other than *Lycyaena* of Swartkrans and Sterkfontein. *Proc. Zool. Soc. Lond.* 124:815–37.

———. 1955*b*. The Lycyaenas of Sterkfontein and Swartkrans, together with some general considerations of the Transvaal fossil hyaenids. *Proc. Zool. Soc. Lond.* 124:839–57.

———. 1955*c*. The fossil carnivores of the Transvaal caves: Machairodontinae. *Proc. Zool. Soc. Lond.* 125 (3–4): 587–615.

———. 1956*a*. The fossil carnivores of the Transvaal caves: Felinae. *Proc. Zool. Soc. Lond.* 126 (1): 83–95.

———. 1956*b*. The fossil carnivores of the Transvaal caves: Canidae. *Proc. Zool. Soc. Lond.* 126 (1): 97–119.

———. 1956*c*. The fossil carnivores of the Transvaal caves: Two new viverrids, together with some general considerations. *Proc. Zool. Soc. Lond.* 126:259–74.

———. 1956*d*. Some fossil carnivores from the Makapansgat valley. *Palaeont. Afr.* 4:57–67.

———. 1956*e*. The fossil suids of the Transvaal caves. *Proc. Zool. Soc. Lond.* 127:527–44.

———. 1958*a*. The fossil Suidae of Makapansgat. *Proc. Zool. Soc. Lond.* 130:329–72.

———. 1958*b*. Adaptive features in the skulls of African Suidae. *Proc. Zool. Soc. Lond.* 131 (1): 135–55.

———. 1965. Large Carnivora. In *Olduvai Gorge 1951–1961,* ed. L. S. B. Leakey. Cambridge: University Press.

———. 1967. The fossil hyaenids of Africa—A reappraisal. In *Background to evolution in Africa,* ed. W. W. Bishop and J. D. Clark. Chicago: University of Chicago Press.

———. 1973. *The carnivores*. The World Naturalist. London: Weidenfeld and Nicolson.

Exton, H. 1899. Presidential address, Geological Society of South Africa, 22d February, 1898. *Trans. Geol. Soc. S. Afr.* 4:6–10.

Frames, M. E. 1898. Remarks on Prof. Prister's paper on Glacial phenomena at Pretoria and on the Rand. *Trans. Geol. Soc. S. Afr.* 3:91–95.

Freedman, L. 1957. The fossil Cercopithecoidea of South Africa. *Ann. Transv. Mus.* 23:121–262.

———. 1965. Fossil and subfossil primates from the

limestone deposits at Taung, Bolt's Farm and Witkrans, South Africa. *Palaeont. Afr.* 9:19–48.

———. 1970. A new checklist of fossil Cercopithecoidea of South Africa. *Palaeont. Afr.* 13:109–10.

Freedman, L., and Brain, C. K. 1972. Fossil cercopithecoid remains from the Kromdraai australopithecine site (Mammalia: Primates). *Ann. Transv. Mus.* 28 (1): 1–16.

———. 1977. A re-examination of the cercopithecoid fossils from Swartkrans (Mammalia: Cercopithecidae). *Ann. Transv. Mus.* 30 (18): 211–18.

Freedman, L., and Stenhouse, N. S. 1972. The *Parapapio* species of Sterkfontein, Transvaal, South Africa. *Palaeont. Afr.* 14:93–111.

Freeman, L. G., and Butzer, K. W. 1966. The Acheulean station at Torralba (Spain): A progress report. *Quaternaria* 8:9–21.

Un frère Mariste de Johannesburg. 1898. Les grottes de Sterkfontein. *Cosmos* 679:133–35.

Frison, G. C., ed. 1974. *The Casper site.* New York: Academic Press.

Galton, F. 1889. *Narrative of an explorer in tropical South Africa, being an account of a visit to Damaraland in 1851.* London: Henry Frowde.

Gargett, V. 1965. Death of a black eagle. *Honeyguide* 50:9–10.

———. 1967. Black eagle experiment. *Bokmakarie* 19:88–90.

———. 1970*a*. Black eagle experiment, no. 2. *Bokmakarie* 22:32–35.

———. 1970*b*. Black eagle survey, Rhodes Matopos National Park: A population study 1964–1968. *Ostrich,* suppl. 8:397–414.

———. 1971. Some observations on the black eagles in the Matopos, Rhodesia. *Ostrich,* suppl. 9:91–124.

———. 1972*a*. Observations at a black eagle nest in the Matopos, Rhodesia. *Ostrich,* 43 (2): 77–108.

———. 1972*b*. Black eagle *Aquila verreauxi* population dynamics. *Ostrich* 43 (3): 177–78.

———. 1975. The spacing of black eagles in the Matopos, Rhodesia. *Ostrich* 46 (1): 1–44.

———. 1976. Dead or Alive? *Afr. Wild Life* 30 (2): 40–41.

———. 1977. A thirteen-year population study of the black eagles in the Matopos, Rhodesia, 1964–1976. *Ostrich* 48:17–27.

———. 1978. Mackinder's eagle owl, *Bubo capensis mackinderi,* in the Matopos, Rhodesia. In *Proceedings of the Symposium on African Predatory Birds,* pp. 46–61. Pretoria: Northern Transvaal Ornithological Society.

Gargett, V., and Grobler, J. H. 1976. Prey of the Cape eagle owl, *Bubo capensis mackinderi* Sharpe 1899, in the Matopos, Rhodesia. *Arnoldia* (Rhodesia) 8 (7): 1–7.

Gentry, A. W. 1965. New evidence on the systematic position of *Hippotragus niro* Hopwood 1936 (Mammalia). *Ann. Mag. Nat. Hist.* ser. 13, 8:335–38.

———. 1966. Fossil Antilopini of East Africa. *Bull. Brit. Mus. Nat. Hist. (Geol.)* 12:45–106.

———. 1970*a*. The Langebaanweg Bovidae. Appendix in A review of the geology and palaeontology of the Plio/Pleistocene deposits at Langebaanweg, Cape Province, ed. Q. B. Hendey. *Ann. S. Afr. Mus.* 56 (2): 114–17.

———. 1970*b*. Revised classification for *Makapania broomi* Wells and Cooke (Bovidae, Mammalia). *Palaeont. Afr.* 13:63–67.

———. 1971*a*. Genus *Gazella.* In *The mammals of Africa: An identification manual,* ed. J. Meester and H. W. Setzer. Washington D.C.: Smithsonian Institution Press.

———. 1971*b*. The earliest goats and other antelopes from the Samos *Hipparion* fauna. *Bull. Brit. Mus. Nat. Hist. (Geol.)* 20 (6): 231–96.

———. 1974. A new genus and species of Pliocene boselaphine (Bovidae, Mammalia) from South Africa. *Ann. S. Afr. Mus.* 65 (5): 145–88.

———. 1976. Bovidae of the Omo Group deposits. In *Earliest man and environments in the Lake Rudolf basin,* ed. Y. Coppens, F. C. Howell, G. Ll. Isaac, and R. E. F. Leakey, pp. 275–92. Chicago: University of Chicago Press.

Gingerich, P. D. 1974. *Proteles cristatus* Sparrman from the Pleistocene of South Africa, with a note on tooth replacement in the aardwolf (Mammalia: Hyaenidae). *Ann. Transv. Mus.* 29 (4): 49–54.

Glover, P. E.; Glover, E. C.; Trump, E. C.; and Wateridge, L. E. D. 1964. The lava caves of Mount Suswa, Kenya, with particular reference to their ecological role. *Studies in Speleology* 1 (1): 51–66.

Gow, C. E. 1973. Habitual sheltering in an extensive cave system by baboons near Bredasdorp, South Africa. *S. Afr. J. Sci.* 69:182.

Graham, C. 1953. Bold leopards. *Afr. Wild Life* 7 (3): 244–45.

Greenwood, M. 1955. Fossil Hystricoidea from the Makapan Valley, Transvaal. *Palaeont. Afr.* 3:77–85.

———. 1958. Fossil Hystricoidea from the Makapan Valley, Transvaal: *Hystrix makapanensis* nom. nov. for *Hystrix major* Greenwood. *Ann. Mag. Nat. Hist.,* ser. 13 (5): 365.

Grindley, J. R., and Nel, E. 1970. Red water and mussel poisoning at Elands Bay, December 1966. *Fisheries Bull. S. Afr.* 6:36–55.

Grindley, J. R.; Speed, E.; and Maggs, T. 1970. The age of the Bonteberg shelter deposits, Cape Peninsula. *S. Afr. Archaeol. Bull.* 25 (97): 24.

Guérin, G. 1928. *La vie des chouettes: Régime et croissance de l'effraye commune* Tyto alba alba *(L.) en vendée.* Paris: P. Lechevalier.

Hanney, P. 1962. Observations on the food of the barn owl *(Tyto alba)* in Southern Nyasaland. *Ann. Mag. Nat. Hist.,* ser. 13 (6): 305–13.

Harris, J. M. 1974. Orientation and variability in the ossicones of African Sivatheriinae (Mammalia: Giraffidae). *Ann. S. Afr. Mus.* 65 (6): 189–98.

———. 1976*a*. Bovidae from the East Rudolf succession. In *Earliest man and environments in the Lake Rudolf basin,* ed. Y. Coppens, F. C. Howell, G. Ll. Isaac, and R. E. F. Leakey, pp. 293–301. Chicago: University of Chicago Press.

———. 1976*b*. Pleistocene Giraffidae (Mammalia, Artiodactyla) from East Rudolf, Kenya. In *Fossil vertebrates of Africa,* ed. R. J. G. Savage and S. C. Coryndon, 4:283–332. London: Academic Press.

———. 1976*c*. Pliocene Giraffoidea (Mammalia, Artiodactyla) from the Cape Province. *Ann. S. Afr. Mus.* 69 (12): 325–53.

Haughton, S. H. 1922. A note on some fossils from the Vaal River gravels. *Trans. Geol. Soc. S. Afr.* 24:11–16.

———. 1932. The fossil Equidae of South Africa. *Ann. S. Afr. Mus.* 28:407–27.

Hemmer, H. 1965. Zur Nomenklatur und Verbreitung des Genus *Dinofelis* Zdansky, 1924 (*Therailurus* Piveteau,

1948). *Palaeont. Afr.* 9:75–89.
Hendey, Q. B. 1968. The Melkbos site: An upper Pleistocene fossil occurrence in the south-western Cape Province. *Ann. S. Afr. Mus.* 52:89–119.
———. 1973. Carnivore remains from the Kromdraai australopithecine site (Mammalia: Carnivora). *Ann. Transv. Mus.* 28:99–112.
———. 1974*a*. The Late Cenozoic Carnivora of the south-sestern Cape Province. *Ann. S. Afr. Mus.* 63:1–369.
———. 1974*b*. New fossil carnivores from the Swartkrans australopithecine site (Mammalia: carnivora). *Ann. Transv. Mus.* 29 (3): 27–48.
———. 1976. The Pliocene fossil occurrences in "E" Quarry, Langebaanweg, South Africa. *Ann. S. Afr. Mus.* 69 (9): 215–47.
Hendey, Q. B., and Hendey, H. 1968. New Quaternary fossil sites near Swartklip, Cape Province. *Ann. S. Afr. Mus.* 52:43–73.
Hendey, Q. B., and Singer, R. 1965. Part III. The faunal assemblages from the Gamtoos valley shelters. *S. Afr. Arch. Bull.* 20 (80): 206–13.
Hewitt, J. 1921. On several implements and ornaments from Strandloper sites in the Eastern Province. *S. Afr. J. Sci.* 18:454–67.
———. 1931. Artefacts from Melkhoutboom. *S. Afr. J. Sci.* 28: 540–48.
Hill, A. 1975. Taphonomy of contemporary and late Cenozoic East African vertebrates. Ph.D. diss. University of London.
———. 1978. Hyaenas, bones and fossil man. *Kenya, Past and Present* 9:9–14.
Hill, A., and Walker, A. 1972. Procedures in vertebrate taphonomy: Notes on a Uganda Miocene fossil locality. *J. Geol. Soc. Lond.* 128:399–406.
Hoernlé, A. 1923. South West Africa as a primitive cultural area. *S. Afr. Geogr. J.* 6:14–28.
Hoesch, W. 1936. *Ornithologische Monatsberichte* 44:6.
Hoffman, A. C. 1953. The fossil alcelaphines of South Africa—genera *Peloroceras, Lunatoceras* and *Alcelaphus. Navors. Nas. Mus. Bloemfontein,* 1 (3): 41–56.
Holloway, R. L. 1970*a*. Australopithecine endocast (Taung specimen, 1924): A new volume determination. *Science* 1968:966–68.
———. 1970*b*. New endocranial values for the australopithecines. *Nature* (London) 227:199–200.
———. 1972. New australopithecine endocast, SK 1585, from Swartkrans, South Africa. *Am. J. Phys. Anthrop.* 37 (2): 173–86.
Honer, M. R. 1963. Observations on the barn owl *(Tyto alba guttata)* in the Netherlands in relation to its ecological population fluctuations. *Ardea* 51:158–95.
Hooijer, D. A. 1975. Miocene to Pleistocene hipparions of Kenya, Tanzania and Ethiopia. *Zool. Verh.* 142:1–80.
———. 1976. The late Pliocene Equidae of Langebaanweg, Cape Province, South Africa. *Zool. Verh.* 148:1–39.
Hopwood, A. T. 1936. New and little-known fossil mammals from the Pleistocene of Kenya Colony and Tanganyika Territory. *Ann. Mag. Nat. Hist.*, ser. 10, 17:636–41.
Howell, F. C. 1961. Ismila: A palaeolithic site in Africa. *Sci. Am.* 205 (4): 118–29.
———. 1966. Observations on the earlier phases of the European Lower Palaeolithic. *Am. Anthrop.* 68 (2): 88–201.
Howell, F. C. 1978. Hominidae. In *Evolution of African mammals,* ed. V. J. Maglio and H. B. S. Cooke, pp. 154–248. Cambridge: Harvard University Press.
Howell, F. C.; Cole, G. H.; and Kleindienst, M. R. 1962. Isimila: An Acheulian occupation site in the Iringa Highlands, Southern Highlands Province, Tanganyika. *Ann. Mus. Roy. Afr. Cent.* 40:43–80.
Howell, F. C., and Coppens, Y. 1974. Inventory of remains of Hominidae from Pliocene/Pleistocene formations of the lower Omo basin, Ethiopia (1967–1972). *Am. J. Phys. Anthrop.* 40:1–16.
———. 1976. An overview of Hominidae from the Omo succession, Ethiopia. In *Earliest man and environments in the Lake Rudolf basin,* ed. Y. Coppens, F. C. Howell, G. Ll. Isaac, and R. E. F. Leakey, pp. 522–32. Chicago: University of Chicago Press.
Howell, F. C., and Petter, G. 1976. Carnivora from Omo Group formations, Southern Ethiopia. In *Earliest man and environments in the Lake Rudolf basin,* ed. Y. Coppens, F. C. Howell, G. Ll. Isaac, and R. E. F. Leakey, pp. 314–31. Chicago: University of Chicago Press.
Howell, F. C.; Fichter, L. S.; and Eck, G. 1969. Vertebrate assemblages from the Usno Formation, White Sands and Brown Sands localities, lower Omo basin, Ethiopia. *Quaternaria* 2:65–88.
Hughes, A. R. 1954*a*. Hyaenas versus australopithecines as agents of bone accumulations. *Am. J. Phys. Anthrop.*, n.s., 12 (4): 467–86.
———. 1954*b*. Habits of hyaenas. *S. Afr. J. Sci.* 51:156–58.
———. 1958. Some ancient and recent observations on hyaenas. *Koedoe* 1:105–14.
———. 1961. Further notes on the habits of hyaenas and bone gathering by porcupines. *Zool. Soc. S. Afr. News Bull.* 3 (1): 35–37.
Hughes, A. R., and Tobias, P. V. 1977. A fossil skull probably of the genus *Homo* from Sterkfontein, Transvaal. *Nature* (London) 265:310–12.
Ilani, G. 1975. Hyenas in Israel. *Israel—Land and Nature,* October 1975, pp. 10–18. From *Teva Va'aretz* 16 (4).
Isaac, G. Ll. 1967*a*. The stratigraphy of the Peninj Group: Early Middle Pleistocene formations west of Lake Natron, Tanzania. In *Background to evolution in Africa,* ed. W. W. Bishop and J. D. Clark, pp. 229–58. Chicago: University of Chicago Press.
———. 1967*b*. Towards the interpretation of occupational debris: Some experiments and observations. *Kroeber Anthrop. Soc. Papers* 37:31–55.
———. 1968. Traces of Pleistocene hunters: An East African example. In *Man the hunter,* ed. R. B. Lee and I. De Vore., pp. 253–61. Chicago: Aldine.
———. 1969. Studies of early culture in East Africa. *World Archaeology* 1:1–28.
———. 1971. The diet of early man: Aspects of archaeological evidence from Lower and Middle Pleistocene sites in Africa. *World Archaeology* 2 (3): 278–99.
———. 1976. The activities of early African hominids: A review of the archaeological evidence for the time span two and a half to one million years ago. In *Human origins: Louis Leakey and the East African evidence,* ed. G. Ll. Isaac and E. R. McCown, pp. 482–514.

Menlo Park, Calif.: Staples Press.

Isaac, G. Ll.; Leakey, R. E. F.; and Behrensmeyer, A. K. 1971. Archaeological traces of early hominid activities, east of Lake Rudolf, Kenya. *Science* 173:1129–34.

Jackson, H. D. 1973*a*. The Cape eagle owl, *Bubo capensis* in Mozambique. *Bull. Brit. Orn. Club* 93:10.

———. 1973*b*. Records of some birds and mammals in the central Chimanimani Mountains of Mozambique and Rhodesia. *Durban Mus. Novit.* 9 (20): 291–305.

Jenkins, T., and Brain, C. K. 1967. The peoples of the lower Kuiseb valley, South West Africa. *Sci. Papers Namib Desert Res. Station* 35:1–24.

Johanson, D. C. 1976. Ethiopia yields first "family" of early man. *Natl. Geogr. Mag.* 150 (6): 791–811.

Johanson, D. C., and Taieb, M. 1976. Plio-Pleistocene hominid discoveries in Hadar, Ethiopia. *Nature* (London) 260:293–97.

Johanson, D. C., and White, T. D. 1979. A systematic assessment of early African hominids. *Science* 203 (4378): 321–30.

Johanson, D. C.; White, T. D.; and Coppens, Y. 1978. A new species of the genus *Australopithecus* (Primates: Hominidae) from the Pliocene of Eastern Africa. *Kirtlandia* 28:1–14.

Jolly, C. J. 1966. The evolution of the baboons. In *The baboon in medical research,* ed. H. Vogtborg, 2:23–50. Austin: University of Texas Press.

———. 1970. The large African monkeys as an adaptive array. In *Old World monkeys,* ed. J. Napier and P. Napier, pp. 139–74. London: Academic Press.

———. 1972. The classification and natural history of *Theropithecus (Simopithecus)* (Andrews), 1916 baboons of the African Pleistocene. *Bull. Brit. Mus. Nat. Hist. (Geol.)* 22 (1): 1–123.

Jones, T. R. 1937. A new fossil primate from Sterkfontein, Krugersdorp, Transvaal. *S. Afr. J. Sci.* 33:709–28.

———. 1978. *The skull and mandible of the South African baboon: A morphological study.* Johannesburg: Witwatersrand University Press.

Kirby, P. R. 1940. Robert Knox and his South African research. *S. Afr. Med. J.* 14:13.

Kitching, J. W. 1953. A new species of fossil baboon from Potgietersrust. *S. Afr. J. Sci.* 50:66–69.

———. 1963. *Bone, tooth and horn tools of Palaeolithic man.* Manchester: Manchester University Press.

———. 1965. A new giant hyracoid from the Limeworks Quarry, Makapansgat, Potgietersrust. *Palaeont. Afr.* 9:91–96.

Klein, R. G. 1972*a*. Preliminary report on the July through September 1970 excavation at Nelson Bay Cave, Plettenberg Bay (Cape Prov., South Africa). In *Palaeoecology of Africa, 1969–71,* ed. E. M. van Zinderen Bakker, pp. 177–208. Cape Town: Balkema.

———. 1972*b*. The late Quaternary mammalian fauna of Nelson Bay Cave (Cape Province, South Africa): Its implications for megafaunal extinctions and environmental and culture change. *Quaternary Res.* 2 (2): 135–42.

———. 1974*a*. A provisional statement on terminal Pleistocene mammalian extinctions in the Cape biotic zone (Southern Cape Province, South Africa). *S. Afr. Arch. Soc.,* Goodwin ser., no. 2, pp. 39–45.

———. 1974*b*. Environment and subsistence of prehistoric man in the southern Cape Province, South Africa. *World Archaeol.* 5 (3): 249–84.

———. 1975*a*. Palaeoanthropological implications of the nonarchaeological bone assemblage from Swartklip 1, south-western Cape Province, South Africa. *Quaternary Res.* 5:275–88.

———. 1975*b*. Middle Stone Age man–animal relationships in southern Africa: Evidence from Die Kelders and Klasies River Mouth. *Science* 190:265–67.

———. 1975*c*. Ecology of Stone Age man at the southern tip of Africa. *Archaeology* 28 (4): 238–47.

———. 1976*a*. The mammalian fauna of the Klasies River Mouth sites, southern Cape Province, South Africa. *S. Afr. Archaeol. Bull.* 31:75–98.

———. 1976*b*. A preliminary report on the "Middle Stone Age" open-air site of Duinefontein 2 (Melkbosstrand, south-western Cape Province, South Africa). *S. Afr. Archaeol. Bull.* 31:12–20.

———. 1976*c*. The fossil history of *Raphicerus* H. Smith, 1827 (Bovidae, Mammalia) in the Cape biotic zone. *Ann. S. Afr. Mus.* 71:169–91.

———. 1977. The ecology of early man in southern Africa. *Science* 197:115–26.

———. N.d. Results of the preliminary analysis of the mammalian fauna from the Redcliff Stone Age site, Rhodesia. Typescript.

Klein, R. G., and Scott, K. 1974. The fauna of Scott's Cave, Gamtoos valley, south-eastern Cape Province. *S. Afr. J. Sci.* 70:186–87.

Knox, R. 1822. Notice relative to the habits of the hyaenas in southern Africa. *Trans. Wernerian Nat. Hist. Soc., Edinburgh* 4:383.

Kolbe, F. F. 1946. The case for the barn owl. *Afr. Wild Life* 1:69–73.

Kruuk, H. 1966. Clan system and feeding habits of spotted hyaenas (*Crocuta crocuta* Erxleben). *Nature* (London) 209:1257–58.

———. 1968. Hyaenas: The hunters nobody knows. *Natl. Geogr. Mag.* 134:44–57.

———. 1972. *The spotted hyaena: A study of predation and social behavior.* Chicago: University of Chicago Press.

———. 1976. Feeding and social behaviour of the striped hyaena (*Hyaena vulgaris* Desmarest). *E. Afr. Wildl. J.* 14:91–111.

Kruuk, H., and Turner, M. 1967. Comparative notes on predation by lion, leopard, cheetah and wild dog in the Serengeti area, E. Africa. *Mammalia '67* 31:1–27.

Kuhn, H. J. 1967. Zur Systematik des Cercopithecidae. In *Progress in primatology,* ed. D. Starck, R. Schneider, and H. J. Kuhn, pp. 25–46. Stuttgart: Fisher.

Kurtén, B. 1956. The status and affinities of *Hyaena sinensis* Owen and *Hyaena ultima* Matsumoto. *Amer. Mus. Novit.,* no. 1764, pp. 1–48.

———. 1957*a*. The bears and hyaenas of the interglacials. *Quaternaria* 4:1–13.

———. 1957*b*. Mammal migrations, Cenozoic stratigraphy and the age of Peking man and the australopithecines. *J. Palaeont.* 31:215–27.

———. 1968. Pleistocene mammals of Europe. London: Weidenfeld and Nicolson.

Laubscher, N. F.; Steffens, F. E.; and Vrba, E. S. 1972. Statistical evaluation of the taxonomic status of a fossil member of the Bovidae (Mammalia: Artiodactyla). *Ann. Transv. Mus.* 28 (2): 17–26.

Lavocat, R. 1956. La faune de rongeurs des grottes à

australopithèques. *Palaeont. Afr.* 4:69–75.

Leakey, L. S. B. 1943. New fossil Suidae from Shungura, Omo. *J. East Afr. Nat. Hist. Soc.* 17:45–61.

———. 1965. *Olduvai Gorge, 1951–61*. Cambridge: Cambridge University Press.

Leakey, L. S. B., and Leakey, M. D. 1964. Recent discoveries of fossil hominids in Tanganyika: At Olduvai and near Lake Natron. *Nature* (London) 202:5–7.

Leakey, L. S. B.; Tobias, P. V.; and Napier, J. R. 1964. A new species of the genus *Homo* from Olduvai Gorge. *Nature* (London) 202:7–9.

Leakey, M. D. 1970. Stone artefacts from Swartkrans. *Nature* (London) 225:1222–25.

———. 1971. *Olduvai Gorge*. Vol. 3. *Excavations in Beds I and II, 1960–1963*. Cambridge: Cambridge University Press.

Leakey, M. G. 1976. Carnivora of the East Rudolf succession. In *Earliest man and environments in the Lake Rudolf basin,* ed. Y. Coppens, F. C. Howell, G. Ll. Isaac, and R. E. F. Leakey, pp. 302–13. Chicago: University of Chicago Press.

Leakey, R. E. F. 1973. Further evidence of Lower Pleistocene hominids from East Rudolf, North Kenya, 1972. *Nature* (London) 242:170–73.

———. 1974. Further evidence of Lower Pleistocene hominids from East Rudolf, North Kenya, 1973. *Nature* (London) 248:653–56.

———. 1976*a*. Hominids in Africa. *Am. Scient.* 64 (2): 174–78.

———. 1976*b*. An overview of the Hominidae from East Rudolf, Kenya. In *Earliest man and environments in the Lake Rudolf basin,* ed. Y. Coppens, F. C. Howell, G. Ll. Isaac, and R. E. F. Leakey, pp. 476–83. Chicago: University of Chicago Press.

Leakey, R. E. F., and Walker, A. C. 1976. *Australopithecus, Homo erectus* and the single species hypothesis. *Nature* (London) 261:572–74.

Lee, R. B. 1968. What hunters do for a living; or, How to make out on scarce resources. In *Man the hunter,* ed. R. B. Lee and I. Devore, pp. 30–48. Chicago: Aldine.

———. 1972. The !Kung bushmen of Botswana. In *Hunters and gatherers today,* ed. M. G. Bicchieri, pp. 327–68. New York: Holt, Rinehart and Winston.

Lee, R. B., and Devore, I. 1968. Problems in the study of hunters and gatherers. In *Man the hunter,* ed. R. B. Lee and I. Devore, pp. 3–12. Chicago: Aldine.

Leechman, D. 1951. Bone grease. *Am. Antiq.* 16 (4): 355–56.

Louw, A. W. 1969. Bushman Rock Shelter, Ohrigstad, eastern Transvaal: A preliminary investigation, 1965. *S. Afr. Archaeol. Bull.* 24 (94): 39–51.

Lubbock, J. 1865. *Prehistoric times*. London: Williams and Norgate.

MacArthur, R. H., and Pianka, E. R. 1966. An optimal use of a patchy environment. *Am. Nat.* 100 (916):603–9.

MacCalman, H. R. 1967. The zoo park elephant site, Windhoek. In *Palaeoecology of Africa, 1964–65,* ed. E. M. van Zinderen Bakker, pp. 102–3. Cape Town: Balkema.

McHenry, H. 1975*a*. Fossil hominid body weight and brain size. *Nature* (London) 254:686–88.

———. 1975*b*. Fossils and the mosaic nature of human evolution. *Science* 190:425–31.

McLachlan, G. R., and Liversidge, R. 1970. *Roberts birds of South Africa*. Cape Town: Central News Agency.

Maggs, T. M. O'C. 1971. Some observations on the size of human groups during the Late Stone Age. In *Rock paintings of South Africa,* ed. M. Schoonraad, pp. 49–53. *S. Afr. J. Sci.,* special issue no. 2.

Maggs, T., and Speed, E. 1967. Bonteberg Shelter. *S. Afr. Archaeol. Bull.* 22:80–93.

Maglio, V. J. 1970. Early Elephantidae of Africa and a tentative correlation of African Plio-Pleistocene deposits. *Nature* (London) 225:328–32.

———. 1973. Origin and evolution of the Elephantidae. *Trans. Amer. Phil. Soc.* 63:1–149.

Maguire, J. 1976. A taxonomic and ecological study of the living and fossil Hystricidae with particular reference to southern Africa. Ph.D. diss., Department of Geology, University of the Witwatersrand.

Maier, W. 1970*a*. Neue Ergebnisse der Systematik und der Stammesgeschichte der Cercopithecoidea. *Z. Säugetierk.* 35:193–214.

———. 1970*b*. New fossil Cercopithecoidea from the Lower Pleistocene cave deposits of the Makapansgat Limeworks, South Africa. *Palaeont. Afr.* 3:69–107.

———. 1973. Palaookologie und zeitliche Einordnung der sudafrikanischen Australopithecinen. *Z. Morph. Anthrop.* 65:70–105.

Malan, B. D. 1959. Early references to the Sterkfontein caves. *S. Afr. J. Sci.* 55: 321–24.

———. 1960. A history of Sterkfontein with a comment by B. D. Malan. *S. Afr. J. Sci.* 56 (5): 110.

Mann, A. E. 1968. The palaeodemography of *Australopithecus*. Ph.D. diss., University of California, Berkeley.

———. 1975. *Paleodemographic aspects of the South African australopithecines*. University of Pennsylvania Publications in Anthropology, no. 1. Philadelphia: University of Pennsylvania.

Martyn, J., and Tobias, P. V. 1967. Pleistocene deposits and new fossil localities in Kenya. *Nature* (London) 215 (5100): 476–80.

Mason, R. J. 1958. Bone tools at the Kalkbank Middle Stone Age site and the Makapansgat australopithecine locality, central Transvaal. Part 1. The Kalkbank site. *S. Afr. Archaeol. Bull.* 13 (51): 94–116.

———. 1962*a*. *Prehistory of the Transvaal: A record of human activity*. Johannesburg: Witwatersrand University Press.

———. 1962*b*. The Sterkfontein stone artefacts and their maker. *S. Afr. Archaeol. Bull.* 17 (66): 109–25.

———. 1964. Iron Age bone artefacts. *S. Afr. Archaeol. Bull.* 19 (74): 38.

———. 1969. Tentative interpretations of new radiocarbon dates for stone artefact assemblages from Rose Cottage Cave, O.F.S. and Bushman Rock Shelter, Tvl. *S. Afr. Archaeol. Bull.* 24 (94): 57–59.

Mason, R. J.; Friede, H.; and Pienaar, J. N. 1974. Kruger Cave. *S. Afr. J. Sci.* 70:375.

Matthews, L. H. 1939. The bionomics of the spotted hyaena, *Crocuta crocuta* Erxl. *Proc. Zool. Soc. Lond.,* ser. A, 109:43–56.

Mayhew, D. F. 1977. Avian predators as accumulators of fossil mammal material. *Boreas* 6:25–31.

Meester, J. 1955. Fossil shrews of South Africa. *Ann. Transv. Mus.* 22:271–78.

———. 1958. Variation in the shrew genus *Myosorex* in southern Africa. *J. Mammal.* 39:325–39.

———. 1963. *A systematic revision of the shrew genus* Crocidura *in southern Africa. Transvaal Museum*

Memoirs, no. 13.

———. 1971. Family Chrysochloridae. In *The mammals of Africa: An identification manual,* ed. J. Meester and H. W. Setzer. Washington, D.C.: Smithsonian Institution Press.

Meester, J.; Davis, D. H. S.; and Coetzee, C. F. 1964. *An interim classification of southern African mammals: Cyclostyled.* King Williams Town: Zoological Society of South Africa and South African Council for Scientific and Industrial Research.

Meester, J., and Dippenaar, N. J. 1978. A new species of *Myosorex* from Knysna. *Ann. Transv. Mus.* 31 (4): 29–42.

Meester,J.; Davis, D. H. S.; and Coetzee, C. G. 1964. *An* malia: Soricidae) from Southern Africa. *Ann. Transv. Mus.* 27:269–77.

Meyer, G. E. 1978. Hyracoidea. In *Evolution of African mammals,* ed. V. J. Maglio and H. B. S. Cooke, pp. 284–314. Cambridge: Harvard University Press.

Mills, M. G. L. 1973. The brown hyaena. *Afr. Wildl.* 27 (4):150–53.

———. 1974. Brown hyaena in the Kalahari Gemsbok National Park: Ecology. *Wld. Wildl. Yb., 1973–4,* pp. 140–48.

———. 1976. Ecology and behaviour of the brown hyaena in the Kalahari, with some suggestions for management. *Proceedings of the Symposium on the Endangered Wildlife Trust,* Pretoria, July 1976, pp. 36–42.

Mills, M. G. L., and Mills, M. E. J. 1977. An analysis of bones collected at hyaena breeding dens in the Gemsbok National Parks (Mammalia: Carnivora. *Ann. Transv. Mus.* 30 (14): 145–55.

Misonne, X. 1969. African and Indo-Australian Muridae: Evolutionary trends. *Ann. Mus. Afr. Cent.,* ser. 8, 172:1–129.

———. 1971. Order Rodentia. In *The mammals of Africa: An identification manual,* ed. J Meester and H. W. Setzer, Part 6. Washington, D.C.: Smithsonian Institution.

Mitchell, B. L.; Shenton, J. B.; and Uys, J. C. M. 1965. Predation on large mammals in the Kafue National Park, Zambia. *Zoologica Africana* 1 (1): 297–318.

Mogg, A. O. D., 1975. Important plants of Sterkfontein: An illustrated guide. Johannesburg: University of the Witwatersrand.

Mohr, E. 1967. *Der Blaubock,* Hippotragus leucophaeus *(Pallas, 1766): Eine Dokumentation. Mammalia Depicta.* Hamburg: Paul Parey.

Mollett, O. 1947. Fossil mammals from the Makapan Valley, Potgietersrust. I. Primates. *S. Afr. J. Sci.* 43:295–303.

Montgomery, T. H. 1899. Observations on owls, with particular reference to their feeding habits. *Am. Nat.* 33:563–72.

Mouritz, L. B. 1915. Notes on the ornithology of the Matopos district, Southern Rhodesia. *Ibis,* 10th ser. 3 (2): 185–216.

Mundy, P. J., and Ledger, J. A. 1976. Griffon vultures, carnivores and bones. *S. Afr. J. Sci.* 72:106–10.

Napier, J., and Napier, P. 1967. *A handbook of living primates.* London: Academic Press.

Nel, J. A. J. 1969. The prey of owls in the Namib Desert. 1. The spotted eagle owl, *Bubo africanus,* at Sossus Vlei. *Sci. Pap. Namib Desert Res. Sta.* 4 (37–53):55–58.

Nel, J. A. J., and Nolte, H. 1965. Notes on the prey of owls in the Kalahari Gemsbok National Park. *Koedoe* 8:75–81.

Noddle, B. 1974. Ages of epiphyseal closure in feral and domestic goats and ages of dental eruption. *J. Archaeol. Sci.* 1:195–204.

Oakley, K. P. 1954*a*. The dating of the Australopithecinae of Africa. *Am J. Phys. Anthrop.,* n.s., 12 (1): 9–28.

———. 1954*b*. Dawn of man in South Africa. *Proc. Roy. Inst. Gt. Brit.* 35 (3):641–56.

———. 1956. The earliest fire-makers. *Antiquity* 30:102–7.

———. 1960. The history of Sterkfontein with a comment by B. D. Malan. *S. Afr. J. Sci.* 56:110.

Olson, T. R. 1974. Taxonomy of the Taung skull. *Nature* (London) 252:85–86.

Pager, H. 1971. The rock art of the Ndedema Gorge and neighbouring valleys, Natal Drakensberg. In *Rock paintings of South Africa,* ed. M. Schoonraad, pp. 27–33. *S. Afr. J. Sci.,* Special issue no. 2.

Parkington, J. E. 1972. Seasonal mobility in the Late Stone Age. *Afr. Stud.* 31 (4): 223–43.

———. 1976. Coastal settlement between the mouths of the Berg and Olifants rivers, Cape Province. *S. Afr. Archaeol. Bull.* 31:127–40.

Parkington, J. E., and Poggenpoel, C. 1971. Excavations at De Hangen, 1968. *S. Afr. Archaeol. Bull.* 26:3–36.

Partridge, T. C. 1973. Geomorphological dating of cave opening at Makapansgat, Sterkfontein, Swartkrans and Taung. *Nature* (London) 246:75–79.

———. 1975. Stratigraphic, geomorphological and palaeoenvironmental studies of the Makapansgat Limeworks and Sterkfontein hominid sites: A progress report on research carried out between 1965 and 1975. Manuscript.

Partridge, T. C., aand Talma, A. S. N.d. Carbon and oxygen isotope measurements at Makapansgat and Sterkfontein, and their palaeoenvironmental significance. In Press.

Patterson, B. 1968. The extinct baboon *Parapapio jonesi* in the early Pleistocene of north western Kenya. *Breviora* 282:1–4.

Patterson, B., and Howells, W. W. 1967. Hominid humeral fragment from early Pleistocene of Northwestern Kenya. *Science* 156 (3771): 64–66.

Patterson, B.; Behrensmeyer, A. K.; and Sill, W. D. 1970. Geology and fauna of a new Pliocene locality in north western Kenya. *Nature* (London) 226:918–21.

Peabody, F. E. 1954. Travertines and cave deposits of the Kaap Escarpment of South Africa, and the type locality of *Australopithecus africanus* Dart. *Bull. Geol. Soc. Am.* 65:671–706.

Pearson, K. 1914. *The life, letters and labours of Francis Galton.* Vol. 1. Cambridge: Cambridge University Press.

Percival, A. B. 1924. *A game ranger's note book.* London: Nisbet.

Peringuey, L. 1911. The Stone Ages of South Africa as represented in the collection of the South African Museum. *Ann. S. Afr. Mus.* 8:1–218.

Perkins, D. 1973. A critique on the methods of quantifying faunal remains from archaeological sites. In *Domestikationsforschung und Geschichte der Haustiere,* ed. J. Matolcsi, pp. 367–69. Budapest: Akadémiai Kiadó.

Perkins, D., and Daly, P. 1968. A hunters' village in

Neolithic Turkey. *Sci. Am.* 219 (5): 96–106.

Petter, F. 1971*a*. Order Lagomorpha. In *The mammals of Africa: An identification manual,* ed. J. Meester and H. W. Setzer, part 5. Washington, D.C.: Smithsonian Institution Press.

———. 1971*b*. Subfamily Gerbillinae. In *The mammals of Africa: An identification manual,* ed. J. Meester and H. W. Setzer, part 6.4. Washington, D.C.: Smithsonian Institution Press.

Petter, G. 1973. Carnivores Pleistocènes du Ravin d'Olduvai (Tanzanie). In *Fossil vertebrates of Africa,* ed. L. S. B. Leakey, R. J. Savage, and S. C. Coryndon, 3:43–100. London: Academic Press.

Pienaar, U. de V. 1969. Predator-prey relationships amongst the larger mammals in the Kruger National Park. *Koedoe* 12:108–76.

Plug, I. 1978. Collecting patterns of six species of vultures. *Ann. Transv. Mus.* 31 (6): 51–63.

Pocock, R. I. 1932. The leopards of Africa. *Proc. Zool. Soc. Lond.* 1:543–91.

———. 1941. *The fauna of British India, including Ceylon and Burma.* Vol. 2. *Mammalia.* London: Taylor and Francis.

Pocock, T. N. 1969. Micro-fauna provisionally identified from sieved material recovered thus far from Dump 8 at Sterkfontein, 1967–1969. Appendix. *S. Afr. Arch. Bull.* 24:168–69.

———. 1970. Pleistocene bird fossils from Kromdraai and Sterkfontein. *Ostrich,* suppl. 8, pp. 1–6.

———. 1976. Pliocene mammalian microfauna from Langebaanweg: A new fossil genus linking the Otomyinae with the Murinae. *S. Afr. J. Sci.* 72:58–60.

Prister, A. 1898. Preliminary notes on glacial phenomena at Pretoria and on the Rand. *Trans. Geol. Soc. S. Afr.* 3:70–82.

Raczynski, J., and Ruprecht, A. L. 1974. The effect of digestion on the osteological composition of owl pellets. *Acta Orn., Warsz.* 14 (2): 1–13.

Rau, C. 1876. *Early man in Europe.* New York: Harper Bros.

Rau, R. E. 1974. Revised list of the preserved material of the extinct Cape Colony quagga, *Equus quagga quagga* (Gmelin). *Ann. S. Afr. Mus.* 65 (2): 41–87.

Reed, C. I., and Reed, B. P. 1928. The mechanism of pellet formation in the great horned owl *(Bubo virginiatus). Science* 68:359–60.

Repenning, C. A. 1965. An extinct shrew from the early Pleistocene of South Africa. *J. Mammal.* 46 (2): 189–96.

Roberts, A. 1951. *The mammals of South Africa.* Johannesburg: Central News Agency, South Africa.

Robinson, J. T. 1953*a*. Meganthropus, australopithecines and hominids. *Am. J. Phys. Anthrop.,* n.s., 11 (1): 1–38.

———. 1953*b*. *Telanthropus* and its phylogenetic significance. *Am. J. Phys. Anthrop.,* n.s., 11:445–502.

———. 1954. The genera and species of the Australopithecinae. *Am. J. Phys. Anthrop.,* n.s., 12:181–200.

———. 1956. *The dentition of the Australopithecinae.* Transvaal Museum Memoirs, no. 9.

———. 1957. Occurrence of stone artefacts with *Australopithecus* at Sterkfontein. *Nature* (London) 180:521–24.

———. 1958. The Sterkfontein tool-maker. *Leech* 28:94–100.

———. 1959. A bone implement from Sterkfontein. *Nature* (London) 184:583–85.

———. 1961. The australopithecines and their bearing on the origin of man and of stone tool-making. *S. Afr. J. Sci.* 57:3–13.

———. 1962. Sterkfontein stratigraphy and the significance of the extension site. *S. Afr. Archaeol. Bull.* 17:87–107.

———. 1963. Adaptive radiation in the australopithecines and the origin of man. In *African ecology and human evolution,* ed. F. C. Howell and F. Bourliere, pp. 385–416. Viking Fund Publications in Anthropology, no. 36. New York: Wenner-Gren Foundation.

———. 1966. The distinctiveness of *Homo habilis. Nature* (London) 209:957–60.

———. 1972. *Early hominid posture and locomotion.* Chicago: University of Chicago Press.

Robinson, J. T., and Mason, R. J. 1957. Occurrence of stone artefacts with *Australopithecus* at Sterkfontein. *Nature* (London) 180:521–24.

Rowe, E. G. 1947. The breeding biology of *Aquila verreauxi* Lesson. Part 1. *Ibis* 89:387–410; part 2, *Ibis* 89:576–606.

Ryan, M. 1961. Man-eating leopard. *Afr. Wild Life* 15 (1): 67–71.

Sale, J. B. 1965. The feeding behaviour of rock hyraxes. (genera *Procavia* and *Heterohyrax*) in Kenya. *E. Afr. Wildl. J.* 3:1–18.

———. 1966. Daily food consumption and mode of ingestion in the hyrax. *J. E. Afr. Nat. Hist. Soc.* 25 (3): 214–24.

Sandelowsky, B. H. 1974. Archaeological investigations at Mirabib Hill Rock Shelter. *S. Afr. Archaeol. Soc.,* Goodwin ser. 2:65–72.

Schaller, G. B. 1967. *The deer and the tiger: A study of wildlife in India.* Chicago: University of Chicago Press.

———. 1972. *The Serengeti lion: A study in predator-prey relations.* Chicago: University of Chicago Press.

Schütt, G. 1974. Die Carnivoren von Würzburg–Schalksberg: Mit einem Beitrag zur biostratigraphischen und zoogeographischen Stellung der altpleistozänen Wirbeltier-faunen vom Mittelmain (unterfranken). *N. Jb. Geol. Paläont. Abh.* 147 (1): 61–90.

Selous, F. C. 1908. *African nature notes and reminiscences.* London: Macmillan.

Sessions, P. H. B. 1966. Notes on the birds of Lengetia farm, Mau Narok. *J. E. Afr. Nat. Hist. Soc.* 26 (1): 18–48.

———. 1972. Observations on Mackinder's eagle owl *Bubo capensis mackinderi* Sharpe on a Kenya farm. *J. E. Afr. Nat. Hist. Soc., Nat. Mus.* 138:1–19.

Shackleton, N. J. 1973. Oxygen isotope analysis as a means of determining season of occupation of prehistoric midden sites. *Archaeometry* 15:133–41.

Shapiro, M. M. J. 1943. Fossil mammalian remains from Bankies, Kroonstad district, O.F.S. *S. Afr. J. Sci.* 39:176–81.

Shaw, E. M.; Woolley, P. L.; and Rae, F. A. 1963. Bushman arrow poisons. *Cimbebasia* 7:2–41.

Shaw, J. C. M. 1937. Evidence concerning a large fossil hyrax. *J. Dent. Res.* no. 1 (16): 37.

———. 1938. The teeth of the South African fossil pig *(Notochoerus capensis* syn. *meadowsi)* and their geological significance. *Trans. R. Soc. S. Afr.* 26:25–37.

Shaw, J. C. Middleton, and Cooke, H. B. S. 1941. New fossil pig remains from the Vaal River gravels. *Trans. R. Soc. S. Afr.* 28:293–99.

Shipman, P., and Phillips, J. E. 1976. On scavenging by hominids and other carnivores. *Current Anthrop.* 17 (1): 170–72.

Shortridge, E. C. 1934. *The mammals of south west Africa.* London: William Heineman.

Siegfried, W. R. 1965. On the food habits of the spotted eagle owl. *Ostrich* 36:146.

———. 1968. Breeding season, clutch and brood sizes in Verreaux's eagle. *Ostrich* 39:139–45.

Silberbauer, G. B. 1972. The G/wi Bushmen. In *Hunters and gatherers today,* ed. M. G. Bicchieri. New York: Holt, Rinehart and Winston.

Silver, I. A. 1969. The ageing of domestic animals. In *Science in Archaeology,* ed. D. Brothwell and E. Higgs, pp. 283–302. Leipzig: Thames and Hudson.

Simons, E. 1972. *Primate evolution: An introduction to man's place in nature.* New York: Macmillan.

Simons, E. L., and Delson, E. 1978. Cercopithecidae and Parapithecidae. In *Evolution of African mammals,* ed. V. J. Maglio and H. B. S. Cooke, pp. 100–119. Cambridge: Harvard University Press.

Simons, J. W. 1966. The presence of leopard and a study of the food debris in the leopard lairs of the Mount Suswa Caves, Kenya. *Bull. Cave Exploration Group E. Afr.* 1:51–69.

Singer, R., and Boné, E. L. 1960. Modern giraffes and the fossil giraffids of Africa. *Ann. S. Afr. Mus.* 45 (4): 375–548.

———. 1966. Hipparion in Africa. *Quaternaria* 8:187–91.

Skead, C. J. 1956. Prey of the barn owl. *Ostrich* 27 (3):148.

———. 1963. Contents of pellets of barn owl, *Tyto alba,* at a rural roost. *Ostrich* 34 (3): 171–72.

Skinner, J. D. 1976. Ecology of the brown hyaena, *Hyaena brunnea* in the Transvaal, with a distribution map for southern Africa. *S. Afr. J. Sci.* 72:262–69.

———. N.d. The striped *Hyaena hyaena* of the Judean and Negev deserts and a comparison with the brown hyaena, *H. brunnea. Israel J. Zool.* In press.

Smart, C. 1976. The Lothagam 1 fauna: Its phylogenetic, ecological, and biogeographic significance. In *Earliest man and environments in the Lake Rudolf basin,* ed. Y. Coppens, F. C. Howell, G. Ll. Isaac, and R. E. F. Leakey, pp. 361–69. Chicago: University of Chicago Press.

Smith, R. M. 1977. Movement patterns and feeding behaviour of leopard in the Rhodes Matopos National Park, Rhodesia. *Arnoldia* (Rhodesia) 8 (13): 1–16.

Smithers, R. H. N. 1971*a*. *The mammals of Botswana.* National Museum of Rhodesia Memoirs, no. 4.

———. 1971*b*. Family Felidae. In *The mammals of Africa: An identification manual,* ed. J. Meester and H. W. Setzer, part 8:1, pp. 1–10. Washington, D.C.: Smithsonian Institution Press.

Southern, H. N. 1954. Tawny owls and their prey. *Ibis* 96:348–410.

Speth, J. D., and Davis, D. D. 1975. Seasonal variability in early hominid predation. Paper presented at the symposium Archeology in Anthropology: Broadening Subject Matter. Seventy-fourth annual meeting of the American Anthropological Association, 1975.

Stevenson-Hamilton, J. 1934. The low-veld: Its wildlife and its people. London: Cassell.

———. 1947. *Wild life in South Africa.* London: Cassell.

Stiles, D. N.; Hay, R. L.; and O'Neil, J. R. 1974. The MNK chert factory site, Olduvai Gorge, Tanzania. *World Archaeology* 5:285–308.

Stoltz, L. P. 1977. The population dynamics of baboons *Papio ursinus* Kerr 1792 in the Transvaal. D.Sc. (wildlife management) thesis, University of Pretoria.

Sutcliffe, A. J. 1969. Adaptations of spotted hyaenas to living in the British Isles. *Mamm. Soc. Bull.* 31:1–4.

———. 1970. Spotted hyaena: Crusher, gnawer, digester and collector of bones. *Nature* (London) 246 (5433): 1110–13.

———. 1973*a*. Caves of the East African Rift Valley. *Trans. Cave Res. Group of Gr. Brit.* 15 (1): 41–65.

———. 1973*b*. Similarity of bones and antlers gnawed by deer to human artefacts. *Nature* (London) 246 (5433):-428–30.

———. 1977. Further notes on bones and antlers chewed by deer and other ungulates. *Deer* 4 (2): 73–82.

Swayne, H. G. C. 1899. The leopard in Somaliland. In *Great and small game of Africa,* ed. H. A. Bryden, pp. 575–79. London: Roland Ward.

Sydow, W. 1961. Fund eines Elefantenskeletts: Windhoek-Zoo (fossil). *Mitt. S. W. Afrika Wiss. Ges.* 2 (1): 2–3.

———. 1963. Elefanten in Zoo. *Mitt. S. W. Afrika Wiss. Ges.* 4 (9): 5–6.

Tappen, N. C. 1969. The relationship of weathering cracks to split-line orientation in bone. *Am. J. Phys. Anthrop.,* n.s., 31 (2): 191–97.

———. 1970. Main patterns and individual differences in baboon skull split-lines and theories of causes of split-line orientation in bone. *Am. J. Phys. Anthrop.,* n.s., 33 (1): 61–71.

Thenius, E. 1966. Zur Stammesgeschichte der Hyänen (Carnivora, Mammalia). *Z. Säugetierk.* 31:293–300.

Tobias, P. V. 1965. *Australopithecus, Homo habilis,* tool-using and tool-making. *S. Afr. Archaeol. Bull.* 20:167–92.

———. 1967. *Olduvai Gorge.* Vol 2. *The cranium of* Australopithecus (Zinjanthropus) boisei. Cambridge: Cambridge University Press.

———. 1968*a*. The taxonomy and phylogeny of the australopithecines. In *Taxonomy and phylogeny of Old World primates with reference to the origin of man,* ed. B. Chiarelli. Turin: Rosenberg and Sellier.

———. 1968*b*. The age of death among the australopithecines. *Anthropologist* (Delhi), special volume, pp. 23–28.

———. 1971. *The brain in hominid evolution.* New York: Columbia University Press.

———. 1973*a*. Implications of the new age estimates of the early South African hominids. *Nature* (London) 246:79–83.

———. 1973*b*. A new chapter in the history of the Sterkfontein early hominid site. *J. S. Afr. Biol. Soc.* 14:30–44.

———. 1974. Aspects of pathology and death among early hominids. *Leech* 44 (3): 119–24.

———. 1975. Long or short hominid phylogenies? Palaeontological and molecular evidences. In *The role of natural selection in human evolution,* ed. F. M. Salzano. Amsterdam: North Holland.

Tobias, P. V., and Hughes, A. R. 1969. The new Witwatersrand University excavation at Sterkfontein. *S. Afr. Archaeol. Bull.* 24 (3–4): 158–69.

Tobias, P. V., and von Koenigswald, G. H. R. 1964. Comparison between the Olduvai hominines and those of Java and some implications for hominid phylogeny.

Nature (London) 204:515–18.
Toerien, M. J. 1952. The fossil hyaenas of the Makapansgat valley. *S. Afr. J. Sci.* 48:293–300.
———. 1955. A sabre-tooth cat from the Makapansgat valley. *Palaeont. Afr.* 3:43–46.
Turnbull-Kemp, P. 1967. *The leopard.* Cape Town: Howard Timmins.
van Hoepen, E. C. N. 1930. Fossiele perde van Cornelia, O.V. S. *Paleont. Navors. Nas. Mus. Bloemfontein* 2 (2): 12–24.
———. 1932*a*. Die stamlyn van die Sebras. *Paleont. Navors. Nas. Mus. Bloemfontein* 2 (3): 25–37.
———. 1932*b*. Voorlopige beskrywing van Vrystaatse soogdiere. *Paleont. Navors. Nas. Mus. Bloemfontein* 2 (5): 63–65.
Van Hoepen, E. C. N., and van Hoepen, H. E. 1932. Vrystaatse wilde varke. *Paleont. Navors. Nas. Mus. Bloemfontein* 2:39–62.
Van Lawick-Goodall, H., and van Lawick-Goodall, J. 1970. *Innocent killers.* London: Collins.
van Riet Lowe, C. 1947. Die ontdekking van die Sterkfontein grotte. *S. Afr. Sci.* 1 (4): 85–86.
Vernon, C. J. 1965. The 1964 black eagle survey in the Matopos, Rhodesia. *Arnoldia* (Rhodesia) 2 (6): 1–9.
———. 1971. Owl foods and other notes from a trip to South West Africa. *Ostrich* 42 (2): 153–54.
———. 1972. An analysis of owl pellets collected in southern Africa. *Ostrich* 43:109–24.
Visser, J. 1963. The black eagles of Zuurhoek, Jansenville. *Afr. Wild Life* 17:191–94.
Vogel, J. C. 1969. Radiocarbon dating of Bushman Rock Shelter, Ohrigstad District. *S. Afr. Archaeol. Bull.* 24 (94): 56.
Voigt, E. A. 1973. Stone Age molluscan utilisation at Klasies River Mouth caves. *S. Afr. J. Sci.* 69:306–9.
———. 1975. Studies of marine mollusca from archaeological sites: Dietary preferences, environmental reconstructions and ethnological parallels. In *Archaeozoological studies,* ed. A. T. Clason, pp. 87–98. Amsterdam: North Holland.
von Koenigswald, G. H. R. 1953. The Australopithecinae and *Pithecanthropus. Proc. K. Ned. Akad. Wet.,* ser. B., 56:403–13.
Voorhies, M. R. 1969. Taphonomy and population dynamics of an early Pliocene vertebrate fauna, Knox County, Nebraska. *Univ. Wyoming Contrib. Geol.,* special paper, 1:1–69.
Vrba, E. S. 1971. A new fossil alcelaphine (Artiodactyla: Bovidae) from Swartkrans. *Ann. Transv. Mus.* 27 (5): 59–82.
———. 1973. Two species of *Antidorcas* Sundevall at Swartkrans (Mammalia: Bovidae). *Ann. Transv. Mus.* 28 (15): 287–352.
———. 1974. Chronological and ecological implications of the fossil Bovidae at the Sterkfontein australopithecine site. *Nature* (London) 250 (5461): 19–23.
———. 1975. Some evidence of chronology and palaeoecology of Sterkfontein, Swartkrans and Kromdraai from the fossil Bovidae. *Nature* (London) 254 (5498): 301–4.
———. 1976*a*. *The fossil Bovidae of Sterkfontein, Swartkrans and Kromdraai.* Mem. Transvaal Museum, no. 21.
———. 1976*b*. The significance of bovid remains as indicators of environment and predation patterns. In *Fossils in the making,* ed. A. K. Behrensmeyer and A. P. Hill. Chicago: University of Chicago Press.
———. 1976*c*. Chronology and palaeoecology of the South African early hominids: Some evidence from the fossil Bovidae of Sterkfontein, Swartkrans and Kromdraai. Paper presented at the Ninth Congress, U.I.S.P.P., Nice, 1976.
———. 1977. New species of *Parmularius* Hopwood and *Damaliscus* Sclater and Thomas (Alcelaphini: Bovidae: Mammalia) from Makapansgat. *Palaeont. Afr.* 20:137–51.
———. 1978. Problematical alcelaphine fossils from the Kromdraai Faunal Site (Mammalia: Bovidae). *Ann. Transv. Mus.* 31 (3): 21–28.
Wallace, J. G. 1948. The barn owl in Michigan. *Tech. Bull. Mich. (St. Coll.) Agric. Exp. Sta.,* no. 208.
Washburn, S. L. 1957. Australopithecines: The hunters or the hunted? *Am. Anthrop.* 59 (4): 612–14.
Washburn, S. L., and Patterson, B. 1951. Evolutionary importance of the South African "Man-apes." *Nature* (London) 167:650–51.
Waterhouse, G., ed. 1932. *Simon van der Stel's journal of his expedition to Namaqualand, 1685–6.* Dublin: Dublin University Press.
Weinert, H. 1950. Über die Neuen Vor- und Frühmenschenfunde aus Afrika, Java, China und Frankreich. *Z. Morphol. Anthropol.* 42:113–48.
———. 1951. Über die Vielgestaltigkeit der Summoprimaten vor den Menschwerdung. *Z. Morphol. Anthropol.* 43:73–103.
Welbourne, R. G. 1973. Prey of the spotted eagle owl. *Witwatersrand Bird Club News Sheet* 83:17–18.
Wells, L. H. 1951. A large fossil klipspringer from Potgietersrust. *S. Afr. J. Sci.* 47:167–68.
———. 1959. The nomenclature of South African fossil equids. *S. Afr. J. Sci.* 55:64–66.
———. 1967. Antelopes in the Pleistocene of southern Africa. In *Background to evolution in Africa,* ed. W. W. Bishop and J. D. Clark. Chicago: University of Chicago Press.
———. 1969*a*. Faunal subdivision of the Quaternary in southern Africa. *S. Afr. Archaeol. Bull.* 24:93–95.
———. 1969*b*. Generic position of *"Phenacotragus" van hoepeni. S. Afr. J. Sci.* 65:162.
———. 1970. A Late Pleistocene faunal assemblage from Driefontein, Cradock District, C.P. *S. Afr. J. Sci.* 66:59–61.
Wells, L. H., and Cooke, H. B. S. 1942. The associated fauna and culture of Vlakkraal thermal springs, O.F.S. 3. The faunal remains. *Trans. Roy. Soc. S. Afr.* 29:214–32.
———. 1956. Fossil Bovidae from the Limeworks Quarry, Makapansgat, Potgietersrust. *Palaeont. Afr.* 4:1–55.
Wender, L. 1948. *Animal encyclopedia: Mammals.* London: George Allen and Unwin.
Wendt, W. E. 1972. Preliminary report on an archaeological research programme in South West Africa. *Cimbebasia,* ser. B, 2 (1): 1–61.
Wheat, J. B. 1967. A paleo-Indian bison kill. *Sci. Am.* 216 (1): 44–52.
White, T. D., and Harris, J. M. 1977. Suid evolution and correlation of African hominid localities. *Science* 198 (4312): 13–21.
White, T. E. 1953*a*. A method of calculating the dietary percentage of various food animals utilized by aboriginal peoples. *Am. Antiq.* 18 (4): 396–98.

———. 1953*b*. Observations on the butchering technique of some aboriginal peoples. No. 2. *Am. Antiq.* 19 (2): 160–64.

———. 1954*a*. Observations on the butchering technique of some aboriginal peoples. No. 4. *Am. Antiq.* 19:257–59.

———. 1954*b*. Observations on the butchering technique of some aboriginal peoples. No. 5. *Am. Antiq.* 19:259–62.

Wilkinson, M. J. 1973. Sterkfontein cave system: Evolution of a karst form. M.A. thesis, Department of Geography, Witwatersrand University.

Wilson, V. J. 1969. The large mammals of the Matopos National Park. *Arnoldia* (Rhodesia) 4 (13): 1–32.

———. 1970. Notes on the breeding and feeding habits of a pair of barn owls, *Tyto alba* (Scopoli), in Rhodesia. *Arnoldia* (Rhodesia) 34 (4): 1–8.

Wilson, V. J., and Child, G. 1966. Notes on development and behaviour of two captive leopards. *Zoologische Garten* 32:67–70.

Wilson, V. J., and Grobler, J. H. 1972. Food of the leopard, *Panthera pardus* (Linn.) in the Rhodes Matopos National Park, Rhodesia, as determined by faecal analysis. *Arnoldia* (Rhodesia) 5 (35): 1–10.

Winterbottom, J. M. 1966. Notes on the food of *Tyto alba* in the Etosha Pan. *Ostrich* 37 (2): 139.

Wolhuter, H. 1948. *Memories of a game ranger*. Johannesburg: Wild Life Protection Society, South Africa.

Woodburn, J. 1968*a*. An introduction to Hadza ecology. In *Man the hunter*, ed. R. B. Lee and I. DeVore, pp. 49–50. Chicago: Aldine.

———. 1968*b*. Stability and flexibility in Hadza residential groupings. In *Man the hunter*, ed. R. B. Lee and I. DeVore, pp. 103–10. Chicago: Aldine.

Yellen, J. E. 1977. Cultural patterning in faunal remains: Evidence from the !Kung bushmen. In *Experimental archaeology*, ed. D. W. Ingersoll. New York: Columbia University Press.

Zapfe, H. 1939. (Quoted by Sutcliffe 1970.) *Palaeobiologica* 7:111.

Zihlman, A., and Tanner, N. N.d. Gathering and the hominid adaptation. In *Female hierarchies*, ed. L. Tiger. Chicago: Aldine. In press.

Index